The Biology of
HEAT SHOCK PROTEINS
and
MOLECULAR CHAPERONES

COLD SPRING HARBOR
MONOGRAPH SERIES

The Biology of
HEAT SHOCK PROTEINS
and
MOLECULAR CHAPERONES

Edited by

Richard I. Morimoto
Northwestern University

Alfred Tissières
University of Geneva

Costa Georgopoulos
University of Geneva

COLD SPRING HARBOR LABORATORY PRESS
1994

QP
552
.H43
B56
1994
c.1

The editors express their appreciation to the authors who enthusiastically supported the efforts to see this book come to fruition. We are grateful to the staff of Cold Spring Harbor Laboratory Press: John Inglis for his support and encouragement and Nancy Ford for overseeing the entire process. A special thanks goes to Dorothy Brown for her conscientious editing and Mary Cozza for her cheerful patience with the editors and authors.

29564889

THE BIOLOGY OF HEAT SHOCK PROTEINS AND MOLECULAR CHAPERONES

Monograph 26
Copyright 1994 by Cold Spring Harbor Laboratory Press
All rights reserved
Printed in the United States of America
Book design by Emily Harste

Quote (p. 2) from Barbara McClintock's Nobel lecture is used, with permission, from the Nobel Foundation (1984).

Library of Congress Cataloging-in-Publication Data

The Biology of heat shock proteins and molecular chaperones / edited by
    Richard I. Morimoto, Alfred Tissières, Costa Georgopoulos.
    p. cm. — (Cold Spring Harbor monograph series; 26)
    Includes bibliographical references.
    ISBN 0-87969-427-0
    ISSN 0270-1847
    1. Heat shock proteins   I. Morimoto, Richard I., 1952-   II. Tissières. Alfred.
III. Georgopoulos, Costa.   IV. Series.
QP552.H43B56   1994
574.87'6—dc20
                                          94-46368
                                              CIP

All Cold Spring Harbor Laboratory Press publications may be ordered directly from Cold Spring Harbor Laboratory Press, 10 Skyline Drive, Plainview, New York 11803-2500. Phone: 1-800-843-4388 in Continental U.S. and Canada. All other locations: (516) 349-1.41930. FAX: (516) 349-1946.

Contents

1

Progress and Perspectives on the Biology of Heat Shock Proteins and Molecular Chaperones

Richard I. Morimoto,[1] Alfred Tissières,[2]
and Costa Georgopoulos[3]

[1]Department of Biochemistry
Molecular Biology, and Cell Biology
Northwestern University, Evanston, Illinois 60208

[2]Department of Molecular Biology
University of Geneva, Geneva 1211 Switzerland

[3]Department of Medical Biochemistry
University of Geneva, Geneva 1211 Switzerland

The Biology of Heat Shock Proteins and Molecular Chaperones
©1994 Cold Spring Harbor Laboratory Press 0-87969-427-0/94 $5 + .00

1

In contrast to such "shocks" for which the genome is unprepared, are those a genome must face repeatedly, and for which it is prepared to respond in a programmed manner. Examples are the "heat shock" responses in eukaryotic organisms, and the "SOS" responses in bacteria. Each of these initiates a highly programmed sequence of events within the cell that serves to cushion the effects of the shock. Some sensing mechanism must be present in these instances to alert the cell to imminent danger, and to set in motion the orderly sequence of events that will mitigate this danger. The responses of genomes to unanticipated challenges are not so precisely programmed. Nevertheless, these are sensed, and the genome responds in a discernible but initially unforeseen manner.

Barbara McClintock–Nobel lecture, 8 December 1983

These prophetic words by Barbara McClintock eloquently capture the essence of the heat shock response. At the time these words were written, it was known that all organisms shared a common response to physiological stress. Some of the genes encoding heat shock proteins had just been cloned and the basis of heat shock gene regulation was in the early stages of investigation. However, the function of the heat shock response and the role of heat shock proteins were still a mystery. This mystery slowly began to unravel, and by the 1980s, some information on the function of the heat shock proteins as proteases or unfolded polypeptide-binding proteins had already accumulated. These early stages of elucidation of the biochemical functions of heat shock proteins and the attempts to elucidate the mechanism of autoregulation of the heat shock response were reviewed in the previous volume, *Stress Proteins in Biology and Medicine* (Morimoto et al. 1990b). A historical perspective on the heat shock response with reference to the key discoveries in the early studies on *Drosophila* is described in the introductory chapter (Morimoto et al. 1990a). As incubation at elevated temperatures was the initial stress used to detect the heat shock response, an understanding of how vertebrates regulate their body temperature and use thermal adaptation in physiology is described by Huey and Bennett (1990). A change in body temperature can be imposed either by the environment or internally by adjusting the thermoregulatory set-point. The well-known inducers of the febrile response, such as endotoxins or exogenous pyrogens including the interleukin-1 (α and β), offer important insights into disease states that may be related to the heat shock response (Kluger 1990). Separate from the effects of heat shock on specific cellular or molecular parameters of the cell, it was well known that exposure of tumors to elevated temperatures or hyperthermia, either alone or in combination with other cancer treatments, had beneficial therapeutic effects. The chapter by Hahn and Li (1990) on thermotolerance, thermoresistance, and thermosensitization of cancer cells describes some of the properties of mammalian cells as they undergo hyperthermic treatment. The effectiveness of hyperthermia as a form of cancer treatment by itself or in conjunction with other cancer treatment modalities is discussed by Abe and Hiraoka (1990) and Dewhirst (1990).

The purpose of this volume, *The Biology of Heat Shock Proteins and Molecular Chaperones*, is to update progress and to introduce new fields of investigation with an emphasis in four areas, namely: (1) the biochemical properties of heat shock proteins and molecular chaperones in protein biogenesis, (2) the role of heat shock proteins in thermotolerance and cytoprotection against cellular insults and tissue trauma, (3) the transcriptional control of heat shock genes in prokaryotes and eukaryotes, and (4) the expression and function of the heat shock response in immunology and disease. Although these areas represent the highlights of the field, the editors refer the reader to many excellent recent reviews for additional discussion of related topics (Ellis and van der Vies 1991; Georgopoulos 1992; Gething and Sambrook 1992; Georgopoulos and Welch 1993; Hendrick and Hartl 1993; Welch 1993).

The purpose of this introductory chapter is to provide an overview of highlights of progress during the past 4 years on the properties and regulation of the heat shock proteins and molecular chaperones with reference to specific chapters for additional discussion. The nomenclature for heat shock proteins and molecular chaperones remains complex and perhaps confusing to the new initiate. However, instead of attempting to propose the use of a single common nomenclature throughout this volume, we have chosen to retain the original nomenclature for its didactic value in indicating the unique convergence of observations from studies in such diverse fields as bacteriophage λ biology, immunoglobulin secretion, and mitochondrial protein import. This diverse nomenclature has been summarized in the previous volume (Morimoto et al. 1990). For example, the terms GroEL, hsp60, chaperonin 60, cpn60 all refer to a single protein family whose function and amino acid sequence have been conserved throughout evolution.

We now have an understanding of how cells sense stress and the events involved in the regulation of the heat shock transcriptional response. Perhaps the most significant area of progress has been the recent elucidation of the biochemical properties of heat shock proteins and the recognition that the majority of the heat shock proteins have a fundamental role as molecular chaperones or components of proteolytic systems. Studies on the properties of heat shock proteins and molecular chaperones have revealed new fundamentals on the regulation of protein folding, protein translocation, and assembly and disassembly of protein complexes and have provided a powerful tool for investigating inducible gene regulation.

I. THE FAMILY OF HEAT SHOCK PROTEINS AND MOLECULAR CHAPERONES

In vivo, there is an acute problem of protein aggregation primarily because of the extremely high intracellular protein concentration in all cellular compartments (100–150 mg/ml) and the presence of hydrophobic interactions. Aggregation, due to hydrophobic interactions, becomes more acute as the temperature rises. As a first approximation, it appears as if all organisms have evolved to live in a particular niche so that the overall structure and stability of their proteins are maintained within the normal temperature range encountered by them. Outside this temperature range, and especially at the high end, massive protein aggregation of both newly synthesized and pre-existing folded proteins occurs. To deal with the problem of high protein concentration and potential protein aggregation, a set of universally conserved proteins has evolved, collectively referred to as molecular chaperones (Ellis and van der Vies 1991; Georgopoulos 1992; Georgopoulos and Welch 1993). The overall purpose of these proteins appears to be the minimization of protein aggregation, thus ensuring proper protein folding and transport. In the case of *Escherichia coli* at 46°C, a temperature at which bacterial growth almost ceases, more than 20% of all cellular polypeptides belong to this class of proteins (Herendeen et al. 1979). Such a phenomenal accumulation most likely reflects the extra need for molecular chaperones to deal with the increased load of protein misfolding and aggregation that occurs at this relatively high temperature range. It should be emphasized here that the term "heat shock" or "stress proteins" is somewhat misleading, inasmuch as most of these proteins are essential for cell growth at all temperature ranges.

Many of the heat shock proteins and molecular chaperones are represented as large gene families with functional homologs in each cellular compartment, including the cytosol, nucleus, and specialized organelles, such as the endoplasmic reticulum (ER), mitochondria, and chloroplasts. Although the problem of protein aggregation and misfolding increases with temperature, necessitating a corresponding increase in the intracellular levels of chaperones, it is clear that not all molecular chaperones belong to the heat shock class of proteins. Conversely, not all heat shock proteins function as molecular chaperones. What is clear, however, is that without adequate levels of chaperone proteins, many of the intracellular proteins would aggregate in vivo. This was nicely shown by the results of Gragerov et al. (1992) with *E. coli* demonstrating wholesale protein aggregation in the absence of sufficient "chaperone power" and the ability of either the DnaK (hsp70) or GroEL (hsp60) chaperone machines to deal successfully with such aggregation when overproduced.

II. THE EMERGENCE OF THE CHAPERONE MACHINES

In the past few years, it has become apparent that the major intracellular chaperone proteins such as hsp60 (GroEL), hsp70 (DnaK), and hsp90 do not function in isolation. Rather, to ensure their proper and efficient function, a number of so-called "cohort" proteins have also evolved (Georgopoulos 1992). In some cases, these cohort proteins themselves are bona fide chaperones, e.g., the DnaJ (hsp40) protein of *E. coli* that works synergistically with the major chaperone member DnaK; in other cases, the sole role of the cohorts appears to be to ensure the efficient function and recycling of the major chaperone protein, e.g., the *E. coli* GroES (hsp10) and GrpE proteins, working with the GroEL (hsp60) and DnaK (hsp70) chaperones, respectively. In addition to the evolution of chaperone machines, an inter-chaperone functional cooperation has also evolved, e.g., hsp60/hsp70 in protein folding in the mitochondria, chloroplasts, and bacteria (see Frydman and Hartl; Langer and Neupert; both this volume) and hsp70/hsp90 in glucocorticoid receptor function in the cytosol (Pratt 1993; see Bohen and Yamamoto, this volume).

III. OVERVIEW OF THE PROPERTIES OF THE VARIOUS HEAT SHOCK
PROTEINS AND MOLECULAR CHAPERONES

As will become obvious shortly, progress on deciphering the biological function of many of the heat shock proteins, including those that serve as molecular chaperones, has been moving at a breathtaking pace. However, judging from the important issues still remaining to be answered, the scientific community working on heat shock proteins and molecular chaperones will be kept busy, clearly well into the 21st century.

A. The hsp100 Class of Proteins

The discovery of this family of proteins has been relatively recent. Its biology is summarized by Parsell and Lindquist (this volume). The protein family has been universally conserved from eukaryotes to bacteria, with the curious exception of *Drosophila*. The hsp104 of yeast and ClpB of *E. coli* family members are required for thermotolerance at extreme temperatures (Sanchez and Lindquist 1990; Squires et al. 1991). Interestingly, the hsp104 protein of yeast can protect cells against high temperatures and high concentrations of ethanol, but it does not protect cells from damage by cadmium (Sanchez et al. 1992). Furthermore, in yeast, a partial functional interchangeability with the hsp70 family of proteins has been observed, suggesting that both protein families may be performing at least some complementary biological functions (Sanchez et al. 1993). The significance of this class of proteins is highlighted by the recent discovery of a homolog protein, also under heat shock regulation, that is imported into mitochondria (Leonhardt et al. 1993). At least some family members are actively involved in the proteolytic pathway, since the *E. coli* ClpA homolog has been shown to constitute an active protease system capable of degrading certain polypeptides in conjunction with the ClpP catalytic protease (for review, see Gottesman and Maurizi 1992; see Georgopoulos et al., this volume).

Important major questions about the hsp100 family members include:

1. Are any of the members bona fide chaperone proteins, able to prevent the aggregation or capable of disaggregating certain protein aggregates?
2. Do they have a role in protein proteolysis by "presenting" substrates to various catalytic subunits?
3. How can hsp104 discriminate polypeptide damage caused by heat as opposed to heavy metals?
4. What is the molecular basis of the common mechanism of action that allows the functional substitution of the hsp104 and hsp70 family of proteins?
5. What is the exact role of ATP hydrolysis in the function of these proteins? In particular, why are two ATP-binding sites required for their biological function?

B. The hsp90 Class of Proteins

The abundance and essentiality for cell growth for this class of proteins, found in the cytosol and ER, suggest that they perform important biological functions in eukaryotes (Pratt 1993). Yet, the gene in *E. coli* that codes for the HtpG homolog protein can be deleted with almost negligible biological consequences (Bardwell and Craig 1988). The precise role of hsp90 in cellular physiology remains a relative enigma, since it has only been seen mostly in association with certain protein kinases, e.g., pp60^{src}, and transcription factors, e.g., the glucocorticoid receptors (see Bohen and Yamamoto; Parsell and Lindquist; both this volume). By binding to these molecules, hsp90 suppresses their function, until it escorts them to their proper final destination, as with pp60^{src}, or until the arrival of the hormone, as with the glucocorticoid receptor. In addition to this regulatory role, it appears that hsp90 has an important role in helping some proteins achieve their functional form, since in its absence, glucocorticoid receptors do not function adequately (Picard et al. 1990), and the activity of the pp60^{src} kinase is not expressed (Xu and Lindquist 1992). Consistent with this, Wiech et al. (1992) have shown that purified hsp90 acts like a bona fide chaperone, inasmuch as it can prevent the aggregation of some denatured polypeptides, thus resulting in increased yields of a folded enzyme.

An increasing number of cohort proteins are found to be associated with hsp90, their varying presence depending on the particular substrate polypeptide being complexed with hsp90. Some of these cohort proteins include hsp56, a heat shock protein that belongs to the immunophilin family of proteins (capable of *cis-trans* prolyl isomerase activity), and two proteins of 50 and 23 kD, respectively (p50 and p23). In addition, a 63-kD protein, also under heat shock regulation, has been identified as part of an intermediate in hsp90/glucocorticoid receptor complex (Smith et al. 1993). hsp70 is also found in hsp90 complexes but is present in substoichiometric quantities, and recent evidence suggests it may function transiently, facilitating the interaction between hsp90 and its substrates (for summary, see Pratt 1993; Bohen and Yamamoto; both this volume).

The outstanding questions with this family of proteins include:

1. What is the range of polypeptide substrates for hsp90?
2. How does hsp90 suppress their function and yet promote their folding?
3. To what degree does hsp90 functionally cooperate with hsp70?
4. What role does hsp90 have in maintaining the cytoskeleton and in polypeptide trafficking within the cell?
5. What are the roles, if any, that covalent modifications, such as phosphorylation and ADP ribosylation, have in hsp90 biology?
6. What is the precise role that the cohort proteins play in assisting hsp90 action?
7. Is there a role of hsp90 in antigen presentation, as the recent work of Li and Srivastava (1993) suggests?

C. The hsp70 (DnaK) Chaperone Machine

The genetic, biochemical, and structural studies of this class of chaperone proteins are at a more advanced level than that of the previous two classes (for details, see Craig et al.; Frydman and Hartl; Langer and Neupert; Hightower et al.; Georgopoulos et al.; McKay et al.; all this volume). There are three chaperone machine members, consisting of a major chaperone, hsp70 (DnaK), and two cohorts DnaJ (hsp40) and GrpE. The DnaK and DnaJ members appear to be bona fide chaperones; DnaK prefers to bind unfolded polypeptides when present in the "stretched" conformation, and DnaJ binds to the more compact intermediate, termed "molten globule" (Landry et al. 1992; Langer et al. 1992). In some instances, the DnaK/DnaJ proteins appear to bind synergistically to an unfolded polypeptide substrate and "hand it" off to the GroEL (hsp60) machine for its final maturation (Langer et al. 1992; see Frydman and Hartl, this volume). Recent studies have shown that the DnaJ protein acts catalytically to allow DnaK to bind to some, but not all, of its polypeptide substrates (for summary, see Georgopoulos et al., this volume). The DnaJ and GrpE cohorts dramatically regulate the otherwise feeble ATPase activity of the DnaK partner, with DnaJ specifically accelerating the rate of hydrolysis of the DnaK-bound ATP, and GrpE accelerating the ADP/ATP exchange by causing the release of DnaK-bound nucleotide (Liberek et al. 1991). One of the important biological roles of the hsp70 (DnaK) machine appears to be the negative autoregulation of the heat shock response, specifically by interfering with the function of the heat shock factor responsible for heat shock gene transcription (see below).

Some of the outstanding questions that need answers in this field include:

1. How many conformational states exist in hsp70, how are they influenced by nucleotides and cohort proteins, and how are they involved in regulating substrate binding and release?
2. To what extent do the various hsp70 homologs vary in their ability to carry out their various chores, such as preventing protein aggregation, maintaining the unfolded state, disaggregating protein aggregates, and autoregulating the heat shock response?
3. How widespread is the ability of the hsp70 chaperone machine to control protein function, once a polypeptide has "properly" folded? For example, can it control protein activity by manipulating their oligomeric states (as in the case of the P1 RepA protein; Wickner et al. 1991) or by gently "massaging" their conformation (as in the case of the p53 protein; Hupp et al. 1992)?

4. How does the hsp70 chaperone machine function to autoregulate the heat shock response?
5. What are the roles, if any, of covalent modifications, such as phosphorylation, in hsp70 function?
6. Is there a role of the hsp70 family in antigen presentation, as the recent work of Udono and Srivastava (1993) suggests?

D. The hsp60 (GroEL) Chaperone Machine

The biochemical knowledge of the functioning of the hsp60 chaperone machine, consisting of the hsp60 (GroEL) and hsp10 (GroES) family members, has vastly increased during the last few years. The presence of this chaperone machine is restricted to eubacteria, mitochondria, and chloroplasts (Ellis and van der Vies 1991). Frydman and Hartl (this volume) summarize the recent biochemical developments, including an elegant model of the mechanism of protein folding by the GroEL chaperone machine and the roles of GroES and bound substrate (Martin et al. 1993b). For additional information, see chapters by Georgopoulos et al., Langer and Neupert, and Parsell and Lindquist (all this volume).

The GroEL chaperone protein appears to bind unfolded polypeptide substrates when present in the compact "molten globule" state, which corresponds to a postulated folding intermediate which still exhibits hydrophobic groups. Substrate folding appears to take place within the central cavity of GroEL (Frydman et al. 1992; Braig et al. 1993). The GroES protein has a pivotal role in substrate maturation by coordinating the ATPase activity of the GroEL subunits. GroES also binds ATP, although it does not hydrolyze it. It has been suggested that this feature of GroES is important in facilitating ATP binding to GroEL (Martin et al. 1993a). A "mobile loop" has been identified in GroES that may directly contact GroEL (Landry et al. 1993). In addition to its classical role of preventing protein disaggregation, the GroEL machine has been shown to be capable of dissolving some protein aggregates, both in vivo and in vitro (Martin et al. 1992; Zieminowicz et al. 1993). The GroEL and DnaK chaperone machines have been shown to act synergistically in promoting the folding of some denatured polypeptides (Langer et al. 1992; see Frydman and Hartl, this volume). Recently, the genes coding for mammalian, plant, and yeast homologs of GroES have been identified, opening the way to studying their molecular biology in eukaryotes (Rospert et al. 1993; see Georgopoulos et al., this volume).

Some of the outstanding questions in the GroEL chaperone field include:
1. What is the substrate-binding domain of GroEL and how is its activity

precisely regulated by conformations imposed by nucleotide binding and hydrolysis, as well as by GroES?

2. How does GroES coordinate ATP binding and hydrolysis by the GroEL subunits? In particular, what is the significance of its ability to bind ATP?

3. How is cooperation between the subunits, each of which has both an ATP-binding site and a substrate-binding site, regulated?

4. What are the domains of GroES that interact with GroEL and vice versa?

5. Does GroES interact with the substrate polypeptide directly, as the work of Bochkareva and Girshovich (1992) suggests, and if so, to what extent does it directly contribute to substrate folding?

E. The hsp27/28 Family of Proteins

The biology of this heat shock family of proteins is summarized by Arrigo and Landry (this volume). An interesting aspect of this class of proteins is its structural and functional similarity to that of α-crystallins. However, although the hsp27/28 class of heat shock proteins was among the initial group of proteins to be identified, very little is known about the actual function of these proteins. In yeast, the genes encoding these proteins can be deleted and the resulting cells have no discernible phenotype (Petko and Lindquist 1986). Recent progress includes the demonstration that both the hsp27/28 family of proteins and some of the crystallin family members exhibit chaperone function in vitro, mainly by preventing polypeptide aggregation (the so-called "junior" chaperones; Jacob et al. 1993; Merck et al. 1993), and undergo phosphorylation in response to stress and mitogenic signals (Arrigo and Landry, this volume). In addition, there is evidence that this class of proteins is involved in cellular thermotolerance.

Important questions include:

1. What is the molecular mechanism by which these proteins exert their chaperone function and thermotolerance?

2. What is the biological role of phosphorylation with respect to their biological function?

3. What are the biochemical activities of these proteins?

F. Many of the Heat Shock Proteins Are Proteases

One of the most interesting developments has been the recent demonstration that many of the heat shock proteins are either bona fide proteases or make up components of a protease system. In eukaryotes, the bulk of ATP-dependent proteolysis is carried out by the ubiquitin system (for review, see Jentsch 1992). In this system, polypeptides to be degraded are covalently attached to ubiquitin, which is itself an extremely conserved heat shock protein. Many of the ubiquitin-conjugating enzymes are also under heat shock or stress regulation. An interesting recent finding in the field is that some of these ubiquitin-conjugating enzymes can somehow distinguish whether polypeptides have been damaged as a result of elevated temperatures or heavy metal treatment (Jungmann et al. 1993).

The work with *E. coli* is summarized by Georgopoulos et al. (this volume). The Lon protease of *E. coli* is a classic example of a heat shock protein that is an ATP-dependent protease capable of degrading a variety of unfolded polypeptides (for review, see Gottesman and Maurizi 1992). Recently, a Lon homolog has been identified in human mitochondria (Wang et al. 1993). Particular interest in this field stems from the fact that the *E. coli* heat shock factor equivalent protein, the σ^{32} transcription factor, is an extremely unstable protein, with a half-life on the order of 1 minute. The newly discovered FtsH/HflB heat shock protein either is a protease or controls the activity of a protease that has a pivotal role in σ^{32} proteolysis (Herman et al. 1993 and in prep.; see Georgopoulos et al., this volume). Thus, at least one of the heat-inducible protease systems negatively autoregulates the heat shock response. Another interesting development is the demonstration that the heat shock catalytic protease, ClpP, can function in conjunction with an array of "cohort" proteins including ClpA and ClpX, whose function might be the "presentation" of the unfolded polypeptide substrate (Gottesman and Maurizi 1992; Gottesman et al. 1993; Wojkowiak et al. 1993; see Georgopoulos et al., this volume).

Important questions in the field include:

1. What is the molecular basis of substrate specificity exhibited by the various ubiquitin-activating enzymes in eukaryotes or by the ClpA, ClpX components of *E. coli*?
2. What is the relationship between unfolded polypeptide degradation and its binding to one or more of the chaperone systems?
3. Which one (or more) of these protease systems is responsible for the rapid intracellular digestion of *E. coli*'s σ^{32}?
4. Are the protease "cohort" proteins, such as ClpA and ClpX, bona fide chaperone proteins that maintain the unfolded state of a polypeptide, thus aiding its degradation?
5. Do either the ClpA or ClpX "cohort" proteins or the major chaperone machines, such as hsp70 (DnaK), participate in the "unfolding" of polypeptides, to varying degrees, thus ensuring and accelerating their proteolysis?

6. What physiological properties of the cell and what features of the substrate determine whether unfolded proteins will be degraded or refolded?

G. "Dedicated" Chaperones

There seem to be several types of "dedicated" chaperones. The first is exemplified by the SecB chaperone of *E. coli*, which ensures the export of polypeptides across the inner membrane of *E. coli*. This specialized capacity of SecB is due to its interaction with the SecA protein, which in turn can engage the secretion machinery by interacting with the SecY membrane protein (for review, see Pugsley 1993; Randall et al., this volume). Wild et al. (1992) have shown that the overproduction of the DnaK/DnaJ chaperone machine can compensate for the lack of SecB in polypeptide export, a dramatic demonstration of the promiscuity of chaperones in substrate binding, especially because SecB has been shown to recognize positively charged side chains preferentially (Randall 1992).

Another example of a dedicated chaperone system is PapD, the prototype of chaperones in the periplasmic space of gram-negative bacteria (for review, see Hultgren et al. 1993). This class of proteins is required for the proper assembly of some pili structures, which serve as virulence factors, by helping adherence to host tissues. What is interesting here is the apparent specialization of the PapD chaperone for its few substrates, on the one hand, and the absence of an ATP requirement to carry out its chaperone function, on the other hand (consistent with the lack of ATP in the periplasmic space). The PapD structure has been solved (Holmgren and Branden 1989), as has the structure of its complex with one of its substrates (Kuehn et al. 1993). The PapD-substrate crystal structure has illustrated a general mechanism by which chaperones may interact with their polypeptide substrates (for discussion, see Kuehn et al. 1993).

Perhaps the ultimate example of a dedicated chaperone is exemplified by the case of intramolecular chaperones, present in several secreted proteases (for review, see Shinde and Inouye 1993). It appears that a long propeptide region has evolved, which is cleaved off following secretion, and may serve a dual function. The first is a chaperone-like function that promotes the proper folding and transport of the mature protease domain (Winther and Sorensen 1991; Baker et al. 1992a; Kobayashi and Inouye 1992). In the case of the α-lytic protease, this requirement has been traced to a lowering of the free energy of a late folding transition state of the mature protease domain (Baker et al. 1992a). The second function is to act as a high-affinity inhibitor of the activity of the mature enzyme, thus ensuring that no damage is done to the producing cell (Baker et al. 1992b). The covalent cross-linking of a chaperone to its target protein has the clear thermodynamic advantage of being equivalent to extremely high chaperone concentrations. Other potential examples of dedicated chaperones that may function for the folding and/or transport of a single polypeptide substrate are the LimA protein of *Pseudomonas* (Hobson et al. 1993), the ExbB protein of *E. coli* (Karlsson et al. 1993), the SycE protein of *Yersinia* (Wattiau and Cornelis 1993), and hsp47, which is a member of the serine protease inhibitor superfamily and binds to the nascent procollagen, presumably to ensure the proper synthesis and folding of the collagen triple-helical protein structure (Nakai et al. 1992).

The interesting conclusion from the above-mentioned studies is that some chaperones have evolved to ensure the proper functioning of a limited number of proteins. Do some of these dedicated chaperones represent the evolution of an otherwise promiscuous chaperone to handle a specific substrate better or do they represent the de novo emergence of a chaperone protein tailored to handle a unique substrate?

H. The TCP-1 Chaperone Machine

One of the more satisfying and latest developments in the field of chaperone protein biology is the discovery of the TCP-1 chaperone system that operates in the cytosol of eukaryotic cells, and which replaces the GroEL (hsp60) machine in certain archaebacteria. The biology of this interesting chaperone machine is summarized by Willison and Kubota and Langer and Hartl (both this volume). The structure of the TCP-1 chaperone machine (TF55 in archaebacteria) resembles overall that of hsp60, although its rotational symmetry appears to be higher. Although the prokaroytic TF55 appears to be an oligomer composed of one type of subunit, the TCP-1 structure must be much more complex, since it is composed of up to nine different subunits, each coded by a related, yet separate, gene. The TCP-1 machine has been shown to participate in the folding of various polypeptides, including actin, tubulin, and luciferase (Frydman et al. 1992; Gao et al. 1992, 1993; Melki et al. 1993). Like certain other bona fide chaperones, it possesses an ATPase activity that is stimulated in the presence of unfolded polypeptide substrates (Trent et al. 1991), and must perform an irreplaceable function since it is essential for yeast viability (Ursic and Culbertson 1991).

Important questions in this emerging field of chaperone biology include:
1. How many types of TCP-1 molecules can be assembled from all these functionally related polypeptide subunits?

2. What is the minimum number of subunits that can constitute an active TCP-1 chaperone machine?
3. How promiscuous is the TCP-1 machine vis-à-vis its polypeptide substrate range?
4. Does it functionally interact with other cytosolic chaperone machines, i.e., hsp70 and hsp90, to carry out its biological chores?
5. Are there cohort proteins that help TCP-1 in the binding or release of polypeptide chains or even in its selection of polypeptide substrates?

I. The Emerging Story of Calnexin

Calnexin is the latest addition to the growing list of promiscuous chaperone proteins. It is a surprising addition, because it is an integral membrane protein, largely located in the lumen of the ER, whose apparent role is to retain incompletely assembled proteins in the ER and to facilitate their proper association on the basis of the exact state of their glycosylation. Calnexin was previously known to be required for the oligomeric assembly of the class I proteins of the major histocompatibility complex (MHC) locus (Degen and Williams 1991). Ou et al. (1993) extended these studies to show that most newly synthesized glycoproteins are bound transiently by calnexin. Treatment with tunicamycin, which blocks the addition of core oligosaccharides, simultaneously prevented binding to calnexin. Consistent with this, polypeptides that normally do not become glycosylated were not bound by calnexin. Hammond and Helenius (1993) have recently synthesized these data, as well as their own, in proposing a specific model of calnexin action. The proposed model emphasizes how polypeptide folding and glycosylation trimming are related to each other and to their retainment by calnexin. It is not known to what extent calnexin acts synergistically with the other chaperones located in the ER, e.g., BiP (hsp70) or Grp94 (hsp90), in assisting proper polypeptide folding.

J. Peptidyl Prolyl *Cis-Trans* Isomerases

One of the slow steps in protein folding is *cis-trans* isomerization of certain proline residues. In eukaryotes, two families of structurally unrelated proteins, usually referred to as immunophilins, can catalyze this slow step. They are defined primarily on the basis of which immunosuppressant drug inhibits their activity. One family of proteins is inhibited by cyclosporin and the second by FK506 (for review, see Gething and Sambrook 1992; Kunz and Hall 1993). The PPIases are ubiquitous and abundant proteins present in the cytosol, the ER, and the mitochondria.

The PPIases are usually soluble proteins and promiscuous in substrate selectivity, the ninaA-encoded PPIase of *Drosophila melanogaster*
demonstrating that this is not necessarily always the case. It turns out that
the ninaA-encoded PPIase is an ER integral membrane protein, expressed in a cell-specific manner and required for the correct folding of a
specific subset of the rhodopsin homologs (Stamnes et al. 1991).

The PPIases have been shown to accelerate the process of protein
folding in vitro and to work together with the protein disulfide isomerase
(PDI) class of proteins (see below) to improve their efficiency as
catalysts of protein folding (Schonbrunner and Schmid 1992). This and
other results raise the interesting question of how PPIases recognize their
target peptide sequences. Perhaps, similar to the hsp70 (DnaK) class,
PPIases recognize an unstructured peptide backbone in the "stretched"
configuration. McNew et al. (1993) have shown that a yeast PPIase can
bind to a short polypeptide, even when such a peptide is devoid of
proline residues! Interestingly, such peptide binding was competed by
cyclosporin, suggesting that the peptide-binding site on PPIases may
overlap that of cyclosporin.

Yeast possesses at least eight different immunophilin genes whose
products are located in various cellular compartments, many of which
appear to have a redundant role in physiology and cell growth (for
review, see Kunz and Hall 1993). Recently, Sykes et al. (1993) have
shown that two PPIases of *Saccharomyces cerevisiae*, one localized in
the ER and the other cytosolic, are bona fide heat shock proteins and that
the presence of either of them is necessary for maximal survival of yeast
after exposure to heat stress.

In the case of *E. coli*, there are two PPIases, one cytosolic and the
other periplasmic, which thus may contribute to the folding of exported
proteins (Compton et al. 1992). Surprisingly, the *Legionella pneumophila mip* gene codes for a cell-surface PPIase homolog that is required
for bacterial virulence (Fischer et al. 1992). Perhaps the *mip*-encoded
PPIase is necessary for the proper folding of host-encoded proteins or
bacterial virulence factors expressed on the cell surface or secreted. The
possible involvement of PPIases in microbial pathogenicity has been
recently summarized by Hacker and Fischer (1993).

K. The PDI Class of Proteins, Involved in Disulfide
   Bond Formation In Vivo

In vivo disulfide bond formation in eukaryotic cells occurs in the ER, because the redox state is much more oxidizing than in the cytosol (Hwang
et al. 1992). Consistent with this, the prototype enzyme affecting di-

sulfide bond formation, PDI, is an abundant protein localized in the lumen of the ER. PDI has been shown to catalyze the isomerization and oxidation of intramolecular disulfide bonds and in so doing to accelerate and promote the correct folding of proteins in vitro (for review, see Noiva and Lennarz 1992). PDI acts like a generalized chaperone by binding nonspecifically to an exposed polypeptide backbone, the presence of a cysteine residue increasing its affinity (Noiva and Lennarz 1992). Consistent with a generalized chaperone role, PDI has been shown to be a subunit of several ER enzyme complexes, involved in the folding or modification of translocated proteins, including prolyl hydroxylase (Noiva and Lennarz 1992).

The PDI equivalent protein of *S. cerevisiae* is also an ER resident protein, essential for yeast viability and whose depletion results in a defect in the transport and maturation of disulfide-containing proteins (for review, see LeMantia and Lennarz 1993). Paradoxically, although the PDI-encoding gene of yeast is essential, its disulfide isomerase activity is not (LaMantia and Lennarz 1993). This was shown through an elimination of the thioredoxin-like active domains of PDI by site-directed mutagenesis. Such yeast mutants were viable, although disulfide bond formation and transport of carboxypeptidase Y were retarded. These results suggest that it is the peptide-binding site of PDI that is necessary for yeast viability and/or its association with other ER-localized proteins.

In gram-negative bacteria like *E. coli*, the periplasmic space is the only oxidizing environment and, in some ways, has a role analogous to that of the ER in eukaryotes. Elegant genetic studies have resulted in the characterization of the DsbA protein, a soluble periplasmic protein involved in both disulfide bond catalysis and disulfide bond exchange reactions (for review, see Bardwell and Beckwith 1993; Creighton and Freedman 1993). In analogy with PDI, the DsbA protein can catalyze and reshuffle disulfide bonds in the periplasmic space. Recently, the DsbB protein of *E. coli* has been described, which is localized in the inner membrane and which is also necessary for wholesale disulfide oxidation reactions in the periplasm (Bardwell et al. 1993; Dailey and Berg 1993; Missiakas et al. 1993). The fact that in *dsbB* mutants, the DsbA protein is found mostly in the reduced form has led to the proposal that the DsbB protein directly or indirectly reoxidizes DsbA (Bardwell et al. 1993). The DsbB protein, in turn, could be re-oxidized by the electron transport system localized in the inner membrane or from the bacterial cytoplasm (Creighton and Freedman 1993). The importance of the catalysis of disulfide bond formation in *E. coli*'s periplasmic space is exemplified by the recent discovery of the DsbC protein, a periplasmic soluble protein, whose absence also results in a significant decrease in the rate of disulfide bond formation in several disulfide-containing periplasmic proteins (Missiakas et al. 1994; Shevchik et al. 1994). The facts (1) that overproduction of DsbC can compensate for the lack of DsbA and vice versa and (2) that DsbA is found in the oxidized state in *dsbC* mutant bacteria argue that the DsbA- and DsbC-controlled disulfide-bond-forming systems represent two parallel pathways that operate independently of each other.

IV. HEAT SHOCK TRANSCRIPTION FACTORS AND THE REGULATION OF HEAT SHOCK GENE TRANSCRIPTION

A. Eukaryotic Cells

Since the seminal observations on heat shock induction of salivary gland chromosomal puffs, a challenging question has been to understand the basis of the temperature or physiological effects on heat shock gene transcription. Although heat shock gene expression is also known to be regulated posttranscriptionally through effects on message stability or at the level of translation initiation, the primary form of regulation in most cell types is at the level of transcription. Therefore, the recent efforts on heat shock gene transcription in eukaryotes have been to decipher how the heat shock transcription factor (HSF) responds to physiological stress and to establish the other events that occur at heat shock promoters leading to the burst in transcriptional activity (Lis and Wu 1993; Morimoto 1993). Substantial progress has been made on the regulatory properties of HSFs through the cloning of HSF genes and specific antibody reagents.

Our understanding of heat shock gene transcription has acquired an additional level of complexity with the discovery that plants and vertebrates encode a family of HSFs, whereas in yeast and *Drosophila*, there is a single HSF-encoding gene (Scharf et al. 1990; Rabindran et al. 1991; Sarge et al. 1991; Schuetz et al. 1991; Nakai and Morimoto 1993; Wu et al.; Morimoto et al.; both this volume). All HSFs share common features including an amino-terminal-localized DNA-binding domain, an adjacent cluster of hydrophobic amino acids organized into heptad repeats (leucine zippers), and distally located heptad repeats located near the extreme carboxyl terminus of the protein (Nieto-Sotelo et al. 1990; Chen et al. 1993; Rabindran et al. 1993). The vertebrate HSFs (HSF1, HSF2, and HSF3) are simultaneously expressed in most but not all cells; however, the DNA-binding properties of each factor is negatively regulated in most cell types. Although there has been much progress in examining the role of the heptad repeats in intermolecular and intramolecular interactions, the mechanistic basis of the events that lead to the conversion of the non-DNA-binding monomer to the active DNA-binding trimeric state remains an area of active investigation (see Wu et al; Morimoto et al; both this volume).

The family of HSFs provides an explanation to the growing number of observations that heat shock gene expression is induced during specific stages of development and differentiation and by a variety of different physiological conditions. HSF1 appears to be a general stress responsive factor and is activated (acquires DNA-binding activity) in response to elevated temperatures, heavy metals, amino acid analogs, and oxidative stress. HSF1 is also activated in adrenals by adrenocorticotropic hormone (ACTH) (Sarge et al. 1993; Baler et al. 1993; see Holbrook and Udelsman, this volume), an observation which reveals that neurohormonal "stress" and other forms of physiological stress that induce the heat shock response converge. In contrast, HSF2 appears to be a developmental HSF and is activated during hemin-induced differentiation of erythroleukemic cells during mouse spermatogenesis and in early mouse embryogenesis (Sistonen et al. 1992, 1994; Sarge et al. 1994). HSF3, cloned from the chicken genome, appears to be a cell-type-specific HSF that responds to heat shock with delayed kinetics relative to HSF1. A critical feature of these HSFs is that the signals that activate the DNA-binding properties of each factor are apparently specific since HSF2 DNA-binding activity is not activated following stress and, likewise, HSF1 DNA-binding activity is not induced during erythroid differentiation or spermatogenesis.

In unstressed cells of higher eukaryotes, HSF1 exists in a monomeric, non-DNA-binding state in both the cytoplasmic and nuclear compartments that undergoes a stress-dependent oligomerization to a trimeric state while translocating and relocalizing within the nucleus (Westwood et al. 1991; Baler et al. 1993; Sarge et al. 1993; Westwood and Wu 1993; Sistonen et al. 1994). Heat shock proteins such as hsp70 have been implicated in regulating the monomeric state of HSF1 as well as in forming stable complexes with the trimeric form of HSF1. The appearance of stable HSF1-hsp70 complexes correlates with the attenuation of the heat shock transcriptional response (Abravaya et al. 1992). In addition, HSF1 exhibits a stress-dependent phosphorylation whose role is not well understood. Neither the conversion of HSF1 monomers to trimers nor the binding of HSF1 trimers to DNA requires phosphorylation. Yet, some of the available data, particularly in yeast, indicate that heat-shock-inducible phosphorylation may be related to HSF transcriptional activity. How these signals are translated into the selective and differential activation of each HSF remains a puzzle.

Among the questions that remain to be answered include:

1. What are the intracellular stress-induced signals that lead to the activation of HSF?
2. How are the heptad repeats utilized for the interconversion of the monomeric to trimeric state of HSF1?
3. How do heat shock proteins alter heat shock factor activities?
4. How are the different HSFs activated in response to different signals?

B. Prokaryotic Cells

In *E. coli*, there is a major heat shock response pathway under the transcriptional control of the σ^{32} factor, as well as minor pathways under the transcriptional control of other factors (Gross et al. 1990; see Georgopoulos et al., this volume). The σ^{32} factor complexes with the RNA polymerase (RNAP) core (E) to constitute the $E\sigma^{32}$ holoenzyme capable of transcribing uniquely σ^{32}-dependent promoters. The regulation of the heat shock response in *E. coli* is very complicated and revolves around three strategies: The first is the posttranslational control of *rpoH* mRNA (encoding the σ^{32} factor), involving repression at low temperature and a sudden and transient derepression at high temperature followed by repression again. The second strategy is the modulation of the extremely unstable half-life of σ^{32}, on the order of 1 minute. Many of the σ^{32}-induced proteins are either proteases or components that control protease activity, and thus may negatively autoregulate the heat shock response. The third strategy is the negative autoregulation by the DnaK/DnaJ/GrpE chaperone machine at a variety of levels, including translational repression, modulation of σ^{32} half-life, sequestration of the σ^{32} factor away from the RNAP core, and prevention of open-complex formation by the σ^{32} holoenzyme. Both the DnaK and DnaJ chaperone proteins can complex with the σ^{32} factor, either alone or synergistically in the presence of ATP (for a detailed review, see Georgopoulos et al., this volume). In addition to the σ^{32}-dependent heat shock response, *E. coli* possesses another stress or heat shock operon under the transcriptional control of newly discovered σE ($\sigma 24$) factor (Mescas et al. 1993; see Georgopoulos et al., this volume). In many other eubacteria, especially the gram-positive ones, the heat shock response appears to be regulated by a mechanism distinct from that of *E. coli* (Wetzstein et al. 1992; see Georgopoulos et al., this volume).

V. HEAT SHOCK PROTEINS AND DISEASE

As the biological role of the heat shock proteins and molecular chaperones in a multitude of cellular processes unfolds, it is of little surprise that they should be implicated in human disease. Hyperthermia has its roots in the annals of ancient medicine. Indeed, the therapeutic effect of heat persists today as the beneficial effects of a sauna in Finland or by the soothing effect of hot tubs in Japan. One wonders whether the positive attributes of heat shock in cytoprotection or thermotolerance are the basis of the complex and pleiotropic protective effects of subtoxic exposures to stress conditions against subsequent exposures to conditions that lead to acute or chronic cell and tissue damage. Yet, for many of the disease-related observations, the questions are whether heat shock proteins provide a useful indicator to detect stress and tissue trauma and, related to this question, whether chronic stress leads to the aberrant synthesis and accumulation of heat shock proteins which itself may be detrimental.

The scientific literature is replete with observations that underscore the potential significance of the link between the aberrant expression of stress proteins and disease states. The abnormal expression of stress proteins has been widely observed in a number of diseases including oxidant injury, ischemia, cardiac hypertrophy, fever, inflammation, metabolic diseases, infection, cell and tissue trauma, and aging (see Holbrook and Udelsman; Nowak and Abe; Benjamin and Williams; Kaufmann and Schoel; all this volume). Serum antibodies against cellular heat shock proteins have been reported for many infectious diseases and autoimmune diseases including rheumatoid arthritis and insulin-dependent diabetes (Heufelder et al. 1991, 1992). Many of these conditions result from or cause cell death and tissue dysfunction. Although the acute response to stress may be important for long-term survival of the affected tissues, the chronic expression of hsp70 and other heat shock proteins and molecular chaperones may have negative effects on protein biogenesis and therefore a variety of cellular activities. The consequence of this could lead to a loss of cell growth potential (see Parsell and Lindquist, this volume) as well as to hamper the ability of the cell to respond to subsequent bouts of physiological stress. Since the ability of a cell to survive stress is dependent on the rapidity and duration of the response, any change in the ability to sense or respond to stress could have profound consequences; e.g., there is growing evidence that "old" cells, either maintained in culture or from aged animals, do not respond with the same vigor to physiological stress as measured by a reduction of heat shock gene expression (see Holbrook and Udelsman, this volume).

A. Heat Shock Proteins and Ischemia

Ischemia is a condition that results from decreased blood flow to tissues, the consequences of which are decreased oxygen and nutrient supply. The increased synthesis of heat shock proteins during conditions of ischemia and reoxygenation during reperfusion represents a well-studied pathophysiological state that impacts the brain and the heart. During postischemic reperfusion, heat shock gene mRNA expression is rapidly activated and corresponding heat shock protein synthesis is elevated in the brain, kidney, heart, and liver tissues (see Benjamin and Williams; Nowak and Abe, both this volume). The postischemic damage may be due to the generation of oxygen-free radicals that have also been implicated in the cellular response to a wide variety of environmental agents including xenobiotics and aromatic hydrocarbons. The ability to detect and respond to such oxidant-damaged proteins may be directly relevant to preventing extensive damage such as that seen following a stroke. An understanding of the regulation and function of heat shock proteins in the brain may also provide information relevant to ischemic injury, neurotransmitter toxicity, and perhaps protection against the accumulation of malfolded proteins (see Nowak and Abe, this volume). Likewise, the induction of hsp70 following aortic constriction or work-overload-induced cardiac hypertrophy could reflect the response to the aberrant synthesis, accumulation, or degradation of proteins, or alternatively, it could reflect a response to events that occur during the partial reentry of the myocardial cell into the growth cycle (see Benjamin and Williams, this volume). In this case, does activation of the heat shock response reflect cell-specific protein damage and does this reveal a potential approach to control this process?

B. Heat Shock Proteins, Aging, and Endocrine Responses

Studies using tissue-culture cells and animal model systems have shown that cells which have attained their replicative potential respond poorly to heat shock relative to early passage cells or cells from young animals (Liu et al. 1989; Choi et al. 1990; Fargnoli et al. 1990; Heydari et al. 1993). These data suggest an age-related decline in the response to physiological stress. The basis for this restriction in heat shock gene activation appears to be at the transcriptional level (Holbrook and Udelsman, this volume). These observations have direct relevance to the cell's ability to sustain damage and to survive such environmental or physiological stresses.

A fascinating observation on the activation of heat shock genes in whole animals was made by Blake et al. (1991). These authors demonstrated that the stress imposed by physical restraint of the rat was sufficient to induce the expression of select heat shock genes in the adrenals and vasculature. More recent data now indicate that restraint stress induces HSF1, the same transcription factor that is induced by heat shock. This "stress" response is endocrine-regulated by ACTH. Hypophysectomized rats did not exhibit this restraint stress heat shock gene induction, whereas the acute treatment of hypophysectomized rats with ACTH induced the expression of hsp70 in the adrenal glands (Blake et al. 1991). The observations on ACTH reveal a link between the neuroendocrine pathways and the models proposed for HSF activation by the appearance of unfolded proteins. How these pathways intersect remains unknown.

C. Heat Shock Proteins and Infectious Diseases

A substantial literature has accumulated on the topic of heat shock proteins as major dominant antigens of a wide range of infectious agents, ranging from bacteria, viruses, and protozoa, to helminths (see Young et al. 1990; Kaufmann and Schoel, this volume). These observations have led to numerous speculations ranging from immune surveillance of damaged or infected cells to the suggestion that the dominant antigenicity of pathogen heat shock proteins may lead to autoimmunity. Expression of heat shock proteins by the pathogen ensures survival in host cells; likewise, both the pathogen and host cell undergo a stress response during phagocytosis. The elevated chronic expression of heat shock proteins in the host cell may subsequently lead to susceptibility to the anti-heat-shock-protein immune surveillance system if the epitope recognized by the specific T-cell population is shared between the pathogen and host. However, it is unlikely that a direct link exists between the dominant antigenicity of heat shock proteins encoded by pathogens and autoimmune disease, as the major antigenic peptide of heat shock proteins encoded by pathogens is typically distinct from the sequence of the corresponding host heat shock protein. Recent progress on the identity of the processed peptides in MHC molecules have identified heat shock protein peptides, as well as peptides corresponding to other proteins, associated with the MHC class I and class II molecules (Newcomb and Cresswell 1993). These data indicate that self-heat-shock-protein peptides are recognized by T cells. Another exciting development in the field is the recent discovery that either the GroEL (hsp60) or DnaK (hsp70) heat shock proteins of *Mycobacterium tuberculosis* or *E. coli* dramatically increase the antigenicity of peptides or oligosaccharides conjugated to them (Barrios et al. 1992; G. Del Giudice, pers. comm). The use of these heat shock proteins should aid in the design of better vaccines in combating infections.

D. A Rationale for Modulating Heat Shock Proteins in Disease

On the basis of the available evidence, it should be possible to modulate the heat shock response in tissues to induce a "thermotolerant" protective state against subsequent traumas. Yet, heat shock treatment is somewhat impractical, in particular for homeotherms. In addition, systemic heat shock is likely to have some negative effects that reduce its attractiveness as a treatment modality. Perhaps it would be more useful to consider other treatments that induce the heat shock response such as the pharmacologically active drugs, salicylates and indomethacin, both of which are potent activators of HSF, enabling it to function at substantially lower temperatures (see Morimoto et al., this volume). The pharmacological manipulation of the heat shock response may provide an alternative modulate heat shock expression in certain tissues as potential intervention or therapy.

Another consideration on the relationship between heat shock proteins and disease is whether the heat shock response or heat shock proteins have a role in cell growth control and, as a consequence, whether deregulation of heat shock expression leads to the neoplastic state. Perhaps the most relevant recent observation is that mutant forms of p53 have been shown to form stable complexes with hsc70, the constitutively expressed member of the hsp70 family (Pinhasi-Kimhi et al. 1986; Hinds et al. 1987). Although the significance of these interactions on the biochemical properties of mutant p53 is not entirely understood, there is some recent evidence to indicate that bacterially expressed p53 does not bind DNA unless incubated with DnaK and ATP (Hupp et al. 1992).

The implication of these observations is that proteins such as p53, which undergo distinct conformational changes that control the active and inert states, may be directly affected by the intracellular concentrations of one or more of the heat shock proteins and molecular chaperones. This may be of some importance as heat shock proteins such as hsp70 are known to be growth-regulated (Wu and Morimoto 1985; Milarski and Morimoto 1986). Shaknovich et al. (1992) have shown that another heat shock protein, hsp90, can also convert the MyoD1 transcription factor from an inactive to an active DNA-binding form in an ATP-independent process. These observations raise the possibility that some of the heat shock proteins and molecular chaperones may exert global regulatory effects by continuously modulating the conformations of various polypeptides, thus influencing their enzymatic activities or oligomerization properties.

ACKNOWLEDGMENT

We thank Susan Lindquist for a critical reading of the manuscript.

REFERENCES

Abe, M. and M. Hiraoka. 1990. Hyperthermia in combination with radiation in the treatment of cancers. In *Stress proteins in biology and medicine* (ed. R.I. Morimoto et al.), pp. 117–130. Cold Spring Harbor Laboratory Press, Cold Spring Harbor, New York.

Abravaya, K., M.P. Myers, S.P. Murphy, and R.I. Morimoto. 1992. The human heat shock protein hsp70 interacts with HSF, the transcription factor that regulates heat shock gene transcription. *Genes Dev.* **6:** 1153–1164.

Baker, D., J.L. Silen, and D.A. Agard. 1992a. Protease pro region required for folding is a potent inhibitor of the mature enzyme. *Protein Sci.* **12:** 339–344.

Baker, D., J.L. Sohl, and D.A. Agard. 1992b. A protein-folding reaction under kinetic control. *Nature* **356:** 263–265.

Baler, R., G. Dahl, and R. Voellmy. 1993. Activation of human heat shock genes is accompanied by oligomerization, modification, and rapid translocation of heat shock transcription factor HSF1. *Mol. Cell. Biol.* **13:** 2486–2496.

Bardwell, J.C.A. and J. Beckwith. 1993. The bonds that tie: Catalyzed disulfide bond formation. *Cell* **74:** 771–779.

Bardwell, J.C.A. and E.A. Craig. 1988. Ancient heat shock gene is dispensable. *J. Bacteriol.* **170:** 2977–2983.

Bardwell, J.C.A., J.O. Lee, G. Jander, N. Martin, D. Belin, and J. Beckwith. 1993. A pathway for disulfide bond formation in vivo. *Proc. Natl. Acad. Sci.* **90:** 1038–1042.

Barrios, C., A.R. Lussow, J. Van Embden, R. Van der Zee, R. Rappuoli, P. Costantino, J.A. Louis, P.-H. Lambert, and G. Del Giudice. 1992. Mycobacterial heat-shock proteins as carrier molecules. II. The use of the 70-kDa mycobacterial heat-shock protein as carrier for conjugated vaccines can circumvent the need for adjuvants and Bacillus Calmette Guerin priming. *Eur. J. Immunol.* **22:** 1365–1372.

Blake, M.J., R. Udelsman, G.J. Feulner, D.D. Norton, and N.J. Holbrook. 1991. Stress-induced heat shock protein 70 expression in adrenal cortex: An adrenocorticotropic hormone-sensitive, age-dependent response. *Proc. Natl. Acad. Sci.* **88:** 9873–9877.

Bochkareva, E.S. and A.S. Girshovich. 1992. A newly synthesized protein interacts with GroES on the surface of chaperonin GroEL. *J. Biol. Chem.* **267:** 25672–25675.

Braig, K., M. Simon, F. Furuya, J. Hainfeld, and A.L. Horwich. 1993. A polypeptide bound by the chaperonin groEL is localized within a central cavity. *Proc. Natl. Acad. Sci.* **90:** 3978–3982.

Chen, Y., N.A. Barlev, O. Westergaard, and B.K. Jakobsen. 1993. Identification of the C-terminal activator domain in yeast heat shock factor: Independent control of transient and sustained transcriptional activity. *EMBO J.* **12:** 5007–5018.

Choi, H.S., Z. Lin, B. Li, and A.Y.-C. Liu. 1990. Age-dependent decrease in the heat-inducible DNA sequence-specific binding activity in human diploid fibroblasts. *J. Biol. Chem.* **265:** 18005–18011.

Compton, L.A., J.M. Davis, J.R. MacDonald, and H.P. Bächinger. 1992. Structural and functional characterization of *Escherichia coli* peptidyl-prolyl *cis-trans* isomerase. *Eur. J. Biochem.* **206:** 927–934.

Creighton, T.E. and R.B. Freedman. 1993. A model catalyst of protein disulphide bond formation. *Curr. Biol.* **3:** 790–793.

Dailey, F.E. and H.C. Berg. 1993. Mutants in disulfide bond formation that disrupt flagellar assembly in *Escherichia coli. Proc. Natl. Acad. Sci.* **90:** 1043–1047.

Degen, E. and D.B. Williams. 1991. Participation of a novel 88-kD protein in the biogenesis of murine class I histocompatibility molecules. *J. Cell Biol.* **112:** 1099–1115.

Dewhirst, M.W. 1990. Hyperthermia in cancer therapy: Model systems. In *Stress proteins in biology and medine* (ed. R.I. Morimoto et al.), pp. 101–116. Cold Spring Harbor Laboratory Press, Cold Spring Harbor, New York.

Ellis, R.J. and S.M. van der Vies. 1991. Molecular chaperones. *Annu. Rev. Biochem.* **60:** 321–347.

Fargnoli, J., T. Kunisada, A.J. Fornace, Jr., E.L. Schneider, and N.J. Holbrook. 1990. Decreased expression of heat shock protein 70 mRNA and protein after heat treatment in cells of aged rats. *Proc. Natl. Acad. Sci.* **87:** 846–850.

Fisher, G., H. Bang, B. Ludwig, K. Mann, and J. Hacker. 1992. Mip protein of *Legionella pneumophila* exhibits peptidyl-prolyl-cis/trans isomerase (PPIase) activity. *Mol. Microbiol.* **6:** 1375–1383.

Frydman, J., E. Nimmesgern, B.H. Erdjument, J.S. Wall, P. Tempst, and F.-U. Hartl. 1992. Function in protein folding of TRiC, a cytosolic ring complex containing TCP-1 and structurally related subunits. *EMBO J.* **11:** 4767–4778.

Gao, Y., I.E. Vainberg, R.L. Chow, and N.J. Cowan. 1993. Two cofactors and cytoplasmic chaperonin are required for the folding of α- and β-tubulin. *Mol. Cell Biol.* **13:** 2478–2485.

Gao, Y., J.O. Thomas, R.L. Chow, G.H. Lee, and N.J. Cowan. 1992. A cytoplasmic chaperonin that catalyzes β-actin folding. *Cell* **69:** 1043–1050.

Georgopoulos, C. 1992. The emergence of the chaperone machines. *Trends Biochem. Sci.* **17:** 295–299.

Georgopoulos, C. and W.J. Welch. 1993. Role of major heat shock proteins as molecular chaperones. *Annu. Rev. Cell Biol.* **9:** 601–635.

Gething, M.-J. and J. Sambrook. 1992. Protein folding in the cell. *Nature* **355:** 33–45.

Gottesman, S. and M.R. Maurizi. 1992. Regulation by proteolysis: Energy-dependent proteases and their targets. *Microbiol. Rev.* **56:** 592–621.

Gottesman, S., W.P. Clark, V. de Crecy-Lagard, and M.R. Maurizi. 1993. ClpX, an alternative subunit for the ATP-dependent Clp protease of *Escherichia coli*. *J. Biol. Chem.* **268:** 22618–22626.

Gragerov, A.I., E. Martin, M.A. Krupenko, M.V. Kashlev, and V.G. Nikiforov. 1992. Protein aggregation and inclusion body formation in *Escherichia coli rpoH* mutant defective in heat shock protein induction. *FEBS Lett.* **291:** 222–224.

Gross, C.A., D.B. Straus, J.W. Erickson, and T. Yura. 1990. The function and regulation of heat shock proteins in *Escherichia coli*. In *Stress proteins in biology and medicine* (ed. R.I. Morimoto et al.), pp. 167–189. Cold Spring Harbor Laboratory Press, Cold Spring Harbor, New York.

Hacker, J. and G. Fischer. 1993. Immunophilins: Structure-function relationship and possible role in microbial pathogenicity. *Mol. Microbiol.* **10:** 445–456.

Hahn, G.M. and G.C. Li. 1990. Thermotolerance, thermoresistance, and thermosensitization. In *Stress Proteins in biology and medicine* (ed. R.I. Morimoto et al.), pp. 79–100. Cold Spring Harbor Laboratory Press, Cold Spring Harbor, New York.

Hammond, C. and A. Helenius. 1993. A chaperone with a sweet tooth. *Curr. Biol.* **3:** 884–886.

Hendrick, J.P. and F.-U. Hartl. 1993. Molecular chaperone functions of heat-shock proteins. *Annu. Rev. Biochem.* **62:** 349–384.

Herendeen, S.L., R.A. VanBogelen, and F.C. Neidhardt. 1979. Levels of major proteins of *Escherichia coli* during growth at different temperatures. *J. Bacteriol.* **139:** 185–194.

Herman, C., T. Ogura, T. Tomoyasu, S. Hiraga, Y. Akiyama, K. Ito, R. Thomas, R. D'Ari, and P. Bouloc. 1993. Cell growth and λ phage development controlled by the same essential *Escherichia coli* gene, *ftsH/hflB*. *Proc. Natl. Acad. Sci.* **90:**

10861–10865.

Heufelder, A.E., J.R. Goellner, B.E. Wenzel, and R.S. Bahn. 1992. Immunohistochemical detection and localization of a 72-kilodalton heat shock protein in autoimmune thyroid disease. *J. Clin. Endocrinol. Metab.* **74:** 724–731.

Heufelder, A.E., B.E. Wenzel, C.A. Forman, and R.S. Bahn. 1991. Detection, cellular localization, and modulation of heat shock proteins in cultured fibroblasts from patients with extrathyroid manifestations of Graves' disease. *J. Clin. Endocrinol. Metab.* **73:** 739–745.

Heydari, A.R., B. Wu, R. Takahashi, R. Strong, and A. Richardson. 1993. Expression of heat shock protein 70 is altered by age and diet at the level of transcription. *Mol. Cell. Biol.* **13:** 2909–2918.

Hinds, P.W., C.A. Finlay, A.B. Frey, and A.J. Levine. 1987. Immunological evidence for the association of p53 with a heat shock protein, hsc70, in p53-plus-ras-transformed cell lines. *Mol. Cell. Biol.* **7:** 2863–2869.

Hobson, A.H., C.M. Buckley, J.L. Aamand, S.T. Jorgensen, B. Diderichsen, and D.J. McConnell. 1993. Activation of a bacterial lipase by its chaperone. *Proc. Natl. Acad. Sci.* **90:** 5682–5686.

Holmgren, A. and C.-I. Branden. 1989. Crystal structure of chaperone protein PapD reveals an immunoglobulin fold. *Nature* **342:** 248–251.

Huey, R.B. and A.F. Bennett. 1990. Physiological adjustments to fluctuating thermal environments: An ecological and evolutionary perspective. In *Stress proteins in biology and medicine* (ed. R.I. Morimoto et al.), pp. 37–60. Cold Spring Harbor Laboratory Press, Cold Spring Harbor, New York.

Hultgren, S.J., S. Abraham, M. Caparon, P. Falk, J.W. St.Geme III, and S. Normark. 1993. Pilus and nonpilus bacterial adhesins: Assembly and function in cell recognition. *Cell* **73:** 887–901.

Hupp, T.R., D.W. Meek, C.A. Midgley, and D.P. Lane. 1992. Regulation of the specific DNA binding function of p53. *Cell* **71:** 875–886.

Hwang, C., A.J. Sinskey, and H.F. Lodish. 1992. Oxidized redox state of glutathionine in the endoplasmic reticulum. *Science* **257:** 1496–1502.

Jakob, U., M. Gaestel, K. Engel, and J. Buchner. 1993. Small heat shock proteins are molecular chaperones. *J. Biol. Chem.* **268:** 1517–1520.

Jentsch, S. 1992. The ubiquitin-conjugation system. *Annu. Rev. Genet.* **26:** 179–207.

Jungmann, J., H.-A. Reins, C. Schobert, and S. Jentsch. 1993. Resistance to cadmium mediated by ubiquitin-dependent proteolysis. *Nature* **361:** 369–371.

Karlsson, M., K. Hannavy, and C.F. Higgins. 1993. ExbB acts as a chaperone-like protein to stabilize TonB in the cytoplasm. *Mol. Microbiol.* **8:** 389–396.

Kluger, M.J. 1990. The febrile response. In *Stress proteins in biology and medicine* (ed. R.I. Morimoto et al.), pp. 61–79. Cold Spring Harbor Laboratory Press, Cold Spring Harbor, New York.

Kobayashi, T. and M. Inouye. 1992. Functional analysis of the intramolecular chaperone. Mutational hot spots in the subtilisin pro-peptide and a second-site suppressor mutation within the subtilisin molecule. *J. Mol. Biol.* **226:** 931–933.

Kuehn, M.J., D.J. Ogg, J. Kihlberg, L.N. Slonim, K. Flemmer, T. Bergfors, and S.J. Hultgren. 1993. Structural basis of pilus subunit recognition by the PapD chaperone. *Science* **262:** 1234–1240.

Kunz, J. and M.N. Hall. 1993. Cyclosporin A, FK506 and rapamycin: More than just immunosuppression. *Trends Biochem. Sci.* **18:** 334–338.

LaMantia, M. and W.J. Lennarz. 1993. The essential function of yeast protein disulfide isomerase does not reside in its isomerase activity. *Cell* **74:** 899–908.

Landry, S.J., R. Jordan, R. McMacken, and L.M. Gierasch. 1992. Different conformations of the same polypeptide bound to chaperone DnaK and GroEL. *Nature* **355:** 455–457.

Landry, S.J., J. Zeirstra-Ryalls, O. Fayet, C. Georgopoulos, and L.M. Gierasch. 1993. Characterization of a functionally important mobile domain of GroES. *Nature* **364:** 255–258.

Langer, T., C. Lu, H. Echols, J. Flanagan, M.K. Hayer, and F.-U. Hartl. 1992. Successive action of DnaK, DnaJ and GroEL along the pathway of chaperone-mediated protein folding. *Nature* **356:** 683–689.

Leonhardt, S.A., K. Fearon, P.N. Danese, and T.L. Mason. 1993. HSP78 encodes a yeast mitochondrial heat shock protein in the Clp family of ATP-dependent proteases. *Mol. Cell. Biol.* **13:** 6304–6313.

Li, Z. and P.K. Srivastava. 1993. Tumor rejection antigen gp96/grp94 is an ATPase: Implications for protein folding and antigen presentation. *EMBO J.* **12:** 3143–3151.

Liberek, K., J. Marszalek, D. Ang, C. Georgopoulos, and M. Zylicz. 1991. The *Escherichia coli* DnaJ and GrpE heat shock proteins jointly stimulate DnaK's ATPase activity. *Proc. Natl. Acad. Sci.* **88:** 2874–2878.

Lis, J. and C. Wu. 1993. Protein traffic on the heat shock promoter: Parking, stalling and trucking along. *Cell* **74:** 1–4.

Liu, A.Y.-C., Z. Lin, H.S. Choi, F. Sorhage, and B. Li. 1989. Attenuated induction of heat shock gene expression in aging diploid fibroblasts. *J. Biol. Chem.* **164:** 12037–12045.

Martin, J., A.L. Horwich, and F.-U. Hartl. 1992. Prevention of protein denaturation under heat stress by the chaperonin Hsp60. *Science* **258:** 995–998.

Martin, J., S. Geromanos, P. Tempst, and F.-U. Hartl. 1993a. Identification of nucleotide-binding regions in the chaperonin proteins GroEL and GroES. *Nature* **366:** 279–282.

Martin, J., M. Mayhew, T. Langer, and F.-U. Hartl. 1993b. The reaction cycle of GroEL and GroES in chaperonin-assisted protein folding. *Nature* **366:** 228–233.

McNew, J.A., K. Sykes, and J.M. Goodman. 1993. Specific cross-linking of the proline isomerase cyclophilin to a non-proline-containing peptide. *Mol. Biol. Cell* **4:** 223–232.

Melki, R., I.E. Vainberg, R.L. Chow, and N.J. Cowan. 1993. Chaperonin-mediated folding of vertebrate actin-related protein and γ-tubulin. *J. Cell Biol.* **122:** 1301–1310.

Merck, K.B., P.J. Groenen, C.E. Voorter, H.H.W.A. De, J. Horwitz, H. Bloemendal, and J.W.W. De. 1993. Structural and functional similarities of bovine α-crystallin and mouse small heat-shock protein. A family of chaperones. *J. Biol. Chem.* **268:** 1046–1052.

Mescas, J., P.E. Rouviere, J.W. Erickson, T.J. Donohue, and C.A. Gross. 1993. The activity of σE, an *Escherichia coli* heat-inducible σ-factor, is modulated by expression of outer membrane proteins. *Genes Dev.* **7:** 2618–2628.

Milarski, K. and R.I. Morimoto. 1986. Expression of human HSP70 during the synthetic phase of the cell cycle. *Proc. Natl. Acad. Sci.* **83:** 9517–9521.

Missiakas, D., C. Georgopoulos, and S. Raina. 1993. Identification and characterization of a new *Escherichia coli* gene dsbB, whose product is involved in the formation of disulfide bonds in vivo. *Proc. Natl. Acad. Sci.* **90:** 7084–7088.

———. 1994. The *Escherichia coli* dsbC (*xprA*) gene encodes a periplasmic protein involved in disulfide bond formation. *EMBO J.* (in press).

Morimoto, R.I. 1993. Cells in stress: Transcriptional activation of heat shock genes. *Science* **259:** 1409–1410.

Morimoto, R., A. Tissières, and C. Georgopoulos. 1990a. The stress response, function of protein, and perspectives. In *Stress proteins in biology and medicine* (ed. R.I. Morimoto et al.), pp. 1–36. Cold Spring Harbor Laboratory Press, Cold Spring Harbor, New York.

―――, eds. 1990b. *Stress proteins in biology and medicine.* Cold Spring Harbor Laboratory Press, Cold Spring Harbor, New York.

Nakai, A. and R.I. Morimoto. 1993. Characterization of a novel chicken heat shock transcription factor, HSF3, suggests a new regulatory pathway. *Mol. Cell. Biol.* **13:** 1983–1997.

Nakai, A., M. Satoh, K. Hirayoshi, and K. Nagata. 1992. Involvement of the stress proein HSP47 in procollagen processing in the endoplasmic reticulum. *J. Cell Biol.* **117:** 903–914.

Newcomb, J.R. and P. Cresswell. 1993. Characterization of endogenous peptide bound to purified HLA-DR molecules and their absence from invariant chain-associated α/β dimer. *J. Immunol.* **150:** 499–507.

Nieto-Sotelo, J., G. Wiederrecht, A. Okuda, and C.S. Parker. 1990. The yeast heat shock transcription factor contains a transcriptional activation domain whose activity is repressed under nonshock conditions. *Cell* **62:** 807–817.

Noiva, R. and W.J. Lennarz. 1992. Protein disulfide isomerase. *J. Biol. Chem.* **267:** 3553–3556.

Ou, W.-J., P.H. Cameron, D.Y. Thomas, and J.J.M. Bergeron. 1993. Association of folding intermediates of glycoproteins with calnexin during protein maturation. *Nature* **364:** 771–776.

Petko, L. and S. Lindquist. 1986. Hsp26 is not required for growth at high temperatures, nor for thermotolerance, spore development, or germination. *Cell* **45:** 885–894.

Picard, D., B. Khursheed, M.J. Garabedian, M.G. Fortin, S. Lindquist, and K.R. Yamamoto. 1990. Reduced levels of hsp90 compromise steroid receptor action in vivo. *Nature* **348:** 166–168.

Pinhasi-Kimhi, O., D. Michalovitz, A. Ben-Zeev, and M. Oren. 1986. Specific interactions between the p53 cellular tumour antigen and major heat shock proteins. *Nature* **320:** 182–184.

Pratt, W.B. 1993. The role of heat shock proteins in regulating the function, folding, and trafficking of the glucocorticoid receptor. *J. Biol. Chem.* **268:** 21455–21458.

Pugsley, A.P. 1993. The complete general secretory pathway in gram-negative bacteria. *Microbiol. Rev.* **57:** 50–108.

Rabindran, S.K., G. Giorgi, J. Clos, and C. Wu. 1991. Molecular cloning and expression of a human heat shock factor, HSF1. *Proc. Natl. Acad. Sci.* **88:** 6906–6910.

Rabindran, S.K., R.I. Haroun, J. Clos, J. Wisniewski, and C. Wu. 1993. Regulation of heat shock factor trimer formation: Role of a conserved leucine zipper. *Science* **259:** 230–234.

Randall, L.L. 1992. Peptide binding by chaperone SecB: Implications for recognition of nonnative structure. *Science* **257:** 241–245.

Rospert, S., B.S. Glick, P. Jeno, G. Schatz, M.J. Todd, G.H. Lorimer, and P.V. Viitanen. 1993. Identification and functional analysis of chaperonin 10, the groES homolog from yeast mitochondria. *Proc. Natl. Acad. Sci.* **90:** 10967–10971.

Sanchez, Y. and S.L. Lindquist. 1990. HSP104 required for induced thermotolerance. *Science* **248:** 1112–1115.

Sanchez, Y.J., K.A. Taulien, K.A. Borkovich, and S. Lindquist. 1992. Hsp104 is required for tolerance to many forms of stress. *EMBO J.* **11:** 2357–2364.

Sanchez, Y., D.A. Parsell, J. Taulien, J.L. Vogel, E.A. Craig, and S. Lindquist. 1993. Genetic evidence for a functional relationship between Hsp104 and Hsp70. *J. Bacteriol.* **175:** 6484–6491.

Sarge, K.D., S.P. Murphy, and R.I. Morimoto. 1993. Activation of heat shock gene transcription by HSF1 involves oligomerization, acquisition of DNA binding activity, and nuclear localization and can occur in the absence of stress. *Mol. Cell. Biol.* **13:** 1392–1407.

Sarge, K.D., O.-K. Park-Sarge, J.D. Kirby, K.E. Mayo, and R.I. Morimoto. 1994. Regulated expression of heat shock factor 2 in mouse testis: Potential role as a regulator of hsp gene expression during spermatogenesis. *Biol. Reprod.* (in press).

Sarge, K.D., V. Zimarino, K. Holm, C. Wu, and R. I. Morimoto. 1991. Cloning and characterization of two mouse heat shock factors with distinct inducible and constitutive DNA-binding ability. *Genes Dev.* **5:** 1902–1911.

Scharf, K.-D., S. Rose, W. Zott, F. Schoff, and L. Nover. 1990. Three tomato genes code for heat stress transcription factors with a remarkable degree of homology to the DNA-binding domain of the yeast HSF. *EMBO J.* **9:** 4495–4501.

Schonbrunner, E.R. and F.X. Schmid. 1992. Peptidyl-prolyl cis-trans isomerase improves the efficiency of protein disulfide isomerase as a catalyst of protein folding. *Proc. Natl. Acad. Sci.* **89:** 4510–4513.

Schuetz, T.J., G.J. Gallo, L. Sheldon, P. Tempst, and R.E. Kingston. 1991. Isolation of a cDNA for HSF2: Evidence for two heat shock factor genes in humans. *Proc. Natl. Acad. Sci.* **88:** 6910–6915.

Shaknovich, R., G. Shue, and D.S. Kohtz. 1992. Conformational activation of a basic helix-loop-helix protein (MyoD1) by the C-terminal region of murine HSP90 (HSP84). *Mol. Cell. Biol.* **12:** 509–568.

Shevchik, V.E., G. Condemine, and J. Robert-Baudony. 1994. Characterization of DsbC, a periplasmic protein of *E. chrysanthemi* and *E. coli* with disulfide isomerase activity. *EMBO J.* (in press).

Shinde, U. and M. Inouye. 1993. Intramolecular chaperones and protein folding. *Trends Biochem. Sci.* **18:** 442–446.

Sistonen, L., K.D. Sarge, and R.I. Morimoto. 1994. Human heat shock factors 1 and 2 are differentially activated and can synergistically induce HSP70 gene transcription. *Mol. Cell. Biol.* (in press).

Sistonen, L., K.D. Sarge, B. Phillips, K. Abravaya, and R. Morimoto. 1992. Activation of heat shock factor 2 during hemin-induced differentiation of human erythroleukemia cells. *Mol. Cell. Biol.* **12:** 4104–4111.

Smith, D.F., W.P. Sullivan, T.N. Marion, K. Zaitsu, B. Madden, D.J. McCormick, and D.O. Toft. 1993. Identification of a 60-kDa stress-related protein, p60, which interacts with hsp90 and hsp70. *Mol. Cell. Biol.* **13:** 869–876.

Squires, C.L., S. Pedersen, B.M. Ross, and C. Squires. 1991. ClpB is the *Escherichia coli* heat shock protein HtpG. *J. Bacteriol.* **173:** 4254–4262.

Stamnes, M.A., B.-H. Shieh, L. Chuman, G.L. Harris, and C.S. Zuker. 1991. The cyclophilin homolog ninaA is a tissue-specific integral membrane protein required for the proper synthesis of a subset of *Drosophila* Rhodopsins. *Cell* **65:** 219–227.

Sykes, K., M.-J. Gething, and J. Sambrook. 1993. Proline isomerases function during heat shock. *Proc. Natl. Acad. Sci.* **90:** 5853–5857.

Trent, J.D., E. Nimmesgern, J.S. Wall, F.-U. Hartl, and A.L. Horwich. 1991. A molecular chaperone from a thermophilic archaebacterium is related to the eukaryotic protein t-complex polypeptide. *Nature* **354:** 490–493.

Udono, H. and P.K. Srivastava. 1993. Heat shock protein 70-associated peptides elicit specific cancer immunity. *J. Exp. Med.* **178:** 1391–1396.

Ursic, D. and M.R. Culbertson. 1991. The yeast homolog to mouse Tcp-1 affects microtubule-mediated processes. *Mol. Cell Biol.* **11:** 2629–2640.

Wang, N., S. Gottesman, M.C. Willingham, M.M. Gottesman, and M.R. Maurizi. 1993. A human mitochondrial ATP-dependent protease that is highly homologous to bacterial Lon protease. *Proc. Natl. Acad. Sci.* **90:** 11247–11251.

Wattiau, P. and G.R. Cornelis. 1993. SycE, a chaperone-like protein of *Yersinia enterocolitica* involved in the secretion of YopE. *Mol. Microbiol.* **8:** 123–131.

Westwood, J.T. and C. Wu. 1993. Activation of *Drosophila* heat shock factor: Conformational change associated with a monomer-to-trimer transition. *Mol. Cell. Biol.* **13:** 3481–3486.

Westwood, J.T., J. Clos, and C. Wu. 1991. Stress-induced oligomerization and chromosomal relocalization of heat-shock factor. *Nature* **353:** 822–827.

Wetzstein, M., U. Volker, J. Dedio, S. Lobau, and U. Zuber. 1992. Cloning, sequencing, and molecular analysis of the *dnaK* locus from *Bacillus subtilis. J. Bacteriol.* **174:** 3300–3310.

Wickner, S., J. Hoskins, and K. McKenney. 1991. Monomerization of RepA dimers by heat shock proteins activates binding to DNA replication origin. *Proc. Natl. Acad. Sci.* **88:** 7903–7907.

Wiech, H., J. Buchner, R. Zimmermann, and U. Jakob. 1992. Hsp90 chaperone protein folding in vitro. *Nature* **358:** 169–170.

Wiech, H., J. Buchner, M. Zimmermann, R. Zimmermann, and U. Jakob. 1993. Hsc70, immunoglobulin heavy chain binding protein, and hsp90 differ in their ability of stimulate transport of precursor proteins into mammalian microsomes. *J. Biol. Chem.* **268:** 7414–7421.

Wild, J., E. Altman, T. Yura, and C.A. Gross. 1992. DnaK and dnaJ heat shock proteins participate in protein export in *Escherichia coli. Genes Dev.* **6:** 1165–1172.

Winther, J.R. and P. Sorensen. 1991. Propeptide of carboxypeptidase Y provides a chaperone-like function as well as inhibition of the enzymatic activity. *Proc. Natl. Acad. Sci.* **88:** 9330–9334.

Wojkowiak, D., C. Georgopoulos, and M. Zylicz. 1993. Isolation and characterization of ClpX, a new ATP-dependent specificity component of the Clp protease of *Escherichia coli. J. Biol. Chem.* **268:** 22609–22617.

Wu, B.J. and R.I. Morimoto. 1985. Transcription of the human HSP70 gene is induced by serum stimulation. *Proc. Natl. Acad. Sci.* **82:** 6070–6074.

Xu, Y. and S. Lindquist. 1993. Heat-shock protein hsp90 governs the activity of pp60[v-src] kinase. *Proc. Natl. Acad. Sci.* **90:** 7074–7078.

Young, D.B., A. Mehlert, and D.F. Smith. 1990. Stress proteins and infectious diseases. In *Stress proteins in biology and medicine* (ed. R.I. Morimoto et al.), pp. 131–166. Cold Spring Harbor Laboratory Press, Cold Spring Harbor, New York.

Ziemienowicz, A., D. Skowyra, J. Zeilstra-Ryalls, O. Fayet, C. Georgopoulos, and M. Zylicz. 1993. Either of the *Escherichia coli* GroEL/GroES and DnaK/DnaJ/GrpE chaperone machines can reactivate heat-treated RNA polymerase: Different mechanisms for the same activity. *J. Biol. Chem.* **268:** 25425–25431.

2

Cytosolic hsp70s of *Saccharomyces cerevisiae*: Roles in Protein Synthesis, Protein Translocation, Proteolysis, and Regulation

Elizabeth A. Craig, Bonnie K. Baxter, Jörg Becker, John Halladay, and Thomas Ziegelhoffer
Department of Biomolecular Chemistry
University of Wisconsin, Madison
Madison, Wisconsin 53706

I. INTRODUCTION

The 70-kD heat shock proteins, or hsp70s, are highly conserved in all organisms studied so far, from bacteria to yeast to humans. Eukaryotes, including the budding yeast *Saccharomyces cerevisiae*, encode multiple hsp70s in their genomes. These related proteins are localized to a variety of cellular compartments, including the cytosol, mitochondria, and endoplasmic reticulum (ER). Functionally, the organellar hsp70s are better understood, having major roles in protein translocation and folding. The functions of the cytosolic hsp70s have been more difficult to define, perhaps because these proteins are involved in multiple processes, including translation, protein translocation, protein folding, and regulation of the heat shock response. This chapter reviews evolutionary analyses and genetic data concerning the roles of these proteins in the cytosol.

The Biology of Heat Shock Proteins and Molecular Chaperones
©1994 Cold Spring Harbor Laboratory Press 0-87969-427-0/94 $5 + .00

II. EVOLUTION OF THE HSP70 MULTIGENE FAMILY

To gain a better understanding of the evolutionary relationships among hsp70s across the biological spectrum, a comparison of 36 hsp70s from 25 diverse genera was conducted (Boorstein et al. 1994). The analysis, carried out by both distance-matrix and character-state methods, showed that the eukaryotic hsp70s comprise four distinct clusters (see Fig. 1). These clusters correspond to the intracellular localization of the proteins: the cytosol, the ER, mitochondria, and chloroplasts.

Analysis of the comparisons revealed that eukaryotic hsp70s appear to have evolved from ancestral genes by two types of mechanisms. Mitochondrial and chloroplast hsp70s appear to be derived from the establishment of an endosymbiotic relationship between a eukaryotic host and bacterial cells. Mitochondrial hsp70s are encoded in the nuclei, but they share the highest degree of identity (60–77%) with the purple bacteria, believed to be the endosymbiotic progenitors of mitochondria (Sagan 1967; Woese 1987; Cegegren et al. 1988). The mitochondrial hsp70s are only 47–53% identical to the eukaryotic hsp70s of the ER or the cytosol.

Divergence following gene duplication is the likely origin of the hsp70s of the cytosol and ER, as these two classes of hsp70s are more closely related to each other than to hsp70s from prokaryotes, chloroplasts, or mitochondria. Interestingly, when the cytosolic hsp70s of *S. cerevisiae* (the *SSA* and *SSB* subfamilies) were compared with the ER-localized hsp70 Kar2p, similarities and differences sorted out by domains of the proteins. The hsp70s have two domains: the amino-terminal ATPase domain and a carboxy-terminal peptide-binding domain (see McKay et al., this volume), connected by a proteolysis-sensitive region (Chappell et al. 1987). Each domain was aligned by matrix and character-based methods (Boorstein et al. 1994). The *SSA* subfamily and *SSB* subfamily ATPase domains were more similar to each other than to the ATPase domain of Kar2p (the amino acid identity of ATPase domains: *SSA1* and *SSB1*, 71.5%; *SSA1* and *KAR2*, 66.8%; *SSB1* and *KAR2*, 63.4%). In contrast, the *SSA* subfamily and Kar2p peptide-binding domains were closer to each other than to the domain of the *SSB* subfamily (*SSA1* and *KAR2*, 63.8%; *SSA1* and *SSB1*, 47.6%; *SSB1* and *KAR2*, 44.7%). This is an intriguing finding in light of the fact that Ssa proteins and Kar2p have both been implicated in the translocation of proteins from the cytosol into the ER (Deshaies et al. 1988; Sanders et al. 1992). Both proteins likely bind to the same polypeptides in assisting this translocation: Ssa proteins on the cytosolic side of the ER membrane and Kar2p on the lumenal side. The similarity in their peptide-binding domains could thus be explained by the need to interact with the same

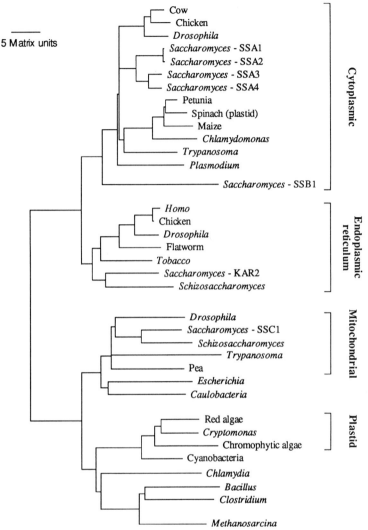

5 Matrix units

Figure 1   Distance-matrix-based phylogenetic trees depicting relationships among 36 hsp70s from 25 different species. The tree was constructed by the distance-matrix methods based on a common progressive alignment of sequences (Fitch and Margoliash 1967; Feng and Doolittle 1990). Intracellular localizations of the eukaryotic hsp70s analyzed are indicated by the square brackets at the right of the figure. (Reprinted, with permission, from Boorstein et al. 1994.)

substrates. In like manner, the similarity in the ATPase domains of the cytosolic *SSA* and *SSB* hsp70 subfamilies may reflect an interaction of both subfamilies with the same or similar regulatory proteins in the cytosol.

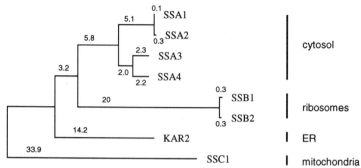

Figure 2 Phylogenetic tree of the *S. cerevisiae* hsp70 subfamilies. The tree was constructed as described in the legend to Fig. 1. (Reprinted, with permission, from Boorstein et al. 1994.)

The comparisons also indicate that the cytosolic subfamilies and Kar2p arose from events that occurred near the time of the eukaryotic/ prokaryotic divergence, estimated to have been from 1.4 to 3.0 billion years ago (Doolittle et al. 1989 and references therein). Cytosolic and organellar hsp70s are known to be important for the transport of proteins across cellular membranes (Deshaies et al. 1988; Kang et al. 1990; Vogel et al. 1990). The critical role of hsp70s in protein translocation into organelles, combined with the similarity in timing of the prokaryotic/ eukaryotic divergence and the divergence of the ER, mitochondria, and plastid hsp70 families, suggests that the ancestral hsp70s may have been important in overcoming the barriers to intracellular protein distribution created by intracellular membranes.

III. CYTOSOLIC HSP70S

Like other eukaryotes, the budding yeast *S. cerevisiae* has cytosolic, mitochondrial, and ER-localized hsp70 proteins (see Fig. 2; for a review, see Craig et al. 1993). We focus here on the cytosolic subfamilies, *SSA* (four members) and *SSB* (two members). The evolutionary analysis described above (Boorstein et al. 1994) revealed that the *SSA* subfamily of cytosolic chaperones in *S. cerevisiae* was more closely related to cytosolic hsp70s from all of the other organisms studied than to the *SSB* subfamily of yeast (see Fig. 1). The *SSB* subfamily is therefore evolutionarily distinct, representing an ancient divergence from the other known cytosolic hsp70s. The *SSA* and *SSB* families are also functionally distinct: Overexpression of an Ssa protein fails to rescue the phenotype of a strain deficient in Ssb proteins, and overexpression of an Ssb protein fails to rescue an Ssa-protein-deficient strain (Craig and Jacobsen 1985).

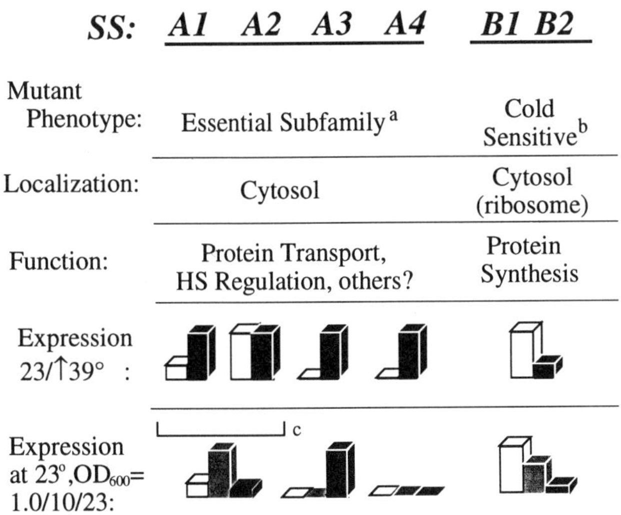

SS: *A1   A2   A3   A4   B1 B2*

Mutant Phenotype:	Essential Subfamily[a]	Cold Sensitive[b]
Localization:	Cytosol	Cytosol (ribosome)
Function:	Protein Transport, HS Regulation, others?	Protein Synthesis

Expression $23/\uparrow39°$:

Expression at $23°$, OD_{600} = 1.0/10/23:

Figure 3 Summary of the *SSA* and *SSB* gene families. Expression at 23ºC, OD_{600} = 1.0/10/23 depicts the relative level of expression of each gene or gene subfamily (in the case of the *SSB* genes) as cells progress from rapid exponential growth (OD_{600} = 1.0, open columns), through diauxic shift (D_{600} = 10, gray columns), to stationary phase (OD_{600} = 23, closed columns). (*a*) See Table 1 for a more complete description of *ssa* phenotypes. (*b*) Strains lacking both *SSB* genes grow poorly at most temperatures but grow progressively more slowly (relative to wild-type) as the temperature is reduced. (*c*) The probe used in these experiments was incapable of distinguishing between the nearly identical *SSA1* and *SSA2* genes.

IV. SSA PROTEINS

The four Ssa proteins comprise an essential subfamily in yeast; at least one of the proteins encoded by the subfamily must be present at high levels for cell viability (Werner-Washburne et al. 1987). Two of the *SSA* genes, *SSA1* and *SSA2*, are expressed under normal growth conditions (see Fig. 3; Table 1). Expression of *SSA1* is further induced by stress, whereas *SSA2* levels remain essentially unchanged. *SSA3* and *SSA4* are not expressed during vegetative growth but are induced by stress. The four Ssa proteins share 84–99% identity at the amino acid level.

Strains in which both *SSA1* and *SSA2* have been inactivated are being used to study the cellular functions of the Ssa proteins. The effects of this double disruption are pleiotropic. *ssa1ssa2* strains grow more slowly than wild-type strains at temperatures ranging from 23ºC to 35ºC and are unable to form colonies at 37ºC (see Table 1) (Craig and Jacobsen 1984). This phenotype has been useful for the isolation of extragenic and multi-copy suppressors (see below). Other phenotypes include constitutive

Table 1 Effect of Various Combinations of *SSA* Subfamily Disruption Mutations On Growth

Genotype	Phenotype
ssa1	temperature sensitive for growth on nonfermentable carbon sources at 37°C[a]
ssa2	indistinguishable from wild type[b]
ssa1ssa2	constitutively thermotolerant; reduced growth rate at permissive temperatures; temperature sensitive for growth at 37°C [b]
ssa3	indistinguishable from wild type[c]
ssa4	indistinguishable from wild type[c]
ssa3ssa4	indistinguishable from wild type[c]
ssa1ssa2ssa3	indistinguishable from *ssa1ssa2*[c]
ssa1ssa2ssa4	inviable[c]
ssa1ssa2ssa3ssa4	inviable[c]

Reprinted, with permission, from Boorstein et al. 1994.
[a]R.J. Nelson and E.A. Craig (unpubl.).
[b]Craig and Jacobsen (1984).
[c]Werner-Washburne et al. (1987).

thermotolerance (i.e., enhanced ability to survive transient exposures to high temperature), increased sensitivity to the amino acid analog canavanine (which is believed to be incorporated into proteins in vivo, resulting in abnormal and misfolded proteins), and elevated constitutive expression of some heat shock genes. *SSA3* and *SSA4* are among the genes whose expression is increased in this strain, although they are not expressed highly enough to bring Ssa protein to wild-type levels. Introduction of *SSA4* on a multicopy plasmid suppresses the temperature sensitivity of the strain, allowing colonies to form at 37°C.

Ssa proteins have previously been implicated in the translocation of proteins into the ER and mitochondria, since depletion of Ssa proteins leads to an accumulation of the precursor form of a mitochondrial protein, the β subunit of the F_1F_0 ATPase, as well as the secreted protein α factor. However, translocation of several other proteins seems to be unaffected (Deshaies et al. 1988; J. Becker and E. Craig, unpubl.). In addition, translocation even of those proteins shown to be affected is not completely blocked. A more severe block has been demonstrated, for example, with alleles of the mitochondrial hsp70 *SSC1*, which produce defective Ssc1p and yet allow cells to survive (Kang et al. 1990). The translocation defect resulting from a depletion of Ssa proteins therefore does not seem to be sufficient to explain the fact that these proteins constitute an essential subfamily. This suggests that other functions of the Ssa proteins may be critical for growth.

A. Regulation of the Heat Shock Response by the Ssa Proteins

One critical function that the *SSA* genes may perform is autoregulation of the heat shock response (see Fernandes et al.; Wu et al.; both this volume). Various groups have implicated cytosolic hsp70s in the negative regulation of heat shock gene expression (DiDomenico et al. 1982; Tilly et al. 1983; Craig and Gross 1991; Morimoto et al. 1993; Mosser et al. 1993). In yeast, overexpression of Ssa1p causes a reduction in the heat-inducible expression from both the *SSA1* and the *SSA4* promoters (Stone and Craig 1990). In addition, *ssa1ssa2* strains exhibit unusually high expression of *SSA4* and some other heat shock genes under optimal growth conditions. The increased expression of *SSA4* appears to be due to activation of the heat shock transcription factor (HSF); the HSF-binding site of the *SSA4* promoter, its heat shock element (HSE), is sufficient to drive high levels of expression from a heterologous promoter in an *ssa1ssa2* strain but not in a wild-type strain (Boorstein and Craig 1990).

Further evidence for a regulatory role of the Ssa proteins came from a search for extragenic suppressors of the temperature-sensitive phenotype of an *ssa1ssa2* strain (Nelson et al. 1992a). A spontaneous mutant was isolated that allows more rapid growth of *ssa1ssa2* strains at 23°C and 30°C; its locus was designated *EXA3*. *EXA3* is very closely linked to *HSF1*, the gene encoding the heat shock transcription factor HSF, suggesting that a mutation in *HSF1* causes suppression. Indeed, at least one mutation is present in the *HSF1* gene from an *EXA3-1* strain as compared to the wild-type strain.

The initial expectation was that a mutation in HSF that suppressed the growth defect of *ssa1ssa2* would lead to increased HSF activity and thus higher levels of the remaining Ssa proteins. When the heat shock response of an *EXA3-1* strain was studied, however, it was found that the increase in the rate of hsp70 expression upon heat shock is slower in this *EXA3* mutant than in wild-type cells, suggesting that the *EXA3-1* mutation may cause a decrease in HSF activity. In addition, introduction of an extra copy of HSF into an *ssa1ssa2* strain is detrimental for growth, further suggesting that HSF is, if anything, too active in these strains and that down-regulation could be advantageous. It is clear that HSF is more active in an *ssa1ssa2* strain than in wild-type, as indicated by the elevated HSE-driven transcription mentioned above. This elevated activity could be explained by assuming an autoregulatory role for Ssa1p and Ssa2p in the heat shock response; the absence of negative feedback from these proteins leads to a constitutively high rate of heat shock gene transcription. This response presumably allows cells to survive by elevating expression of *SSA3* and *SSA4*, but the global overexpression of other heat shock genes may well be detrimental under normal growth conditions. A delicate balance must therefore be struck to allow sufficient hsp70 expression without detrimental overexpression of other heat shock proteins. A down-regulation of HSF activity might suppress the defect of *ssa1ssa2* cells by improving this balance. The suppression of the *ssa1ssa2* growth defect by the *EXA3-1* mutation thus appears to reflect a defect in the global regulation of the heat shock response in this strain, which can be partially rescued by alteration of the heat shock transcription factor itself.

As mentioned above, previous data have implicated the hsp70s in regulation of the heat shock response. In addition, it has been suggested before that an hsp70 mutant could be rescued by altering the level of heat shock gene expression. Bukau and Walker (1990) identified suppressors in *Escherichia coli* of a deletion of the single known hsp70-encoding gene in that organism, *dnaK*. The suppressors allowed improved growth at 30°C, a permissive temperature for this slow-growing, temperature-sensitive strain. One complementation group contained mutants of the σ factor, σ^{32}, which is responsible for directing RNA polymerase to heat shock promoters upon temperature upshift. Interestingly, these σ^{32} mutants appeared to have either decreased activity or decreased stability, resulting in lowered expression of heat shock genes. This result led these researchers to propose that, at least in part, the slow growth of strains deficient in DnaK was due to overexpression of heat shock proteins.

In summary, there are many lines of evidence implicating the cytosolic *SSA* subfamily of hsp70s in autoregulation of the heat shock response. More work remains to be done to clarify the nature of the constitutive heat shock response of *ssa1ssa2* mutants and to elucidate the alterations that the *EXA3-1* mutation causes in HSF activity. The specific interactions responsible for autoregulation remain to be identified in yeast and in other organisms. However, evidence is mounting that autoregulation does occur and that interference with it can be detrimental for the cell.

B. A Possible Interaction between the Ssa Proteins and the Ubiquitin-dependent Proteolysis System

Possible evidence for yet another role of the Ssa proteins came from analysis of a multicopy suppressor of the temperature-sensitive phenotype of *ssa1ssa2* strains. A yeast library constructed on a multicopy plasmid was transformed into an *ssa1ssa2* strain, and plasmids were isolated that allow improved growth at 30–35°C. Upon further analysis, one suppressor plasmid was found to contain *UBP3*, which was isolated by

Baker et al. (1992) as a gene encoding a ubiquitin-processing protease. The isolation of this plasmid-borne suppressor suggested that there might be a functional connection between the Ssa proteins and ubiquitin-dependent proteolysis.

To understand a possible relationship between the hsp70s and the ubiquitin system, a brief review of this proteolysis pathway is called for (see Fig. 4). The ubiquitin system, extremely highly conserved in all eukaryotes studied so far, is a major pathway for selective protein degradation and is believed to be responsible for the degradation of abnormal and short-lived proteins (for reviews, see Hershko and Ciechanover 1992; Hochstrasser 1992; Jentsch 1992). Ubiquitin itself is a 76-amino-acid protein that when attached in a multimeric chain to a protein substrate targets that protein for degradation. In the first step of this pathway, ubiquitin is activated in an ATP-dependent reaction by a ubiquitin-activating enzyme (E1). The activated ubiquitin molecule is then transferred to a ubiquitin-conjugating enzyme (E2). Finally, the activated ubiquitin molecule is transferred to a protein substrate, sometimes with the aid of a ubiquitin-protein ligase or recognin (E3). An isopeptide linkage is formed between the carboxyl terminus of the ubiquitin molecule and an internal lysine residue of the substrate protein. Targeting for degradation requires the presence of a multi-ubiquitin chain on a substrate, in which many ubiquitin molecules are linked together with isopeptide linkages by a ubiquitin-conjugating enzyme. This chain of ubiquitin molecules allows the substrate protein to be recognized and degraded by the 26S protease, a large, multisubunit, ATP-dependent proteolytic complex in the cytosol. The ubiquitin molecules are then released and recycled for reactivation and reuse.

Many of the components of this system have been identified in *S. cerevisiae*, and their genes have been cloned and sequenced. Ubiquitin is encoded by four genes, *UBI1* through *UBI4* (Özkaynak et al. 1987), and each encodes a polyprotein. In the case of *UBI1*, *UBI2*, and *UBI3*, which are expressed under optimal growth conditions and down-regulated after a heat shock, ubiquitin is encoded as a fusion to a ribosomal protein. *UBI4*, which is stress-inducible and is responsible for the production of ubiquitin under stress conditions (Finley et al. 1987), encodes a pentameric head-to-tail ubiquitin repeat. Many of the genes encoding ubiquitin-conjugating enzymes have also been cloned (Hochstrasser 1992). Expression of two of these genes, *UBC4* and *UBC5*, is known to be induced by heat shock (Seufert and Jentsch 1990).

Four ubiquitin-processing proteases have been identified so far in *S. cerevisiae*: *YUH1*, *UBP1*, *UBP2*, and *UBP3* (Miller et al. 1989; Tobias and Varshavsky 1991; Baker et al. 1992). All four were isolated on the

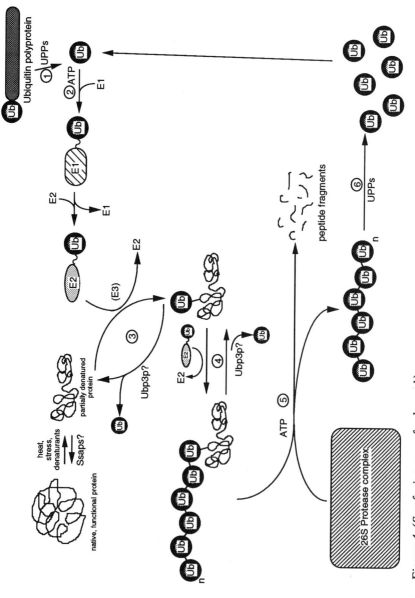

Figure 4 (See facing page for legend.)

basis of their ability to cleave ubiquitin from carboxy-terminal extensions, ranging from short peptides to β-galactosidase in the different screens employed. The in vivo roles of these proteases are not yet clear. There are at least two places in the ubiquitin pathway that require a ubiquitin protease activity: the release of free ubiquitin from its natural carboxy-terminal extensions during or after translation, and the release of ubiquitin from its protein substrates at the 26S protease complex. Either of these activities would presumably facilitate cellular proteolysis by enhancing the efficiency of the ubiquitin system. However, proteases could also be required to serve a proofreading function for the system, removing ubiquitin from incorrect substrates before they can be degraded. This would actually decrease the overall rate of proteolysis while presumably enhancing its precision and preventing errors. Overexpression of a particular ubiquitin-processing protease such as Ubp3p could therefore either increase or decrease the rate of cellular protein degradation, depending on the site of action of the particular protease.

In considering whether higher or lower rates of ubiquitin-dependent proteolysis would be likely to improve the growth of our *ssa1ssa2* mutant strain, two contradictory models were apparent. It is likely that the relative depletion of hsp70 in this strain results in the accumulation of misfolded proteins in the cytosol. An increase of proteolysis might ameliorate this situation by helping to rid the cell of these proteins before they can seriously impair cell growth. Alternatively, it is possible that some of these misfolded proteins could eventually fold correctly and

Figure 4 Model of the suggested roles for the cytosolic *SSA* hsp70 proteins and the ubiquitin-processing proteases (UPPs) in the ubiquitin proteolysis pathway. From top right: (1) Ubiquitin (Ub) is encoded as a polyprotein, which must be processed by a UPP to yield free ubiquitin. (2) Ubiquitin is activated in an ATP-dependent reaction by a ubiquitin-activating enzyme (E1). Activated Ub is then passed to a ubiquitin-conjugating enzyme or (E2). (3) Ubiquitin is attached to a substrate protein via an isopeptide linkage, sometimes with the aid of ubiquitin-protein ligase, recognin, or E3. (4) A multi-ubiquitin chain is built on the substrate by ubiquitin-linked E2s. (5) The 26S protease complex recognizes the multi-ubiquitinated substrate protein and degrades it, releasing the ubiquitin chain. (6) The ubiquitin chain is processed by a UPP to release free ubiquitin, which re-enters the cycle. UPPs may also act as proofreaders of the ubiquitin pathway, counteracting steps 3 and 4 when the substrate is inappropriate. The Ssa proteins may act to help a partially denatured protein refold before it can be recognized by the ubiquitin system at all (*top left*). UPP is used here to represent any of the ubiquitin-processing proteases, whether encoded by *YUH1, UBP1, UBP2, UBP3,* or other as yet undiscovered genes.

function, either by themselves or with the aid of the remaining Ssa proteins, but that they are being targeted for destruction before this can occur. In this case, a decrease in the overall rate of proteolysis could be useful for growth. We attempted to discriminate between these two possibilities by artificially increasing the rate of ubiquitin-dependent proteolysis in our *ssa1ssa2* strain by the introduction of a multicopy plasmid bearing the gene encoding polyubiquitin, *UBI4*. Introduction of this plasmid in *ssa1ssa2* led to an increase in the levels of both free ubiquitin and ubiquitin conjugates, yet failed to suppress the temperature-sensitive growth defect. Indeed, *ssa1ssa2* strains transformed with *UBI4* grew more slowly than the parent strains. This suggested that proteolysis may already be too elevated in *ssa1ssa2*. In addition, a deletion of the *UBP3* gene in either a wild-type or an *ssa1ssa2* background leads to the accumulation of high-molecular-weight ubiquitin conjugates, suggesting that *UBP3* acts to remove ubiquitin from protein conjugates, thus preventing their destruction. Overexpressed Ubp3p therefore appears to improve the growth of *ssa1ssa2* strains by decreasing the rate of protein degradation.

The discovery of *UBP3* as a suppressor of an hsp70 defect is not the first time that a link has been suggested between the heat shock proteins or molecular chaperones and cellular proteolysis. As mentioned above, expression of several of the genes encoding components of the ubiquitin pathway in yeast, *UBI4*, *UBC4*, and *UBC5*, is induced by heat shock. Mutations in these genes lead to defects typical of heat shock protein mutants, including temperature sensitivity and constitutive thermotolerance (Finley et al. 1987; Seufert and Jentsch 1990). In *E. coli*, mutations in the heat shock genes *dnaK*, *dnaJ*, and *grpE* lead to alterations in the proteolysis of several test peptides, as well as altered degradation of the short peptide fragments produced by treatment of the cells with puromycin (Straus et al. 1988; Goldberg 1992; Sherman and Goldberg 1992). Current evidence, however, has not yet clarified the nature of the interactions involved.

In our system, it is possible that the data do not reflect a direct connection between the heat shock proteins and proteolysis. Suppression of *ssa1ssa2* by *UBP3* could be yet another indication of the defect in heat shock regulation in this strain. Since several components of the ubiquitin pathway are known to be induced by heat shock, and at least some heat shock proteins are constitutively induced in an *ssa1ssa2* background, it would be reasonable to assume that *UBI4*, *UBC4*, and *UBC5* are overexpressed in an *ssa1ssa2* background as well, leading to a detrimentally elevated rate of ubiquitin-dependent proteolysis. Overexpression of *UBP3* could cause suppression simply by countering the overexpression of these other genes. However, preliminary analyses by Northern blot do not support this hypothesis: Whereas mRNA levels of *SSA4* and other heat shock genes are drastically elevated in an *ssa1ssa2* background, those of *UBC4* and *UBC5* are not noticeably increased. Total ubiquitin message from the four ubiquitin-encoding genes also remains unchanged. It is possible that these messages are elevated enough to perturb the system without the differences being clear by Northern analysis, but the absence of a clear induction has led us to formulate other hypotheses. We speculate that suppression by *UBP3* does not reflect a regulatory defect in the heat shock response, but instead indicates a direct interaction between the systems of heat shock and proteolysis. This could be explained through a molecular chaperone function of the Ssa proteins.

Molecular chaperones such as the hsp70s have repeatedly been shown to interact with unfolded proteins. A protein that has been partially denatured or misfolded in vivo has two possible fates: It can be targeted for degradation or it can refold and continue to function. Much evidence has been accumulated to indicate that the ubiquitin pathway in eukaryotic cells is responsible for recognizing abnormal or misfolded proteins and destroying them. This function is clearly essential for normal growth. In some cases, however, a misfolded protein may be salvageable, perhaps through the binding and release of molecular chaperones such as the Ssa proteins. A balance would then exist between the degradation and salvage pathways, tuned to allow the rescue of some proteins while preventing the accumulation of others that are nonsalvageable. In the case of a strain lacking Ssa1p and Ssa2p and thus relatively depleted in *SSAs*, this equilibrium would be shifted toward the degradation of more of these temporarily denatured proteins, leading to an increase in proteolysis. This might result in the loss of proteins that could otherwise refold and function, and which would be necessary for optimal growth. Overexpression of *UBP3* in this situation would help to shift the balance back toward normal and thus allow better growth.

Much work remains to be done to elucidate the link between the ubiquitin system and the heat shock proteins. The data so far indicate that such a link exists; these two central systems are clearly interdependent in the cell. The precise interplay between the two, however, and the regulatory mechanisms that allow for balance between salvage and repair remain to be clarified. These early findings should help to point the way for future research.

The Ssa proteins have thus been shown to have a role in several diverse areas of normal growth. They are involved in the translocation of proteins into other cellular compartments such as mitochondria and the ER. They are important for the proper regulation of the global heat shock response. Finally, they may be important in protecting temporarily denatured or misfolded proteins from degradation by the ubiquitin system. Clearly, the *SSA* subfamily of cytosolic hsp70s has a number of vital functions.

V. SSB PROTEINS

The "*SSA* type" of cytosolic hsp70 is the only type of hsp70 identified in other eukaryotes so far; the Ssb proteins have only been studied in *S. cerevisiae*. The two Ssb proteins, encoded by *SSB1* and *SSB2*, are 99.3% identical to each other, and single *ssb* mutants have no phenotype that we have detected (Craig and Jacobsen 1985). The Ssb proteins share significant homology with the other six hsp70s of *S. cerevisiae*, but they are not induced by heat shock. Instead, they are most highly expressed under optimal growth conditions, when they are among the most abundant proteins in the cell. They are thus technically heat shock cognates rather than true heat shock proteins. The *SSB* subfamily is the only hsp70 subfamily in yeast that is not essential: An *ssb1ssb2* double mutant is viable, but slow-growing and cold-sensitive. This phenotype cannot be rescued by overexpression of Ssa1p, just as Ssb1p cannot rescue the *ssa1ssa2* phenotype (Craig and Jacobsen 1985). This lack of functional complementation between the two cytosolic hsp70 families suggests that the two classes of proteins have different roles in the cell.

There is evidence that the Ssb proteins are involved in the process of protein synthesis. Nelson et al. (1992b) reported that the majority of the Ssb protein in a yeast cell is associated with translating ribosomes. A large percentage of Ssb protein comigrated with monosomes (single ribosomes) and polysomes (ribosomes linked together by an mRNA) in a sucrose gradient (see Fig. 5). After incubation with RNase to destroy the mRNA linking the ribosomes, the polysome peaks collapsed to a single monosome peak at 80S. In this case, the peaks of Ssb proteins collapsed as well, so that the majority of Ssb proteins continued to migrate with the 80S ribosomes. These data strongly suggested that Ssb proteins are associated with ribosomes. Ssb proteins appear to bind only to translating ribosomes, however. This has been shown by treatment of cells with azide, which depletes ATP levels and prevents initiation of protein synthesis (Carter et al. 1980). Ribosomes in lysates from cells treated with azide therefore migrated as monosomes, just as those from RNase-treated extracts did. However, the Ssb proteins did not comigrate with the ribosomes in lysates from azide-treated cells but remained at the top of the gradient with soluble proteins.

The interaction between Ssb proteins and translating ribosomes could be due to binding between the Ssb proteins and the emerging nascent

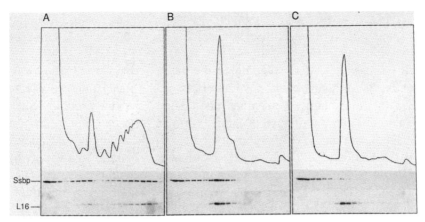

Figure 5 Comigration of ribosomes and Ssb proteins after treatment with RNase and the translation dependence of association of Ssb proteins with ribosomes. Extracts were prepared from wild-type cells; RNase was added to one third of the cells. To another one third of the cells, 10 mM sodium azide was added for 15 min prior to harvest to allow translating ribosomes to run off the mRNA and block reinitiation. The extracts were run on sucrose gradients, and tracings of the optical density at 260 nm were made. Fractions were collected, and the presence of Ssb proteins and ribosomal protein L16 were determined by immunodetection methods. (Reprinted, with permission, from Nelson et al. 1992b.)

polypeptide chain, as similar interactions have been reported for hsp70s from mammalian cells (Beckmann et al. 1990). If such an interaction were occurring, the Ssb proteins should be released from ribosomes by the antibiotic puromycin. Puromycin is an analog of aminoacyl-tRNA which is covalently added onto the growing end of the polypeptide chain, blocking further amino acid incorporation and resulting in the release of the nascent chain from the ribosome. To test the effect of puromycin on the association of Ssb proteins with ribosomes, yeast cell extracts were treated with the antibiotic and analyzed on sucrose gradients. This treatment left the polysomes intact but resulted in a dramatic reduction in the amount of the Ssb protein migrating with polysomes.

The effect of puromycin on the comigration of Ssb protein and ribosomes has more than one possible explanation. This result is consistent with the idea that Ssb proteins bind directly to the emerging polypeptide, perhaps serving to prevent its misfolding in the early stages of its synthesis. This is an appealing idea, because it fits well with the findings from many laboratories that hsp70s bind to unfolded proteins (Cyr et al. 1992; DeLuca-Flaherty et al. 1990; Flynn et al. 1989) and are thought to be important for maintaining a conformation necessary for successful folding of proteins in the cytosol and for translocation into organelles. However, it is possible that Ssb proteins do not associate directly with the nascent chain, but rather associate with translating ribosomes in a manner that is disrupted by either the binding of puromycin or the release of the nascent chain.

It is possible that by binding to the nascent chain or to ribosomes directly, Ssb proteins could have an effect on the translation process itself, rather than simply preventing misfolding of the new protein. To try to determine the importance of the Ssb protein–ribosome interaction on translation, we analyzed the phenotype of strains containing null alleles of both *SSB1* and *SSB2*. The sensitivity of *ssb1ssb2* strains to a variety of inhibitors of protein synthesis when compared to wild-type strains was determined. The *ssb1ssb2* strains were hypersensitive to several drugs, including paromomycin, hygromycin B, and G418, all of which are aminoglycoside antibiotics and act on the 40S ribosomal subunit to inhibit polypeptide chain elongation. The *ssb1ssb2* strain was also hypersensitive to verrucarin A, a member of the trichothecene family of antibiotics that blocks polypeptide bond formation and acts primarily on the 60S ribosomal subunit. However, *ssb1ssb2* cells are not hypersensitive to cycloheximide or anisomycin, both of which are known to act on the 60S subunit to inhibit peptide bond formation. Although not indicative of a particular mechanism, this susceptibility of *ssb1ssb2* strains to a variety of antibiotics strongly suggests that the Ssb class of hsp70 proteins affects the translation process per se, rather than exerting an indirect effect through protein folding.

The idea of a direct link between the Ssb proteins and the translation process was further supported by the results of a multicopy suppressor screen of an *ssb1ssb2* double mutant strain. An *ssb1ssb2* strain was transformed with a library of yeast genomic DNA. Transformants were selected for rapid growth at 23°C, a semipermissive temperature for growth of this strain. Apart from *SSB1* and *SSB2*, only one gene was isolated from the centromere-based library. This gene, named *HBS1* (for hsp70 subfamily B suppressor), was a previously unidentified gene having significant sequence similarity to the translation elongation factor EF-1α. EF-1α is a highly conserved protein found in all eukaryotes. The similarity between EF-1α and *HBS1* extends throughout the entire length of EF-1α, with an overall identity of 33%. Since the sequence similarity extends well beyond the three highly conserved sequence blocks found in many GTP-binding proteins, Hbs1p may well be functionally related to EF-1α and thus directly involved in translation.

In conclusion, the association of Ssb proteins with translating ribosomes, the fact that disruption of the *SSB* genes results in sensitivity to antibiotics known to inhibit translation, and the identification of a suppressor gene encoding a protein related to a translation factor all lead to the conclusion that Ssb proteins are necessary for normal protein synthesis. Although neither the role of Ssb proteins in the translation process nor that of Hbs1p in suppression is yet known, one can speculate based on an understanding of the nature of the interaction of hsp70s with polypeptides and the role of EF-1α in the translation process. Since Ssb proteins are very abundant proteins, and a large percentage are found associated with ribosomes, there is approximately enough hsp70 to allow binding of one hsp70 per nascent chain. One could imagine that an Ssb protein binds to a nascent chain as it emerges from the tunnel of the 60S subunit, preventing intramolecular interactions of the nascent chain with itself or with the surface of the ribosome and thus facilitating the movement of the remainder of the polypeptide through the tunnel as it is synthesized. However, an important question to address is whether a small proportion of the nascent chains are binding several hsp70s or each is binding one.

How might overexpression of Hbs1p suppress the growth defect of deletion of *SSB1* and *SSB2*? Since Hbs1p is related to EF-1α, it is useful to consider the role of EF-1α in translation. During the translocation step of translation, movement of the peptidyl tRNA opens the A site of the ribosome, allowing binding of the aminoacyl tRNA needed for formation of the next peptide bond. EF-1α mediates tRNA binding to the ribosome, releasing the amino acyl tRNA upon the hydrolysis of bound GTP. If Hbs1p is acting as an elongation factor like EF-1α, then overexpression may allow an increase in the rate of peptide bond formation, which may have become limiting in the absence of Ssb proteins. For example, the peptidyl ribosome complex might be distorted due to "backing up" of the nascent chain in the tunnel. On the other hand, Hbs1p may in fact inhibit activity of EF-1α by competing with it for a limiting factor, rather than functioning in the same reaction. In such a scenario, the resultant slowing of the rate of elongation may allow more time for the polypeptide to transit the tunnel of the 60S ribosome, thus alleviating problems at the exit site caused by lack of Ssb proteins. Until the function of Hbs1p and the function of Ssb proteins themselves are better understood, the mode of suppression by overproduction of Hbs1p must remain speculative.

VI. RELATIONSHIP BETWEEN THE SSA AND SSB CYTOSOLIC HSP70S

To date, the data accumulated indicates distinct functions for the two types of cytosolic hsp70s, Ssa and Ssb proteins. Initially, the phenotypes of disruption strains suggested that the Ssa and Ssb proteins might be very similar in function but expressed under different conditions. The *ssa1ssa2* strain was unable to grow at higher temperatures, a condition under which *SSB* expression is relatively low; growth of the *ssb1ssb2* strain was compromised at lower temperatures, a condition under which *SSA* expression is relatively low. However, *SSB1* could not suppress the *ssa1ssa2* growth defect, even when placed under the control of the *SSA2* promoter so that its expression would not decrease at higher growth temperatures. In addition, *SSA1* could not suppress the cold sensitivity of the *ssb1ssb2* mutant even when under the control of the *SSB1* promoter such that the expression would not decrease at lower temperature. This lack of phenotypic suppression clearly indicates that Ssa and Ssb proteins are performing different functions. The differences in function between the two cytosolic hsp70 families may at least in part be due to differences in the polypeptide-binding properties of the two classes of proteins. The first experiments designed to address this question suggest that although Ssa proteins are able to uncoat clathrin-coated vesicles, Ssb proteins have almost no uncoating activity (Gao et al. 1991). In addition, localization may also be a factor. Up to 73% of Ssb protein, but no more that 10% of Ssa protein, has been found to be associated with polysomes.

Many questions remain to be answered concerning the interplay between these two types of hsp70s in the shepherding of a polypeptide from a translating ribosome to its final destination in the cell. Ssb proteins certainly are involved in a very early event at the ribosome, as their absence seems to affect the elongation of the nascent chain. Ssa proteins, on the other hand, are needed for efficient translocation of at least some proteins into mitochondria and the ER and may be involved in rescuing incompletely folded proteins from the proteolytic machinery. It is easy to envision how both Ssa and Ssb proteins might bind to polypeptides at different stages of their maturation. Perhaps Ssb proteins bind to the amino terminus of nascent chains as they emerge from the ribosome, facilitating the translation process, whereas Ssa proteins bind to internal sites either while the protein is still attached to the ribosome or shortly after translation is completed. Alternatively, Ssa and Ssb proteins may bind to different subsets of proteins. For example, Ssa proteins might preferentially bind to proteins destined for translocation into the ER and mitochondria. Binding of Ssa proteins may also prevent degradation either by directly facilitating folding into an active, proteolytic-resistant conformation or by directly preventing access of proteases to the unfolded chains.

VII.   DNAJ-LIKE PROTEINS OF THE CYTOSOL

hsp70s do not function alone. In *E. coli*, DnaK, an hsp70 homolog, acts with two other heat shock proteins, DnaJ and GrpE (see Georgopoulos, this volume). The *dnaJ* gene of *E. coli* is located in an operon with *dnaK*, and considerable evidence indicates that the proteins encoded by these two genes function together in a variety of processes including bacteriophage λ replication and regulation of the heat shock response. Eukaryotes have a number of DnaJ homologs (for review, see Caplan et al. 1993; Silver and Way 1993). The best studied is the *SEC63* gene product (see Brodsky and Shekman, this volume), an integral ER membrane protein of *S. cerevisiae*. *S. cerevisiae* has two identified cytosolic DnaJ homologs, the products of the *SIS1* (Luke et al. 1991) and *YDJ1* (*MAS5*) genes (Caplan and Douglas 1991; Atencio and Yaffe 1992). *SIS1* is an essential gene, whereas strains containing disruptions of *YDJ1* grow very slowly and are inviable at 37°C (Caplan and Douglas 1991; Atencio and Yaffe 1992). A comparison of the phenotypes of the *sis1* and *ydj1* mutants with the phenotypes of the hsp70 mutants would suggest that Ydj1p interacts with Ssa proteins and Sis1p interacts with Ssb proteins. *ydj1* mutants are defective in translocation of proteins into the ER and mitochondria (Atencio and Yaffe 1992; Caplan et al. 1992). Ydj1p stimulates the ATPase activity of Ssa1p, as DnaJ stimulates the ATPase activity of DnaK. In addition, Ydj1p is attached to membranes via a carboxy-terminal farnesyl group (Caplan et al. 1992), thus targeting it to a site where it could logically interact with Ssa proteins during the translocation process. Temperature-sensitive *sis1* mutants are defective in the initiation of translation (Zhong and Arndt 1993). Sis1p is associated mainly with free 40S subunits and small polysomes (Zhong and Arndt 1993). As discussed above, Ssb proteins are associated with all classes of polysomes, but not free subunits or nontranslating monosomes, and may be involved either in the later stages of initiation or in elongation. It is possible that Sis1p is involved in targeting of Ssb proteins to polysomes. However, lest this model of interaction of Sis1p with Ssb proteins and Ydj1p with Ssa proteins seem too simple, it should be pointed out that the regulation of expression of the genes encoding these putative DnaJ-hsp70 pairs is very different. Although some of the *SSA* genes are very heat inducible (Werner-Washburne et al. 1989), expression of *YDJ1* increases only slightly upon heat shock (Atencio and Yaffe 1992). On the other hand, the *SIS1* gene is quite heat inducible (Luke et al. 1991), whereas expression of *SSB* genes is turned off after a heat shock (Werner-Washburne et al. 1989).

VIII. CONCLUDING REMARKS

It is evident that the molecular chaperones play many roles in the eukaryotic cytosol. As discussed in this chapter, the hsp70s have been implicated in translation, folding, and translocation, and an additional role may involve the rescue of misfolded proteins. hsp70s are also clearly involved in regulation of the heat shock response. Nor do these proteins function alone: Many DnaJ-like proteins have been isolated in recent years, and their contributions to the processes of translation and translocation are already apparent. Experiments performed over the next few years should allow a much fuller appreciation of the physiological functions of molecular chaperones in the complex milieu of the cytosol.

REFERENCES

Atencio, D. and M. Yaffe. 1992. *MAS5*, a yeast homolog of DnaJ involved in mitochondrial import. *Mol. Cell. Biol.* **12:** 283–291.

Baker, R., J. Tobias, and A. Varshavsky. 1992. Ubiquitin-specific proteases of *Saccharomyces cerevisiae. J. Biol. Chem.* **267:** 23364–23375.

Beckmann, R.P., L. Mizzen, and W. Welch. 1990. Interaction of Hsp70 with newly synthesized proteins: Implications for protein folding and assembly. *Science* **248:** 850–856.

Boorstein, W.R. and E.A. Craig. 1990. Structure and regulation of the *SSA4* HSP70 gene of *Saccharomyces cerevisiae. J. Biol. Chem.* **265:** 18912–18921.

Boorstein, W., T. Ziegelhoffer, and E.A. Craig. 1994. Molecular evolution of the HSP70 multigene family. *J. Mol. Evol.* **37:** (in press).

Bukau, B. and G. Walker. 1990. Mutations altering heat shock specific subunit of RNA polymerase suppress major cellular defects of *E. coli* mutants lacking the DnaK chaperone. *EMBO J.* **9:** 4027–4036.

Caplan, A. and M. Douglas. 1991. Characterization of *YDJ1:* A yeast homologue of the bacterial *dnaJ* protein. *J. Cell Biol.* **114:** 609–621.

Caplan, A., D. Cyr, and M. Douglas. 1992. YDJ1p facilitates polypeptide translocation across different intracellular membranes by a conserved mechanism. *Cell* **71:** 1143–1155.

―――. 1993. Eukaryotic homologues of *Escherichia coli dnaJ:* A diverse protein family that functions with HSP70 stress proteins. *Mol. Biol. Cell.* **4:** 555–563.

Caplan, A., J. Tsai, P. Casey, and M. Douglas. 1992. Farnesylation of Ydj1p is required for function at elevated growth temperatures in *S. cerevisiae. J. Biol. Chem.* **267:** 18890–18895.

Carter, C.J., M. Cannon, and A. Jimenez, 1980 A trichodermin-resistant mutant of *Saccharomyces cerevisiae* with an abnormal distribution of native subunits. *Eur. J. Biochem.* **107:** 173–183.

Cegegren, R., M.W. Gray, Y. Abel, and D. Sankoff. 1988. The evolutionary relationship among known life forms. *J. Mol. Evol.* **28:** 98–112.

Chappell, T.G., B.B. Konforti, S.L. Schmid, and J.E. Rothman. 1987. The ATPase core of a clathrin uncoating protein. *J. Biol. Chem.* **262:** 746–751.

Craig, E.A. and C.A. Gross. 1991. Is hsp70 the cellular thermometer? *Trends Biochem. Sci.* **16:** 135–140.

Craig, E.A. and K. Jacobsen. 1984. Mutations of the heat-inducible 70 kilodalton genes of yeast confer temperature-sensitive growth. *Cell* **38:** 841–849.

―――. 1985 Mutations in cognate gene of *Saccharomyces cerevisiae* HSP70 result in reduced growth rates at low temperatures. *Mol. Cell. Biol.* **5:** 3517–3524.

Craig, E.A., B.D. Gambill, and R.J. Nelson. 1993. Heat shock proteins: Molecular chaperones of protein biogenesis. *Microbiol. Rev.* **57:** 402–414.

Cyr, D., X. Lu, and M. Douglas. 1992. Regulation of Hsp70 function by a eukaryotic DnaJ homolog. *J. Biol. Chem.* **267:** 20927–20931.

DeLuca-Flaherty, C., B.B. McKay, P. Parham, and B.L. Hill. 1990. Uncoating protein (hsc70) binds a conformationally labile domain of clathrin light chain LCa to stimulate ATP hydrolysis. *Cell* **62:** 875–887.

Deshaies, R., B. Koch, M. Werner-Washburne, E. Craig, and R. Schekman. 1988. A subfamily of stress proteins facilitates translocation of secretory and mitochondrial precursor polypeptides. *Nature* **332:** 800–805.

DiDomenico, B., G. Bugaisky, and S. Lindquist. 1982. Heat shock and recovery are mediated by different translational mechanisms. *Cell* **31:** 593–603.

Doolittle, R., K. Anderson, and D.-F. Feng. 1989. Estimating the prokaryotic-eukaryotic divergence time from protein sequence. In *The hierarchy of life* (ed. B. Fernholm et al.), pp. 73–85. Elsevier, Amsterdam.

Feng, D.-F. and R.F. Doolittle. 1990. Progressive alignment and phylogenetic tree construction of protein sequences. In *Methods in enzymology. Molecular evolution: Computer analysis of protein and nucleic acid sequences* (ed. R.F. Dolittle), pp. 375–387. Academic Press, San Diego.

Finley, D., E. Özkaynak, and A. Varshavsky. 1987. The yeast polyubiquitin gene is essential for resistance to high temperature, starvation and other stresses. *Cell* **49:** 1035–1046.

Fitch, W.M. and E. Margoliash. 1967. Construction of phylogenetic trees. **155:** 279–284.

Flynn, G.C., T.G. Chappell, and J.E. Rothman. 1989. Peptide binding and release by proteins implicated as catalysts of protein assembly. *Science* **245:** 385–390.

Gao, B., J. Biosca, E.A. Craig, L.E. Greene, and E. Eisenberg. 1991. Uncoating of coated vesicles by yeast hsp70 proteins. *J. Biol. Chem.* **266:** 19565–19571.

Goldberg, A. 1992. The mechanism and functions of ATP-dependent proteases in bacterial and animal cells. *Eur. J. Biochem.* **203:** 9–23.

Hershko, A. and A. Ciechanover. 1992. The ubiquitin system for protein degradation. *Annu. Rev. Biochem.* **61:** 761–807.

Hochstrasser, M. 1992. Ubiquitin and intracellular protein degradation. *Curr. Opin. Cell Biol.* **4:** 1024–1031.

Jentsch, S. 1992. The ubiquitin-conjugation system. *Annu. Rev. Genet.* **26:** 179–207.

Kang, P.J., J. Ostermann, J. Shilling, W. Neupert, E.A. Craig, and N. Pfanner. 1990. Hsp70 in the mitochondrial matrix is required for translocation and folding of precursor proteins. *Nature* **348:** 137–143.

Luke, M., A. Suttin, and K. Arndt. 1991. Characterization of *SIS1*, a *Saccharomyces cerevisiae* homologue of bacterial *dnaJ* proteins. *J. Cell Biol.* **114:** 623–638.

Miller, H.I., W.J. Henzel, J.B. Ridgway, W.J. Kuang, V. Chisholm, and C.C. Liu. 1989. Cloning and expression of a yeast ubiquitin-protein cleaving activity in *Escherichia coli. Bio/Technology* **7:** 698–704.

Morimoto, R.I., K.D. Sarge, and K. Abravaya. 1993. Transcriptional regulation of heat shock genes. *J. Biol. Chem.* **267:** 21987–21990.

Mosser, D.D., J. Duchaine, and B. Massie. 1993. The DNA-binding activity of the human heat shock transcription factor is regulated in vivo by hsp70. *Mol. Cell. Biol.* **13:** 5427–5438.

Nelson, R.J., M. Heschl, and E.A. Craig. 1992a. Isolation and characterization of extragenic suppressors of mutations in the *SSA* hsp70 genes of *Saccharomyces cerevisiae*.

Genetics **131**: 277–285.

Nelson, R.J., T. Ziegelhoffer, C. Nicolet, M. Werner-Washburne, and E.A. Craig. 1992b. The translation machinery and seventy kilodalton heat shock protein cooperate in protein synthesis. *Cell* **71**: 97–105.

Özkaynak, E., D. Finley, M.J. Solomon, and A. Varshavsky. 1987. The yeast ubiquitin genes: A family of natural gene fusions. *EMBO J.* **6**: 1429–1439.

Sagan, L. 1967. On the origin of mitosing cells. *J. Theoret. Biol.* **14**: 225–274.

Sanders, S., K. Whitfield, J. Vogel, M. Rose, and R. Schekman. 1992. Sec61p and BiP directly facilitate polypeptide translocation into the ER. *Cell* **69**: 353–366.

Seufert, W. and S. Jentsch. 1990. Ubiquitin-conjugating enzymes UBC4 and UBC5 mediate selective degradation of short-lived and abnormal proteins. *EMBO J.* **9**: 543–550.

Sherman, M. and A. Goldberg. 1992. Involvement of the chaperonin DnaK in the rapid degradation of a mutant protein in *Escherichia coli*. *EMBO J.* **11**: 71–77.

Silver, P.A. and J. Way. 1993. Eukaryotic DnaJ homologs and the specificity of Hsp70 activity. *Cell* **74**: 5–6.

Stone, D.E. and E.A. Craig. 1990. Self regulation of 70 kilodalton heat shock proteins in *Saccharomyces cerevisiae*. *Mol. Cell. Biol.* **10**: 1622–1632.

Straus, D., W. Walter, and C. Gross. 1988. *Escherichia coli* heat shock gene mutants are defective in proteolysis. *Genes Dev.* **2**: 1851–1858.

Tilly, K., N. McKittrick, M. Zylicz, and C. Georgopoulos. 1983. The DnaK protein modulates the heat shock response of *Escherichia coli*. *Cell* **34**: 641–646.

Tobias, J.W. and A. Varshavsky. 1991. Cloning and functional analysis of the ubiquitin-specific protease gene *UBP1* of *Saccharomyces cerevisiae*. *J. Biol. Chem.* **266**: 12021–12028.

Vogel, J.P., L.M. Misra, and M.D. Rose. 1990. Loss of BiP/grp78 function blocks translocation of secretory proteins in yeast. *J. Cell Biol.* **110**: 1885–1895.

Werner-Washburne, M., D.E. Stone, and E.A. Craig. 1987. Complex interactions among members of an essential subfamily of hsp70 genes in *Saccharomyces cerevisiae*. *Mol. Cell. Biol.* **7**: 2568–2577.

Werner-Washburne, M., J. Becker, J. Kosics-Smithers, and E.A. Craig. 1989. Yeast Hsp70 RNA levels change in response to the physiological status of the cell. *J. Bacteriol.* **171**: 2680–2688.

Woese, C.R.. 1987. Bacterial evolution. *Microbiol. Rev.* **51**: 221–271.

Zhong, T. and K.T. Arndt. 1993. The yeast *SIS1* protein, a DnaJ homolog, is required for initiation of translation. *Cell* **73**: 1175–1186.

3

Chaperoning Mitochondrial Biogenesis

Thomas Langer and Walter Neupert
Institut für Physiologische Chemie
München, Germany

I. INTRODUCTION

The parallel development of powerful in vitro systems and of genetic approaches has allowed considerable progress in understanding the mechanisms of protein transport into various cellular compartments. It is becoming ever more evident that transport processes across different cellular membranes are based on similar principles. Polypeptide chains appear to traverse lipid bilayers through proteinaceous pores. Translocation requires a "translocation-competent," rather unfolded conformation. As a consequence, proteins must be partially unfolded or kept in an unfolded conformation prior to the translocation event and must refold after crossing the lipid membrane bilayer. In recent years, increasing evidence was obtained that both represent assisted processes. Molecular chaperones, in many cases originally identified as heat shock proteins, modulate the folding state of polypeptide chains in different cellular compartments.

The Biology of Heat Shock Proteins and Molecular Chaperones
©1994 Cold Spring Harbor Laboratory Press 0-87969-427-0/94 $5 + .00

Mitochondria, which contain heat shock proteins of the hsp70 and hsp60 family, proved to be a useful model system to study the function of chaperone proteins in protein translocation and folding. Although mitochondria contain their own DNA and independent systems for replication and protein synthesis, only a few subunits of the oxidative phosphorylation system and, in some organisms, components mediating splicing and translation of mitochondrial mRNA are encoded by the mitochondrial genome (Grivell 1989). About 95% of the total mass of mitochondrial proteins are encoded in the nucleus. They are synthesized on cytosolic polyribosomes, many of them as precursor molecules with amino-terminal presequences containing the targeting information. Import can occur posttranslationally followed by sorting to the various subcompartments of mitochondria, the outer and inner membranes, the intermembrane, and the matrix space.

In recent years, an increasing number of components have been identified that are involved in the import and sorting of mitochondrial proteins (for reviews, see Glick and Schatz 1991; Segui-Real et al. 1992; Hannavy et al. 1993; Kiebler et al. 1993). In this chapter, we focus on the function of molecular chaperones in import and folding of mitochondrial proteins. In particular, we discuss their roles in maintaining a translocation-competent conformation of mitochondrial precursor molecules in the cytosol, in mediating the translocation process across mitochondrial membranes, and in the folding of matrix-localized proteins.

II. MAINTENANCE OF TRANSLOCATION COMPETENCE IN THE CYTOSOL

A. Conformation of Mitochondrial Proteins during Membrane Translocation

It is now generally agreed that proteins must attain a loosely folded conformation to traverse biological membranes, although with some organelles, in particular peroxisomes and glyoxysomes, the need for unfolding has not been proven. Studies of mitochondrial protein import provided direct experimental evidence for the requirement of a "translocation-competent" conformation of precursor proteins during the translocation process, which differs from the completely folded, native state: (1) Tight folding into a stable tertiary structure, e.g., induced by the presence of substrate analogs or cofactors, was found to prevent the import of precursor proteins into mitochondria (Eilers and Schatz 1986; Chen and Douglas 1987; Rassow et al. 1989; Wienhues et al. 1991). Removal of the ligand restored import competence of the precursor protein. Conversely, destabilization of the native conformation by point mutations results in a more efficient import into mitochondria (Chen and Douglas 1988; Vestweber and Schatz 1988). (2) A nonnative conformation of precursor proteins during the translocation process is suggested by the identification of translocation intermediates spanning the inner and outer membranes (Schleyer and Neupert 1985). The two mitochondrial membranes form a barrier of about 10–12 nm as measured by electron microscopy. Using a set of fusion proteins consisting of aminoterminal parts of cytochrome b_2 of various lengths and dihydrofolate reductase (DHFR), Rassow et al. (1990) showed that about 50 amino acid residues are sufficient to span both mitochondrial membranes. This excludes that precursor proteins traverse membranes in their native conformation and suggests an extended or β-sheet structure of the spanning portion of a polypeptide chain, rather than an α-helical structure.

Up to now, physicochemical data have not been available that describe directly the conformation of translocation competent, mitochondrial precursor proteins. However, in view of the rapid collapse of proteins into a compact conformation after dilution from denaturant in vitro (Kim and Baldwin 1990), a completely unfolded conformation of mitochondrial precursor proteins prior to membrane translocation seems to be very unlikely. Proteins were proposed to traverse membranes in a "molten-globule"-like conformation characterized by the presence of secondary structural elements and a flexible, disordered tertiary structure (Bychkova et al. 1988). At this point, it should be noted that molecular chaperones, whose function in maintaining translocation competence is discussed in the following section, were found to stabilize unfolded proteins in a compact conformation without ordered tertiary structure (Martin et al. 1991b; Langer et al. 1992b).

B. Function of Cytosolic Chaperone Proteins

A nonnative conformation of precursor proteins during membrane translocation implies that their folding must be modulated in the cytosol (Fig. 1). Precursor proteins to be transported across membranes could fold to the native state and become unfolded during membrane translocation (Pfanner et al. 1990; Skerjanc et al. 1990) or their folding is prevented in the cytosol. One obvious possibility would be that the amino-terminal presequence modulates the folding state of precursor molecules. After translocation, specific proteases within mitochondria cleave off the presequence, which would then allow folding to the native structure. However, the presequence is not sufficient to confer prolonged translocation competence. After dilution of mitochondrial precursor proteins from denaturant into in vitro import assays, translocation competence is usual-

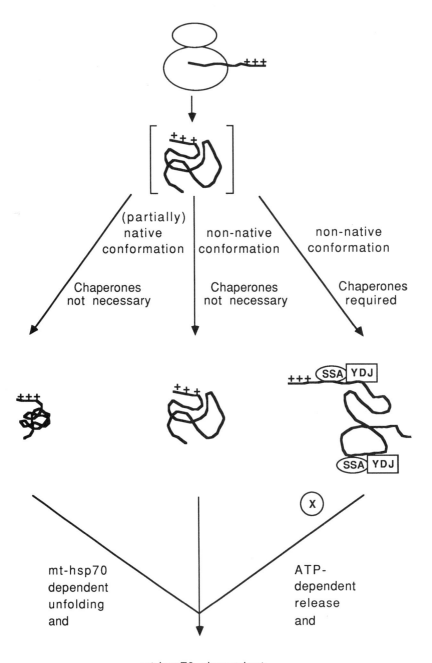

Figure 1 Possible mechanisms of maintenance of translocation competence of precursor proteins in the cytosol. (SSA) Ssa1p/Ssa2p; (YDJ) Ydj1p; (X) NEM-sensitive factor; (mt-hsp70) mitochondrial hsp70.

ly rapidly lost. In contrast, it was shown that precursor proteins are kept transport-competent for long periods in the cytosol, in many cases depending on the presence of ATP (Pfanner et al. 1987, 1990). Indeed, in recent years, several ATP-dependent cytosolic factors have been identified that stabilize various precursor proteins in the cytosol, preventing their folding or aggregation. As discussed in a later section, however, (partial) folding of precursor proteins in the absence of these factors does not necessarily abolish translocation competence. Rather, unfolding in some cases can be promoted by the mitochondrial import machinery, in particular hsp70 (mt-hsp70), in the matrix (Fig. 1; see Section III.D).

Studies in yeast revealed that molecular chaperones of the hsp70 family help to maintain a translocation-competent conformation of mitochondrial precursor proteins as well as proteins targeted to the endoplasmic reticulum, chloroplasts, and the nucleus (Chirico et al. 1988; Deshaies et al. 1988; Murakami et al. 1988; Waegemann et al. 1990; Dingwall and Laskey 1992). It is well established that hsp70 proteins interact with unfolded polypeptide chains in an ATP-dependent manner. Among the six cytosolic hsp70 proteins identified in yeast, evidence for a role in maintaining transport competence exists for Ssa1p and Ssa2p. A yeast strain, in which the *SSA1*, *SSA2*, and *SSA4* genes are deleted, could be rescued by expression of *SSA1* from a galactose-regulated promoter (Deshaies et al. 1988). Genetic depletion of Ssa1p resulted in the accumulation of precursor forms of the mitochondrial inner membrane protein $F_1\beta$ and of α-factor in the cytosol, suggesting a common step in posttranslational protein transport across different membranes. The stimulating effect of Ssa1p/Ssa2p on import of prepro-α-factor into microsomes was also demonstrated biochemically in in vitro transport systems (Chirico et al. 1988). The identification of the temperature-sensitive yeast mutant *mas3*, which maps to the yeast heat shock factor (HSF), provides further evidence for a function of heat shock proteins in protein transport (Smith and Yaffe 1991). At the nonpermissive temperature, in the absence of an induction of *SSA1*, the rate of posttranslationally imported mitochondrial precursor proteins was decreased drastically. Interestingly, overexpression of Ssa1p alone did not relieve this phenotype, indicating that additional heat shock proteins are functioning during early steps of mitochondrial protein import (Smith and Yaffe 1991).

Although a direct physical interaction of Ssa1p/Ssa2p with prepro-α-factor was recently demonstrated by coimmunoprecipitation (Chirico 1992), so far, no stable binary complexes were isolated between cytosolic hsp70 and mitochondrial precursor proteins. Therefore, a detailed description of the mode of action of hsp70 in the cytosol is still not available. Precursor proteins were described to be part of a 200–250-kD protein complex, which contains cytosolic hsp70 (Sheffield et al. 1990). ATP-dependent dissociation of the complex could be prevented by *N*-ethylmaleimide (NEM) treatment of the cytosol. The NEM insensitivity of hsp70 proteins suggests the presence of an additional, NEM-sensitive subunit of the complex. Although this component has not yet been identified, eukaryotic homologs of the *Escherichia coli* heat shock proteins DnaJ and GrpE are attractive candidates. DnaJ and GrpE interact functionally with the *E. coli* hsp70 homolog DnaK (Liberek et al. 1991) and modulate its ATP-dependent interaction with unfolded polypeptide chains (Zylicz et al. 1989; Liberek et al. 1991; Langer et al. 1992b). Indeed, a number of homologs of DnaJ were recently identified localized in various compartments of a eukaryotic cell (Kurihara and Silver 1992; Caplan et al. 1993).

The *YDJ1* gene (also called *MAS5*) was identified by screening a yeast expression library with a polyclonal antiserum raised against a partially purified nuclear fraction (Caplan and Douglas 1991) and independently by screening for yeast mutants displaying a defect in mitochondrial protein import (Atencio and Yaffe 1992). Subsequent biochemical analysis clearly demonstrated that Ydj1p is required for efficient posttranslational protein import into mitochondria (Caplan et al. 1992a). In temperature-sensitive yeast mutant strains at the nonpermissive temperature, precursor proteins of the α, β, and γ subunits of the F_1-ATPase and of citrate synthase accumulate in the cytosol. The dependence of import on intact Ydj1p is obviously more strict at higher temperature. Only minor import defects were observed in a *mas5* deletion mutant at 23°C, whereas cells were not viable at 37°C. Interestingly, Ydj1p is farnesylated in vivo, which is essential for the function of the protein at high temperatures (Caplan et al. 1992b). Upon shift of the temperature to 37°C, the protein was partially relocalized to the membrane fraction dependent on the presence of the farnesyl lipid moiety. However, although enriched at the cytosolic side of the endoplasmic reticulum membrane, Ydj1p was not found in the outer membrane of mitochondria. A specific targeting function of the farnesyl group of Ydj1p during protein transport is therefore still speculative.

How does Ydj1p affect mitochondrial precursor proteins in the cytosol? The prokaryotic homolog DnaJ slightly stimulates the ATPase activity of DnaK (Liberek et al. 1991). This effect is far more pronounced in the presence of another heat shock protein, GrpE. Therefore, it is likely that Ydj1p exerts its effects in collaboration with hsp70 proteins in the cytosol. Indeed, purified Ydj1p functionally interacts with Ssa1p, as it stimulates the ATPase activity of Ssa1p up to ninefold (Cyr et al. 1992). Under these conditions, a permanently unfolded polypeptide chain, carboxymethylated α-lactalbumin, was released from Ssa1p/Ssa2p in vitro. It remains to be determined whether Ydj1p is indeed part of the described cytosolic complex of about 200–250 kD containing Ssa1p/Ssa2p and mitochondrial precursor proteins (Sheffield et al. 1990). Because Ydj1p, as DnaJ, is found to be insensitive to NEM (D. Cyr, pers. comm.; T. Langer, unpubl.), the complex should contain additonal component(s). Although not identified in the cytosol of eukaryotic cells up to now, a protein homologous to the *E. coli* GrpE protein is a likely candidate.

C. Cytosolic Factors with Targeting Function for Mitochondria

In view of the involvement of Ssa1p/Ssa2p and Ydj1p in the import of proteins into both mitochondria and endoplasmic reticulum, interaction of these molecular chaperones with the presequences, if it exists, apparently does not contribute to the specificity of targeting. However, chaperone proteins may stabilize precursor proteins in a conformation that allows interaction of the presequence with specific receptor proteins at the outer surface of mitochondria. Indeed, cytosolic targeting factors seem not to be absolutely required, as efficient import of a chemically pure preprotein into isolated yeast mitochondria was described to occur (Becker et al. 1992). The specific recognition of transport-competent precursor proteins by receptor proteins in the outer mitochondrial membrane is apparently sufficient for correct targeting in vitro. Nevertheless, cytosolic factors that bind specifically to mitochondrial presequences appear to exist, and several such factors have been identified in mammalian cells.

A presequence-binding factor (PBF) was purified from rabbit reticulocyte lysate (Murakami and Mori 1990). Whereas no interaction was observed with mature ornithine transcarbamoyltransferase (OTC), the precursor form was efficiently bound by PBF (Murakami et al. 1992). PBF is a homo-oligomeric protein of 50-kD subunits, with an $s_{20,w}$ value of 5.5S. Depletion of rabbit reticulocyte lysate from PBF prevented import of OTC, aspartate aminotransferase, and malate dehydrogenase into mitochondria. Readdition of purified PBF fully restored import. In contrast, import of 3-oxoacyl-CoA thiolase, which lacks a cleavable presequence, did not depend on PBF. The mode of PBF action has not been understood up to now. Direct evidence for a chaperone-like role of PBF is so far lacking. It has been suggested that PBF might modulate the conformation of precursor proteins synergistically, with hsp70 proteins conferring additional mitochondrion-specific targeting information to the complex (Murakami et al. 1992).

Another cytosolic factor that stimulates mitochondrial protein import was isolated from rat liver (Ono and Tuboi 1988, 1990; Hachiya et al. 1993). This factor, termed mitochondrial-import-stimulating factor (MSF), is composed of two subunits of 30 and 32 kD. In contrast to PBF, MSF exhibits strong ATPase activity in the presence of a transport-incompetent precursor protein (Hachiya et al. 1993). ATP hydrolysis was reported to result in depolymerization of an in-vitro-synthesized mito-chondrial precursor protein. Therefore, MSF may represent a novel chaperone protein specific for mitochondrial precursor proteins with a dual function: On the one hand, it may recognize presequences and con-fer translocation competence to precursor proteins; on the other hand, it may target precursor proteins to mitochondria. Interestingly, the two ac-tivities were affected differently by NEM treatment (Hachiya et al. 1993). Whereas no effect of the alkylating agent was observed on prese-quence binding and the ATPase activity of MSF, the stimulating effect of MSF on mitochondrial import was abolished, suggesting impairment of the release from MSF. NEM exhibited a similar effect on a cytosolic complex containing hsp70 and a mitochondrial precursor protein (Mura-kami et al. 1988; Sheffield et al. 1990). However, in contrast to PBF, stimulation of import by MSF did not depend on cytosolic hsp70.

The relative importance of MSF, PBF, or hsp70 for mitochondrial protein import in vivo remains to be determined. It might well be that a diverse set of factors interact with various parts of a precursor protein, resulting in stabilization of a transport-competent conformation and effi-cient targeting to mitochondria.

III. PROTEIN TRANSLOCATION ACROSS MITOCHONDRIAL MEMBRANES

Translocation-competent precursor proteins are specifically recognized by receptor proteins at the outer surface of mitochondria, which are part of a protein complex in the outer membrane. This receptor complex con-sisting of at least six different proteins mediates binding and insertion into the translocation pore in the outer membrane of mitochondria (for review, see Kiebler et al. 1993). The targeting sequences are then thought to make contact with components of the inner membrane. Translocation of the presequence across the inner membrane into the matrix strictly depends on an energized inner membrane (Gasser et al. 1982; Schleyer and Neupert 1982). The electrical potential ($\Delta\psi$) may exert an elec-trophoretic effect on the positively charged presequence or influence the conformation of an inner membrane component in a manner such that translocation is triggered. This hypothesis is supported by the finding that differences in the positive charge of presequences are reflected in a different sensitivity of import for the uncoupler carbonyl cyanide m-chlorophenylhydrazone (CCCP) (Martin et al. 1991a). Further translocation into the matrix does not require an energized inner membrane, but it does require the hydrolysis of ATP. The role of ATP for mitochondrial protein import was a matter of debate for a long time, mainly because ATP depletion experiments with isolated mitochondria were performed under various conditions, resulting in different ATP levels both outside and inside mitochondria. Two ATP-dependent steps of mitochondrial protein import are now well characterized. (1) In the cytosol, ATP is required to maintain a transport-competent conformation of precursor proteins, as discussed above. (2) The translocation of polypeptide chains across the inner mitochondrial membrane is mediated by a matrix-localized hsp70 protein (mt-hsp70) in an ATP-dependent manner.

A. mt-hsp70-dependent Membrane Translocation

First evidence for a function of mt-hsp70 in the translocation process was obtained upon characterization of the yeast mutant *ssc1-2*, which contains a temperature-sensitive allele of the mt-hsp70 gene *SSC1* (Kang et al. 1990; Ostermann et al. 1990). At the nonpermissive temperature, precursor proteins of $F_1\beta$, hsp60, and Ssc1p (mt-hsp70) itself accumulated in the cytosol in vivo. Consistently, import was impaired in in vitro systems. The mutation in the *SSC1* gene affected import of proteins of the inner membrane (e.g., the Rieske-Fe/S-protein and the ADP/ATP carrier), the intermembrane space (e.g., cytochrome c_1), and the matrix (e.g., the β subunit of the F_1-ATPase). At nonpermissive temperature in the *ssc1-2* mutant, these proteins accumulated at the surface of mitochondria, as assessed by their accessibility to externally added protease. However, the amino-terminal presequences reached the matrix space and were cleaved off by the matrix-processing peptidase. Obviously, the accumulated translocation intermediates were spanning both mitochondrial membranes, indicating that mt-hsp70 acts already during membrane translocation. Indeed, a precursor protein partly translocated into the matrix could be cross-linked to mt-hsp70 (Scherer et al. 1990). In addition, electron microscopic studies revealed a localization of mt-hsp70 near the inner membrane (Carbajal et al. 1993).

On the basis of these results, a model for the translocation of proteins across mitochondrial membranes mediated by mt-hsp70 was proposed (Fig. 2) (Neupert et al. 1990; Neupert and Pfanner 1993). This model predicts cycles of binding of mt-hsp70 to an incoming precursor to provide the driving force for the translocation across the membrane. Spontaneous "breathing" of the polypeptide on the outside would be suf-

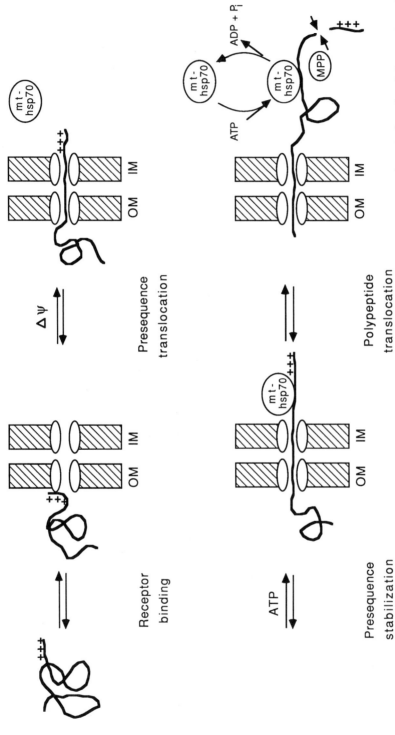

Figure 2   Model of mt-hsp70-mediated membrane translocation of mitochondrial precursor proteins. A functional interaction of proteins homologous to *E. coli* DnaJ and GrpE is conceivable but remains to be demonstrated. (Mt-hsp70) Mitochondrial hsp70; (MPP) matrix processing peptidase; (OM) outer mitochondrial membrane; (IM) inner mitochondrial membrane.

ficient to allow the passage of limited segments of the precursor through the translocation pores in the outer and inner membranes. According to this view, binding of mt-hsp70 to incoming segments of the precursor protein shifts the equilibrium of folded and unfolded state on the outside by trapping the unfolded precursor in a stepwise fashion on the *trans* side of the two mitochondrial membranes. The model would also imply that breakdown of folded domains on the outside is a cooperative effect. After initial unfolding steps, only little energy input is necessary, since then free energy stabilizing the folded conformation is no longer existing as a force preventing complete unfolding. The free energies that further stabilize a folded conformation upon binding of a ligand are usually in the range of a few kcal/mole, thus relatively small. Still, they are sufficient to block import efficiently (Eilers and Schatz 1986; Chen and Douglas 1987; Rassow et al. 1989; Wienhues et al. 1991). This would support the view that advantage is taken of the spontaneous reversible unfolding on the outside by the mt-hsp70-binding system inside. Obviously, the hsp70-binding/ATP hydrolysis system cannot work when spontaneous unfolding outside is strongly impaired by binding of a ligand.

B. Requirement for ATP in the Matrix

To test some predictions of this model, the energetics of membrane translocation was studied in more detail. mt-hsp70 mediates at least one important ATP-dependent step in the mitochondrial matrix during translocation. Therefore, the requirement of ATP in the matrix most likely reflects the function of mt-hsp70. ATP concentrations can be modulated in the matrix under various import conditions (Hwang and Schatz 1989; Stuart et al. 1994). In the absence of substrates for the respiratory chain and by inhibition of the ATP synthase and the ADP/ATP carrier, ATP levels in the matrix can be decreased drastically in vitro. Reduction of the ATP concentration from normal levels of about 1.4 mM to 280 μM did not impair the translocation of matrix-localized proteins or proteins finally localized in the intermembrane space (Stuart et al. 1994). However, at ATP concentrations of about 150 μM, import of matrix-localized proteins like the β subunit of the F_1-ATPase or the Su9(1-69)-DHFR fusion protein was affected. The ATP available to mt-hsp70 under these conditions is extremely low since a considerable amount of total ATP in the matrix is bound to mitochondrial proteins, in particular to F_1-ATPase with affinities in the nanomolar range (Cross and Nalin 1981). Although the binding constant for ATP of mt-hsp70 has not been determined so far, it is expected to be in the micromolar range. DnaK, the *E. coli* hsp70 homolog, has an ATP-binding constant of about 20 μM (Liberek et al. 1991). Therefore, most likely under conditions of extreme ATP depletion, mt-hsp70 in the matrix becomes inactive.

Interestingly, at these extremely low matrix ATP levels, processing of precursor proteins was very inefficient and import-competent proteins accumulated at the surface of the mitochondria (Cyr et al. 1993). The observation of inefficient processing of precursor proteins in the presence of Δψ indicates that the presequence can reach the matrix. However, the membrane potential is not sufficient to translocate presequences across the inner membrane in a stable manner. In addition to Δψ, ATP is required in the matrix. Precursor proteins, accumulated outside the inner membrane of mitochondria in the presence of Δψ, but absence of matrix ATP, could be chased into the matrix by adding ATP. Most likely, the ATP-dependent interaction of mt-hsp70 with the incoming polypeptide chain arrests the presequence on the matrix side of the inner membrane in a topology that allows cleavage by the matrix-processing peptidase (MPP) (Cyr et al. 1993). The observation of only inefficient processing at low ATP concentrations suggests that already the binding of aminoterminal segments of the precursor protein to mt-hsp70 requires the presence of ATP. Indeed, after import in ATP-depleted mitochondria, partly translocated Su9(1-69)-DHFR could only be coimmunoprecipitated with mt-hsp70 shortly after readdition of ATP (Manning-Krieg et al. 1991). Consistent results were obtained when two temperature-sensitive alleles of mt-hsp70, ssc1-2 and ssc1-3, were analyzed (Kang et al. 1990; Gambill et al. 1993). In the temperature-sensitive mutant ssc1-2, carrying a mutation in the putative peptide-binding domain, precursor proteins are bound to mt-hsp70 and processed efficiently at normal ATP levels. However, the release of bound polypeptides is impaired. On the other hand, in the temperature-sensitive mutant ssc1-3, carrying a point mutation near the ATP-binding site, which may prevent binding of ATP to mt-hsp70, binding and efficient processing were not observed.

Taken together, these results indicate that ATP-dependent mt-hsp70 binding is sufficient to arrest the presequence in a stable manner on the matrix side of the inner membrane and allow efficient processing. Complete translocation of matrix-localized proteins into the matrix, however, requires several cycles of ATP-dependent binding and release from mt-hsp70. Even after unfolding of precursor proteins in vitro, import of matrix-targeted precursor proteins did not occur under conditions of extreme ATP depletion or at nonpermissive temperature in ssc1-3 mutant mitochondria (Gambill et al. 1993; Stuart et al. 1994). Interestingly, under these conditions, efficient in vitro import of polypeptides into the matrix was observed in the ssc1-2 mutant, in which binding of precursor proteins to mt-hsp70 is still possible. This suggests that already the ATP-dependent binding of (several) mt-hsp70 by itself to newly imported amino-terminal segments of precursor proteins may be sufficient to drive the translocation, independent of the hydrolysis of ATP.

C. Matrix ATP Requirement for Import of Intermembrane Space Proteins

Import of several proteins localized to the intermembrane space was also found to depend on mt-hsp70 and matrix ATP. Whereas cytochrome b_2 accumulates as a translocation intermediate spanning both mitochondrial membranes in the *ssc1-2* mutant under nonpermissive temperature, no processing was observed in the *ssc1-3* mutant or after ATP depletion of the matrix (Voos et al. 1993; Stuart et al. 1994). A stepwise reduction of ATP levels in the mitochondrial matrix during import revealed a less-stringent ATP requirement for sorting of proteins to the intermembrane space compared to matrix-localized proteins. The hydrophobic part of the bipartite presequences in intermembrane space proteins that contain the sorting information relieves the requirement for the import for matrix ATP/mt-hsp70 (Voos et al. 1993; Stuart et al. 1994). A fusion protein containing the complete presequence of cytochrome b_2 fused to mouse DHFR is transported to the intermembrane space even at very low matrix ATP levels or in the absence of functional mt-hsp70. In contrast, after deletion of the hydrophobic part of the presequence, which results in missorting of the otherwise identical precursor protein into the matrix, import is strictly dependent on the presence of ATP. Similar observations were made studying the import of cytochrome c_1 (Stuart et al. 1994). The efficient, mt-hsp70-independent processing of intermembrane space proteins in the matrix indicates that in this case, stable translocation of the presequence across the inner membrane is achieved by binding to another, not yet identified, protein that may interact with the hydrophobic part of the bipartite presequence.

The less-stringent dependence of intermembrane space proteins on mt-hsp70 does not allow a differentiation between the models presently proposed for the sorting of intermembrane space proteins, namely, the stop-transfer model and the conservative sorting model (Hartl and Neupert 1990; Glick et al. 1992). However, the efficient sorting at low matrix ATP levels of preproteins, loosely folded prior to import and destined to the intermembrane space, indicates that the precursor protein may be present in the matrix only with parts of the entire length at a given time. In frame of the conservative sorting model, this suggests that the polypeptide chain is exported to the intermembrane space in a cotranslocational manner as proposed earlier (Koll et al. 1992). This mechanism could provide the energy for the movement of the polypeptide chain from the matrix to the intermembrane space.

D. mt-hsp70-mediated Unfolding of Precursor Proteins

ATP-dependent binding to mt-hsp70 drives the vectorial movement of a precursor protein across mitochondrial membranes into the matrix. Several lines of evidence indicate that binding of mt-hsp70 to newly imported segments of precursor proteins can indirectly promote the unfolding of domains at the outer surface of mitochondria, another key element of the model for its function in membrane translocation (Fig. 2): (1) The block of complete import of several precursors into $sscl-2$ mitochondria at nonpermissive temperature can be circumvented by urea denaturation of precursor proteins prior to import (Kang et al. 1990; Gambill et al. 1993). (2) In contrast to various fusion proteins containing the presequence of cytochrome b_2, import and sorting of cytochrome b_2 itself to the intermembrane space require matrix ATP and mt-hsp70 (Voos et al. 1993; Stuart et al. 1994). Cytochrome b_2, a lactate dehydrogenase, contains a tightly folded heme-binding domain (cytochrome-b_5-like) followed by a flavin-containing domain. Upon protease treatment of the precursor, the cytochrome b_5 domain is found to form a protease-resistant fragment prior to import (B. Glick; R. Stuart; both pers. comm.). A precursor protein, in which this domain was deleted, did not depend on matrix ATP and mt-hsp70 in its import (Stuart et al. 1994). Consistently, to reach the intermembrane space, fusion proteins containing amino-terminal parts of cytochrome b_2 of various lengths and DHFR only required matrix ATP if the cytochrome b_5 domain was intact. As shown in $sscl-2$ mitochondria, urea denaturation of the precursor protein prior to import circumvented the necessity of ATP in the matrix for the translocation process.

An unfolding reaction on the outside of the mitochondrion mediated by mt-hsp70 implies that folding of precursor proteins in the cytosol does not necessarily prevent efficient import (Fig. 1). Rather, import of folded (partially) proteins depends strictly on the action of mt-hsp70, which promotes unfolding outside. This is consistent with a mechanism by which unfolding at the mitochondrial surface occurs essentially in a spontaneous reaction. As a consequence, cytosolic chaperones may not even be required to maintain a translocation-competent conformation of certain precursor proteins, one such example being cytochrome b_2. However, after stabilization of the folded structure by adding substrate analogs or ligands, e.g., methotrexate for DHFR fusion proteins or heme to a heme-binding domain, the energy provided by binding to mt-hsp70 (followed by ATP-dependent release) seems not to be sufficient to facilitate the unfolding. Under these conditions, translocation intermediates spanning across inner and outer membranes accumulate. From this scenario, it can be predicted that after removal of the ligand, e.g., methotrexate in the case of DHFR fusion proteins, the import of the translocation intermediates requires matrix ATP and mt-hsp70.

Taken together, several key predictions of the current view of the mt-hsp70-mediated membrane translocation process (Fig. 2) received experimental support. mt-hsp70 function could be studied either by ATP depletion of the matrix, preventing binding to mt-hsp70, or by characterizing temperature-sensitive mutants with a defect in binding (*ssc1-3*) or in release of polypeptides (*ssc1-2*). These approaches unraveled several functions of mt-hsp70: Presequences are stabilized in the matrix by ATP-dependent binding to mt-hsp70. The vectorial movement of the complete polypeptide chain across the two mitochondrial membranes requires several cycles of ATP-dependent binding and release from mt-hsp70. These interactions not only provide the energy for the translocation process itself, but can also promote unfolding of precursor proteins outside of mitochondria by shifting the equilibrium of folding to the unfolded state.

E. mt-hsp70-independent Translocation Across the Outer Membrane of Mitochondria

Whereas the complete transport of polypeptide chains across the mitochondrial inner membrane requires mt-hsp70 and the membrane potential, several precursor proteins can be translocated across the outer mitochondrial membrane in a manner independent of mt-hsp70. Cytochrome *c*, a soluble protein of the intermembrane space, follows a quite exceptional import pathway (for review, see Stuart and Neupert 1990; Lill et al. 1992b). Efficient import does not require receptor proteins at the surface of mitochondria or the hydrolysis of ATP. Attachment of the heme group in the intermembrane space, catalyzed by cytochrome *c* heme lyase (CCHL), and subsequent folding are thought to drive membrane translocation (Nicholson et al. 1988). CCHL, on the other hand, is transported via the receptor complex in the outer membrane into the intermembrane space seemingly independent from an external energy source (Lill et al. 1992a). Neither ATP depletion nor destruction of the membrane potential across the inner membrane reduced the import efficiency. It is so far not clear how the energy is provided for the vectorial movement of CCHL across the lipid bilayer. Folding of CCHL or binding to a yet unidentified factor in the intermembrane space could drive the import reaction. Similarly, as in the matrix, a chaperone-like protein might be involved in these processes. However, the observed import of CCHL into isolated outer membrane vesicles argues against the requirement of a soluble factor in the intermembrane space (Mayer et al. 1993). Matrix-localized proteins cannot be imported into these vesicles, most likely because a driving force is missing that in intact mitochondria is provided by the simultaneous, mt-hsp70-dependent translocation across the inner membrane.

IV. FOLDING AND ASSEMBLY OF MITOCHONDRIAL PROTEINS

After membrane translocation, newly imported polypeptides have to attain their native conformation at their site of function. In many cases, this seems to be an assisted process. An increasing number of genes are being characterized whose functions are required for the assembly of protein complexes in the inner membrane, e.g., the F_1F_0-ATPase, the ubiquinol–cytochrome c oxidoreductase, and cytochrome c oxidase (for review, see Grivell 1989; Ackermann and Tzagoloff 1990; Luis et al. 1990; Buchwald et al. 1991). In many cases, the function of the products of these genes seems to be restricted to assisting assembly of a particular protein complex, i.e., they may function as "private" chaperones. In contrast, the molecular chaperone hsp60, localized in the matrix, was shown to mediate folding and assembly of many mitochondrial proteins (Cheng et al. 1989; Martin et al. 1992; Hallberg et al. 1993).

hsp60 belongs to a family of highly conserved proteins, termed chaperonins (cpn60) (Hemmingsen et al. 1988), that occur in prokaryotes (Hendrix 1979; Hohn et al. 1979) and eukaryotes, where it is present in mitochondria (McMullin and Hallberg 1987, 1988; Jindal et al. 1989; Mizzen et al. 1989; Picketts et al. 1989) and in chloroplasts (Barraclough and Ellis 1980; Martel et al. 1990). hsp60 is encoded by an essential gene whose transcription is increased two- to threefold upon temperature shift to 39°C (Reading et al. 1989). Under these conditions, the protein represents about 0.3% of total cell protein. As other chaperonins, hsp60 is a homo-oligomeric protein composed of 14 subunits with a molecular mass of 60 kD. These subunits are arranged in two-stacked heptameric rings, thereby forming the characteristic barrel-like structure (Hutchinson et al. 1989). hsp60 exhibits an ATPase activity that is modulated by a cochaperonin (cpn10) homologous to GroES in *E. coli* (Goloubinoff et al. 1989a). Although so far only identified in mammalian and plant mitochondria (Lubben et al. 1990; Hartman et al. 1992a,b), the ubiquitous occurrence of GroES homologs is very likely. Mitochondrial cpn10 proteins consist of seven identical 10-kD subunits that form a ring-like structure.

A. hsp60-dependent Assembly of Matrix-localized and Inner Membrane Proteins

The yeast *HSP60* gene was originally identified in the mutant *mif4* that lacked enzymatic activity of imported mitochondrial proteins (Cheng et al. 1989). At the same time, hsp60 was found in the yeast genome and its DNA sequence was determined (Johnson et al. 1989; Reading et al. 1989). Subsequent biochemical characterization of the temperature-sensitive mutant *mif4* provided direct evidence for the involvement of hsp60 in the assembly of mitochondrial proteins (Cheng et al. 1989). Import of a number of precursor proteins localized in the matrix or the inner membrane was analyzed and found not to be affected at nonpermissive temperature in vivo. However, assembly of the β subunit of the F_1-ATPase or of ornithine transcarbamoylase and the maturation of the Rieske-Fe/S-protein were impaired. Under these conditions, a large number of matrix proteins, including Mif4p (hsp60), were found as aggregates in the membrane pellet after extraction of mitochondria. This points to a general role of hsp60 in the assembly of mitochondrial matrix proteins. Recently, these observations were further confirmed by genetic depletion of hsp60 (Hallberg et al. 1993). Yeast strains with a disrupted *HSP60* gene were rescued by expression of the wild-type gene from a galactose-inducible promoter. Growth of cells on glucose-containing medium resulted in depletion of hsp60. As in *mif4* mitochondria, proteins were imported normally but remained insoluble. Interestingly, as with other matrix proteins, hsp60 is required for its own assembly (Cheng et al. 1990; Hallberg et al. 1993). In addition, hsp60 also seems to be required for the assembly of some mitochondrially encoded proteins. In plant mitochondria, the newly synthesized α subunit of the F_1-ATPase was found to be associated with hsp60 (Prasad et al. 1990). Taken together, these results demonstrate the requirement of the hsp60 complex for the biogenesis of mitochondrial matrix proteins. However, it was not possible on the basis of these studies to distinguish whether hsp60 affects folding or oligomerization of newly imported proteins.

B. Role of hsp60 for Sorting Proteins to the Intermembrane Space

A function of hsp60 for sorting of proteins to the intermembrane space is currently a matter of debate. At nonpermissive temperature, accumulation of the intermediate form of cytochrome b_2 was observed in the *mif4* strain in vivo (Cheng et al. 1989). Consistently, cytochrome b_2 or hybrid proteins containing various amino-terminal parts of cytochrome b_2 fused to DHFR were found in association with hsp60 in in vitro experiments (Koll et al. 1992). The association with hsp60 was taken as an additional evidence for the conservative sorting model which predicts that cytochrome b_2 traverses the matrix on its sorting pathway to the inter- membrane space. As demonstrated in vitro using the purified *E. coli* hsp60 homolog GroEL, ATP hydrolysis resulted in efficient release of the bound protein only if the hydrophobic part of the bipartite prese- quence of cytochrome b_2 was deleted (Koll et al. 1992). This was inter- pretated to suggest an antifolding effect of hsp60 on intermembrane space proteins with a bipartite presequence. The hydrophobic part of the presequence may promote a prolonged association with hsp60 that keeps the import intermediate in a conformation competent for re-export.

Recently, however, these results were challenged. A different pheno- type was described for import of cytochrome b_2 and cytochrome c_1 into the intermembrane space of mitochondria isolated from the *mif4* strain (Glick et al. 1992). In addition, both proteins were found to be imported with unchanged efficiency after genetic depletion of hsp60 (Hallberg et al. 1993). Although the latter result suggests that hsp60 may not be es- sential for correct sorting of cytochrome b_2 and cytochrome c_1 into the intermembrane space, a kinetic effect of hsp60 was not excluded. hsp60 may stabilize intermediates in an export-competent conformation espe- cially under conditions that favor import into the matrix over re-export into the intermembrane space. This might explain the accumulation of the intermediate form of cytochrome b_2 that was synthesized and im- ported into mitochondria after almost complete depletion of hsp60 (Hallberg et al. 1993).

C. hsp60-mediated Folding of Monomeric Proteins in the Matrix

The demonstration of an impaired assembly of several newly imported proteins in the *mif4* mutant strain at nonpermissive temperature raised the intriguing question of whether folding of monomeric proteins is mediat- ed by molecular chaperones in vivo. An assisted folding reaction might be required to cope with the high protein concentration in the mitochon- drial matrix, which may be as high as 500 mg/ml (Schwerzmann et al. 1986), and thus favor aggregation of newly imported or synthesized polypeptides. A hybrid protein containing DHFR fused to a mitochondri- al targeting domain (amino acids 1–69 of subunit 9 of the ATP synthase;

Su9[1-69]-DHFR) was used to study folding of proteins within mito-chondria (Ostermann et al. 1989). In the native conformation, DHFR exhibits an intrinsic protease resistance, allowing assessment of the folding state of the protein. In *Neurospora crassa*, folding of the DHFR was found to occur with a half-time of about 2 minutes (Ostermann et al. 1989). In contrast, spontaneous refolding of purified DHFR from denaturant in vitro takes place at a considerably faster rate (Touchette et al. 1986). Together with the observed ATP dependence of the folding reaction in mitochondria, these results pointed to a role of hsp60 in mediating the folding of DHFR after import. Indeed, a stable complex of newly imported, unfolded DHFR with hsp60 was isolated from a mito-chondrial matrix extract at reduced ATP levels (Ostermann et al. 1989). Addition of ATP resulted in folding of the DHFR in a protease-resistant conformation. Besides hsp60, an additional factor in the matrix was required for efficient folding, most likely a protein homologous to *E. coli* GroES, in the meantime identified in mitochondria of various organisms (Lubben et al. 1990; Hartman et al. 1992a,b). Additional evidence for an hsp60 function in folding of DHFR was obtained by importing a fusion protein, pOTC-DHFR, into *mif4* mitochondria in vivo (Martin et al. 1992). DHFR could only be extracted in a soluble, enzymatically active conformation at 23°C, whereas at the nonpermissive temperature, most of the protein was recovered in the membrane pellet. Therefore, despite the ability of DHFR to refold in vitro spontaneously after dilution from denaturant, in vivo folding is mediated by hsp60. This is also suggested by the slower kinetics of DHFR folding observed in vivo.

These studies established the role of hsp60 in mediating the folding of newly imported, monomeric proteins. The general function of hsp60 is underlined by the observation that under stress conditions (e.g., high temperature), hsp60 prevents the denaturation of a large number of preexisting mitochondrial proteins as well (Martin et al. 1992). After import into mitochondria in vivo, DHFR, a thermolabile protein, was inactivated at 37°C in the absence of functional hsp60 but was stabilized in an enzymatically active conformation in the presence of hsp60. An ATP-dependent association with hsp60 was only detected at high temperature, conditions that result in denaturation of DHFR. This suggests that under stress conditions in vivo, proteins are stabilized by ATP-dependent association with hsp60. Interestingly, in vivo, mt-hsp70 is not able to compensate for hsp60 in maintaining DHFR enzymatically active at 37°C.

A detailed characterization of the folding activity of chaperonins, including mitochondrial hsp60, was performed by reconstitution of the folding reaction in vitro using purified components. These studies revealed principles of chaperonin action as reviewed elsewhere (Gething and Sambrook 1992; Hendrick and Hartl 1993). Chaperonins of different origins, such as mitochondrial hsp60 and *E. coli* GroEL, can substitute for each other in in vitro folding assays, although with reduced efficiency (Goloubinoff et al. 1989b). This demonstrates the existence of a conserved, ATP-dependent mechanism of chaperonin action. The reduced efficiency of the folding reaction observed when using heterologous components may reflect the parallel evolution of the cochaperonin.

D. Sequential Action of mt-hsp70 and hsp60 in the Mitochondrial Matrix

In addition to hsp60, folding of newly imported proteins in the mitochondrial matrix requires functional mt-hsp70 (Fig. 3). As discussed in an earlier section, mt-hsp70 promotes the translocation of polypeptide chains across mitochondrial membranes by cycles of ATP-dependent binding to the incoming protein. In the yeast mutant *ssc1-2* at the nonpermissive temperature, the block of import could be circumvented by urea denaturation of precursor proteins (Kang et al. 1990; Gambill et al. 1993). Under these conditions, a fusion protein containing DHFR (Su9[1-69]-DHFR) was imported completely into the matrix. However, DHFR remained bound to mt-hsp70 in an unfolded conformation as demonstrated by coimmunoprecipitation and by assessing its protease sensitivity after lysis of mitochondria. Obviously, protein folding requires the ATP-dependent release of newly imported proteins from mt-hsp70 and most likely transfer to hsp60. A sequential interaction of mt-hsp70 and hsp60 with newly imported matrix proteins was already suggested by the observation that in contrast to mt-hsp70, functional inactivation of hsp60 did not affect the import reaction (Cheng et al. 1989). Indeed, upon import in vitro, the precursor of β-MPP, a subunit of the dimeric matrix processing peptidase of yeast, could be coimmunoprecipitated with mt-hsp70 and hsp60 successively (Manning-Krieg et al. 1991). ATP hydrolysis promotes release from mt-hsp70 and binding to hsp60. In addition, newly imported hsp60 was found in a transient complex with mt-hsp70 prior to its assembly (Manning-Krieg et al. 1991). Interestingly, in hsp60-depleted mitochondria, newly imported hsp60 remained associated with mt-hsp70 even in the presence of ATP (Hallberg et al. 1993). Because of the lack of preexisting hsp60 oligomers, which are required for assembly (Cheng et al. 1990), hsp60 subunits remain bound to mt-hsp70.

The cooperation of mt-hsp70 and hsp60 in mediating folding of proteins localized in the mitochondrial matrix raises the intriguing question of how the ATP-dependent transfer of a polypeptide chain from mt-

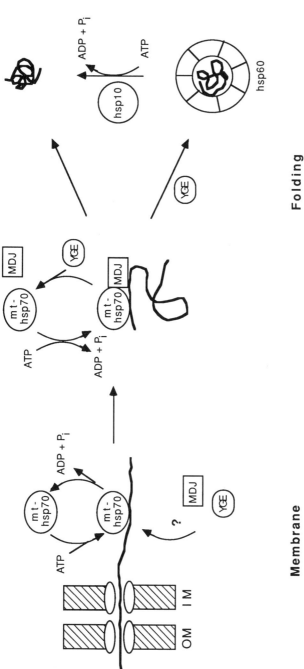

Figure 3 Hypothetical model of the role of mitochondrial chaperone proteins in protein folding in the mitochondrial matrix. The direct functional interaction of Mdj1p and of Yge1p and of Ygelp (see Section IV.D and E) with mt-hsp70 in this process remains to be demonstrated. (Mt-hsp70) Mitochondrial hsp70; (MDJ) mitochondrial DnaJ homolog Mdj1p; (YGE) mitochondrial GrpE-homolog Ygelp; (OM) outer mitochondrial membrane; (IM) inner mitochondrial membrane.

hsp70 to hsp60 is regulated. Reconstitution experiments with purified components allowed further insights into the mechanism of chaperone-mediated protein folding (Langer et al. 1992b). The homologous proteins from *E. coli*, DnaK and GroEL, were used in these studies that share 58% (DnaK and Ssc1) and 54% (GroEL and hsp60) sequence identity with their mitochondrial counterparts (Craig et al. 1989; Reading et al. 1989). The transfer of a protein from DnaK to GroEL was found to be tightly regulated. The ATP-dependent interaction of an unfolded polypeptide chain with DnaK is modulated by two other heat shock proteins of *E. coli*, DnaJ and GrpE (Liberek et al. 1991; Langer et al. 1992b). Binding of DnaJ, which can act as a molecular chaperone on its own, increases the affinity of DnaK for an unfolded protein (Zylicz et al. 1989; Wickner et al. 1991; Langer et al. 1992b). A complex between DnaK and DnaJ is formed that is stabilized by ATP hydrolysis by DnaK. GrpE mediates the ADP release from DnaK, resulting in a decreased substrate affinity of DnaK. Under these conditions, an efficient transfer of the protein to GroEL is observed. Subsequently, folding of the polypeptide chain occurs in association with GroEL in an ATP-dependent manner, most likely within the central cavity of the GroEL cylinder (Martin et al. 1991b; Langer et al. 1992a; Braig et al. 1993). The sequential interaction of the molecular chaperones, DnaK and GroEL, seems to be directed by their binding specificity (Langer et al. 1992b). Whereas DnaK, like various eukaryotic hsp70 proteins (Palleros et al. 1991), exhibits high substrate affinity for polypeptides that are in an unfolded conformation lacking secondary structures, a polypeptide chain in the process of folding to its native state is stabilized by GroEL in a collapsed state characterized by a disordered tertiary structure (Martin et al. 1991b).

The successive interaction of DnaK and GroEL with a polypeptide chain during its folding in vitro may mimic the situation prevailing in mitochondria. Participation in mitochondria of proteins with a function similar to that of *E. coli* DnaJ and GrpE is an attractive possibility. Indeed, a protein homologous to *E. coli* GrpE was recently identified (Yge1p; E. Craig, pers. comm.) that was localized to the mitochondrial matrix. In agreement with the predicted function, the protein is encoded by an essential gene (E. Craig, pers. comm.). On the other hand, a general importance of DnaK-DnaJ-like interactions in eukaryotes is suggested by the identification of proteins homologous to *E. coli* DnaJ in various compartments of a eukaryotic cell; part of them has already been shown to interact functionally with hsp70 proteins (Kurihara and Silver 1992; Caplan et al. 1993). In *Saccharomyces cerevisiae*, *SCJ1* (37% sequence identity to DnaJ) was identified as a gene whose overexpression results in missorting of a nucleus-targeted cytochrome c_1 fusion protein to mitochondria (Blumberg and Silver 1991). However, a mitochondrial localization of Scj1p by cellular subfractionation has not been demonstrated.

E. Identification of the Mitochondrial DnaJ Homolog Mdj1

Recently, during DNA sequencing of an *S. cerevisiae* λ clone library, another gene was identified that turned out to encode a mitochondrial DnaJ homolog and was therefore termed Mdj1p (<u>m</u>itochondrial <u>D</u>na<u>J</u>) (N. Rowley et al., in prep.). The *MDJ1* gene exhibits striking similarity with already known DnaJ homologs. The gene encodes a protein of 511 residues that is 33% identical to *E. coli* DnaJ over the entire length. Moreover, the characteristic sequence motifs found in DnaJ homologs are also present in the *MDJ1* gene. The "J region" of Mdj1p is 50% identical to that in *E. coli* DnaJ and 54% identical to that in Ydj1p, which is located in the cytosol of *S. cerevisiae*. In addition, *MDJ1* contains a glycine-rich region as well as a four times repeated cysteine-containing motif in the central part of Mdj1p, both motifs being characteristic for members of the DnaJ family. In contrast to other known DnaJ homologs, an amino-terminal extension is found in Mdj1p that is rich in basic amino acids, a characteristic feature of mitochondrial presequences. Indeed, Mdj1p is synthesized as a larger presursor protein and imported into isolated mitochondria followed by cleavage of the presequence. The protein was localized to the mitochondrial matrix, more precisely to the inner side of the inner membrane.

To analyze the function of Mdj1p within mitochondria, a gene disruption was carried out ($\Delta mdj1$). As with other DnaJ homologs, Mdj1p is not essential for viability. Disruption of the *MDJ1* gene resulted in a *petite* phenotype in yeast. Whereas normal growth on fermentable carbon sources at 30°C was observed, cells were inviable at 37°C and unable to grow on nonfermentable carbon sources at any temperature. Growth at 37°C could be restored by transformation of the disruptant strain with the complete *MDJ1* gene.

Interestingly, *MDJ1* is required for the maintenance of mitochondrial DNA. No mitochondrial DNA was found in the disruptant strain. In view of the known functions of other DnaJ homologs, in particular their functional interaction with hsp70 proteins, an impaired protein import or folding may account for this effect. Alternatively, the observed ρ^0 phenotype may reflect a function of Mdj1p in mitochondrial DNA replication or in translation, which is required for maintaining mitochondrial DNA. To obtain further evidence for the role of Mdj1p in mitochondrial biogenesis, protein import and folding within mitochondria were studied.

In contrast to an inactivation of mt-hsp70, disruption of *MDJ1* did not affect protein import into different mitochondrial subcompartments. Matrix-localized proteins (e.g., β subunit of the F_1-ATPase), proteins located in the intermembrane space (e.g., cytochrome b_2), as well as proteins of the inner (e.g., ADP/ATP-carrier) and outer membranes (e.g., MOM38) were imported in the absence of Mdj1p with the same efficiencies and kinetics as those observed in wild type. In contrast, Mdj1p seems to participate in the folding of both newly imported and preexisting proteins within mitochondria. After import of DHFR fusion proteins into mitochondria, protease-resistant folded DHFR was formed even in the absence of Mdj1p, however, with reduced efficiency. In the Δ*mdj1* strain, insoluble DHFR was in the pellet fraction after low-speed centrifugation, most likely representing aggregated protein. This indicates a role of Mdj1p in folding of newly imported proteins and, in addition, in folding of preexisting proteins. Completely imported DHFR exhibited a decreased heat stability at 37°C in Δ*mdj1*, suggesting a role of Mdj1p in stabilization of preexisting mitochondrial proteins against heat denaturation.

Taken together, these results demonstrate the importance of Mdj1p for the formation of respiratory-competent mitochondria. Further experiments are required to demonstrate a functional interaction with Ssc1p in mediating import and folding of proteins. Such a cooperation would also point to a protein homologous to *E. coli* GrpE recently identified in yeast (as discussed above). It may seem surprising that deletion of *MDJ1* affects folding of mitochondrial proteins, but not membrane translocation, in view of the participation mt-hsp70 in both processes. Studies of the function of *E. coli* DnaK demonstrated that DnaK binds extended polypeptide chains with high affinity, whereas a stable complex with compact folding intermediates can only be detected in the presence of DnaJ (Langer et al. 1992b). Therefore, different conformational states of precursor proteins during membrane translocation and during subsequent folding may account for the different effects of *MDJ1* deletion. The observed requirement of Mdj1p for maintenance of mitochondrial DNA may suggest a role of Mdj1p in mitochondrial DNA replication or protein synthesis, comparable to DnaJ in the replication of viral DNA in *E. coli* (Georgopoulos et al. 1990) or Sis1p in initiation of translation in *S. cerevisiae* (Zhong and Arndt, 1993).

V. PERSPECTIVES

Molecular chaperones are known to fulfill essential functions during biogenesis of mitochondria. Although general functions are recognized, in many cases, a detailed analysis of the mode of action of the various chaperones is still awaited. This holds particularly true for their role in maintaining a translocation-competent conformation in the cytosol. Questions addressing the composition of cytosolic complexes that contain mitochondrial precursor proteins, the coordination of various chaperone proteins in the cytosol, or the interplay with targeting factors specific for mitochondrial presequences need to be answered.

Future studies should also allow further insights into the function of chaperone proteins within mitochondria. Although a sequential action of mt-hsp70 and hsp60 has been described, the general importance of this pathway is still a matter of debate. The analysis of Mdj1p mutants may provide further clues as to how the cooperation is regulated. Moreover, mitochondrial chaperone proteins may also participate in processes other than protein import and folding, such as DNA replication, DNA recombination, protein synthesis, and degradation. Again, mitochondria may turn out to represent a useful model system to discover novel functions of molecular chaperones.

REFERENCES

Ackermann, S. and A. Tzagoloff. 1990. ATP10, a yeast nuclear gene required for the assembly of the mitochondrial F_1-F_0 complex. *J. Biol. Chem.* **265:** 9952–9995.

Atencio, D.P. and M.P. Yaffe. 1992. MAS5, a yeast homolog of DnaJ involved in mitochondrial protein import. *Mol. Cell. Biol.* **12:** 283–291.

Barraclough, R. and R.J. Ellis. 1980. Protein synthesis in chloroplasts. IX: Assembly of newly-synthesized large subunits into ribulose bisphosphate carboxylase in isolated intact pea chloroplasts. *Biochem. Biophys. Acta* **608:** 19–31.

Becker, K., B. Guiard, J. Rassow, T. Söllner, and K. Pfanner. 1992. Targeting of a chemically pure preprotein to mitochondria does not require the addition of a cytosolic signal recognition factor. *J. Biol. Chem.* **267:** 5637–5643.

Blumberg, H. and P.A. Silver. 1991. A homologue of the bacterial heat-shock gene DnaJ that alters protein sorting in yeast. *Nature* **349:** 627–630.

Braig, K., J. Hainfeld, M. Simon, F. Furuya, and A.L. Horwich. 1993. Polypeptide bound to the chaperonin groEL binds within a central cavity. *Proc. Natl. Acad. Sci.* **90:** 3978–3982.

Buchwald, P., G. Krummeck, and G. Rödel. 1991. Immunological identification of yeast SCO1 protein as a component of the inner mitochondrial membrane. *Mol. Gen. Genet.* **229:** 413–420.

Bychkova, V., R.H. Pain, and O.B. Ptitsyn. 1988. The "molten globule" state is involved in the translocation of proteins across membranes? *FEBS Lett.* **238:** 231–234.

Caplan, A.J. and M.G. Douglas. 1991. Characterization of YDJ1: A yeast homologue of the bacterial dnaJ protein. *J. Cell Biol.* **114:** 609–621.

Caplan, A., D.M. Cyr, and M.G. Douglas. 1992a. YDJ1p facilitates polypeptide translocation across different intracellular membranes by a conserved mechanism. *Cell* **71:** 1143–1155.

———. 1993. Eukaryotic homologues of *Escherichia coli* DnaJ: A diverse protein family that functions with HSP70 stress proteins. *Mol. Biol. Cell* **4:** 555–563.

Caplan, A., J. Tsai, P.J. Casey, and M.G. Douglas. 1992b. Farnesylation of YDJ1p is required for function at elevated growth temperatures in *Saccharomyces cerevisiae. J. Biol. Chem.* **267:** 18890–18895.

Carbajal, E., J.-F. Beaulieu, L.M. Nicole, and R.M. Tanguay. 1993. Intramitochondrial localization of the main 70-kDa heat-shock cognate protein in *Drosophila* cells. *Exp. Cell Res.* **207:** 300–309.

Chen, W.-J. and M. Douglas. 1987. The role of protein structure in the mitochondrial import pathway: Unfolding of mitochondrially bound precursors is required for membrane translocation. *J. Biol. Chem.* **262:** 15605–15609.

————. 1988. An F_1-ATPase β-subunit precursor lacking an internal tetramer-forming domain is imported into mitochondria in the absence of ATP. *J. Biol. Chem.* **263:** 4997–5000.

Cheng, M.Y., F.-U. Hartl, and A.L. Horwich. 1990. The mitochondrial chaperonin hsp60 is required for its own assembly. *Nature* **348:** 455–458.

Cheng, M.Y., F.-U. Hartl, J. Martin, R.A. Pollock, F. Kalousek, W. Neupert, E.M. Hallberg, R.L. Hallberg, and A.L. Horwich. 1989. Mitochondrial heat-shock protein hsp60 is essential for assembly of proteins imported into yeast mitochondria. *Nature* **337:** 620–625.

Chirico, W. 1992. Dissociation of complexes between 70 kDa stress proteins and presecretory proteins is facilitated by a cytosolic factor. *Biochem. Biophys. Res. Commun.* **189:** 1150–1156.

Chirico, W., M. Waters, and G. Blobel. 1988. 70K heat-shock related proteins stimulated protein translocation into microsomes. *Nature* **332:** 805–810.

Craig, E.A., J. Kramer, J. Shilling, W.M. Werner, S. Holmes, S.J. Kosic, and C.M. Nicolet. 1989. SSC1, an essential member of the yeast HSP70 multigene family, encodes a mitochondrial protein. *Mol. Cell. Biol.* **9:** 3000–3008.

Cross, R. and C.M. Nalin. 1981. Adenine nucleotide binding sites on beef heart F_1-ATPase. *J. Biol. Chem.* **257:** 2874–2881.

Cyr, D., X. Lu, and M.G. Douglas. 1992. Regulation of Hsp70 function by a eukaryotic DnaJ homolog. *J. Biol. Chem.* **267:** 20927–20931.

Cyr, D., R.A. Stuart, and W. Neupert. 1993. A matrix-ATP requirement for presequence translocation across the inner membrane of mitochondria. *J. Biol. Chem.* **268:** 23751–23754.

Dingwall, C. and R. Laskey. 1992. The nuclear membrane. *Science* **258:** 942–947.

Deshaies, R., B. Koch, M. Werner-Washburne, E. Craig, and R. Schekman. 1988. A subfamily of stress proteins facilitates translocation of secretory and mitochondrial precursor polypeptides. *Nature* **332:** 800–805.

Eilers, M. and G. Schatz. 1986. Binding of a specific ligand inhibits import of a purified precursor protein into mitochondria. *Nature* **322:** 228–232.

Gambill, B., W. Voos, P.J. Kang, B. Miao, T. Langer, E.A. Craig, and K. Pfanner. 1993. A dual role for mitochondrial heat shock protein 70 in membrane translocation of preproteins. *J. Cell Biol.* **123:** 9–26.

Gasser, S., G. Daum, and G. Schatz. 1982. Import of proteins into mitochondria: Energy-dependent uptake of precursors by isolated mitochondria. *J. Biol. Chem.* **257:** 13034–13041.

Georgopoulos, C., D. Ang, K. Liberek, and M. Zylicz. 1990. Properties of the *Escherichia coli* heat shock proteins and their role in bacteriophage λ growth. In *Stress proteins in biology and medicine* (ed. R.I. Morimoto et al.), pp. 191–222. Cold Spring Harbor Laboratory Press, Cold Spring Harbor, New York.

Gething, M.-J. and J. Sambrook, J. 1992. Protein folding in the cell. *Nature* **355:** 33–45.

Glick, B. and G. Schatz. 1991. Import of proteins into mitochondria. *Annu. Rev. Genet.* **25:** 21–44.

Glick, B., A. Brandt, K. Cunningham, S. Müller, R. Hallberg, and G. Schatz. 1992. Cytochromes c_1 and b_2 are sorted to the intermembrane space of yeast mitochondria by a stop-transfer mechanism. *Cell* **69:** 809–822.

Goloubinoff, P., A.A. Gatenby, and G.H. Lorimer. 1989a. GroE heat-shock proteins promote assembly of foreign prokaryotic ribulose bisphosphate carboxylase oligomers in *Escherichia coli. Nature* **337:** 44–47.

Goloubinoff, P., J.T. Christeller, A.A. Gatenby, and G.H. Lorimer. 1989b. Reconstitution of active dimeric ribulose bisphosphate carboxylase from an unfolded state depends on two chaperonin proteins and MgATP. *Nature* **342:** 884–889.

Grivell, L. 1989. Nucleo-mitochondrial interactions in yeast mitochondrial biogenesis. *Eur. J. Biochem.* **182:** 477–493.

Hachiya, N., R. Alam, Y. Sakasegawa, M. Sakaguchi, N. Mihara, and T. Omura. 1993. A mitochondrial import factor purified from rat liver cytosol is an ATP-dependent conformational modulator for precursor proteins. *EMBO J.* **12:** 1579–1586.

Hallberg, E., Y. Shu, and R.L. Hallberg. 1993. Loss of mitochondrial hsp60 function: Nonequivalent effects on matrix-targeted and intermembrane-targeted proteins. *Mol. Cell. Biol.* **13:** 3050–3057.

Hannavy K., S. Rospert, and G. Schatz. 1993. Protein import into mitochondria: A paradigm for the translocation of polypeptides across membranes. *Curr. Biol.* **5:** 694–700.

Hartl, F.-U. and W. Neupert. 1990. Protein sorting to mitochondria: Evolutionary conservations of folding and assembly. *Science* **247:** 930–938.

Hartman, D., D. Dougan, N.J. Hoogenraad, and P.B. Hoj. 1992a. Heat shock proteins of barley mitochondria and chloroplasts. Identification of organellar hsp10 and 12: Putative chaperonin 10 homologues. *FEBS Lett.* **305:** 147–150.

Hartman, D.J., N.J. Hoogenraad, R. Condron, and P.B. Hoj. 1992b. Identification of a mammalian 10-kDa heat shock protein, a mitochondrial chaperonin 10 homologue essential for assisted folding of trimeric ornithine transcarbamoylase *in vitro. Proc. Natl. Acad. Sci.* **89:** 3394–3398.

Hemmingsen, S.M., C. Woolford, S. van der Vies, K. Tilly, D.T. Dennis, C.P. Georgopoulos, R.W. Hendrix, and R.J. Ellis. 1988. Homologous plant and bacterial proteins chaperone oligomeric protein assembly. *Nature* **333:** 330–334.

Hendrick, J., and F.-U. Hartl. 1993. Molecular chaperone function of heat-shock proteins. *Annu. Rev. Biochem.* **62:** 349–384.

Hendrix, R. 1979. Purification and properties of GroE, a host protein involved in bacteriophage assembly. *J. Mol. Biol.* **129:** 375–392.

Hohn, T., B. Hohn, A. Engel, M. Wortz, and P.R. Smith. 1979. Isolation and characterization of the host protein GroE involved in bacteriophage λ assembly. *J. Mol. Biol.* **129:** 359–373.

Hutchinson, E.G., W. Tichelaar, G. Hofhaus, H. Weiss, and K.R. Leonard. 1989. Identification and electron microsopic analysis of a chaperonin oligomer from *Neurospora crassa. EMBO J.* **8:** 1485–1490.

Hwang, S. and G. Schatz. 1989. Translocation of proteins across the mitochondrial inner membrane, but not into the outer membrane, requires nucleotide triphosphates in the matrix. *Proc. Natl. Acad. Sci.* **86:** 8432–8436.

Jindal, S., A.K. Dudani, B. Singh, C.B. Harley, and R.S. Gupta. 1989. Primary structure of a human mitochondrial protein homologous to the bacterial and plant chaperonins and to the 65-kilodalton mycobacterial antigen. *Mol. Cell. Biol.* **9:** 2279–2283.

Johnson, R.B., K. Fearon, T. Mason, and S. Jindal. 1989. Cloning and characterization of the yeast chaperonin HSP60 gene. *Gene* **84:** 295–302.

Kang, P.-J., J. Ostermann, J. Shilling, W. Neupert, E. Craig, and N. Pfanner. 1990. Requirement for hsp70 in the mitochondrial matrix for translocation and folding of precursor proteins. *Nature* **348:** 137–142.

Kiebler, M., K. Becker, N. Pfanner, and W. Neupert. 1993. Mitochondrial protein import: Specific recognition and membrane translocation of preproteins. *J. Membr. Biol.* **135:** 191–207.

Kim, P. and R.L. Baldwin. 1990. Intermediates in the folding reactions of small proteins. *Annu. Rev. Biochem.* **59:** 631–660.

Koll, H., B. Guiard, J. Rassow, J. Ostermann, A. Horwich, W. Neupert, and F.-U. Hartl. 1992. Antifolding activity of hsp60 couples protein import into the mitochondrial matrix with export to the intermembrane space. *Cell* **68:** 1163–1175.

Kurihara, T. and P.A. Silver. 1992. DnaJ homologs and protein transport. In *Membrane biogenesis and protein targeting* (ed. W. Neupert and R. Lill), pp. 309–384. Elsevier, New York.

Langer, T., G. Pfeifer, J. Martin, W. Baumeister, and F.-U. Hartl. 1992a. Chaperonin-mediated protein folding: GroES binds to one end of the GroEL cylinder, which accommodates the protein substrate within its central cavity. *EMBO J.* **11:** 4757–5765.

Langer, T., C. Lu, H. Echols, J. Flanagan, M. Hayer-Hartl, and F.-U.Hartl. 1992b. Successive action of DnaK, DnaJ and GroEL along the pathway of chaperone-mediated protein folding. *Nature* **356:** 683–689.

Liberek, K., J. Marszalek, D. Ang, C. Georgopoulos, and M. Zylicz. 1991. *Escherichia coli* DnaJ and GrpE heat shock proteins jointly stimulate ATPase activity of DnaK. *Proc. Natl. Acad. Sci.* **88:** 2874–2878.

Lill, R., R. Stuart, M. Drygas, F. Nargang, and W. Neupert. 1992a. Import of cytochrome *c* heme lyase into mitochondria: A novel pathway into the intermembrane space. *EMBO J.* **11:** 449–456.

Lill, R., C. Hergersberg, H. Schneider, T. Söllner, R. Stuart, and W. Neupert. 1992b. General and exceptional pathways of protein import into the sub-mitochondrial compartments. In *Membrane biogenesis and protein targeting* (ed. W. Neupert and R. Lill), pp. 265–278. Elsevier, New York.

Lubben, T.H., A.A. Gatenby, G.K. Donaldson, G.H. Lorimer, and P.V. Viitanen. 1990. Identification of a groES-like chaperonin in mitochondria that facilitates protein folding. *Proc. Natl. Acad. Sci.* **87:** 7683–7687.

Luis, A.M., A. Alconada, and J.M. Cuezva. 1990. The alpha regulatory subunit of the mitochondrial F_1-ATPase complex is a heat-shock protein. Identification of two highly conserved amino acid sequences among the alpha-subunits and molecular chaperones. *J. Biol. Chem.* **265:** 7713–7716.

Manning-Krieg, U., P. Scherer, and G. Schatz. 1991. Sequential action of mitochondrial chaperones in protein import into the matrix. *EMBO J.* **10:** 3273–3280.

Martel, R., L.P. Cloney, L.E. Pelcher, and S.M. Hemmingsen. 1990. Unique composition of plastid chaperonin-60: α and β polypeptide-encoding genes are highly divergent. *Gene* **94:** 181–187.

Martin, J., A.L. Horwich, and F.-U. Hartl. 1992. Prevention of protein denaturation under heat stress by the chaperonin Hsp60. *Science* **258:** 995–998.

Martin, J., K. Mahlke, and N. Pfanner. 1991a. Role of an energized inner membrane in mitochondrial protein import. *J. Biol. Chem.* **266:** 18051–18057.

Martin, J., T. Langer, R. Boteva, A. Schramel, A. Horwich, and F.-U. Hartl. 1991b. Chaperonin-mediated protein folding at the surface of groEL through a "molten globule"-like intermediate. *Nature* **352**: 36–42.

Mayer, A., R. Lill, and W. Neupert. 1993. Translocation and insertion of precursor proteins into isolated outer membranes of mitochondria. *J. Cell Biol.* **121**: 2233–2243.

McMullin, T.W. and R.L. Hallberg. 1987. A normal mitochondrial protein is selectively synthesized and accumulated during heat shock in *Tetrahymena thermophila. Mol. Cell. Biol.* **7**: 4414–4423.

———. 1988. A highly evolutionary conserved mitochondrial protein is structurally related to the protein encoded by the *Escherichia coli* groEL gene. *Mol. Cell. Biol.* **8**: 371–380.

Mizzen, L.A., C. Chang, J.I. Garrels, and W.J. Welch. 1989. Identification, characterization, and purification of two mammalian stress proteins present in mitochondria, grp75, a member of the hsp70 family and hsp58, a homolog of the bacterial GroEL protein. *J. Biol. Chem.* **264**: 20664–20675.

Murakami, H., D. Pain, and G. Blobel. 1988. 70K heat-shock related protein is one of at least two distinct cytosolic factors stimulating protein import into mitochondria. *J. Cell Biol.* **107**: 2051–2057.

Murakami, K. and M. Mori. 1990. Purified presequence binding factor (PBF) forms an import-competent complex with a purified mitochondrial precursor protein. *EMBO J.* **9**: 3201–3208.

Murakami, K., S. Tanase, Y. Morino, and. N. Mori. 1992. Presequence binding factor-dependent and -independent import of proteins into mitochondria. *J. Biol. Chem.* **267**: 13119–13122.

Neupert, W. and N. Pfanner. 1993. Roles of molecular chaperones in protein targeting to mitochondria. *Philos. Trans. R. Soc. Lond.* **339**: 355–362.

Neupert, W., F.-U. Hartl, E.A. Craig, and N. Pfanner. 1990. How do polypeptides cross the mitochondrial membranes? *Cell* **63**: 447–450.

Nicholson, D., C. Hergersberg, and W. Neupert. 1988. Role of cytochrome *c* heme lyase in the import of cytochrome *c* into mitochondria. *J. Biol. Chem.* **263**: 19034–19042.

Ono, H. and S. Tuboi. 1988. The cytosolic factor required for import of precursors of mitochondrial precursor proteins into mitochondria. *J. Biol. Chem.* **263**: 3188–3193.

———. 1990. Purification and identification of a cytosolic factor required for import of precursors of mitochondria proteins into mitochondria. *Arch. Biochem. Biophys.* **280**: 299–304.

Ostermann, J., A. Horwich, W. Neupert, and F.-U. Hartl. 1989. Protein folding in mitochondria requires complex formation with hsp60 and ATP hydrolysis. *Nature* **341**: 125–130.

Ostermann, J., W. Voss, P. Kang, E. Craig, W. Neupert, and N. Pfanner. 1990. Precursor proteins in transit through mitochondrial contact sites interact with hsp70 in the matrix. *FEBS Lett.* **277**: 281–284.

Palleros, D.R., W.J. Welch, and A.L. Fink. 1991. Interaction of hsp70 with unfolded proteins: Effects of temperature and nucleotides on the kinetics of binding. *Proc. Natl. Acad. Sci.* **88**: 5719–5723.

Pfanner, N., M. Tropschug, and W. Neupert. 1987. Mitochondrial protein import: Nucleoside triphosphates are involved in conferring import-competence to precursors. *Cell* **49**: 815–823.

Pfanner, N., J. Rassow, B. Guiard, T. Söllner, F.-U. Hartl, and W. Neupert. 1990. Energy requirements for unfolding and membrane translocation of precursor proteins during import into mitochondria. *J. Biol. Chem.* **265**: 16324–16329.

Picketts, D.J., C.S.K. Mayanil, and R.S. Gupta. 1989. Molecular cloning of a Chinese hamster mitochondrial protein related to the "chaperonin" family of bacterial and plant proteins. *J. Biol. Chem.* **264:** 12001–12008.

Prasad, T.K., E. Hack, and R.L. Hallberg. 1990. Function of the maize mitochondrial chaperonin hsp60: Specific association between hsp60 and newly synthesized F_1-ATPase alpha subunits. *Mol. Cell. Biol.* **10:** 3979–3986.

Rassow, J., F.-U. Hartl, B. Guiard, N. Pfanner, and W. Neupert. 1990. Polypeptides traverse the mitochondrial envelope in an extended state. *FEBS Lett.* **275:** 190–194.

Rassow, J., B. Guiard, U. Wienhues, V. Herzog, F.-U. Hartl, and W. Neupert. 1989. Translocation arrest by reversible folding of a precursor protein imported into mitochondria: A means to quantitate translocation contact sites. *J. Cell Biol.* **109:** 1421–1428.

Reading, D.S., R.L. Hallberg, and A.M. Meyers. 1989. Characterization of the yeast *HSP60* gene coding for a mitochondrial assembly factor. *Nature* **337:** 655–659.

Scherer, P.E., U.C. Krieg, S.T. Hwang, D. Vestweber, and G. Schatz. 1990. A precursor protein partly translocated into yeast mitochondria is bound to a 70 kd mitochondrial stress protein. *EMBO J.* **9:** 4315–4322.

Schleyer, M. and W. Neupert. 1982. Requirement of a membrane potential for the post-translational transfer of proteins into mitochondria. *Eur. J. Biochem.* **125:** 109–116.

―――. 1985. Transport of proteins into mitochondria: Translocation intermediates spanning contact sites between inner and outer membranes. *Cell* **43:** 330–350.

Schwerzmann, K., L.M. Cruz-Orive, R. Eggmen, A. Sänger, and E.R. Weibel. 1986. Molecular architecture of the inner membrane of mitochondria from rat liver: A combined biochemical and stereological study. *J. Cell Biol.* **102:** 97–103.

Segui-Real, B., R. Stuart, and W. Neupert. 1992. Transport of proteins into the various subcompartments of mitochondria. *FEBS Lett.* **313:** 2–7.

Sheffield, W., G. Shore, and S. Randall. 1990. Mitochondrial protein import: Effects of 70-kd hsp70 on polypeptide folding, aggregation and import competence. *J. Biol. Chem.* **265:** 11069–11076.

Skerjanc, I.S., W.P. Sheffield, S.K. Randall, J.R. Silvius, and G. Shore. 1990. Import of precursor proteins into mitochondria: Site of polypeptide unfolding. *J. Biol. Chem.* **265,** 9444–9451.

Smith, B. and M.P. Yaffe. 1991. A mutation in the yeast heat-shock factor gene causes temperature-sensitive defects in both mitochondrial protein import and the cell cycle. *Mol. Cell. Biol.* **11:** 2647–2655.

Stuart, R. and W. Neupert. 1990. Apocyptochrome *c*: An exceptional mitochondrial precursor protein using an exceptional import pathway. *Biochimie* **72:** 115–121.

Stuart, R.A., A. Gruhler, I. van der Klei, B. Guiard, H. Koll, and W. Neupert. 1994. The requirement of matrix ATP for the import of precursor proteins into the mitochondrial matrix and intermembrane space. *Eur. J. Biochem.* (in press).

Touchette, N.A., K.M. Perry, and C.R. Matthews. 1986. Folding of dihydrofolate reductase from *Escherichia coli*. *Biochemistry* **25:** 5445–5452.

Vestweber, D. and G. Schatz. 1988. Point mutations destabilizing a precursor protein enhance its post-translational import into mitochondria. *EMBO J.* **7:** 1147–1151.

Voos, W., B.D. Gambill, B. Guiard, K. Pfanner, and E.A. Craig. 1993. Presequence and mature part of preproteins strongly influence the dependence of mitochondrial protein import on heat shock protein 70 in the matrix. *J. Cell Biol.* **123:** 109–118.

Waegemann, K., H. Paulsen, and J. Soll. 1990. Translocation of proteins into isolated chloroplasts requires cytosolic factors to obtain import competence. *FEBS Lett.* **261:** 89–92.

Wickner, S., J. Hoskins, and K. McKenney. 1991. Function of DnaJ and DnaK as chaperones in origin-specific DNA binding by RepA. *Nature* **350:** 165–167.

Wienhues, U., K. Becker, M. Schleyer, B. Guiard, M. Tropschug, A. Horwich, N. Pfanner, and W. Neupert. 1991. Protein folding causes an arrest of preprotein translocation in mitochondria in vivo. *J. Cell Biol.* **115:** 1601–1609.

Zhong, T. and K.T. Arndt. 1993. The yeast SIS1 protein, a DnaJ homolog, is required for the initiation of translation. *Cell* **73:** 1175–1186.

Zylicz, M., D. Ang, K. Liberek, and C. Georgopoulos. 1989. Initiation of lambda DNA replication with purified host- and bacteriophage-encoded proteins: The role of the dnaK, dnaJ and grpE heat shock proteins. *EMBO J.* **8:** 1601–1608.

4

Heat Shock Cognate Proteins and Polypeptide Translocation Across the Endoplasmic Reticulum Membrane

Jeffrey L. Brodsky and Randy Schekman
Department of Molecular and Cell Biology and
the Howard Hughes Medical Institute
University of California
Berkeley, California 94720

I. INTRODUCTION

Protein translocation into the lumen of the endoplasmic reticulum (ER) is an essential step in the biogenesis of secretory proteins and of some proteins that are destined to reside in intracellular organelles. Translocation requires the hydrolysis of ATP and the function of proteins in the cytosol, in the ER membrane, and in the ER lumen (for review, see Walter and Lingappa 1986; Nunnari and Walter 1992; Rapoport 1992; Sanders and Schekman 1992). Proteins that are destined for translocation contain a signal peptide that targets the nascent chain to the cytoplasmic face of the ER membrane.

Signal peptides generally consist of a continuous stretch of nonpolar amino acids at the amino terminus of the protein (for review, see von Heijne 1990). The signal peptide is removed by signal peptidase and N-linked glycosylation may occur as the translocating polypeptide emerges in the lumen of the ER. The ER membrane is equipped with proteins that act as protein permeases or pores which facilitate the transport of the polypeptide across the ER bilayer (Simon and Blobel 1991; Görlich et al.

1992b; Müsch et al. 1992; Sanders et al. 1992). Once translocation is complete, the protein folds into its native conformation. Protein folding in the ER is aided by lumenal factors that directly catalyze folding and others that prevent unproductive folding pathways (for review, see Gething and Sambrook 1992).

Proteins are translocated across membranes in an unfolded or extended conformation (Verner and Schatz 1988; Langer and Neupert, this volume), as the various structures and large diameter of native proteins would otherwise present too great a challenge to the translocation machinery. Unfolded proteins contain ionic, polar, nonpolar, and hydrophobic moieties that are exposed to the solvent. How does the translocation apparatus handle the heterogeneity in amino acid side chains? This is especially difficult because charged and polar amino acid groups must traverse the hydrophobic lipid bilayer of the ER. Experimental evidence suggests, however, that secretory polypeptides are in an aqueous environment as they cross the ER membrane (Gilmore and Blobel 1985; Crowley et al. 1993). The problem then lies in preventing solvent-exposed hydrophobic patches of amino acids from aggregating before proper folding can occur in the lumen of the ER. The resolution of this issue comes from the observation that ATP hydrolysis is required for protein translocation and from the genetic and biochemical analyses of translocation in both yeast and mammalian cells. These studies have implicated a class of molecules known as heat shock cognate proteins that appear to be essential for protein translocation.

Heat shock proteins are intracellular factors whose synthesis is induced upon cellular stress (Parsell and Lindquist, this volume). Heat shock cognate proteins (hscs) are homologs of heat shock proteins but are expressed constitutively, which implies that the hscs perform a "housekeeping" function in the cell. Among the hscs are those with a molecular mass of 70 kD that possess ATPase activity (hsc70s). BiP, an hsc70 in the ER lumen, was originally isolated because it associated with unfolded immunoglobin heavy chains (Haas and Wabl 1983; for review, see Gething, this volume); the affinity of BiP for heavy chains is lost upon addition of ATP (Bole et al. 1986; Munro and Pelham 1986). Pelham (1986) first proposed that hsc70s bind the exposed hydrophobic regions of misfolded polypeptides and prevent them from forming protein aggregates. Since malfolded proteins are expected to accumulate under conditions of cellular stress, the inducible 70-kD proteins were hypothesized to be involved in the repair of stress-damaged proteins, and the constitutive 70-kD proteins were postulated to have a role in normal folding pathways. In support of a role for the constitutive hsc70s in assisted protein folding, hsc70s bind to short peptides and then release them upon ATP hydrolysis. Furthermore, peptides stimulate the ATPase activity of hsc70s (Flynn et al. 1989). Flynn et al. (1991) demonstrated that BiP binds preferentially to peptides that are composed of aliphatic amino acids. Thus, hsc70s interact with linear arrays of amino acids that are usually found buried in the interior of proteins. Peptides that contain these sequences normally aggregate if presented to an aqueous environment. Therefore, the interaction between BiP and unfolded polypeptides in the cell likely prevents protein aggregation. Because translocated proteins remain unfolded during translocation (Verner and Schatz 1988), hsc70s are expected to facilitate protein translocation.

Escherichia coli contains an hsp70 known as DnaK (for review, see Ang et al. 1991; Georgopoulos, this volume). The *dnaK* gene is found in an operon with a gene that encodes a second heat shock protein, DnaJ. Mutations in DnaK, DnaJ, and a third heat shock protein, GrpE, block bacteriophage λ DNA replication. At the onset of λ DNA synthesis, the DnaB helicase is found in an inert preprimosomal protein complex at the replication origin. The helicase is subsequently activated by removing the λP protein from the preprimosomal complex. Zylicz et al. (1989) demonstrated that this release requires DnaK, DnaJ, GrpE, and ATP. Using purified proteins, Liberek et al. (1991) showed that DnaJ stimulates the hydrolysis of ATP by DnaK, whereas GrpE specifically catalyzes nucleotide release from DnaK. Together, these results demonstrate that the ATPase activity of the 70-kD proteins can be modified and that the hsp70 protein complex (DnaK, DnaJ, and GrpE) can regulate the function of other protein complexes.

In this chapter, we show how the field of protein translocation has intersected with studies on the heat shock proteins. A general theme in the first section of this chapter is that proteins destined for the ER lumen are presented to the translocation machinery in a nonnative conformation. Because hsc70s can bind, stabilize, and release protein segments that would otherwise aggregate in an aqueous environment, they are ideally suited for this chore. It is for this reason that hsc70s are referred to as "molecular chaperones" or "unfoldases" (Ellis 1987; Rothman 1989; Gething and Sambrook 1992; Craig 1993; Hendrick and Hartl 1993). The second part of this review considers how the hsc70s may regulate the activity of the translocation apparatus and provide the energy for translocation. We discuss the cytoplasmic hsc70s that act as chaperones in both mammals and yeast, as well as the essential membrane and lumenal translocation factors in the yeast ER that are DnaK and DnaJ homologs. Finally, we present a model demonstrating how some of the yeast translocation factors might function, and we consider other components that may be required for protein translocation.

II. MAMMALIAN HSC70 AND POSTTRANSLATIONAL TRANSLOCATION

Protein translocation in ER-derived mammalian microsomes mainly proceeds cotranslationally (for review, see Walter and Lingappa 1986; Siegel and Walter 1988). As the signal peptide emerges from the ribosome, it binds to a ribonucleoprotein particle known as the signal recognition particle (SRP). SRP has at least three functions: It arrests or retards translocation, it binds to the signal peptide and prevents it from being buried in the nascent protein, and it targets the ribosome/nascent chain complex to the ER membrane through its affinity for a membrane protein complex of the ER, the docking protein or SRP receptor (SR). The SR complex contains an integral membrane component (the β subunit) and a peripheral membrane protein (the α subunit). Interaction of SRP with SR at the ER membrane triggers the release of the signal peptide from SRP. Cotranslational translocation of the nascent polypeptide commences once SRP is released from the SR and the membrane, in a reaction that requires GTP hydrolysis (Connolly and Gilmore 1989; Connolly et al. 1991). The essential feature of this pathway is that the translocated protein is kept from folding until it is in the ER lumen.

Some proteins can be translocated into mammalian microsomes independently of the SRP-mediated pathway. Zimmermann and colleagues have demonstrated that small (about 60–70-amino-acid) precursors can be translocated posttranslationally into dog pancreas microsomes in the absence of SRP and SR (Watts et al. 1983; Zimmermann and Mollay 1986; Schlenstedt and Zimmermann 1987; Wiech et al. 1987; Schlenstedt et al. 1990, 1992). Because SRP-mediated arrest of translation in vitro yields nascent chains of only 70 amino acids (Walter and Blobel 1981), a precursor such as preprocecropin, which is only 64 amino acids in its entirety, is unlikely to interact with SRP during its synthesis. Posttranslational translocation of these precursors requires ATP as well as factors present in the rabbit reticulocyte lysate in which the precursors are synthesized (Schlenstedt and Zimmermann 1987; Wiech et al 1987; Müller and Zimmermann 1988; Schlenstedt et al. 1990, 1992). When chemically synthesized preprocecropin is added directly to a reaction containing microsomes, reticulocyte lysate, and ATP, translocation is not observed (Klappa et al. 1991). When the peptide is first denatured before being added to the reaction, translocation occurs but without a requirement for any additional cytosolic factors (Klappa et al. 1991). This result implies that posttranslational translocation requires a partially denatured polypeptide or at least a precursor that lacks significant tertiary structure. Previous work on the posttranslational translocation of precursors into isolated mitochondria, into yeast microsomes, and across the inner membrane of *E. coli* led other investigators to reach identical conclusions

(Chirico et al. 1988; Deshaies et al. 1988; Verner and Schatz 1988; Wickner et al. 1991; Langer and Neupert, this volume). Specific cytosolic proteins are required to maintain small precursors in a translocation-competent (i.e., unfolded) conformation. Pelham (1986) suggested that hsc70s are suited to perform this role. To test this hypothesis, the M13 procoat protein was synthesized from an *E. coli* lysate and mixed with pure hsc70 or bovine serum albumin (BSA) (Zimmermann et al. 1988). Only the cytosolic hsc70 increases the protease sensitivity of the peptide. The M13 procoat is translocated into dog pancreas microsomes in incubations that contain pure hsc70 and ATP and limiting amounts of reticulocyte lysate (Zimmermann et al. 1988). These observations suggest that the hsc70-bound precursor is unfolded (i.e., protease-sensitive) and retained in a translocation-competent state. Recently, Wiech et al. (1993) demonstrated that hsc70-stimulated translocation is specific for the cytosolic hsc70, because two other hsc70s, DnaK and BiP, fail to substitute for the cytosolic hsc70 and do not competitively inhibit translocation when they are added to a reaction that already contains hsc70. In accordance with this result, we find that neither DnaK nor BiP substitutes for the yeast hsc70 requirement in an in vitro translocation assay with yeast microsomes (see below; Brodsky et al. 1993). One could conclude from these experiments that DnaK and BiP fail to bind productively to the precursor, a result that is somewhat surprising because DnaK, cytosolic hsc70, and BiP are more than 50% identical (Rose et al. 1989). An additional cytosolic factor is required for translocation that may interact with hsc70 or be required for the proper association/dissociation of hsc70 and the precursor polypeptide.

III. POSTTRANSLATIONAL TRANSLOCATION IN YEAST: INVOLVEMENT OF CYTOSOLIC HSC70

The in vitro assay for protein translocation into yeast microsomes measures the conversion of a yeast mating pheromone precursor, prepro-α factor (ppαf), to its signal-peptide-cleaved and triply glycosylated form (3gpαf). Translocation of ppαf into microsomes occurs posttranslationally and requires the hydrolysis of ATP and a soluble factor(s) that is found in a yeast post-ribosomal supernatant fraction (Hansen et al. 1986; Rothblatt and Meyer 1986; Waters and Blobel 1986; Waters et al. 1986). A soluble factor required for ppαf translocation, Ssa1p/Ssa2p, was identified through combined genetic and biochemical efforts (Chirico et al. 1988; Deshaies et al. 1988).

The yeast *Saccharomyces cerevisiae* contains four genes, *SSA1* through *SSA4*, which encode cytosolic hsc70s that are about 80% identical to each other (for review, see Craig 1990; Craig et al., this volume). The *SSA* genes are characterized by a high degree of functional redundancy and complex genetic interactions. For example, yeast are viable at any temperature when either the *SSA1* or *SSA2* genes are deleted, but the *ssa1ssa2* double mutant is temperature-sensitive for growth (Craig and Jacobsen 1984). Cell growth at the permissive temperature in the *ssa1ssa2* strain is due to the expression of the *SSA4* gene product, which normally is expressed at elevated temperatures but not at room temperature (Werner-Washburne et al. 1987). As expected, the *ssa1ssa2ssa4* triple deletion strain is inviable at room temperature, but it can be rescued by expressing the *SSA3* gene product, a factor that like Ssa4p is synthesized only at elevated temperatures (Werner-Washburne et al. 1987).

Deshaies et al. (1988) used a strain in which the *SSA1 SSA2 SSA4* genes were deleted but which contained a plasmid with the *SSA1* gene subject to a galactose-regulated promoter. When growing cells are shifted from galactose-containing media to glucose-containing media, the cells stop dividing and begin to accumulate untranslocated ppαf in the cytosol. Translocation of a vacuolar protein, carboxypeptidase Y, into the ER is also impaired (Deshaies et al. 1988). Waters et al. (1986) previously found that when ppαf is synthesized in a wheat-germ extract, translocation into yeast microsomes requires an energy source and additional proteins contained in a post-ribosomal supernatant fraction. Using this assay, Deshaies et al. (1988) showed that Ssa1p in the presence of limiting amounts of cytosol also restored translocation activity to wheat-germ-translated ppαf. These authors concluded that Ssa1p, a cytosolic hsc70, is required for protein translocation in yeast.

Chirico et al. (1988) simultaneously confirmed the results of Deshaies et al. (1988) by purifying the proteins from yeast cytosol that enhance the translocation of wheat-germ-translated ppαf into yeast microsomes. They obtained a fraction that contained the pure *SSA1* and *SSA2* gene products (Chirico et al. 1988). The presumed functional redundancy of Ssa1p and Ssa2p is not surprising given that these proteins are predicted to be 97% identical at the amino acid level (Craig and Jacobsen 1984).

How do the Ssa1p/Ssa2p hsc70s stimulate protein translocation? By coupling successive rounds of ATP hydrolysis to the binding and release of hydrophobic domains in the precursor (Flynn et al. 1991), they could "coat" and retain the polypeptide in an unfolded conformation in the cytosol. In support of this hypothesis, Chirico et al. (1988) observed that urea-denatured ppαf was translocated faster than native ppαf in the yeast translocation reaction. When Ssa1p/Ssa2p are also present in these reactions, the translocation rates of both the native and the denatured precursors are further increased. Therefore, the urea-denatured precursor is an even better substrate in a translocation reaction if the hsc70s are present. Interestingly, although the initial rates of translocation of the denatured ppαf (in the absence of hsc70) and the native ppαf (with hsc70) are different, the same yield of translocation is observed in the two reactions. This result implies that the Ssa proteins are as efficient as urea in the ability to form translocation-competent (i.e., partially unfolded) precursors.

Chirico (1992) recently demonstrated a direct interaction between ppαf and cytosolic hsc70. Antibody against hsc70 coprecipitates ppαf when it is added to a translocation reaction in the absence of membranes. The hsc70-ppαf complex is abolished by the combined addition of yeast cytosol and ATP (Chirico 1992). Overall, posttranslational protein translocation in both yeast (Chirico et al. 1988; Deshaies et al. 1988; Brodsky et al. 1993) and probably mammalian microsomes (Zimmermann et al. 1988; Wiech et al. 1993) requires a cytosolic hsc70 that binds directly to the precursor. In both cases, limiting amounts of cytosol also are required, implying that at least one other component is necessary for protein translocation.

The additional cytosolic component(s) that promotes translocation has not been identified. The factor is sensitive to N-ethylmaleimide (NEM) (Chirico et al. 1988; Wiech et al. 1993), but attempts to purify it from either reticulocyte lysate or yeast cytosol have been unsuccessful (Wiech and Zimmermann 1993; B. Koch and R. Schekman, unpubl.). One possibility is that the factor regulates the activity of the cytosolic hsc70. For example, it may free the precursor from the hsc70 and transfer it to the translocation apparatus. A likely candidate for this factor would be a cytosolic eukaryotic homolog of DnaJ or GrpE, two proteins that are known to modulate the activity of the *E. coli* hsc70, DnaK (Liberek et al. 1991).

Yeast contain at least five DnaJ homologs (for review, see Caplan et al. 1993; Silver and Way 1993). Two of these homologs, Sis1p and Ydj1p, are found in yeast cytosol. Sis1p is required for the initiation of protein translation (Zhong and Arndt 1993) and is also present in the nucleus (Luke et al. 1991). Ydj1p is a prenylated protein that shows cytoplasmic and perinuclear ER localization by indirect immunofluorescence (Caplan and Douglas 1991; Caplan et al. 1992b). *YDJ1* is isogenic with *MAS5*, a gene that was identified in a screen for mutants that accumulate untranslocated mitochondrial precursor proteins (Atencio and Yaffe 1992). This observation suggests that the *YDJ1/MAS5* gene product may have a role in mitochondrial and ER translocation, since Ssa1p/Ssa2p is required for the translocation of both mitochondrial and ER precursor proteins (Chirico et al. 1988; Deshaies et al. 1988). Caplan et al. (1992a) confirmed that Ydj1p/Mas5p participates in ER protein translocation, as a yeast strain that contains a temperature-sensitive mutant allele of *YDJ1* (*ydj1-151*) accumulates untranslocated precursors of both a mitochondrial and an ER protein at the nonpermissive temperature (Caplan et al. 1992a). In combination with the results of Deshaies et al. (1988) and Chirico et al. (1988), which show the involvement of the hsc70s in translocation, these results imply that the *SSA1* (the DnaK homolog) and *YDJ1/MAS5* (the DnaJ homolog) gene products act together to promote posttranslational translocation in yeast.

To determine whether the DnaK and DnaJ homologs in yeast cytosol interact in vitro, Cyr et al. (1992) prepared pure Ssa1p and carboxy-methylated α-lactalbumin (CMLA), a protein substrate that is modified so that it maintains an unfolded conformation in the absence of denaturant. A CMLA-Ssa1p complex was isolated by native gel electrophoresis after the pure proteins had been incubated together at 30°C. The CMLA-Ssa1p complex was dissociated by the addition of ATP and Ydj1p but not by either one alone (Cyr et al. 1992). The authors also demonstrated that purified Ydj1p stimulated the ATPase activity of Ssa1p tenfold. The purified protein product of the temperature-sensitive mutant allele of *YDJ1*, *ydj1-151*, is only 16% as efficient as the wild-type protein in activating the ATPase activity of Ssa1p at 30°C (Caplan et al. 1992a). The conclusion is that protein translocation in yeast of at least some substrates requires active Ssa1p/Ssa2p and Ydj1p. Ssa1p/Ssa2p and Ydj1p interact in vitro and presumably interact in the yeast cytosol to facilitate the translocation of certain precursor proteins.

A model for the function of the yeast cytosolic hsc70s has precursor-bound Ssa1p/Ssa2p targeted to the ER membrane by virtue of its affinity for Ydj1p (see Fig. 1). Ydj1p is attached to the membrane by the presence of a farnesyl group at the carboxyl terminus of the protein (Caplan et al. 1992b). Alternatively, because a significant proportion of Ydj1p is cytoplasmic (Caplan and Douglas 1991), a complex containing Ssa1p/Ssa2p, Ydj1p, and protein precursor may form in the cytosol and then be recruited to the ER membrane by some other means. The role of Ydj1p may simply be to stabilize the complex that consists of the hsc70 and the unfolded protein precursor. A factor at the membrane would then release the molecular chaperones from the precursor as translocation begins. The dissociation of the precursor may require the hydrolysis of ATP by hsc70, an event that could be stimulated by a protein in the ER membrane. Conversely, Ydj1p may bind directly to the unfolded precursor, consistent with the observation that DnaJ prevents the aggregation of an unfolded polypeptide upon dilution out of denaturant (Langer et al.

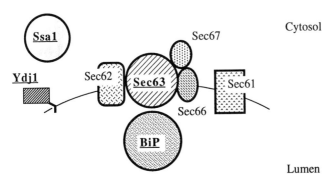

Figure 1 Factors involved in posttranslational translocation in the yeast ER membrane. hscs are written in boldface and are underlined. The arrangement of the Sec proteins reflects the observations of Deshaies et al. (1991) and Brodsky and Schekman (1993). See text for details.

1992). In this scenario, Ssa1p/Ssa2p and/or Ydj1p are released at the ER membrane when the precursor is transferred to the translocation apparatus. The primary feature of these models for posttranslational translocation is that the precursor polypeptide is prevented from folding until it is in the ER lumen.

Yeast cytosolic factors in addition to Ssa1p/Ssa2p and Ydj1p are required for protein translocation in vivo. Genes that encode the *S. cerevisiae* homologs of the 54-kD (*SRP54*) subunit, the 19-kD (*SRP19*) subunit (*SEC65*), and the RNA molecule (scR1) of the mammalian SRP complex have been isolated (Hann et al. 1989; Amaya et al. 1990; Hann and Walter 1991; Stirling and Hewitt 1992), along with the α subunit of SR (Ogg et al. 1992). When the Srp54p or SRα proteins are depleted from growing yeast, the cells accumulate untranslocated precursor proteins in the cytosol (Amaya and Nakano 1991; Hann and Walter 1991; Ogg et al. 1992). Surprisingly, disruption of the genes that encode the Srp54p, Sec65p, and SRα proteins or the scR1 RNA component produces strains that are viable but grow poorly (Felici et al. 1989; Hann and Walter 1991; Stirling and Hewitt 1992).

There is strong biochemical and genetic evidence that yeast Srp54p, Sec65p, and scR1 form a complex (Hann and Walter 1991; Hann et al. 1992; Stirling and Hewett 1992). Assuming that the yeast SRP complex, like the mammalian SRP complex, mediates cotranslational protein translocation, why are the genes that encode its components dispensable? One possibility is that protein precursors may be translocated either co- or posttranslationally in yeast. In the absence of the components required for the cotranslational pathway for protein translocation, the factors in the posttranslational pathway (e.g., Ssa1p/Ssa2p and Ydj1p) may substitute functionally for SRP and SR. It is also possible that the cytosolic hsc70s and Ydj1p in yeast facilitate the cotranslational or SRP-mediated pathway for protein translocation. For example, cytosolic hsp70 associates with newly synthesized proteins and nascent polypeptide chains in HeLa cells (Beckmann et al. 1990). The reconstitution of protein translocation using purified cytosolic components ultimately will determine how all of these factors interact.

IV. ER MEMBRANE AND LUMENAL PROTEINS REQUIRED FOR PROTEIN TRANSLOCATION IN YEAST: IDENTIFICATION OF TWO HEAT SHOCK PROTEIN HOMOLOGS

Five essential genes in yeast are required for protein translocation in vivo and encode membrane or lumenal proteins of the ER: *SEC61, SEC62, SEC63, KAR2,* and *SSS1. SEC61, SEC62,* and *SEC63* were obtained through genetic selections designed to identify mutant yeast strains that accumulate untranslocated protein precursors at the nonpermissive temperature (Deshaies and Schekman 1987; Toyn et al. 1988; Rothblatt et al. 1989; Stirling et al. 1992). The *KAR2* gene was identified both because it is homologous to the 70-kD mammalian protein BiP (Normington et al. 1989) and because *kar2* mutant strains are defective for karyogamy (nuclear fusion) (Rose et al. 1989). Sec61p, Sec62p, and Sec63p are ER membrane proteins (Deshaies and Schekman 1990; Feldheim et al. 1992; Stirling et al. 1992). Sec61p is predicted to span the ER bilayer eight to ten times and is similar to *E. coli* SecY and mammalian Sec61, proteins that are thought to be part of the protein translocation pore (Görlich et al. 1992b; Rapoport 1992; Stirling et al. 1992). Consistent with this prediction, a chemical cross-linking agent caused Sec61p to be coupled to a precursor protein that was partially translocated into the ER lumen of yeast microsomes (Müsch et al. 1992; Sanders et al. 1992). Sec62p spans the ER membrane twice, and microsomes prepared from a *sec62* strain are defective for translocation in vitro (Deshaies and Schekman 1989, 1990). However, the function of Sec62p during protein translocation remains unknown.

The *KAR2* gene encodes the yeast homolog of the mammalian ER lumenal hsc70, BiP. Kar2p is essential for vegetative growth and is 50% identical to the *E. coli* hsp70 homolog, DnaK (Normington et al. 1989; Rose et al. 1989). Yeast strains with certain mutations in the *KAR2* gene (known as class I mutants) accumulate untranslocated protein precursors in the cytosol and stop dividing at the nonpermissive temperature (Vogel et al. 1990). A similar phenotype is seen when a galactose-regulated promoter is used to deplete BiP from growing wild-type yeast (Vogel et al. 1990; Nguyen et al. 1991). Microsomes prepared from strains with class I mutations in *kar2* display a temperature-sensitive translocation defect in an in vitro translocation assay (Sanders et al. 1992).

Because a DnaK homolog in the yeast ER is required for protein translocation both in vivo and in vitro, one might expect that an essential DnaJ homolog also exists in this compartment; Sec63p appears to fill this role. Microsomes prepared from the *sec63-1* mutant strain show a temperature-sensitive translocation defect in vitro (Rothblatt et al. 1989). Sec63p spans the ER membrane three times and contains a domain in the lumen of the ER that is 43% identical to DnaJ (Sadler et al. 1989; Feldheim et al. 1992). The DnaJ domain of Sec63p is oriented such that it is in the same compartment as BiP, leading to the prediction that the two proteins physically interact (Normington et al. 1989; Rose et al. 1989). Furthermore, because both proteins are required for protein translocation and because the bacterial versions of these proteins (DnaK and DnaJ) cooperate to regulate several biochemical processes (Ang et al. 1991), Sec63p-BiP interaction similarly may regulate some aspect of protein translocation.

To examine how the components of the yeast translocation apparatus operate (see Fig. 1), we have reconstituted the protein translocation reaction in vitro (Brodsky et al. 1993). Yeast microsomes are solubilized with octylglucoside, and the proteins are reintroduced into a synthetic phospholipid bilayer upon removal of the detergent. The resulting proteoliposomes translocate ppαf in a reaction that is dependent on ATP hydrolysis and wild-type *SEC* gene products, mimicking the microsome-based translocation reaction (Rothblatt et al. 1989). Intravesicular BiP is also necessary for translocation in the reconstituted proteoliposomes; neither Ssa1p nor DnaK substitutes for BiP in promoting translocation into the vesicles (Brodsky et al. 1993). In addition, neither BiP nor DnaK replaces the cytosolic Ssa1p requirement to facilitate posttranslational translocation (Brodsky et al. 1993). Similarly, posttranslational transloca-tion assayed with mammalian microsomes is stimulated by hsc70 but not by BiP (see above; Wiech et al. 1993). Together, these results suggest that specific hsc70s are topologically oriented to fulfill unique roles during protein translocation. In the cytosol, Ssa1p probably interacts with a DnaJ homolog, Ydj1p, during translocation (see above). In the ER lumen, BiP may have to interact specifically with Sec63p to promote translocation, whereas other hsc70s may be unable to bind to Sec63p or to interact functionally with the translocation complex. Alternatively, BiP may bind directly to the translocating polypeptide in the lumen of the ER (see below) in a manner that cannot be satisfied by cytosolic hsc70s.

Deshaies et al. (1991) investigated molecular interactions among members of the translocation Sec protein family. Radioactively labeled yeast microsomes were solubilized, treated with cross-linker, and incubated with antibodies to either Sec62p or Sec63p. The resulting immunoprecipitate contained Sec61p, Sec62p, Sec63p, and two proteins with molecular masses of 31.5 kD and 23 kD. In native immunoprecipitations, Sec62p, Sec63p, and the 31.5- and 23-kD proteins cosediment (Deshaies et al. 1991), implying that the *SEC* gene products form a complex in the membrane. Because Sec61p associates with the complex only in the presence of cross-linker, its association may be transient or less tight. The existence of this complex was predicted on the basis of genetic interaction in the form of synthetic lethality (Rothblatt et al. 1989). Strains that contain pairwise combinations of temperature-sensitive alleles of *sec61*, *sec62*, and *sec63* are unable to grow at temperatures that normally are permissive for the single mutant alleles (Rothblatt et al. 1989). In other instances, synthetic lethality indicates that the corresponding wild-type gene products physically interact or are required for a common step in a biochemical pathway (Huffaker et al. 1987; Salminen and Novick 1987; Kaiser and Schekman 1990).

The 31.5- and 23-kD proteins in the translocation complex have been cloned and sequenced (Feldheim et al. 1993; Kurihara and Silver 1993; D. Feldheim and R. Schekman, unpubl.). The genes encoding these proteins are not essential, but strains that have deletions in either gene accumulate untranslocated protein precursors (Feldheim et al. 1993; Kurihara and Silver 1993; D. Feldheim and R. Schekman, unpubl.). The corresponding gene products are now known as Sec66p (the 31.5-kD protein) and Sec67p (the 23-kD protein). Microsomes and reconstituted vesicles prepared from the *sec66* or *sec67* deleted strains are defective for translocation in vitro (Brodsky and Schekman 1993). *SEC66* and *SEC67* are probably identical to the *SEC71* and *SEC72* genes, respectively (N. Green, unpubl.), which were identified previously in a genetic screen for mutants that are defective in the insertion of a membrane fusion protein (Green et al. 1992).

Another component of the translocation complex was recently identified in a screen for multicopy suppressors of the *sec61* temperature-sensitive phenotype (Esnault et al. 1993). The gene, *SSS1*, encodes a 9-kD ER protein and is essential. Depletion of the protein results in translocation defects, although its overproduction restores translocation in a *sec61* strain. Chemical cross-linking indicates that Sss1p and Sec61p interact directly (D. Feldheim, unpubl.).

The absence of BiP from the solubilized, cross-linked Sec protein complex in yeast microsomes (Deshaies et al. 1991) was unexpected. Hence, either the BiP-Sec63p association is labile or the two proteins may not make direct contact in vivo. We addressed this issue by developing a scheme for the purification of functional Sec63p from octylglucoside-solubilized membranes (Brodsky and Schekman 1993). After purification by DEAE-Sepharose, Superose-6, and hydroxyapatite chromatography, we obtained a protein complex that contains Sec63p, Sec66p, Sec67p, and BiP. BiP remains in the complex if the final step of the purification is performed in buffer or buffer supplemented with ATP, but BiP dissociates from the complex if the nonhydrolyzable ATP analog, ATPγS, is included; therefore, ATP binding and hydrolysis regulate BiP's association. BiP also separates from the complex if membranes from the *sec63-1* strain are used as the starting material for the purification (Brodsky and Schekman 1993). The point mutation in *sec63-1* converts a conserved alanine in the DnaJ domain to a threonine (Nelson et al. 1993); thus, it seems probable that the DnaJ domain mediates the interaction between BiP and Sec63p.

A complementary genetic approach was taken to examine the interaction between Sec63p and BiP. Scidmore et al. (1993) isolated dominant suppressors in the *sec63-1* temperature-sensitive mutant strain that localized to point mutations in *KAR2*. In addition, the *sec63-1* mutation and some *kar2* mutant alleles display synthetic lethal interactions. The genetic data support the conclusion that Sec63p and BiP interact.

There is a large excess of BiP relative to Sec63p in the yeast ER; therefore, the majority of BiP is not associated with the Sec63p complex (Brodsky and Schekman 1993), suggesting that BiP has additional physiological roles in the ER. Yeast BiP can be expected to bind to solvent-exposed hydrophobic domains of denatured or unfolded proteins, as is the case for mammalian BiP (Flynn et al. 1991). These substrates accumulate upon cellular stress, such as heat shock. In fact, transcription of the *KAR2* gene and BiP levels are increased by heat shock or by the accumulation of misfolded proteins in the ER (Normington et al. 1989; Rose et al. 1989; Tokunaga et al. 1992; Kohno et al. 1993). During protein translocation, polypeptides emerge in the intravesicular space in an unfolded conformation (Verner and Schatz 1988). Mitochondrial and ER hsc70s may bind to emerging precursor polypeptides during translocation. Scherer et al. (1990) demonstrated that the yeast mitochondrial matrix hsc70, Ssc1p, became associated with a modified precursor protein arrested during protein translocation into mitochondria. Sanders et al. (1992) subsequently showed that yeast BiP became associated with nascent polypeptides in the lumen of the ER, as BiP could be chemically

cross-linked to a translocation-arrested precursor protein in yeast microsomes. Cross-linking requires ATP and is less efficient if microsomes
contain a class I *kar2* mutation (Sanders et al. 1992). Overall, association
of the precursors with the chaperones prevents unproductive folding
pathways (i.e., aggregation) from occurring and thus stabilizes the
polypeptide until translocation is complete. Proteins attain their native
structures after they are released from hsc70 (Gething and Sambrook
1992).

To define further the role that BiP plays in the ER, J. Vogel et al.
(pers. comm.) isolated dominant lethal mutations in BiP. There are many
ways to envision how a dominant mutant form of BiP could act. The
protein may bind but fail to release the translocating polypeptide in the
ER or may have an altered rate of ATPase activity (J. Vogel, pers.
comm.). If the dominant BiP mutants are unable to exchange ATP for
ADP, they would be bound constitutively to Sec63p and keep the translocation machine in a permanently "on" or "off" configuration (Brodsky
and Schekman 1993). The net effect is that the cell dies as precursor
proteins accumulate in the cytosol or jam the translocation complex.

V. MODELS FOR BIP FUNCTION DURING PROTEIN TRANSLOCATION IN YEAST

What is the driving force for translocation or, more specifically, how is
protein translocation across the ER membrane coupled to the hydrolysis
of ATP? To date, only two factors involved in the translocation reaction
have been identified as ATPases: Ssa1p/Ssa2p and BiP (see Fig. 1). Only
BiP has been shown to associate with the translocation complex
(Brodsky and Schekman 1993). How does BiP use ATP binding and
hydrolysis to activate the translocation complex or to drive the reaction
forward? To answer these questions, the overall translocation reaction
must be dissected into a stepwise process.

The following steps in yeast protein translocation can be deduced
from the cross-linking experiments of Sanders et al. (1992) and Müsch et
al. (1992): (1) Protein precursors can bind to a member of the translocation complex (Sec62p; Deshaies et al. 1991) in the absence of an energy
source; (2) ATP, Sec62p, and Sec63p are required to transfer the precursor to Sec61p; and (3) BiP function is required for at least two steps in
protein translocation: association of the precursor with Sec61p and translocation of the polypeptide across the ER membrane.

In Figure 2, we depict four stages of protein translocation across the
yeast ER membrane. In panel A, the nascent polypeptide engages an acceptor protein complex at the ER membrane. As discussed above, the
precursor is transferred to the acceptor after it is released from cytosolic

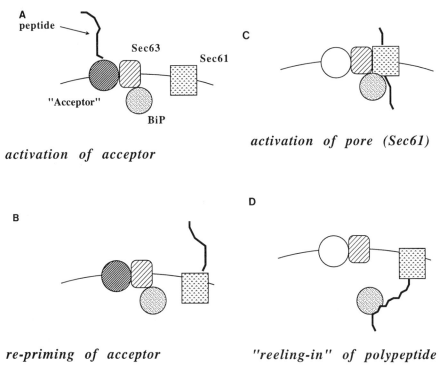

Figure 2 Models for BiP activity during protein translocation in the yeast ER.

chaperones, such as Ssa1p/Ssa2p. Sec62p is predicted to be a component of the acceptor complex, and energy is not required for its association with the precursor (Müsch et al. 1992). Next, the precursor contacts Sec61p in a reaction that requires ATP and Sec63p (Fig. 2B) (Müsch et al. 1992; Sanders et al. 1992). BiP is also required at some point for cross-linking the precursor protein to Sec61p (Sanders et al. 1992), which then becomes intimately associated with the translocating polypeptide in the ER membrane (Fig. 2C). Finally, in the lumen of the ER, the translocated polypeptide interacts with BiP (Fig. 2D) (Sanders et al. 1992) before it is released from the translocation complex and attains its native conformation.

There are two ways to envision how BiP facilitates the interaction of precursor with Sec61p: It may activate or assemble the acceptor complex or it may recycle the acceptor to ready it for another round of transloca- tion. These possibilities are depicted in Figure 2, A and B. How does BiP transduce a signal through the ER membrane to the acceptor complex? BiP forms a complex with Sec63p, Sec66p, and Sec67p (Brodsky and Schekman 1993). There is some evidence to suggest that Sec66p and

Sec67p are involved in signal peptide recognition of protein precursors (D. Feldheim and R. Schekman, unpubl.). Through its interaction with Sec63p, BiP is indirectly associated with Sec62p, Sec66p, and Sec67p, proteins that are probably constituents of the acceptor complex. The acceptor complex may be regulated by a conformational change that could be mediated by BiP binding to ATP and/or ATP hydrolysis. Sec63p may be the transducer of this signal, because BiP binds to the acceptor complex via the DnaJ domain of Sec63p (Brodsky and Schekman 1993). Overall, this model is similar to that proposed for the activation of bacteriophage λ DNA replication by DnaK and DnaJ (Ang et al. 1991). In this case, the DnaK-DnaJ complex uses ATP to remove the λP protein from the preprimosomal protein complex, which in turn stimulates the DnaB helicase. Thus, ATP and DnaK indirectly activate an enzyme in a preassembled complex. Similarly, our model employs ATP and the lumenal DnaK homolog, BiP, to drive protein translocation into the ER.

BiP is also required for a step in translocation after the binding of Sec61p to the precursor polypeptide (Sanders et al. 1992). In Figure 2, C and D, we present two possible roles that BiP could play to complete the translocation of a protein. As discussed above, the BiP-Sec63p complex may activate the translocation complex. One facet of BiP's role may include an indirect interaction with Sec61p to increase the fidelity of translocation (Fig. 2C). The role for BiP as a driving force for protein translocation may be essential because the translocation pore in both mammalian microsomes (Garcia et al. 1988; Ooi and Weiss 1992; Nicchitta and Blobel 1993) and *E. coli* (Schiebel et al. 1991) has been shown to allow bidirectional "slipping" of the translocating polypeptide. Thus, once Sec61p engages the precursor polypeptide, BiP may act to ensure translocation directionality by acting on the Sec61p-precursor complex.

BiP interaction with the precursor in the lumen of the ER may also promote the directionality of translocation (Sanders et al. 1992). Multiple BiP contacts with successively longer stretches of the translocating polypeptide may drive import, or "reel-in" the precursor (Fig. 2D). Because the binding of nascent precursors to BiP requires ATP (Sanders et al. 1992), the hydrolysis of ATP by BiP may be the only energy that is necessary for translocation. Lumenal ATP is achieved by an ATP transporter whose activity is required for translocation (Mayinger and Meyer 1993). In thermodynamic terms, the BiP-precursor protein complex has a lower free energy than the extended polypeptide as it is translocated across the ER membrane. Therefore, it may be the binding energy of the BiP-nascent polypeptide complex that drives translocation forward. Once translocation is complete, ATP hydrolysis would sequentially release the polypeptide and allow folding to begin. In this model, ATP hydrolysis by

BiP must be regulated; Sec63p or perhaps a soluble DnaJ homolog in the ER may also be required. Because the translocating polypeptide on both sides of the ER membrane is bound to an hsc70, the release of both the cytosolic and lumenal factors must be coordinated with translocation. This process could be regulated by membrane-bound DnaJ homologs on both sides of the membrane (Ydj1p and Sec63p).

VI. MAMMALIAN BIP AND TRANSLOCATION

What is the function of BiP in the mammalian microsome during poly-peptide translocation? The available data imply that BiP may not be re-quired for translocation or may be necessary only at low levels or only in special cases. Bulleid and Freedman (1988) and Paver et al. (1989) showed that dog pancreas microsomes washed in a pH 9 buffer were capable of translocating a protein precursor. Under these conditions, the authors demonstrated that BiP was extracted from the washed micro-somes. It is possible that the secretory precursor proteins used in these studies do not require BiP or that low, undetected levels of BiP are suffi-cient to translocate the protein. Similarly, Nicchitta and Blobel (1990) reconstituted the protein translocation reaction from cholate-solubilized dog pancreas microsomes and demonstrated that the reconstituted vesicles were capable of translocating a protein precursor even though BiP was absent. Later, using a different procedure, Nicchitta and Blobel (1993) showed that lumenal components were required to complete translocation into emptied microsomes. It is not clear from these experi-ments whether BiP or some other lumenal protein is rate-limiting.

Recently, Görlich and Rapoport (1993) reconstituted protein trans-location from purified mammalian ER proteins. They found that proteinaceous factors were not required inside the reconstituted vesicles; only the known ER membrane components SR, TRAM (Görlich et al. 1992a), and a complex containing mammalian Sec61p (Görlich et al. 1992b) were necessary to translocate ppαf. The energy required to push the translocating polypeptide through the membrane must come from the ribosome during protein synthesis. The variance between this result and that of Nicchitta and Blobel (1993) with regard to the requirement for lumenal components may be that some previously required factors are no longer necessary when translocation is reconstituted with purified com-ponents. A simpler set of protein requirements is obtained when the *E. coli* translocation reaction is reconstituted with pure proteins (Brundage et al. 1990; Driessen and Wickner 1990).

Is there an inherent difference between protein translocation across the ER membrane in yeast and higher eukaryotes? In mammalian ER translocation, it is possible that BiP still facilitates translocation but is not absolutely required. This issue may be clarified by examining a variety of protein precursors in these systems. It is also possible that in higher eukaryotes, another molecule or set of proteins have substituted for the function of the Sec63p-BiP complex. To test this hypothesis, cross-species mixing experiments can be performed with purified proteins in the yeast and mammalian reconstituted systems (Nicchitta and Blobel 1990; Brodsky et al. 1993). Finally, it is important to emphasize that even though the in vitro mammalian translocation reactions are cotrans-lational, whereas the yeast translocation reactions are posttranslational, mutations in *kar2* also affect precursors that are translocated cotransla-tionally in vivo (Vogel et al. 1990). Therefore, the requirement for BiP in yeast translocation appears to be general. The variance between yeast and mammalian translocation in terms of this requirement has yet to be resolved.

VII. PERSPECTIVES

The study of protein translocation has progressed from the use of crude membrane preparations, to the genetic and biochemical identification of required components, and finally to defined reconstituted systems using some of these purified components. A subset of the factors necessary for protein translocation in the ER are eukaryotic homologs of the *E. coli* proteins DnaJ and DnaK. The thorough biochemical characterization of bacteriophage λ DNA replication by C. Georgopoulos and R. McMacken and their colleagues has contributed to the dissection of the ER protein translocation pathway.

GrpE is another protein involved in λ DNA replication whose homolog has not yet emerged in studies on protein translocation in eukaryotes. GrpE stimulates the nucleotide exchange activity of DnaK (Liberek et al. 1991) and along with ATP catalyzes the release of an un-folded protein from the DnaK-DnaJ complex (Langer et al. 1992). As discussed above, DnaK and DnaJ homologs in both the cytosol and the ER lumen interact with each other. ATP regulates the interaction of BiP and Sec63p (Brodsky and Schekman 1993), and Ydj1p and ATP releases peptide bound to Ssa1p (Cyr et al. 1992). A novel GrpE could act specif-ically with either Ydj1p/Ssa1p or Sec63p/BiP. Prokaryotic GrpE homo-logs at best have a sequence similarity of 50% (Wetzstein and Schumann 1990), which makes the direct cloning of a eukaryotic version prob-lematic. Alternatively, a genetic screen could be designed to reveal a eukaryotic GrpE homolog. Of course, it remains possible that Sec63p and BiP interact and regulate the protein translocation complex in lieu of a GrpE-like nucleotide exchange activity. Nevertheless, the degree of conservation between the protein machines responsible for DNA replication (Ang et al. 1991), protein folding (Langer et al. 1992), and protein translocation favors the existence of a eukaryotic GrpE.

The discovery of accessory molecules involved in protein translocation is probably far from over. There are many unanswered mechanistic questions in protein translocation. The stoichiometry and assembly of the translocation complex have not been investigated sufficiently. In addition, the translocation of multispanning membrane proteins into the ER may require additional factors. A biophysical analysis ultimately will be required to understand the regulation of the protein translocation machine and to obtain a dynamic view of its operation.

ACKNOWLEDGMENTS

The manuscript was greatly improved by the generous comments of D. Feldheim, S. Lyman, and K. Römisch. We also thank J. Vogel (M. Rose laboratory), A. Caplan (M. Douglas laboratory), and H. Wiech (this laboratory) for their many helpful discussions. J.L.B. was supported by a postdoctoral research grant from the American Cancer Society.

REFERENCES

Amaya, Y. and A. Nakano. 1991. SRH1 protein, the yeast homolog of the 54kDa subunit of signal recognition particle, is involved in ER translocation of secretory proteins. *FEBS Lett.* **283**: 325–328.

Amaya, Y., A. Nakano, K. Ito, and M. Mori. 1990. Isolation of a yeast gene, *SRH1*, that encodes a homolog of the 54K subunit of mammalian signal recognition particle. *J. Biochem.* **107**: 457–463.

Ang, D., K. Liberek, D. Skowyra, M. Zylicz, and C. Georgopoulos. 1991. Biological role and regulation of the universally conserved heat shock proteins. *J. Biol. Chem.* **266**: 24233–24236.

Attencio, D.P. and M.P. Yaffe. 1992. *MAS5*, a yeast homolog of dnaJ involved in mitochondrial protein import. *Mol. Cell. Biol.* **12**: 283–291.

Beckmann, R.P., L.A. Mizzen, and W.J. Welch. 1990. Interaction of hsp70 with newly synthesized proteins: Implications for protein folding and assembly. *Science* **248**: 850–854.

Bole, D.G., L.M. Hendershot, and J.F. Kearney. 1986. Posttranslational association of immunoglobin heavy chains in nonsecreting and secreting hybridomas. *J. Cell Biol.* **102**: 1558–1566.

Brodsky, J.L. and R. Schekman. 1993. A Sec63p-BiP complex from yeast is required for protein translocation in a reconstituted proteoliposome. *J. Cell Biol.* **123**: 1355–1363.

Brodsky, J.L., S. Hamamoto, D. Feldheim, and R. Schekman. 1993. Reconstitution of protein translocation from solubilized yeast membranes reveals topologically distinct roles for BiP and cytosolic hsc70. *J. Cell Biol.* **120**: 95–102.

Brundage, L., J.P. Hendrick, E. Sciebel, A.J.M. Driessen, and W. Wickner. 1990. The purified integral membrane protein SecY/E is sufficient for reconstitution of SecA-dependent precursor protein translocation. *Cell* **62:** 649–657.

Bulleid, N.J., and R.B. Freedman. 1988. Defective co-translational formation of di-sulphide bonds in protein disulphide-isomerase-deficient microsomes. *Nature* **335:** 649–651.

Caplan, A.J. and M.G. Douglas. 1991. Characterization of *YDJ1:* A yeast homolog of the *E. coli* dnaJ gene. *J. Cell Biol.* **114:** 609–622.

Caplan, A.J., D.M. Cyr, and M.G. Douglas. 1992a. YDJ1 facilitates polypeptide trans-location across different intercellular membranes by a conserved mechanism. *Cell* **71:** 1143–1155.

———. 1993. Eukaryotic homologs of *Escherichia coli* dnaJ: A diverse protein family that functions with hsp70 stress proteins. *Mol. Biol. Cell* **4:** 555–563.

Caplan, A.J., J. Tsai, P.J. Casey, and M.G. Douglas. 1992b. Farnesylation of YDJ1p is re-quired for function at elevated temperatures in *S. cerevisiae. J. Biol. Chem.* **267:** 18890–18895.

Chirico, W. 1992. Dissociation of complexes between 70Kd. stress proteins and pre-secretory proteins is facilitated by a cytosolic factor. *Biochem. Biophys. Res. Commun.* **189:** 1150–1156.

Chirico, W.J., M.G. Waters, and G. Blobel. 1988. 70K heat shock related proteins stimu-late protein translocation into microsomes. *Nature* **332:** 805–810.

Connolly, T. and R. Gilmore. 1989. The signal recognition particle receptor mediates the GTP-dependent displacement of SRP from the signal sequence of the nascent polypep-tide. *Cell* **42:** 497–505.

Connolly, T., P.J. Rapiejko, and R. Gilmore. 1991. Requirement of GTP hydrolysis for dissociation of the signal recognition particle from its receptor. *Science* **252:** 1171–1173.

Craig, E.A. 1990. Role of hsp70 in translocation of proteins across membranes. In *Stress proteins in biology and medicine* (ed. R.I. Morimoto et al.), pp. 279–286. Cold Spring Harbor Laboratory Press, Cold Spring Harbor, New York.

———. 1993. Chaperones: Helpers along the pathways to protein folding. *Science* **260:** 1902–1903.

Craig, E.A. and K. Jacobsen. 1984. Mutations of the heat inducible 70 kilodalton genes of yeast confer temperature sensitive growth. *Cell* **38:** 841–849.

Crowley, K.S., G.D. Reinhart, and A.E. Johnson. 1993. The signal sequence moves through a ribosomal tunnel into a noncytoplasmic aqueous environment at the ER membrane early in translocation. *Cell* **73:** 1101–1116.

Cyr, D.M., X. Lu, and M.G. Douglas. 1992. Regulation of eucaryotic hsp70 function by a dnaJ homolog. *J. Biol. Chem.* **267:** 20927–20931.

Deshaies, R.D. and R. Schekman. 1987. A yeast mutant defective at an early stage in im-port of secretory precursors into the endoplasmic reticulum. *J. Cell Biol.* **105:** 633–645.

———. 1989. *SEC62* encodes a putative membrane protein required for protein trans-location into the yeast endoplasmic reticulum. *J. Cell Biol.* **109:** 2653–2664.

———. 1990. Structural and functional dissection of Sec62p, a membrane bound com-ponent of the yeast endoplasmic reticulum protein import machinery. *Mol. Cell. Biol.* **10:** 6024–6035.

Deshaies, R.D., S. Sanders, D.A. Feldheim, and R. Schekman. 1991. Assembly of yeast Sec proteins involved in translocation into the endoplasmic reticulum into a membrane-bound multisubunit complex. *Nature* **349:** 806–808.

Deshaies, R.D., B.D. Koch, M. Werner-Washburne, E.A. Craig, and R. Schekman. 1988.

A subfamily of stress proteins facilitates translocation of secretory and mitochondrial precursor proteins. *Nature* **332:** 800–805.

Driessen, A.J.M. and W. Wickner. 1990. Solubilization and functional reconstitution of the protein-translocation enzymes of *Escherichia coli. Proc. Natl. Acad. Sci.* **87:** 3107–3111.

Ellis, R.J. 1987. Proteins as molecular chaperones. *Nature* **328:** 378–379.

Esnault, Y., M. Blondel, R.D. Deshaies, R. Schekman, and F. Kepes. 1993. The yeast *SSS1* gene is essential for secretory protein translocation, and encodes a highly conserved protein of the endoplasmic reticulum. *EMBO J.* **12:** 4083–4093.

Feldheim, D., J. Rothblatt, and R. Schekman. 1992. Topology and functional domains of Sec63p, an endoplasmic reticulum membrane protein required for secretory protein translocation. *Mol. Cell. Biol.* **12:** 3288–3296.

Feldheim, D., K. Yoshimura, A. Admon, and R. Schekman. 1993. Structural and functional characterization of Sec66p, a new subunit of the polypeptide translocation apparatus in the yeast ER. *Mol. Biol. Cell* **4:** 931–939.

Felici, F., G. Cesareni, and J.M.X. Hughes. 1989. The most abundant small cytoplasmic RNA of *Saccharomyces cerevisiae* has an important function required for cell growth. *Mol. Cell. Biol.* **9:** 3260–3268.

Flynn, G.C., T.C. Chappell, and J.E. Rothman. 1989. Peptide binding and release by proteins implicated as catalysts of protein assembly. *Science* **245:** 385–390.

Flynn, G.C., J. Pohl, M.T. Flocco, and J.E. Rothman. 1991. Peptide binding specificity of the molecular chaperone BiP. *Nature* **353:** 726–730.

Garcia, P.D., J.H. Ou, W.J. Rutter, and P. Walter. 1988. Targeting of the hepatitis B virus precore protein to the endoplasmic reticulum membrane: After signal peptide cleavage translocation can be aborted and the product released into the cytoplasm. *J. Cell Biol.* **106:** 1093–1104.

Gething, M.J., and J. Sambrook. 1992. Protein folding in the cell. *Nature* **355:** 33–45.

Gilmore, R. and G. Blobel. 1985. Translocation of secretory proteins across the microsomal membrane occurs through an environment accessible to aqueous perturbants. *Cell* **42:** 497–505.

Görlich, D. and T.A. Rapoport. 1993. Protein translocation into proteoliposomes reconstituted from purified components of the endoplasmic reticulum membrane. *Cell* **75:** 615–630.

Görlich, D., E. Hartmann, S. Prehn, and T. Rapoport. 1992a. A protein of the endoplasmic reticulum involved early in polypeptide translocation. *Nature* **357:** 47–52.

Görlich, D., S. Prehn, E. Hartmann, K.U. Kalies, and T.A. Rapoport. 1992b. A mammalian homolog of SEC61p and SECY is associated with ribosomes and nascent polypeptides during translocation. *Cell* **71:** 489–503.

Green, N., H. Fang, and P. Walter. 1992. Mutants in three novel complementation groups inhibit membrane protein insertion into and soluble protein translocation across the endoplasmic reticulum membrane of *Saccharomyces cerevisiae. J. Cell Biol.* **116:** 597–604.

Haas, I.G. and M. Wabl. 1983. Immunoglobin heavy chain binding protein. *Nature* **306:** 387–389.

Hann, B.C. and P. Walter. 1991. The signal recognition particle in yeast. *Cell* **67:** 131–143.

Hann, B.C., M.A. Poritz, and P. Walter. 1989. *Saccharomyces cerevisiae* and *Schizosaccharomyces pombe* contain a homolog to the 54-kD subunit of the signal recognition particle that in *S. cerevisiae* is essential for growth. *J. Cell Biol.* **109:** 3223–3230.

Hann, B.C., C. Stirling, and P. Walter. 1992. *SEC65* gene product is a subunit of the yeast signal recognition particle required for its integrity. *Nature* **356:** 532–533.

Hansen, W., P.D. Garcia, and P. Walter. 1986. In vitro protein translocation across the yeast endoplasmic reticulum: ATP-dependent post-translational translocation of the prepro-α-factor. *Cell* **45:** 397–406.

Hendrick, J.P. and F.U. Hartl. 1993. Molecular chaperoning functions of heat shock proteins. *Annu. Rev. Biochem.* **62:** 349–384.

Huffaker, T.C., M.A. Hoyt, and D. Botstein. 1987. Genetic analysis of the yeast cytoskeleton. *Annu. Rev. Genet.* **21:** 259–284.

Kaiser, C. and R. Schekman. 1990. Distinct sets of SEC genes govern transport vesicle formation and fusion early in the secretory pathway. *Cell* **61:** 723–733.

Klappa, P., P. Mayinger, R. Pipkorn, M. Zimmermann, and R. Zimmermann. 1991. A microsomal protein is involved in ATP-dependent transport of presecretory proteins into mammalian microsomes. *EMBO J.* **10:** 2795–2803.

Kohno, K., K. Normington, J. Sambrook, M.J. Gething, and K. Mori. 1993. The promoter region of the yeast *KAR2* (BiP) gene contains a regulatory domain that responds to the presence of unfolded proteins in the endoplasmic reticulum. *Mol. Cell. Biol.* **13:** 877–890.

Kurihara, T. and P. Silver. 1993. Suppression of a *sec63* mutation identifies a novel component of the yeast ER translocation apparatus. *Mol. Biol. Cell* **4:** 919–930.

Langer, T., C. Lu, H. Echols, J. Flanagan, M.K. Hayer, and F.U. Hartl. 1992. Successive action of dnaK, dnaJ and GroEL along the pathway of chaperone-mediated folding. *Nature* **356:** 683–689.

Liberek, K., J. Marszalek, D. Ang, C. Georgopoulos, and M. Zylicz. 1991. *Escherichia coli* dnaJ and grpE heat shock proteins jointly stimulate ATPase activity of dnaK. *Proc. Natl. Acad. Sci.* **88:** 2874–2878.

Luke, M.M., A. Sutton, and K.A. Arndt. 1991. Characterization of *SIS1*, a *Saccharomyces cerevisiae* homolog of bacterial dnaJ proteins. *J. Cell Biol.* **114:** 623–638.

Mayinger, P. and D.I. Meyer. 1993. An ATP transporter is required for protein translocation into the yeast endoplasmic reticulum. *EMBO J.* **12:** 659–666.

Müller, G. and R. Zimmermann. 1988. Import of honeybee prepromelittin into the endoplasmic reticulum: Energy requirements for membrane insertion. *EMBO J.* **7:** 639–648.

Munro, S. and H.R.B. Pelham. 1986. An hsp70 like protein in the ER: Identity of the 78-kd glucose regulated protein and immunoglobin heavy chain binding protein. *Cell* **46:** 291–300.

Müsch, A., M. Wiedmann, and T.A. Rapoport. 1992. Yeast Sec proteins interact with polypeptides traversing the endoplasmic reticulum membrane. *Cell* **69:** 343–352.

Nelson, M.K., T. Kurihara, and P.A. Silver. 1993. Extragenic suppressors of mutations in the cytoplasmic C terminus of *SEC63* define five genes in *Saccharomyces cerevisiae*. *Genetics* **134:** 159–173.

Nguyen, T.H., D.T.S. Law, and D.B. Williams. 1991. Binding protein BiP is required for translocation of secretory proteins into the endoplasmic reticulum in *Saccharomyces cerevisiae*. *Proc. Natl. Acad. Sci.* **88:** 1565–1569.

Nicchitta, C.V. and G. Blobel. 1990. Assembly of translocation-competent proteoliposomes from detergent-solubilized rough microsomes. *Cell* **60:** 259–269.

———. 1993. Lumenal proteins of the mammalian endoplasmic reticulum are required to complete protein translocation. *Cell* **73:** 989–998.

Normington, K., K. Kohno, Y. Kozutsumi, M.J. Gething, and J. Sambrook. 1989. *S. cerevisiae* encodes an essential protein in sequence and function to mammalian BiP. *Cell* **57:** 1223–1236.

Nunnari, J. and P. Walter. 1992. Protein targeting to and translocation across the membrane of the endoplasmic reticulum. *Curr. Opin. Cell Biol.* **4:** 573–580.

Ogg, S.C., M.A. Poritz, and P. Walter. 1992. Signal recognition particle receptor is important for cell growth and protein secretion in *Saccharomyces cerevisiae. Mol. Biol. Cell* **3:** 895–911.

Ooi, C.E. and J. Weiss. 1992. Bidirectional movement of a nascent polypeptide across microsomal membranes reveals requirements for vectorial translocation of proteins. *Cell* **71:** 87–96.

Paver, J.L., H.C. Hawkins, and R.B. Freedman. 1989. Preparation and characterization of dog pancreas microsomal membranes specifically depleted of protein disulphide-isomerase. *Biochem. J.* **257:** 657–663.

Pelham, H.R.B. 1986. Speculations on the functions of the major heat shock and glucose regulated proteins. *Cell* **46:** 959–961.

Rapoport, T.A. 1992. Transport of proteins across the endoplasmic reticulum membrane. *Science* **258:** 931–936.

Rose, M.D., L.M. Misra, and J.P. Vogel. 1989. *KAR2*, a karyogamy gene, is the yeast homolog of the mammalian BiP/GRP78 gene. *Cell* **57:** 1211–1221.

Rothblatt, J.A. and D.I. Meyer. 1986. Secretion in yeast: Reconstitution of the translocation and glycosylation of α-factor and invertase in a homologous cell-free system. *Cell* **44:** 619–628.

Rothblatt, J.A., R.J. Deshaies, S.L. Sanders, G. Daum, and R. Schekman. 1989. Multiple genes are required for proper insertion of secretory proteins into the endoplasmic reticulum in yeast. *J. Cell Biol.* **109:** 2641–2652.

Rothman, J.E. 1989. Polypeptide chain binding proteins: Catalysts of protein folding and related processes in cells. *Cell* **59:** 591–601.

Sadler, I., A. Chiang, T. Kurihara, J. Rothblatt, J. Way, and P. Silver. 1989. A yeast gene important for protein assembly into the endoplasmic reticulum and the nucleus has homology to dnaJ, an *Escherichia coli* heat shock protein. *J. Cell Biol.* **109:** 2665–2675.

Salminen, A. and P. Novick. 1987. A ras-like protein is required for a post-golgi event in yeast secretion. *Cell* **49:** 527–538.

Sanders, S.L. and R. Schekman. 1992. Polypeptide translocation across the endoplasmic membrane. *J. Biol. Chem.* **267:** 13791–13794.

Sanders, S.L., K.M. Whitfield, J.P. Vogel, M.D. Rose, and R.W. Schekman. 1992. Sec61p and BiP directly facilitate polypeptide translocation into the ER. *Cell* **69:** 353–365.

Scherer, P.E., U.C. Krieg, S.T. Hwang, D. Vestweber, and G. Schatz. 1990. A protein precursor partly translocated into yeast mitochondria is bound to a 70 kd mitochondrial stress protein. *EMBO J.* **9:** 4315–4322.

Schiebel, E., A.J.M. Driessen, F.U. Hartl, and W. Wickner. 1991. Δμ+ and ATP function at different steps of the catalytic cycle of preprotein translocase. *Cell* **64:** 927–939.

Schlenstedt, G. and R. Zimmermann. 1987. Import of frog prepropeptide GLa into microsomes requires ATP but does not involve docking protein or ribosomes. *EMBO J.* **6:** 699–703.

Schlenstedt, G., G.H. Gudmundsson, H.G. Boman, and R. Zimmermann. 1990. A large presecretory protein translocates both cotranslationally, using signal recognition particle and ribosome, and posttranslationally, without these ribonucleoparticles, when synthesized in the presence of mammalian microsomes. *J. Biol. Chem.* **265:** 13960–13968.

———. 1992. Structural requirements for transport of preprocecropinA and related presecretory proteins into mammalian microsomes. *J. Biol. Chem.* **267:** 24328–24332.

Scidmore, M., H.H. Okamura, and M.D. Rose. 1993. Genetic interactions between *KAR2* and *SEC63*, encoding eukaryotic homologues of DnaK and DnaJ in the endoplasmic reticulum. *Mol. Biol. Cell* **4:** 1145–1159.

Siegel, V. and P. Walter. 1988. Functional dissection of the signal recognition particle. *Trends Biol. Sci.* **13:** 314–316.

Silver, P.A. and J. Way. 1993. Eukaryotic DnaJ homologs and the specificity of hsp70 activity. *Cell* **74:** 5–6.

Simon, S.M. and G. Blobel. 1991. A protein conducting channel in the endoplasmic reticulum. *Cell* **65:** 371–380.

Stirling, C.J. and E.W. Hewitt. 1992. The *S. cerevisiae SEC65* gene encodes a component of yeast signal recognition particle with homology to human SRP19. *Nature* **356:** 534–537.

Stirling, C.J., J. Rothblatt, M. Hosobuchi, R. Deshaies, and R. Schekman. 1992. Protein translocation mutants defective in the insertion of integral membrane proteins into the endoplasmic reticulum. *Mol. Biol. Cell* **3:** 129–142.

Tokunaga, M., A. Kawamura, and K. Kohno. 1992. Purification and characterization of BiP/Kar2 protein from *Saccharomyces cerevisiae. J. Biol. Chem.* **267:** 17553–17559.

Toyn, J., A.R. Hibbs, P. Sanz, J. Crowe, and D.I. Meyer. 1988. In vivo and in vitro analysis of *ptl1*, a yeast ts mutant with a membrane-associated defect in protein translocation. *EMBO J.* **7:** 4347–4353.

Verner, K. and G. Schatz. 1988. Protein translocation across membranes. *Science* **241:** 1307–1313.

Vogel, J.P., L.M. Misra, and M.D. Rose. 1990. Loss of BiP/GRP78 function blocks translocation of secretory proteins in yeast. *J. Cell Biol.* **110:** 1885–1895.

von Heijne, G. 1990. The signal peptide. *J. Membr. Biol.* **115:** 195–201.

Walter, P. and G. Blobel. 1981. Translocation of proteins across the endoplasmic reticulum. III. Signal recognition particle causes signal sequence dependent and site specific arrest of chain elongation that is released by microsomal membranes. *J. Cell Biol.* **91:** 557–561.

Walter, P. and V.R. Lingappa. 1986. Mechanism of protein translocation across the endoplasmic reticulum membrane. *Annu. Rev. Cell Biol.* **2:** 499–516.

Waters, M.G. and G. Blobel. 1986. Secretory protein translocation in a yeast cell-free system can occur post-translationally and requires ATP hydrolysis. *J. Cell Biol.* **102:** 1543–1550.

Waters, M.G., W.J. Chirico, and G. Blobel. 1986. Protein translocation across the yeast microsomal membrane is stimulated by a soluble factor. *J. Cell Biol.* **103:** 2629–2636.

Watts, C., W. Wickner, and R. Zimmerman. 1983. M13 procoat and pre-immunoglobulin share processing specificity but use different membrane receptor mechanisms. *Proc. Natl. Acad. Sci.* **80:** 2809–2813.

Werner-Washburne, M., D.E. Stone, and E.A. Craig. 1987. Complex interactions among members of an essential subfamily of hsp70 genes in *Saccharomyces cerevisiae. Mol. Cell. Biol.* **7:** 2568–2577.

Wetzstein, M. and W. Schumann. 1990. Nucleotide sequence of a *Bacillus subtilis* gene homologous to the grpE gene of *E. coli* located immediately upstream of the dnaK gene. *Nucleic Acids Res.* **18:** 1289.

Wickner, W., A.J.M. Driessen, and F.U. Hartl. 1991. The enzymology of protein translocation across the *Escherichia coli* plasma membrane. *Annu. Rev. Biochem.* **60:** 101–124.

Wiech, H. and R. Zimmermann. 1993. Role of molecular chaperones in transport of proteins into the mammalian endoplasmic reticulum. In *Protein folding and recovery* (ed. J.L. Cleland and J. King), pp 84–101. American Chemical Society, Washington, D.C.

Wiech, H., M. Sagstetter, G. Muller, and R. Zimmermann. 1987. The ATP-requiring step in assembly of M13 procoat protein into microsomes is related to preservation of transport competence of the precursor protein. *EMBO J.* **6:** 1011–1016.

Wiech, H., J. Buchner, M. Zimmermann, R. Zimmermann, and U. Jakob. 1993. Hsc70, immunoglobulin heavy chain binding protein, and hsp90 differ in their ability to stimulate transport of precursor proteins into mammalian microsomes. *J. Biol. Chem.* **268:** 7414–7421.

Zimmermann, R. and C. Mollay. 1986. Import of honeybee prepromelittin into the endoplasmic reticulum. *J. Biol. Chem.* **261:** 12889–12986.

Zimmermann, R., M. Sagstetter, M.J. Lewis, and H.R.B. Pelham. 1988. Seventy-kilodalton heat shock proteins and an additional component from reticulocyte lysate stimulate import of M13 procoat protein into microsomes. *EMBO J.* **7:** 2875–2880.

Zhong, T. and K.T. Arndt. 1993. The yeast SIS1 protein, a DnaJ homolog, is required for the initiation of translation. *Cell* **73:** 1175–1186.

Zylicz, M., D. Ang, K. Liberek, and C. Georgopoulos. 1989. Initiation of λ DNA replication with purified host and bacteriophage encoded proteins: The role of the dnaK, dnaJ, and grpE heat shock proteins. *EMBO J.* **8:** 1601–1608.

5

Structure, Function, and Regulation of the Endoplasmic Reticulum Chaperone, BiP

Mary-Jane Gething,[1,2] Sylvie Blond-Elguindi,[1,2] Kazutoshi Mori,[1,3] and Joseph F. Sambrook[1,4]
[1]Department of Biochemistry,
[2]Howard Hughes Medical Institute, and [4]McDermott Center
University of Texas Southwestern Medical Center,
Dallas, Texas 75235-9050

I. Introduction
II. Structure of BiP and Role of Posttranslational Modification
III. Role of BiP in the Endoplasmic Reticulum
IV. (Poly)peptide Recognition by BiP
V. Regulation of BiP Transcription: The Unfolded Protein Response
VI. Concluding Remarks

I. INTRODUCTION

Within the lumen of the endoplasmic reticulum (ER), a variety of resident ER proteins assist newly translocated nascent polypeptides to fold into their correct tertiary and quaternary structures. These resident proteins include both molecular chaperones that recognize and stabilize partially folded intermediates during polypeptide folding and assembly and enzymes that catalyze rate-determining steps in folding, such as protein disulfide isomerase and peptidyl prolyl isomerases (for review, see Gething and Sambrook 1992). Prominent among the molecular chaperones, both in its relative abundance in the ER lumen and as a focus of recent experimental attention, is BiP, the sole ER-located member of the stress-70 (hsp70) protein family. In mammalian cells, this protein was originally described independently as the immunoglobulin heavy-chain binding protein, found in noncovalent association with heavy chains in myeloma cells that do not synthesize immunoglobulin light chains (Haas and Wabl 1983), and as the glucose-regulated protein, Grp78 (Pouyssegur et al. 1977), one of a set of highly abundant ER

---

[3]Present address: HSP Research Institute, Kyoto Research Park, Kyoto 660, Japan.

The Biology of Heat Shock Proteins and Molecular Chaperones
©1994 Cold Spring Harbor Laboratory Press 0-87969-427-0/94 $5 + .00

proteins whose rate of synthesis is increased by glucose starvation or a variety of other cellular stress conditions (Lee 1987). In *Saccharomyces cerevisiae,* BiP is the product of the *KAR2* gene (Normington et al. 1989; Rose et al. 1989; Nicholson et al. 1990), one of a class of genes originally defined as being involved in nuclear fusion following mating of yeast cells (Polaina and Conde 1982). BiP is now known to associate transiently with a wide variety of newly synthesized wild-type exocytotic proteins and more permanently with malfolded or unassembled proteins whose transport from the ER is blocked (for review, see Gething and Sambrook 1992). These observations, together with the finding that BiP is required for the assembly of toxin oligomers in the yeast ER (Schonberger et al. 1991), indicate that BiP has a role in the folding and assembly of newly synthesized proteins in the ER lumen. Like other stress-70 proteins, BiP is thought to function by recognizing unfolded polypeptides and, by inhibiting intra- or intermolecular aggregation, maintaining them in a state competent for subsequent folding and oligomerization (Gething and Sambrook 1992; Craig et al. 1993). In addition to modulating protein folding in the ER lumen, BiP is directly or indirectly involved in the translocation of secretory precursors across the ER membrane (Vogel et al. 1990; Nguyen et al. 1991; Sanders et al. 1992). This chapter reviews current knowledge about the structure, function, and regulation of BiP.

II. STRUCTURE OF BIP AND ROLE OF POSTTRANSLATIONAL MODIFICATION

The nucleotide sequences of BiP cDNAs cloned from mammalian species (Munro and Pelham 1986; Kozutsumi et al. 1989) predict a protein of 654 amino acids that includes an 18-residue amino-terminal signal sequence and a carboxy-terminal tetrapeptide (Lys-Asp-Glu-Leu) required for retrieval of BiP molecules that leave the ER (Pelham 1989). The BiP protein from the yeast *S. cerevisiae* (Kar2p) has 682 residues, including a longer (42-amino-acid) signal sequence and a slightly different (His-Asp-Glu-Leu) carboxy-terminal retrieval signal (Normington et al. 1989; Rose et al. 1989; Nicholson et al. 1990). The sequences encoding BiP proteins from insect (Rubin et al. 1993) and plant (Fontes et al. 1991) species as well as other lower eukaryotes (Heschl and Baillie 1989; Kumar and Zheng 1992; Bangs et al. 1993) have been reported. BiP proteins are highly conserved, there being 99% amino acid sequence identity between the mature proteins from human and rodent cells and 67% identity between the proteins from mouse cells and from yeast. Comparison of the sequences of BiP proteins with those of other members of the stress-70 family reveals that the amino-terminal two-thirds (~450 amino acids) are much more highly conserved than the carboxy-terminal portions, suggesting a conserved domain followed by a more variable region.

Although the three-dimensional structure of BiP has not been determined, David McKay and his colleagues crystallized an amino-terminal 44-kD fragment of bovine hsc70, which retains the ATPase activity of the molecule, and determined its structure to a resolution of 2.2 Å revealing two structural lobes with the nucleotide bound at the base of a deep cleft between them (Flaherty et al. 1990, 1991; see McKay et al., this volume). The amino acid sequences of the corresponding amino-terminal domains of the BiP proteins, all of which share greater than 60% identity and 80% similarity with that of the cytosolic hsc70 protein, can be mapped without difficulty onto the backbone of the hsc70 structure, suggesting that the proteins have essentially identical conformations (D. McKay, pers. comm.). The structure of the carboxy-terminal domain, which includes the binding site for unfolded polypeptides, has not been determined for BiP or any other stress-70 protein. However, two groups (Flajnik et al. 1991; Rippmann et al. 1991) have proposed that the structure of a major portion of this domain closely resembles the $\alpha_1\alpha_2$ peptide-binding superdomain of the human major histocompatibility complex (MHC) class I antigen HLA. Recently, evidence has been obtained that the peptide-binding domain of hsc70 is included in a 159-amino-acid stretch that immediately follows the site of cleavage of the 44-kD ATPase domain (Wang et al. 1993). This fragment essentially corresponds to the HLA-homology domain, lacking only a few residues of the long α-helix of the α_2 domain. The remaining approximately 70-residue region that lies between the "HLA-homology domain" and the carboxy-terminal ER retrieval signal is the least conserved region of the BiP molecule.

Like other stress-70 proteins, BiP binds ATP with high affinity and displays a weak ATPase activity that is stimulated by binding of unfolded proteins and some, but not all, synthetic peptides (Flynn et al. 1989, 1991; Blond-Elguindi et al. 1993b) (see below). Binding of adenine nucleotides causes conformational changes in the protein that result in altered sensitivity to proteases: ATP protects an approximately 60-kD fragment and ADP protects an approximately 48-kD fragment (Kassenbrock and Kelly 1989). This effect of ATP does not require hydrolysis since ATPγS can substitute. In contrast, addition of ATP, but not of nonhydrolyzable analogs, to cell extracts causes dissociation of complexes between BiP and protein substrates such as immunoglobulin heavy chains (Munro and Pelham 1986). It was originally proposed that ATP hydrolysis, which takes place in vitro with a turnover time of about 5 minutes, might provide a timed mechanism of release of BiP from its substrate, freeing the polypeptide to continue the folding process (Rothman 1989). However, recent studies with *Escherichia coli* and mammalian hsp70 proteins indicate that in the presence of physiological concentrations of K^+ ions, dissociation of polypeptide substrates from hsp70 molecules occurs rapidly following ATP binding and precedes ATP hydrolysis by several minutes (Palleros et al. 1993). Furthermore, rates of ATP hydrolysis observed in vitro do not necessarily reflect those occurring in vivo, since physiological Ca^{++} concentrations inhibit the ATPase activity of BiP (Kassenbrock and Kelly 1989), and co-chaperones present in the ER may significantly stimulate its activity. In the case of *E. coli* DnaK, the simultaneous presence of the co-chaperones DnaJ and GrpE increase the ATPase activity of DnaK by up to 50-fold (Liberek et al. 1991). Many eukaryotic homologs of DnaJ have been identified in recent years (Caplan et al. 1993), including two in the ER of *S. cerevisiae*. Sec63p, an ER membrane protein that is a component of the translocation apparatus (Rothblatt et al. 1989; Sadler et al. 1989), exposes to the ER lumen a domain homologous to DnaJ (Feldheim et al. 1992), and genetic (Scidmore et al. 1993) and biochemical evidence (Brodsky and Schekman 1993) supports a direct interaction between this domain and the yeast BiP protein (see Brodsky and Schekman, this volume). A second DnaJ-related protein, Scj1p, was initially thought to be located within mitochondria (Blumberg and Silver 1991). However, recent evidence (A. Hundal, P. Silver, J.F. Sambrook, and M.-J. Gething, unpubl.) suggests that it is an ER-resident protein whose synthesis (like that of BiP) is under the control of an ER-to-nucleus (ERN) signaling pathway that responds to the levels of unfolded proteins in the lumen of the ER (Mori et al. 1993). GrpE-related proteins are beginning to be identified in eukaryotic cells (E. Craig; K. Palter; both pers. comm.), although there is as yet no evidence for a homolog located in the ER.

In vivo in mammalian cells, BiP exists in interconvertible monomeric and oligomeric forms (Hendershot et al. 1988; Carlino et al. 1992; Freiden et al. 1992; Blond-Elguindi et al. 1993a; Toledo et al. 1993) and can be posttranslationally modified by phosphorylation on serine and threonine residues (Welch et al. 1983; Hendershot et al. 1988; Leustek et al. 1991) and by adenylation (Carlsson and Lazarides 1983; Leno and Ledford 1990). In vitro BiP, like several other hsp70 family members, has a Ca^{++}-dependent autophosphorylation activity (Leustek et al. 1991, 1992) whose substrate is a threonine residue located in the ATP-binding cleft (Gaut and Hendershot 1993). This threonine (residue 207 in the mature, signal-cleaved mouse BiP molecule or residue 229 in the full-length precursor) is apparently not a detectable site of phosphorylation in vivo (Gaut and Hendershot 1993), and its modification may occur as a side reaction occurring during ATP hydrolysis. To what extent the degree of posttranslational modification affects the basal ATPase activity of the BiP molecules or their propensity to oligomerize is not known. However, conditions that increase the levels of unfolded polypeptides in the ER lumen cause a decrease in the extent of modification of BiP (Carlsson and Lazarides 1983; Hendershot et al. 1988; Leno and Ledford 1990; Leustek et al. 1991) and an increase in the proportion of monomeric species (Freiden et al. 1992). Only unmodified, monomeric BiP molecules are found in complexes with unfolded or unassembled polypeptides (Hendershot et al. 1988; Freiden et al. 1992). These observations led Hendershot and co-workers to propose that the dimeric, posttranslationally modified forms of BiP constitute a storage pool of inactive protein that can be mobilized in response to an increase in the concentration of unfolded or unassembled polypeptide substrates (Fig. 1). Results of in vitro studies are consistent with this hypothesis, since synthetic peptides that are capable of binding BiP can cause the conversion of a complex mixture of oligomeric and monomeric BiP species to a single monomeric form which has a higher ATPase activity (Blond-Elguindi et al. 1993a). Peptides can bind to both monomeric and oligomeric forms of BiP (Blond-Elguindi et al. 1993a), ruling out the possibility that only the monomeric species is active in complexing protein substrates (Freiden et al. 1992).

III. ROLE OF BIP IN THE ENDOPLASMIC RETICULUM

In mammalian cells, BiP associates transiently with many if not all newly synthesized wild-type transmembrane and secretory proteins until they fold or assemble in the ER (Bole et al. 1986; Dorner et al. 1987; Ng et al. 1989; Blount and Merlie 1991; Earl et al. 1991; Kim et al. 1992; Knittler and Haas 1992) and more permanently with misfolded, underglycosylated or unassembled proteins whose transport from the ER is blocked (Bole et al. 1986; Gething et al. 1986; Hurtley et al. 1989; Machamer et al. 1990; Navarro et al. 1991). BiP does not interact with native polypeptides. Experiments performed in yeast also indicate a role for BiP during protein folding and assembly in the ER since cells expressing the *kar2-1* mutant allele, which has sustained a single-amino-acid substitution in the carboxy-terminal peptide-binding domain of BiP (J. Vogel and M. Rose, pers. comm.), fail to support the assembly into pentamers of the B subunit of a heat-labile enterotoxin (Schonberger et al. 1991). Other studies in yeast indicate an additional role for BiP in translocation of newly synthesized polypeptides across the ER membrane. Thus, cells contain-

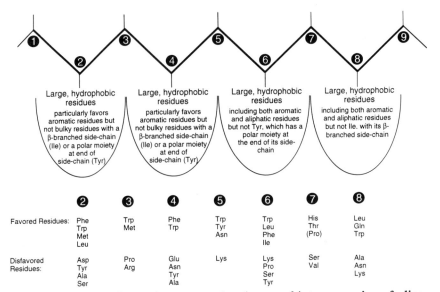

Figure 1 A model illustrating proposed pathways of interconversion of oligomeric and monomeric species of BiP and their interactions with unfolded polypeptide substrates. (*Anti-clockwise from top*) In quiescent, unstressed cells, the majority of BiP molecules are present as oligomeric species that are posttranslationally modified by phosphorylation and/or ADP ribosylation (Welch et al. 1983; Hendershot et al. 1988; Carlsson and Lazarides 1983; Leno and Ledford 1990; Leustek et al. 1991). Conditions that result in an increase in the concentration of unfolded polypeptides in the ER promote the reversal of modifications (Carlsson and Lazarides 1983; Hendershot et al. 1988; Leno and Ledford 1990; Leustek et al. 1991) and the dissociation of BiP oligomers (Freiden et al. 1992). In vitro, both the oligomeric and monomeric species are capable of binding peptides, and peptide-induced stimulation of BiP's ATPase activity reflects the conversion of inactive or less active oligomers into active monomers (Blond-Elguindi et al. 1993a). When peptide is bound, exchange of ADP for ATP is favored. High concentrations of ATP can also cause dissociation of BiP oligomers in vitro (Carlino et al. 1992; Toledo et al. 1993). Binding of ATP by BiP then promotes release of the (poly)peptide from the complex (Palleros et al. 1993). Hydrolysis of ATP returns BiP to the ADP-bound form having highest affinity for unfolded polypeptide substrates. In the continued presence of unfolded polypeptides and ATP, cycles of polypeptide binding, ATP binding, polypeptide release, and ATP hydrolysis continue. When the concentration of unfolded polypeptides in the ER decreases, free BiP molecules become modified and oligomer formation is favored. This scheme was adapted from that shown in Fig. 5 of Blond-Elguindi et al. (1993a) on the basis of new data concerning the order of peptide release and ATP hydrolysis by hsp70 proteins (Palleros et al. 1993).

ing temperature-sensitive *kar2* alleles with amino acid substitutions in the amino-terminal ATPase domain (J. Vogel and M. Rose, pers. comm.) accumulate secretory precursors in the cytosol (Vogel et al. 1990), as do cells in which the concentration of wild-type BiP is depleted below approximately 10% of its normal level (Vogel et al. 1990; Nguyen et al. 1991). In vitro experiments that reconstitute the translocation reaction using yeast microsomes demonstrated that incorporation of purified yeast BiP into microsomes prepared from a translocation-defective *kar2* strain restored translocation of prepro-α-factor (Brodsky et al. 1993). A direct association between BiP and the translocating polypeptide could be detected using a modified prepro-α-factor fused to avidin to construct a translocation intermediate that is trapped spanning the microsome membrane (Sanders et al. 1992). As part of this study, some mutant *kar2* alleles were shown to decrease the association of the translocating chain with Sec61p, which together with three other transmembrane proteins (Sec62p, Sec63p, and Sec66p) and one peripherally associated protein (Sec67p) are components of the translocation machinery that can be isolated as a complex by cross-linking and coimmunoprecipitation (Deshaies et al. 1991; see also Brodsky and Schekman, this volume). Finally, BiP can be isolated in a smaller complex that includes Sec63p, Sec66p, and Sec67p (Brodsky and Schekman 1993). The association of BiP with the complex is mediated through a nucleotide-dependent interaction that is compromised by a single amino-acid substitution of a conserved residue in the DnaJ-like domain of Sec63p.

On the other side of the ER membrane, stress-70 proteins are also required for translocation of secretory precursors. Thus, the import of M13 procoat protein into dog pancreas microsomes requires functional hsp70 protein (Zimmermann et al. 1988), whereas in yeast, the constitutively expressed cytosolic hsp70 proteins, Ssa1p and Ssa2p, are required for ER translocation in vivo (Deshaies et al. 1988) and in vitro (Chirico et al. 1988). BiP and cytosolic hsp70 proteins may therefore be performing similar functions in the ER lumen and in the cytosol, raising the possibility that the various members of the hsp70 family might be able to substitute for each other if placed in the appropriate location. However, in vitro experiments indicate that this is not the case, at least when substituted at roughly equivalent concentrations. Bovine BiP cannot replace bovine hsc70 in stimulating transport of precursor proteins into mammalian microsomes (Wiech et al. 1993), and yeast BiP and Ssa1p do not have interchangeable roles in supporting protein translocation into reconstituted yeast microsomes (Brodsky et al. 1993). Since stress-70 proteins have quite similar patterns of peptide-binding specificity (see next section), it seems likely that their lack of interchangeability may result from their failure to interact efficiently with co-chaperones other than their own specific partners.

IV. (POLY)PEPTIDE RECOGNITION BY BIP

BiP's role as a general chaperone during protein folding in the ER lumen depends on its ability to recognize a wide variety of nascent polypeptides that share no obvious sequence similarity while accurately discriminating between properly folded and unfolded structures. In vitro, BiP can interact with short synthetic peptides whose binding stimulates its ATPase activity (Flynn et al. 1989) and alters its oligomeric state (Blond-Elguindi et al. 1993a). Experiments using a small set of randomly chosen synthetic peptides showed that their affinities for BiP varied over a 1000-fold range (Flynn et al. 1989). Analysis of the BiP-binding capacity of a collection of synthetic peptides of defined length but random sequence (Flynn et al. 1991) demonstrated that the minimum length for binding was seven to eight residues. Bulk sequencing of the population of water-soluble heptameric peptides that bound to BiP revealed a positive selection for amino acids with aliphatic side chains, particularly at positions 3 through 6. Unfortunately, this approach could not identify the sequences of individual selected peptides whose affinities for BiP could be directly measured.

The characteristics of peptides that bind to BiP have also been investigated using affinity screening of large, highly diverse libraries of peptides displayed by fusion to the amino termini of the four to five copies of the pIII protein present at one tip of bacteriophage fd particles (Blond-Elguindi et al. 1993b). This powerful technique (for review, see Scott and Smith 1990; Smith 1991; Dower 1992; Scott 1992) is based on the ability of filamentous bacteriophages to display foreign peptides on their outer surfaces with little effect on their viability and involves the specific screening and affinity purification of bacteriophages displaying peptides that are ligands for a particular protein using a method called "biopanning" (Parmley and Smith 1988). Bacteriophage display has several advantages over studies with random synthetic peptides. First, the peptides are attached to the bacteriophage proteins, minimizing peptide solubility problems. Second, the sequences of individual peptides can be easily deduced by DNA sequencing of the appropriate coding region of the bacteriophage genome. This approach avoids technical problems encountered during amino acid sequence analysis, such as the inability to analyze certain residues or positions of the peptide. Finally, peptides corresponding to individual bacteriophage-displayed sequences can be synthesized and their biological activity assayed directly.

Two bacteriophage libraries with random octapeptides or dodecapeptides displayed by the pIII protein were utilized to study the peptide-
binding specificity of murine BiP (Blond-Elguindi et al. 1993b). Fewer
than 1 in 1000 of the bacteriophages from the original unpanned libraries
showed any BiP-binding activity. The peptides displayed by BiP-binding
bacteriophages selected by affinity panning show extensive sequence
diversity, consistent with the observed "promiscuity" of BiP's interaction
with a wide variety of unrelated nascent polypeptides, and usually exhibit marked hydrophobicity, consistent with the likelihood that BiP interacts with sequences normally located in the interior of a fully folded
protein. In confirmation of the fidelity of the selection system, synthetic
peptides with sequences corresponding to those on selected bacteriophages bind to BiP and stimulate its ATPase activity, whereas peptides corresponding to those displayed by bacteriophages randomly
chosen from the original library do not. Comparison of 114 BiP-binding
sequences and an identical number of nonbinding sequences showed that
tryptophan, phenylalanine, and leucine, and to a lesser extent methionine
and isoleucine, were increased in abundance in the peptides displayed by
BiP-binding bacteriophages. Shifting of the frame of many of the oc-
tameric sequences by one position revealed a heptameric motif best described as Hy(W/X)HyXHyXHy, where Hy is a bulky aromatic or hydrophobic residue (most frequently tryptophan, phenylalanine, or
leucine, but also methionine and isoleucine), W is tryptophan, and X is
any amino acid. The alternating pattern of Hy residues in the binding
motif is compatible with peptides being bound to BiP in an extended
conformation, with the bulky aromatic/hydrophobic side chains lying on
one side of the strand and pointing into a binding cleft on the BiP
molecule (Fig. 2). This model is consistent with nuclear magnetic
resonance studies of the binding of peptides to BiP (S. Blond-Elguindi et
al., unpubl.) and to its homolog in *E. coli*, DnaK (Landry et al. 1992),
which indicate that peptides are bound in an extended conformation.
Taken together, the data indicate that the peptide-binding site on the BiP
molecule accommodates a linear sequence of seven amino acids and includes four major pockets that can accommodate large hydrophobic/aro-
matic residues. Occupancy of at least two of the pockets by preferred
residues is sufficient to promote stable binding between the peptide and
the BiP molecule. The four pockets in the binding cleft have similar
overall preferences for large, hydrophobic or aromatic side chains, although there is some variation between them in the degree to which individual residues are preferred or excluded (see Fig. 2). For example, the
pockets that bind the side chains of residues 2, 4, and 8 exclude aliphatic
residues (such as isoleucine) that have a β-branched side chain, whereas

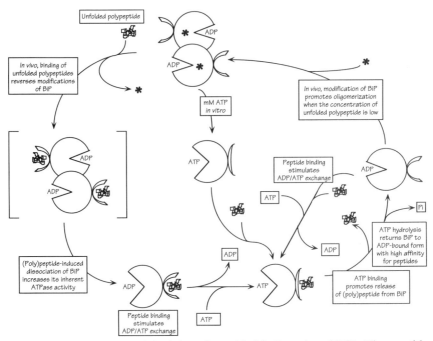

Figure 2 Schematic model of the of peptide-binding site of BiP. The peptide backbone is shown as an extended chain. The side chains of the even-numbered residues extend into four deep pockets in the peptide-binding site that have similar overall preferences for large, hydrophobic, or aromatic side chains. However, there is some variation between the pockets in the degree to which individual residues are preferred or excluded. There may in addition be one or two minor pockets on the other side of the cleft, or residues positioned at the top of the cleft, whose side chains can interact with those of the odd-numbered residues of the peptide. For each residue of the heptameric recognition sequence (numbered 2 through 8 in this representation to underscore the fact that the motif need not be at the amino terminus) are listed those amino acids whose presence at this position is either highly favored or disfavored in BiP-binding octapeptides selected by bacteriophage panning (Blond-Elguindi et al. 1993b). Proline is favored at position 7 only when it is flanked by bulky hydrophobic residue(s) at position 6 and/or 8. (Adapted from Blond-Elguindi et al. 1993b.)

only the latter pocket will tolerate tyrosine, whose structure, except for the polar moiety at the end of the side chain, is identical to that of phenylalanine (which is favored or tolerated in all four pockets). There may in addition be one or two minor pockets on the other side of the cleft, or residues at the top of the cleft, whose side chains can interact with those of the odd-numbered residues of the peptide.

A detailed comparison of the abundance of each of the 20 amino acids at each position in the BiP-binding and nonbinding heptapeptides yielded a scoring system that codifies the observed positional preferences (Blond-Elguindi et al. 1993b). Heptameric sequences that conform to the general consensus motif and have high "BiP scores" were present within BiP-binding dodecapeptides displayed on bacteriophages independently selected from the second library and within synthetic peptides known to bind to BiP (Flynn et al. 1989; Oblas et al. 1990). A computer program that calculates the scores for all possible heptapeptides in a polypeptide sequence can be used to predict BiP-binding sites in natural proteins (Blond-Elguindi et al. 1993b). Preliminary analyses of the positions of predicted binding sites in proteins of known structure indicate that in the native protein, these sites are located either in the interior of a folded domain or at the interface between subunits (S. Blond-Elguindi, J. Buchner, and M.-J. Gething, unpubl.), consistent with BiP's proposed role in stabilizing partially folded or partially assembled polypeptides in the ER.

Because of the high degree of homology in sequence and function between all of the members of the stress-70 protein family (Gething and Sambrook 1992), it is likely that there will be significant similarity among the structures of the peptide-binding sites of all these proteins. Many synthetic peptides are bound equally well by bovine BiP, bovine hsc70, and DnaK (A. Fourie et al., unpubl.), indicating that there is significant overlap in the peptide-binding specificity of these family members. However, some peptides are recognized by one family member but not by others (our unpublished results). This pattern of overall similarity but specific differences in peptide recognition could result from the hsp70 family members differing in having one or more amino acid substitutions that affect the shape, depth, or charge density of one of the binding pockets. Such a change would not affect the binding of peptides that do not utilize this pocket to accommodate the hydrophobic residues that stabilize binding, as long as these peptides do not contain a residue whose side chain cannot fit into the altered pocket or is of the wrong charge.

As discussed earlier, two groups have proposed that the three-dimensional structure of the carboxy-terminal peptide-binding domains of hsp70 proteins resembles that of the peptide-binding $\alpha_1\alpha_2$ superdomain of the human MHC class I antigen, HLA (Flajnik et al. 1991; Rippmann et al. 1991). Peptides bind to MHC molecules in an extended conformation in a binding groove that contains pockets or subsites which vary in their relative positioning, depth, and size (Saper et al. 1991; Fremont et al. 1992; Madden et al. 1992; Matsumura et al. 1992; Brown et al. 1993). In the class I molecule, two deep pockets accommodate the amino and carboxyl termini of the peptide, which is usually an octamer or nonamer. A third deep pocket found in different positions in different alleles has a major role in determining allele-specific peptide binding by providing chemical and structural complementarity for a particular "anchor" residue (Rammensee et al. 1993). Although the structure of the class I peptide-binding pocket itself does not vary significantly when different peptides are bound, peptides bound to the same class I allele may take up quite dissimilar conformations, particularly in their central portions (Madden et al. 1993). Class II molecules bind longer peptides (12–24 residues) in an extended conformation stabilized by hydrogen bonding between main chain peptide atoms and some conserved residues in the peptide-binding cleft. The cleft contains one deep pocket lined with nonpolar residues that is likely to accommodate the dominant aromatic anchor residue of the bound peptide, as well as three other pockets likely to bind the side chains of a secondary anchor residue and two allele-specific residues positioned three, five, and six or eight amino acids further along the peptide (Brown et al. 1993; Hammer et al. 1993). Even if the overall structures of the peptide-binding domains of MHC and hsp70 molecules are similar, it is likely that the details of the binding interactions will vary. First, unlike class I, BiP and other stress-70 molecules bind sequences that are embedded in long polypeptide chains so that free amino and carboxyl termini are not available for hydrogen bonding. In this respect, stress-70 proteins more closely resemble class II molecules, which bind longer peptides that project out of both ends of the binding cleft (Brown et al. 1993). Second, the fact that the BiP-binding motif involves hydrophobic and aromatic residues (but not tyrosine) suggests that hydrogen bonds to atoms in the peptide side chains do not contribute significantly to the interaction between hsp70 molecules and their (poly)peptide ligands. Third, peptides that bind BiP do not contain a single dominant anchor residue. Rather, the more generalized requirement for the binding of large hydrophobic residues in any two or more of the pockets facilitates recognition of an even greater variety of peptides than is observed for MHC molecules while maintaining the ability to discriminate between sequences likely to be exposed by properly folded and unfolded or unassembled structures.

Although available evidence is generally consistent with the possibility that the structure of the peptide-binding domain of hsp70 molecules resembles that of MHC proteins, the recently published structure of the complex of the *E. coli* periplasmic chaperone PapD and one of its peptide ligands (Kuehn et al. 1993) provides another model of how a chaperone can bind an extended polypeptide chain. PapD, which is required for the assembly of pilus subunits into P pili, is composed of two β-barrel globular domains positioned such that the overall shape of the molecule resembles a boomerang with a cleft between the two domains

(Holmgren and Branden 1989). Peptides corresponding to the carboxyl termini of pilus subunits block pilus assembly in vitro. Analysis of the structure of the complex of PapD and the carboxy-terminal peptide of PapG revealed the peptide forming a parallel β-strand interaction that extends the β-sheet of the amino-terminal domain of the PapD molecule (Kuehn et al. 1993). The carboxyl terminus of the peptide is anchored in the interdomain cleft. Interestingly in the context of the preference of BiP for peptides containing bulky hydrophobic residues in alternate positions, the pilus peptide residues that interact with the PapD β-sheet, i.e., Phe$^{2'}$, Leu$^{4'}$, Met$^{6'}$, and Met$^{8'}$, are part of a pattern of alternating hydrophobic and hydrophilic residues that is conserved at the carboxyl termini of pilus subunits. These residues are likely to contribute significantly to both the specificity and strength of the interaction between PapD and the pilus peptides (Kuehn et al. 1993).

V. REGULATION OF BIP TRANSCRIPTION: THE UNFOLDED PROTEIN RESPONSE

Under normal growth conditions, BiP is synthesized constitutively and abundantly, comprising approximately 5% of the lumenal content of the ER. In mammalian cells, its synthesis can be further induced by the accumulation of mutant proteins in the ER or by a variety of stress conditions, including glucose starvation and treatment with calcium ionophores, amino acid analogs, or drugs that inhibit glycosylation (for review, see Lee 1987, 1992), whose common denominator is believed to be the accumulation in the ER of unfolded polypeptides (Kozutsumi et al. 1988; Nakaki et al. 1989). Similarly, in yeast cells, the amount of *KAR2* (BiP) mRNA can be increased from its constitutive level either by raising the concentration of secretory protein precursors in the ER by incubating *sec* mutants at nonpermissive temperatures (Normington et al. 1989; Rose et al. 1989), or by accumulating nonglycosylated polypeptides in the ER following treatment of cells with tunicamycin (Normington et al. 1989; Rose et al. 1989), or by accumulating an aggregated proform of murine α-amylase in the ER lumen (Tokunaga et al. 1992). In yeast, BiP and other members of the stress-70 family respond in a compartment-specific manner to the presence of unfolded proteins (Normington et al. 1989). Thus, the synthesis of BiP is induced only by the accumulation of secretory precursors within the ER, whereas synthesis of cytosolic hsp70 protein(s) increases in response to the accumulation of unfolded precursors in the cytoplasm.

The promoters of mammalian BiP (*grp78*) genes contain *cis*-acting regulatory elements required for high-basal-level expression and for induction by malfolded proteins, glycosylation block, or the calcium ionophore A23187 (Wooden et al. 1991; Lee 1992). These include (1) a 36-bp "Grp core" element involved in stress inducibility that includes a 28-bp sequence that is highly conserved in the promoters of mammalian and avian genes that encode Grp78 and another stress-responsive ER resident protein, Grp94 (Chang et al. 1989; Liu and Lee 1991); (2) a series of CCAAT or CCAAT-like motifs, the most proximal of which is required to mediate the effects of the upstream regulatory elements and which binds the transcription factor CTF/NFI (Wooden et al. 1991); (3) a second stress-responsive element that provides partial stress inducibility when the Grp core and proximal CCAAT motif are deleted (Wooden et al. 1991); and (4) a cAMP response element (CRE) that functions as a major basal level regulatory element and is also necessary to maintain high promoter activity under stress-induced conditions (Alexandre et al. 1991).

Transcription of the yeast *KAR2* (BiP) gene is regulated by three independent *cis*-acting elements: (1) a functional heat shock element (HSE) containing four repeats of the 5-bp modular units defined by (Lis et al. 1990) [aGAAccTTCtgGAAatTTCa]; (2) a pyrimidine-rich region that is similar in sequence to the consensus element for binding of the mammalian transcription factor Sp1 and that contributes to the high level of basal expression; and (3) a 22-bp element (UPRE) that shows significant homology with the consensus sequence found in the promoters of the mammalian glucose-regulated genes (Resendez et al. 1985; Chang et al. 1989) and is required for the induction of BiP mRNA by unfolded proteins (Mori et al. 1992; Kohno et al. 1993). Internal deletions of 10-bp segments from the HSE and from the UPRE independently abolished the heat shock response and the unfolded protein response, whereas a 10-bp deletion of the Sp1-like element significantly decreased the basal level of expression of the *KAR2* gene (Mori et al. 1992). Transplantation of the 22-bp UPRE into the heterologous *CYC1* promoter causes the promoter to respond to the presence of unfolded proteins in the ER (Mori et al. 1992). Sequences closely related to the *KAR2* UPRE are present in the promoter regions of other unfolded protein-responsive yeast genes including *EUG1*, which encodes an ER resident protein related to protein disulfide isomerase (Tachibana and Stevens 1992); *FKB2*, which encodes the ER-located member of the FKB family of peptidyl prolyl isomerases (Partaledis and Berlin 1993); and *SCJ1*, which encodes the ER homolog of DnaJ (Blumberg and Silver 1991). Extracts of unstressed yeast cells contain a protein (UPRF) that binds specifically to the UPRE (Mori et al. 1992). Stressing of the yeast cells by tunicamycin treatment before extracts were prepared had no effect on the pattern or extent of binding activity, indicating that, as is the case with the yeast heat shock transcription factor (HSF; Sorger et al. 1987), UPRF activates transcription of the *KAR2* gene by virtue of posttranslational modifications that cannot be detected by shifts in the mobility of the specific DNA-protein bands.

Evidence from studies in both yeast and mammalian cells suggests that the proximal signal for the induction of UPRE-controlled genes is not an increase in the concentration of unfolded or unassembled proteins or of BiP-protein complexes in the ER, but rather the consequent decrease in the concentration of free BiP (for review, see Kohno et al. 1993). The pathway of transduction of the unfolded protein response across the ER membrane and to the nucleus must at minimum involve (1) a sensor to measure the concentration of free BiP in the ER lumen, (2) a mechanism to recognize the signal and transduce it across the lipid bilayer to the cytoplasmic face of the ER membrane, where (3) modification (perhaps by phosphorylation) of a messenger (perhaps the UPRF itself or an intermediary kinase) facilitates transmission of the activation signal to the *KAR2* promoter in the nucleus. To identify the participants in the unfolded protein response pathway in *S. cerevisiae*, Mori et al. (1993) constructed a reporter plasmid in which the *lacZ* gene encoding β-galactosidase is controlled by a modified *KAR2* promoter that contains a functional UPRE but lacks the HSE and Sp1-like elements. *sec53* cells containing this reporter were mutagenized and screened for mutants that no longer induce *lacZ* transcription when unfolded secretory precursors accumulate following growth at the nonpermissive temperature. Subsequently, Cox et al. (1993) synthesized a 38-bp DNA that includes the UPRE (Mori et al. 1992), fused it to a transcriptionally silent *CYC1* promoter upstream of the *lacZ* reporter gene, and introduced the construct into wild-type yeast cells. Following mutagenesis, mutants were identified that failed to synthesize β-galactosidase activity following growth in the presence of the glycosylation inhibitor, tunicamycin. Both screens identified mutants defective in the same gene, termed *ERN1* (Mori et al. 1993) or *IRE1* (Nikawa and Yamashita 1992; Cox et al. 1993). *ern1* mutants, which are recessive, display wild-type levels of basal transcription from the UPRE in the promoter of the *KAR2* (BiP) gene but are impaired in the UPR response. The heat shock response is unaffected. The signaling defect affects not only transcription from the *KAR2* promoter, but also transcription of the coregulated *EUG1* gene, which encodes a protein structurally and functionally related to protein disulfide isomerase (Tachibana and Stevens 1992). Gene disruption experiments demonstrated that the *ERN1* gene product is not essential for vegetative growth but protects cells against certain types of stress by maintaining sufficient levels of functional BiP (and possibly other coregulated ER proteins), which is essential for viability. *ern1* cells grow normally and can utilize sucrose as sole carbon source and mate efficiently, indicating that, in these mutant cells, secretion of invertase and α factor are unimpaired.

The single-copy *ERN1* gene was cloned from a library of yeast genomic DNA by complementation of the defect in the UPR response of the *ern1-1* mutant allele (Mori et al. 1993). *ERN1* mRNA (a single 3.9-kb species) is present at very low levels, and its weak promoter (transcriptional activity ~2% of that of the uninduced *KAR2* promoter) is not induced by conditions that lead to induction of the *KAR2* and *EUG1* genes, i.e., heat shock or accumulation of unfolded proteins in the ER (Mori et al. 1993). Sequencing of the *ERN1* gene revealed that it contains an open reading frame of 3345 bp, which encodes a protein of 1115 amino acids with a predicted molecular weight of 126,974. The authenticity of the cloned gene was confirmed by also cloning and sequencing the *ern1-1* and *ern1-2* mutant alleles (Mori et al. 1993). These contain single point mutations, causing truncations near the amino terminus and carboxyl terminus, respectively, and have no complementing activity. Comparison of the sequence of the wild-type gene with others deposited in the Gen-Bank database revealed that the gene is essentially identical (four nucleotide differences, three amino acid differences) to *IRE1*, a gene required for inositol phototrophy in *S. cerevisiae* (Nikawa and Yamashita 1992). Severe *ern1* mutants are indeed inositol auxotrophs; however, a site-directed mutant that retains only about 4% of the UPR response (Mori et al. 1993) does not require inositol for growth (K. Mori and R. McMillan, unpubl.), indicating that the two phenotypes are not tightly linked.

The polypeptide encoded by the *ERN1* gene (Fig. 3) is predicted to contain an amino-terminal signal sequence and a membrane-spanning domain of 16 hydrophobic residues located approximately in the middle of the polypeptide chain (residues 541–556) (Mori et al. 1993). Antibodies raised against amino-terminal fragments of Ern1p expressed in *E. coli* recognize a polypeptide of approximately 136 kD that is present at very low abundance in *ERN*+ cells and at high abundance in cells overexpressing the *ERN1* gene from a multicopy plasmid (Mori et al. 1993). Consistent with the predictions from the deduced amino acid sequence, immunoblotting showed that Ern1p behaved as an integral membrane protein following extraction of yeast microsomes with high salt, sodium carbonate, or urea but was released following detergent solubilization. A decrease of about 6 kD in molecular mass following treatment with endo H was consistent with the presence of four potential glycosyla-

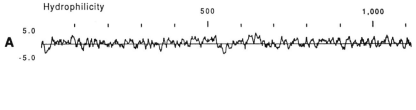

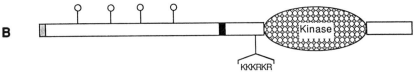

Figure 3 Domain structure of Ern1p. Hydropathy analysis of the amino acid sequence of Ern1p (Mori et al. 1993) shows that the polypeptide has two stretches of significantly hydrophobic residues. One lies near the amino terminus and has a sequence compatible with function as a signal peptide to facilitate translocation of the nascent polypeptide into the ER. The second lies at the center of the polypeptide and includes an uninterrupted stretch of 16 hydrophobic amino acids (residues 541–556), which is long enough to span the lipid bilayer once. The amino acid sequence that lies between these two motifs is hydrophilic in character and displays no sequence similarity to any protein in the database (other than Ire1p; Nikawa and Yamashita 1992). This domain contains four cysteine residues and four consensus sequences (Asn.Xaa.Ser/Thr) for N-linked glycosylation. Following the potential transmembrane domain is a stretch of 117 residues that also has no counterpart in the database but contains a short, highly basic sequence that includes 6 consecutive lysines and arginines embedded in a sequence of 18 residues of which 12 are basic amino acids. The carboxy-terminal half of the molecule contains a region (residues 673–980) that shows significant sequence similarity to the catalytic domains of protein kinases (Hanks and Quinn 1991). In particular, Ern1p is most closely related to the CDC28/cdc2+ protein kinase family (Hunter 1991), the amino acid sequence of its putative catalytic domain displaying 27% identity and 46% similarity to that of the *CDC28* gene product. Finally, following the kinase homology domain is a stretch of 135 residues that again show no similarity to other proteins in the database.

tion sites (Asn-X-Ser/Thr) in the amino-terminal half of the sequence. There was no evidence of hypermannosylation of the oligosaccharide side chains, which is characteristic of glycoproteins that move to or through the yeast Golgi apparatus, consistent with the localization of the protein in ER membranes. Finally, protease treatment of microsomes in the absence and presence of detergent followed by immunoblotting demonstrated that the amino-terminal domain is protected from protease in intact microsomes, indicating that the amino-terminal domain is located in the lumen of the ER and that the carboxy-terminal half of the molecule is located in the cytosol (Mori et al. 1993). The amino-terminal 672 amino acids of Ern1p show no sequence similarity to any protein in the database other than Ire1p. Because of its localization, it is presumed to be the ligand-binding domain of the molecule. As discussed above, previous studies indicate that the proximal signal for initiation of the UPR response is the level of free BiP in the ER, but it is not known whether the ligand for the receptor might be BiP itself or another molecule such as Ca^{++} that might be released when BiP binds polypeptide substrates (BiP is a high-capacity, low-affinity Ca^{++}-binding protein; Macer and Koch 1989). Residues 673–980 display significant similarity to the catalytic domains of protein kinases, being most closely related to the $CDC28/cdc2^+$ family of serine/threonine kinases. The 12 conserved subdomains characteristic of all known protein kinases are present in the Ern1p sequence, and 11 of the 12 invariant residues are perfectly conserved. Mutation of two of the invariant residues either abolished (K702A) or significantly diminished (K702R, D828A) the ability of the *ERN1* gene to complement the defect in the UPR response in *ern1-1* mutant cells, indicating that the kinase activity of Ern1p is essential for its function.

Characterization of the Ern1p transmembrane receptor protein kinase has identified in one molecule at least two of the essential components of the UPR pathway: the lumenal sensor and the mechanism for transducing the signal across the ER membrane. However, we do not yet know the identity of the ligand for the receptor, nor whether the UPRF transcription factor is the immediate substrate of the kinase or whether there are intermediate messengers, such as one or more additional kinases, between Ern1p and UPRF. The genetic screens described above each yielded, in addition to the *ern1/ire1* mutants, an additional mutant defective in an as yet uncharacterized gene required for UPR signaling (Cox et al. 1993; Mori et al. 1993). Recent studies have also identified an *ern* mutant that displays a constitutively induced UPR response (I. Elguindi et al., unpubl.). Before too long, therefore, all the components of the pathway should be identified and characterized.

VI. CONCLUDING REMARKS

Although the work of many laboratories in recent years has demonstrated the involvement of BiP in protein translocation and folding in the ER and provided a preliminary description of the mechanisms by which this chaperone functions and is regulated, much remains to be understood. Definition of the structure of the carboxy-terminal peptide-binding domain of this or any other hsp70 protein is perhaps the "Holy Grail." With the availability of a complete three-dimensional structure and with knowledge of the peptide-binding specificity of the molecule, we should be able to comprehend in detail how BiP interacts with its polypeptide substrates and how nucleotide binding and hydrolysis modulate their association. Important also will be the characterization of the cast of co-chaperones that function with BiP in the ER and the determination of how they interact to modulate polypeptide recognition or dissociation. An intriguing question is whether BiP functions in the same manner in promoting polypeptide translocation and in facilitating protein folding and assembly, or whether different mechanisms are involved in these two processes. Finally, definition of all the components involved in transmitting the unfolded protein response from the lumen of the ER to the nucleus will provide the first description of an intracellular signaling pathway.

REFERENCES

Alexandre, S., T. Nakaki, L. Vanhamme, and A.S. Lee. 1991. A binding site for the cyclic adenosine 3',5'-monophosphate-response element-binding protein as a regulatory element in the grp78 promoter. *Mol. Endocrin.* **5:** 1862–1872.

Bangs, J.D., L. Uyetake, M.J. Brickman, A.E. Balber, and J.C. Boothroyd. 1993. Molecular cloning and cellular localization of a BiP homologue in *Trypanosoma brucei. J. Cell Sci.* **105:** 1101–1113.

Blond-Elguindi, S., A.M. Fourie, J.F. Sambrook, and M.-J.H. Gething. 1993a. Peptide-dependent stimulation of the ATPase activity of the molecular chaperone BiP is the result of conversion of oligomers to active monomers. *J. Biol. Chem.* **268:** 12730–12735.

Blond-Elguindi, S., S.E. Cwirla, W.J. Dower, R.J. Lipshutz, S.R. Sprang, J.F. Sambrook, and M.-J. Gething. 1993b. Affinity panning of a library of peptides displayed on bacteriophages reveals the binding specificity of BiP. *Cell* **75:** 717–728.

Blount, P. and J.P. Merlie. 1991. BiP associates with newly synthesized subunits of the mouse muscle nicotinic receptor. *J. Cell Biol.* **113:** 1125–1132.

Blumberg, H. and P.A. Silver. 1991. A homologue of the bacterial heat-shock gene DnaJ that alters protein sorting in yeast. *Nature* **349:** 627–630.

Bole, D.G., L.M. Hendershot, and J.F. Kearney. 1986. Posttranslational association of immunoglobulin heavy chain binding protein with nascent heavy chains in nonsecreting and secreting hybridomas. *J. Cell Biol.* **102:** 1558–1566.

Brodsky, J.L. and R. Schekman. 1993. A Sec63p-BiP complex from yeast is required for protein translocation in a reconstituted proteoliposome. *J. Cell Biol.* **123:** 1355–1363.

Brodsky, J.L., S. Hamamoto, D. Feldheim, and R. Schekman. 1993. Reconstitution of protein translocation from solubilized yeast membranes reveals topologically distinct roles for BiP and cytosolic Hsc70. *J. Cell Biol.* **120:** 95–102.

Brown, J.H., T.S. Jardetzky, J.C. Gorga, L.J. Stern, R.G. Urban, J.L. Strominger, and D. C. Wiley. 1993. Three-dimensional structure of the human class II histocompatibility antigen HLA-DR1. *Nature* **364:** 33–39.

Caplan, A.J., D.M. Cyr, and M.G. Douglas. 1993. Eukaryotic homologues of *Escherichia coli* dnaJ: A diverse protein family that functions with HSP70 stress proteins. *Mol. Biol. Cell* **4:** 555–563.

Carlino, A., H. Toledo, D. Skaleris, R. DeLisio, H. Weissbach, and N. Brot. 1992. Interactions of liver Grp78 and *Escherichia coli* recombinant Grp78 with ATP: Multiple species and disaggregation. *Proc. Natl. Acad. Sci.* **89:** 2081–2085.

Carlsson, L. and E. Lazarides. 1983. ADP-ribosylation of the M_r 83,000 stress-inducible and glucose-regulated protein in avian and mammalian cells: Modulation by heat shock and glucose starvation. *Proc. Natl. Acad. Sci.* **80:** 4664–4668.

Chang, S.C., A.E. Erwin, and A.S. Lee. 1989. Glucose-regulated protein (GRP94 and GRP78) genes share common regulatory domains and are coordinately regulated by common trans-acting factors. *Mol. Cell. Biol.* **9:** 2153–2162.

Chirico, W.J., M.G. Waters, and G. Blobel. 1988. 70K heat shock related proteins stimulate protein translocation into microsomes. *Nature* **332:** 805–809.

Cox, J.S., C.E. Shamu, and P. Walter. 1993. Transcriptional induction of genes encoding endoplasmic reticulum resident proteins requires a transmembrane protein kinase. *Cell* **73:** 1197–1206.

Craig, E.A., B.D. Gambill, and R.J. Nelson. 1993. Heat shock proteins: Molecular chaperones of protein biogenesis. *Microbiol. Rev.* **57:** 402–414.

Deshaies, R.J., S.L. Sanders, D.A. Feldheim, and R. Schekman. 1991. Assembly of yeast Sec proteins involved in translocation into the endoplasmic reticulum into a membrane-bound multisubunit complex. *Nature* **349:** 806–808.

Deshaies, R.J., B.D. Koch, M. Werner-Washburne, E.A. Craig, and R. Schekman. 1988. A subfamily of stress proteins facilitates translocation of secretory and mitochondrial precursor polypeptides. *Nature* **332:** 800–805.

Dorner, A.J., D.G. Bole, and R.J. Kaufman. 1987. The relationship of N-linked glycosylation and heavy chain-binding protein association with the secretion of glycoproteins. *J. Cell Biol.* **105:** 2665–2674.

Dower, W.J. 1992. Phage power. *Curr. Biol.* **2:** 251–253.

Earl, P.L., B. Moss, and R.W. Doms. 1991. Folding, interaction with GRP78-BiP, assembly, and transport of the human immunodeficiency virus type 1 envelope protein. *J. Virol.* **65:** 2047–2055.

Feldheim, D., J. Rothblatt, and R. Schekman. 1992. Topology and functional domains of sec63p, an endoplasmic reticulum membrane protein required for secretory protein translocation. *Mol. Cell. Biol.* **12:** 3288–3296.

Flaherty, K.M., C. DeLuca-Flaherty, and D.B. McKay. 1990. Three-dimensional structure of the ATPase fragment of a 70K heat-shock cognate protein. *Nature* **346:** 623–628.

Flaherty, K.M., D.B. McKay, W. Kabash, and K.C. Holmes. 1991. Similarity of the three-dimensional of actin and the ATPase fragment of a 70-kDa heat shock cognate protein. *Proc. Natl. Acad. Sci.* **88:** 5041–5045.

Flajnik, M.F., C. Canel, J. Kramer, and M. Kasahara. 1991. Which came first, MHC class I or class II? *Immunogenetics* **33:** 295–300.

Flynn, G.C., T.G. Chappell, and J.E. Rothman. 1989. Peptide binding and release by proteins implicated as catalysts of protein assembly. *Science* **245:** 385–390.

Flynn, G.C., J. Pohl, M.T. Flocco, and J.E. Rothman. 1991. Peptide-binding specificity of the molecular chaperone BiP. *Nature* **353:** 726–730.

Fontes, E.B.P., B.B. Shank, R.L. Wrobel, S.P. Moose, G.R. O'Brian, E.T. Wurtzel, and R.S. Boston. 1991. Characterization of an immunoglobulin-binding protein analog in the maize *floury*-2 endosperm mutant. *Plant Cell* **3:** 483–496.

Freiden, P.J., J.R. Gaut, and L.M. Hendershot. 1992. Interconversion of three differently modified and assembled forms of BiP. *EMBO J.* **11:** 63–70.

Fremont, D.H., M. Matsumura, E.A. Stura, P.A. Peterson, and I.A. Wilson. 1992. Crystal structures of two viral peptides in complex with murine MHC class I H-2Kb. *Science* **257:** 919–927.

Gaut, J.R. and L.M. Hendershot. 1993. The immunoglobulin-binding protein in vitro autophosphorylation site maps to a threonine within the ATP binding cleft but is not a detectable site of in vivo phosphorylation. *J. Biol. Chem.* **268:** 12691–12698.

Gething, M.-J. and J.F. Sambrook. 1992. Protein folding in the cell. *Nature* **355:** 33–45.

Gething, M.-J., K. McCammon, and J. Sambrook. 1986. Expression of wild-type and mutant forms of influenza hemagglutinin: The role of folding in intracellular transport. *Cell* **46:** 939–950.

Haas, I.G. and M. Wabl. 1983. Immunoglobulin heavy chain binding protein. *Nature* **306:** 387–389.

Hammer, J., P. Valsasnini, K. Tolba, D. Bolin, J. Higelin, B. Takacs, and F. Sinigaglia. 1993. Promiscuous and allele-specific anchors in HLA-DR-binding peptides. *Cell* **74:** 197–203.

Hanks, S.K. and A.M. Quinn. 1991. Protein kinase catalytic domain sequence database: Identification of conserved features of primary structure and classification of family members. *Methods Enzymol.* **200:** 38–62.

Hendershot, L.M., J. Ting, and A.S. Lee. 1988. Identity of the immunoglobulin heavy-chain-binding protein with the 78,000-dalton glucose-regulated protein and the role of posttranslational modifications in its binding function. *Mol. Cell. Biol.* **8:** 4250–4256.

Heschl, M.F.P. and D.L. Baillie. 1989. Characterization of the hsp70 multigene family of *Caenorhabditis elegans*. *DNA* **8:** 233–243.

Holmgren, A. and C.-I. Branden. 1989. Crystal structure of chaperone protein PapD reveals an immunoglobulin fold. *Nature* **342:** 248–251.

Hunter, T. 1991. Protein kinase classification. *Methods Enzymol.* **200:** 3–37.

Hurtley, S.M., D.G. Bole, H. Hoover-Litty, A. Helenius, and C.S. Copeland. 1989. Interactions of misfolded influenza virus hemagglutinin with binding protein (BiP). *J. Cell Biol.* **108:** 2117–2126.

Kassenbrock, C.K. and R.B. Kelly. 1989. Interaction of heavy chain binding protein (BIP/GRP78) with adenine nucleotides. *EMBO J.* **8:** 1461–1467.

Kim, P.S., D. Bole, and P. Arvan. 1992. Transient aggregation of nascent thyroglobulin in the endoplasmic reticulum: Relationship to the molecular chaperone, BiP. *J. Cell Biol.* **118:** 541–549.

Knittler, M.R. and I.G. Haas. 1992. Interaction of BiP with newly synthesized immunoglobulin light chain molecules: Cycles of sequential binding and release. *EMBO J.* **11:** 1573–1581.

Kohno, K., K. Normington, J. Sambrook, M.-J. Gething, and K. Mori. 1993. The promoter region of the yeast KAR2 (BiP) gene contains a regulatory domain that responds to the presence of unfolded proteins in the endoplasmic reticulum. *Mol. Cell. Biol.* **13:** 877–890.

Kozutsumi, Y., M. Segal, K. Normington, M.-J. Gething, and J. Sambrook. 1988. The presence of malfolded proteins in the endoplasmic reticulum signals the induction of glucose regulated proteins. *Nature* **332:** 462–464.

Kozutsumi, Y., K. Normington, E. Press, C. Slaughter, J. Sambrook, and M.-J. Gething. 1989. Identification of immunoglobulin heavy chain binding protein as glucose regulated protein 78 on the basis of amino acid sequence, immunological crossreactivity and functional activity. *J. Cell Sci.* (suppl.) **11:** 115–137.

Kuehn, M.J., D.J. Ogg, J. Kihlberg, L.N. Slonim, K. Flemmer, T. Bergfors, and S.J. Hultgren. 1993. Structural basis of polus subunit recognition by the PapD chaperone. *Science* **262:** 1234–1241.

Kumar, N. and H. Zheng. 1992. Nucleotide sequence of a *Plasmodium falciparum* stress protein with similarity to mammalian 78 kDa glucose-regulated protein. *Mol. Biochem. Parasitol.* **56:** 353–356.

Landry, S.J., R. Jordan, R. McMacken, and L.M. Gierasch. 1992. Different conformations for the same polypeptide bound to chaperones DnaK and GroEL. *Nature* **355:** 455–457.

Lee, A.S. 1987. Coordinated regulation of a set of genes by glucose and calcium ionophores in mammalian cells. *Trends Biol. Sci.* **12:** 20–23.

———. 1992. Mammalian stress response: Induction of the glucose-regulated protein family. *Curr. Opin. Cell Biol.* **4:** 267–273.

Leno, G.H. and B.E. Ledford. 1990. Reversible ADP-ribosylation of the 78 kDa glucose-regulated protein. *FEBS Lett.* **276:** 29–33.

Leustek, T., H. Toledo, N. Brot, and H. Weissbach. 1991. Calcium-dependent autophosphorylation of the glucose-regulated protein, grp78. *Arch. Biochem. Biophys.* **289:** 256–261.

Leustek, T., D. Amir-Shapira, H. Toledo, N. Brot, and H. Weissbach. 1992. Autophosphorylation of 70 kDa heat shock proteins. *Cell. Mol. Biol.* **38:** 1–10.

Liberek, K., J. Marszalek, D. Ang, C. Georgopoulos, and M. Zylicz. 1991. *Escherichia coli* DnaJ and GrpE heat shock proteins jointly stimulate ATPase activity of DnaK. *Proc. Natl. Acad. Sci.* **88:** 2874–2878.

Lis, J.T., H. Xiao, and O. Perisic. 1990. Modular units of heat shock regulatory regions: Structure and function. In *Stress proteins in biology and medicine* (ed. R.I. Morimoto et al.), pp. 411–428. Cold Spring Harbor Laboratory Press, Cold Spring, Harbor, New York.

Liu, E.S. and A.S. Lee. 1991. Common sets of nuclear factors binding to the conserved promoter sequence motif of two coordinately regulated ER protein genes, GRP78 and GRP94. *Nucleic Acids Res.* **19:** 5425–5431.

Macer, D.R.J. and G.L.E. Koch. 1989. Identification of a set of calcium proteins in reticuloplasm, the luminal contents of the endoplasmic reticulum. *J. Cell Sci.* **91:** 61–70.

Machamer, C.E., R.W. Doms, D.G. Bole, A. Helenius, and J.K. Rose. 1990. Heavy chain binding protein recognizes incompletely disulfide-bonded forms of Vesicular Stomatitis Virus G protein. *J. Biol. Chem.* **265:** 6879–6883.

Madden, D.R., D.N. Garboczi, and D.C. Wiley. 1993. The antigenic identity of peptide-MHC complexes: A comparison of the conformations of five viral peptides presented by HLA-A2. *Cell* **75:** 693–708.

Madden, D.R., J.C. Gorga, J.L. Strominger, and D.C. Wiley. 1992. The three-dimensional structure of HLA-B27 at 2.1 Å resolution suggests a general mechanism for tight peptide binding to MHC. *Cell* **70:** 1035–1048.

Matsumura, M., D.H. Fremont, P.A. Peterson, and I.A. Wilson. 1992. Emerging principles for the recognition of peptide antigens by MHC class I molecules. *Science* **257:** 927–934.

Mori, K., W. Ma, M.-J. Gething, and J. Sambrook. 1993. A transmembrane protein with a cdc2+/CDC28-related kinase activity is required for signaling from the ER to the nucleus. *Cell* **74:** 743–756.

Mori, K., A. Sant, K. Kohno, K. Normington, M.-J. Gething, and J.F. Sambrook. 1992. A 22-bp cis-acting element is necessary and sufficient for the induction of the yeast

KAR2 (BiP) gene by unfolded proteins. *EMBO J.* **11:** 2583–2593.

Munro, S. and H.R.B. Pelham. 1986. An Hsp70-like protein in the ER: Identity with the 78 kd glucose-regulated protein and immunoglobulin heavy chain binding protein. *Cell* **46:** 291–300.

Nakaki, T., R.J. Deans, and A.S. Lee. 1989. Enhanced transcription of the 78,000-dalton glucose-regulated protein (GRP78) gene and association of GRP78 with immunoglobulin light chains in a nonsecreting B-cell myeloma line (NS-1). *Mol. Cell. Biol.* **9:** 2233–2238.

Navarro, D., I. Qadri, and L. Pereira. 1991. A mutation in the ectodomain of herpes simplex virus 1 glycoprotein B causes defective processing and retention in the endoplasmic reticulum. *Virology* **184:** 253–264.

Ng, D.T.W., R.E. Randall, and R.A. Lamb. 1989. Intracellular maturation and transport of the SV5 type II glycoprotein hemagglutinin-neuraminidase: Specific and transient association with GRP78-BiP in the endoplasmic reticulum and extensive internalization from the cell surface. *J. Cell Biol.* **109:** 3273–3289.

Nguyen, T.H., D.T.S. Law, and D.B. Williams. 1991. Binding protein BiP is required for translocation of secretory proteins into the endoplasmic reticulum in *Saccharomyces cerevisiae*. *Proc. Natl. Acad. Sci.* **88:** 1565–1569.

Nicholson, R.C., D.B. Williams, and L.A. Moran. 1990. An essential member of the HSP70 gene family of *Saccharomyces cerevisiae* is homologous to immunoglobulin heavy chain binding protein. *Proc. Natl. Acad. Sci.* **86:** 1159–1163.

Nikawa, J. and S. Yamashita. 1992. *IRE1* encodes a putative protein kinase containing a membrane-spanning domain and is required for inositol phototrophy in *Saccharomyces cerevisiae*. *Mol. Microbiol.* **6:** 1441–1446.

Normington, K., K. Kohno, Y. Kozutsumi, M.-J. Gething, and J. Sambrook. 1989. *S. cerevisiae* encodes an essential protein homologous in sequence and function to mammalian BiP. *Cell* **57:** 1223–1236.

Oblas, B., N.D. Boyd, J. Luber-Narod, V.E. Reyes, and S.E. Leeman. 1990. Isolation and identification of a polypeptide in the Hsp 70 family that binds substance P. *Biochem. Biophys. Res. Commun.* **166:** 978–983.

Palleros, D.R., K. Reid, L. Shi, W.J. Welch, and A.L. Fink. 1993. ATP-induced protein-Hsp70 complex dissociation requires K+ but not ATP hydrolysis. *Nature* **365:** 664–666.

Parmley, S.F. and G.P. Smith. 1988. Antibody-selectable filamentous fd phage vectors: Affinity purification of target genes. *Gene* **73:** 305–318.

Partaledis, J.A. and V. Berlin. 1993. The *FKB2* gene of *Saccharomyces cerevisiae*, encoding the immunosuppressant-binding protein FKBP-13, is regulated in response to accumulation of unfolded proteins in the endoplasmic reticulum. *Proc. Natl. Acad. Sci.* **90:** 5450–5454.

Pelham, H.R.B. 1989. Control of protein exit from the endoplasmic reticulum. *Annu. Rev. Cell Biol.* **5:** 1–23.

Polaina, J. and J. Conde. 1982. Genes involved in the control of nuclear fusion during the sexual cyle of *Saccharomyces cerevisiae*. *Mol. Gen. Genet.* **186:** 253–258.

Pouyssegur, J., R.P.C. Shiu, and I. Pastan. 1977. Induction of two transformation-sensitive membrane polypeptides in normal fibroblasts by a block in glycoprotein synthesis or glucose deprivation. *Cell* **11:** 941–947.

Rammensee, H.-G., K. Falk, and O. Rotzschke. 1993. Peptides naturally presented by MHC class I molecules. *Annu. Rev. Immunol.* **11:** 213–244.

Resendez, I., Jr., J.W. Attenello, A. Grafsky, C.S. Chang, and A.S. Lee. 1985. Calcium ionophore A23187 induces expression of glucose-regulated genes and their heterologous fusion genes. *Mol. Cell. Biol.* **5:** 1212–1219.

Rippmann, F., W.R. Taylor, J.B. Rothbard, and N.M. Green. 1991. A hypothetical model for the peptide binding domain of hsp70 based on the peptide binding domain of HLA. *EMBO J.* **10:** 1053–1059.

Rose, M.D., L.M. Misra, and J.P. Vogel. 1989. *KAR2*, a karyogamy gene, is the yeast homologue of mammalian BiP/GRP78. *Cell* **57:** 1211–1221.

Rothblatt, J.A., R.J. Deshaies, S.L. Sanders, G. Daum, and R. Schekman. 1989. Multiple genes are required for proper insertion of secretory proteins into the endoplasmic reticulum in yeast. *J. Cell Biol.* **109:** 2641–2652.

Rothman, J.E. 1989. Polypeptide chain binding proteins: Catalysts of protein folding and related processes in cells. *Cell* **59:** 591–601.

Rubin, D.M., A.D. Mehta, J. Zhu, S. Shoham, X. Chen, Q.R. Wells, and K.B. Palter. 1993. Genomic structure and sequence analysis of *Drosophila melanogaster* HSC70 genes. *Gene* **128:** 155–163.

Sadler, I., A. Chiang, T. Kurihara, J. Rothblatt, J. Way, and P. Silver. 1989. A yeast gene important for protein assembly into the endoplasmic reticulum and the nucleus has homology to DnaJ, an *Escherichia coli* heat shock protein. *J. Cell Biol.* **109:** 2665–2675.

Sanders, S.L., K.M. Whitfield, J.P. Vogel, M.D. Rose, and R.W. Schekman. 1992. Sec61p and BiP directly facilitate polypeptide translocation into the ER. *Cell* **69:** 353–365.

Saper, M.A., P.J. Bjorkman, and D.C. Wiley. 1991. Refined structure of the human histocompatibility antigen HLA-A2 at 2.6 Å resolution. *J. Mol. Biol.* **219:** 277–319.

Schonberger, O., T.R. Hirst, and O. Pines. 1991. Targeting and assembly of an oligomeric bacterial entertoxoid in the endoplasmic reticulum of *Saccharomyces cerevisiae*. *Mol. Microbiol.* **11:** 2663–2671.

Scidmore, M.A., H.H. Okamura, and M.D. Rose. 1993. Genetic interactions between *KAR2* and *SEC63*, encoding eukaryotic homologues of DnaK and DnaJ in the endoplasmic reticulum. *Mol. Biol. Cell* **4:** 1145–1159.

Scott, J.K. 1992. Discovering peptide ligands using epitope libraries. *Trends Biol. Sci.* **17:** 241–245.

Scott, J.K. and G.P. Smith. 1990. Searching for peptide ligands with an epitope library. *Science* **249:** 386–390.

Smith, G.P. 1991. Surface presentation of protein epitopes using bacteriophage expression systems. *Curr. Opin. Biotech.* **2:** 668–673.

Sorger, P.K., M.J. Lewis, and H.R.B. Pelham. 1987. Heat shock factor is regulated differently in yeast and HeLa cells. *Nature* **329:** 81–84.

Tachibana, C. and T.H. Stevens. 1992. The yeast EUG1 gene encodes an endoplasmic reticulum protein that is functionally related to protein disulfide isomerase. *Mol. Cell. Biol.* **12:** 4601–4611.

Tokunaga, M., A. Kawamura, and K. Kohno. 1992. Purification and characterization of BiP/Kar2 protein from *Saccharomyces cerevisiae*. *J. Biol. Chem.* **267:** 17553–17559.

Toledo, H., A. Carlino, V. Vidal, B. Redfield, M.Y. Nettleton, J.P. Kochen, N. Brot, and H. Weissbach. 1993. Dissociation of glucose-regulated protein Grp78 and Grp78-IgE Fc complexes by ATP. *Proc. Natl. Acad. Sci.* **90:** 2505–2508.

Vogel, J.P., L.M. Misra, and M.D. Rose. 1990. Loss of Bip/GRP78 function blocks translocation of secretory proteins in yeast. *J. Cell Biol.* **110:** 1885–1895.

Wang, T.-F., J. Chang, and C. Wang. 1993. Identificaiton of the peptide-binding domain of hsc70. *J. Biol. Chem.* **268:** 26049–26051.

Welch, W.J., J.I. Garrels, G.P. Thomas, J.J.-C. Lin, and J.R. Feramisco. 1983. Biochemical characterization of the mammalian stress proteins and identification of two stress proteins as glucose- and Ca^{2+}-ionophore-regulated proteins. *J. Biol. Chem.* **258:** 7102–7111.

Wiech, H., J. Buchner, M. Zimmermann, R. Zimmermann, and U. Jakob. 1993. Hsc70, immunoglobulin heavy chain binding protein, and hsp90 differ in their ability to stimulate transport of precursor proteins into mammalian microsomes. *J. Biol. Chem.* **268:** 7414–7421.

Wooden, S.K., L.J. Li, D. Navarro, I. Qadri, L. Pereira, and A.S. Lee. 1991. Transactivation of the grp78 promoter by malfolded proteins, glycosylation block, and calcium ionophore is mediated through a proximal region containing a CCAAT motif which interacts with CTF/N-I. *Mol. Cell. Biol.* **11:** 5612–5623.

Zimmermann, R., M. Sagstetter, M.J. Lewis, and H.R.B. Pelham. 1988. Seventy-kilodalton heat shock proteins and an additional component from reticulocyte lysate stimulate import of M13 procoat protein into microsomes. *EMBO J.* **7:** 2875–2880.

6

Heat Shock 70-kD Proteins and Lysosomal Proteolysis

J. Fred Dice, Fernando Agarraberes, Melissa Kirven-Brooks, Laura J. Terlecky,[1] and Stanley R. Terlecky[1]
Department of Physiology
Tufts University School of Medicine
Boston, Massachusetts 02111

I. INTRODUCTION

There are multiple pathways of intracellular proteolysis in eukaryotic cells (Dice 1987; Hershko and Ciechanover 1992; Olson et al. 1992). The best studied cytosolic pathway is dependent on ATP and ubiquitin and is responsible for the degradation of many abnormal proteins as well as many short-lived normal proteins (Rechsteiner 1991; Hershko and Ciechanover 1992). Other cytosolic proteolytic pathways are independent of ubiquitin, and still others are not cytosolic but are contained within organelles such as the endoplasmic reticulum and mitochondrion (Olson et al. 1992).

It is also well established that lysosomes have an important role in overall proteolysis (Mortimore 1987), and lysosomes appear to be able to internalize intracellular proteins in a variety of ways. For example, degradation of many plasma membrane proteins, as well as certain other intracellular membrane proteins, is through endocytosis and delivery to lysosomes (Fig. 1) (Hare 1990). In well-nourished cells, lysosomes appear to be able to internalize cytosolic proteins by a poorly understood process called microautophagy (Fig. 1) in which the lysosomal membrane invaginates at multiple locations (Ahlberg et al. 1982; Dice 1987).

Present address: [1]Department of Biology, 0322 Bonner Hall, University of California at San Diego, La Jolla, California 92093-0322.

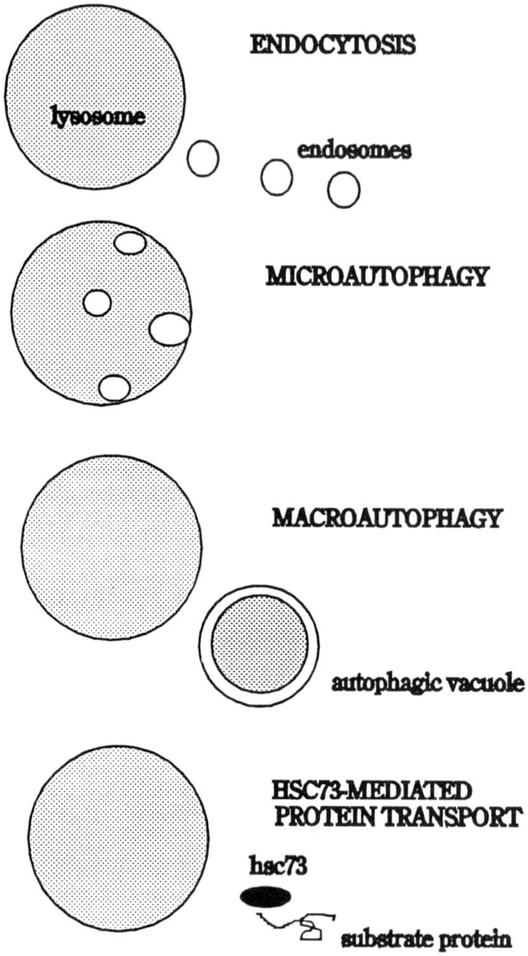

Figure 1 Schematic representations of pathways of lysosomal proteolysis. For descriptions of the pathways, see the text.

These intralysosomal vesicles presumably decompose, whereupon their contents can be digested by lysosomal hydrolases. When cultured cells reach confluence and in certain tissues of fasted animals, macroautophagy (Fig. 1) is stimulated (Cockle and Dean 1982; Knecht et al. 1984; Mortimore 1987). Macroautophagy begins with the formation of double-membraned autophagic vacuoles that sequester areas of cytoplasm. The membranes of these autophagic vacuoles are derived from ribosome-free areas of the rough endoplasmic reticulum and from smooth endoplasmic reticulum (Dunn 1990a,b; Ueno et al. 1991). These vacuoles then acidify and acquire lysosomal hydrolases to form autophagosomes. Both micro-autophagy (Ahlberg et al. 1982; Marzella and Glaumann 1987) and mac-roautophagy (Kominami et al. 1983; Kopitz et al. 1990) appear to be nonselective in that many different proteins are sequestered at approximately the same rates.

An additional pathway of lysosomal proteolysis is activated in confluent cell monolayers in response to withdrawal of serum growth factors. This pathway is restricted to cytosolic proteins that contain peptide sequences biochemically related to Lys-Phe-Glu-Arg-Gln (KFERQ). The mechanism by which proteins with KFERQ-like peptide regions are targeted to lysosomes for degradation is similar in many respects to the import of newly synthesized proteins into organelles. Therefore, proteins with KFERQ-like peptide regions may enter lysosomes by crossing a membrane bilayer rather than by vesicular pathways (Fig. 1).

II. RIBONUCLEASE A AS A PROBE FOR SELECTIVE LYSOSOMAL PROTEOLYSIS

Red-cell-mediated microinjection has been used by several laboratories to introduce specific radioactively labeled proteins into the cytosol of cultured cells (Schlegel and Rechsteiner 1978; Kulka and Loyter 1979; McElligott and Dice 1984). In confluent cultures of human fibroblasts, radioactively labeled ribonuclease A (RNase A[1]) was degraded with a half-life of approximately 100 hours in serum-supplemented cells and 50 hours in serum-deprived cells (Backer et al., 1983). The half-lives of certain microinjected proteins, but not others, were also reduced in response to serum withdrawal (Fig. 2).

Several lines of evidence indicated that microinjected RNase A was degraded within lysosomes: (1) A small amount of microinjected RNase A, but not other proteins, fractionated with lysosomes (McElligott et al. 1985); (2) degradation of RNase A was partially inhibited by ammonium chloride (McElligott et al. 1985); (3) degradation products of [³H]raffinose-RNase A accumulated only in lysosomes (McElligott et al. 1985); and (4) the same peptides derived from RNase A were released from cells after microinjection and after lysosomal hydrolysis following endocytosis (Isenman and Dice 1989).

The amino-terminal 20 amino acids of RNase A (RNase S-peptide; Fig. 3) were shown to be required for the enhanced degradation of RNase

---

[1]The abbreviations used are RNase A, bovine pancreatic ribonuclease A; RNase S-peptide, residues 1–20 of RNase A; RNase S-protein, residues 21–124 of RNase A; hsp70, heat shock protein of 70 kD; prp73, peptide recognition protein of 73 kD; hsc73, heat shock cognate protein of 73 kD; Grp78, glucose-regulated protein of 78 kD; DnaK, the product of the *E. coli dnak* gene; Ssa1, the product of the *Saccharomyces cerevisiae SSA1* gene; GAPDH, glyceraldehyde-3-phosphate dehydrogenase; MAb, monoclonal antibody.

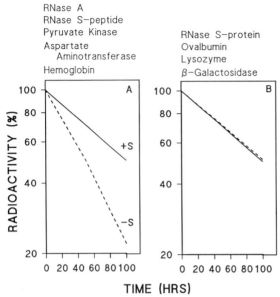

Figure 2 Selectivity in the enhanced degradation of proteins microinjected into cultured cells. Radioactively labeled proteins were microinjected into cultured human fibroblasts or HeLa cells, and time-dependent loss of radioactivity from cells was followed in the presence (+S) and absence (−S) of serum in the medium. This loss reflects degradation of the microinjected protein to free amino acids and peptides (McElligott and Dice 1984).

A in response to serum withdrawal (Backer et al. 1983). Covalent attachment of this peptide to heterologous proteins whose degradation rates were not serum-regulated caused their degradation rates to increase in response to serum deprivation (Table 1) (Backer and Dice 1986). Amino acids 7–11 within the RNase S-peptide (KFERQ) was identified as the region essential for enhanced degradation (Dice et al. 1986). Coinjection of [125I]RNase A and increasing amounts of unlabeled KFERQ blocked the enhanced degradation of RNase A in response to serum deprivation without affecting its degradation rate in the presence of serum (Chiang and Dice 1988).

Recent results showed that in human fibroblasts, *Escherichia coli* β-galactosidase introduced into the cytosol was degraded at the same rate in the presence and absence of serum. However, an RNase S-peptide–β-galactosidase fusion protein was degraded more rapidly in the absence of serum (Table 1) (L. Jeffreys-Terlecky et al., unpubl.). The effects of mutations in each of the codons of the KFERQ region are currently being examined to determine experimentally the types of peptides that will lead to enhanced degradation in response to serum withdrawal.

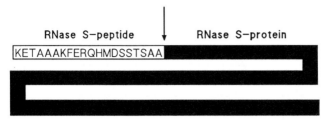

RNase S-peptide   RNase S-protein

KETAAAKFERQHMDSSTSAA

Figure 3 Schematic representation of RNase A. Amino-terminal 20 amino acids (RNase S-peptide) are shown. The arrow indicates the subtilisin cleavage site.

III. GENERALITY OF THIS PATHWAY OF PROTEIN DEGRADATION

To examine whether peptide regions similar to KFERQ exist in intracellular proteins, polyclonal antibodies to KFERQ were raised and IgGs specifically directed toward the pentapeptide were isolated (Chiang and Dice 1988). These antibodies immunoprecipitated 30% of [³H]-leucine-labeled cytosolic proteins from human fibroblasts. Loss of radioactivity from immunoreactive and nonimmunoreactive cytosolic proteins was followed by incubating radioactively labeled cells in unlabeled medium containing excess leucine to prevent reutilization of the isotope. The immunoreactive proteins were preferentially degraded in response to serum withdrawal, whereas nonimmunoreactive proteins were degraded at the same rate in the presence and absence of serum.

A similar pathway of proteolysis was shown to be active in liver, kidney, and heart of fasted rats (Chiang and Dice 1988; Wing et al. 1991). In liver, this pathway of proteolysis was activated with prolonged fasting (Wing et al. 1991). Apparently, nonselective macroautophagy is activated first in response to fasting in order to supply amino acids for continued protein synthesis, gluconeogenesis, and use as an energy source. During a prolonged fast, however, continued nonselective degradation of

Table 1 Effect of RNase S-Peptide Sequences on Serum-regulated Degradation of Test Proteins

	$T_{1/2}$ (h)	
Protein	+ serum	– serum
Lysozyme	59	59
RNase S-peptide–lysozyme	47	21
Insulin A-chain	104	104
RNase S-peptide–insulin A-chain	100	44
β-Galactosidase[a]	134	154
RNase S-peptide–β-galactosidase[a]	178	78

Half-life values can be found in Backer and Dice (1986).
[a]Data from L. Jeffreys-Terlecky et al. (unpubl.).

cellular proteins might lead to detrimental depletion of critical enzymes or regulatory proteins. Instead, the selective lysosomal degradation pathway is activated that is specific for proteins containing KFERQ-like peptide sequences. These proteins have presumably evolved to contain such peptide regions because they are dispensable during a prolonged fast.

IV. ROLES OF HEAT SHOCK 70-KD PROTEINS IN SELECTIVE LYSOSOMAL PROTEOLYSIS

A possible mechanism for selectivity in lysosomal proteolysis would be the recognition by an intracellular protein of the KFERQ-like peptide regions in proteins that are degraded more rapidly in response to serum withdrawal. A protein of 73 kD, purified on an RNase S-peptide affinity column, was designated prp73 for peptide recognition protein of 73 kD (Chiang et al. 1989). This protein bound to RNase S-peptide with a K_d of 8 μM (Terlecky et al. 1992).

A monoclonal antibody that recognizes many members of the 70-kD heat shock protein (hsp70) family (Kurtz et al. 1986) reacted with prp73 purified from human fibroblast cytosol. In addition, sequence data obtained from purified prp73 tentatively identified it as the constitutively expressed heat shock cognate protein of 73 kD (hsc73) (Chiang et al. 1989). This was an exciting discovery since hsp70 family members had been implicated in the transport of precursor proteins into mitochondria and endoplasmic reticulum (Chirico et al. 1988; Deshaies et al. 1988).

hsc73 and prp73 were functionally equivalent in several assays. Both proteins bound to RNase A, RNase S-peptide, KFERQ, aspartate aminotransferase, and pyruvate kinase, proteins and peptides that contain KFERQ-like peptide motifs. Neither hsc73 nor prp73 binds to ovalbumin, lysozyme, ubiquitin, or β-galactosidase, proteins that lack KFERQ-like peptide motifs (Terlecky et al. 1992).

Such selectivity in substrate binding of hsc73 is somewhat controversial. Some evidence suggests that the binding is rather nonspecific in that hsc73 transiently associates with many different nascent chains as they emerge from polysomes (Beckmann et al. 1990), and hsc73 binds to several denatured proteins but not to their native counterparts (Palleros et al. 1991). On the other hand, Flynn et al. (1989) demonstrated that the ability of two different peptides to interact with hsc73 differed by 14-fold, and preferential binding of hsc73 to certain peptides has been shown by other investigators (DeLuca-Flaherty et al. 1990; Hightower et al., this volume). Perhaps hsc73 is able to interact weakly with many different polypeptides, and even nonpeptide molecules (Alvares et al. 1990), but higher-affinity binding may require specific peptide characteristics.

V. SELECTIVE DEGRADATION OF PROTEINS BY ISOLATED LYSOSOMES

To try to elucidate the mechanism of degradation of KFERQ motif-containing proteins, an in vitro assay using lysosomes isolated from human fibroblasts was developed (Chiang et al. 1989; Terlecky et al. 1992; Terlecky and Dice 1993). Maximal degradation of [^{3}H]RNase S-peptide and [^{3}H]RNase A by isolated lysosomes required both ATP and hsc73 (Chiang et al. 1989; Terlecky et al. 1992). hsc73 and prp73 functioned identically in stimulating degradation of [^{3}H]RNase S-peptide by isolated lysosomes under conditions where three other members of the hsp70 family (DnaK, Ssa1, and Grp78) had no activity (Terlecky et al. 1992). This degradation was selective because [^{3}H]RNase S-protein and [^{3}H]dihydrofolate reductase, proteins that do not contain KFERQ motifs, were degraded little, if at all, under the same conditions (Terlecky and Dice 1993).

Degradation of [^{3}H]RNase S-peptide could be inhibited by reducing the temperature, and degradation appeared to occur within intact lysosomes because it could be inhibited by ammonium chloride (Chiang et al. 1989) and leupeptin (Terlecky and Dice 1993). Hydrolases released from damaged lysosomes could not account for the proteolysis because the incubation buffer maintained the pH at 7.2, and there is no detectable RNase S-peptide hydrolyzing activity at this pH (Terlecky and Dice 1993).

Lysosomal uptake of [^{3}H]RNase S-peptide was saturable (K_m = 5 μM). At 4°C and in the presence of hsc73, [^{3}H]RNase S-peptide specifically bound to lysosomal membranes, and this binding was reduced by prior mild trypsinization (Terlecky and Dice 1993). No such binding was observed for RNase S-protein. Presumably, this RNase-S-peptide-binding component is a receptor or a polypeptide transport channel. We have shown that a 39-kD protein within the lysosomal membrane specifically binds to RNase S-peptide, and we are currently determining its identity (R. Skurat and J.F. Dice, unpubl.).

Lysosomes isolated from serum-deprived cells were twice as active in protein uptake in vitro as were lysosomes derived from serum-supplemented cells. Correlated with this increased activity was an increased amount of hsc73 in the lysosomal fraction (Terlecky and Dice 1993). Some of this hsc73 appeared to be associated with the lysosome surface because it could be removed with trypsin. However, most of the lysosomal hsc73 appeared to be in the lumen of the lysosome since it was not digested by trypsin unless the lysosomal membrane was disrupted. Indirect immunofluorescence using laser scanning confocal microscopy confirmed the colocalization of a fraction of hsc73 with lysosomal marker enzymes (S.R. Terlecky et al., unpubl.). These studies also demonstrated that certain lysosomes did not contain hsc73 and that serum withdrawal caused a marked morphological change in hsc73-containing lysosomes; they fused to become a tubular network. The functional significance of these morphological changes is not yet clear.

Lysosomes isolated from rat liver have recently been reported to take up selectively and degrade glyceraldehyde-3-phosphate dehydrogenase (GAPDH; Aniento et al. 1993). This process appeared to be very similar to the uptake and degradation of RNase A by fibroblast lysosomes in that it was selective and stimulated by hsc73 and ATP (A.M. Cuervo et al., unpubl.). Indeed, uptake of GAPDH by rat liver lysosomes could be competed with RNase A or RNase S-peptide, and uptake of RNase A could be competed by GAPDH. RNase S-protein and ovalbumin showed no competition in these assays. Interestingly, in this rat liver lysosome system, an import intermediate of RNase A was detected in which most of the molecule had entered the lysosome while a small portion remained outside (A.M. Cuervo et al., unpubl.).

The hsp70 in the lumen of fibroblast and rat liver lysosomes was identified as hsc73 using a monoclonal antibody (MAb), 13D3 (Maekawa et al. 1989), that was shown to be specific for hsc73 (Terlecky et al. 1992; S.R. Terlecky et al., unpubl.). Figure 4 compares the degree of recognition of denatured hsp70s in immunoblots for the hsc73-specific MAb 13D3 and for the more broadly reacting MAb 7.10 (Kurtz et al. 1986). Other experiments showed that MAb 7.10 recognized only denatured proteins, whereas MAb 13D3 had the unusual property for a monoclonal antibody that it also immunoprecipitated native hsc73. Other researchers have also reported hsp70s within lysosomes (Mayer et al. 1991; Domanico et al. 1993).

Cells were allowed to endocytose large amounts of MAb 13D3 in order to neutralize the intralysosomal hsc73. This treatment had no effect on the intralysosomal degradation of endocytosed [^{3}H]RNase A (S.R. Terlecky et al., unpubl.). However, when cells were metabolically labeled with [^{3}H]leucine and then protein degradation followed, the endocytosed 13D3 was found to block completely the enhanced protein degradation in response to serum withdrawal without altering proteolysis in the presence of serum (Fig. 5). Endocytosis of MAb P32, a control IgM, had no effect on degradation, and endocytosis of 13D3 in the presence of an equimolar amount of hsc73 also had no effect on proteolysis (Fig. 5) (S.R. Terlecky et al., unpubl.). The intralysosomal hsc73 is likely to be required to pull the substrate proteins across the lipid bilayer, a role that has been shown for other hsp70 family members within mitochondria and the endoplasmic reticulum (Kang et al. 1990; Vogel et al. 1990; Nicchita and Blobel 1993).

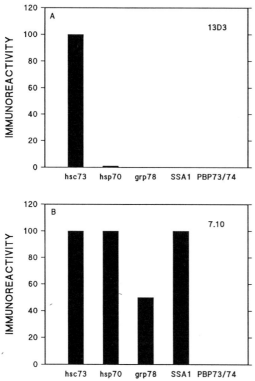

Figure 4 Comparison of abilities of MAbs 13D3 and 7.10 to recognize hsp70s
in immunoblots. hsp70s were separated by SDS-PAGE and immunoblotted
using standard techniques. MAb 7.10 was generously provided by Dr. Susan
Lindquist (University of Chicago), and MAb 13D3 was generously provided by
Edward Eddy (National Institute of Environmental Health Sciences) and by
Joseph Chandler (Maine Biotechnology Services, Inc.). The immunoreactivity
of hsc73 is taken as 100% for each monoclonal antibody. hsp70 refers to the
major heat-inducible family member from human fibroblasts (Terlecky et al.
1992). The glucose-regulated protein (Grp78) is also commonly referred to as
BiP and was provided by Seth Sadis and Larry Hightower (University of Connecticut) or identified on immunoblots of isolated microsomes. SSA1 refers to
the product of the *SSA1* gene from *S. cerevisiae* and was provided by Bruce
Koch and Randy Sheckman (University of California at Berkeley). PBP73/74 is
the hsp70 family member described by Domanico et al. (1993) and was provided by Drs. Diane DeNagel and Susan Pierce (Northwestern University).

Our results suggest the following model (Fig. 6) for steps in the selective entry of proteins into lysosomes: (1) binding of hsc73 to cytosolic
proteins containing KFERQ motifs, (2) binding of the substrate protein
to a lysosomal membrane protein, (3) import of the substrate protein by

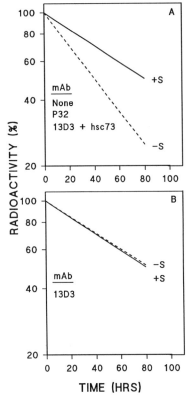

Figure 5 Effect of endocytosed IgMs on intracellular proteolysis. Confluent cultures of human fibroblasts were metabolically labeled with [³H]leucine for 2 days. Prior to beginning degradation measurements, the cells were allowed to endocytose IgMs for 16 hr. (*A*) Enhanced degradation in the absence (−S) compared to the presence (+S) of serum was evident for cells that endocytosed no IgMs, a control IgM (P32), or 13D3 with an equimolar amount of hsc73. (*B*) Enhanced degradation in the absence of serum was blocked by endocytosis of 13D3.

the action of hsc73 within the lysosomal lumen, and (4) intralysosomal degradation of the protein.

VI. FUTURE DIRECTIONS

The mechanisms by which hsc73 stimulates lysosomal degradation of proteins containing KFERQ-like peptide regions are not yet known. hsc73 stimulates import of precursor proteins into mitochondria at least in part by preventing misfolding into a transport-incompetent conformation (Hendrick and Hartl 1993). Such a role seems unlikely for the stimulation of lysosomal uptake of mature, folded proteins such as RNase A.

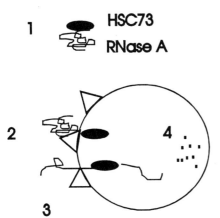

LYSOSOME

Figure 6   A possible pathway for the selective degradation of proteins by lysosomes. For description, see text.

Perhaps hsc73 unfolds proteins prior to lysosomal import. Alternatively, hsc73 may participate in binding of the substrate proteins to the receptor or protein transporters in the lysosomal membrane. Finally, it is possible that hsc73 itself can act as a protein transporter (Fig. 6).

How hsc73 enters the lysosome remains to be established. It may enter along with substrate proteins, but hsc73 could itself be a substrate for this selective lysosomal uptake pathway since two KFERQ motifs are contained within the hsc73 sequence (Terlecky et al. 1992). hsc73 could also enter lysosomes by nonselective microautophagy or macroautophagy. hsc73 would then have to be relatively resistant to intralysosomal hydrolysis since it is more abundant in lysosomes than are other more abundant cytosolic proteins such as hsp90 (see Terlecky and Dice 1993).

Another important issue to resolve is how hsc73 actions are regulated. This protein appears to facilitate protein transport into mitochondria, endoplasmic reticulum, nuclei, and peroxisomes (Chirico et al. 1988; Deshaies et al. 1988; Dingwall and Laskey 1992; Hendrick and Hartl 1993; Walton et al. 1993), as well as facilitating the lysosomal degradation pathway described here. It initially appeared that hsc73 redistributed from membranes into the cytosol and that total cellular levels of hsc73 increased in response to serum withdrawal (Chiang et al. 1989). Subsequent studies have not consistently revealed this redistribution to the cytosol, and total intracellular levels of hsc73 are not greatly changed in response to serum withdrawal (T.S. Olson et al., unpubl.). The earlier results (Chiang et al. 1989) were obtained following subcellular fractionation and may have been due to incomplete recovery of cellular hsc73 from cells grown in the presence of serum.

Other possible ways to regulate hsc73 functions include posttranslational modifications and/or binding of cofactors that could alter peptide-binding properties or subcellular localization. However, it is also possible that components other than hsc73 in the various polypeptide import pathways are regulated. For example, activation of the lysosomal degradation pathway may be due to changes in the substrate proteins that expose the KFERQ-like regions and/or due to activation of the putative receptor or peptide transporter on the lysosome surface. Answering these questions is likely to uncover new aspects of the roles of hsp70s as molecular chaperones.

VII. SUMMARY

Lysosomes are able to internalize cellular proteins in a variety of ways. One pathway is selective for cytosolic proteins containing peptide sequences biochemically related to KFERQ. This pathway is activated in confluent monolayers of cultured cells in response to deprivation of serum growth factors and applies to approximately 30% of cytosolic proteins. Animals also activate this proteolytic pathway in tissues such as liver, kidney, and heart in response to fasting.

This lysosomal degradation pathway has been reconstituted in vitro using highly purified lysosomes. Uptake and degradation of substrate proteins are stimulated by ATP and hsc73. This pathway is selective and saturable, and at 4°C, substrate proteins bind to a protein component of lysosomal membranes. A fraction of cellular hsc73 is associated with lysosomes, both with the lysosomal membrane and within the lysosomal lumen. hsc73 within the lysosomal lumen is required for polypeptide import into the organelle. These results suggest that the lysosomal polypeptide import process is similar in many respects to those for import of proteins for residence in other organelles.

ACKNOWLEDGMENTS

We thank former and current members of the laboratory who contributed to this work, especially Hui-Ling Chiang, Jonathan Backer, Stephen A. Goff, Sharla R. Short, and Ronald Skurat. Research in the authors' laboratory was supported by National Institutes of Health grant AG-06116.

REFERENCES

Ahlberg, J., L. Marzella, and H. Glaumann. 1982. Uptake and degradation of proteins by isolated rat liver lysosomes. Suggestion of a microautophagic pathway of proteolysis. *Lab. Invest.* **47:** 523–532.

Alvares, K., A. Carrillo, P.M. Yuan, H. Kawang, R.I. Morimoto, and J.K. Reddy. 1990. Identification of a cytosolic peroxisome proliferator binding protein as a member of the heat shock protein HSP70 family. *Proc. Natl. Acad. Sci.* **87:** 5293–5297.

Aniento, F., E. Roche, A.M. Cuervo, and E. Knecht. 1993. Uptake and degradation of glyceraldehyde-3-phosphate dehydrogenase by rat liver lysosomes. *J. Biol. Chem.* **268:** 19463–19470.

Backer, J.M. and J.F. Dice. 1986. Covalent linkage of ribonuclease S-peptide to micro-injected proteins causes their intracellular degradation to be enhanced during serum withdrawal. *Proc. Natl. Acad. Sci.* **83:** 5830–5834.

Backer, J.M., L. Bourret, and J.F. Dice. 1983. Regulation of catabolism of microinjected ribonuclease A requires the amino terminal twenty amino acids. *Proc. Natl. Acad. Sci.* **80:** 2166–2170.

Beckmann, R.P., L.A. Mizzen, and W.J. Welsh. 1990. Interaction of hsp70 with newly synthesized proteins: Implications for protein folding and assembly. *Science* **248:** 850–854.

Chiang, H.-L. and J.F. Dice. 1988. Peptide sequences that target proteins for enhanced degradation during serum withdrawal. *J. Biol. Chem.* **263:** 6797–6805.

Chiang, H.-L., S.R. Terlecky, C.P. Plant, and J.F. Dice. 1989. A role for a 70-kilodalton heat shock protein in lysosomal proteolysis of intracellular proteins. *Science* **246:** 282–285.

Chirico, W.J., M.G. Waters, and G. Blobel. 1988. 70K heat shock related proteins stimulate protein translocation into microsomes. *Nature* **332:** 805–810.

Cockle, S.M. and R.T. Dean. 1982. The regulation of proteolysis in normal fibroblasts as they approach confluence. Evidence for participation of the lysosomal system. *Biochem. J.* **208:** 243–249.

DeLuca-Flaherty, C., D.B. McKay, P. Parnum, and B.L. Hill. 1990. Uncoating protein (hsc70) binds a conformationally labile domain of clathrin light chain LC_a to stimulate ATP hydrolysis. *Cell* **62:** 875–887.

Deshaies, R.J., B.D. Koch, M. Werner-Washburne, E.A. Craig, and R. Schekman. 1988. 70 kD stress protein homologues facilitate translocation of secretory and mitochondrial precursor polypeptides. *Nature* **332:** 800–805.

Dice, J.F. 1987. Molecular determinants of protein half-lives in eukaryotic cells. *FASEB J.* **1:** 349–357.

Dice, J.F., H.-L. Chiang, E.P. Spenser, and J.M. Backer. 1986. Regulation of catabolism of microinjected ribonuclease A: Identification of residues 7-11 as the essential pentapeptide. *J. Biol. Chem.* **262:** 6853–6859.

Dingwall, C. and R. Laskey. 1992. The nuclear membrane. *Science* **258:** 942–947.

Domanico, S.Z., D.C. DeNagel, J.N. Dahlseid, J.M. Green, and S.K. Pierce. 1993. Cloning of the gene encoding peptide-binding protein 74 shows that it is a new member of the heat shock protein 70 family. *Mol. Cell. Biol.* **13:** 3598–3610.

Dunn, W.A. 1990a. Studies on the mechanisms of autophagy: Formation of the autophagic vacuole. *J. Cell. Biol.* **110:** 1923–1933.

———. 1990b. Studies on the mechanisms of autophagy: Maturation of the autophagic vacuole. *J. Cell. Biol.* **110:** 1935–1945.

Flynn, G.C., T.G. Chappell, and J.E. Rothman. 1989. Peptide binding and release by proteins implicated as catalysts in protein assembly. *Science* **245:** 385–390.

Hare, J.F. 1990. Mechanisms of membrane protein turnover. *Biochim. Biophys. Acta* **1031:** 71–90.

Hendrick, J.P. and F.-U. Hartl. 1993. Molecular chaperone functions of heat shock proteins. *Annu. Rev. Biochem.* **62:** 349–384.

Hershko, A. and A. Ciechanover. 1992. The ubiquitin system for protein degradation. *Annu. Rev. Biochem.* **61:** 761–807.

Isenman, L.D. and J.F. Dice. 1989. Secretion of intact proteins and peptide fragments by lysosomal pathways of protein degradation. *J. Biol. Chem.* **264:** 21591–21596.

Kang, P.-J., J. Ostermann, J. Shilling, W. Neupert, E.A. Craig, and N. Pfanner. 1990. Requirement for hsp70 in the mitochondrial matrix for translocation and folding of precursor proteins. *Nature* **348:** 137–143.

Knecht, E., J. Hernandez-Yago, and S. Grisolia. 1984. Regulation of lysosomal autophagy in transformed and non-transformed mouse fibroblasts under several growth conditions. *Exp. Cell. Res.* **154:** 224–232.

Kominami, E., E. Hashida, E. A. Khairallah, and N. Katunuma. 1983. Sequestration of cytoplasmic enzymes in an autophagic vacuole-lysosomal system induced by injection of leupeptin. *J. Biol. Chem.* **258:** 6093–6100.

Kopitz, J., G.O. Kisen, P.B. Gordon, P. Bohley, and P.O. Seglen. 1990. Non-selective autophagy of cytosolic enzymes in isolated rat hepatocytes. *J. Cell Biol.* **111:** 941–954.

Kulka, R.G. and A. Loyter. 1979. The use of fusion methods for the microinjection of animal cells. In *Current topics of membranes and transport* (ed. F. Bronner and A Kleinzeller), pp. 365–430. Academic Press, New York.

Kurtz, S., J. Rossi, L. Petko, and S. Lindquist. 1986. An ancient developmental induction: Heat shock proteins induced in sporulation and oogenesis. *Science* **231:** 1154–1157.

Maekawa, M., D.A. O'Brien, R.L. Allen, and E.M. Eddy. 1989. Heat-shock cognate protein (hsc71) and related proteins of mouse spermatogenic cells. *Biol. Repro.* **40:** 843–852.

Marzella, L. and H. Glaumann. 1987. Autophagy, microautophagy, and crinophagy as mechanisms for protein degradation. In *Lysosomes: Their role in protein breakdown* (ed. H. Glaumann and F.J. Ballard), pp. 319–367, Academic Press, New York.

Mayer, R.J., J. Lowe, M. Landon, H. McDermott, J. Tuckwell, F. Doherty, and L. Lazlo. 1991. Ubiquitin and the lysosomal system: Molecular pathological and experimental findings. In *Heat shock* (ed. B. Maresca and S. Lindquist), pp. 299–314. Springer-Verlag, Heidelberg.

McElligott, M.A. and J.F. Dice. 1984. Microinjection of cultured cells using red cell-mediated fusion and osmotic lysis of pinosomes: A review of methods and applications. *Biosci. Rep.* **4:** 451–466.

McElligott, M.A., P. Miao, and J.F. Dice. 1985. Lysosomal degradation of ribonuclease A and ribonuclease S-protein microinjected into human fibroblasts. *J. Biol. Chem.* **260:** 11986–11993.

Mortimore, G.E. 1987. Mechanism and regulation of induced and basal protein degradation in liver. In *Lysosomes: Their role in protein breakdown* (ed. H. Glaumann and F.J. Ballard), pp. 415–444, Academic Press, New York.

Nicchitta, C.V. and G. Blobel. 1993. Lumenal proteins of the mammalian endoplasmic reticulum are required to complete protein translocation. *Cell* **73:** 989–998.

Olson, T.S., S.R. Terlecky, and J.F. Dice. 1992. Pathways of intracellular protein degradation in eukaryotic cells. In *Stability of protein pharmaceuticals: In vivo pathways of degradation and strategies for protein stabilization* (ed. T.J. Ahern and M.C. Manning), pp. 89–118, Plenum Publishing, New York.

Palleros, D.R., W.J. Welsh, and A.L. Fink. 1991. Interaction of hsp70 with unfolded proteins: Effects of temperature and nucleotides on the kinetics of binding. *Proc. Natl. Acad. Sci.* **88:** 5719–5723.

Rechsteiner, M. 1991. Natural substrates of the ubiquitin proteolytic pathway. *Cell* **66:** 615–618.

Schlegel, R.A. and M. Rechsteiner. 1978. Red cell-mediated microinjection of macromolecules into mammalian cells. *Methods Cell Biol.* **20:** 341–354.

Terlecky, S.R. and J.F. Dice. 1993. Polypeptide import and degradation by isolated lysosomes. *J. Biol. Chem.* **268:** 23490–23495.

Terlecky, S.R., H.-L. Chiang, T.S. Olson, and J.F. Dice. 1992. Protein and peptide binding and stimulation of *in vitro* lysosomal proteolysis by the 73-kDa heat shock cognate protein. *J. Biol. Chem.* **267:** 9202–9209.

Ueno, T., D. Muno, and E. Kominami. 1991. Membrane markers of endoplasmic reticulum preserved in autophagic vacuolar membranes from leupeptin-administered rat liver. *J. Biol. Chem.* **266:** 18995–18999.

Vogel, J.P., L.M. Misra, and M.D. Rose. 1990. Loss of BiP/GRP78 function blocks translocation of secretory proteins in yeast. *J. Cell Biol.* **110:** 1855–1895.

Walton, P.A., J.P. Morello, and W.J. Welch. 1993. Inhibition of the peroxisomal import of a microinjected protein by coinjection of antibodies to members of the 70 kD heat shock protein family. *J. Cell. Biochem.* (suppl.) **17C:** 22.

Wing, S.S., H.-L. Chiang, A.L. Goldberg, and J.F. Dice. 1991. Proteins containing peptide sequences related to KFERQ are selectively depleted in liver and heart, but not skeletal muscle, of fasted rats. *Biochem. J.* **275:** 165–169.

7

Stress-70 Proteins and Their Interaction with Nucleotides

David B. McKay, Sigurd M. Wilbanks,
Kevin M. Flaherty, Jeung-Hoi Ha,
Melanie C. O'Brien, and Lauren L. Shirvanee
Beckman Laboratories for Structural Biology
Department of Cell Biology
Stanford University School of Medicine
Stanford, California 94305

I. INTRODUCTION

The first members of the 70-kD heat shock protein family, alias "stress-70" family, were identified in the 1970s. In 1974, Tissières and colleagues observed that the level of synthesis of some *Drosophila* cell proteins was dramatically enhanced by heat shock (Tissières et al. 1974). In addition, studies during the 1970s on bacteriophage λ yielded an *Escherichia coli* mutant that failed to support λ replication and was temperature-sensitive for growth at 42°C (Georgopoulos 1977). The product of the mutant gene was identified as the DnaK protein; the mutant became known as *dnaK756* (Georgopoulos et al. 1979).

In 1984, sequencing of the *E. coli* DnaK protein revealed that it is homologous to the previously sequenced *Drosophila* 70-kD heat shock protein (Ingolia et al. 1980; Bardwell and Craig 1984). The high degree of sequence conservation of stress-70 genes within eukaryotes allowed the *Drosophila* gene to be used as a hybridization probe in cloning cDNAs of other stress-70 proteins (see, e.g., Hunt and Morimoto 1985;

The Biology of Heat Shock Proteins and Molecular Chaperones
©1994 Cold Spring Harbor Laboratory Press 0-87969-427-0/94 $5 + .00

Voellmy et al. 1985; Wu et al. 1985; Munro and Pelham 1986). Purification of stress-70 proteins, including those that were present at low levels in cells, became relatively facile once it was recognized that they bound ATP tightly and could be affinity-purified on ATP-agarose (Welch and Feramisco 1985). As the number of sequences and purified proteins grew, it became apparent that many members of this family of proteins are present in cells under normal conditions; their presence is not strictly stress-related (for review, see Pelham 1986; Gething and Sambrook 1992). Expression of some stress-70 proteins in response to heat shock or other forms of cell stress is only one of the collective functions of the larger family of proteins.

In efforts unrelated to heat shock, Rothman and colleagues, in their studies of cellular trafficking of proteins, characterized in vitro a "clathrin uncoating ATPase" that disassembled the clathrin cages of coated vesicles in an ATP-dependent reaction (Schlossman et al. 1984). Surprisingly, these authors subsequently found that the clathrin-uncoating ATPase was a constitutively expressed member of the stress-70 protein family (Chappell et al. 1986). Two years later, it was reported that this protein also facilitates transmembrane targeting of proteins from the cytoplasm to mitochondria (Chirico et al. 1988) and the endoplasmic reticulum (Murakami et al. 1988). The proteins of the stress-70 family appear to participate in a diverse set of activities, including, but not restricted to, facilitation of initiation of λ replication by disassembly of a specific preinitiation complex (Zylicz et al. 1988; Alfano and McMacken 1989), disassembly of clathrin cages, facilitation of transmembrane translocation of polypeptides, participation in protein (re)folding during translation (or during transmembrane translocation) (Beckmann et al. 1990; Scherer et al. 1990; Langer et al. 1992), and participation in "repair" of damaged proteins after heat shock. A crucial open question in this system is can all of these functions be reconciled with a single mechanism of activity at the biochemical level? In this chapter, we discuss what is currently known about the ATPase activity of stress-70 proteins; a companion chapter by Hightower and colleagues in this volume discusses interaction of stress-70 proteins with peptides and unfolded proteins.

II. PRIMARY STRUCTURE

More than 60 complete amino acid sequences of stress-70 proteins are available (Swiss Protein Data Bank, Release 25.0), mostly translated from nucleic acid sequences. Apparent homology in the primary structure suggests a pattern of evolution that reflects the pattern of cellular compartmentalization of different stress-70 proteins (Fig. 1). Prokaryotes

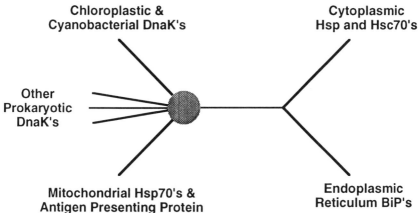

Figure 1   A parsimonious tree of stress-70 protein sequences. The gray circle denotes uncertainty in tree topology. This tree was constructed with Protpars (in the program suite PHYLIP 3.4; J. Felsenstein, University of Washington), from an alignment prepared using a simultaneous alignment and phylogeny algorithm (Hein 1990). The Swiss Protein Data Bank accession numbers and species names of the sequences used were as follows. **Prokaryotes**: P27542, *Chlamydia pneumoniae*; P04475, *Escherichia coli*; P19993, *Mycobacterium leprae*; Q00488, *Mycobacterium paratuberculosis*; Q01100, *Haloarcula marismortui*. **Chloroplast and cyanobacteria**: P30723, *Porphyra umbilicalis*; P22358, *Synechocystis* sp. (strain PCC 6803). **Mitochondrial**: P20583, *Trypanosoma cruzi*; P12398, *Saccharomyces cerevisiae*. **Antigen Presenting**: L11066 (this is the Genbank accession number for the nucleic acid sequence), human. **Cytoplasmic**: P25840, *Chlamydomonas reinhardtii*; P17804, *Leishmania donovani*; P11143, *Zea mays*; P05456, *Trypanosoma cruzi*; P10591, *S. cerevisiae*; P11142, human; P22202, *S. cerevisiae* (SSB1). **BiPs**: P11021, human; P16474, *S. cerevisiae*.

have a single stress-70 protein, DnaK, which is typically present at approximately 1% of total cellular protein under normal conditions and whose rate of synthesis is increased in response to heat shock. DnaKs from eubacteria and archaebacteria share approximately 50% identity. Eukaryotes typically have both constitutive (hscs) and stress-inducible (hsps) cytoplasmic/nuclear representatives that segregate as a phylogenetic class distinct from bacterial DnaKs. The eukaryotic BiPs, localized in the endoplasmic reticulum, also segregate as a class distinct from both DnaKs and hsp70s; BiPs typically share 60–70% amino acid sequence identity with their cytoplasmic counterparts. The bacterial DnaKs have about 45–50% sequence identity with the eukaryotic hsp70s, hsc70s, and BiPs. The mitochondrial and chloroplast stress-70 proteins appear to form distinct groups that are more closely related to prokaryotic DnaKs

Position in HSC70	Insertions and Deletions Relative to HSC70						Final Sequences	
	Leader	80-117	190-191	278	383-386	509	610-645	Terminal
Cytosolic	none	0	+0,1,2	0	0	0	(GGMP)n	(V,I)EEVD
BiP	+ 23-47	+ 0-1	- 1	+ 1	- 2	0	D,E	(H,K,D)DEL
Mitochondrial	+ 24-49	- 2-3	- 2	+ 5	- 4	- 1	G,S,N	KQ
Chloroplastic	none	- 3	- 2	+ 5	- 4	- 1	G,A,P	D,E,K
Bacterial dnaK	none	- 3 or -27	- 1,2	+ 5	- 4	- 1	G,A,P	D,E,K

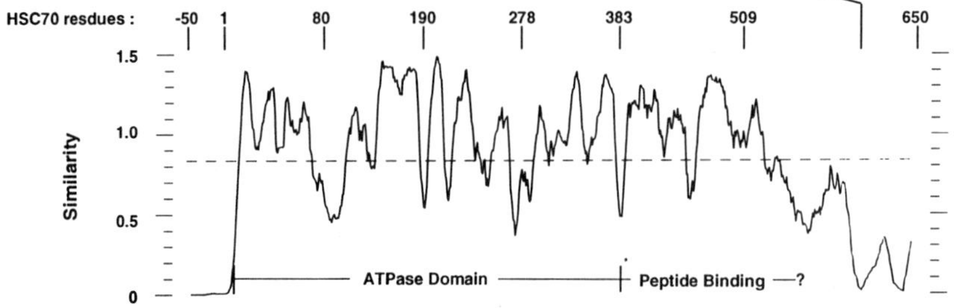

Figure 2 Comparison of primary sequences. (*Top*) The sizes of amino acid insertions (+) and deletions (−) relative to bovine hsc70, which characterize different branches of the hsp70 family, are tabulated. Listed at the far right are the carboxy-terminal sequence motifs of each family, including the conserved terminal sequence (or composition) and the typical preterminal composition (or in the case of the cytosolic members, the predominant repeat). (*Bottom solid line*) Similarity among hsc70 sequences, with position given by human hsc70 numbering. The alignment was prepared by the program PlotSimilarity and PileUp (Genetics Computer Group, Madison, Wisconsin) from 61 hsp70 and DnaK sequences from Swiss Protein Data Bank release 25.0, and two antigen-presenting proteins (Domanico et al. 1993). Each position was scored from 1.5 (identity) to 0 (no similarity), with conservative substitutions receiving intermediate scores, and the score was averaged over ten residues. (*Dashed line*) Average across the entire sequence. The apparent functional domains of hsc70 are shown at the bottom.

than to other eukaryotic proteins. A recently characterized stress-70 protein that has a role in antigen presentation and is localized in vesicles and on the surface of cells (Domanico et al. 1993) is most closely related in sequence to mitochondrial stress-70 proteins, sharing approximately 60% amino acid identity with them. Additionally, aligned sequences show that the prokaryotic DnaKs, and the mitochondrial, chloroplast, and antigen-presenting stress-70 proteins, share several short deletions that distinguish them from their eukaryotic counterparts.

Generally, the stress-70 proteins are most highly conserved in the first approximately 530 amino acids, with substantially less conservation in the range of residues 530–600, followed by highly variable sequences in the carboxy-terminal 30–50 amino acids (Fig. 2). The hsp70 and hsc70 proteins have a stretch of about 10–30 uncharged amino acids that is rich in glycine, alanine, proline, and hydrophobic residues; in many cases, there is a distinctive (GGMP) motif repeated five to ten times. No specific function has yet been ascribed to this sequence motif. The BiP proteins have a 10–15-residue G,A,S,P-rich region, followed by a short acidic region and an endoplasmic reticulum retention signal, (K/H/D)DEL. The DnaK proteins have a 10–20-residue G,A,P-rich segment leading into a highly charged, predominantly acidic termination. In this case also, no specialized function has been found for the carboxy-terminal sequence motif.

An early indication of functional organization within the primary structure of stress-70 proteins was given by Rothman and colleagues, who showed that digestion of bovine hsc70 with chymotrypsin yielded an amino-terminal fragment of about 44 kD that bound ATP agarose and retained ATPase activity but did not bind clathrin and that by implication had lost its peptide-binding activity (see Chappell et al. 1987). This demonstrated that the ATP-binding activity resided in the amino-terminal portion of the molecule and further suggested that an essential part of the peptide-binding function must reside in the carboxy-terminal portion. Additional evidence has subsequently accumulated to demonstrate unequivocally that the ATPase activity of stress-70 proteins is encompassed within the first 380–390 residues of the primary structure (Milarski and Morimoto 1989; Flaherty et al. 1990). It is presumed that the peptide-binding activity of the molecule lies within the remainder of the primary structure, although its boundaries have not been delineated precisely; theoretical models that would place it within the region of residues 385–540 are discussed by Hightower and colleagues (this volume).

III. TERTIARY STRUCTURE

To date, no full-length stress-70 protein has proven amenable to crystallization and structure determination. The X-ray crystallographic structure (Fig. 3) of the ATPase fragment has been solved and refined to 2.2 Å resolution (Flaherty et al. 1990). It has four structural domains that form two lobes with a cleft between them. Nucleotide is bound at the base of the cleft in a binding pocket formed primarily by the two lower domains of the structure. The high level of sequence conservation in the stress-70 protein family implies that the ATPase fragments of other members will have a similar tertiary structure. The pattern of insertions and deletions within the stress-70 family is consistent with this. The first and largest deletion in aligned sequences of DnaKs relative to hsc70 (after residue 80; see Fig. 2) is in the upper right domain of the structure; it is interest-

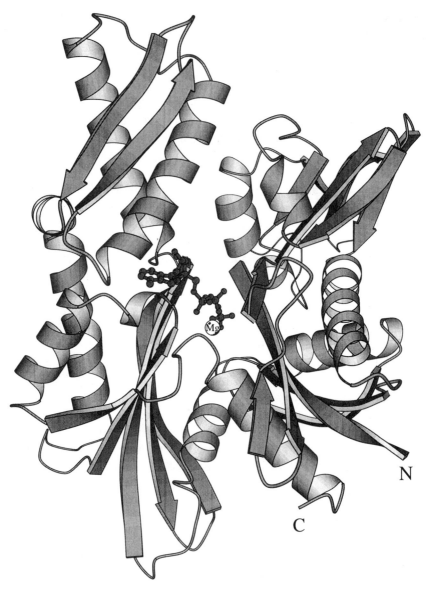

Figure 3 Three-dimensional structure of the ATPase fragment of hsc70, with ADP, P_i, and the Mg^{++} ion included. Water molecules are not included. Molscript was used to make this figure (Kraulis 1991).

ing to note that this is the same place that actin has a foreshortened, methylhistidine-containing loop relative to hsc70. The subsequent small deletions in other stress-70 proteins occur at domain boundaries (position 190, between the two lobes, and position 383, at the end of the ATPase domain). The insertion at residue 278 is in a surface loop that connects a β strand and an α helix in the upper left domain. Consequently, none of the insertions or deletions would be expected to disrupt the core tertiary structure of the ATPase fragment.

The ATPase fragment is in overall tertiary structure strikingly similar to that of actin, despite negligible similarity between these proteins at the amino acid sequence level (Kabsch et al. 1990; Flaherty et al. 1991). Additionally, the nucleotide-binding fold (the two lower domains) resembles that of hexokinase (Fletterick et al. 1975), regardless of the overall dissimilarity in the remainder of the structure (the upper domains). The observed similarity in structure between the hsc70 ATPase fragment, actin, and hexokinase clearly establishes these proteins as members of a common structural superfamily. The sequences of the β strands that bind the Mg^{++} ion and β phosphate of MgADP in both actin and hsc70 identify a sequence fingerprint that is characteristic of this family, (I/L/V)X (I/L/V/C)DXG(T/S/G)(T/S/G)XX(R/K/C). On the basis of this fingerprint, the prediction that glycerol kinase and its close relatives belong to the same structural family has recently been confirmed by the three-dimensional structure of glycerol kinase (Hurley et al. 1993). Additionally, it has been suggested that some prokaryotic cell cycle proteins share this nucleotide-binding motif (Bork et al. 1992; Gupta and Singh 1992).

Refinement of the wild-type ATPase fragment structure has revealed details of the active site region (Fig. 4), including the interaction of the protein with MgADP and P_i (K.M. Flaherty et al., unpubl.). The Mg^{++} ion is octahedrally coordinated; its first coordination shell is formed by oxygens from the P_i and the P_β of ADP, plus four water molecules; the protein does not contact Mg^{++} directly. Amino acid side chains of several residues are observed to interact with the metal-nucleotide complex in the active site, including the carboxylates of D10, E175, D199, and D206, the ε-amino group of K71, and the hydroxyl of T204 (residues numbered according to bovine hsc70 sequence). These residues are rigorously conserved throughout the heat shock protein family and can be suggested as candidates that may influence the ATPase activity of the protein; results from mutagenesis of many of these residues are discussed later in this chapter.

VI. TEMPERATURE-DEPENDENT STRUCTURAL TRANSITIONS

Temperature-dependent structural transitions have been monitored in both *E. coli* DnaK and bovine hsc70 by fluorescence and circular dichroism, and further by Fourier transform infrared spectroscopy and scanning calorimetry in DnaK. The results are summarized in Table 1. A point of general agreement is that for both DnaK and hsc70, an initial

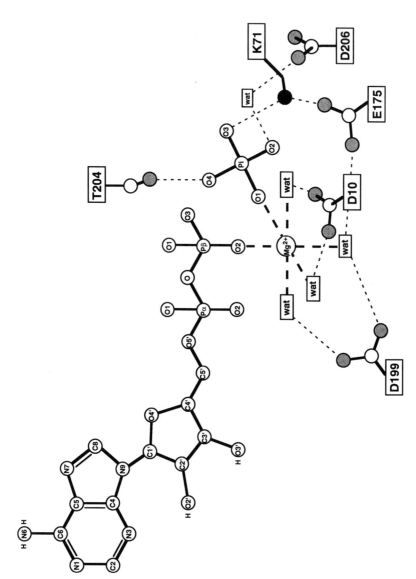

Figure 4 Schematic drawing of interactions of residues in the active site of hsc70 with MgADP and P$_i$. (*Solid lines*) Covalent bonds; (*dashed lines*) apparent hydrogen bond.

Table 1 Apparent Midpoints (°C) of Thermal Transitions of DnaK and Bovine hsc70 with Various Nucleotides

	No nucleotide	Mg:ADP	ATP	Mg:ATP	Mg:AMPPCP
DnaK					
FTIR[1]	55	—	—	60	55
ATPase[2,3]	—	—	—	60	—
Fluorescence[4]	42	—	55	60	—
CD[4]	41, 71	59, 71	52, 69	59, 71	—
Calorimetry[6]	45.2, 58.0, 73.3	51.8, 59.0, 72.9	—	—	
amino-terminal proteolytic	47.5, 79.4				
carboxy-terminal proteolytic	50.4, 58.2, 70.6				
hsc70					
Fluorescence[4]	42, 65	—	—	—	
CD[4,5]		57, n.d.[a]	—	57, n.d.[a]	

Sources of data: (1) Banecki et al. 1992; (2) Liberek et al. 1991a; (3) McCarty and Walker 1991; (4) Palleros et al. 1992; (5) Palleros et al. 1991; (6) Montgomery et al. 1993.

[a] n.d. indicates not determined

transition in the temperature range 40–42°C occurs in the absence of nucleotides, and that bound nucleotide—either MgADP or MgATP— shifts the first transition to a higher temperature in the range 52–59°C. Circular dichroism shows at least one further transition, approximately 70°C for DnaK and 65°C for hsc70, whose behavior is independent of nucleotides. The first transition appears to be reversible for DnaK but leads irreversibly to large aggregates of hsc70 (Palleros et al. 1992). The spectroscopic data have been interpreted within a three-state model involving cooperative transitions of the entire polypeptide chain, first from the native state to a molten globule and then to a fully denatured state (Palleros et al. 1991, 1992). Evidence presented in support of this interpretation includes the relatively modest increase in Stokes' radius of the intermediate state (43 Å), as compared to the native state (38 Å), and a larger change in shifting to the fully denatured state (80 Å), measured by gel filtration.

Calorimetry experiments on full-length DnaK in absence of nucleotide detect three thermal transitions, consistent with a four-state model for the protein (Montgomery et al. 1993). Calorimetry experiments were also done separately on the amino-terminal ATPase fragment and a carboxy-terminal fragment of DnaK; they showed two and three apparent transitions, respectively. These data were interpreted with a four-state model in which different folding domains denature independently and noncooperatively at different temperatures, in contrast to the cooperative model described above. Comparison of the calorimetric transition temperatures of the fragments with those of full-length DnaK suggests that the first transition is intrinsic to the amino-terminal ATPase region of the protein. The increase in the lowest transition temperature upon binding of nucleotide, observed by all the techniques, is consistent with nucleotide binding stabilizing the ATPase region against thermal unfolding.

Regardless of differences of interpretation of the temperature-dependent unfolding behavior of DnaK and hsc70, it is clear that under intracellular conditions ([ATP] ≥1 mM), where stress-70 proteins would be saturated with nucleotide (K_m[ATP] ≤1 μM; K_d[ADP] ~0.1 μM), the major temperature-dependent structural changes occur at temperatures (≥52°C) outside the normal range of heat shock response in both *E. coli* (≤45°C) and mammals (≤40°C).

V. THE ATPASE ACTIVITY

Stress-70 proteins bind denatured proteins and short peptides and release them in response to binding and hydrolysis of ATP (possibly with subsequent release of P_i or ADP) (see, e.g., Kassenbrock et al. 1988; Hurtley et al. 1989; de Silva et al. 1990; Hendershot 1990; Palleros et al. 1991). Conversely, binding of peptides or denatured proteins to stress-70 proteins raises the ATPase rate above its basal level. Peptide-dependent enhancement of ATPase activity above basal level is often used as a convenient assay for stress-70 protein activity. The activities can be formally subdivided into (1) ATP binding and hydrolysis, followed by product release (P_i and ADP); (2) peptide binding and release; and (3) a mechanism of coupling peptide binding/release cycle with the ATPase cycle.

The ATPase turnover rates of isolated stress-70 proteins are relatively slow, typically about 0.1 $(min)^{-1}$ in the absence of peptides or denatured proteins and severalfold higher in the presence of stimulatory peptides (Table 2). K_m values for the ATPase activity are reported in the range 0.7–1.4 μM for bovine hsc70 and 0.1–0.4 μM for mammalian BiPs; values of K_m are not affected significantly by peptide binding. It has been demonstrated that ADP is a strong product inhibitor of the ATPase activity, with a K_I of approximately 0.1 μM (Palleros et al. 1991; Gao et al. 1993; O'Brien and McKay 1993); it has become apparent that product inhibition may affect the values of kinetic parameters derived from data taken under assay conditions where ADP binding is significant.

It is notable that the peptide-independent turnover rate of the hsc70 ATPase fragment, 0.8 $(min)^{-1}$, is greater than the turnover rates reported for full-length hsc70 under similar assay conditions, suggesting that the peptide-binding domain of the full-length protein attenuates the intrinsic ATPase activity of the ATPase fragment of the protein and that peptide binding relieves the attenuation.

Differences between the basal ATPase rates of natural and recombinant proteins have been observed for both BiP (Blond-Elguindi et al. 1993) and hsc70 (Sadis and Hightower 1992). In the case of bovine BiP, it has been shown that purified recombinant protein is completely monomeric, although a substantial fraction of natural BiP purified from bovine liver is multimeric (Blond-Elguindi et al. 1993). Addition of BiP-binding peptides dissociates the multimers. Furthermore, at high peptide concentration, the ATPase turnover rates of natural and recombinant BiPs are equal. Gething and co-workers consequently suggest that aggregated forms of BiP may have little or no ATPase activity and that the twofold difference of ATPase activity between basal and peptide-stimulated protein is the intrinsic difference between basal and maximum activity of fully active, monomeric protein.

The ATPase activity of stress-70 proteins shows a strong temperature dependence. The basal ATPase rate of DnaK rises steeply with temperature, increasing 70-fold from 20°C to 53°C (McCarty and Walker 1991).

Table 2 Recent Values for Kinetic Parameters of ATPase Activity Derived from Data in the Literature

Protein	Basal or peptide-stimulated?	K_m (μM)	k_{cat} (mole ATP) (mole enzyme)$^{-1}$ (min)$^{-1}$	Conditions
E. coli DnaK[1]	basal	n.d.[a]	0.15	37°C 100 mM HEPES (pH 8.1) 35 mM KCl 5 mM MgCl$_2$ 70 mM ATP
DnaK[2]	basal	n.d.	0.14–0.27	30°C 30 mM HEPES (pH 8.8)
DnaK756	basal	n.d.	~18	40 mM KCl 50 mM NaCl 7 mM Mg acetate 0.29 mg/ml BSA 100 μM ATP
Bovine BiP[3]	basal	n.d.	0.026	37°C 20 mM HEPES (pH 7.0) 20 mM NaCl 2 mM MgCl$_2$ 8 μM ATP react for 10 min
	peptide-stimulated	n.d.	0.08–0.16	

				Conditions
Bovine BiP[4]				37°C
natural	basal	—	0.03	20 mM HEPES (pH 7.0)
				20 mM KCl
recombinant	basal	0.4	0.09	2 mM MgCl$_2$
				10 mM (NH$_4$)$_2$SO$_4$
natural and	peptide-stimulated	—	0.13	0.5 mM DTT$_E$
recombinant				0.1–8 μM ATP
Canine BiP[5]	basal	0.1	0.34	37°C
				50 mM HEPES (pH 6.0)
				25 mM NaCl
				1 mM MgCl$_2$
				1 mg/ml BSA
				10–50 μM ATP
Bovine hsc70[3]	peptide-stimulated	n.d.	0.23–0.24	37°C
				20 mM HEPES (pH 7.0)
				20 mM NaCl
				2 mM MgCl$_2$
				8 μM ATP
				react for 10 min
Bovine hsc70[6]	peptide-stimulated	n.d.	1.0	37°C
natural				40 mM HEPES (pH 7.0)
				25 mM KCl
				4.5 mM Mg acetate
				550 μM ATP

(Continued on following page.)

Table 2 (Continued.)

Protein	Basal or peptide-stimulated?	K_m (μM)	k_{cat} (mole ATP) (mole enzyme)$^{-1}$ (min)$^{-1}$	Conditions
Bovine hsc70[7] natural	basal	1.37	0.15 per hsc70 dimer	25 mM KCl 10 mM $(NH_4)_2SO_4$ 2 mM Mg acetate 0.1 mM EDTA 1 mM DTT 50 μM ATP 25°C
	ApoC-stimulated	1.44	0.38 per hsc70 dimer	
Bovine hsc70[8]	basal	n.d.	0.05	37°C
Bovine hsc70[9] recombinant	peptide-stimulated	0.7 ± 0.5	0.2 ± 0.1	40 mM HEPES (pH 7.0) 25 mM KCl
recombinant ATPase fragment	peptide-independent	0.7 ± 0.1	0.8 ± 0.1	4.5 mM Mg acetate [ATP] range 0.125 μM to 25 μM
Rat hsc70[10] recombinant	peptide-stimulated	n.d.	0.2–0.4	37°C
recombinant ATPase fragment	peptide-independent	n.d.	0.2–0.4	

Sources: (1) McCarty and Walker 1991; (2) Liberek et al. 1991a; (3) Flynn et al. 1989; (4) Blond-Elguindi et al. 1993; (5) Kassenbrock and Kelly 1989; (6) DeLuca-Flaherty et al. 1990; (7) Sadis and Hightower 1992; (8) Gao et al. 1992; (9) O'Brien and McKay 1993; (10) Huang et al. 1993.
[a]n.d. indicates not determined.

Above 53°C, the activity drops precipitously, possibly as a consequence of the structural transition observed near this temperature by both circular dichroism and calorimetry. The smooth increase in activity from 30°C to 50°C is further evidence that the thermal transition observed at 42°C in the absence of nucleotide is not of functional significance.

The activity of stress-70 proteins can also be influenced by accessory proteins; the ATPase rate of DnaK increases about 50-fold above its basal level by the concerted effect of the *E. coli* GrpE and DnaJ proteins, although neither protein acting alone has such a substantial effect (Liberek et al. 1991a). Data have been presented to support a model in which the two proteins affect different steps of the ATPase cycle; specifically, DnaJ accelerates ATP hydrolysis and GrpE accelerates nucleotide exchange by DnaK. The role of DnaJ and GrpE homologs in eukaryotes has not yet been elucidated, but they may also modulate the activity of stress-70 proteins. The role of DnaJ and GrpE in DnaK activity is discussed further by Hightower and colleagues (this volume).

The modulation of the ATPase activity by accessory proteins, its slow turnover rate, and strong product inhibition by ADP, together with its coupling to peptides, hint that regulation of peptide binding is the primary role for nucleotide binding and hydrolysis. It has been suggested that the turnover time of seconds to minutes may be tuned to the time constants of the cellular processes in which the stress-70 proteins participate, such as polypeptide synthesis and translocation (Rothman 1989; Flynn et al. 1991). If this were the case, then it would be expected that mutant stress-70 proteins whose overall ATPase rates were either substantially lower *or* substantially higher than those of wild-type could have deleterious effects on the cellular metabolism and the heat shock response; an example of the latter case is the DnaK756 mutant, which is aberrant in heat shock response and deficient in supporting bacteriophage λ replication, and in which the protein has a basal ATPase turnover rate that is two orders of magnitude greater than that of wild type (Liberek et al. 1991a).

Questions that arise in this context include: What step of the ATPase cycle is rate-limiting, and how does the cycle couple to peptide binding? The rate of ADP release has been measured for hsc70. Under conditions where a chase nucleotide is added in molar excess to prevent rebinding of released nucleotide, the rate of ADP release has been reported to be 0.16 $(min)^{-1}$ at 37°C (Sadis and Hightower 1992) and about 0.3 $(min)^{-1}$ at 25°C (estimated from published graphical data) (Gao et al. 1993). The K_d for ADP is approximately 0.1 μM; this implies that k_{on} of ADP binding is on the order of about 10^5 (M sec)$^{-1}$, substantially slower than a diffusion-limited process. Hightower and colleagues have demonstrated that binding a representative denatured protein substrate, apocytochrome *c*, doubles the release rate of ADP. Furthermore, in both the absence and the presence of apocytochrome *c*, the steady-state ATPase turnover rates of hsc70 are essentially equal to the ADP release rates (Sadis and Hightower 1992). They have suggested on the basis of these data that unfolded polypeptides stimulate the ATPase rate of hsc70 by accelerating ADP/ATP exchange.

Eisenberg and colleagues have observed that in the presence of saturating ATP (maintained with an ATP regenerating system), the ratio of ATP:ADP bound to hsc70 is approximately 0.85:0.15 (Gao et al. 1993). They conclude that the hydrolytic step in the ATPase cycle is rate-limiting; their measured values of 0.05 $(min)^{-1}$ for the ATPase rate, compared to approximately 0.3 $(min)^{-1}$ for the ADP release rate, are consistent with this suggestion. They also have shown that at nucleotide concentrations as high as 200 μM, a maximum of one nucleotide is bound per hsc70 protomer; hsc70 has a single nucleotide-binding site, rather than two distinct sites, which had been suggested by other investigators in earlier work on the clathrin uncoating reaction (Schmid and Rothman 1985a,b; Schmid et al. 1985; Rothman and Schmid 1986).

VI. MUTANTS AND MUTAGENESIS

DnaK mutants have been selected in *E. coli* by screening for high constitutive expression of heat shock proteins (Wild et al. 1992). The mutants that were isolated fall into two classes. Class I mutants are recessive, permit replication of bacteriophage λ, and survive at temperatures up to 40°C. The mutations in this class are T154S and A174T; both of these residues are remote from the ATP-binding site of DnaK. Class II mutants are dominant, do not support growth of λ, and are temperature-sensitive for cell growth above 34°C. All of these mutants have point mutations in residues that interact with the nucleotide in such a manner that they can be expected to affect ATP binding (G229D and G341D) or the ATPase activity (D201N, equivalent to D206 in hsc70; E171K, equivalent to E175 in hsc70; and the triple mutation G198D + E230K + a carboxy-terminal mutation). The fact that the class II mutations localize to the nucleotide-binding site underscores the importance of the ATPase activity in DnaK function in vivo.

Site-directed mutagenesis of residues in the nucleotide-binding site of BiP—specifically, the mutations T37G, T229G, and E201G (corresponding to T13, T204, and E175 of hsc70)—produced proteins whose ATPase activity is reduced to less than 4% of wild-type activity under the conditions employed (Gaut and Hendershot 1993b). The mutant proteins still bind ATP and immunoglobulin heavy chains; however, they are deficient in heavy-chain release.

In recombinant rat hsc70, the mutation D10N yielded a protein with substantially lowered ATPase activity in both the ATPase fragment and full-length protein (Huang et al. 1993). In a recombinant bovine hsc70 ATPase fragment, mutations D199N and D199S result in proteins whose K_m values are essentially unchanged from those of wild-type but whose k_{cat} values are reduced approximately two orders of magnitude. Mutations D10N and D10S also reduce the turnover approximately two orders of magnitude, but with a concomitant order of magnitude increase in K_m. E175Q and E175S increase K_m values two orders of magnitude, with a tenfold decrease of k_{cat}. D206N and D206S reduce k_{cat} values approximately tenfold without significantly affecting K_m values (S.M. Wilbanks et al., unpubl.).

Overall, results from site-specific mutagenesis indicate that (1) no single mutation leads to a complete knockout of the ATPase activity, and no single residue has yet been identified as catalytically indispensable; (2) for each case examined, changes in residues that interact with the nucleotide have a substantial effect on the overall ATPase activity, either through k_{cat} or K_m or both; and (3) the residues whose modification results in the greatest decrease of k_{cat} are those whose side chains ligate the H_2O molecules in the first coordination shell of the Mg^{++} ion.

The mutant DnaK756 protein is somewhat of an enigma in this context. Originally, the *dnaK756* mutant strain was selected for its deficiency in supporting bacteriophage λ growth; it also overproduces heat shock proteins at 30°C and fails to turn off the heat shock response at 42°C (Georgopoulos 1977; Georgopoulos et al. 1979). The DnaK756 protein has an abnormally high basal ATPase activity (Table 2). Recently, the *dnaK756* gene has been sequenced; the protein has three point mutations: G32D, G455D, and G468D (Miyazaki et al. 1992). Two of these are in the carboxy-terminal region of the protein, and the one in the amino-terminal domain at G32 is remote from the nucleotide-binding site. The mechanism by which these mutations exert a dramatic effect on the steady-state ATPase activity is still unclear. However, the ability of the carboxyl terminus to attenuate the ATPase rate suggests a rationalization for the effect of carboxy-terminal mutations on the ATPase rate.

VII.　CALCIUM-DEPENDENT AUTOPHOSPHORYLATION

199 of DnaK (which is equivalent to T204 of hsc70) is the site of Ca^{++}-dependent in vitro autophosphorylation (Zylicz et al. 1983; McCarty and Walker 1991). The equivalent threonine of BiP, T229, is also autophos-

phorylated in vitro with CaATP (Leustek et al. 1991; Gaut and Hendershot 1993a); similar calcium-dependent autophosphorylation has been observed in mitochondrial stress-70 proteins (Mizzen et al. 1991). Autophosphorylation is not observed with MgATP; the activity has a divalent cation requirement for Ca^{++}. Mutation of this threonine residue in DnaK or BiP eliminates the autophosphorylation activity (except when it is replaced with serine, which retains about 12% of the original autokinase activity).

In the structure of the hsc70 ATPase fragment with ATP or non-hydrolyzable analogs bound, the hydroxyl of this threonine (T204 in hsc70; see Fig. 4) is positioned such that nucleophilic attack on the γ phosphate, with subsequent formation of a phosphoprotein intermediate, would be a sterically feasible step in ATP hydrolysis (O'Brien and McKay 1993). The BiP mutant proteins T229G, T229S, and T229D retain Mg^{++}-dependent ATPase activity that ranges approximately 6–22% of wild type (Gaut and Hendershot 1993b); the ATPase activity of DnaK mutant proteins T199D, T199A, and T199V is reduced to less than 1% of wild type (McCarty and Walker 1991). Mutagenesis of T204 to valine or glutamic acid in hsc70 results in proteins whose k_{cat} values are similar to that of wild-type protein, but whose K_m values are increased by two orders of magnitude (O'Brien and McKay 1993). These data argue against the formation of a phosphoprotein intermediate at T204 during the normal course of the ATP hydrolysis.

Determination of the structure of the hsc70 ATPase fragment with CaAMPPNP bound reveals a weakly bound Ca^{++} ion that is displaced from the Mg^{++}-binding site in a manner that could interfere with normal hydrolysis and P_i binding and could enhance the likelihood of phosphorylation of T204 (K.M. Flaherty et al., unpubl.). Taken in sum, the mutagenesis and structural data suggest that the CaATP-dependent autophosphorylation activity observed in vitro is a nonproductive side reaction resulting from the manner in which the Ca^{++} ion binds in the active site; it is unlikely to be a regulatory modification of stress-70 proteins in vivo.

VIII. CONFORMATIONAL CHANGES AND OLIGOMERIZATION

Evidence from several sources indicates that stress-70 proteins have at least two distinct conformations, one of which has high affinity for peptides and denatured proteins and the second of which has a substantially lower affinity. Two different sets of experiments provide evidence for a conformational change in DnaK. Fink and colleagues (Palleros et al.

1992) have monitored the intrinsic tryptophan fluorescence of DnaK and have found both a change in intensity and a shift in the wavelength of peak intensity between DnaK with MgADP present (λ_{max} = 333 nm; greater peak fluorescence intensity) and DnaK with MgATP present (λ_{max} = 327 nm; lesser peak intensity). DnaK complexed with the non-hydrolyzable analog Mg[γ-S]ATP showed a spectrum indistinguishable from that of DnaK with MgADP, suggesting that it is hydrolysis of the nucleoside triphosphate (or a later step), rather than nucleotide binding, that is required to shift the protein into a "ATP-bound" conformation.

Schematically similar conclusions were reached by Georgopoulos and colleagues (Liberek et al. 1991b), using protease digestion patterns as a probe of conformation for DnaK. Distinctly different digestion patterns were observed for DnaK in the presence of MgADP versus in the presence of MgATP. Consistent with the fluorescence experiments of Fink and colleagues, MgATP hydrolysis appeared to be necessary to shift DnaK into the "ATP-bound" conformation; proteolytic digestion in the presence of MgAMPPNP, a nonhydrolyzable analog of MgATP, resulted in a digestion pattern similar to that observed in the presence of MgADP. They further demonstrated that the "ADP-bound" conformation of DnaK had a relatively high affinity for denatured proteins (specifically, for reduced, carboxy-methylated bovine pancreatic trypsin inhibitor [RCM-BPTI]) and that addition of MgATP to DnaK in the "ADP-bound" conformation complexed with RCM-BPTI resulted in both a release of RCM-BPTI and a shift to the "ATP-bound" conformation. Interestingly, the mutant DnaK756 protein appears to be locked in the "ATP-bound" conformation; it also has a substantially lower affinity for RCM-BPTI than the wild-type DnaK protein. Taken together, these data support a two-state model for DnaK in absence of peptide, with the transition between states resulting from ATP hydrolysis. With a leap of faith, one can suggest that these two states that are observed in the absence of bound peptide approximate the high-peptide-affinity and low-peptide-affinity states of DnaK, although there is evidence that peptide binding has some effect on the conformation of the proteins (DeLuca-Flaherty et al. 1990; Park et al. 1993). This relatively simple two-state model, which can be offered as a paradigm that may generalize to other stress-70 proteins, is consistent with observations that stress-70 proteins release bound peptides in response to ATP binding and hydrolysis.

In the absence of peptides or denatured proteins, stress-70 proteins self-associate to form oligomers of low order, primarily dimers and trimers (Schlossman et al. 1984; Freiden et al. 1992). Incubation of these oligomers with ATP results in dissociation to monomers. In addition, it has been shown in the specific case of bovine BiP that addition of peptide results in dissociation of the oligomers (Blond-Elguindi et al. 1993). Furthermore, stress-70 proteins complexed to denatured polypeptides separate on gel filtration as single protomers rather than as multimers (Palleros et al. 1991). A question that arises is whether the reversible oligomerization has any functional significance. One model of self-association that can be suggested, and is consistent with these data, is that in absence of target peptides, stress-70 proteins bind "self-peptides"—relatively unstructured segments of polypeptide within the protein itself—and that this association is broken up either by conversion of the protein to the low-peptide-affinity form by ATP or by addition of competing peptides. Such self-association could serve as a mechanism for modulating the concentrations of active stress-70 proteins capable of binding (and rebinding) target polypeptides under cellular conditions where there is a dearth of such substrates.

IX. PHOSPHORYLATION AND ADP-RIBOSYLATION

In vivo phosphorylation of two hsp70s of *Dictyostelium*, primarily (~70–90%) on threonine residues and to a lesser extent (~10–30%) on serine residues, was reported in 1982 (Loomis et al. 1982). A year later it was reported that in vivo labeling of chick embryo fibroblasts and rat embryo cell lines with [^{3}H]adenosine, as well as in vitro labeling of fibroblast whole-cell lysates with [^{32}P]NAD^{+}, resulted in labeling of a protein with a relative molecular weight of 83,000, thought to be the BiP protein, suggesting a possible ADP ribosylation of BiP (Carlsson and Lazarides 1983). The labeling decreased under conditions of heat shock and glucose starvation, conditions that increase the synthesis of BiP in cells. Since these early observations, in vivo phosphorylation and ADP ribosylation of BiP have been further characterized. In particular, Hendershot and colleagues have shown that the modified BiP protein is found exclusively in the multimer pool; monomeric peptide-free BiP and BiP complexed with immunoglobulin heavy chains are not significantly modified (Freiden et al. 1992).

In vivo, DnaK is found to be phosphorylated on serine and threonine residues; the phosphorylation is primarily on serine under normal growth conditions but shifts predominantly to threonine when *E. coli* is infected with bacteriophage M13 (Rieul et al. 1987). BiP is phosphorylated in vivo on both serine and threonine residues; however, T229, the site of in vitro phosphorylation, is not phosphorylated in vivo (Gaut and Hendershot 1993a). The specific target residues of in vivo phosphorylation have not yet been determined for either BiP or DnaK.

The pattern of phosphorylation, and additionally ADP ribosylation in the case of BiP, is suggestive of regulatory modifications that might modulate the activity of stress-70 proteins in vivo. However, at this time, there is no case in which a modifying enzyme has been purified and its effect on a target stress-70 protein characterized. Hence, the functional significance of posttranslational modification of stress-70 proteins remains an open question.

X. CONCLUDING REMARKS

Stress-70 proteins were first appreciated for their participation in the disassembly of specific supramolecular complexes. Only more recently have they been recognized for their widespread participation in preventing misfolding and/or aggregation of proteins, a recognition that has earned them the title of "molecular chaperone." Although the duties of a social chaperone are generally well-defined ("one delegated to ensure proper behavior"), the duties of molecular chaperones in a biochemical milieu are still somewhat nebulous. Nevertheless, consensus is beginning to emerge concerning certain aspects of the biochemistry of stress-70 proteins. We now have an extensive database on the primary structures of the stress-70 family as well as the tertiary structure of the ATPase fragment, allowing the identification of crucial residues and conserved motifs. Substantial information is accumulating on the ATPase cycle and mechanism; data from several members of the family suggest that nucleotide binding is tight (K_m ~1 μM), and the ATPase is slow (k_{cat} ~0.1 min^{-1}) and is subject to complex regulation by nucleotides, peptides, unfolded proteins, and accessory proteins (e.g., GrpE and DnaJ). Several active site residues and a bound magnesium ion contribute to catalysis, but no one residue has been shown to be absolutely required. Physical studies indicate that members of the stress-70 family undergo a large conformational shift following ATP hydrolysis.

The basis of the peptide specificity and the biochemical mechanism coupling the ATPase cycle and the peptide-binding/release cycle are less well understood. Significant open questions include identification both of the slow step in hydrolysis and of the step that is coupled to peptide release, as well as determination of the significance of posttranslational modifications. We may hope that further studies will provide a clear understanding of the biochemical mechanism of stress-70 proteins in vitro, and how it relates to their biological functions in vivo, allowing us to describe the duties of a molecular chaperone in unambiguous biochemical terms.

ACKNOWLEDGMENTS

This work has been supported by grant GM-39928 to D.B.M. and National Research Service Award GM-15141 to M.C.O. from the National Institues of Health.

REFERENCES

Alfano, C. and R. McMacken. 1989. Heat shock protein-mediated disassembly of nucleoprotein structures is required for the initiation of bacteriophage lambda DNA replication. *J. Biol. Chem.* **264:** 10709–10718.

Banecki, B., M. Zylicz, E. Bertoli, and F. Tanfani. 1992. Structural and functional relationships in DnaK and DnaK756 heat-shock proteins from *Escherichia coli. J. Biol. Chem.* **267:** 25051–25058.

Bardwell, J.C. and E.A. Craig. 1984. Major heat shock gene of *Drosophila* and the *Escherichia coli* heat-inducible *dnaK* gene are homologous. *Proc. Natl. Acad. Sci.* **81:** 848–852.

Beckmann, R.P., L.A. Mizzen, and W.J. Welch. 1990. Interaction of Hsp 70 with newly synthesized proteins: Implications for protein folding and assembly. *Science* **248:** 850–854.

Blond-Elguindi, S., A.M. Fourie, J.F. Sambrook, and M.J. Gething. 1993. Peptide-dependent stimulation of the ATPase activity of the molecular chaperone BiP is the result of conversion of oligomers to active monomers. *J. Biol. Chem.* **268:** 12730–12735.

Bork, P., C. Sander, and A. Valencia. 1992. An ATPase domain common to prokaryotic cell cycle proteins, sugar kinases, actin, and hsp70 heat shock proteins. *Proc. Natl. Acad. Sci.* **89:** 7290–7294.

Carlsson, L. and E. Lazarides. 1983. ADP-ribosylation of the M_r 83,000 stress-inducible and glucose-regulated protein in avian and mammalian cells: Modulation by heat shock and glucose starvation. *Proc. Natl. Acad. Sci.* **80:** 4664–4668.

Chappell, T.G., B.B. Konforti, S.L. Schmid, and J.E. Rothman. 1987. The ATPase core of a clathrin uncoating protein. *J. Biol. Chem.* **262:** 746–751.

Chappell, T.G., W.J. Welch, D.M. Schlossman, K.B. Palter, M.J. Schlesinger, and J.E. Rothman. 1986. Uncoating ATPase is a member of the 70 kilodalton family of stress proteins. *Cell* **45:** 3–13.

Chirico, W.J., M.G. Waters, and G. Blobel. 1988. 70K Heat shock related proteins stimulate protein translocation into microsomes. *Nature* **332:** 805–810.

DeLuca-Flaherty, C., D.B. McKay, P. Parham, and B.L. Hill. 1990. Uncoating protein (hsc70) binds a conformationally labile domain of clathrin light chain LC_a to stimulate ATP hydrolysis. *Cell* **62:** 875–887.

de Silva, A.M., W.E. Balch, and A. Helenius. 1990. Quality control in the endoplasmic reticulum: Folding and misfolding of vesicular stomatitis virus G protein in cells and *in vitro. J. Cell Biol.* **111:** 857–866.

Domanico, S.Z., D.C. DeNagel, J.N. Dahlseid, J.M. Green, and S.K. Pierce. 1993. Cloning of the gene encoding peptide-binding protein 74 shows that it is a new member of the heat shock protein 70 family. *Mol. Cell. Biol.* **13:** 3598–3610.

Flaherty, K.M., C. DeLuca-Flaherty, and D.B. McKay. 1990. Three dimensional structure of the ATPase fragment of a 70 kilodalton heat shock cognate protein. *Nature* **346:** 623–628.

Flaherty, K.M., D.B. McKay, W. Kabsch, and K.C. Holmes. 1991. Similarity of the three-dimensional structures of actin and the ATPase fragment of a 70-kDa heat shock cognate protein. *Proc. Natl. Acad. Sci.* **88:** 5041–5045.

Fletterick, R.J., D.J. Bates, and T.A. Steitz. 1975. The structure of a yeast hexokinase monomer and its complexes with substrates at 2.7 Å resolution. *Proc. Natl. Acad. Sci.* **72:** 38–42.

Flynn, G.C., T.G. Chappell, and J.E. Rothman. 1989. Peptide binding and release by proteins implicated as catalysts of protein assembly. *Science* **245:** 385–390.

Flynn, G.C., J. Pohl, M.T. Flocco, and J.E. Rothman. 1991. Peptide-binding specificity of the molecular chaperone BiP. *Nature* **353:** 726–730.

Freiden, P.J., J.R. Gaut, and L.M. Hendershot. 1992. Interconversion of three differentially modified and assembled forms of BiP. *EMBO J.* **11:** 63–70.

Gao, B., E. Eisenberg, and L.E. Greene. 1992. Binding of nucleotides to the bovine brain uncoating ATPase. *Mol. Biol. Cell* **3S:** 58a.

Gao, B., Y. Emoto, L. Greene, and E. Eisenberg. 1993. Nucleotide binding properties of bovine brain uncoating ATPase. *J. Biol. Chem.* **268:** 8507–8513.

Gaut, J.R. and L.M. Hendershot. 1993a. The immunoglobulin-binding protein *in vitro* autophosphorylation site maps to a threonine within the ATP binding cleft but is not a detectable site of *in vivo* phosphorylation. *J. Biol. Chem.* **268:** 12691–12698.

―――. 1993b. Mutations within the nucleotide binding site of immunoglobulin-binding protein inhibit ATPase activity and interfere with release of immunoglobulin heavy chain. *J. Biol. Chem.* **268:** 7248–7255.

Georgopoulos, C.P. 1977. A new bacterial gene (*groPC*) which affects λ DNA replication. *Mol. Gen. Genet.* **151:** 35–39.

Georgopoulos, C.P., B. Lam, A. Lundquist-Heil, C.F. Rudolph, J. Yochem, and M. Feiss. 1979. Identification of the *E. coli dnaK* (*groPC756*) gene product. *Mol. Gen. Genet.* **172:** 143–149.

Gething, M.J. and J. Sambrook. 1992. Protein folding in the cell. *Nature* **355:** 33–45.

Gupta, R.S. and B. Singh. 1992. Cloning of the HSP70 gene from *Halobacterium marismortui*: Relatedness of archaebacterial HSP70 to its 24 eubacterial homologs and a model for the evolution of the HSP70 gene. *J. Bacteriol.* **174:** 4594–4605.

Hein, J. 1990. Unified approach to alignment and phylogenies. *Methods Enzymol.* **183:** 626–645.

Hendershot, L.M. 1990. Immunoglobulin heavy chain and binding protein complexes are dissociated in vivo by light chain addition. *J. Cell. Biol.* **111:** 829–837.

Huang, S., M.-Y. Tsai, Y.-M. Tzou, W. Wu, and C. Wang. 1993. Aspartyl residue 10 is essential for ATPase activity of rat hsc70. *J. Biol. Chem.* **268:** 2063–2068.

Hunt, C. and R.I. Morimoto. 1985. Conserved features of eukaryotic hsp70 genes revealed by comparison with the nucleotide sequence of human hsp70. *Proc. Natl. Acad. Sci.* **82:** 6455–6459.

Hurley, J.H., H.R. Faber, D. Worthylake, N.D. Meadow, S. Roseman, D.W. Pettigrew, and S.J. Remington. 1993. Structure of the regulatory complex of *Escherichia coli* IIIG1c with glycerol kinase. *Science* **259:** 673–677.

Hurtley, S.M., D.G. Bole, H. Hoover-Litty, A. Helenius, and C.S. Copeland. 1989. Interactions of misfolded influenza virus hemagglutinin with binding protein (BiP). *J. Cell. Biol.* **108:** 2117–2126.

Ingolia, T.D., E.A. Craig, and B.J. McCarthy. 1980. Sequence of three copies of the gene for the major *Drosophila* heat shock induced protein and their flanking regions. *Cell* **21:** 669–679.

Kabsch, W., H.G. Mannherz, D. Suck, E.F. Pai, and K.C. Holmes. 1990. Atomic structure of the actin: DNase I complex. *Nature* **347**: 37–44.

Kassenbrock, C.K. and R.B. Kelly. 1989. Interaction of heavy chain binding protein (BiP/GRP78) with adenine nucleotides. *EMBO J.* **8**: 1461–1467.

Kassenbrock, C.K., P.D. Garcia, P. Walter, and R.B. Kelly. 1988. Heavy-chain binding protein recognizes aberrant polypeptides translocated in vitro. *Nature* **333**: 90–93.

Kraulis, P. 1991. MOLSCRIPT: A program to produce both detailed and schematic plots of protein structures. *J. Appl. Crystallogr.* **24**: 946–950.

Langer, T., C. Lu, H. Echols, J. Flanagan, M.K. Hayer, and F.-U. Hartl. 1992. Successive action of DnaK, DnaJ and GroEL along the pathway of chaperone-mediated protein folding. *Nature* **356**: 683–689.

Leustek, T., H. Toledo, N. Brot, and H. Weissbach. 1991. Calcium-dependent autophosphorylation of the glucose-regulated protein, Grp78. *Arch. Biochem. Biophys.* **289**: 256–261.

Liberek, K., J. Marszalek, D. Ang, C. Georgopoulos, and M. Zylicz. 1991a. *Escherichia coli* DnaJ and GrpE heat shock proteins jointly stimulate ATPase activity of DnaK. *Proc. Natl. Acad. Sci.* **88**: 2874–2878.

Liberek, K., D. Skowyra, M. Zylicz, C. Johnson, and C. Georgopoulos. 1991b. The *Escherichia coli* DnaK chaperone, the 70-kDa heat shock protein eukaryotic equivalent, changes conformation upon ATP hydrolysis, thus triggering its dissociation from a bound target protein. *J. Biol. Chem.* **266**: 14491–14496.

Loomis, W., S. Wheeler, and J. Schmidt. 1982. Phosphorylation of the major heat shock protein of *Dictyostelium discoideum*. *Mol. Cell. Biol.* **2**: 484–489.

McCarty, J.S. and G.C. Walker. 1991. DnaK as a thermometer: Threonine-199 is site of autophosphorylation and is critical for ATPase activity. *Proc. Natl. Acad. Sci.* **88**: 9513–9517.

Milarski, K.L. and R.I. Morimoto. 1989. Mutational analysis of the human HSP70 protein: Distinct domains for nucleolar localization and adenosine triphosphate binding. *J. Cell. Biol.* **109**: 1947–1962.

Miyazaki, T., S. Tanaka, H. Fujita, and H. Itikawa. 1992. DNA sequence analysis of the dnaK gene of *Escherichia coli* B and of two *dnaK* genes carrying the temperature-sensitive mutations *dnaK7*(Ts) and *dnaK756*(Ts). *J. Bacteriol.* **174**: 3715–3722.

Mizzen, L.A., A.N. Kabiling, and W.J. Welch. 1991. The two mammalian mitochondrial stress proteins, grp 75 and hsp 58, transiently interact with newly synthesized mitochondrial proteins. *Cell Reg.* **2**: 165–179.

Montgomery, D., R. Jordan, R. McMacken, and E. Freire. 1993. Thermodynamic and structural analysis of the folding/unfolding transitions of the *Escherichia coli* molecular chaperone dnaK. *J. Mol. Biol.* **232**: 680–692.

Munro, S. and H.R. Pelham. 1986. An Hsp70-like protein in the ER: Identity with the 78 kd glucose-regulated protein and immunoglobulin heavy chain binding protein. *Cell* **46**: 291–300.

Murakami, H., D. Pain, and G. Blobel. 1988. 70-kD heat shock-related protein is one of at least two distinct cytosolic factors stimulating protein import into mitochondria. *J. Cell Biol.* **107**: 2051–2057.

O'Brien, M.C. and D.B. McKay. 1993. Threonine 204 of the chaperone protein hsc70 influences the structure of the active site but is not essential for ATP hydrolysis. *J. Biol. Chem.* **268**: 24323–24329.

Palleros, D.R., W.J. Welch, and A.L. Fink. 1991. Interaction of hsp70 with unfolded proteins: Effects of temperature and nucleotides on the kinetics of binding. *Proc. Natl. Acad. Sci.* **88**: 5719–5723.

Palleros, D.R., K.L. Reid, J.S. McCarty, G.C. Walker, and A.L. Fink. 1992. DnaK, hsp73, and their molten globules. Two different ways heat shock proteins respond to heat. *J. Biol. Chem.* **267:** 5279–5285.

Park, K., G.C. Flynn, J.E. Rothman, and G.D. Fasman. 1993. Conformational change of chaperone Hsc70 upon binding to a decapeptide: A circular dichroism study. *Protein Sci.* **2:** 325–330.

Pelham, H.R.B. 1986. Speculations on the functions of the major heat shock and glucose-regulated proteins. *Cell* **46:** 959–961.

Rieul, C., J.C. Cortay, F. Bleicher, and A.J. Cozzone. 1987. Effect of bacteriophage M13 infection on phosphorylation of dnaK protein and other *Escherichia coli* proteins. *Eur. J. Biochem.* **168:** 621–627.

Rothman, J.E. 1989. Polypeptide chain binding proteins: Catalysts of protein folding and related processes in cells. *Cell* **59:** 591–601.

Rothman, J.E. and S.L. Schmid. 1986. Enzymatic recycling of clathrin from coated vesicles. *Cell* **46:** 5–9.

Sadis, S. and L.E. Hightower. 1992. Unfolded proteins stimulate molecular chaperone Hsc70 ATPase by accelerating ADP/ATP exchange. *Biochemistry* **31:** 9406–9412.

Scherer, P.E., U.C. Krieg, S.T. Hwang, D. Vestweber, and G. Schatz. 1990. A precursor protein partly translocated into yeast mitochondria is bound to a 70 kd mitochondrial stress protein. *EMBO J.* **9:** 4315–4322.

Schlossman, D.M., S.L. Schmid, W.A. Braell, and J.E. Rothman. 1984. An enzyme that removes clathrin coats: Purification of an uncoating ATPase. *J. Cell Biol.* **99:** 723–733.

Schmid, S.L. and J.E. Rothman. 1985a. Enzymatic dissociation of clathrin cages in a two-stage process. *J. Biol. Chem.* **260:** 10044–10049.

———. 1985b. Two classes of binding sites for uncoating protein in clathrin triskelions. *J. Biol. Chem.* **260:** 10050–10056.

Schmid, S.L., W.A. Braell, and J.E. Rothman. 1985. ATP catalyzes the sequestration of clathrin during enzymatic uncoating. *J. Biol. Chem.* **260:** 10057–10062.

Tissières, A., H.K. Mitchell, and U.M. Tracy. 1974. Protein synthesis in salivary glands of *Drosophila melanogaster*: Relation to chromosome puffs. *J. Mol. Biol.* **84:** 389–398.

Voellmy, R., A. Ahmed, P. Schiller, P. Bromley, and D. Rungger. 1985. Isolation and functional analysis of a human 70,000-dalton heat shock protein gene segment. *Proc. Natl. Acad. Sci.* **82:** 4949–4953.

Welch, W.J. and J.R. Feramisco. 1985. Rapid purification of mammalian 70,000-dalton stress proteins: Affinity of the proteins for nucleotides. *Mol. Cell Biol.* **5:** 1229–1237.

Wild, J., L.A. Kamath, E. Ziegelhoffer, M. Lonetto, Y. Kawasaki, and C.A. Gross. 1992. Partial loss of function mutations in DnaK, the *Escherichia coli* homologue of the 70-kDa heat shock proteins, affect highly conserved amino acids implicated in ATP binding and hydrolysis. *Proc. Natl. Acad. Sci.* **89:** 7139–7143.

Wu, B., C. Hunt, and R. Morimoto. 1985. Structure and expression of the human gene encoding major heat shock protein HSP70. *Mol. Cell. Biol.* **5:** 330–341.

Zylicz, M., J.H. LeBowitz, R. McMacken, and C. Georgopoulos. 1983. The dnaK protein of *Escherichia coli* possesses an ATPase and autophosphorylating activity and is essential in an in vitro DNA replication system. *Proc. Natl. Acad. Sci.* **80:** 6431–6435.

Zylicz, M., D. Ang, K. Liberek, T. Yamamoto, and C. Georgopoulos. 1988. Initiation of lambda DNA replication reconstituted with purified lambda and *Escherichia coli* replication proteins. *Biochim. Biophys. Acta* **951:** 344–350.

8

Interactions of Vertebrate hsc70 and hsp70 with Unfolded Proteins and Peptides

Lawrence E. Hightower and Seth E. Sadis[1]
Department of Molecular and Cell Biology
The University of Connecticut
Storrs, Connecticut 06269-3044

Ivone M. Takenaka
Bristol-Myers Squibb Pharmaceutical Research Institute
Princeton, New Jersey 08543

I. INTRODUCTION

This chapter is the companion to McKay's chapter in this volume on the ATP-binding domain of the vertebrate hsp70 protein family. We have attempted to pull together what is known about the interactions of the

---

[1]Present Address: Department of Cell Biology, Harvard Medical School, Boston, Massachusetts 02115.

The Biology of Heat Shock Proteins and Molecular Chaperones
©1994 Cold Spring Harbor Laboratory Press 0-87969-427-0/94 $5 + .00

carboxy-terminal domains of vertebrate hsc70 and hsp70 with unfolded proteins and peptides and about the interactions between the peptide-binding and ATPase domains. The molecular structure of the crystallized ATP-binding domain has been determined by X-ray diffraction (Flaherty et al. 1990); however, the three-dimensional structure of the peptide-binding domain is not known. Therefore, computerized structural predictions have been used to model the carboxy-terminal domain.

Because of the manner in which information on the functions of this protein family was initially developed, primarily through serendipitous discoveries of roles in normal cellular processes, much more is known about the constitutive or cognate members such as hsc70 than the stress-inducible members (hsp70). Thus, evidence of "molecular chaperone" functions comes primarily from studies of the cognates. Molecular chaperones are proteins that facilitate the folding or assembly and dis-assembly of other proteins but are not part of the finished structure (for recent reviews, see Ang et al. 1991; Craig et al. 1993; Georgopoulos and Welch 1993; Hendrick and Hartl 1993). The current hypotheses regarding chaperoning mechanisms for the hsp70 family include (1) binding to nascent polypeptides to delay folding until synthesis of the polypeptide chain is completed and all of the information needed for folding is present, (2) delaying folding until the polypeptide reaches the appropriate cellular compartment, (3) blocking nonproductive folding interactions such as aggregation, (4) translocation of polypeptides from sites of synthesis to sites of membrane translocation, (5) holding subunits in an assembly-competent form as in the assembly of steroid hormone receptors, and (6) active participation in disassembly of clathrin cages, where hsc70 is thought to bind to a flexible, exposed segment of clathrin light chain. Hypotheses regarding the role of inducible hsp70 in stressed cells are basically extrapolations of the molecular chaperoning concept and include (1) augmentation of the same functions carried by hsc70, (2) solubilization of denatured protein aggregates, (3) facilitation of renaturation of denatured proteins to restore function, and (4) transloca-tion of irreversibly damaged proteins to degradative organelles and proteasomes.

Several years ago, Milton Schlesinger wrote an amusing review of in-itial efforts to uncover the cellular functions of the hsp70, a process that he aptly likened to the pursuit of the elusive creature in Lewis Carroll's tale "*The Hunting of the Snark: An Agony in Eight Fits*" (Schlesinger 1986). Milton echoed the sentiment of many of us when he hoped that the search for function would not be a "trivial pursuit." With the snark's territory now in sight, the hope seems justified. In normal cells, hsc70 functions in the final step in the "central dogma" of molecular biology, which in the past was written DNA←—→RNA→Protein, but which is increasingly seen now as DNA←—→RNA→Polypeptide→Folded Protein. In addition, in stressed cells, hsp70 appears to be involved in the restoration of protein homeostasis following proteotoxic insults.

II. DISCUSSION

A. A Working Model of a Peptide-stimulated ATPase

1. Interactions with Nucleotides and Peptide Sequences

All of the known members of the hsp70 family can be described operationally as peptide- or unfolded-protein-stimulated ATPases. Most hypotheses concerning the role of ATP hydrolysis have invoked conformational changes in hsp70 driven by the energy of ATP hydrolysis to explain the release of unfolded proteins. This was primarily based on the fact that the slowly hydrolyzable ATP analog ATPγS does not support the release of polypeptides from hsp70. Now Tony Fink and co-workers report that release of unfolded proteins precedes ATP hydrolysis for DnaK, bovine hsc70, and human hsp72 using as a model substrate reduced carboxymethylated α-lactalbumin (RCMLA), a permanently unfolded form of α-lactalbumin (Palleros et al. 1993). These authors find that binding of MgATP to hsp70 is sufficient to cause the conformational change needed for dissociation of RCMLA and that this process is K+-dependent.

This and other information have been collected in Figure 1 to produce a model of the cycle. Since the K_m for ATP of bovine hsc70 is 1–2 μm in the presence or absence of unfolded proteins and the ATP concentration in cells is in the millimolar range, nucleotides should not be limiting in vivo, and virtually all cellular hsc70s will have either ADP or ATP bound. In step 1, an unfolded protein (UP) binds hsc70-ADP to form a ternary complex hsc70-ADP-UP. Evidence from several studies indicates that this complex is relatively stable (Palleros et al. 1991; Sadis and Hightower 1992). A change in the shape of the peptide-binding site (Fig. 1, hatched area) is proposed upon UP binding since Park et al. (1993) using circular dichroism spectroscopy found evidence of a conformational change in hsc70 upon binding to a decapeptide, and a new protease-protected fragment of hsc70 is found upon peptide binding (DeLuca-Flaherty et al. 1990).

We propose that a conformational change in the peptide-binding domain causes a conformational change in the nucleotide-binding domain since the model UP apocytochrome *c* stimulates a twofold increase in the rate of ADP release from hsc70 in the presence of ATP

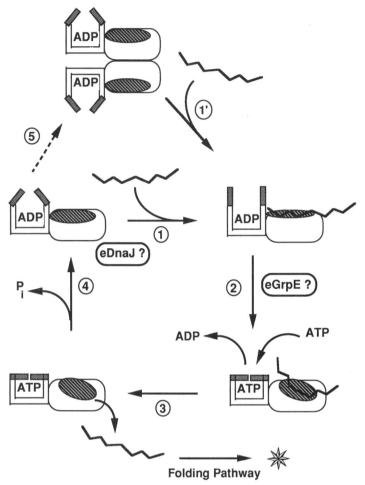

Figure 1 A working model of the reaction cycle of a peptide-stimulated ATP-ase. The model is described in detail in the text.

(Sadis and Hightower 1992). In the same study, rate constants of the nucleotide exchange reaction measured in the absence and presence of apocytochrome c (0.16 and 0.34 min^{-1}, respectively) closely matched the k_{cat} values derived from ATP hydrolysis measurements (0.15 and 0.38 min^{-1}, respectively). This suggests that the rate-limiting step in the ATP-ase reaction under our in vitro conditions is ADP release and that UPs stimulate ATP hydrolysis by accelerating the rate of ADP/ATP exchange. Using different conditions that included an ATP-regenerating system, Eisenberg and co-workers showed that the catalytic step is rate-limiting under their in vitro conditions (Gao et al. 1993). These authors also provided evidence of only one ATP-binding site per hsc70 monomer. Since the rate-limiting step in the presence of accessory proteins has not been determined and regulation under these conditions is likely to be more complicated, the most significant aspect of the in vitro work with hsc70 is that both the nucleotide exchange step and the catalytic step are slow and provide opportunities for regulation of the hydrolytic cycle. For bovine hsc70, addition of apocytochrome c in the presence of ATP increases the V_{max} of ATP hydrolysis about threefold with no significant change in the K_m. ADP binds to hsc70 with about a sixfold higher affinity than ATP, and for DnaK, an accessory protein GrpE stimulates ADP release (Liberek et al. 1991a). A sequence apparently encoding a eukaryotic homolog of GrpE has recently been cloned in K. Palter's laboratory, and given the similarities in the hsp70 family, the existence of such a protein, designated eGrpE in Figure 1, is not surprising.

Fink and co-workers observed a change in tryptophan fluorescence emission spectra of DnaK in the presence of K^+/MgATP, indicating a conformational change in the nucleotide-binding domain where the single tryptophan residue is located (Palleros et al. 1993). This is indicated by a hypothetical closing of the binding pocket in Figure 1, step 2. We speculate that the conformational change in the nucleotide-binding domain causes a conformational change in the peptide-binding domain, reducing its affinity for the polypeptide substrate, which is released in step 3.

In step 4, ATP hydrolysis and release of P_i produce another conformational change in the nucleotide-binding domain that causes a conformational change in the peptide-binding domain, returning it to a higher-affinity site. It has been shown for DnaK that ATP hydrolysis and release of peptides are accompanied by a conformational change as evidenced by a change in the pattern of protease-protected fragments of DnaK (Liberek et al. 1991b). This model postulates significant conformational flexibility for both domains of hsc70. McKay and colleagues noted that the topological similarity between hexokinase and the hsc70 ATPase domain suggests the possibility that the ATPase domain may undergo substrate-induced conformational changes similar to that observed with hexokinase (Flaherty et al. 1990). Conformational flexibility of the carboxy-terminal domain may explain the difficulties that have been encountered in attempts to crystallize the intact hsc70 protein. The model also includes allosteric interactions between the domains. There is already evidence that the carboxyl terminus of hsc70 affects ATP hydrolysis, based on the properties of a 60-kD chymotryptic fragment missing its carboxyl terminus, and removal of the entire peptide-binding domain actually stimulates the rate of ATP hydrolysis of the remaining amino-terminal 44-kD fragment, suggesting a close association of the domains (Chappell et al. 1987).

2. Interactions with Accessory Proteins

In *Escherichia coli*, the accessory protein DnaJ stimulates ATPase activity and release of protein substrates from DnaK. Eukaryotic homologs of DnaJ have been discovered recently and are designated eDnaJ in Figure 1 along with a question mark to indicate that we do not know exactly where in the cycle they act (Caplan et al. 1993; Silver and Way 1993). Of direct interest for hsc70 function is hsp40, a stress-inducible basic protein isolated from HeLa cells that has amino acid sequence homology with bacterial DnaJ (Hattori et al. 1993). This protein localizes to the nucleus and particularly the nucleolus of heat-shocked cells and returns to the cytoplasm during recovery; i.e., it has the same intracellular pattern as hsp70 and hsc70. Several hypotheses have been advanced regarding the relationship of hsp70 family and DnaJ family proteins: (1) DnaJ-like proteins may target hsp70 family proteins to specific locations such as ribosomes and membrane translocation sites; (2) they may serve to concentrate hsp70 family proteins and increase the efficiency with which they work; (3) since DnaJ appears to be a chaperone in its own right, it may help determine the kinds of substrates to which an associated hsp70 family protein may bind; (4) DnaJ-like proteins may stabilize substrate binding to hsp70 family proteins as in step 1 of Figure 1, and there is some experimental support for this possibility (Langer et al. 1992); and (5) they may be required to stimulate the dissociation of certain hsp70-ATP-UP complexes that are unusually stable, i.e., stimulate step 3. In support of this possibility, a cytosolic factor was recently discovered that facilitates dissociation of yeast hsp70 bound to prepro-α factor, a complex that is stable in the presence of ATP alone (Chirico 1992), and dissociation of a complex between yeast hsp70 and RCMLA requires both the eukaryotic DnaJ homolog Ydj1 and ATP (Cyr et al. 1992). Whereas reticulocyte hsc70 can slow the folding of precursor proteins destined for mitochondrial import and can prevent precursor aggregation, hsc70 alone does not support import. Another protein appears to be required for efficient dissociation of the precursor-hsc70 complex (see Sheffield et al. 1990).

The fact that hsc70 may bind to a polypeptide at one physical location in the cell and dissociate from it in another location while maintaining a stable complex during the translocation process suggests that different combinations of accessory proteins may be needed to promote binding, stabilization, and dissociation at different locations along the chaperoning pathways. Evidence has recently been marshalled in support of a hypothesis that hsp70 family proteins may serve to cross-link proteins to microfilament networks (Tsang 1993), and this represents another potential opportunity for regulation of interactions by accessory proteins.

3. The Role of Oligomers

Preparations of purified hsp70 family proteins routinely contain mixtures of monomers, dimers, and higher-order oligomers. A major unanswered question in the field is what role if any does oligomerization have in the regulation of hsp70 function? Oligomeric forms of hsc70 are converted to monomers in the presence of ATP, and when $[\gamma$-$^{32}P]$ATP is included as a tracer, radioactive ATP is recovered only on monomeric hsc70. However, the conversion to monomers has not been proven rigorously using hydrodynamic methods. When preparations are depleted of ATP, oligomers reform (Kim et al. 1992). Recently, heterodimers consisting of mammalian hsp70 and hsc70 have been detected by immunoprecipitation from heat-shocked cell lysates (Brown et al. 1993). The dimers are thought to be converted to monomers in the presence of ATP, but purified hsp70 and hsc70 did not form heterodimers even in the absence of ATP, again suggesting a role for accessory proteins.

Mammalian hsc70 undergoes thermally induced aggregation in the mammalian heat shock temperature range (40–47°C), and it has been suggested that a conformational change causes the protein to self-associate (Sadis et al. 1990b; Palleros et al. 1991). Unlike the dimers formed at 20–37°C, the thermally induced oligomers cannot be dissociated with Mg^{++}/ATP and presumably represent a different type of protein-protein interaction. Here, we have used the term aggregation to denote essentially irreversible associations, reserving the term oligomerization for reversible associations.

Gething and co-workers (Elguindi et al. 1993) have incorporated oligomer dissociation into their model of the peptide-stimulated ATPase cycle of Grp78. They propose that oligomeric forms bind amino acid sequences and consequently are converted to monomers (Fig. 1, step 1), which is the form that would undergo ATP/ADP exchange and ATP hydrolysis. When the concentration of unfolded proteins is lowered, Grp78 would return to its oligomeric forms (Fig. 1, step 5). They note in their paper that similar observations have been made for bovine hsc70, and there is evidence for both hsc70 and Grp78 that both monomers and oligomers can bind unfolded proteins (Elguindi et al. 1993). Another question remaining to be answered is what role the posttranslational phosphorylation and adenine nucleotide modifications have in the Grp78 cycle, particularly in oligomerization, since such modifications have not been detected on hsc70.

B. Computer-assisted Predictions of the Structure of the Carboxy-terminal Portion of hsc70

1. Secondary Structure Predictions

The abundance of secondary structure elements in mammalian hsc70 has been calculated on the basis of analyses of far-UV circular dichroism (CD) spectra (Sadis et al. 1990a; Park et al. 1993). Analyses of CD spectra based on two different methods (Yang et al. 1986; Manavalan and Johnson 1987) yielded similar estimates of α helix (~41%) but differed considerably in their estimations of β-strand and β-turn, neither of which agreed with the secondary structure predictions obtained by the Chou-Fasman method (Chou and Fasman 1974) or the Garnier-Osguthorpe-Robson (GOR) method (Garnier et al. 1978). Analyses of CD spectra by the LINCOMB method (Perczel et al. 1992) yielded values of 24% β-strand and 24% β-turn that are in closer agreement with Chou-Fasman predictions but gave estimates of α-helical content that are considerably lower than those of other methods. For the secondary structure predictions for the carboxy-terminal domain of rat hsc70 (Fig. 2), the GOR method was used, which predicts for the entire protein 40% α helix, 24% β-strand, 13% β-turn, and 23% aperiodic structures. After comparing the results of the different methods, we concluded that the GOR method gave reasonable consensus predictions of α helix and β-strand for hsc70 but may have underpredicted β-turns (13%→24%) in favor of aperiodic structures (23%→11%).

In Figure 2, residue 387 was taken as the beginning of the carboxy-terminal domain, since residue 386 marks the end of the 44-kD amino-terminal fragment crystallized by McKay and colleagues. The existence of the predicted helix $\alpha_{366-380}$ has been confirmed in the three-dimensional structure of the 44-kD fragment. The carboxy-terminal domain of hsc70 can be divided into three distinct regions on the basis of secondary structure predictions.

The first region (amino acids 392–511) contains β-strand and β-turn predictions primarily. The second region (amino acids 512–605) is predicted to be extensively α-helical. This region in hsc70 has an unusually high content of charged and polar amino acids: 51% of the residues in this region are glutamic acid, lysine, aspartic acid, or glutamine. The third region, beginning at amino acid 606 and extending to the carboxyl terminus, is predicted to contain mostly alternating aperiodic structures

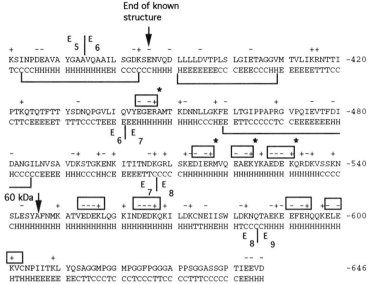

Figure 2 Amino acid sequence of the carboxy-terminal portion of rat hsc70 (O'Malley et al. 1985). Below the sequence is the GOR predicted secondary structure (H = α helix, E = β strand, T = β turn, C = aperiodic structure). The AXAB motifs are boxed, and asterisks mark the ones conserved in Grp78. The brackets mark the most highly conserved segments between hsc70 and Grp78. Exon junctions are labeled with an E and a subscript designating the exon number. The arrow marked "end of known structure" at residue 386 denotes the end of the 44-kD crystallized fragment. The estimated end of the 60-kD chymotryptic cleavage fragment used in some of Rothman's studies is marked.

and β-turns. Within this segment are four repeats of the sequence Gly-Gly-X-Pro, where X is methionine, phenylalanine, or alanine. A similar sequence is found at the amino terminus of mammalian and avian elastins, but its function in this protein is unknown. The carboxy-terminal tetrapeptide is the acidic sequence Glu-Glu-Val-Asp, which is conserved in cytoplasmic members of the eukaryotic hsp70 and hsp90 families.

The strength of the α-helical predictions can be assessed by plotting the α-helix conformational parameters generated by the GOR algorithm (Fig. 3A). The α-helix conformational parameters assigned to different amino acid residues are a statistical reflection of the occurrence of each amino acid residue within α-helical secondary structures taken from the known protein structure database, normalized by amino acid composition within the database. The amino acid sequence between residues 512 and 605 of rat hsc70 generates an extensive run of high-magnitude α-helix conformational parameters. In fact, the length and the intensity of the α-

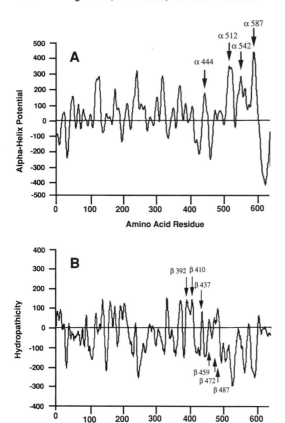

Figure 3 (*A*) Plot of the α helix conformational parameters generated by the GOR algorithm using the rat hsc70 amino acid sequence. Positive regions indicate amino acid sequences that are predicted to form α-helical secondary structures, whereas negative regions have a poor tendency for α-helix formation. (*B*) Plot of the hydropathicity of the rat hsc70 sequence generated by the Kyte-Doolittle algorithm using a seven-residue window size. These data were then smoothed by taking a running average of every seven residues. Positive regions indicate areas that are rich in hydrophobic residues, whereas negative regions indicate areas rich in hydrophilic residues.

helix conformational parameters generated by this segment are greater than those in any other region within hsc70. Three long contiguous α helices are predicted to reside within this region. The GOR prediction of the number of residues in helical secondary structure in the peptide-binding domain (amino acids 387–646) is 110 residues, and this number can be compared to the number determined using CD data and the number of residues known to be in helical structures in the amino-terminal

ATPase, which is 154 residues (D. McKay, pers. comm.). Our estimate from CD analysis of 41% helix for the entire hsc70 protein yields a total of 265 residues (0.41 x 646). Subtracting 154 from 265 gives 111 residues in helical structures in the peptide-binding domain, in close agreement with the 110 residues from the GOR prediction.

During inspection of this segment, we noted the presence of several copies of an element defined by the sequence E/D-x-E/D-K/R/H (acidic-x-acidic-basic). This AXAB motif is identified in Figure 2 as boxed charge symbols above the amino acid sequence. Nine copies of the AXAB motif are present in the hsc70 amino acid sequence. Seven of these are clustered within the 90-amino-acid segment beginning at amino acid 514. In the three predicted α helices nearest the carboxyl terminus, the clusters of AXAB motifs repeat every seven residues, as shown in Figure 4B for the helix beginning at amino acid 512. The spacing enables each amino acid side chain within the motif unit to extend from the same side of the predicted α helix. This effect is most evident when these sequences are diagrammed as extended α-helical wheels (Fig. 4A). In addition, the spacing of the few hydrophobic residues present in these predicted α helices confines these residues to one side of the extended α helix, marked as positions "a" and "d" by convention.

The hsc70 sequence was analyzed by the Kyte-Doolittle hydropathicity algorithm (Kyte and Doolittle 1982) to assess the distribution of hydrophilic and hydrophobic residues (Fig. 3B). The carboxy-terminal portion of hsc70 contains only a few areas enriched in hydrophobic residues, and these areas are found between residues 390 and 480 in predicted β strands. Interestingly, these hydrophobic regions are among the most highly conserved sequences in the carboxy-terminal domain of the hsp70 family, as marked by the brackets in Figure 2, based on a comparison of rat Grp78 and hsc70 (Wooden and Lee 1992). Some of the AXAB motifs (Fig. 2, asterisks) are also conserved between these two hsp70 family members. After about residue 540, the amino acid sequence identity declines considerably, giving each member of the hsp70 family a distinctive segment (see Fig. 6). It is not known whether this sequence divergence represents a structurally or functionally less constrained portion of the molecule or whether different family members have evolved different functions for this segment.

Recent analyses of other proteins such as the human class I and class II major histocompatibility complex (MHC) antigens that bind diverse polypeptide sequences have revealed binding domains formed by α helices. On the basis of the strong α-helix predictions in the carboxy-terminal domain of hsc70 and these known structures, we proposed that a channel or groove-like domain formed by α helices in hsp70 family

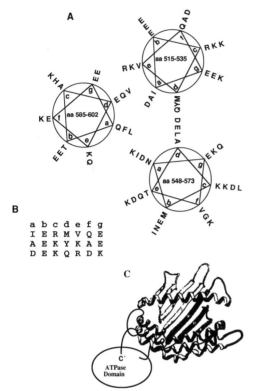

Figure 4 (*A*) Extended α-helical wheel projections of the three predicted helices in the carboxy-terminal region of hsc70. The wheels have been positioned roughly to accommodate salt bridges and hydrophobic interactions. (*B*) The three heptads of helix $\alpha_{512-536}$. (*C*) The "fancy clam" model of the peptide-binding domain showing the proposed triple-helical bundle and the flexible region at the carboxyl terminus extending to bind to the ATPase domain, represented by an oval. Only the C′ in the oval represents the carboxyl terminus of the polypeptide.

proteins may restrict binding to extended or disordered polypeptides (Sadis et al. 1990b). Our model is reminiscent of that proposed for the binding domain for signal sequences in the 54-kD signal recognition protein as well. Such a domain may account in part for the ability of hsp70 family proteins to distinguish denatured or misfolded proteins from native structures and to bind to nascent polypeptide chains. hsc70 might have greatest affinity for peptide ligands that include charged side chains, since the predicted α helices available in the binding domain contain mainly charged and polar residues. The possibility of hydrogen bonding to the main-chain amides and carbonyl oxygens of peptide ligands was also suggested. As discussed in greater detail below, there is now experimental support for these two hypotheses along with a requirement for at least two large aliphatic residues in the core of peptide ligands that bind hsp70 family proteins.

2. Tertiary Structure Predictions

Computer modeling based on fine secondary structural predictions requires an additional leap of faith since current algorithms are only about 70% accurate in predicting known structures. Two studies (Flajnik et al. 1991; Rippmann et al. 1991) took the leap and presented several tantalizing models of the putative peptide-binding domain of hsp70 family proteins superimposed upon the known three-dimensional structure of the $\alpha_1\alpha_2$ domains of HLA-A2 (Bjorkman et al. 1987). These two domains interact to form a peptide-binding cleft in which the base or floor of the cleft is formed by eight β strands on top of which are positioned two antiparallel α helices. The fine secondary structure GOR predictions from Figure 2 are shown diagrammatically in Figure 5 to illustrate the parts of hsc70 used in these models. The region including amino acids 392–536 can be described as tandem $\beta_4\alpha$ patterns, and the predicted $\beta_4\alpha\beta_4\alpha$ domain fits with an HLA-type fold. Helix $\alpha_{444-454}$, which forms one wall of the predicted binding groove, is a particularly dicey call because it is a relatively weak prediction, as shown in Figure 3A, but Rippman and colleagues pointed out that the known helix in the HLA protein is also poorly predicted by the secondary structure algorithms. The helix that forms the other wall $\alpha_{512-536}$ has the intriguing potential for aligning residues of the same charge along one side of the helix (Fig. 4A). Interestingly, three of the predicted β strands, $\beta_{392-398}$, $\beta_{459-463}$, and $\beta_{472-478}$, are in the most highly conserved regions (Fig. 2, brackets) and form the central, relatively hydrophobic floor of the predicted binding groove. In comparing the two superimposed structures, Flajnik and co-workers noted that the amino acids whose side chains are predicted to extend into the hsc70 peptide-binding groove are either polar or charged, which is different from the residues that extend into the HLA groove. The majority of the amino acids that are conserved between the two structures are in regions that might be expected to confer structural stability. They also noted that the absence of a stabilizing disulfide bond found in HLA but not in the putative peptide-binding domain of hsc70 may provide more flexibility, which would be desirable in light of recent data suggesting conformational changes upon peptide and nucleotide binding (see discussion of Fig. 1). Recently, the 18-kD segment (amino acids 384–543) of rat hsc70 immediately following the 44-kD fragment was expressed in *E. coli* and purified (Wang et al. 1993). This segment

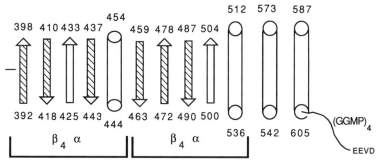

Figure 5 Diagrammatic representation of the secondary structure predictions shown in Fig. 2. The α helices are represented by cylinders and β strands are indicated by arrows (arrows with diagonal fill are predicted hydropathic strands from Fig. 3). The numbers indicate the amino acid residues predicted to begin and end each element. The putative HLA-like fold is indicated by $\beta_4\alpha\beta_4\alpha$. The flexible segment contains four repeats of the GGXP motif, and the conserved tetrapeptide EEVD marks the carboxyl terminus.

has the same high affinity for peptides as intact hsc70, lending experimental support to the positioning of the peptide-binding domain in the model described above.

The hsp70 sequences used in both modeling studies ended between $\alpha_{512-536}$ and $\alpha_{542-573}$ in a putative aperiodic region thought to contain a chymotryptic cleavage site used by Rothman and co-workers to generate a 60-kD fragment that retains clathrin-binding activity (Chappell et al. 1987). The remainder of the carboxy-terminal segment contains the very strong α-helical predictions between residues 542 and 605. In the spirit of stimulating further thought, we offer an embellishment of the HLA-based model that we call the "fancy clam" model (Fig. 4C). In this model, extended helix $\alpha_{542-573}$ runs antiparallel to extended helix $\alpha_{512-536}$ and extended helix $\alpha_{587-605}$ runs parallel to helix $\alpha_{512-536}$, forming a triple-helical bundle, shown as an axial projection in Figure 4A. An interesting comparison is the triple-helical coiled coil, which involves a heptad repeat similar to that shown in Figure 4B, i.e., positions a and d are mainly neutral residues and positions e and g have mainly basic and acidic residues, respectively, that participate in ionic interactions between helices. In the proposed hsc70 helical bundle, ionic interactions between acidic and basic residues on the helices would help position the helices and stabilize the bundle. The predominantly neutral a and d faces would constitute the interior residues of the bundle in which the helices would be extended but not coiled.

Finally, there is what appears to be a flexible, "sticky" appendage at

Table 1 Competition Binding Assay

A. Hexamers	K_i (mM)[a]
V R K K S R	8.8
G I K K K A	2.1
A G I K K K	>1
B. Heptamers	K_i (µM)
N I V R K K K	2.5
N I V I L K K	40
N Y I K K K A	200
C. Heptamers	
N I V A K K K	16
N I V A A K K	24
N I V A A A K	45
N I V A A A A	68
D. Control Heptamers	K_i (mM)
Y V E D E S G	>0.6
N I V E D E S	>1

The components of the assay were a ^{3}H-labeled cytochrome *c* fragment (residues 81–104) at 8 µM, purified bovine hsc70 at 1 µM, and increasing amounts of competing peptide. Bound and unbound radioactive fragments were separated on Sephadex G50 columns and assayed by liquid scintillation counting.

[a]The K_i was defined as the concentration of competing peptide that gave 50% inhibition of radioactive fragment binding.

the very carboxyl terminus composed of the GGXP repeats and the acidic EEVD terminal tetrapeptide. Judging from the properties of the 60-kD chymotryptic fragment (discussed in more detail below), an intact carboxyl terminus is needed for a productive interaction between the ATPase and peptide-binding domains, since the 60-kD fragment can hydrolyze ATP but cannot release from clathrin. The flexible segment and/or the two helices preceding it may associate with the amino-terminal ATPase domain with conserved EEVD participating. Interestingly, on the basis of comparisons with peptides that bind with high affinities, the sequence of this flexible appendage seems to be designed to preclude interaction with the peptide-binding site (Table 1). The positioning of helix $\alpha_{587-605}$ parallel to and contacting helix $\alpha_{512-536}$ that forms a side of the putative peptide-binding groove in the fancy clam model may allow the flexible finger to reach out and touch the ATPase domain and possibly affect the conformation of the peptide-binding groove via interactions between these helices, thus contributing to inter-domain regulation (Fig. 4C).

3. Evolution of Peptide-binding Domains

Flajnik et al. (1991) proposed an evolutionary scheme for the emergence of MHC molecules that included transfer of nucleotide sequences encoding peptide-binding domains from an ancestral hsp70 family gene into an ancestral class I gene. It is interesting in this context to examine the exon junctions of the gene encoding hsc70 (Fig. 2, $E_{subscript}$), since such a transfer might have involved "exon shuffling." Wooden and Lee (1992) observed that the intron-exon boundaries are not highly conserved between the mammalian *hsc70* and *grp78* genes but do occur at similar positions. In hsc70, exons 6 and 7 encode a predicted pattern of $\alpha\beta_4$, and the helix $\alpha_{512-536}$ that comprises a wall of the predicted binding groove is part of exon 8. Exon 8 contains two strong predictions of long helices, and the exon boundaries are neatly positioned in predicted aperiodic regions. The exon junctions of Grp78 fit the concept of a transferable peptide-binding domain better in that exon 7 encodes the predicted $\beta_4\alpha$ pattern and the second $\beta_4\alpha$ pattern is part of exon 8, which includes the remainder of the carboxy-terminal coding sequence. An ancestral sequence encoding $\beta_4\alpha$ might have been transferred and then duplicated later.

C. Interactions of hsc70 with Folded Proteins, Peptides, and Unfolded Proteins

1. hsc70 Can Distinguish between Folded and Unfolded Forms of the Same Protein

The results of two studies have validated a broadly held hypothesis in the field by showing that hsc70 can indeed distinguish between folded and unfolded forms of the same protein, binding only to the latter. Using gel filtration high-performance liquid chromatography (HPLC), Palleros and co-workers showed that bovine hsc70 and human hsp72/73 form a stable complex with RCMLA and with a mutant form of staphylococcal nuclease but not with native α-lactalbumin and native staphylococcal nuclease (Palleros et al. 1991). Sadis and co-workers showed that apocytochrome *c* but not native cytochrome *c* binds hsc70 and stimulates its ATPase activity (Sadis et al. 1990b; Sadis and Hightower 1992). Determination of the characteristics of the binding site(s) on unfolded proteins recognized by hsp70 family members is still in the early stages. However, it is clear from the data at hand that the basis of recognition is not a specific amino acid sequence. It is also apparent that there are rules governing the interactions and that the selectivity is considerable, probably on the same order as other targeting motifs such as the signal sequences.

2. Interactions with Folded Proteins

Bovine hsc70 has been a major focus of biochemical and biophysical studies of the vertebrate hsp70 family, and calf brain has been a relatively abundant source of the protein (~10 mg purified hsc70 per brain). Bovine brain is also a good source of clathrin, and the purification and study of this protein led to the biochemical characterization of hsc70 as a clathrin uncoating ATPase (Ungewickell 1985; Chappell et al. 1986). In addition to providing an example of the interaction of hsc70 with a folded protein, clathrin, this system has shed light on the location of the protein/peptide-binding site of hsp70 family proteins. Rothman and colleagues characterized proteolytic fragments of hsc70 generated by chymotrypsin for their ability to associate with clathrin coats and hydrolyze ATP (Chappell et al. 1987). Using an antipeptide antibody specific for the extreme amino terminus of hsc70, it was demonstrated that chymotrypsin digestion begins at the carboxyl terminus of hsc70 and proceeds toward the amino terminus. Two relatively stable intermediates are produced, a 60-kD and a 44-kD species. The 60-kD fragment binds clathrin cages (empty baskets of polymerized triskelions) but is defective in coupling ATP hydrolysis to clathrin release. In contrast, the 44-kD fragment has completely lost the ability to bind clathrin cages. Both fragments are able to bind and hydrolyze ATP, and it is the 44-kD amino-terminal fragment that McKay and co-workers crystallized and determined the ATP-binding site in the three-dimensional structure. Using a molecular genetic approach, Milarski and Morimoto (1989) provided additional evidence that the peptide-binding domain is located in the carboxy-terminal portion of hsp70. They identified discrete functional domains in hsp70 by constructing a series of deletions in the human hsp70 gene and analyzing the properties of the encoded proteins following transfection of the mutant constructs into CV1 (monkey kidney) cells. Their evidence also indicates that the hsp70 protein may be roughly divided into two separate domains, an amino-terminal region responsible for ATP binding and a carboxy-terminal region responsible for nucleolar localization after heat shock, i.e., protein binding. Mutations in the amino terminus of hsp70 that blocked ATP binding did not inhibit stress-induced nucleolar localization, whereas deletion of amino acids 437–617 blocked nucleolar localization after heat shock. More extensive analyses of deletion mutants are currently under way in the Morimoto laboratory for hsp70 and in Linda Hendershot's laboratory for Grp78, and these studies should yield a more precise localization of the peptide-binding site(s).

Studies of clathrin binding have provided a model for how hsp70 family proteins interact with potentially conformationally flexible regions of otherwise folded proteins, specifically clathrin light chains (LC_a and LC_b). To identify the light-chain region involved in binding, DeLuca-Flaherty et al. (1990) tried 15 synthetic peptides derived from human clathrin light chains (LC_a and LC_b) in hsc70-stimulated ATPase assays, and only one peptide gave substantial stimulation even at millimolar concentrations. This peptide was based on a proline- and glycine-rich region of LC_a that underwent reversible conformational changes and has the sequence FAILDGGAPGPQPHGEPPGGPDAVD. The enrichment in glycine and proline residues suggests that this region in LC_a is not part of a periodic secondary structure such as an α helix or a β-sheet, but instead perhaps forms a connecting loop or turn structure in the native protein. Another example is the binding of hsc70 to mutant p53 proteins. Lam and Calderwood (1992) identified a sequence PLSQETFS GLWKLLPPEDG, derived from the p53 tumor antigen, that blocks the binding of mutant p53 to human hsp70. The sequences from p53 and CL_a are both enriched for glycine and proline residues, which may allow a region in an otherwise folded protein to assume an accessible aperiodic structure. This sequence also binds Grp78, and the hydrophobic residues underlined in the sequence above are considered to be a motif of three or four hydrophobic residues in alternating positions characteristic of Grp78-binding peptides selected from a phage display library (Elguindi et al. 1993). However, this sequence is not the only one in p53 that interacts with hsc70. Hainaut and Milner (1992) reported that complex formation between mammalian hsc70 and mutant p53 requires the carboxy-terminal 28 amino acids of p53: SSHLKSKKGQSTSRHKKLMFKTEG PDSD. As we discuss later, peptides selected from a phage display library for hsc70 binding contain multiple basic residues as well. It is possible that hsc70 interacts with folded proteins such as clathrin light chain, SV40 large T antigen, and adenovirus E1A by the putative α-helical channel structure seeking out relatively extended loop or turn regions on the surfaces of these native proteins. The story with mutant p53 may be more complex, possibly involving a region of the protein containing an aperiodic loop and another region that is exposed in mutants but not in wild-type p53.

3. How Much Functional Diversity Exists within the hsp70 Family?

Before proceeding further, it may be useful to discuss one of the paradigms in the heat shock field in the context of protein interactions. The hsp70 family is considered one of the evolutionarily most highly conserved protein families based on nucleotide and amino acid sequence comparisons, which show 50% amino acid sequence identity between

DnaK and *Drosophila* hsp70, for example (Bardwell and Craig 1984), and on similarities in the physical properties of the proteins such as almost superimposable CD spectra for DnaK and bovine hsc70 (Sadis et al. 1990b). In stressed mammalian cells, hsc70 and hsp70 colocalize in nucleoli and both proteins return to the cytoplasm during recovery where they associate with nascent polypeptides (Brown et al. 1993). Thus, it is likely that many of the major properties of hsp70 family proteins will be held in common. However, several recent studies indicate that not all hsp70 family members are functionally interchangeable. Neither Grp78 nor DnaK can substitute for hsc70 in facilitating protein translocation into mammalian microsomes in vitro (Wiech et al. 1993). Two yeast hsp70 family proteins, Ssa1p and Ssa2p, which have only a 3% difference in their amino acid sequences, nevertheless have markedly different activities in uncoating clathrin-coated vesicles (Gao et al. 1991). Furthermore, evidence was presented suggesting that isozymes with lower uncoating activity may inhibit the activity of isozymes with higher activity in a mixture. In addition, it has recently been shown that even closely related vertebrate species can have dramatically different hsp70 isoform patterns (White et al. 1994). By analogy with HLA molecules, only a few amino acid substitutions in an otherwise conserved hsp70 peptide-binding site could alter its selectivity. However, a comparison of the amino acid sequences of several vertebrate and yeast hsp70 family proteins suggests a different intepretation of this diversity (Fig. 6). The sequences of Ssa1p and Ssa2p, for example, show no significant sequence differences in their putative peptide-binding regions; however, these proteins have differences in the carboxy-terminal region of the putative aperiodic sequence. Likewise, nonconservative amino acid substitutions are mainly in the segment of vertebrate hsp70 family proteins after residue 560. Thus, it seems more likely that variation in hsp70 family proteins may affect other properties instead of the selectivity of peptide binding, such as protein stability, interdomain regulation, and oligomerization.

4. Interactions with Peptides

Three other peptide sequences have been reported that interact with hsc70. They are (1) Pc 81–104, IFAGIKKKSERVDLIAYLKDATSK, a fragment of cytochrome *c* that binds to Pbp74, a constitutive hsp70 family member thought to be involved in antigen presentation on cell surfaces (Vanbuskirk et al. 1989; Domanico et al. 1993), and also to hsc70 (S.E. Sadis and L.E. Hightower, unpubl.), (2) KRQIYTDLEMNRLGK and also (3) KLIGVLSSLFRPK, peptides from vesicular stomatitis virus

```
KSINPDEAVA  YGAAVQAAIL  SGDKSENVQD  LLLLDVTPLS  LGIETAGGVM  TVLIKRNTTI -420
..........  ..........  ..........  ..........  ..........  ..........
.........G  ..........  M.........  ......A...  ..L.......  .A.....S..
R.........  ..........  T..E.SKT..  ......A...  ..........  .K..P..S..
R.........  ..........  T..E.SKT..  ......A...  ..........  .K..P..S..
TCCCCHHHHH  HHHHHHHHEH  CCCCCCHHHH  HEEEEEEECC  CEEECCCHHE  EEEEETTTCC
                       - -+
PTKQTQTFTT  YSDNQPGVLI  QVYEGERAMT  KDNNLLGKFE  LTGIPPAPRG  VPQIEVTFDI -480
..........  ..........  ..........  ..........  ..........  ..........
......I...  ..........  ..........  .......R..  .S.......-.  ..........
S..KFEI.S.  .A........  ..F....K.   ..........  .S........  .........V
...KSEV.S.  .A........  ..F....K.   ..........  .S........  .........V
CTTCEEEEET  TTTCCCTEEE  EEEHHHHHHH  HHHHCCCHEE  ETTCCCCCTC  CEEEEEEEHH
                       - -+        - -+ ---    +
DANGILNVSA  VDKSTGKENK  ITITNDKGRL  SKEDIERMVQ  EAEKYKAEDE  KQRDKVSSKN -540
.....M....  A.........  ..........  ..........  ......C..D  V.........
........T.  T......A..  ..........  ...E......  ......A..E  ...ER..A..
.S........  .E.G...S..  ..........  ......K..A  ....F.E...  .ESQRIA...
.S........  .E.G...S..  ..........  ......K..A  ....F.E...  .ESQRIA...
HCCCCCEEEE  HHHCCCHHCE  EEEETTCCCC  CHHHHHHHHH  HHHHHHHHHH  HHHHHHCCCC
  ↓60kDa   ---+        -- -+                               - -+  - -
SLESYAFNMK  ATVEDEKLQG  KINDE-DKQKILDKCN------EIISW LDKNQTAEKE  EFEHQQKELE -600
..........  S.........  ..S..-..T...E...------.V.G..........  .Y..H.....
A.........  SA......K.  ..SEA-..K.V....Q------.V...  ..A.TL...D  ...KR....
Q...I.YSL.  N.--------  -.SEAG..LEQA..DTVTKKAE.T...  ..S.T..S..  ..DDKL...Q
Q...I.YSL.  N.------.--  -.SEAG..LEQA..DAVTKKAE.T.A.  ..S.T..T..  ..DD.L...Q
CHHHHHHHHH  HHHHHHHHHH  HHHHH-HHHHHHHHTT------HHEHH HTCCCCHHHH  HHHHHHHHHH
+
KVCNPIITKL  YQSAGGMPGG  MP----GGFPG-GGA  PPSGGASSGP  TIEEVD      rat hsc70  -646
..........  ..G.......  ..EGMA.....A...  A.G..G....  ......      trout hsc70
.......SG.  ..G...-..-  -.----...-.AQ.-  -.K..SG...  ......      human hsp70
DIA...MS..  ..-...A...  AAGGAP.....--..  ..APE.E-..  .V....      yeast SSA1
E.A...MS..  ..-...A.E.  AA---P......--..  ..APE.E-..  .V....      yeast SSA2
HTHHHEEEEE  EECCTTCCCTC  CC----TCCCT-TCC  CCTTTCCCCC  CEEHHH
```

Figure 6 Amino acid sequence alignment of several vertebrate and yeast hsp70 family proteins. The vertebrate sequence alignments follow those of Zafarullah et al. (1992) and the yeast sequences were taken from Slater and Craig (1989). The periods denote identity and the dashes are gaps introduced into the hand-aligned sequences. The last line contains the secondary structure predictions for rat hsc70 as described in Fig. 2.

with K_m values (concentrations of peptides that give half-maximal velocities of ATP hydrolysis) of 770 μM and 55 μM, respectively (Flynn et al. 1989). This last peptide, also known as vsv-C, was used in the only direct analysis to date of the conformation of a peptide bound to an hsp70 family protein, which in this study was bacterial DnaK (Landry et al. 1992). By analyzing transferred nuclear Overhauser effects in two-dimensional nuclear magnetic resonance (NMR) spectra, Gierasch and co-workers showed that the bound peptide is in an extended conformation. They concluded that the region of the peptide including Ile-3 and Val-5 has the strongest interactions with DnaK and that peptide binding is probably due to hydrogen bonding between DnaK and the peptide backbone along with hydrophobic contacts and possible salt bridges with the side chains. Consistent with these observations, Gragerov et al. (1994) screened a peptide display library using DnaK and found a peptide, NRLLLTG, with a high affinity for this heat shock protein. On the basis of an analysis of a set of peptides, they concluded that peptides containing internal hydrophobic residues are preferred substrates, that inclusion of a basic residue in the peptide is beneficial, and that peptides with acidic residues have poor affinities for DnaK.

The observation that the bound peptide is in an extended conformation is of particular interest when combined with recent studies which suggest that hsc/hsp70 may bind to nascent polypeptide chains as they emerge from ribosomes. One of the first indications that stress proteins were involved in translation came from biochemical and ultrastructural studies of mammalian hsp70 in cells recovering from thermal and chemical stress in which hsp70 appeared to colocalize with translationally active cytoplasmic ribosomes (Welch and Suhan 1986). Subsequent studies showed that hsc70 (called hsp73 in these studies) and hsp70 (called hsp72) bound to nascent polypeptide chains in polysomes in normal and stressed cells, respectively (Beckman et al. 1990). Furthermore, these interactions are transient and ATP-dependent in normal cells and relatively stable in stressed cells that have accumulated damaged proteins incapable of folding properly (Beckman et al. 1992). Recently, in vitro translation systems have been used to model these interactions. For example, translation systems primed with Sindbis virus mRNA produce an autoprotease that folds and functions cotranslationally to process viral polypeptides. In the presence of exogenously added hsp70 family proteins, autoprotease activity was inhibited, indicating that the chaperones interact with nascent polypeptides by blocking either the folding event that forms the autoprotease or the cleavage step (Ryan et al. 1992). In addition, it has been proposed from studies of yeast hsp70 that this chaperone aids the passage of nascent polypeptide chains through the ribosomal channel and into the cytoplasm by interacting with the nascent chain as it emerges from the channel (Nelson et al. 1992). A prediction from these models that levels of molecular chaperones in cells should increase with increasing protein synthetic capacity has recently found experimental support. Transforming growth factor β1 regulates protein synthesis in cultured chicken embryo cells, and increased overall rates of protein synthesis are preceded by increased synthesis of hsp70 and hsp90 family chaperones (Takenaka and Hightower 1992, 1993).

hsc70 binding may occur even before the emerging polypeptide chain gains secondary structure. Since the formation of secondary structure during protein folding in vitro has been detected within milliseconds (Matthews 1993), chaperones may be present in polysomes to catch nascent polypeptide chains before folding begins. The segments of nascent chains involved in binding may differ from those in proteins like clathrin light chains, where the presentation of the binding site may involve a conformational change. Enrichment in residues like proline and glycine in the vicinity of the residues directly engaged in binding may be needed to disrupt secondary structures and to facilitate conformational changes in folded proteins but may not be necessary for nascent polypeptides.

Rothman and co-workers have determined the chain-length dependence of the peptide-binding site of Grp78 (Flynn et al. 1991). They found maximum peptide-stimulated ATPase activity at a peptide length of seven residues. Their approach involved allowing Grp78 to bind peptides from a pool of random sequences and then sequencing the mixture of bound peptides. A statistical profile of the amino acid specificity of each position in the binding site on Grp78 was determined based on the degree of enrichment or exclusion (relative to the original pool of random peptides) of each amino acid at a given position in the bound peptide. Aliphatic residues were preferred at all positions but particularly at positions 3–6, and there was a small enrichment for arginine and lysine at position 7. But in general, charged residues were excluded as were polar uncharged amino acids, aromatic residues, and proline. The authors concluded that the peptide-binding site of Grp78 holds aliphatic residues in an environment that is equivalent energetically to the hydrophobic interior of folded proteins. They further concluded that only 1.6 aliphatic residues per span of 7 amino acids is sufficient for productive binding and that a few polar and charged residues per span of 7 amino acids are tolerated in the binding site. This led to the hypothesis that binding sites for hsp70 family proteins are likely to be very common in unfolded proteins.

An advantage to blocking folding of nascent chains until their synthesis is complete is that all of the information needed for folding is made available for use when the completed polypeptide is released. There are two hypotheses regarding the mechanism of release. In one hypothesis, the placement in polypeptides of binding sites for chaperones with different dissociation constants could program the release of different parts of the chain and, in effect, direct the folding process to some extent (Rothman 1989). A second hypothesis accommodates the cooperative nature of protein folding by proposing that all portions of a polypeptide chain involved in folding of a domain are released from their chaperone(s) at once. This would be most easily accomplished if only one chaperone-binding site existed per independently folding domain, many fewer than predicted by Flynn and colleagues. As soon as more than one chaperone-binding site per independently folding domain is proposed, the problem of coordinating the release of all of the chaperones to free the polypeptide for folding arises for the second hypothesis. These two hypotheses embody quite different roles for hsp70 family chaperones. In the first hypothesis, chaperones direct folding by determining the order in which segments are allowed to fold, i.e., chaperones contribute information to the folding process. In the second hypothesis, chaperones delay folding until synthesis of the polypeptide is completed and block nonproductive interactions that lead to aggregation, i.e., chaperones facilitate folding and control when it starts but do not contribute information to the process once folding begins. Future studies that determine the number and placement of binding sites on polypeptides will be needed to evaluate these possibilities. It may emerge that cytoplasmic proteins that fold rapidly have minimal numbers of binding sites. In contrast, multiple binding sites may be useful for molecules whose folding must be delayed longer than that of cytoplasmic proteins, such as those destined for translocation. Here, relatively uncoordinated binding and release of chaperones from multiple binding sites may assure that polypeptides remain "chaperoned" until translocation sites are reached. Alternatively, accessory proteins like eDnaJ and eGrpE could have a major role in the timing and coordination of release.

Currently, very few peptides are known that bind to hsp70 family proteins, too few to reach final conclusions on the relative contributions of the various properties to the selectivity of binding. It is obvious that more systematic methods to identify peptides that bind to hsp70 family proteins are needed, and several such studies are under way using phage display libraries (Scott 1992). One of us (I.M.T., in prep.) has used this approach to select peptides that bind to bovine hsc70. A 15-mer phage display random peptide library was constructed by inserting degenerate oligonucleotides at the amino-terminal coding region of the pIII gene of an fd-tet derivative filamentous phage DNA (Devlin et al. 1990; Scott and Smith 1990). This library was composed of 5.7×10^8 clones and 2.5×10^{11} transducing units per milliliter. About 5×10^9 trasducing units were incubated with bovine hsc70 (>95% purity). Of 100 clones selected for binding specifically to hsc70, 15 different DNA sequences coding for a consensus of two to four basic amino acids, predominantly lysines, were obtained. On the basis of this consensus sequence, 21 different peptides were synthesized and tested in vitro for binding to purified hsc70. The relative affinities of the peptides for hsc70 were estimated using the competition binding assay (Table 1). The value K_i was defined as the concentration of competing peptide that gave 50% inhibition of binding of a radioactively labeled fragment of cytochrome c known to bind hsc70. The K_i values ranged over three orders of magnitude. In accord with the findings of Flynn and co-workers for Grp78, hexamers (Table 1, group A) were 1000-fold weaker binders than the best heptamer, which was the putative consensus peptide NIVRKKK with a K_i of 2.5 μM. Interestingly, hsc70 has recently been implicated in the nuclear import of karyophilic proteins, which have basic nuclear localization sequences such as PKKKRKV of SV40 T antigen. Anti-DDDED antibodies inhibit protein translocation into nuclei in vivo, and these antibodies also bind to hsc70 (Imamoto et al. 1992). One wonders if the binding site for these nuclear localization sequences is one of the negatively charged clusters on α helices of hsc70 predicted in Figure 4A. A less likely but formally possible alternative site might be the EEVD sequence at the carboxyl terminus. Recently, the region on the class I MHC molecule that interacts with the molecular chaperone p88 in the lumen of the endoplasmic reticulum has been identified (Margolese et al. 1993). The major interaction site mapped to the transmembrane spanning segment and three flanking amino acids that in one construct were KRR on the lumenal side. The combination of neutral residues in the membrane spanning segment followed by charged residues suggests the possibility that p88 and hsc70 may have similar binding motifs.

Comparison of the heptamers in group B suggests that the two large aliphatic residues at positions 2 and 3 make a major contribution to binding and that arginine at position 4 is also an important contributor. Since positively charged residues were enriched substantially in the selected peptides, it is important to note that the original library contained arginine and lysine at reasonable frequencies of 8.4% and 2.8%, respectively. To assess the contribution of the positively charged residues systematically, alanine was substituted for arginine and lysine successively in NIVRKKK to generate the group-C peptides. Loss of each positive residue weakened binding, with the loss of arginine at position 4 having the greatest effect. Interestingly, replacement of all of the positive residues resulted in a peptide that still exhibited relatively good binding with a K_i of 68 μM. In contrast, the presence of negatively charged residues (group D) resulted in relatively weak binding. These findings are consistent with those of Gragerov et al. (1994) for peptides that bind DnaK.

III. CONCLUDING REMARKS

The results of these initial studies to characterize the selectivity of peptide binding to hsc70 are in several features similar to those for Grp78: (1) Binding affinities improve dramatically in heptamers compared to hexamers, so the minimum length requirement of the binding site is the same, and (2) there is a requirement for at least two large neutral residues in the peptide core. Several years ago, Pelham (1986) suggested that a likely distinguishing characteristic between native and thermally denatured proteins is the surface display of neutral amino acid residues on denatured proteins and proposed hydrophobic interactions between hsp70 and its protein targets. The idea that the hsp70 chaperones may distinguish unfolded from folded proteins based on the display of aliphatic residues on the surface of unfolded proteins that are normally buried in folded proteins now has some experimental support.

On the basis of the study summarized in Table 1, hsc70 has a preference for one or more positively charged residues, which was also noted in passing in the study by Flynn et al. (1991) for Grp78 and was found for peptides binding to DnaK by Gragerov et al. (1994). We propose a two-step interaction between hsc70 and extended polypeptide chains in which ionic forces provide the initial attraction between a negatively charged cluster on an α helix in or near the peptide-binding site of hsc70 and positively charged residues in target amino acid sequences. Ionic interactions are relatively weak in aqueous environments with bond strengths of 1–3 kcal/mole; however, ionic attraction is omnidirectional and is exerted over a greater distance than any other chemical bond. In the second step, the binding of the peptide triggers a conformational change in the binding domain of hsc70, bringing the neutral residues of the peptide into a hydrophobic environment as envisioned by Rothman and colleagues. The contributions to binding affinity from this second step, i.e., an induced fit, would thus be shorter-range van der Waals attractive forces and hydrogen bonding with the peptide backbone. Evidence of the latter interaction was found in Landry and Gierasch's NMR study. Another interesting possibility raised by a motif that consists of an aliphatic core followed by an ionic segment is that hsc70 might monitor the "edges" of proteins, i.e., the transition segments of polypeptides that go from the interior to the surface of the folded protein. This may be a particularly effective way for hsc70 to function in cells as a "folding cop" since initial interactions with positively charged regions would periodically bring hsc70 into contact with a protein "under surveillance," but detection of exposed, adjacent aliphatic residues would be required to make the "arrest."

ACKNOWLEDGMENTS

We thank our colleagues Al Phillips, Debra Kendall, and Judy Kelly for many helpful conversations about proteins and peptides. A special thanks is due K. Raghavendra for his collaboration on CD spectroscopy. We thank Mary Brown and Carol White for critiques of the manuscript, Sau-Mei Leung and John Ryan for the computer drawing, and Tony Fink along with Sashe Gragerov for providing preprints of their papers. The authors benefited from grants from the National Science Foundation and the National Institutes of Health.

REFERENCES

Ang, D., K. Liberek, D. Skowyra, M. Zylicz, and C. Georgopoulos. 1991. Biological role and regulation of the universally conserved heat shock proteins. *J. Biol. Chem.* **266:** 24233–24236.

Bardwell, J.C.A. and E.A. Craig. 1984. Major heat shock gene of *Drosophila* and the *Escherichia coli* heat inducible dnaK gene are homologous. *Proc. Natl. Acad. Sci.* **79:** 525–529.

Beckman, R.P., M. Lovett, and W.J. Welch. 1992. Examining the function and regulation of hsp-70 in cells subjected to metabolic stress. *J. Cell. Biol.* **117:** 1137–1150.

Beckman, R.P., L.A. Mizzen, and W.J. Welch. 1990. Interaction of hsp70 with newly synthesized proteins: Implications for protein folding and assembly. *Science* **248:** 850–854.

Bjorkman, P.J., M.A. Saper, B. Samraoui, W.S. Bennet, J.L. Strominger, and D.C. Wiley. 1987. Structure of the human class I histocompatibility antigen, HLA-A2. *Nature* **329:** 506–512.

Brown, C.R., R.L. Martin, W.J. Hansen, R.P. Beckmann, and W.J. Welch. 1993. The constitutive and stress inducible forms of hsp-70 exhibit functional similarities and interact with one another in an ATP-dependent fashion. *J. Cell Biol.* **120:** 1101–1112.

Caplan, A.J., D.M. Cyr, and M.G. Douglas. 1993. Enkaryotic homologues of *Escherichia coli* dnaJ: A diverse protein family that functions with hsp70 stress proteins. *Mol. Biol. Cell* **4:** 555–563.

Chappell, T.G., B.B. Konforti, S.L. Schmid, and J.E. Rothman. 1987. The ATPase core of a clathrin uncoating protein. *J. Biol. Chem.* **262:** 746–751.

Chappell, T.G., W.J. Welch, D.M. Schlossman, K.B. Palter, M.J. Schlesinger, and J.E. Rothman. 1986. Uncoating ATPase is a member of the 70 kilodalton family of stress proteins. *Cell* **45:** 3–13.

Chirico, W.J. 1992. Dissociation of complexes between 70 kDa stress proteins and presecretory proteins is facilitated by a cytosolic factor. *Biochem. Biophys. Res. Commun.* **189:** 1150–1156.

Chou, P.Y., and G.D. Fasman. 1974. Prediction of protein conformation. *Biochemistry* **13:** 222–245.

Craig, E.A., B.D. Gambill, and R.J. Nelson. 1993. Heat shock proteins; molecular chaperones of protein biogenesis. *Microbiol. Rev.* **57:** 402–414.

Cyr, D.M., X.Y. Lu, and M.G. Douglas. 1992. Regulation of Hsp70 function by a eukaryotic DnaJ homolog. *J. Biol. Chem.* **267:** 20927–20931.

DeLuca-Flaherty, C., D.B. McKay, P. Parham, and B.L. Hill. 1990. Uncoating protein (hsc70) binds a conformationally labile domain of clathrin light chain LCa to stimulate ATP hydrolysis. *Cell* **62:** 875–887.

Devlin, J.J., L.C. Paganiban, and P.E. Devlin. 1990. Random peptide libraries: A source of specific protein binding molecules. *Science* **249:** 404–406.

Domanico, S.Z., D.C. Denagel, J.N. Dahlseid, J.M. Green, and S.K. Pierce. 1993. Cloning of the gene encoding peptide-binding protein-74 shows that it is a new member of the heat shock protein-70 family. *Mol. Cell. Biol.* **13:** 3598–3610.

Elguindi, S., A.M. Fourie, J.F. Sambrook, and M.J.H. Gething. 1993. Peptide-dependent stimulation of the ATPase activity of the molecular chaperone BiP is the result of conversion of oligomers to active monomers. *J. Biol. Chem.* **268:** 12730–12735.

Flaherty, K.M., C. DeLuca-Flaherty, and D.B. McKay. 1990. Three-dimensional structure of the ATPase fragment of a 70K heat-shock cognate protein. *Nature* **346:** 623–628.

Flajnik, M.F., C. Canel, J. Kramer, and M. Kasahara. 1991. Which came first, MHC class-I or class-II? *Immunogenetics* **33:** 295–300.

Flynn, G.C., T.G. Chappell, and J.E. Rothman. 1989. Peptide binding and release by proteins implicated as catalysts of protein assembly. *Science* **245:** 385–390.

Flynn, G.C., J. Pohl, M.T. Flocco, and J.E. Rothman. 1991. Peptide-binding specificity of the molecular chaperone BiP. *Nature* **353:** 726–730.

Gao, B., Y. Emoto, L. Greene, and E. Eisenberg. 1993. Nucleotide binding properties of bovine brain uncoating ATPase. *J. Biol. Chem.* **268:** 8507–8513.

Gao, B., J. Biosca, E.A. Craig, L.E. Greene, and E. Eisenberg. 1991. Uncoating of coated vesicles by yeast hsp70 proteins. *J. Biol. Chem.* **266:** 19565–19571.

Garnier, J., D.J. Osguthorye, and B. Robson. 1978. Analysis of the accuracy and implications of simple methods for predicting the secondary structure of globular proteins. *J. Mol. Biol.* **120:** 97–120.

Georgopoulos, C. and W.J. Welch. 1993. Role of major heat-shock proteins as molecular chaperones. *Annu. Rev. Cell Biol.* **9:** 601–634.

Gragerov, A., Z. Li, Z. Xun, W. Burkholder, and M.E. Gottesman. 1994. Specificity of DnaK-peptide binding. *J. Mol. Biol.* **235:** (in press).

Hainaut, P. and J. Milner. 1992. Interaction of heat-shock protein 70 with p53 translated in vitro: Evidence for interaction with dimeric p53 and for a role in the regulation of p53 conformation. *EMBO J.* **11:** 3513–3520.

Hattori, H., T. Kaneda, B. Lokeshwar, A. Laszlo, and K. Ohtsuka. 1993. A stress-inducible 40-kDa protein (hsp40)-purification by modified 2-dimensional gel electrophoresis and co-localization with hsc70(p73) in heat-shocked HeLa cells. *J. Cell Sci.* **104:** 629–638.

Hendrick, J.P. and F.U. Hartl. 1993. Molecular chaperone functions of heat-shock proteins. *Annu. Rev. Biochem.* **62:** 349–384.

Imamoto, N., Y. Matsuoka, T. Kurihara, K. Kohno, M. Miyagi, F. Sakiyama, Y. Okada, S. Tsunasawa, and Y. Yoneda. 1992. Antibodies against 70-kD heat shock cognate protein inhibit mediated nuclear import of karyophilic proteins. *J. Cell. Biol.* **119:** 1047–1061.

Kim, D.H., Y.J. Lee, and P.M. Corry. 1992. Constitutive hsp70-oligomerization and its dependence on ATP binding. *J. Cell. Physiol.* **153:** 353–361.

Kyte, J. and R.F. Doolittle. 1982. A simple method for displaying the hydropathic character of a protein. *J. Mol. Biol.* **157:** 105–132.

Lam, K.T. and S.K. Calderwood. 1992. Hsp70 binds specifically to a peptide derived from the highly conserved domain (1) region of p53. *Biochem. Biophys. Res. Commun.* **184:** 167–174.

Landry, S.J., R. Jordan, R. McMacken, and L.M. Gierasch. 1992. Different conformations for the same polypeptide bound to chaperones DnaK and GroEL. *Nature* **355:** 455–457.

Langer, T., C. Lu, H. Echols, J. Flanagan, M.K. Hayer, and F.-U. Hartl. 1992. Successive action of dnaK, dnaJ and GroEL along the pathway of chaperone-mediated protein folding. *Nature* **356:** 683–689.

Liberek, K., J. Marszalek, D. Ang, C. Georgopoulos, and M. Zylicz. 1991a. *Escherichia coli* DnaJ and GrpE heat shock proteins jointly stimulate ATPase activity of DnaK. *Proc. Natl. Acad. Sci.* **88**: 2874–2878.

Liberek, K., D. Skowyra, M. Zylicz, C. Johnson, and C. Georgopoulos. 1991b. The *Escherichia coli* DnaK chaperone, the 70-kDa heat shock protein eukaryotic equivalent, changes conformation upon ATP hydrolysis, thus triggering its dissociation from a bound target protein. *J. Biol. Chem.* **266**: 14491–14496.

Manavalan, P. and W.C. Johnson, Jr. 1987. Variable selection method improves the prediction of protein secondary structure from circular dichroism spectra. *Anal. Biochem.* **167**: 76–85.

Margolese, L., G.L. Waneck, C.K. Suzuki, E. Degen, R.A. Flavell, and D.B. Williams. 1993. Identification of the region on the class I histocompatibility molecule that interacts with the molecular chaperone, p88 (calnexin, IP90). *J. Biol. Chem.* **268**: 17959–17966.

Matthews, C.R. 1993. Pathways of protein folding. *Annu. Rev. Biochem.* **62**: 653–683.

Milarski, K.L. and R.I. Morimoto. 1989. Mutational analysis of the human hsp70 protein: Distinct domains for nucleolar localization and adenosine triphosphate binding. *J. Cell Biol.* **109**: 1947–1962.

Nelson, R.J., T. Ziegelhoffer, C. Nicolet, M. Werner-Washburne, and E.A. Craig. 1992. The translational machinery and 70 kDa heat shock protein cooperate in protein synthesis. *Cell* **71**: 97–105.

O'Malley, K., A. Mauron, J.D. Barchas, and L. Kedes. 1985. Constitutively expressed rat mRNA encoding a 70-kilodalton heat-shock-like protein. *Mol. Cell. Biol.* **5**: 3476–3483.

Palleros, D., W. Welch, and A. Fink. 1991. Interaction of Hsp70 with unfolded proteins: Effects of temperature and nucleotides on the kinetics of binding. *Proc. Natl. Acad. Sci.* **88**: 5719–5723.

Palleros, D.R., K.L. Reid, L. Shi, W.J. Welch, and A.L Fink. 1993. ATP-induced protein-hsp70 complex dissociation requires K+ and does not involve ATP hydrolysis. Analogy to G proteins. *Nature* **365**: 664–666.

Park, K., G.C. Flynn, J.E. Rothman, and G.D. Fasman. 1993. Conformational change of chaperone Hsc70 upon binding to a decapeptide: A circular dichroism study. *Protein Sci.* **2**: 325–330.

Pelham, H.R.B. 1986. Speculations on the functions of the major heat shock and glucose-regulated proteins. *Cell* **46**: 959–961.

Perczel, A., K. Park, and G.D. Fasman. 1992. Analysis of the circular dichroism spectrum of proteins using the convex constraint algorithm: A practical guide. *Anal. Biochem.* **203**: 83–93.

Rippmann, F., W.R. Taylor, J.B. Rothbard, and N.M. Green. 1991. A hypothetical model for the peptide binding domain of hsp70 based on the peptide binding domain of HLA. *EMBO J.* **10**: 1053–1059.

Rothman, J.E. 1989. Polypeptide chain binding proteins: Catalysts of protein folding and related processes in cells. *Cell* **59**: 591–601.

Ryan, C., T.H. Stevens, and M.J. Schlesinger. 1992. Inhibitory effects of hsp70 chaperones on nascent polypeptides. *Protein Sci.* **1**: 980–985.

Sadis, S. and L.E. Hightower. 1992. Unfolded proteins stimulate molecular chaperone Hsc70 ATPase by accelerating ADP/ATP exchange. *Biochemistry* **31**: 9406–9412.

Sadis, S., K. Raghavendra, and L.E. Hightower. 1990a. Secondary structure of the mammalian 70-kilodalton heat shock cognate protein analyzed by circular dichroism spectroscopy and secondary structure prediction. *Biochemistry* **29**: 8199–8206.

Sadis, S., K. Raghavendra, T.M. Schuster, and L.E. Hightower. 1990b. Biochemical and biophysical comparison of bacterial DnaK and mammalian hsc73, two members of a ancient stress protein family. In *Current research in protein chemistry* (ed. J.J. Villafranca), pp. 339–347. Academic Press, New York.

Schlesinger, M. 1986. Heat shock proteins: The search for functions. *J. Cell. Biol.* **103**: 321–325.

Scott, J.K. 1992. Discovering peptide ligands using epitope libraries. *Trends Biochem. Sci.* **17**: 241–245.

Scott, J.K. and G.P. Smith. 1990. Searching for peptide ligands with an epitope library. *Science* **249**: 386–390.

Sheffield, W.P., G.C. Shore, and S.K. Randall. 1990. Mitochondrial precursor protein: Effects of 70-kilodalton heat shock protein on polypeptide folding, aggregation, and import competence. *J. Biol. Chem.* **265**: 11069–11076.

Silver, P.A. and J.C. Way. 1993. Eukaryotic DnaJ homologs and the specificity of hsp70 activity. *Cell* **74**: 5–6.

Slater, M.R. and E.A. Craig. 1989. The SSA1 and SSA2 genes of yeast *Saccharomyces cerevisiae*. *Nucleic Acids Res.* **17**: 805–806.

Takenaka, I.M. and L.E. Hightower. 1992. Transforming growth factor-beta1 rapidly induces Hsp70 and Hsp90 molecular chaperones in cultured chicken embryo cells. *J. Cell. Physiol.* **152**: 568–577.

———. 1993. Regulation of chicken Hsp70 and Hsp90 family gene expression by transforming growth factor-beta-1. *J. Cell. Physiol.* **155**: 54–62.

Tsang, T.C. 1993. New model for 70 kDa heat-shock proteins' potential mechanisms of function. *FEBS Lett.* **323**: 1–3.

Ungewickell, E. 1985. The 70-kd mammalian heat shock proteins are structurally and functionally related to the uncoating protein that releases clathrin triskelion from coated vesicles. *EMBO J.* **4**: 3385–3391.

Vanbuskirk, A., B.L. Crump, E. Margoliash, and S.K. Pierce. 1989. A peptide binding protein having a role in antigen presentation is a member of the hsp70 heat shock family. *J. Exp. Med.* **170**: 1799–1809.

Wang, T.-F., J. Chang, and C. Wang. 1993. Identification of the peptide-binding domain of hsc70. *J. Biol. Chem.* **268**: 26049–26051.

Welch, W.J. and J.P. Suhan. 1986. Cellular and biochemical events in mammalian cells during and after recovery from physiological stress. *J. Cell Biol.* **103**: 2035–2052.

White, C.N., L.E. Hightower, and R.J. Schultz. 1994. Variation in heat shock proteins among species of desert fishes (Poeciliidae, *Poeciliopsis*). *Mol. Biol. Evol.* **11**: 106–119.

Wiech, H., J. Buchner, M. Zimmerman, R. Zimmerman, and U. Jakob. 1993. Hsc70, immunoglobulin heavy chain binding protein, and Hsp90 differ in their ability to stimulate transport of precursor proteins into mammalian microsomes. *J. Biol. Chem.* **268**: 7414–7421.

Wooden, S.K. and A.S. Lee. 1992. Comparison of the genomic organizations of the rat grp78 and hsc73 gene and their evolutionary implications. *DNA Sequence* **3**: 41–48.

Yang, J.T., C.-S.C. Wu, and H.M. Martinez. 1986. Calculation of protein conformation from circular dichroism. *Methods Enzymol.* **130**: 208–269.

Zafarullah, M., J. Wisniewski, N.W. Shworak, S. Schieman, S. Misra, and L. Gedamu. 1992. Molecular cloning and characterization of a constitutively expressed heat shock cognate hsc71 gene from rainbow trout. *Eur. J. Biochem.* **204**: 893–900.

9

Properties of the Heat Shock Proteins of *Escherichia coli* and the Autoregulation of the Heat Shock Response

Costa Georgopoulos,[1] Krzysztof Liberek,[1,2] Maciej Zylicz,[2] and Debbie Ang[1]
[1]Département de Biochimie Médicale
Centre Médical Universitaire
University of Geneva
1211 Geneva 4, Switzerland
[2]Department of Molecular Biology
University of Gdansk
80-822 Kladki, Gdansk, Poland

The Biology of Heat Shock Proteins and Molecular Chaperones
©1994 Cold Spring Harbor Laboratory Press 0-87969-427-0/94 $5 + .00

I.   INTRODUCTION

The studies of bacteriophage λ/*Escherichia coli* interactions, initiated more than 25 years ago, have resulted in the identification of a plethora of interesting and useful host functions (for summary, see Friedman et al. 1984; Georgopoulos et al. 1990). These bacterial functions are essential not only for bacteriophage λ growth and development, but for bacterial growth as well. The ability to study both λ-specific and host-specific phenotypes at nonpermissive temperatures was instrumental in the elucidation of the function of these bacterial genes. In many instances, these genetic studies identified a series of *E. coli* genes that are required for a specific step in λ development, strongly suggesting that their gene products act synergistically to carry out the particular λ developmental step. The availability of these genes on plasmids plus the large number of mutations in them was the starting block for the elucidation of the molecular mechanism of the action of their gene products.

The best studied of these *E. coli* functions, namely, the NusA/NusB transcriptional antitermination system and the DnaK and GroEL chaperones, all turned out to be supramolecular machines made up of intimately interacting parts (Friedman et al. 1984; Das 1992; Georgopoulos 1992; Georgopoulos and Welch 1993). For the purpose of this particular volume, the most interesting *E. coli* functions are the DnaK chaperone machine, composed of the DnaK (the hsp70 homolog), DnaJ (the hsp40 homolog), and GrpE heat shock proteins, and the GroEL chaperone machine, composed of the GroES (the hsp10 homolog) and GroEL (the hsp60 homolog) proteins. The DnaK chaperone machine was originally shown to be indispensable for λ DNA replication, its main function being to liberate the DnaB helicase from its complex with the λP protein (for review, see Georgopoulos et al. 1990). The GroEL chaperone machine was originally shown to be indispensable for the morphogenesis of the heads of bacteriophages λ and T4 and the morphogenesis of the bacteriophage T5 tail (for review, see Georgopoulos et al. 1990).

Eukaryotic counterparts for all five heat shock protein members have been identified in most organisms, underscoring the universal conservation of both the function and the structure of these proteins. Biochemical studies with purified proteins have clearly shown that although each individual chaperone machine may function slightly differently, to a large extent, the information derived from studies of one organism's chaperone machines finds immediate application to those of other organisms.

II.   THE HTPG PROTEIN, THE HSP90 HOMOLOG

The work of Bardwell and Craig (1987) resulted in the identification of the *htpG* gene of *E. coli* as the hsp90 eukaryotic homolog, and whose deletion did not affect bacterial growth except at extremely high temperatures (Bardwell and Craig 1988). Perhaps its function is dispensable because other *E. coli* chaperones can normally substitute for it, except at very high temperatures when chaperone function could be limiting. Some support for such a possible chaperone function for HtpG comes from the work of Ueguchi and Ito (1992). These workers selected for genes that, when cloned on multicopy plasmids, can restore bacterial growth as well as suppress export defects caused by temperature-sensitive mutations in the *secY* or *secA* genes. With this selection, they identified the *groE* operon as a multicopy suppressor of the *secA51* mutation and the *htpG* gene as a multicopy suppressor of the *secY24* mutation. The overproduction of HtpG was shown to accelerate markedly the processing (export) of the outer membrane OmpA porin protein. Whether HtpG acts like a true chaperone by maintaining preOmpA in a translocation-competent form or by "correcting" the conformation of the mutant SecY24 protein is not known.

III. THE DNAK CHAPERONE MACHINE

The various details of the biochemical properties of the DnaK chaperone machine in polypeptide binding and release and the existence of various conformational domains and states of DnaK are high-lighted by Frydman and Hartl, Craig et al., Langer and Neupert, High-tower et al., and McKay et al. (all this volume). Here, we emphasize those functions that are of special interest and are not covered in detail elsewhere.

One of the reasons why the DnaK, DnaJ, and GrpE proteins are referred to as a "machine" is that they work together to carry out a variety of biochemical functions. Figure 1 depicts the requirements for DnaK in the "resurrection" of the activity of heat-inactivated RNA polymerase (RNAP). In the absence of the DnaJ and GrpE cohorts, more than 100 molecules of DnaK chaperone protein are needed to resurrect one molecule of aggregated RNAP, whereas in the presence of DnaJ or DnaJ/ GrpE, progressively fewer molecules of DnaK are needed (Ziemieno-wicz et al. 1993). This experiment exemplifies the synergistic functioning of the three heat shock proteins in protein disaggregation.

Liberek et al. (1991a) showed that the weak ATPase of DnaK can be stimulated 50-fold in the joint presence of DnaJ and GrpE. The DnaJ protein specifically accelerates the rate of hydrolysis of DnaK-bound ATP, whereas GrpE induces the release of all DnaK-bound nucleotides (Fig. 2). As a consequence of accelerating the ATPase activity of DnaK, the presence of DnaJ would favor the DnaK-ADP form, thought to be the one that binds substrate best (see Frydman and Hartl, this volume). In

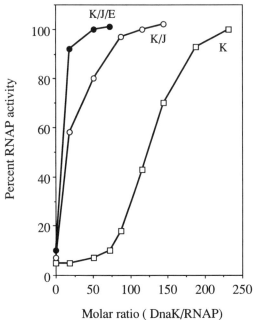

DnaK-dependent reactivation of
heat-inactivated RNAP

Figure 1 DnaK, DnaJ, and GrpE proteins function synergistically. The influence of DnaJ and GrpE on the reactivation of heat-inactivated RNAP by
DnaK and ATP (4 mM) is shown. The final concentrations of RNAP, DnaJ, and
GrpE in the reactivation assay were 0.12, 0.12, and 2 μM, respectively. (*Open
box*) DnaK alone; (*open circle*) DnaK and DnaJ; (*closed circle*) DnaK, DnaJ,
and GrpE. For further details, see the text and Ziemienowicz et al. (1993).
(Reprinted, with permission, from Ziemienowicz et al. 1993.)

this respect, GrpE should play the opposite role, since it causes the
release of ADP, and thus may weaken the DnaK-substrate interaction.

There is little doubt that the DnaJ and GrpE homologs in eukaryotes
will act in an analogous manner with the appropriate hsp70 homologs to
assist in substrate binding and release, although each particular system
may exhibit its own peculiarities. For example, the *Saccharomyces
cerevisiae* Ydj1 homolog interacts with the Ssa1 homolog to activate
substantially its ATPase (tenfold) and simultaneously cause the release
of an Ssa1-bound substrate polypeptide (Cyr et al. 1992). In this system,
DnaJ of *E. coli* can substitute in vivo for the missing Ydj1 function, although the details of its action in a purified system have not yet been
studied (Caplan et al. 1992). In contrast to this promiscuity of interaction
of the bacterially encoded DnaJ protein with a yeast-encoded hsp70

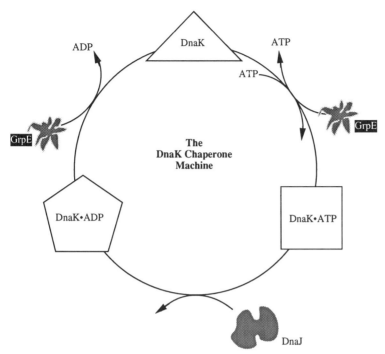

Figure 2 Molecular mechanism of action of the DnaJ and GrpE cohorts in the ATPase of DnaK. The DnaJ protein specifically catalyzes the hydrolysis of DnaK-bound ATP, thus playing a role equivalent to that of a GTPase-activating protein (GAP) with Ras. The GrpE protein catalyzes the release of either ADP or ATP bound to DnaK, a role resembling that of a guanine nucleotide release protein (GNRP) with Ras (Boguski and McCormick 1993).

homolog, Brodsky et al. (1993) (see also Brodsky and Shekman, this volume) have emphasized that the various hsp70 proteins may perform specialized functions perhaps because they interact effectively with only certain DnaJ homologs. No evidence for the involvement of a GrpE-like protein in Ssa1/Ydj1 function has been collected as yet.

In some instances, individual members of the DnaK chaperone machine are seen to function independently of the other members or to require the assistance of only one of the other two members. These individual examples will be taken up separately in the discussion of each gene.

A. The Biology of the *dnaK* Gene

The *dnaK* gene of *E. coli* can be deleted only in certain genetic backgrounds, and even then, *dnaK* null mutants grow extremely slowly, accumulating extragenic suppressors readily (Bukau and Walker 1990; D.

Ang and C. Georgopoulos, unpubl.). However, the *dnaK* gene can be deleted in all *E. coli* genetic backgrounds tested provided the bacteria are maintained in minimal media (D. Ang and C. Georgopoulos, unpubl.). This is also true for four out of five missense *dnaK* mutations tested, i.e., only one, *dnaK756*, can be transduced freely in bacteria grown on rich (LB) media (C. Georgopoulos, unpubl.). When these *dnaK* mutants are transferred from minimal media to LB media, they filament extensively and die (survivors can be obtained readily at a frequency of 10^{-4}). This behavior is reminiscent of a *secB* null mutant, the *secB* gene normally coding for a major chaperone that specializes in the protein export pathway (see Randall et al., this volume). Interestingly, Wild et al. (1992) showed that overproduction of both DnaK and DnaJ allows *secB* null mutant bacteria to grow on LB media and corrects some of the export defects associated with a lack of SecB as well. This is a perfect example of chaperone machines substituting for one another, provided they are in excess, in the export of some polypeptides. It is not known whether overproduction of SecB allows *dnaK* null mutant bacteria to grow on LB media.

Work from a variety of laboratories had led to the conclusion that ATP hydrolysis affects the conformation of DnaK and that this correlates with the release of bound polypeptide substrates. This conclusion was based mostly on the fact that nonhydrolyzable ATP analogs, such as ATPγS, do not release the bound polypeptide. The recent studies of Palleros et al. (1993) and Schmid et al. (1994) showed that this conclusion is not neccessarily correct and added further insights to this seemingly complex process. Palleros et al. (1993) showed that binding of ATP (but not binding of nonhydrolyzable ATP analogs) was sufficient to result in the rapid dissociation of a DnaK-bound polypeptide substrate in the presence of K^+ ions. The fact that the DnaK T199A missense mutant, which is crippled in both the ATPase and autophosphorylating activities (McCarthy and Walker 1991), releases substrate upon ATP binding argues in favor of such a conclusion (Palleros et al. 1993).

Schmid et al. (1994) used a substrate peptide tagged with a fluorophore moiety to probe in detail the kinetics of DnaK-substrate binding and release. Their results indicated the following: (1) Although the overall affinity of DnaK for the peptide ligand is substantially reduced in the presence of ATP, the means by which this end result is achieved was completely unexpected. Apparently, ATP causes a large acceleration in the rate of DnaK-peptide formation, but it causes an even greater acceleration in the rate of DnaK-peptide dissociation. (2) Nonhydrolyzable ATP analogs have no effect on the overall affinity for peptide ligand, although they substantially accelerate both the binding and release reaction. It is possible that these differences reflect a variation in the "quality" of ATP binding to DnaK, as opposed to binding by nonhydrolyzable analogs. (3) ADP has no effect on the peptide binding and release reaction. It appears then that the binding and release of this particular substrate by DnaK are much faster in the presence of ATP, occurring in approximately the same time range as polypeptide chain elongation.

All of these results highlight the complexity of the process of substrate binding and release by DnaK and question the validity of attempts to generalize from specific examples. Most likely, each case of substrate binding must be treated individually. For example, whereas the DnaK-λP or DnaK-σ^{32} complexes contain bound nucleotide, the DnaK-σ^{32}-DnaJ complex does not (K. Liberek and D. Wall, unpubl.). The quality of the substrate-DnaK complex may affect the ability of GrpE to influence release of the substrate; e.g., GrpE accelerates the release of rhodanese from the DnaK-rhodanese-DnaJ complex (Langer et al. 1992b), but neither the formation nor the apparent stability of the DnaK-σ^{32}-DnaJ complex is compromised by the presence of GrpE (Liberek and Georgopoulos 1993). It could be that the DnaK-rhodanese-DnaJ complex is stabilized by ADP and that GrpE destabilizes it by releasing the bound ADP (see Frydman and Hartl, this volume), whereas the DnaK-σ^{32}-DnaJ complex is qualitatively different since it does not contain ADP and hence perhaps cannot be destabilized by GrpE. All of these substrate-binding modes may have evolved to augment DnaK's binding capacity range and may reflect a differential role in its ability to promote the folding or degradation of the bound polypeptide substrate. A recent article by T.-F. Wang et al. (1993) showed that an 18-kD polypeptide, spanning amino acid residues 384–543 of hsc70, constitutes the peptide-binding domain. The peptide-binding properties of this 18-kD polypeptide are indistinguishable from those of native hsc70, suggesting that it may contain the binding site in its entirety.

Another question often asked is the biological role of the observed calcium-dependent autophosphorylation of the DnaK protein at T199 (equivalent to T204 in hsc70). This is discussed in some detail by McKay et al. (this volume). The conclusion reached, following mutagenesis and enzyme kinetic studies, is that the observed autophosphorylation reaction is most likely a nonproductive side reaction. To complicate the situation, Sherman and Goldberg (1993) recently reported that heat shock increases the binding of DnaK to some of its polypeptide substrates by promoting its phosphorylation at an as yet undetermined site, but different from T199. The biological importance of this phosphorylation reaction is clouded by its limited molecular analysis and the small percentage of DnaK molecules that it affects.

One important function of the DnaK chaperone, as pointed out by the work of Hupp et al. (1992), is to change the conformation of seemingly folded polypeptides. The case reported by Hupp and colleagues is that of the p53 tumor suppressor protein which, when purified from *E. coli*, does not bind to its proper DNA target. Incubation with DnaK and ATP somehow "massages" p53 into a form that can readily bind to its target DNA sequence. This "activation" by DnaK and ATP also has been observed with some purified mutant forms of the p53 protein (Hupp et al. 1993). The p53 example is reminiscent of the activation of the F factor RepE protein by the DnaJ chaperone, allowing RepE to bind to its own target sequence (Kawasaki et al. 1992). These results suggest that chaperones not only prevent aggregation of polypeptides and assist their folding, but also modulate the conformation of certain proteins even after they have assumed a seemingly folded state.

B. The Biology of the *dnaJ* Gene

The *dnaJ* gene in bacteria is usually located immediately downstream from the *dnaK* gene and is always cotranscribed with it under heat shock regulation. The *dnaJ* gene is not absolutely essential for *E. coli* viability, since it can be deleted in all genetic backgrounds tested at temperatures up to 42–43°C. However, the *dnaJ*-deleted strains grow slower at all temperatures and readily accumulate extragenic suppressors (Sell et al. 1990). At temperatures above 43°C, the *dnaJ* null bacteria do not form colonies and actually die. Perhaps what is most surprising about *dnaJ*, given the pivotal role that its product has in the biology of the DnaK chaperone machine and the regulation of the heat shock response, is that it can be deleted at all. It is possible that *E. coli* possesses another DnaJ-like function that can substitute for DnaJ to some extent. Alternatively, it could be that the need for DnaJ protein becomes particularly acute at temperatures above 43°C, when many polypeptide chains may begin to unfold and aggregate.

During the last 2 years, the DnaJ protein has emerged as a chaperone in its own right. It binds tightly to some denatured proteins such as rhodanese and luciferase and in so doing prevents their aggregation (Langer et al. 1992b; Schröder et al. 1993). In addition, it has been shown recently that DnaJ can bind to a nascent polypeptide emerging from the ribosome and thus prevent its folding (Hendrick et al. 1993). It has been proposed that the DnaK chaperone binds to this DnaJ-substrate complex and further stabilizes it. The addition of GrpE causes the release of the substrate (see Frydman and Hartl, this volume). The DnaJ protein will also bind to some seemingly folded polypeptides, such as the σ^{32}

and bacteriophage P1 RepA proteins, thus facilitating the binding of DnaK to them (Wickner et al. 1991; Gamer et al. 1992; Liberek and Georgopoulos 1993). Finally, DnaJ has been shown to activate the F factor RepE protein for binding to its target sequence, whereas DnaK and GrpE have no effect on this reaction (Kawasaki et al. 1992).

Although such solitary actions of DnaJ exist and may be biologically relevant, its most important function is to act synergistically with DnaK to (1) suppress polypeptide aggregation, thus promoting protein folding (Langer et al. 1992b), (2) "resurrect" the activity of some mildly aggregated enzymes (Ziemienowicz et al. 1993; Schröder et al. 1993), and (3) negatively autoregulate the σ^{32}-directed heat shock response (Gamer et al. 1992; Liberek and Georgopoulos 1993; see below).

The DnaJ protein has been shown to possess several highly conserved features (Fig. 3), including a "J" domain (see below), a G-rich domain that usually follows the J domain, putative zinc-binding domains, and a less well-conserved carboxy-terminal segment that may be involved in polypeptide substrate binding.

The 70 or so amino-terminal residues of DnaJ of *E. coli* constitute the most highly conserved domain of the protein. This domain is used as a "fingerprint" to identify all DnaJ family members (for review, see Bork et al. 1992; Silver and Way 1993; Caplan et al. 1993). There is ample evidence that this domain is of paramount importance to the function of DnaJ and is specifically used to interact with the DnaK protein. Some of the evidence for this conclusion derives from the following studies: (1) The "classical"*dnaJ*259 mutation was sequenced and shown to result in a H33Q change, located in the highly conserved HPD tripeptide sequence (Fig. 2) (Wall et al. 1994). The mutant DnaJ259 protein was purified and shown to be incapable of interacting with DnaK, as judged from the fact that it does not accelerate the hydrolysis of DnaK-bound ATP (Wall et al. 1994). Although DnaJ259 can form a complex with σ^{32}, it is incapable of supporting the formation of a DnaJ-σ^{32}-DnaK complex in the presence of ATP (K. Liberek and D. Wall, unpubl.). These results suggest that DnaJ259 can bind the σ^{32} substrate, but it is uniquely defective in its ability to interact with DnaK. (2) The complementary situation was seen with DnaJ12, a truncated version of DnaJ that comprises the first 108 amino-terminal amino acids. DnaJ12 can interact with DnaK in two ways. First, it accelerates the hydrolysis of DnaK-bound ATP, almost to the same extent as full-length DnaJ (Wall et al. 1994). Second, although DnaJ12 does not itself complex with σ^{32}, it is capable of stimulating the formation of a DnaK-σ^{32} complex in the presence of ATP. This ability of DnaJ12 to stimulate the formation of a DnaK-substrate complex is selective, inasmuch as it stimulates the formation of the DnaK-σ^{32} complex

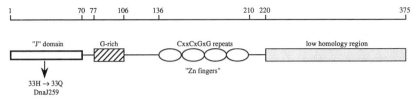

Figure 3 Structure/function of the DnaJ protein and its relationship to other family members. The 375-amino-acid-residue DnaJ protein of *E. coli* is organized into four distinct domains conserved among the family members. The approximately 70 amino-terminal amino acid residue region makes up the "J" domain, the only one conserved in all family members. A mutation in this domain (e.g., DnaJ259) specifically interferes with interaction with DnaK but not with substrate binding (Wall et al. 1994). A G-rich domain follows, whose inframe deletion drastically affects J protein activity (D. Wall, pers. comm.). The four $CxxCxGxG(x)_{6-14}$ repeats are conserved in some of the family members. They are called "Zn fingers" because purified DnaJ protein contains two molecules of zinc (K. Liberek, unpubl.). Deletion of this region results in a DnaJ protein with a temperature-sensitive phenotype for bacteriophage λ growth (D. Wall, pers. comm.). The carboxy-terminal part of DnaJ is the least conserved among family members and is most likely involved in substrate binding. The dimerization domain of DnaJ is not known.

(which is devoid of any bound nucleotide) but not the DnaK-λP complex in the presence of ATP (K. Liberek and D. Wall, unpubl.). Thus, it appears that DnaK has at least two binding modes for substrate polypeptides and that DnaJ can influence the binding to substrate in one mode but not in the other. In some respects, DnaK is "transformed" by DnaJ or DnaJ12 in the presence of ATP into a new conformation capable of binding tightly only to a selective group of substrates such as σ^{32}. Clearly, DnaK is a multifaceted protein, the complete understanding of whose structure/function, interaction with the DnaJ/GrpE cohorts, and various substrates will keep molecular biologists busy for a while longer.

The DnaJ12 peptide has been shown to perform another function exhibited by the full-length protein, namely, the regulation of the heat shock response. In a *dnaJ* deletion mutant background, the heat shock response is overexpressed as expected from previous studies (Straus et al. 1990; Sell et al. 1990). The overproduction of heat shock proteins is substantially down-regulated in these mutants if the DnaJ12 protein is produced from a plasmid (D. Wall, pers. comm.). This result is interesting because DnaJ12 itself cannot bind to σ^{32} and can only exert its effects through a modulation of DnaK's substrate-binding properties and its ATPase activity. It also suggests that DnaK's binding to σ^{32}, in a mode dictated by DnaJ12, is sufficient to regulate the heat shock response properly. Perhaps *E. coli* regulates the heat shock response by utilizing more than one mechanism, thus ensuring the proper level of heat shock response appropriate for each occasion.

C. The Biology of the *grpE* Gene

The *grpE* gene of *E. coli* is monocistronic, under the control of both $E\sigma^{70}$ and $E\sigma^{32}$, and is located at 56 minutes (for review, see Georgopoulos et al. 1990). The classical *grpE*280 mutation was isolated on the basis of blocking bacteriophage λ growth at 42°C and subsequently shown to confer a temperature-sensitive phenotype for bacterial growth (for review, see Georgopoulos et al. 1990). The mutation was sequenced and shown to result in an amino acid substitution at position 122 (L. Baird, unpubl.). The GrpE280 mutant protein does not interact with DnaK as judged from the lack of formation of a stable DnaK-GrpE280 complex (Ang 1988; Johnson et al. 1989) and failure to accelerate the ATPase of DnaK (Liberek et al. 1991a). The GrpE280 protein, however, is not totally devoid of enzymatic activity since its overproduction leads to suppression of the bacterial temperature-sensitive phenotype and ability to propagate bacteriophage λ (Ang 1988). The same *grpE*280 mutation was re-isolated following a massive hunt for other *grpE* mutations that also block bacteriophage λ growth. In addition, mutations that result in the alteration of amino acids at positions 53, 102, 127, and 177 were identified (B. Wu, pers. comm.). Some of these mutations cause a temperature-sensitive phenotype for bacteriophage λ, allowing growth at 30°C but not at 42°C. The biochemical analysis of all these new GrpE mutant proteins will help to elucidate the biology of this interesting protein.

As for its biological role, there is no evidence that the GrpE protein possesses any function other than to simply assist DnaK to carry out its own biological function effectively. This auxiliary role is critical to DnaK's chaperone biology such that the *grpE* gene cannot be deleted under all conditions in all wild-type genetic backgrounds tested (Ang and Georgopoulos 1989). However, *grpE* can be deleted in *E. coli* hosts that were previously adapted to the loss of the *dnaK* gene through the accumulation of extragenic suppressors (Ang and Georgopoulos 1989).

The only protein that GrpE is known to interact with is DnaK, and the resulting DnaK-GrpE complex is resistant to 2 M KCl and is disrupted in the presence of either ATP or the nonhydrolyzable analog, ATPγS (Zylicz et al. 1987; A. Wawrzynow, pers. comm.). These results show that the DnaK/GrpE interaction can be disrupted differently from the DnaK/substrate interaction (Liberek et al. 1991b; Palleros et al. 1993).

The GrpE protein causes the release of nucleotide bound to DnaK, thus in essence acting like an ADP/ATP exchange factor. The binding site for GrpE on DnaK has been suggested to be around amino acid 32 by B. Bukau's laboratory. One of the three mutations associated with the classical *dnaK756* mutation maps at position 32 (Miyazaki et al. 1992), causing a G→D substitution, and has been shown to interfere with the binding of GrpE to DnaK756. In addition, the deletion of amino acid residues 28–33 in the wild-type DnaK sequence does not interfere with its ATPase- or substrate-binding properties, but rather specifically interferes with binding of GrpE (B. Bukau, pers. comm.). This putative GrpE-binding region of DnaK is predicted to be in the immediate vicinity of the ATP-binding site, which helps to explain the release of nucleotide upon GrpE binding.

GrpE also acts as a "discrimination" factor to target DnaK to some of its substrates. For example, although GrpE causes the release of λP from its complex with DnaK (Liberek et al. 1990), it stabilizes this complex in the presence of DnaJ (Osipiuk et al. 1993). Thus, GrpE may exert two major effects on DnaK: (1) It causes the release of bound nucleotide and thus may destabilize the binding of some substrates and (2) it causes an additional conformational change that may further destabilize or stabilize the DnaK-substrate complex, depending on the nature of the complex and/or the presence of DnaJ. GrpE does not act as a competitor for substrate binding, since a λP-DnaK-GrpE complex can be readily isolated following cross-linking with glutaraldehyde (Osipiuk et al. 1993).

Another biological situation where DnaK-GrpE is seen to act independently of DnaJ is the conversion of the "inactive" form of the DnaA5 mutant protein to an "active" form. This conversion cannot be carried out by DnaK alone and requires the presence of GrpE (Hupp and Kaguni 1993). This "massage" of a protein's structure is not facilitated by the presence of DnaJ and is in fact inhibited by it (J. Kaguni, pers. comm.). Skowyra and Wickner (1993) have also found conditions under which GrpE is needed to assist the DnaK-DnaJ complex to activate the DNA-binding properties of the P1 RepA protein. Under these conditions (1 mM Mg^{++}), the DnaK and DnaJ proteins, in the absence of GrpE, are completely unable to activate RepA by themselves.

To understand the biological role of GrpE, *dnaK* mutants were selected that have a GrpE-independent phenotype. Two missense *dnaK* mutants, *dnaK325* and *dnaK332*, were isolated by plating *E. coli grpE280* bacteria at 42°C and selecting for survivors (A. Maddock and D. Ang, unpubl.). Subsequent genetic analyses indicated that (1) bacteria carrying either the *dnaK325* or *dnaK332* allele cannot tolerate the presence of high levels of wild-type GrpE protein and (2) the *dnaK325* mutation allows the deletion of the *grpE* gene at temperatures up to 42°C, whereas the *dnaK332* mutation allows the deletion only up to 30°C. The *dnaK332* mutation was sequenced and shown to result in a substitution at amino acid position 71 (A. Maddock, pers. comm.). The mutant DnaK332 protein was purified and shown to be capable of performing some of its biological roles in the absence of GrpE, including the in vitro replication of bacteriophage λ DNA and the spontaneous release of bound ADP, a reaction usually mediated by GrpE (A. Maddock, B. Banecki, and A. Wawrzynow, pers. comm.). The *dnaK325* mutation was also sequenced and shown to result in a substitution at amino acid position 10. Despite the fact that the amino acids mutated in *dnaK332* and *dnaK325* are separated by 61 amino acid residues in the primary sequence of DnaK, they are within 8 Å of each other in the native protein, assuming that the hsc70 crystal structure reflects that of DnaK as well (see McKay et al., this volume). Amino acid residue G10 of DnaK corresponds to G12 of hsc70 and R71 corresponds to R72 of hsc70. The neighboring residues of hsc70, D10 and K71, are thought to be directly involved in MgATP binding (see McKay et al., this volume). Hence, perhaps it is not surprising that the DnaK332 mutant protein spontaneously releases ADP.

In a complementary study, it was also found that expression of carboxy-terminal truncations of the *dnaK* gene, when present on a multicopy vector, can also bypass the requirement for GrpE provided they do not exceed amino acid position 490 (A. Maddock, pers. comm.). This observation suggests that the peptide-binding site of DnaK may be closer to the ATPase site than previously suspected. Interestingly, all carboxy-terminally deleted DnaK proteins exhibit a much higher ATPase activity than wild type (A. Maddock and B. Banecki, pers. comm.). Perhaps the extreme carboxy-terminal region of DnaK can somehow interact with the ATPase domain of the protein to down-regulate its activity. In its absence, the ATPase activity of DnaK becomes deregulated, thus bypassing the requirement for GrpE.

D. Yeast GrpE and DnaJ Mitochondrial Homologs

Recently, the *S. cerevisiae* GrpE homolog was identified in mitochondria and its corresponding gene was cloned (Bolliger et al. 1994). The GrpE homolog was purified on the basis of its ability to bind to an Ssc1 (the mitochondrial hsp70 homolog)-affinity column and be released with ATP, the same scheme previously used to purify the *E. coli* GrpE protein (Zylicz et al. 1987). The corresponding gene was identified in parallel studies by hybridization and subsequently cloned and sequenced. The predicted amino acid sequence is approximately 30% identical to *E. coli*'s GrpE and contains all partial amino acid sequences obtained with the purified protein, indicating that the appropriate gene was cloned. The yeast GrpE protein contains a mitochondrial target sequence that is cleaved upon entry into mitochondria. Yeast that carry a disrupted grpe gene are inviable, a phenotype similar to that seen with the *grpE* gene of *E. coli*, demonstrating its importance in Ssc1 protein function. The availability of abundant amounts of both Ssc1 and yeast GrpE proteins should help to decipher the exact role that GrpE has in Ssc1 biology (perhaps as an ADP/ATP exchange factor in analogy with GrpE of *E. coli*). The laboratory of E.A. Craig has independently carried out such genetic studies (E.A. Craig, pers. comm.).

A DnaJ homolog that resides in yeast mitochondria and functions in the folding of newly imported and preexisting proteins has also been recently identified and studied (W. Neupert, pers. comm.; see Langer and Neupert, this volume).

IV. THE GROEL CHAPERONE MACHINE

The original genetic studies that led to the identification of the *groES groEL* operon have been reviewed in detail in previous publications (Georgopoulos et al. 1990; Zeilstra-Ryalls et al. 1991; Georgopoulos and Welch 1993). The biochemical properties of the GroEL chaperone machine, especially those necessary in protein folding and assembly, are reviewed in detail by Frydman and Hartl (this volume). Here, we concentrate on supplementary biological aspects of this machine that are not dealt with in detail elsewhere in this volume.

A. The GroEL Chaperone Machine Can Also Disaggregate Some Protein Aggregates

Previous work had shown that the DnaK chaperone alone can protect *E. coli* RNA polymerase (RNAP) from heat inactivation. In addition, DnaK and ATP are capable of disaggregating aggregates of RNAP formed upon heat treatment (Skowyra et al. 1990). Since then, it has been shown that the GroEL machine is also capable of protecting various polypeptides from thermal aggregation and inactivation (for review, see Frydman and Hartl, this volume). In an effort to compare the chaperone powers of the DnaK and GroEL machines, we examined in detail their ability to protect RNAP from heat inactivation, as well as to "resurrect" heat-inactivated, aggregated RNAP. We found that the GroEL protein alone can protect RNAP from heat inactivation when present in a 10:1 molar ratio of native GroEL to RNAP (Ziemienowicz et al. 1993). To obtain equivalent protection, a ratio of 140:1 DnaK to RNAP is required (Skowyra et al. 1990). The presence of GroES further improves protection of RNAP by wild-type GroEL but not by the GroEL673 mutant protein. The inability of wild-type GroES to enhance protection of RNAP by the GroEL673 mutant suggests that the two GroE proteins do not functionally interact (Ziemienowicz et al. 1993).

The GroEL chaperone machine was also tested for its ability to "resurrect" RNAP activity from heat-inactivated aggregates. It was found that the GroEL chaperone machine, using energy derived from ATP hydrolysis, reactivates RNAP by dissolving the large protein aggregates formed during incubation at high temperature. The GroEL chaperone machine is very efficient in this "resurrection" reaction, requiring only ten chaperone machine molecules per heat-treated RNAP holoenzyme molecule. Resurrection of RNAP activity is largely accelerated when heat-inactivated RNAP is preincubated for 10 minutes at 30°C with only GroEL, followed by the addition of GroES and ATP. This result suggests that the presence of GroES may interfere with the efficient binding of GroEL to the heat-inactivated RNAP. The mutant GroEL673 protein shows no ability to reactivate heat-inactivated RNAP. This result correlates with the above mentioned observation that addition of wild-type GroES does not improve the ability of GroEL673 to protect RNAP from heat inactivation. These in vitro findings with purified proteins are in complete agreement with the in vivo findings of Gragerov et al. (1992), who showed that the overproduction of either the GroEL/GroES or DnaK/DnaJ systems can prevent wholesale protein aggregation.

It is likely that the demonstrated overlap in function of the DnaK and GroEL chaperone machines ensures that important molecules, such as RNAP, will stay functional under adverse physiological conditions, such as those encountered under heat stress, thus increasing the probability of bacterial survival. In this context, Sherman and Goldberg (1992b) have reported that the GroEL protein becomes phosphorylated following heat shock and that this phosphorylation enables GroEL to release its substrate without the assistance of GroES. The importance of this finding is obscured by the relatively minor amount of GroEL that purportedly becomes phosphorylated and the limited biochemical analysis of this modification.

B. Possible Sites of GroES/GroEL Interaction

It has been known for some time that the GroES and GroEL proteins physically interact and that this interaction is important in the mediation of protein folding (for review, see Georgopoulos et al. 1990; Zeilstra et al. 1991; Georgopoulos and Welch 1993). A possible site of GroES has been recently identified that is an excellent candidate for promoting interaction with GroEL (Landry et al. 1993). The one-dimensional nuclear magnetic resonance (NMR) spectrum of purified GroES indicated the presence of a region that is substantially more mobile than the rest of the molecule. Two-dimensional NMR analysis pinpointed this to the amino acid 17–32 region of GroES (referred to as the "mobile loop" of GroES). Additional support for the existence of this unstructured mobile loop came from the fact that partial trypsin proteolysis of GroES results in its preferential cleavage. However, both susceptibility to trypsin and the unstructured state of the mobile loop are suppressed when GroES is complexed with GroEL.

Among the 17 *groE* mutant strains originally isolated by their inability to support the growth of bacteriophage λ, 8 were assigned to the *groES* group (Georgopoulos et al. 1990; Zeilstra-Ryalls et al. 1991). All 8 independently isolated *groES* mutations were sequenced and assigned to this mobile loop region (6/8 result in a G24D change, 1 results in a G23D change, and 1 results in an A31V change). All altered amino acids represent highly conserved positions. Consistent with the idea that the mobile loop is responsible for binding to GroEL, a 20-residue peptide spanning amino acids 13–32 was synthesized and shown to be capable of binding to GroEL. In contrast, a mutant peptide that carries the G24D change not only interacts less well with GroEL, but does so in an overall different qualitative fashion. The wild-type synthetic loop peptide does not compete with a typical "substrate" synthetic peptide for binding to GroEL (Landry et al. 1993).

Why are all *groES* mutations located in the region of GroES that is responsible for interacting with GroEL? The answer must lie in the selection exerted for the isolation of these mutations. The original selection was for mutations that block the lytic pathway of bacteriophage λ, yet still allow bacterial growth at 30–37°C (Georgopoulos et al. 1973). Since we now know that both the *groES* and *groEL* genes are essential for bacterial growth under all conditions examined (Fayet et al. 1989), a gross distortion of the function of either could not have been tolerated in the original selection scheme. But why do the *groE* mutants block specifically λ as opposed to bacterial growth? The answer may be due to the fact that bacteriophage λ has an extra need for chaperone power, especially at the later stages of its infectious cycle when it attempts to assemble the structural virion components. There are approximately 1000 λ-encoded polypeptides present in a λ head structure, and assuming the presence of 200 heads per infected cell, 200,000 head-related polypeptides (not to mention those required for tails) must be assembled into capsids in a relatively short period of time. Any diminution in GroEL/GroES chaperone power at this stage could lead to irreversible aggregation, resulting in the abnormal head structures seen under the electron microscope (Georgopoulos et al. 1973). Since the mutant bacterial host can "limp" along without its full complement of GroEL chaperone power, it can still form a colony. However, at high temperatures, there is greater need for chaperone power, and in addition, the activity of some mutant GroE proteins could be further compromised, thus resulting in the observed bacterial temperature-sensitive phenotype (Georgopoulos et al. 1973).

Two genetic approaches have been used to determine the GroES-binding region of GroEL, both involving direct selection for colony-forming ability. The first was initiated by Tilly and Georgopoulos (1982) and involved the isolation of extragenic suppressors of *groES* temperature-sensitive mutations. Mutant *groES* bacteria were spread on LB plates and placed at the nonpermissive temperature. Colonies that came up were tested against bacteriophage T4 to identify those extragenic suppressors that map in the *groEL* gene. Those putative *groEL** suppressor mutations that block bacteriophage T4 growth were sequenced and shown to result in amino acid changes at positions 174, 190, and 375 (J. Zeilstra and O. Fayet, pers. comm.). The observed allele-specific pattern of suppression with two of the *groES* mutations suggests that amino acid residues 174–190 may be part of the corresponding region of GroEL that interacts with GroES.

The second genetic approach took advantage of a previous finding, namely, that the overexpression of both *groES* and *groEL* genes can restore colony-forming ability to mutant *dnaA*46 bacteria at 42°C (Fayet et al. 1986). The plasmid pOF15 codes for a GroES protein that lacks amino acids 36 and 37. As a consequence, pOF15 will not suppress the growth defect of *dnaA*46 mutant bacteria at 42°C (Fayet et al. 1986). Two *groEL* mutations were identified, resulting in amino acid changes at either positions 169 or 173, that restore suppression of the *dnaA*46 temperature-sensitive phenotype (J. Zeilstra and O. Fayet, pers. comm.). Amino acids 169 and 173 are in the region of GroEL previously identified as suppressing missense *groES* mutations (see above), further strengthening the suggestion that this segment of GroEL interacts with GroES. Support for this possibility comes from three other observations. The first line of evidence is that the *groES*619 *groEL**174 mutant bacteria, besides acquiring the ability to grow at high temperature, simultaneously become sensitive to infection by bacteriophage λ, implying again that a workable GroES/GroEL interaction has been restored. The second line of evidence comes from the fact that one of the two original mutations present in *groEL*673 results in a change at codon 173 which confers by itself a bacteriophage-λ-resistant phenotype (Zeilstra-Ryalls et al. 1993). It is likely that the purified GroEL673 mutant protein cannot interact with GroES, as evidenced by the fact that GroES does not stimulate the ability of GroEL673 to protect RNAP from heat inactivation, and that GroEL673 cannot reactivate heat-inactivated RNAP, a function that absolutely requires a workable GroES/GroEL interaction (Ziemienowicz et al. 1993; see above). The third line of evidence comes from the work of Baneyx and Gatenby (1992). These authors showed that the GroEL140 mutant protein, altered at amino acid residue 201 (Zeilstra-Ryalls et al. 1993), is substantially defective in its interaction with GroES. In summary, all of these results, taken together, suggest that the region of GroEL in the vicinity of amino acid residues 169–201 is important for its proper interaction with GroES.

In contrast to this preliminary conclusion derived from genetic and biochemical studies, the presence of GroES protects the 16 carboxy-terminal amino acid residues of GroEL from proteolytic cleavage (Langer et al. 1992a; see Frydman and Hartl, this volume). This result suggests that the carboxy-terminal portion of GroEL may also participate in its interaction with GroES. Interestingly, McLennon et al. (1993) have shown that this part of GroEL is dispensable for its biological function.

C. The Bacteriophage T4 Encodes a GroES Homolog

It was shown more than 20 years ago that the virulent bacteriophage T4 also requires the GroEL function for proper assembly of its head (for review, see Zeilstra-Ryalls et al. 1991). Compensatory bacteriophage T4-encoded mutations could be readily isolated, all of which were shown to map in gene *31* (Keppel et al. 1990). In all respects, the phenotype of wild-type T4 infecting a mutant *groEL* host, such as *groEL*44, is equivalent to a T4 gene *31⁻* mutant infecting a wild-type *E. coli* host. Some of these *31* mutations behave in an allele-specific manner, indicating genetically that the GroEL and gp31 proteins interact. These results, taken together, suggested that the gp31 protein may replace or supplement the GroES function in bacteriophage T4 head assembly (Keppel et al. 1990).

The following biochemical and genetic data of S.M. van der Vies et al. (in prep.) demonstrate that the T4-encoded gp31 indeed functions as a cochaperone of GroEL: (1) When gene *31* is cloned under an inducible promoter and its product overproduced, three distinct phenotypes of *groES* mutants are suppressed, i.e., the ability to propagate bacteriophage λ, the ability to propagate bacteriophage T5 (whose growth cycle on *groES* mutant hosts is blocked at the level of tail morphogenesis;

Georgopoulos et al. 1990; Zeilstra-Ryalls et al. 1991), and the ability of the mutant bacteria to grow at the nonpermissive temperature (albeit not to the same extent as wild-type *E. coli*); (2) overproduction of gp31 also restores the ability of mutant *groES* bacteria to assemble the decahexameric form of cyanobacterial Rubisco enzyme; and (3) the gp31 protein was purified to homogeneity and shown to possess the following cochaperone properties in vitro: (i) the oligomeric gp31 protein (~90 kD) binds to GroEL in the presence of MgATP. In the absence of MgATP, gp31 does not coelute with GroEL, (ii) gp31 inhibits the ATPase of GroEL to the same degree as GroES, and (iii) GroEL and gp31 function together to promote the folding of denatured dimeric Rubisco from *Rhodospirillum rubrum*, with the same rate and final yield as those of GroEL and GroES (S.M. van der Vies et al., in prep.). In summary, all available evidence demonstrates that gp31 is a functional analog of GroES, although no obvious amino acid sequence homology between the two proteins exists.

The reason bacteriophage T4 encodes a cochaperone of GroEL may be related to its "do-or-die" lytic life cycle. Since T4 is a virulent bacteriophage, it is to its advantage to grow fast and infect all neighboring bacteria. But to do so, it must assemble a larger number of head structures in a shorter time interval than the temperate bacteriophage λ. The gp31 protein may have evolved as a cochaperone (having been "borrowed" from another organism by T4?) in order to handle efficiently the morphogenetic load in a timely fashion. Or perhaps the GroEL/gp31 system is another example of a chaperone system that evolved to fold a specific substrate, since in the absence of gp31 function, the gp23 capsid protein is found in large protein aggregates (for review, see Zeilstra-Ryalls et al. 1991).

D. Yeast GroES Mitochondrial Homolog

Recently, an *S. cerevisiae* hsp10 (GroES homolog) protein has been purified from mitochondria and extensively characterized (Rospert et al. 1993b). The hsp60^{yeast}/hsp10^{yeast} and the heterologous GroEL$^{E.coli}$/hsp10^{yeast} combinations can support the in vitro folding of Rubisco protein, whereas the hsp60^{yeast}/GroES$^{E.coli}$ combination cannot. Interestingly, whereas both the hsp10^{yeast} and GroES$^{E.coli}$ homologs can inhibit the ATPase of GroEL$^{E.coli}$, neither can inhibit the ATPase of hsp60^{yeast}. The hsp10^{yeast}-coding gene has been cloned and shown to be an essential gene at all temperatures (Rospert et al. 1993a; G. Schatz, pers. comm.). The rat gene that codes for a mitochondrially located hsp10 homolog of GroES has also been identified and its product has been characterized (Hartman et al. 1993).

V.   OTHER INTERESTING HEAT SHOCK GENES AND THEIR PRODUCTS

A.   Proteases

Most of the intracellular proteases in *E. coli* are ATP-dependent, as judged from the fact that depletion of ATP inhibits protein degradation (for review, see Gottesman and Maurizi 1992). During the past few years, an increasing number of heat shock proteins have been shown either to be proteases or to participate in protease action. At some level, the chaperone and protease systems converge to coordinate their action, to assure the rapid and timely destruction of permanently unfolded proteins. On the one hand, proteases resemble chaperones in the sense that they must recognize the unfolded, denatured state in a protein, so that they do not interfere with the function of properly folded proteins. On the other hand, chaperones, by maintaining proteins in an unfolded state, may aid their accessibility to proteases and hence increase their rate of degradation. C. Gross' laboratory has shown that mutations in all major chaperone genes of *E. coli* (*dnaK, dnaJ, grpE, groES*, and *groEL*) lead to the hypodegradation of abnormal polypeptides, such as puromycyl peptides (Straus et al. 1988). The interpretation is that both the DnaK and GroE chaperone systems are needed to best "present" these substrates to proteases. In the absence of a fully functional major chaperone machine, the proteases may have limited access to their polypeptide substrates.

1.   The Lon Protease

The Lon (or La) protease was one of the first heat shock proteins to be identified, purified, and characterized. Its biology and biochemistry have been reviewed recently by Gottesman and Maurizi (1992). The only relevant information to be provided here is that Lon has been observed to associate with some abnormal polypeptides, such as PhoA61, in a complex that includes DnaK and GrpE, but not DnaJ (Sherman and Goldberg 1992a). The association of the PhoA61 protein with DnaK correlates with its degradation, especially because in a *dnaK* deletion background, the mutant PhoA61 protein is completely stable. Interestingly, in a *dnaK756* mutant background, the proteolysis of the PhoA61 polypeptide is accelerated. One interpretation is that the DnaK756 polypeptide can bind to PhoA61 and present it to proteases but cannot readily release it, thus increasing its overall accessibility to proteases. In support of this interpretation, Liberek et al. (1991b) have actually shown that DnaK756 is defective in the release of a bound polypeptide substrate.

Recently, N. Wang et al. (1993) have reported the discovery of a human mitochondrial protein that has high homology with the bacterial Lon protease, demonstrating the dramatic conservation of this ATP-dependent proteolytic system in nature. A gene coding for a mitochondrially located Lon protease homolog has also been identified and characterized in *S. cerevisiae* (G. Schatz, pers. comm.).

2. The Clp Proteases

In recent years, another ATP-dependent protease system of *E. coli* has been identified, purified, and extensively characterized, namely, Clp. The Clp (or Ti) protease, unlike Lon, has a bipartite nature, being composed of two types of subunits, ClpP and ClpA (for a detailed review, see Gottesman and Maurizi 1992). Briefly, the ClpP component is made up of two hexameric rings, and itself possesses an ATP-independent protease activity for very short peptides but not for longer unstructured polypeptides. The ClpP protein acquires the ability to degrade abnormal proteins only in conjunction with ClpA, itself a hexameric protein. The $(ClpP)_{12}(ClpA)_6$ complex requires ATP for its formation and is extremely stable. The ClpA protein has a basal ATPase activity that is enhanced in the presence of appropriate polypeptide substrates. Kroh and Simon (1990) showed that the *clpP* gene, coding for heat shock protein F21.5, is under σ^{32}-dependent regulation. Although a first report indicated that ClpA is not under heat shock regulation, the data by Chuang et al. (1993) suggest that it may in fact be under σ^{32} regulation. Thus, both components of the ClpP/ClpA protease are heat shock proteins.

Recently, it has been shown that ClpP can also functionally participate in the degradation of some polypeptides independently of ClpA, by working in conjunction with another subunit, called ClpX (Gottesman et al. 1993; Wojtkowiak et al. 1993). This work started with a search for *E. coli* proteases that specifically destabilize the bacteriophage λO protein, the origin-binding protein that triggers the series of molecular events leading to λ DNA replication (for review, see Georgopoulos et al. 1990). A bipartite protease system was purified and found to consist of ClpP and ClpX. The ClpX component has a substrate-stimulated ATPase (A. Wawrzynow and D. Wojtkowiak, pers. comm.) that is encoded by a gene cotranscribed with and distal to *clpP* (Gottesman et al. 1993). Consistent with this in vitro result, inactivation of the *clpX* gene led to a dramatic stabilization of the λO protein in vivo, yet it did not affect the degradation of ClpA/ClpP substrate polypeptides (Gottesman et al. 1993).

An interpretation of all the data is that the ClpP component plays the role of a "master" protease that is "attracted" to various polypeptide substrates presented by different specificity factors, such as ClpA and ClpX (Gottesman et al. 1993; Wotjkowiak et al. 1993). It could be that the newly discovered *clpB* gene, which is under σ^{32}-dependent heat shock regulation and codes for a polypeptide that has homology with ClpA (for review, see Squires and Squires 1992; see below), can also complex with ClpP to give rise to yet another ATP-dependent protease system. Consistent with this, it has been recently shown that the ATPase of purified ClpB is stimulated in the presence of certain polypeptides (Woo et al. 1992). Of course, the possibility exists that, in addition to ClpP, other "master" proteases exist in *E. coli* that similarly interact with these or other regulatory subunits to ensure efficient intracellular proteolysis.

The *clpB* gene encodes two proteins highly homologous to that of hsp104 in yeast (Kitigawa et al. 1991; Squires et al. 1991; Squires and Squires 1992; see Parsell and Lindquist, this volume). It can be translated to give rise to two inframe 93- and 79-kD polypeptides (the F84.1 and F68.5 heat shock proteins; Squires et al. 1991), both of which associate with purified anaerobic nucleotide triphosphate reductase, suggesting that they perform chaperone functions (Eliasson et al. 1990). Park et al. (1993) have recently shown that the ATPase activity of the 93-kD form of ClpB is activated by certain protein substrates. In contrast, although the 79-kD form of ClpB also possesses an ATPase activity, this is not stimulated by protein substrates and furthermore inhibits in *trans* the stimulation by protein substrates of the ATPase of the 93-kD form of ClpB. The significance of this elaborate modulation of the ATPase activity of ClpB in its biological function remains to be seen. A null mutation in *clpB* exhibits a mild growth defect at extreme temperatures and reduced thermotolerance, as is the case with its *S. cerevisiae* homolog (Squires et al. 1991; Sanchez et al. 1993; see Parsell and Lindquist, this volume). It is possible that its putative chaperone and protease functions become important to the survival of *E. coli* only at extremely elevated temperatures because of the high intracellular demands for protein protection, disaggregation, or degradation that necessitate the intact presence of all chaperone/protease systems.

Recently, a new heat shock gene, *HSP78*, has been identified in *S. cerevisiae* that codes for a ClpB homolog which is imported into mitochondria (Leonhardt et al. 1993). No phenotype was found associated with the disruption of the *HSP78* gene.

3. The Putative FtsH/HflB Protease

In a recent paper, Herman et al. (1993) synthesized the combined efforts of various laboratories in showing that the cell division gene *ftsH* is identical to the *hflB* locus of *E. coli*, originally defined as a protease system that partly controls λcII intracellular degradation. It turns out that *ftsH/hflB* is an essential gene under σ^{32}-dependent heat shock regulation that codes for an integral membrane protein (Herman et al. 1993 and in prep.). More importantly for this section is the possibility that the FtsH/HflB protein either is a protease or controls the activity of a protease that degrades the σ^{32} polypeptide, since in an *ftsH*1 temperature-sensitive background, the σ^{32} polypeptide is dramatically stabilized under nonpermissive conditions (C. Herman et al., in prep.; see below). For example, it could be that FtsH/HflB is a chaperone that binds σ^{32} and presents it to an as yet unidentified protease. It is still possible that the *ftsH*1-mediated stabilization of σ^{32} is an indirect effect, the cells being "stressed" because of the lack of FtsH function or the presence of the FtsH1 mutant protein.

4. The DegP (HtrA) Periplasmic Protease

The *degP* (*htrA*) gene has been shown to be under heat shock regulation (Lipinska et al. 1988), exclusively under the transcriptional control of the σ^{24} factor, which appears to control an independent heat shock regulon (Erickson and Gross 1989). The *degP (htrA)* gene is essential for *E. coli* growth at temperatures above 40°C (Lipinska et al. 1989). The DegP (HtrA) protein has been shown to be a protease, localized in the periplasmic space (Strauch and Beckwith 1988; Lipinska et al. 1989, 1990; Strauch et al. 1989). Its substrates include abnormal periplasmic proteins (Strauch and Beckwith 1988; Strauch et al. 1989), the colicin A lysis protein (Cavard et al. 1989), and casein (Lipinska et al. 1990). Most likely, the DegP (HtrA) protease is identical to the previously purified Do protease of *E. coli* (Seal et al. 1991).

Baird et al. (1991) characterized the *sohB* gene as a multicopy suppressor of the temperature-sensitive phenotype of *degP(htrA)* null mutations. Interestingly, SohB is an exported protein that exhibits substantial sequence similarity to SppA (protease IV) protein, a signal peptide peptidase found in the cytoplasmic membrane that digests cleaved signal peptides (Suzuki et al. 1987; Baird et al. 1991). On the basis of these findings, it is tempting to speculate that the SohB protein is a periplasmic protease, whose substrate specificity overlaps that of DegP (HtrA). A recent finding that underscores the biological significance of the *degP(htrA)* equivalent gene of *Salmonella typhimurium* is that it is necessary for its pathogenicity (Johnson et al. 1991).

B. The *htrC* and *htpY* Genes May Control the Heat Shock Response

Two additional *E. coli* genes have been identified, both under σ^{32} regulation, that directly or indirectly modulate the heat shock response but in opposite directions. The *htrC* gene was identified because its insertional inactivation resulted in a temperature-sensitive phenotype for bacterial growth (Raina and Georgopoulos 1990). The *htrC* null mutant exhibits a number of pleiotropic phenotypes, even at the permissive temperature of 30°C, including (1) overproduction of the $E\sigma^{32}$-regulated heat shock proteins, (2) deregulation in the rate of synthesis of a few additional polypeptides, and (3) an overall defect in cellular proteolysis as judged from the reduced rate of puromycyl-containing polypeptide degradation. It could be that *htrC* is an inhibitor of the $E\sigma^{32}$-regulated heat shock response or simply that the cell is "stressed" in the absence of HtrC function. The *htrC* gene is under $E\sigma^{32}$ regulation, being maximally transcribed at 50°C (Raina and Georgopoulos 1990).

The *htpY* gene is located 700 bp upstream of *dnaK* and is transcribed under the control of the $E\sigma^{32}$ holoenzyme in the same clockwise direction with respect to the *E. coli* chromosome as *dnaK* (Missiakas et al. 1993). The gene can be deleted without noticeable effects on bacterial growth in most genetic backgrounds within the normal growth temperature range of *E. coli*. Although *htpY* null bacteria are viable, the expression from various $E\sigma^{32}$-directed heat shock promoters is significantly decreased, suggesting that HtpY somehow has a positive role in the regulation of the heat shock response. Consistent with this interpretation, overproduction of the HtpY protein results in a generalized increase in transcription from $E\sigma^{32}$-regulated promoters in *E. coli*. The molecular mechanism by which *htpY* positively regulates the $E\sigma^{32}$ response is not known.

C. The Identification of Additional Heat Shock Genes

Chuang ct al. (1993) prepared cDNA from mRNA extracted from cultures under various experimental treatments and used this to probe DNA dot blots made from λ transducing bacteriophages spanning the genome of *E. coli*. From the hybridization patterns, they were able to identify those regions of the *E. coli* genome whose transcription increases under each experimental protocol.

Chuang et al. (1993) extended this study in order to identify as many heat shock genes as possible. Transducing bacteriophages carrying those regions of the *E. coli* chromosome that are preferentially transcribed at high temperature were used to infect UV-irradiated *E. coli* (thus suppressing endogenous protein synthesis) at 50°C and to identify the synthesized proteins by labeling with [^{35}S]methionine. DNA was also extracted from these bacteriophages and used to program an in vitro *E.*

coli transcription/translation system. At least 16 new *E. coli* genes under
σ^{32} control were identified because their expression increased upon sup-
plementation with exogenous $E\sigma^{32}$ holoenzyme. Two genes, called *hslS*
and *hslT*, mapping at 83 minutes, are most likely identical to the heat
shock genes *ibpA* and *ibpB* of Allen et al. (1992). The IbpA and IbpB
heat shock proteins were shown to be present in inclusion bodies formed
by certain heterologous protein production (Allen et al. 1992). One possibility is that IbpA and IbpB may be chaperone proteins that become
"trapped" during inclusion body formation. The IbpA and IbpB proteins
are 52% identical to each other and share some homology with the small
heat shock proteins of eukaryotes. Another potentially interesting finding
of this study is that three genes, *hflX, hflK,* and *hflC,* which map at 95
minutes and whose products may constitute the HflA protease of *E. coli*
(Noble et al. 1993), also seem to be under σ^{32} regulation. In addition, a
gene corresponding to *clpA* and mapping at 19 minutes aiso appears to be
under heat shock regulation. Thus, *lon, clpP, clpA, clpB, clpX, ftsH/hflB,
hflX, hflK,* and *hflC,* whose products are either proteases or polypeptides
that assist protease action, are all under $E\sigma^{32}$ transcriptional regulation.

VI. REGULATION OF THE σ^{32}-PROMOTED HEAT SHOCK
RESPONSE IN *E. COLI*

From pioneering work in the laboratories of T. Yura, C. Gross, and F.C.
Neidhardt, the "classical" *E. coli* heat shock response is known to be under the transcriptional regulation of the *rpoH (htpR; hin)* gene product.
The *rpoH* gene product is the σ^{32} transcription factor that complexes
with the RNAP core (E) to constitute the $E\sigma^{32}$ holoenzyme. $E\sigma^{32}$ recognizes the classical heat shock promoters located upstream of heat shock
genes such as *dnaK, groES, lon,* and others (Gross et al. 1990). There is
no known functional overlap between $E\sigma^{32}$- and $E\sigma^{70}$-directed promoters, each set being transcribed exclusively by their corresponding holoen-
zymes. The bulk of the genetic and biochemical work on the regulation
and function of the *rpoH* gene has been done in the laboratories of T.
Yura and C. Gross during the past few years. For a summary of the early
history and contributions from these laboratories, see the review article
published in the first volume of this series (Gross et al. 1990), and for a
review of more recent studies, see Yura et al. (1993). In this section, we
highlight briefly the transcriptional and translational regulations of the
rpoH gene and then discuss the various mechanisms by which the DnaK
chaperone machine can negatively regulate both the intracellular levels
and activity of the σ^{32} polypeptide.

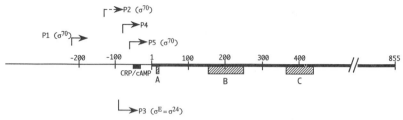

Figure 4 Regulation of *rpoH* gene expression. Transcriptional regulation from the various promoters is discussed in the text. The numbering of nucleotides is in reference to the structural gene (which begins at +1 and ends at 855). Mutations that disrupt putative base pairing of the mRNA secondary structure involving regions A and B lead to higher translational rates at low temperature. Compensatory second-site mutations that restore base pairing also restore translational repression at low temperature. Consistent with the mRNA secondary structure being a critical element in repression of translation, compensatory mutations that replace a putative A:U pair with a C:G pair decrease the rate of translation at low temperature and the extent of heat inducibility at high temperatures. All of these studies (elegantly summarized by Yura et al. 1993) show that effects of temperature on *rpoH* gene expression are mediated, at least partly, by its mRNA secondary structure.

A. Transcriptional Regulation of the *rpoH* Gene

Very little additional information is available today about the transcriptional regulation of the *rpoH* gene that was not known 4 years ago (Gross et al. 1990). Figure 4 highlights many interesting features of *rpoH* gene regulation that operate either at the transcriptional level or at the translational level.

There are at least four promoters that transcribe the *rpoH* gene: P1, P4, and P5, which are transcribed by $E\sigma^{70}$ (the P2 promoter is strain-specific), and P3, which is transcribed by the $E\sigma^E$ (σ^{24}) holoenzyme. Under normal physiological conditions, the P1 promoter contributes the bulk of *rpoH* gene transcription, and the rest contribute varying minor amounts (Gross et al. 1990; Yura et al. 1993). The P3 promoter usage pattern is very interesting because it gradually increases with a corresponding increase in temperature within the normal growth range of *E. coli*, but at extremely high temperatures, such as 51°C, P3 constitutes the sole operating transcription system of the *rpoH* gene (Erickson and Gross 1989). At such extreme temperatures, $E\sigma^{70}$-directed transcription ceases, whereas $E\sigma^E$ transcription continues unabatedly. If transcription of the *rpoH* gene were to cease, the extremely short half-life of its product (see below) would lead to the rapid depletion of its intracellular levels. The continuous presence of σ^{32} at high temperatures assures the continuous transcription of heat shock genes and the corresponding accumulation of heat shock proteins, such as the DnaK and GroEL chaperone machines, whose protein protective and disaggregating activities are especially needed at the high temperatures. The work of Gragerov et al. (1992) has nicely documented that in the absence of high levels of either the DnaK or GroEL chaperone machines, such wholesale intracellular aggregation does indeed take place, especially at high temperatures.

The gene encoding σ^E has been cloned, following intensive work in a number of laboratories over the past years (S. Raina; C. Gross; both pers. comm.). Besides the *rpoH* P3 promoter, the promoter of the *degP* (*htrA*) gene of *E. coli* is the only other known promoter to be recognized by $E\sigma^E$ (Lipinska et al. 1988; Erickson and Gross 1989). However, as opposed to the multiple promoters of the *rpoH* gene, the $E\sigma^E$-directed promoter of *degP* (*htrA*) is the only known promoter of the gene. Since *degP* (*htrA*) null mutants are inviable at temperatures above 42°C (Lipinska et al. 1989), it is expected that the σ^E-coding gene should be an essential *E. coli* gene above 42°C.

The only negative control known to be exerted on *rpoH* gene transcription is that by the DnaA replication protein of *E. coli* (Wang and Kaguni 1989a). These authors showed that in vitro, purified DnaA protein can bind to the two DnaA boxes present in the *rpoH* promoter region, shown in Figure 3, and thus attenuate transcription from the P3 and P4 promoters (the P5 promoter was not tested in these experiments). The same result was found in vivo following overproduction of the DnaA protein. Since transcription from the major *rpoH* promoter, P1, is not affected by DnaA, it is likely that this negative regulation serves to "fine-tune" *rpoH* gene transcription in the cell by the "master-regulator" protein, DnaA (Georgopoulos 1989).

B. Posttranslational Control of the *rpoH* mRNA

Although there is a modest accumulation of *rpoH* gene transcripts following a shift from 30°C to 42°C, the levels of σ^{32} increase dramatically primarily as a consequence of posttranslational regulation. The large transient increase in the σ^{32} levels is mostly due to two factors: (1) increased translation of the *rpoH* mRNA at 42°C, suggesting the existence of thermoregulation at the translational level, and (2) a transient stabilization of the half-life of σ^{32} (Straus et al. 1987; for summary, see Gross et al. 1990; Yura et al. 1993). The extensive use of *rpoH-lacZ* protein fusions and their deletion derivatives enabled the localization of three important *cis*-acting regulatory regions of the *rpoH* gene, shown in Figure 3. The first region, called A, is a small one corresponding to nucleotides

6–20 in the structural gene. It is a *cis*-acting positive element that ensures a high rate of translation by most likely interacting with a complementary region present in 16S rRNA. Its deletion results in a 15-fold decrease in the rate of translation of the *rpoH* mRNA (Nagai et al. 1991; Yura et al. 1993).

The second *cis*-acting region, B, has been localized to nucleotides 153–247 and acts as a negative element of translation, since its elimination leads to high translational expression at low temperatures (Kamath-Loeb and Gross 1991; Nagai et al. 1991). The fact that some point mutations in region A also relieve the temperature-dependent translation repression led to the proposal of the existence of a secondary structure that includes nucleotides 1–210 of the *rpoH* mRNA transcript. In this putative regulatory structure, the AUG initiation codon and region A can base pair with complementary sequences present in region B, resulting in the efficient sequestration of the initiation codon. The model is strongly supported by the existence of compensatory mutations that simultaneously restore putative base pairing and thermoregulation of the *rpoH* mRNA (for summary, see Yura et al. 1993). The fact that some of the compensatory mutations (which substitute a G:C for the original A:T pair) result in a super-repressed translational phenotype adds further credence to the idea that the *rpoH* mRNA secondary structure is responsible for the observed thermoregulation of translation (Yuzawa et al. 1993). It is not known whether a *trans*-acting factor can bind to and stabilize the mRNA secondary structure in a manner analogous to that of the Rom (Rop) protein encoded by the ColE1 family of plasmids that stabilizes the otherwise transient and unstable interaction between the inhibitory RNA and primer RNA molecules (Tomizawa and Som 1984). If such a putative protein factor exists, its activity should be thermosensitive, since the rate of translation of the *rpoH* mRNA increases almost instantaneously at 42°C.

The third *cis*-acting region, C, has been localized to nucleotides 364–433 and appears to be responsible for the translational repression of the *rpoH* mRNA by the DnaK chaperone machine. It was previously observed that when *rpoH* mRNA is induced at 42°C under the control of the λP_L promoter, the kinetics of *rpoH* mRNA synthesis and σ^{32} polypeptide synthesis are different. Although the mRNA levels are found, as expected, at the same high levels, the rate of σ^{32} synthesis is dramatically repressed following a period of acute synthesis (Grossman et al. 1987). Furthermore, this repression of *rpoH* mRNA translation does not occur in either *dnaK*, *dnaJ*, or *grpE* mutant bacteria (Straus et al. 1990). Yura et al. (1993), utilizing a set of inframe deletion derivatives of an *rpoH-lacZ* protein fusion construct, were able to pinpoint the region responsible for this translational repression control to nucleotides 364–433 (which code for amino acids 122–144 of σ^{32}). The use of a frameshift mutation that altered the amino acid sequence at that region abolished translational repression, suggesting that it is somehow mediated by the sequence of the 122–144-amino-acid polypeptide segment of the σ^{32} protein (Yura et al. 1993). It could be that one or more of the DnaK chaperone machine members can bind to this particular segment of σ^{32} as it emerges from ribosomes and arrest σ^{32} translation directly (by acting as a signal recognition particle) or indirectly, e.g., by somehow making the unfinished σ^{32} polypeptide accessible to proteases (see below).

C. Instability of the σ^{32} Polypeptide

The σ^{32} polypeptide has been shown to be a highly unstable protein in vivo, its half-life being on the order of 45–60 seconds between 30ºC and 42ºC, and substantially longer at 22ºC (Straus et al. 1987, 1990; Tilly et al. 1989). The half-life of the full-length σ^{32} polypeptide has been shown to be stabilized under certain experimental conditions: (1) Following a temperature shift to 42ºC, the half-life is stabilized for a short, transient period of time, after which its extreme instability is restored (Straus et al. 1987, 1990), and (2) in *dnaK*, *dnaJ*, or *grpE* mutant backgrounds, the half-life of σ^{32} is stabilized at all temperatures tested (Tilly et al. 1989; Straus et al. 1990), suggesting that these three heat shock proteins directly or indirectly participate in σ^{32} degradation.

D. Mechanisms of Autoregulation of the Heat Shock Response

At the time of writing, a number of observations, both old and very recent, suggested that the heat shock response of *E. coli* is autoregulated at a remarkable variety of levels by the DnaK chaperone machine. All of the evidence are consistent with the idea that the level of unfolded or aggregated proteins in the cell controls the extent of heat shock gene expression. Most likely, these proteins exert this control through the titration of the DnaK chaperone machine (for discussion, see Craig and Gross 1991; Wild et al. 1993; Yura et al. 1993). These regulatory checkpoints include (1) the specific blockage of Eσ^{32} holoenzyme function following binding to its promoter, (2) sequestration of the σ^{32} polypeptide away from the RNAP core, thus preventing the formation of the Eσ^{32} holoenzyme, (3) arrest of σ^{32} translation, and (4) proteolysis of σ^{32}. Below, we briefly describe the existing data suggesting that control over σ^{32} function and synthesis is most likely applied at all these levels.

1. It has been shown that the presence of the DnaK/DnaJ/GrpE proteins specifically interferes with $E\sigma^{32}$-directed transcription in vitro (B. Bukau, pers. comm.; K. Liberek, unpubl.). B. Bukau's group has also shown that the DnaJ heat shock protein alone is capable of blocking the formation of an open complex (which is absolutely needed for the initiation of transcription) by the $E\sigma^{32}$ holoenzyme but not that by the $E\sigma^{70}$ holoenzyme.

2. Both the DnaK and DnaJ proteins have been shown to bind to σ^{32} in vitro. Gamer et al. (1992) showed that when σ^{32} is immobilized on a nickel-affinity column (because of the presence of a hexa-histidine tag), all members of the DnaK chaperone machine specifically bind to it. The DnaK and GrpE members are eluted from the column by ATP, whereas DnaJ remains bound to σ^{32} under these conditions. When the σ^{32} polypeptide was overproduced and purified, it was noted that approximately half of it was associated with the DnaK chaperone, even at late stages of the purification procedure (Liberek et al. 1992). The DnaK-σ^{32} interaction was confirmed using purified protein components. It was shown that (i) the DnaK-σ^{32} complex is an inherently weak one. To drive most of σ^{32} into a complex, approximately eight- to tenfold molar excess of DnaK is needed. The DnaK-σ^{32} complex is a typical DnaK-substrate complex, inasmuch as it is destroyed in the presence of hydrolyzable ATP, (ii) the mutant DnaK756 protein does not bind substantially to σ^{32}, since no stable complex was observed. This result was satisfying because $dnaK756$ bacteria were previously shown to overexpress the heat shock proteins at all temperatures (Tilly et al. 1983), and (iii) the purified σ^{32} protein is perfectly functional since it can bind to the RNAP core to constitute an active $E\sigma^{32}$ holoenzyme (Liberek et al. 1992). It is not known what feature of σ^{32} is recognized by DnaK.

The DnaJ heat shock protein binds purified σ^{32} much stronger than DnaK, a DnaJ-σ^{32} complex being readily observable at a 1:1 molar ratio. The DnaJ-σ^{32} complex is not influenced by ATP (Gamer et al. 1992; Liberek and Georgopoulos 1993). Surprisingly, a very stable complex of DnaK-σ^{32}-DnaJ forms exclusively in the presence of ATP. Nonhydrolyzable ATP analogs do not support the formation of such a DnaK-σ^{32}-DnaJ complex. As a control, neither DnaK nor DnaJ is capable of forming a complex with σ^{70}, the major σ factor of *E. coli* (Liberek and Georgopoulos 1993).

Finally, the DnaK/DnaJ chaperone machine has been shown to be capable of "stripping off" σ^{32} from a preformed $E\sigma^{32}$ holoenzyme complex in an ATP-dependent reaction, but only in the presence of σ^{70}. In the complete absence of σ^{70} (an unlikely in vivo event), the

DnaK chaperone machine cannot strip off σ^{32} from the $E\sigma^{32}$ holoenzyme (Liberek and Georgopoulos 1993).

3. Arrest of σ^{32} translation. As noted above, this conclusion comes from in vivo studies from the laboratories of C. Gross and T. Yura. Briefly, the evidence is that in either *dnaK, dnaJ,* or *grpE* mutants, the arrest of σ^{32} translation, normally seen in wild-type bacteria following the initial burst of σ^{32} translation at 42°C, is not observed. A likely scenario is the following: When the levels of the DnaK/DnaJ chaperone machine are high and simultaneously there is minimum need for its services (i.e., the heat shock damage has been repaired, and/or subsided), the now available DnaK/DnaJ proteins bind with high affinity to the 122–144-amino-acid region of σ^{32} as it emerges from the ribosome and either block translation per se (by acting analogously to a signal recognition particle) or make the growing σ^{32} polypeptide chain readily accessible to proteases (for discussion, see Yura et al. 1993). In the latter case, "translational" arrest could be synonymous to σ^{32} breakdown before the full σ^{32} polypeptide is even translated.

4. Presentation of full-length σ^{32} polypeptide to proteases. The evidence for this possibility comes primarily from the fact that the half-life of finished σ^{32} polypeptide chains is dramatically stabilized in a *dnaK, dnaJ,* or *grpE* mutant background (Tilly et al. 1989; Straus et al. 1990). However, the molecular mechanism by which this is brought about is not at all understood. The half-life of σ^{32} in a crude in vitro system is on the order of 60 minutes, far longer than that observed in vivo. The half-life is still of the same order of magnitude when a DnaK-σ^{32}-DnaJ preformed complex is added to this crude in vitro system (K. Liberek, unpubl.). Furthermore, σ^{32} is also stable when freshly synthesized in an S30 transcription/translation system, even in the presence of excess DnaK, DnaJ, and GrpE. Clearly, either a component is missing from the in vitro crude system or the in vivo reaction conditions have not been adequately reproduced.

Genetic experiments have shown that neither the Lon- nor ClpP-dependent proteolytic systems have a major role in influencing σ^{32} stability, since mutations in the corresponding heat shock genes, *lon* and *clpP*, do not affect σ^{32} half-life (Gottesman and Maurizi 1992). Of course, if the σ^{32} polypeptide is being degraded in vivo by many different proteolytic systems, the absence of any one of them would appear to have little or no effect. Recently, a serendipitous breakthrough may have occurred in understanding σ^{32} intracellular breakdown. A recent publication has shown that the *ftsH* gene, an essential gene previously thought to participate exclusively in the bacterial cell division process, is identical to the *hflB* gene (Herman et al. 1993). The *hflB* gene was originally identified by a single *E. coli* mutation that favored the lysogenic pathway in a bacteriophage λcIII mutant (Friedman et al. 1984) and subsequently shown to code for an in vivo proteolytic system that helps to degrade the λcII protein (Banuett et al. 1986).

It has been shown that in an *ftsH1* temperature-sensitive mutant, the half-life of σ^{32} is dramatically stabilized at the nonpermissive temperature (from 40 sec to 9 min; C. Herman et al., in prep.). Thus, FtsH/HflB could be part of a proteolytic system involved in the degradation not only of λcII, but also of σ^{32}. However, it is not known at this time whether the FtsH/HflB protein is itself a protease capable of degrading σ^{32}, whether it acts by "presenting" σ^{32} to the actual catalytic protease, or whether σ^{32} stabilization in this genetic background is an even more indirect effect. FtsH/HflB is an integral membrane protein whose bulk is predicted to lie in the interior of the cell. In this context, it is also interesting that the *ftsH* gene is under $E\sigma^{32}$ transcriptional control, since its transcripts accumulate following a temperature shift to 42°C or overproduction of σ^{32} (C. Herman et al., in prep.).

VII.   THE EXISTENCE OF OTHER HEAT SHOCK REGULONS

The σ^E factor was originally identified in the laboratories of C. Gross and J. Kaguni as a 24-kD protein present in RNAP preparations. The σ^E polypeptide was excised from a gel following SDS-PAGE and was shown to impact on the RNAP core the ability to initiate transcription at the P3 promoter of the *rpoH* gene (Erickson and Gross 1989; Wang and Kaguni 1989b). Erickson and Gross (1989) also showed that $E\sigma^E$ transcribes the promoter of the *degP* (*htrA*) gene of *E. coli*. Since it was previously shown that both the P3 promoter of *rpoH* and the promoter of *degP* (*htrA*) are more active at higher temperatures (Erickson et al. 1987; Lipinska et al. 1988), the σ^E transcription factor must represent a second heat shock regulon in *E. coli*, whose other gene members remain to be identified. Mescas et al. (1993) have recently suggested that the $E\sigma^E$ regulon constitutes a complementary stress system that responds to changes in the periplasm and outer membrane. The recent cloning of the gene that encodes σ^E should enable the rapid elucidation of its function (S. Raina; C. Gross; both pers. comm.).

The laboratory of P. Model has also presented evidence for the existence of yet another heat shock regulon under the control of the $E\sigma^{54}$ holoenzyme (Weiner et al. 1991). The σ^{54} polypeptide is the product of the *rpoN* gene and helps transcribe a diverse array of genes, including those involved in nitrogen metabolism (Kustu et al. 1989). The transcription of the four-gene *psp* operon is induced at high temperatures as well as following other stresses such as osmotic shock and infection by filamentous bacteriophage f1 (Weiner et al. 1991). Interestingly, Kleerebezem and Tommassen (1993) have shown that intracellular conditions that block the export pathway in *E. coli* also result in the induction of the *psp* operon.

Fischer et al. (1993) showed that one of five *groE* operons in *Bradyrhizobium japonicum* is coregulated with symbiotic nitrogen fixation genes under the control of the σ^{54} polypeptide. It could be that this regulation reflects a more specialized chaperone role in nitrogen fixation for this particular GroEL/GroES system or by a need of the host bacterium to modulate its chaperone levels in response to specific extracellular conditions and physiological needs.

VIII.   HEAT SHOCK GENE EXPRESSION IN OTHER BACTERIA

The use of σ^{32}-like transcription factors to promote heat shock gene transcription is not universal and may not even be operating in all gram-negative bacteria. Studies from various laboratories over the past few years have shown that in gram-positive bacteria, the corresponding *dnaK* and *groEL* genes are indeed under heat shock regulation. However, no evidence for specific heat shock promoter sequences exists, although many heat shock genes have been cloned and sequenced. In all cases studied, heat-inducible transcriptional start sites match those at low temperatures and are preceded exclusively by vegetative promoters. A unifying feature of such heat shock gene induction seems to be the presence of a nine-nucleotide inverted repeat separated by nine nucleotides and located somewhere between the transcriptional start site and the beginning of the structural gene (Wetzstein et al. 1992). It is possible that a repressor molecule, functioning at either the DNA or RNA level, normally limits heat shock gene expression by binding to this inverted repeat element. This putative repressor could be the sensor that responds to the presence of heat (by undergoing a conformational change?), thus allowing higher levels of transcription or translation (increased mRNA translation will also lead to higher intracellular mRNA levels, through increased stabilization of the mRNA). The same nine-nucleotide inverted repeat structure was also seen upstream of the *groE* operon of the gram-negative bacterium *Agrobacterium tumefaciens* (Segal and Ron 1993), exemplifying the lack of universality of the σ^{32}-promoted response among bacteria.

IX. CONCLUSION

The deciphering of the biological function of many of the heat shock proteins discussed in this chapter exemplifies the enhanced power of genetics when coupled to biochemistry. Many of these studies were initiated more than 25 years ago in an attempt to identify *E. coli* functions necessary for bacteriophage λ growth (for review, see Friedman et al. 1984; Georgopoulos 1992). Many of these *E. coli* genes were subsequently shown to encode heat shock proteins that perform essential host functions. Parallel studies identified additional heat shock proteins, many of which appear to be part of proteolytic systems. The purification of all of these heat shock proteins and the use of appropriate in vitro biochemical systems have helped to decipher their fundamental role in protein folding, polypeptide transport, protein disaggregation, and regulation of the heat shock response. Future studies will undoubtedly clarify more details of the biochemical properties of these fascinating proteins, including the mechanisms of interaction among individual members of each chaperone machine and the cooperation among the various chaperone and proteolysis machines in carrying out their biological function. The elucidation of these mechanisms of function will clearly have an impact on our understanding of gene regulation as well as the proper assembly and function of all cellular proteins and structures.

ACKNOWLEDGMENTS

We thank our many colleagues who have contributed to these studies over the years and for permission to quote their unpublished experiments. These studies were supported throughout the years by grants from the National Institutes of Health, the National Science Foundation, the Polish Academy of Sciences, the National Swiss Foundation, and the Canton of Geneva.

REFERENCES

Allen, S.P., J.O. Polazzi, J.K. Gierse, and A.M. Easton. 1992. Two novel heat shock genes encoding proteins produced in response to heterologous protein expression in *Escherichia coli. J. Bacteriol.* **174:** 6938–6947.

Ang, D. 1988. "The role of the *Escherichia coli* heat shock protein, grpE, in *Escherichia coli* growth and λ DNA replication." Ph.D. thesis, University of Utah, Salt Lake City.

Ang, D. and C. Georgopoulos. 1989. The grpE heat shock protein is essential for *Escherichia coli* viability at all temperatures but is dispensable in certain mutant backgrounds. *J. Bacteriol.* **171:** 2748–2755.

Baird, L., B. Lipinska, S. Raina, and C. Georgopoulos. 1991. Identification of the *Escherichia coli sohB* gene, a multicopy suppressor of the HtrA (DegP) null phenotype. *J.*

Bacteriol. **173:** 5763–5770.

Baneyx, F. and A.A. Gatenby. 1992. A mutation in GroEL interferes with protein folding by reducing the rate of discharge of sequestered polypeptides. *J. Biol. Chem.* **267:** 11637–11644.

Bardwell, J.C.A. and E.A. Craig. 1987. Eukaryotic M_r 83,000 heat shock protein has a homologue in *Escherichia coli*. *Proc. Natl. Acad. Sci.* **84:** 5177–5181.

———. 1988. Ancient heat shock gene is dispensable. *J. Bacteriol.* **170:** 2977–2983.

Banuett, F., M.A. Hoyt, L. McFarlane, H. Echols, and I. Herskowitz. 1986. *hflB*, a new *Escherichia coli* locus regulating lysogeny and the level of bacteriophage lambda cII protein. *J. Mol. Biol.* **187:** 213–224.

Boguski, M.S. and F. McCormick. 1993. Proteins regulating Ras and its relatives. *Nature* **366:** 643–654.

Bolliger, L., O. Deloche, B.S. Glick, C. Georgopoulos, P. Jenö, N. Kronidou, M. Horst, N. Morishima, and G. Schatz. 1994. A mitochondrial homolog of bacterial GrpE interacts with mitochondrial hsp70 and is essential for viability. *EMBO J.* (in press).

Bork, P., C. Sander, A. Valencia, and B. Bukau. 1992. A module of the dnaJ heat shock proteins found in malaria parasites. *Trends Biochem. Sci.* **17:** 129.

Braig, K., M. Simon, F. Furuya, J. Hainfeld, and A.L. Horwich. 1993. A polypeptide bound by the chaperonin groEL is localized within a central cavity. *Proc. Natl. Acad. Sci.* **90:** 3978–3982.

Brodsky, J.L., S. Hamamoto, D. Feldheim, and R. Schekman. 1993. Reconstitution of protein translocation from solubilized yeast membranes reveals topologically distinct roles for Bip and cytosolic hsp70. *J. Cell Biol.* **120:** 95–102.

Bukau, B. and G.C. Walker. 1990. Mutations altering heat shock specific subunit of RNA polymerase suppress major cellular defects of *E. coli* mutants lacking the DnaK chaperone. *EMBO J.* **9:** 4027–4036.

Caplan, A.J., D.M. Cyr, and M.G. Douglas. 1992. YDJ1p facilitates polypeptide translocation across different intracellular membranes by a conserved mechanism. *Cell* **71:** 1143–1155.

———. 1993. Eukaryotic homologues of *Escherichia coli* dnaJ: A diverse protein family that functions with HSP70 stress proteins. *Mol. Biol. Cell* **4:** 555–563.

Cavard, D., C. Lazdunski, and S.P. Howard. 1989. The acylated precursor form of the colicin A lysis protein is a natural substrate of the *degP* protease. *J. Bacteriol.* **171:** 6316–6322.

Chuang, S.-E., D.L. Daniels, and F.R. Blattner. 1993. Global regulation of gene expression in *Escherichia coli*. *J. Bacteriol.* **175:** 2026–2036.

Craig, E.A. and C.A. Gross. 1991. Is hsp70 the cellular thermometer? *Trends Biochem. Sci.* **16:** 135–140.

Cyr, D.M., X. Lu, and M.G. Douglas. 1992. Regulation of eukaryotic hsp70 function by a dnaJ homolog. *J. Biol. Chem.* **267:** 20927–20931.

Das, A. 1992. How the phage lambda *N* gene product suppresses transcription termination: Communication of RNA polymerase with regulatory proteins mediated by signals in nascent RNA. *J. Bacteriol.* **174:** 6711–6716.

Eliasson, R., M. Fontecave, H. Jörnvall, M. Krook, E. Pontis, and P. Reichard. 1990. The anaerobic ribonucleoside triphosphate reductase from *Escherichia coli* requires S-adenosylmethionine as a cofactor. *Proc. Natl. Acad. Sci.* **87:** 3314–3318.

Ellis, R.J. and S.M. van der Vies. 1991. Molecular chaperones. *Annu. Rev. Biochem.* **60:** 321–347.

Erickson, J.W. and C.A. Gross. 1989. Identification of the σ^E subunit of *Escherichia coli* RNA polymerase: A second alternate σ factor involved in high-temperature gene expression. *Genes Dev.* **3:** 1462–1471.

Erickson, J.W., V. Vaughn, W.A. Walter, F.C. Neidhardt, and C.A. Gross. 1987. Regulation of the promoters and transcripts of *rpoH*, the *Escherichia coli* heat shock regulatory gene. *Genes Dev.* **1:** 419–432.

Fayet, O., J.-M. Louarn, and C. Georgopoulos. 1986. Suppression of the *E. coli dnaA*46 mutation by amplification of the *groES* and *groEL* genes. *Mol. Gen. Genet.* **202:** 435–445.

Fayet, O., T. Ziegelhoffer, and C. Georgopoulos. 1989. The *groES* and *groEL* heat shock genes of *Escherichia coli* are essential for bacterial growth at all temperatures. *J. Bacteriol.* **171:** 1379–1385.

Fischer, H.M., M. Babst, T. Kaspar, G. Acuna, F. Arigoni, and H. Hennecke. 1993. One member of a *groESL*-like chaperonin multigene family in *Bradyrhizobium japonicum* is co-regulated with symbiotic nitrogen fixation genes. *EMBO J.* **12:** 2901–2912.

Friedman, D.I., E.R. Olson, K. Tilly, C. Georgopoulos, I. Herskowitz, and F. Banuett. 1984. Interactions of bacteriophage λ and host macromolecules in the growth of bacteriophage l. *Microbiol. Rev.* **48:** 299–325.

Gamer, J., H. Bujard, and B. Bukau. 1992. Physical interaction between heat shock proteins DnaK, DnaJ, and GrpE and the bacterial heat shock transcription factor σ^{32}. *Cell* **69:** 833–842.

Georgopoulos, C. 1989. The *Escherichia coli* dnaA initiation protein: A protein for all seasons. *Trends Genet.* **5:** 319–321.

———. 1992. The emergence of the chaperone machines. *Trends Biochem. Sci.* **17:** 295–299.

Georgopoulos, C. and W.J. Welch. 1993. Role of major heat shock proteins as molecular chaperones. *Annu. Rev. Cell Biol.* **9:** 601–635.

Georgopoulos, C., D. Ang, K. Liberek, and M. Zylicz. 1990. Properties of the *E. coli* heat shock proteins and their role in bacteriophage λ growth. In *Stress proteins in biology and medicine* (ed. R.I. Morimoto et al.), pp. 191–221. Cold Spring Harbor Laboratory Press, Cold Spring Harbor, New York.

Georgopoulos, C., R.W. Hendrix, S.R. Casjens, and A.D. Kaiser. 1973. Host participation in bacteriophage λ head assembly. *J. Mol. Biol.* **76:** 45–60.

Gottesman, S. and M.R. Maurizi. 1992. Regulation by proteolysis: Energy-dependent proteases and their targets. *Microbiol. Rev.* **56:** 592–621.

Gottesman, S., W.P. Clark, V. de Crecy-Lagard, and M.R. Maurizi. 1993. ClpX, an alternative subunit for the ATP-dependent Clp protease of *Escherichia coli*. *J. Biol. Chem.* **268:** 1–9.

Gragerov, A., E. Nudler, N. Komissarova, G.A. Gaitanaris, M.E. Gottesman, and V. Nikiforov. 1992. Cooperation of GroEL/GroES and DnaK/DnaJ heat shock proteins in preventing protein misfolding in *E. coli*. *Proc. Natl. Acad. Sci.* **89:** 10341–10344.

Gross, C.A., D.B. Straus, J.W. Erickson, and T. Yura. 1990. The function and regulation of heat shock proteins in *Escherichia coli*. In *Stress proteins in biology and medicine* (ed. R.I. Morimoto et al.), pp. 167–189. Cold Spring Harbor Laboratory Press, Cold Spring Harbor, New York.

Grossman, A.D., D.B. Straus, W.A. Walter, and C.A. Gross. 1987. σ^{32} synthesis can regulate the synthesis of heat shock proteins in *Escherichia coli*. *Genes Dev.* **1:** 179–184.

Hartman, D.J., N.J. Hoogenraad, R. Condron, and P.B. Høj. 1993. The complete primary structure of rat chaperonin 10 reveals a putative βαβ nucleotide-binding domain with homology to p21^{ras}. *Biochim. Biophys. Acta* **1164:** 219–222.

Hendrick, J.P., T. Langer, T.A. Davis, F.-U. Hartl, and M. Wiedman. 1993. Control of folding and membrane translocation by binding of the chaperone DnaJ to nascent polypeptides. *Proc. Natl. Acad. Sci.* **90:** 10216–10220.

Herman, C., T. Ogura, T. Tomoyasu, S. Hiraga, Y. Akiyama, K. Ito, R. Thomas, R. D'Ari, and P. Bouloc. 1993. Cell growth and λ phage development controlled by the same essential *Escherichia coli gene, ftsH/hflB. Proc. Natl. Acad. Sci.* **90:** 10861–10865.

Hupp, T.R. and J.M. Kaguni. 1993. Activation of DnaA5 protein by GrpE and DnaK heat shock proteins in initiation of DNA replication in *Escherichia coli. J. Biol. Chem.* **268:** 13137–13142.

Hupp, T.R., D.W. Meek, C.A. Midgley, and D.P. Lane. 1992. Regulation of the specific DNA binding function of p53. *Cell* **71:** 875–886.

———. 1993. Activation of the cryptic DNA binding function of mutant forms of p53. *Nucleic Acids Res.* **21:** 3167–3174.

Johnson, C., G.N. Chandrasekhar, and C. Georgopoulos. 1989. The dnaK and grpE heat shock proteins of *Escherichia coli* interact both *in vivo* and *in vitro. J. Bacteriol.* **171:** 1590–1596.

Johnson, K., I. Charles, G. Dougan, D. Pickard, P. O'Gaota, T. Ali, I. Miller, and C. Hormaecha. 1991. The role of a stress-response protein in *Salmonella typhimurium* virulence. *Mol. Microbiol.* **5:** 401–407.

Kamath-Loeb, A.S. and C.A. Gross. 1991. Translational regulation of σ^{32} synthesis: Requirement for an internal control element. *J. Bacteriol.* **173:** 3904–3906.

Kawasaki, Y., C. Wada, and T. Yura. 1992. Binding of RepE initiator protein to mini-F DNA origin (*ori2*). *J. Biol. Chem.* **267:** 11520–11524.

Keppel, F., B. Lipinska, D. Ang, and C. Georgopoulos. 1990. Mutational analysis of the phage T4 morphogenetic gene *31*, whose product interacts with the *E. coli* groEL protein. *Gene* **90:** 19–25.

Kitigawa, M., C. Wada, S. Yoshioka, and T. Yura. 1991. Expression of ClpB, an analog of the ATP-dependent protease regulatory subunit in *Escherichia coli*, is controlled by a heat shock σ factor (σ^{32}). *J. Bacteriol.* **173:** 4247–4253.

Kleerebezem, M. and J. Tommassen. 1993. Expression of the *pspA* gene stimulates efficient protein export in *Escherichia coli. Mol. Microbiol.* **7:** 947–956.

Kroh, H.E. and L.D. Simon. 1990. The ClpP component of Clp protease is the σ^{32}-dependent heat shock protein F21.5. *J. Bacteriol.* **172:** 6026–6034.

Kustu, S., E. Santero, J. Keener, D. Popham, and D. Weiss. 1989. Expression of σ^{54} (*ntrA*-) dependent genes is probably united by a common mechanism. *Microbiol. Rev.* **53:** 367–376.

Landry, S.J., J. Zeilstra-Ryalls, O. Fayet, C. Georgopoulos, and L.M. Gierasch. 1993. Characterization of a functionally important mobile domain of GroES. *Nature* **364:** 255–258.

Langer, T., G. Pfeifer, J. Martin, W. Baumeister, and F.-U. Hartl. 1992a. Chaperonin-mediated protein folding: GroES binds to one end of the GroEL cylinder, which accommodates the protein substrate within its central cavity. *EMBO J.* **11:** 4757–4765.

Langer, T., C. Lu, H. Echols, J. Flanagan, M.K. Hayer, and F.-U. Hartl. 1992b. Successive action of DnaK, DnaJ and GroEL along the pathway of chaperone-mediated protein folding. *Nature* **356:** 683–689.

Leonhardt, S.A., K. Fearon, P.N. Danese, and T.L. Mason. 1993. *HSP78* encodes a yeast mitochondrial heat shock protein in the Clp family of ATP-dependent proteases. *Mol. Cell. Biol.* **13:** 6304–6313.

Liberek, K. and C. Georgopoulos. 1993. Autoregulation of the *Escherichia coli* heat shock response by the DnaK and DnaJ heat shock proteins. *Proc. Natl. Acad. Sci.* **90:**

11019–11023.

Liberek, K., T.P. Galitski, M. Zylicz, and C. Georgopoulos. 1992. The DnaK chaperone modulates the heat shock response of *Escherichia coli* by binding to the σ^{32} transcription factor. *Proc. Natl. Acad. Sci.* **89:** 3516–3520.

Liberek, K., J. Marszalek, D. Ang, C. Georgopoulos, and M. Zylicz. 1991a. *Escherichia coli* DnaJ and GrpE heat shock proteins jointly stimulate ATPase activity of DnaK. *Proc. Natl. Acad. Sci.* **88:** 2874–2878.

Liberek, K., D. Skowyra, M. Zylicz, C. Johnson, and C. Georgopoulos. 1991b. The *Escherichia coli* DnaK chaperone protein, the Hsp70 eukaryotic equivalent, changes its conformation upon ATP hydrolysis, thus triggering its dissociation from a bound target protein. *J. Biol. Chem.* **266:** 14491–14496.

Liberek, K., J. Osipiuk, M. Zylicz, D. Ang, J. Skorko, and C. Georgopoulos. 1990. Physical interactions among bacteriophage and *Escherichia coli* proteins required for initiation of λ DNA replication. *J. Biol. Chem.* **265:** 3022–3029.

Lipinska, B., S. Sharma, and C. Georgopoulos. 1988. Sequence analysis and transcriptional regulation of the *htrA* gene of *Escherichia coli:* A σ^{32}-independent mechanism of heat-inducible transcription. *Nucleic Acids Res.* **16:** 10053–10067.

Lipinska, B., M. Zylicz, and C. Georgopoulos. 1990. The HtrA (DegP) protein, essential for *Escherichia coli* growth at high temperatures, is an endopeptidase. *J. Bacteriol.* **172:** 1791–1797.

Lipinska, B., O. Fayet, L. Baird, and C. Georgopoulos. 1989. Identification, characterization and mapping of the *Escherichia coli htrA* gene, whose product is essential for bacterial viability only at elevated temperatures. *J. Bacteriol.* **171:** 1574–1584.

Martin, J., A.L. Horwich, and F.-U. Hartl. 1992. Prevention of protein denaturation under heat stress by the chaperonin Hsp60. *Science* **258:** 995–998.

Martin, J., S. Geromanos, P. Tempst, and F.-U. Hartl. 1993. Identification of nucleotide-binding regions in the chaperonin proteins GroEL and GroES. *Nature* **366:** 279–282.

McCarthy, J.S., and G.C. Walker. 1991. DnaK as a thermometer: Threonine-199 is site of autophosphorylation and is critical for ATPase activity. *Proc. Natl. Acad. Sci.* **88:** 513–517.

McLennan, N.F., A.S. Girshovich, N.M. Lissin, Y. Charters, and M. Masters. 1993. The strongly conserved carboxyl-terminus glycine-methionine motif of the *Escherichia coli* GroEL chaperonin is dispensable. *Mol. Microbiol.* **7:** 49–58.

Mecsas, J., P.E. Rouviere, J.W. Erickson, T.J. Donohue, and C.A. Gross. 1993. The activity of σ^E, an *Escherichia coli* heat-inducible σ-factor, is modulated by expression of outer membrane proteins. *Genes Dev.* **7:**2618–2628.

Missiakas, D., C. Georgopoulos, and S. Raina. 1993. The *Escherichia coli* heat shock gene *htpY*: Mutational analysis, cloning, sequencing, and transcriptional regulation. *J. Bacteriol.* **175:** 2613–2624.

Miyazaki, T., S. Tanaka, H. Fujita, and H. Itikawa. 1992. DNA sequence analysis of the *dnaK* gene of *Escherichia coli* B and of two *dnaK* genes carrying the temperature-sensitive mutations *dnaK7*(Ts) and *dnaK756*(Ts). *J. Bacteriol.* **174:** 3715–3722.

Nagai, H., H. Yuzawa, and T. Yura. 1991. Interplay of two *cis*-acting mRNA regions in translational control of σ^{32} synthesis during the heat shock response of *Escherichia coli*. *Proc. Natl. Acad. Sci.* **88:** 10515–10519.

Noble, J.A., M.A. Innis, E.V. Koonin, K.E. Rudd, F. Banuett, and I. Herskowitz. 1993. The *Escherichia coli hflA* locus encodes a putative GTP-binding protein and two membrane proteins, oen of which contains a protease-like domain. *Proc. Natl. Acad. Sci.* **90:** 10866–10870.

Osipiuk, J., C. Georgopoulos, and M. Zylicz. 1993. Initiation of λ DNA replication: The

Escherichia coli small heat-shock proteins, DnaJ and GrpE, increase DnaK's affinity for the λP protein. *J. Biol. Chem.* **268:** 4821–4827.

Palleros, D.R., K.L. Reid, L. Shi, W.J. Welch, and A.L. Fink. 1993. ATP-induced protein-Hsp70 complex dissociation requires K^+ but not ATP hydrolysis. *Nature* **365:** 664–666.

Park, S.K., K.I. Kim, K.M. Woo, J.H. Seol, K. Tanaka, A. Ichihara, D.B. Ha, and C.H Chung. 1993. Site-directed mutagenesis of the dual translational initiation sites of the *clpB* gene of *Escherichia coli* and characterization of its gene products. *J. Biol. Chem.* **268:** 20170–20174.

Raina, S. and C. Georgopoulos. 1990. The identification and characterization of a new heat shock gene, *htrC*, whose product is essential for *Escherichia coli* viability at high temperatures. *J. Bacteriol.* **172:** 3417–3426.

Rospert, S., T. Junne, B.S. Glick, and G. Schatz. 1993a. Cloning and disruption of the gene encoding yeast mitochondrial chaperonin 10, the homolog of *E. coli* groES. *FEBS Lett.* **335:** 358–360.

Rospert, S., B.S. Glick, P. Jenö, G. Schatz, M.J. Todd, G.H. Lorimer, and P.V. Viitanen. 1993b. Identification and functional analysis of chaperonin 10, the groES homolog from yeast mitochondria. *Proc. Natl. Acad. Sci.* **90:** 10967–10971.

Sanchez, Y., D.A. Parsell, J. Taulien, J.L. Vogel, E.A. Craig, and S. Lindquist. 1993. Genetic evidence for a functional relationship between Hsp104 and Hsp70. *J. Bacteriol.* **175:** 6484–6491.

Schmid, D., H. Gehring, and P. Christen. 1994. Kinetics of molecular chaperone action. *Science* (in press).

Schröder, H., T. Langer, F.-U. Hartl, and B. Bukau. 1993. DnaK, DnaJ and GrpE form a cellular chaperone machinery capable of repairing heat-induced protein damage. *EMBO J.* **12:** 4137–4144.

Segal, G. and E.Z. Ron. 1993. Heat shock transcription of the *groESL* operon of *Agrobacterium tumefaciens* may involve a hairpin-loop structure. *J. Bacteriol.* **175:** 3083–3088.

Sell, S.M., C. Eisen, D. Ang, M. Zylicz, and C. Georgopoulos. 1990. The isolation and characterization of *dnaJ* null mutants of *Escherichia coli. J. Bacteriol.* **62:** 939–944.

Seol, J.H., S.K. Woo, E.M. Jung, S.J. Yoo, C.S. Lec, K. Kim, K. Tanaka, A. Ichihara, D.B. Ha, and C.H. Chung. 1991. Protease Do is essential for survival of *Escherichia coli* at high temperatures: Its identity with the *htrA* gene product. *Biochem. Biophys. Res. Commun.* **176:** 730–736.

Sherman, M.Y. and A.L. Goldberg. 1992a. Involvement of the chaperonin DnaK in the rapid degradation of a mutant protein in *Escherichia coli. EMBO J.* **11:** 71–77.

———. 1992b. Heat shock in *Escherichia coli* alters the protein-binding properties of the chaperonin GroEL by inducing its phosphorylation. *Nature* **257:** 167–169.

———. 1993. Heat shock of *Escherichia coli* increases binding of dnaK (the hsp70 homolog) to polypeptides by promoting its phosphorylation. *Proc. Natl. Acad. Sci.* **90:** 8648–8652.

Silver, P.A. and J.C. Way. 1993. Eukaryotic DnaJ homologs and the specificity of Hsp70 activity. *Cell* **74:** 5–6.

Skowyra, D. and S. Wickner. 1993. The interplay of the GrpE heat shock protein and Mg^{2+} in RepA monomerization by DnaJ and DnaK. *J. Biol. Chem.* **268:** 25296–25301.

Skowyra, D., C. Georgopoulos, and M. Zylicz. 1990. The *Escherichia coli* dnaK protein, the hsp70 homologue, can reactivate heat-inactivated RNA polymerase in an ATP hydrolysis-dependent reaction. *Cell* **62:** 939–944.

Squires, C. and C.L. Squires. 1992. The Clp proteins: Proteolysis regulators or molecular chaperones? *J. Bacteriol.* **174:** 1081–1085.

Squires, C.L., S. Pedersen, B.M. Ross, and C. Squires. 1991. ClpB is the *Escherichia coli* heat shock protein HtpG. *J. Bacteriol.* **173:** 4254–4262.

Strauch, K.L. and J. Beckwith. 1988. An *Escherichia coli* mutation preventing degradation of abnormal periplasmic proteins. *Proc. Natl. Acad. Sci.* **85:** 1576–1580.

Strauch, K.L., K. Johnson, and J. Beckwith. 1989. Characterization of *degP*, a gene required for proteolysis in the cell envelope and essential for growth of *Escherichia coli* at high temperature. *J. Bacteriol.* **171:** 2689–2696.

Straus, D., W. Walter, and C.A. Gross. 1987. The heat shock response of *E. coli* is regulated by changes in the concentration of σ^{32}. *Nature* **329:** 348–351.

————. 1988. *Escherichia coli* heat shock gene mutants are defective in proteolysis. *Genes Dev.* **2:** 1851–1858.

————. 1990. DnaK, DnaJ, and GrpE heat shock proteins negatively regulate heat shock gene expression by controlling the synthesis and stability of σ^{32}. *Genes Dev.* **4:** 2202–2209.

Suzuki, T., A. Itoh, S. Ichihara, and S. Mizushima. 1987. Characterization of the *sppA* gene coding for protease IV, a signal peptide peptidase of *Escherichia coli*. *J. Bacteriol.* **169:** 2523–2528.

Tilly, K. and C. Georgopoulos. 1982. The *groEL* and *groES* morphogenetic gene products of *Escherichia coli* interact *in vivo*. *J. Bacteriol.* **149:** 1082–1088.

Tilly, K., J. Spence, and C. Georgopoulos. 1989. Modulation of the stability of *Escherichia coli* heat shock regulatory factor σ^{32}. *J. Bacteriol.* **171:** 1585–1589.

Tilly, K., N. McKittrick, M. Zylicz, and C. Georgopoulos. 1983. The dnaK protein modulates the heat-shock response of *Escherichia coli*. *Cell* **34:** 641–646.

Tomizawa, J. and T. Som. 1984. Control of ColE1 plasmid replication: Enhancement of binding of RNA I to the primer transcript by the rom protein. *Cell* **38:** 871–878.

Ueguchi, C. and K. Ito. 1992. Multicopy suppression: An approach to understanding intracellular functioning of the protein export system. *J. Bacteriol.* **174:** 1454–1461.

Wall, D., M. Zylicz, and C. Georgopoulos. 1994. Identification of the conserved domain of the *Escherichia coli* DnaJ protein that stimulates DnaK's ATPase activity and is required for λ DNA replication. *J. Biol. Chem.* (in press).

Wang, N., S. Gottesman, M.C. Willingham, M.M. Gottesman, and M.R. Maurizi. 1993. A human mitochondrial ATP-dependent protease that is highly homologous to bacterial Lon protease. *Proc. Natl. Acad. Sci.* **90:** 11247–11251.

Wang, Q. and J.M. Kaguni. 1989a. dnaA protein regulates transcription of the *rpoH* gene of *Escherichia coli*. *J. Biol. Chem.* **264:** 7338–7344.

————. 1989b. A novel sigma factor is involved in expression of the *rpoH* genes of *Escherichia coli*. *J. Bacteriol.* **171:** 4248–4253.

Wang, T.-F., J.-H. Chang, and C. Wang. 1993. Identification of the peptide binding domain of hsc70. *J. Biol. Chem.* **268:** 26049–26051.

Weiner, L., J.L. Brissette, and P. Model. 1991. Stress-induced expression of the *Escherichia coli* phage shock protein operon is dependent on σ^{54} and modulated by positive and negative feedback mechanisms. *Genes Dev.* **5:** 1912–1923.

Wetzstein, M., U. Völker, J. Dedio, S. Löbau, and U. Zuber. 1992. Cloning, sequencing, and molecular analysis of the *dnaK* locus from *Bacillus subtilis*. *J. Bacteriol.* **174:** 3300–3310.

Wickner, S., J. Hoskins, and K. McKenney. 1991. Monomerization of RepA dimers by heat shock proteins activates binding to DNA replication origin. *Proc. Natl. Acad. Sci.* **88:** 7903–7907.

Wild, J., E. Altman, T. Yura, and C.A. Gross. 1992. DnaK and dnaJ heat shock proteins participate in protein export in *Escherichia coli. Genes Dev.* **6**: 1165–1172.

Wild, J., W.A. Walter, C.A. Gross, and E. Altman. 1993. Accumulation of secretory protein precursors in *Escherichia coli* induces the heat shock response. *J. Bacteriol.* **175**: 3992–3997.

Wojtkowiak, D., C. Georgopoulos, and M. Zylicz. 1993. Isolation and characterization of ClpX, a new ATP-dependent specificity component of the Clp protease of *Escherichia coli. J. Biol. Chem.* **268**: 22609–22617.

Woo, K.M., K.I. Kim, A.L. Goldberg, D.B. Ha, and C.H. Chung. 1992. The heat-shock protein ClpB in *Escherichia coli* is a protein-activated ATPase. *J. Biol. Chem.* **267**: 20429–20434.

Yura, T., H. Nagai, and H. Mori. 1993. Regulation of the heat-shock response in bacteria. *Annu. Rev. Microbiol.* **47**: 321–350.

Yuzawa, H., H. Nagai, H. Mori, and T. Yura. 1993. Heat induction of σ^{32} synthesis mediated by mRNA secondary structure: A primary step of the heat shock response in *Escherichia coli. Nucleic Acids Res.* **21**: 5449–5455.

Zeilstra-Ryalls, J., O. Fayet, and C. Georgopoulos. 1991. The universally-conserved GroE chaperonins. *Annu. Rev. Microbiol.* **45**: 301–325.

Zeilstra-Ryalls, J., O. Fayet, L. Baird, and C. Georgopoulos. 1993. Sequence analysis and phenotypic characterization of *groE* mutations that block λ and T4 bacteriophage growth. *J. Bacteriol.* **175**: 1134–1143.

Ziemienowicz, A., D. Skowyra, J. Zeilstra-Ryalls, O. Fayet, C. Georgopoulos, and M. Zylicz. 1993. Either of the *Escherichia coli* GroEL/GroES and DnaK/DnaJ/GrpE chaperone machines can reactivate heat-treated RNA polymerase: Different mechanisms for the same activity. *J. Biol. Chem.* **268**: 25425–25431.

Zylicz, M., D. Ang, and C. Georgopoulos. 1987. The grpE protein of *Escherichia coli*: Purification and properties. *J. Biol. Chem.* **262**: 17437–17442.

10

Molecular Chaperone Functions of hsp70 and hsp60 in Protein Folding

Judith Frydman and Franz-Ulrich Hartl
Cellular Biochemistry and Biophysics Program
Memorial Sloan-Kettering Cancer Center
New York, New York 10021

I. INTRODUCTION

Molecular chaperones have the general property of interacting with other proteins in their nonnative conformations. The chaperones of the hsp70 and hsp60 families function by preventing aggregation of newly synthesized polypeptides, and they then mediate their folding to the native state in an ATP-dependent process. Both classes of components are functionally distinct but can cooperate in a sequential folding pathway that has been reconstituted in vitro. Intensive functional studies of these proteins have led to a revision of the long-held view that protein folding in the cell is a spontaneous process. In this chapter, we discuss the mechanistic principles of hsp70 and hsp60 action in protein folding.

The Biology of Heat Shock Proteins and Molecular Chaperones
©1994 Cold Spring Harbor Laboratory Press 0-87969-427-0/94 $5 + .00

II. THE PROBLEM OF IN VIVO PROTEIN FOLDING

All of the information necessary for a polypeptide chain to fold to its native state is contained within its amino acid sequence (Anfinsen 1973). How the information present in the linear sequence of amino acid building blocks is transformed into a unique three-dimensional structure is not yet clear. However, considerable progress has been made by biophysicists in understanding the pathways through which a defined, thermodynamically stable structure arises from the astronomically large number of conformations the polypeptide chain adopts in the unfolded state (Jaenicke 1987). Many examples exist, especially for small proteins, where folding occurs spontaneously in vitro after the unfolded polypeptide is diluted from denaturant into aqueous solution. The pathway followed by globular proteins involves the restriction of conformational space by the very rapid collapse (within milliseconds) of hydrophobic residues to the interior of the molecule. This process gives rise to an ensemble of compact folding intermediates or molten globule-like states that contain elements of secondary structure but lack the stable tertiary structure interactions of the native state (Christensen and Pain 1991). Since these folding intermediates still expose hydrophobic surfaces to the solvent, they tend to engage in unproductive intermolecular interactions that result in the formation of insoluble aggregates. This tendency to aggregate usually necessitates that in vitro refolding experiments be performed at low concentrations of protein and at low temperature, conditions that minimize aggregation. How then is the efficient de novo folding of proteins realized within the cellular environment? In vivo, the high concentration of both total protein (20–30%) and in particular unfolded polypeptides (30–50 µM nascent chains in the *Escherichia coli* cytosol) would strongly favor unproductive interactions over the correct folding pathway. Another important consideration is that in the cell, proteins emerge from ribosomes, or at the *trans*-side of subcellular membranes, with their amino termini first and are therefore not available for folding until synthesis is completed. In vitro folding studies have shown that due to the cooperative nature of the folding process, the formation of stable tertiary structure requires the presence of a complete polypeptide or at least a complete folding domain (usually 100–200 amino acid residues) (Jaenicke 1987; Flanagan et al. 1991). Until such a chain length is reached, ribosome-bound polypeptides would exist as partially folded chains with a strong tendency to undergo aberrant interactions.

In recent years, it has become clear how cells solve the problem of in vivo protein folding. Newly synthesized proteins are shepherded to their correct native structure by several structurally unrelated families of proteins, now generally termed "molecular chaperones" (Gething and

Sambrook 1992; Hendrick and Hartl 1993). Molecular chaperones are ubiquitous. They bind to and stabilize the nonnative conformations of other proteins and subsequently facilitate their correct folding by releasing them in a controlled manner. The common property shared by molecular chaperones is the ability to recognize structural elements exposed in unfolded or partially denatured proteins, such as hydrophobic surfaces (Gething and Sambrook 1992; Hendrick and Hartl 1993). They do not bind to native proteins, and they are capable of interacting with many different polypeptide chains without exhibiting an apparent sequence preference. Many molecular chaperones are also classified as stress or heat shock proteins as their expression can be induced under a variety of cellular stresses. It should be noted, however, that most "stress proteins" also have essential functions under normal cellular conditions. The molecular chaperones described initially were nucleoplasmin (Laskey et al. 1978) and the chloroplast ribulose bisphosphate carboxylase (Rubisco)-binding protein, based on their role in promoting the oligomeric assembly of nucleosomes and Rubisco, respectively (Ellis 1987). The definition of molecular chaperone has subsequently been extended to include proteins, such as the members of the hsp70 and hsp60 families (Table 1), which mediate the folding of monomeric polypeptide chains in ATP-dependent reactions.

Cells appear to have developed two strategies to prevent the incorrect intermolecular association of polypeptide chains and to promote their proper folding (Hendrick and Hartl 1993). First, chaperones, such as the members of the hsp70 family, shield hydrophobic surfaces to prevent aggregation at a stage of biosynthesis, when productive folding of the polypeptide chain is not yet possible. This is the case during translation or membrane translocation of proteins. The hsp70s act as monomers or dimers and function in cooperation with DnaJ-like chaperones. Second, a complete but as yet unfolded protein (or a protein domain) is sequestered from the cellular milieu to prevent aggregation and at the same time allow folding to the native state to proceed. This function is carried out by the hsp60s (or "chaperonins") whose unique double-ring structure creates a propitious environment that facilitates the folding of proteins under the conditions prevailing in the cytosol. The members of the hsp70/DnaJ and hsp60 families, respectively, can act together in a sequential pathway of chaperoned folding.

III. CHAPERONES OF THE HSP70 FAMILY

The members of the 70-kD family of stress proteins (Table 1) were discovered through their rapid induction during exposure of cells to heat or

Table 1 Components of the hsp70, hsp60, and TRiC Families of Molecular Chaperones with Their Functions

Subcellular localization	Organism	Chaperone	Cooperating factors	Activity
hsp70 FAMILY				
Prokaryotes cytosol	*E. coli*	DnaK	DnaJ, GrpE	stabilizes newly made proteins in vivo; preserves folding competence of polypeptides; promotes oligomer assembly/disassembly; reactivates thermally inactivated proteins in vivo and in vitro; facilitates protein degradation; controls heat shock response
Eukaryotes cytosol	*S. cerevisiae*	Ssa1–4p	Ydj1p (DnaJ homolog)	stimulates protein transport across organellar membranes; binds to nascent polypeptides; dissociates clathrin from clathrin coats; promotes lysosomal degradation of cytosolic proteins
		Ssb1,2p	GrpE homolog not identified	
	mammals	Hsc73	NEM-sensitive factor, hsp40(?)	
Endoplasmic reticulum	*S. cerevisiae*	Kar2p	Sec63p (has limited homology with DnaJ)	promotes protein translocation into ER
	mammals	BiP/Grp78	none identified	binds unassembled subunits of multisubunit ER proteins
Mitochondria	*S. cerevisiae*	Ssc1p	DnaJ, GrpE homologs	promotes protein translocation into mitochondria and subsequent folding

Location	Organism	Protein	Co-chaperone	Function
Chloroplasts	P. sativum, E. gracilis	cthsp70	none identified	promotes insertion of light-harvesting complex into thylakoid membrane
hsp60 FAMILY				
Prokaryotes cytosol	E. coli	GroEL	GroES	promotes protein folding; required for phage assembly
Eukaryotes mitochondria	S. cerevisiae, N. crassa	hsp60	hsp10 (GroES homolog)	promotes folding and assembly of imported proteins; binds heat-denatured mitochondrial proteins, preventing aggregation
	mammals	hsp58	hsp10	
Chloroplasts	plants	cpn60 (Rubisco-binding protein)	cpn10	promotes folding and assembly of imported proteins
TRiC FAMILY				
Archaebacteria cytosol	Sulfolobus, Pyrodictium	TF55, thermosome	none identified, none identified	binds thermally denatured proteins and mediates their ATP-dependent folding
Eukaryotes cytosol	mammals	TRiC (TCP-1 ring complex)	two factors for tubulin folding	promotes folding of actin, tubulin, and firefly luciferase

For references, see text and review articles by Gething and Sambrook (1992) and Hendrick and Hartl (1993).

other forms of stress (Tissières et al. 1974). The association of these proteins with denatured polypeptides was proposed on the basis of the accumulation of hsp70 in the nucleoli of cells (Pelham 1986). The hsp70s are now known to be involved in a large variety of cellular functions related to almost every aspect of what can be described as the "life cycle" of proteins. In the following sections, we summarize the roles of hsp70 in different cellular processes and try to understand them in the context of a general molecular mechanism of action.

The generic name hsp70 is used for the members of a family of highly conserved proteins that are present in prokaryotic and eukaryotic cells under normal growth conditions and under stress. They possess weak ATPase activity and bind both to peptides and to extended polypeptide chains in the absence of ATP; release of the bound substrate requires hydrolyzable ATP. The hsp70 structure presents little variation from *E. coli* to mammals. It consists of two domains; the amino-terminal domain of approximately 40 kD contains the ATP-binding site and is more conserved than the carboxy-terminal 25-kD domain that carries the substrate-binding function (for review, see Gething and Sambrook 1992; Hendrick and Hartl 1993). Elucidation of the crystal structure of the amino-terminal domain revealed that although unrelated in its sequence, it has the same folding motif as G-actin and hexokinase (Flaherty et al. 1990). Changes in the nucleotide status of these proteins trigger major conformational alterations, which is probably also the case with hsp70. For example, ATP binding changes the fragment pattern obtained after trypsin treatment of DnaK, the hsp70 homolog of *E. coli* (Liberek et al. 1991b). The conformational changes induced by ATP binding result in a reduced affinity of the chaperone for the substrate polypeptide.

Using rapid gel-filtration techniques, Palleros et al. (1993) have shown recently that the dissociation of the bound protein from DnaK is triggered by the binding of ATP. This surprising finding revises the widely held view that ATP hydrolysis is required for peptide or polypeptide release from hsp70. hsp70 can alternate between the ADP and ATP states with high and low substrate affinities, respectively. At physiological concentrations of ATP and K^+, the affinity of hsp70 for proteins is low. On the other hand, binding of ADP strengthens the interaction of hsp70 with substrate polypeptides (Palleros et al. 1991). hsp70 has a higher affinity for ADP than for ATP (Palleros et al. 1991), and therefore substrates will bind stably to ADP-hsp70 until the nucleotide is exchanged for ATP. As we discuss below, the interconversion between the two states is catalyzed by additional factors that regulate the binding and release of unfolded polypeptide by hsp70.

A. Substrate Binding Specificity of hsp70

What elements in a polypeptide chain are recognized by hsp70? The greater structural variability in the carboxy-terminal substrate-binding domains among the members of the hsp70 family suggests differences in their substrate-binding specificities. It is known that different hsp70s are not functionally interchangeable (Brodsky et al. 1993). Although there is no detailed information about the structure of the carboxy-terminal domain, it has been proposed to resemble the peptide-binding site of major histocompatibility complex class I (MHC-I) molecules on the basis of secondary structure prediction and computer modeling (Rippmann et al. 1991).

The MHC-I molecule presents nonapeptides at the cell surface, and in *Xenopus*, it shows limited sequence homology with hsp70 (Flajnik et al. 1991). This may reflect the evolutionary origin of the antigen-presenting system. The analysis of the substrate-binding specificity of BiP, the resident hsp70 of the endoplasmic reticulum (ER), was examined by two independent approaches. One selection scheme evaluated the binding of peptides of increasing length by their effect on the ATPase activity of BiP (Flynn et al. 1991). This study revealed that this hsp70 binds with optimal affinity to heptameric peptides and that hydrophobic amino acids were favored at every position. The other approach characterized the binding preference of BiP by affinity panning of libraries of bacteriophages that display random peptide sequences (Blond-Elguindi et al. 1993).

This study showed that BiP binds peptides containing a subset of aromatic and hydrophobic residues at alternating positions. This would be consistent with the peptide binding in an extended conformation, with the side chains of alternating residues pointing into a cleft on the BiP molecule. Binding to BiP requires the presence of only two hydrophobic residues in the correct spacing. This motif is redundant enough to occur frequently in most proteins. hsp70 also binds to unfolded proteins such as reduced carboxymethylated α-lactalbumin, a protein with little or no secondary structure. Presumably, these proteins expose stretches of hydrophobic amino acids that would normally be buried within the interior of the protein and are recognized by hsp70. Nuclear magnetic resonance (NMR) analysis of the conformation of a peptide bound to DnaK indicates that it is indeed maintained in an extended state (Landry et al. 1992). It has been proposed that hsp70 provides the bound segment of an unfolded chain with an environment energetically similar to that of the hydrophobic core of a folded protein (Flynn et al. 1991).

B. Role of hsp70 in Membrane Translocation of Proteins

Stably folded proteins are generally unable to traverse the membranes of subcellular organelles such as mitochondria, chloroplasts, or ER. Precursor proteins destined for posttranslational uptake into these organelles must be stabilized by molecular chaperones in a loosely folded, translocation-competent state while in the cytosol (for review, see Gething and Sambrook 1992; Hendrick and Hartl 1993). Genetic analysis in yeast has indicated an important role for the Ssa proteins (cytosolic hsp70) in the import of proteins into mitochondria and the ER (Deshaies et al. 1988). Biochemical experiments have demonstrated that hsp70 is involved in keeping precursor proteins import-competent, although purified hsp70 alone is not sufficient to carry out this function (Chirico et al. 1988; Murakami et al. 1988; Sheffield et al. 1990). In the case of import into mitochondria, the precursor protein has been detected in a 250-kD complex containing hsp70 and other as yet uncharacterized cytosolic components (Sheffield et al. 1990). A requirement for hsp70 has also been observed for import of proteins into the nucleus.

C. Cooperation of hsp70s with the Chaperone DnaJ

The hsp70 homolog of *E. coli*, DnaK, is one of the most studied members of the hsp70 family. Most cellular functions of DnaK require the cooperation of two additional factors, the chaperone DnaJ and the nucleotide exchange factor GrpE (Georgopolous 1992). The recent identification of multiple DnaJ homologs in eukaryotic cells suggests that the mechanisms by which DnaK is regulated are of general significance (Caplan et al. 1993). DnaJ is a constitutively expressed stress protein of 45 kD that is encoded in the same operon as DnaK. DnaJ usually functions as a dimer and was initially characterized through its functional cooperation with DnaK and GrpE in bacteriophage λ DNA replication (Georgopolous 1992). For λ DNA synthesis to proceed, the λP protein must dissociate from the replication complex at the origin of replication. Dissociation begins when DnaJ binds to λP and targets DnaK to the complex, which then dissociates by addition of the 20-kD regulator GrpE in an ATP-dependent manner (Alfano and McMacken 1989; Zylicz et al. 1989). The DnaK system has a similar role in the replication of phages P1 and P7 DNA. In this case, the replication protein RepA exists as an inactive dimer that does not bind DNA (Wickner et al. 1991; Lindquist 1992). Dissociation of the RepA dimer into active monomers is achieved by the sequential binding of DnaJ and DnaK followed by ATP hydrolysis. The rate at which dissociation occurs is substantially enhanced in the presence of GrpE. Interestingly, DnaK does not bind to either λP or RepA in the absence of DnaJ.

Another cellular process in which DnaK, DnaJ, and GrpE cooperate is the regulation of the transcription of heat shock genes as a response to various forms of cellular stress. Under normal growth conditions, the heat shock transcription factor σ^{32} is bound to DnaK, DnaJ, and GrpE (Gamer et al. 1992; Liberek et al. 1992) and is very rapidly degraded with a half-life of minutes. It is thought that upon exposure to heat stress, σ^{32} is displaced from DnaK and DnaJ by the increased amount of denatured protein and is then able to interact with the transcriptional machinery. It was found that σ^{32} is dissociated from its complex with DnaK and GrpE in the presence of ATP (Gamer et al. 1992). Notably, in the examples discussed above, DnaK and DnaJ bind to folded proteins. Presumably, these substrates transiently expose structural features typically found in unfolded proteins, such as hydrophobic surfaces, that are recognized by these chaperones. Release from DnaK and DnaJ is likely to result in a conformational change in these substrate proteins upon which they lose their binding affinity for the chaperone. The capacity of DnaJ to function as a chaperone has been demonstrated with unfolded model proteins in vitro (Langer et al. 1992b). For example, purified DnaJ efficiently binds to unfolded rhodanese and luciferase, preventing their aggregation. The elements recognized by DnaJ in unfolded proteins are probably different from those recognized by DnaK. Unlike the latter, DnaJ does not bind the extended conformation of reduced and carboxymethylated α-lactalbumin. In contrast to DnaK, DnaJ binds casein, a protein that exposes hydrophobic regions (Langer et al. 1992b).

What is the effect of DnaJ and GrpE on DnaK? Both proteins interact with DnaK in vitro. DnaJ alone activates the ATPase of DnaK approximately twofold by stimulating the hydrolysis of DnaK-bound ATP, whereas GrpE promotes exchange of the DnaK-bound ADP for ATP. In the presence of both DnaJ and GrpE, the ATPase activity of DnaK can be stimulated up to 50-fold (Liberek et al. 1991a). As mentioned above, the ADP-bound state of DnaK has a high affinity for substrate protein, and the ATP-bound state has a low substrate affinity. In Section G, we discuss how the coordinated action of DnaJ and GrpE could regulate the interconversion of these two functional states of DnaK in a manner reminiscent of the regulation of GTP-binding proteins.

D. The Eukaryotic Homologs of DnaJ

A variety of DnaJ homologs have been identified in different eukaryotic cell compartments including the cytosol, ER, mitochondria, and nucleus (Table 1), whereas a eukaryotic GrpE homolog has so far only been described for the mitochondrial matrix. The yeast protein Ydj1p/Mas5p is a cytosolic DnaJ homolog that is required for the posttranslational import of proteins into mitochondria (Caplan and Douglas 1991; Atencio and Yaffe 1992). It is likely to cooperate in this process with cytosolic hsp70. Another yeast DnaJ homolog, Sis1p, was identified as a suppressor of the Sit4p kinase and is localized in the cytosol and nucleus (Luke et al. 1991). In the cytosol, this protein participates in regulating the initiation of translation (Zhong and Arndt 1993). hsp40 has recently been characterized as a mammalian protein that has homology with Sis1p (Cheetham et al. 1992; Hattori et al. 1992). A further type of DnaJ homologous protein, Scj1p, is involved in the transport of proteins to the nucleus. Its subcellular localization is not known exactly (Blumberg and Silver 1991). The ER contains an integral membrane protein, Sec63p, which is structurally related to DnaJ and cooperates with the hsp70 BiP in protein translocation (see Feldheim et al. 1992). Similarly, the mitochondrial homolog of DnaJ, Mdj1p, has a role in mediating the folding of imported proteins. However, in contrast to the mitochondrial hsp70, it does not appear to be required for their translocation across the membrane (W. Neupert, pers. commun.). All of these proteins share a 70-amino-acid sequence called the "J-domain" which is thought to interact with the hsp70 proteins (Caplan et al. 1993). Two other domains characteristic of DnaJ are a sequence rich in glycine and phenylalanine and a cysteine-rich domain that might be a protein-binding zinc finger. These domains are conserved in some, but not all, DnaJ homologs (Caplan et al. 1993). The direct interaction between unfolded polypeptides and eukaryotic DnaJ homologs has not yet been demonstrated. It seems possible that the different proteins function together with specific forms of hsp70 regulating their function. For example, Ydj1p has been shown to stimulate the ATPase activity of the hsp70 Ssa proteins (Cyr et al. 1992) but not of the Ssb proteins (M. Douglas, pers. comm.).

E. Role of hsp70 in Thermal Stress

hsp70 has a role in protecting preexisting proteins against denaturation under conditions of cellular stress. Although such a function had been assumed early on (Pelham 1986), direct evidence for it has been presented only recently. DnaK can prevent the heat inactivation of *E. coli* RNA polymerase when present at a high molar excess, and in vitro, it is even able to renature the heat-inactivated protein (Skowyra et al. 1990). The function of DnaK in the prevention of heat denaturation of proteins has also been shown in vivo using firefly luciferase as a model protein (Schroder et al. 1993). Upon expression in the *E. coli* cytosol, luciferase

is rapidly inactivated when the cells are incubated at 42°C but can renature after a temperature down-shift to 30°C. This reactivation does not
occur in either *dnaK*, *dnaJ*, or *grpE* mutants, indicating that the
DnaK/DnaJ/GrpE chaperone team is required for this process. These observations have been reproduced in vitro illustrating the functional interaction of the three proteins (Schroder et al. 1993). When exposed to
42°C, luciferase denatures and aggregates. The presence of DnaJ during
incubation at 42°C does not prevent inactivation of the enzyme but suppresses its aggregation. Luciferase can then be reactivated by the addition of DnaK, GrpE, and Mg/ATP. The combination of DnaK together
with DnaJ during the thermal denaturation is more effective in preventing aggregation, reflecting the functional cooperation of both these
chaperones.

F. Role of hsp70 in De Novo Protein Folding

Mutations in the mitochondrial hsp70 of yeast, Ssc1p, are defective in
protein import into mitochondria (Kang et al. 1990). The physical association of mitochondrial hsp70 with translocating polypeptide chains
emerging from the inner surface of the membrane has been demonstrated
(Scherer et al. 1990). From these and other findings, it has been concluded that binding of the translocating polypeptide by mitochondrial
hsp70, and perhaps cooperating chaperones, is necessary to drive the import process and determine its unidirectionality (for details, see Langer
and Neupert, this volume). In addition to its role in translocation,
mitochondrial hsp70 is also required for the subsequent folding of newly
imported polypeptides that traverse the mitochondrial membranes as extended chains. In many cases, however, mitochondrial hsp70 is not sufficient to guarantee folding to the native state, and the newly imported
polypeptides must be transferred to the mitochondrial GroEL homolog
hsp60 for final folding (Ostermann et al. 1989; Manning et al. 1991).
Analogous to the situation in mitochondria, the hsp70 of the ER, BiP, is
required for the translocation of secretory precursors into the ER lumen
and for their subsequent folding and assembly (Vogel et al. 1990;
Nguyen et al. 1991). BiP cooperates in this process with the DnaJ-related
Sec63p (Sanders et al. 1992). After completion of translocation, the
proteins fold with the cooperation of additional folding catalysts and
chaperones such as protein disulfide isomerase and calnexin (Gething
and Sambrook 1992).

The high degree of conservation between the chaperone components
in the mitochondrial matrix and the *E. coli* cytosol suggests that the reactions of protein folding in these compartments follow similar principles.

This is also supported by the recent identification of DnaJ and GrpE homologs in mitochondria (W. Neupert and E. Craig, pers. comm.). The sequential action of chaperones of the hsp70 and hsp60 classes has been reconstituted in vitro using the *E. coli* DnaK (hsp70), DnaJ, and GrpE as well as GroEL and GroES (Langer et al. 1992b). Mitochondrial rhodanese, a 35-kD monomeric protein, served as a model substrate. Rhodanese has a low tendency to refold spontaneously when diluted from denaturant, but rather aggregates even at high protein dilution. These experiments revealed details about the function of the hsp70 system. Similar to the observations made with thermally unfolded luciferase (Schroder et al. 1993), the presence of DnaK alone did not prevent the aggregation of guanidinium-denatured rhodanese except when added at a very high molar excess. On the other hand, DnaJ was more efficient in preventing rhodanese aggregation, and this effect was synergistically potentiated by the presence of DnaK. Surprisingly, a tight ternary complex between unfolded rhodanese, DnaK, and DnaJ was formed in the presence of Mg/ATP (Langer et al. 1992b). In contrast, the weaker interaction between DnaK and rhodanese was resolved upon addition of Mg/ATP, indicating that DnaJ changes the interaction of the unfolded protein with DnaK. The protein bound to DnaJ and DnaK forms a 230-kD complex and has the conformation of a partially folded intermediate based on its intrinsic tryptophan fluorescence and anilino-naphthalene sulfonate (ANS)-binding properties. When GrpE and GroEL/GroES were added to the rhodanese/DnaK/DnaJ complex, efficient Mg/ATP-dependent refolding of rhodanese was observed. In the absence of GroEL/GroES, rhodanese was stably maintained in a nonaggregated, folding-competent state even when Mg/ATP and GrpE were present (Langer et al. 1992b).

More recently, the mode of action of DnaK, DnaJ, and GrpE in the folding of denatured proteins was investigated further. Firefly luciferase has been useful in these studies because its refolding does not require GroEL/GroES (A. Szabo et al., in prep.). DnaJ, DnaK, and GrpE allow the efficient ATP-dependent reactivation of denaturant-unfolded luciferase, whereas in the absence of chaperone, the protein aggregates. As in the case of unfolded rhodanese, the formation of a 250-kD complex between luciferase, DnaK, and DnaJ requires the presence of Mg^{++} and ATP. Reisolation of this complex showed that its DnaK has only ADP bound. Addition of GrpE then causes the dissociation of ADP but no change in the state of luciferase. In contrast, similar experiments with rhodanese showed that this protein is transferred to GroEL following the GrpE-dependent dissociation of ADP. Interestingly, in the case of unfolded luciferase, only the addition of either Mg/ATP or the non-hydrolyzable analog AMP-PNP resulted in the dissociation of the ternary complex with DnaK and DnaJ (A. Szabo et al., in prep.). In the presence of Mg/ATP, most of the luciferase was efficiently refolded, whereas addition of AMP-PNP allowed the folding of only a fraction (~30%) of the luciferase molecules present. Significantly, another 30% of the luciferase aggregated. These results are consistent with recent findings (Palleros et al. 1993) that ATP binding induces the low-affinity state of DnaK for unfolded polypeptide and is in itself sufficient to trigger the release of bound substrate. However, a single cycle of binding and release is not sufficient for full reactivation of luciferase. ATP hydrolysis is necessary for the reformation of the ternary complex between substrate protein, DnaK, and DnaJ to allow multiple rounds of interaction (A. Szabo et al., in prep.). It should be noted that no principal distinction could be established in these reconstitution studies between the interaction of rhodanese and luciferase with the hsp70 system. The different folding propensities of the two proteins are the reason luciferase is not dependent on GroEL/GroES to reach its native state. If, however, GroEL is included in the reaction, luciferase will be transferred from the complex with DnaK/DnaJ to GroEL as is the case for rhodanese.

G. The hsp70 Reaction Cycle of *E. coli* DnaK, DnaJ, and GrpE

On the basis of the observations described above and the known properties of hsp70, the following model can be proposed for the mechanism of DnaK, DnaJ, and GrpE in protein folding (Fig. 1): (1) In the absence of protein substrate, there is only a weak interaction between DnaK and DnaJ. Under physiological conditions, DnaK is in an ATP-bound form with low affinity for unfolded polypeptides and peptides. Unfolded protein is first bound by DnaJ. (2) The DnaJ complex then activates the ATP-hydrolytic activity of DnaK and stabilizes the ADP-bound state of DnaK that binds the DnaJ-stabilized protein strongly. In this way, DnaK is recruited into a ternary complex with DnaJ and substrate polypeptide. (3) The action of GrpE releases the bound ADP from DnaK and labilizes the interaction of DnaK with the protein. For some proteins, such as rhodanese, this may be sufficient to allow transfer to GroEL. (4) The net result of GrpE action is ATP binding to DnaK, which induces the low-affinity state of DnaK resulting in the release of the bound protein. The released protein may then fold to the native state, be transferred to a "downstream" chaperone system such as GroEL/GroES, or rebind to DnaJ and DnaK.

This reaction cycle would be consistent with the action of DnaK, DnaJ, and GrpE described in other systems, such as the regulation of

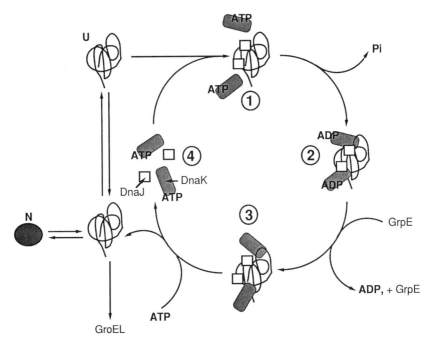

Figure 1 Model for the reaction cycle of DnaK, DnaJ, and GrpE in protein fold-ing. (*1*) Unfolded protein (U) interacts with DnaJ; (*2*) DnaK is recruited by DnaJ and hydrolyzes bound ATP to ADP, which results in the formation of a tight ternary complex; (*3*) GrpE causes the dissociation of ADP from DnaK, which weakens the interaction between DnaK and U allowing protein transfer to GroEL in the case of rhodanese; (*4*) upon ATP binding to DnaK, the ternary complex dissociates. U may fold to the native state (N), rebind to DnaJ/DnaK, or be transferred to the hsp60, GroEL, dependent on the folding properties of the substrate protein.

DNA replication and of the heat shock response. In these cases, other-wise folded substrate proteins (e.g., λP and RepA) may interact with DnaJ and DnaK by exposing chaperone recognition elements typically present in unfolded polypeptides. Binding of DnaJ and DnaK could then result in conformational changes of the substrate, such as disassembly of an oligomeric complex. In this way, the more general chaperone func-tions of the hsp70 system in protein folding could be used for a multitude of specific regulatory purposes.

H. Interaction of hsp70 and DnaJ with Nascent Polypeptides

Since the acquisition of stable tertiary structure requires that a complete polypeptide domain be available, folding will not occur cotranslationally in a residue by residue fashion as the amino terminus of the polypeptide chain emerges from the ribosome. As a consequence, nascent (i.e., ribosome-bound) chains will remain unfolded or partially folded. A function in shielding the hydrophobic surfaces of nascent chains in order to prevent misfolding and aggregation is attributed to hsp70. The association of hsp70 with a large spectrum of ribosome-bound chains has indeed been demonstrated in coimmunoprecipitation experiments using anti-hsp70 antibodies (Beckmann et al. 1990). This interaction is disrupted by Mg/ATP. Whether hsp70 is required for efficient folding has not yet been addressed. Yeast strains with a loss of function in the hsp70s of the Ssb class show a reduced efficiency of translation, which is suppressed by a protein that has homology with elongation factor EF-1α (Nelson et al. 1992). The Ssb protein is associated with translating ribosomes and released upon puromycin treatment. It seems possible that the binding of hsp70 to the nascent chains facilitates their efficient passage through the ribosomal exit channel (Nelson et al. 1992). This could be analogous to the role of mitochondrial hsp70 and BiP in membrane translocation of proteins.

Using cell-free translation extracts, Hendrick et al. (1993) have recently shown that *E. coli* DnaJ binds cotranslationally to nascent chains of firefly luciferase and chloramphenicol acetyltransferase very early in translation. Binding of DnaJ allows the co- or posttranslational binding of DnaK. This interaction is productive for folding upon addition of GrpE. The DnaJ homologs of yeast, Sis1p and Ydj1p, are respectively associated with ribosomes or have a role in maintaining precursor proteins competent for membrane translocation. Together, these findings suggest that the hsp70s cooperate with DnaJ homologs in protecting nascent and newly synthesized polypeptides. It is conceivable that DnaJ binds initially and then targets hsp70 to the substrate. DnaJ would thus have an important function in selecting polypeptides for hsp70 binding. Future research will have to test this hypothesis.

IV. THE CHAPERONINS: HSP60 AND TRiC FAMILIES

The "chaperonin" class of molecular chaperones comprises two protein families that share important structural and functional properties (Table 1). In prokaryotic cells and in organelles of endosymbiotic origin (mitochondria and chloroplasts), the chaperonins are represented by the highly conserved hsp60 proteins. In the eukaryotic cytosol and in archaebacterial cells, the function of the hsp60s is apparently fulfilled by the recently identified members of the TRiC family (TRiC = TCP-1 ring complex).

A. The hsp60s: Function in Protein Folding and Assembly

The hsp60s (also known as chaperonin 60 or cpn60; Hemmingsen et al. 1988) are large oligomeric complexes of 60-kD subunits arranged as two stacked heptameric rings with a central cavity (Hendrix 1979; Hohn et al. 1979). Although stress inducible in bacteria, mitochondria, and presumably also in chloroplasts, the hsp60s are constitutively expressed and are essential for growth under all conditions. The different members of the family show a high degree of sequence homology (Gething and Sambrook 1992; Hendrick and Hartl 1993). The hsp60s cooperate with a smaller stress-inducible protein, hsp10 (also termed chaperonin 10 or cpn10). These "cochaperonins" are also highly conserved. They have a subunit size of approximately 10 kD and form single heptameric rings that bind to hsp60 (Lubben et al. 1990; Hartman et al. 1992). The hsp60 of *E. coli*, GroEL, and its hsp10, GroES, are encoded in a single operon (Tilly et al. 1981; Chadrasekhar et al. 1986).

The bacterial chaperonin GroEL was initially identified as a host protein necessary for the assembly of the heads of bacteriophage λ and the tails of phage T5 (Georgopoulos et al. 1973). The chloroplast chaperonin Rubisco subunit-binding protein was discovered due to its association with large subunits of ribulose bisphosphate carboxylase (Hemmingsen and Ellis 1986). This complex was dissociated in the presence of Mg/ATP and proved to be an intermediate in the assembly of the chloroplast enzyme from eight large and eight small subunits. The analysis of a temperature-sensitive yeast mutant strain defective in the gene encoding mitochondrial hsp60 demonstrated that chaperonin function is indeed essential for the assembly of oligomeric enzyme complexes (Cheng et al. 1989, 1990). Subsequently, evidence was provided that the basic function of the chaperonins is the folding of monomeric polypeptide chains. When the enzyme dihydrofolate reductase (DHFR) was imported into mitochondria as a fusion protein carrying a cleavable presequence, it was found in a complex with hsp60. Mg/ATP was required to release the protein from this complex in a folded form (Ostermann et al. 1989). It is noteworthy that DHFR folds spontaneously in vitro but in vivo requires molecular chaperones for correct folding.

The general role of chaperonins in the folding of monomeric proteins is underscored by the fact that, in vitro, GroEL can assist the folding of a large number of guanidinium unfolded enzymes that would otherwise aggregate when diluted from denaturant. This indicates that GroEL must be able to recognize certain structural features, such as hydrophobic surfaces, that are generally exposed by unfolded proteins. Indeed, a large fraction (~50%) of the proteins of *E. coli* are able to bind to the chaperonin when added from denaturant and are released in the presence of Mg/ATP (Viitanen et al. 1992). The requirement of GroEL for the assembly of oligomeric complexes appears to be a consequence of its role in folding the individual subunits to an "assembly-competent" state, which could then be stabilized by the chaperonin until the subunits have engaged in stable interactions with their partner molecules. For example, GroEL and GroES mediate the folding of subunits of ornithine transcarbamylase (OTC) which will then associate spontaneously to the active OTC trimer (Zheng et al. 1993).

The general role of GroEL in the folding of cytosolic proteins in vivo has recently been confirmed by the generation and analysis of a temperature-sensitive lethal mutant in the *groEL* gene (Horwich et al. 1993). Shift of the mutant cells to the nonpermissive temperature (37°C) reduced the rate of translation and impaired phage assembly. Significantly, two test proteins, the monomeric maltose-binding protein (expressed without signal sequence for export) and the trimeric OTC, were synthesized but failed to fold to their active forms (Horwich et al. 1993). The general fate of newly synthesized cytoplasmic proteins was analyzed in pulse-labeling experiments followed by cell fractionation and two-dimensional gel electrophoresis. After shift to the nonpermissive temperature, a large number of cytoplasmic proteins were absent from the fraction of soluble protein. Most of these proteins were detected in aggregates, although some appeared to be degraded. Notably, a subset of proteins appeared to fold normally in the *groEL* mutant strain (Horwich et al. 1993). These proteins may depend on other chaperone proteins, such as DnaK, DnaJ, and GrpE, for their folding and assembly.

B. hsp60 as a Stress Protein

The members of the hsp60/hsp10 chaperonin family are induced by forms of cellular stress that can cause protein denaturation (Hightower 1991). When mitochondria are exposed to heat stress, a large variety of polypeptides associate with the mitochondrial hsp60 (Martin et al. 1992). Functional hsp60 is required in vivo to prevent the thermal inactivation of DHFR imported into mitochondria. These findings indicate that members of the hsp60 family indeed function in stabilizing preexisting proteins under stress conditions. In doing so, they might cooperate with members of the hsp70 family (see below). At elevated temperatures, the de novo folding of proteins also requires the presence of higher levels of hsp60/hsp10. Overproduction of both GroEL and GroES, but not of each protein separately, can protect newly synthesized proteins from aggregating in *E. coli* cells lacking a normal heat shock response (Gragerov et al. 1992).

The capacity of the chaperonins to stabilize proteins under heat stress has been reconstituted in vitro. GroEL is able to prevent the thermal aggregation of a variety of enzymes including DHFR, rhodanese, and α-glucosidase by forming a binary complex with the heat-denatured proteins (Buchner et al. 1991; Holl et al. 1991; Martin et al. 1992). Importantly, GroEL must be present during thermal unfolding prior to the formation of aggregates in order to be effective. In the case of DHFR, the protease sensitivity and spectral characteristics of the protein are the same whether binding to the chaperonin occurred during thermal unfolding or upon dilution from denaturant (Martin et al. 1992). When the temperature is lowered, the thermally unfolded protein can be renatured from the binary complex with GroEL upon incubation with Mg/ATP and GroES.

C. Mechanism of Protein Folding by GroEL/GroES

1. GroEL Stabilizes the Collapsed State of Folding Proteins

GroEL binds to a large variety of unfolded proteins but not to their native counterparts. What is the nature of the structural elements recognized by the chaperonins? When substrate proteins such as Rubisco, DHFR, or rhodanese are bound to GroEL, they are maintained in a protease-sensitive conformation with tryptophan fluorescence properties intermediate between those of the unfolded and the native states (Martin et al. 1991; van derVies et al. 1992). Furthermore, the GroEL-associated proteins have a high affinity for the fluorescent dye ANS, a diagnostic probe of the so-called "molten globule" state that binds to exposed hydrophobic surfaces. This has led to the proposal that GroEL binds to and stabilizes the collapsed folding intermediates that form early during folding and that are particularly prone to aggregation (Martin et al. 1991). Unlike hsp70, GroEL does not recognize short peptides or polypeptides in extended conformations with high affinity. Conformational analysis by NMR has shown that a 13-mer peptide derived from rhodanese was weakly bound by GroEL and stabilized in an α-helical conformation (Landry and Gierasch 1991; Landry et al. 1992). In contrast, DnaK (hsp70) interacts with the same peptide in an extended form (Landry et al. 1992). Since GroEL also binds all β-sheet proteins (Schmidt and Buchner 1992), it seems likely that GroEL recognizes structural motifs other than secondary structure elements, possibly the exposure of hydrophobic residues or surfaces. This would be consistent with the results of a recent analysis of the binding specificity of GroEL for a series of conformers of the 15-kD protein α-lactalbumin (M. Hayer-Hartl, pers. comm.). These forms ranged from the reduced carboxymethylated α-lactalbumin (RCM-LA), which is fully extended, to the native protein (Ewbank and Creighton 1993). Neither the native nor the fully extended forms of α-lactalbumin are bound by GroEL. However, disruption of the tertiary structure of the protein, by removal of the stabilizing Ca^{++} and reduction of one of the four disulfide bonds (3SS-lactalbumin), creates the molten globule state in which the remaining three disulfides rearrange freely. A subset of the large number of possible 3SS-rearranged forms bind efficiently to GroEL, whereby the binding affinity appears to increase with the exposure of hydrophobic surface as measured by the adsorption of ANS. It will be of interest to determine whether the structural elements recognized by GroEL are present in a linear segment of the protein or whether they are composed of regions distant in the primary sequence.

Only one to two protein molecules can bind per GroEL double toroid and only one bound molecule is released and folded (Martin et al. 1991; Bochkareva et al. 1992). Electron microscopy has shown that the unfolded protein is bound within the central cavity of the cylinder created by the two rings (Langer et al. 1992a; Braig et al. 1993). This space is about 6 nm in diameter, and it appears that each ring cavity could accommodate a compact folding intermediate for a protein of approximately 70 kD (Braig et al. 1993). Folding of larger proteins could proceed domainwise, perhaps in concert with other chaperone systems such as DnaK and DnaJ.

2. GroES Is a Regulator of GroEL Function

Unfolded rhodanese, Rubisco, and a number of other proteins that bind tightly to GroEL have been shown to require both GroES and ATP hydrolysis for their folding (Goloubinoff et al. 1989; Badcoe et al. 1991; Buchner et al. 1991; Martin et al. 1991; Mendoza et al. 1991). Proteins that associate less stably with the chaperonin, such as DHFR, can be released for folding in the absence of GroES just by the binding of ATP or a nonhydrolyzable analog to GroEL (Laminet et al. 1990; Martin et al. 1991; Viitanen et al. 1991). These latter proteins are characterized by the ability to reach the native state spontaneously in vitro. Presumably, they engage in fewer contacts with the ring subunits of GroEL. ATP binding lowers the substrate affinity of GroEL sufficiently to set these proteins free for spontaneous folding. However, in vivo productive protein folding by GroEL may generally require GroES. Only in the presence of GroES does the folding reaction occur in a shielded environment, thus effectively preventing protein aggregation. In the absence of GroES, rhodanese is only slowly released from GroEL, even in the presence of

Mg/ATP. The released, partially folded protein rapidly rebinds to the chaperonin but can be displaced by the loosely folded protein casein, which itself is a substrate for GroEL binding (Martin et al. 1991). This displacement results in the aggregation of the partially folded rhodanese. However, in the presence of GroES, casein is unable to interfere with the productive folding of rhodanese that occurs upon addition of Mg/ATP. The shielding of the folding protein from solution appears to be achieved by GroES regulating the binding and release of substrate by GroEL. Since the substrate protein is contained in the cavity of the chaperonin cylinder, folding could then occur under conditions analogous to those in a test tube at infinite dilution of unfolded protein.

What are the structural and functional consequences of GroES binding to GroEL? GroES forms a single heptameric toroid that associates asymmetrically to one end surface of the GroEL double toroid (Langer et al. 1992a). Electron microscopic image analysis revealed that GroES induces conformational changes both in the GroES-bound and in the opposite ring of GroEL (Langer et al. 1992a). This allosteric effect may prevent another GroES ring from stably associating with the free end of the GroEL cylinder. Binding to GroEL also results in the stabilization of a flexible loop in GroES extending between residues 17 and 32 (Landry et al. 1993). The weak potassium-dependent ATPase of GroEL is at least 50% inhibited by binding of GroES (Chadrasekhar et al. 1986; Viitanen et al. 1990; Martin et al. 1991; Todd et al. 1993). GroEL has 14 nucleotide-binding sites (1 per subunit) that bind ATP cooperatively (Gray and Fersht 1991; Bochkareva et al. 1992). Importantly, GroES increases the cooperativity of ATP binding and hydrolysis (Gray and Fersht 1991; Bochkareva et al. 1992) and also increases dramatically the affinity of GroEL for ADP (Jackson et al. 1993; Martin et al. 1993b; Todd et al. 1993). This effect is most pronounced in the interacting GroEL ring. Conversely, nucleotide binding to GroEL is necessary for the formation of a stable GroEL/GroES complex (Bochkareva et al. 1992; Langer et al. 1992a). Azido-ATP has been cross-linked to Tyr-477 in the carboxy-terminal third of the GroEL subunits, indicating that the purine ring of the nucleotide can be in close proximity to this region (Martin et al. 1993a). Tyr-477 is located within a protease-stable approximately 40-kD domain of the GroEL subunits that participates in forming the outer ring layers of the GroEL complex. GroES binding protects the seven interacting GroEL subunits from proteolytic cleavage of their 16 carboxy-terminal residues (residues 531–547). Interestingly, GroES also binds ATP, although it does not hydrolyze it, and this capacity may be important in facilitating the cooperative binding of ATP to GroEL (Martin et al. 1993a).

3. The GroEL-GroES Reaction Cycle in Protein Folding

Three affinity states of GroEL for substrate binding can be distinguished: The nucleotide-free form of GroEL appears to bind unfolded substrate protein most tightly, whereas the ATP state of GroEL has the lowest substrate affinity (Bochkareva et al. 1992; Jackson et al. 1993; Martin et al. 1993b). The ADP form of GroEL associates with the substrate polypeptide with an affinity closer to that of the nucleotide-free state than to that of the ATP state. These forms probably represent different conformational states of GroEL that are interconvertible by nucleotide exchange and ATP hydrolysis (Bochkareva et al. 1992; Jackson et al. 1993; Martin et al. 1993b). Since GroEL and GroES are both present within the cell, an unfolded polypeptide is likely to encounter the chaperonin as a complex with GroES that is in the ADP-bound state. How does this result in productive folding? Insight into this question has been gained by the finding that the polypeptide substrate and GroES counteract each others' effects on GroEL (Martin et al. 1993b). Substrate binding results in the dissociation of the tightly bound ADP from GroEL, and as a consequence, GroES is transiently released. The dissociation of ADP will be followed by ATP binding. At this point, GroES will reassociate with GroEL to trigger a round of highly cooperative ATP hydrolysis. Reassociation of GroES may occur at the substrate-bearing ring of GroEL (Martin et al. 1993b). The concerted action of the GroEL subunits will result in the release of the bound protein from its multiple attachment sites into the GroEL cavity. The protein could then fold through the folding pathway determined by its primary sequence. If, upon a single release event, the protein has not yet completed folding and still exposes GroEL recognition sites, it will rebind to GroEL triggering another reaction cycle. The hydrolysis of approximately 100 ATP measured per rhodanese molecule folded indicates that multiple rounds of interaction indeed occur. Thus, the underlying principle of GroEL-mediated protein folding hinges on alternating cycles of binding and release of GroES and substrate until the protein has lost its affinity for GroEL (Martin et al. 1993b). Association of GroES with only one of the GroEL rings imparts the asymmetry required for this process and ensures cooperative ATP hydrolysis at the level of the heptameric rings. The following sequence of steps can be proposed for the GroEL-GroES reaction cycle (Fig. 2):

1. Under physiological conditions, GroEL is in a complex with GroES. ADP is tightly bound to the subunits in the ring adjacent to GroES and with lower affinity in the opposite ring.
2. Unfolded protein binds to the GroEL ring opposite to GroES and trig-

gers ADP dissociation from the GroEL subunits. This in turn results in the release of GroES.

3. ADP-ATP exchange weakens the affinity for the bound protein.
4. GroES rebinds to GroEL in the ATP state and may cover the ring that contains the bound substrate. Cooperative ATP hydrolysis releases the substrate protein for folding in the ring cavity.
5. GroES binding becomes stabilized in the regained ADP state and partially folded protein may reassociate for another round of interaction.

D. The TRiC Family of Chaperonins in the Eukaryotic Cytosol

It appears that the chaperonins of the GroEL/hsp60 family provide a sequestered environment allowing proteins to pursue their folding potential within the dense cellular milieu. Assuming that this principle of chaperone function is of general significance for protein folding, components equivalent to hsp60 could be expected to exist in the eukaryotic cytosol as well. No direct homologs of hsp60 were found in this compartment. However, recent studies of the heat shock response of thermophilic archaebacteria led to the discovery of a new family of chaperonins that have homology with TCP-1, a known eukaryotic protein (Table 1) (Trent et al. 1991). The gene for TCP-1 (for *t*-complex *p*olypeptide) was originally identified as a constituent of the mouse t-complex, a chromosomal region linked to the phenomenon of male transmission ratio distortion. The TCP-1 protein is highly expressed in developing spermatids but is also present in the cytosol of most eukaryotic cells, including yeast, *Drosophila*, and plants (Hendrick and Hartl 1993). The yeast gene is essential (Ursic and Culbertson 1991). Interestingly, yeast carrying a temperature-sensitive TCP-1 protein are defective in mitotic spindle assembly and are hypersensitive to microtubule affecting drugs (Ursic and Culbertson 1991).

E. TRiC Structure and Function

TCP-1 has been purified from mouse and bovine testis and from rabbit reticulocytes as a component of a 970-kD complex that shares characteristic properties with the chaperonins of the hsp60 family (Frydman et al. 1992; Gao et al. 1992; Lewis et al. 1992). This *TCP-1 ring complex* (TRiC) is hetero-oligomeric containing eight to ten different polypeptides that range in size from 50 to 65 kD (Frydman et al. 1992; Lewis et al. 1992; also see Willison and Kubota, this volume). This complex composition distinguishes TRiC from its archaebacterial homologs TF55 of *Sulfolobus* and the so-called "thermosome" of *Pyrodictium* (Phipps et al.

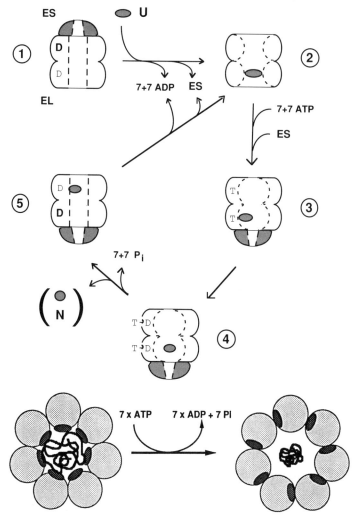

Figure 2 (*Top*) Reaction cycle of GroEL and GroES in protein folding. (D, boldface) High-affinity ADP-binding state of a heptameric GroEL ring; (D, lightface) low ADP-affinity binding state; (T) ATP-binding state; (U,N) unfolded and native protein, respectively. In this model, the GroEL/GroES complex is favored under physiological conditions (*1*). Upon binding of unfolded protein to GroEL, ADP and consequently GroES dissociate (*2*). ATP binding weakens the interaction between GroEL and substrate protein and causes GroES to rebind (*3*). ATP hydrolysis causes release of the protein within the GroEL cavity, allowing it to fold (*4*). Generation of the ADP state increases affinity of GroEL for GroES. Folding to the native state is either completed, or the protein is rebound by GroEL in a partially folded form (*5*), reentering the cycle at step 2. (*Bottom*) Model for ATP-dependent release of substrate protein from multiple attachment sites on GroEL into the central cavity for folding. (*Dark shaded areas*) Putative polypeptide-binding sites of the GroEL subunits.

1991; Trent et al. 1991). Several or perhaps all of these polypeptides are structurally related to TCP-1, indicating that they are members of a new protein family (Frydman et al. 1992). Consistent with this, several genes encoding TCP-1 homologous proteins have been identified in yeast, mice, and humans. These proteins present a weak sequence homology with hsp60/GroEL, which appears to be more pronounced in defined regions that represent a putative nucleotide-binding site (Gupta 1990; Frydman et al. 1992). Electron micrographs show that TRiC is composed of two stacked rings that define a central cavity (Frydman et al. 1992; Gao et al. 1992). The three-dimensional structure of the *Pyrodictium* thermosome has been reconstructed from electron micrographs (Phipps et al. 1993). It differs from GroEL in that the double toroid has eightfold symmetry, resulting in a larger cavity.

TRiC has a weak ATPase activity and has been shown to bind a number of different unfolded proteins upon dilution from denaturant. ATP-hydrolysis-dependent refolding by TRiC has been demonstrated for firefly luciferase, recombinant actin, and α, β, and γ tubulins (Frydman et al. 1992; Gao et al. 1992, 1993; Melki et al. 1993). Interestingly, bovine TRiC is able to mediate the refolding/assembly of α and β tubulins purified from bovine brain without assistance from additional components. This reaction depends on the presence of assembled tubulin dimers. In contrast, the release of tubulin from GroEL requires GroES, suggesting that TRiC might not depend on a GroES-like cochaperonin (Frydman et al. 1992). However, the release of recombinant α and β (but not of the homologous γ) tubulins from TRiC requires additional factors (Gao et al. 1993; Melki et al. 1993). The role of these factors is not yet clear, but they do not appear to be equivalent to GroES. When tubulin is translated in reticulocyte lysate, it forms a complex with TRiC that can be partially purified (Yaffe et al. 1992). Upon addition of Mg/ATP, the bound protein folds and assembles by exchanging with subunits in added tubulin dimers.

Although initially proposed to be specialized for the folding and assembly of cytoskeletal proteins, it is now becoming clear that TRiC has a more general chaperonin function. TRiC also appears to be involved in the assembly of hepatitis B capsids (W. Lingappa and W. Welch, pers. comm.) and interacts with a spectrum of ^{35}S-labeled unfolded proteins as well as with newly translated DHFR, luciferase, and chloramphenicol acetyltransferase. Moreover, a TCP-1-related chaperonin appears to be functional in the cytosol of plant cells (Mummers et al. 1993). TRiC is not inducible by heat stress and its mode of action might be different from that of GroEL. One attractive possibility is that distinct subunits of TRiC confer a specificity for certain sets of proteins. The large number of different subunits that have been identified suggests that more than one defined chaperonin complex may exist. It is interesting in this context that mutants in the yeast TCP-1 gene cause disturbances in tubulin metabolism (Ursic and Culbertson 1991), whereas recently identified mutants in a yeast gene encoding another TRiC subunit are impaired in the assembly of the actin cytoskeleton (D. Drubin et al., pers. comm.).

V. PERSPECTIVES: A GENERAL PATHWAY OF CHAPERONE-MEDIATED PROTEIN FOLDING?

The differential binding capabilities of the hsp70 and hsp60 chaperone systems suggest that these proteins could function in a sequential reaction accompanying a newly synthesized polypeptide along its folding pathway. Such a reaction occurs for the folding of mitochondrial proteins. On the basis of the reconstitution of this pathway using the homologous components of *E. coli*, it has been proposed that the folding of proteins in the bacterial cytosol may follow similar reactions (Fig. 3). This view has received support from the demonstration that DnaK (hsp70) and DnaJ can be found in association with ribosome-bound polypeptide chains, whereas GroEL appears to interact posttranslationally (A. Gragerov, pers. comm.). Indeed, addition of GroEL and GroES increases the specific activity of in-vitro-translated rhodanese (Tsalkova et al. 1993). Furthermore, the efficient folding of proteins at elevated temperature has been shown in vivo to require both the DnaK/DnaJ/GrpE and the GroEL/GroES chaperone systems (Gragerov et al. 1992). The recent identification of TRiC suggests that the principle of sequential action of hsp70 and hsp60 chaperones may be of importance for protein folding in the eukaryotic cytosol as well. hsp70 and TRiC have indeed been demonstrated to interact with nascent chains of luciferase in a large chaperone complex, which shields the elongating polypeptide very efficiently (J. Frydman and F.-U. Hartl, unpubl.).

The pathways by which newly synthesized polypeptides fold in the eukaryotic cytosol may well be of considerable complexity. For example, the chaperone hsp90, which is present in this compartment at high concentrations, has a role in the folding of a certain subset of proteins (for review, see Pratt 1993; Gething and Sambrook 1992; see also Bohen and Yamamoto, this volume). Whether these proteins interact first with TRiC or bypass the chaperonin is not clear. Thus, protein traffic in the cell may be segregated into different routes that are governed by different chaperones. This would pose interesting questions of how folding polypeptide chains are directed to these pathways. Another area of interest will be to understand the functional role of chaperones in determining

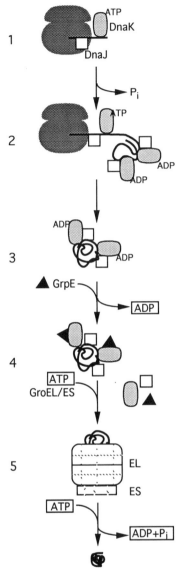

Figure 3 Model for the pathway of chaperone-mediated protein folding in the cytosol of bacteria (modified from Langer et al. 1992b). (*1*) DnaK and DnaJ associate with the polypeptide chain as it emerges from the ribosome; (*2*) interaction of DnaJ with DnaK stabilizes DnaK in its ADP state; (*3*) a tight ternary complex of DnaJ, DnaK, and unfolded polypeptide is formed; (*4*) GrpE dissociates ADP from DnaK; (*5*) ATP binding to DnaK releases the unfolded protein, permitting its transfer to GroEL for folding to the native state. A similar pathway may exist in the eukaryotic cytosol in which TRiC fulfills the role of GroEL.

whether a protein folds or is degraded. One might also speculate about the existence of chaperone dysfunctions, which could cause disease by reducing the efficiency of protein folding in certain cell types or organelles.

ACKNOWLEDGMENTS

We thank numerous colleagues in the molecular chaperone and protein folding fields for stimulating discussions, and the Deutsche Forschungsgemeinschaft and the National Institutes of Health for support. J.F is a recipient of a postdoctoral National Institutes of Health fellowship.

REFERENCES

Alfano, C. and R. McMacken. 1989. Ordered assembly of nucleoprotein structures at the bacteriophage lambda replication origin during the initiation of DNA replication. *J. Biol. Chem.* **264:** 10699–10708.

Anfinsen, C.B. 1973. Principles that govern the folding of protein chains. *Science* **181:** 223–230.

Atencio, D.P. and M.P. Yaffe. 1992. MAS5, a yeast homolog of DnaJ involved in mitochondrial protein import. *Mol. Cell. Biol.* **12:** 283–291.

Badcoe, I.G., C.J. Smith, S. Wood, D.J. Halsall, J. Holbrook, P. Lund, and A.R. Clarke. 1991. Binding of a chaperonin to the folding intermediates of lactate dehydrogenase. *Biochemistry* **30:** 9195–9200.

Beckmann, R.P., L.A. Mizzen, and W.J. Welch. 1990. Interaction of Hsp 70 with newly synthesized proteins: Implications for protein folding and assembly. *Science* **248:** 850–854.

Blond-Elguindi, S., S.E. Cwirla, W.J. Dower, R.J. Lipshutz, S.R. Sprang, J.F. Sambrook, and M.-J. Gething. 1993. Affinity panning of a library of peptides displayed on bacteriophages reveals the binding specificity of BiP. *Cell* **75:** 717–729.

Blumberg, H. and P. Silver. 1991. A homolog of the bacterial heat-shock gene *dnaJ* that alters protein sorting in yeast. *Nature* **349:** 627–630.

Bochkareva, E.S., N.M. Lissin, G.C. Flynn, J.E. Rothman, and A.S. Girshovich. 1992. Positive cooperativity in the functioning of molecular chaperone GroEL. *J. Biol. Chem.* **267:** 6796–6800.

Braig, K., F. Furuya, J. Hainfeld, and A.L. Horwich. 1993. Gold-labeled DHFR binds in the center of GroEL. *Proc. Natl. Acad. Sci.* **90:** 3978–3982.

Brodsky, J.L., S. Hamamoto, D. Feldheim, and R. Schekman. 1993. Reconstitution of protein translocation from solubilized yeast membranes reveals topologically distinct roles for BiP and cytosolic Hsc70. *J. Cell Biol.* **120:** 95–102.

Buchner, J., M. Schmidt, M. Fuchs, R. Jaenicke, R. Rudolph, F.X. Schmid, and T. Kiefhaber. 1991. GroE facilitates refolding of citrate synthase by suppressing aggregation. *Biochemistry* **30:** 1586–1591.

Caplan, A.J. and M.G. Douglas. 1991. Characterization of YDJ1: A yeast homologue of the bacterial dnaJ protein. *J. Cell Biol.* **114:** 609–621.

Caplan, A.J., D. Cyr, and M.G. Douglas. 1993. Eukaryotic homologues of *Escherichia coli* dnaJ: A diverse protein family that functions with Hsp70 stress proteins. *Mol. Biol.*

Cell **4:** 555–563.

Chadrasekhar, G.N., K. Tilly, C. Woolford, R. Hendrix, and C. Georgopoulos. 1986. Purification and properties of the groES morphogenetic protein of *Escherichia coli. J. Biol. Chem.* **261:** 12414–12419.

Cheetham, M.E., J.P. Brion, and B.H. Anderton. 1992. Human homologues of the bacterial heat-shock protein DnaJ are preferentially expressed in neurons. *Biochem. J.* **284:** 469–476.

Cheng, M.Y., F.-U. Hartl, and A.L. Horwich. 1990. The mitochondrial chaperonin hsp60 is required for its own assembly. *Nature* **348:** 455–458.

Cheng, M.Y., F.-U. Hartl, J. Martin, R.A. Pollock, F. Kalousek, W. Neupert, E.M. Hallberg, R.L. Hallberg, and A.L. Horwich. 1989. Mitochondrial heat-shock protein hsp60 is essential for assembly of proteins imported into yeast mitochondria. *Nature* **337:** 620–625.

Chirico, W.J., M.G. Waters, and G. Blobel. 1988. 70K heat shock related proteins stimulate protein translocation into microsomes. *Nature* **332:** 805–810.

Christensen, H. and R.H. Pain. 1991. Molten globule intermediates and protein folding. *Eur. Biophys. J.* **19:** 221–229.

Cyr, D.M., X. Lu, and M.G. Douglas. 1992. Regulation of hsp70 function by a eukaryotic DnaJ homolog. *J. Biol. Chem.* **267:** 20927–20931.

Deshaies, R.J., B.D. Koch, M. Werner-Washburne, E.A. Craig, and R. Schekman. 1988. A subfamily of stress proteins facilitates translocation of secretory and mitochondrial precursor polypeptides. *Nature* **332:** 800–805.

Ellis, R.J. 1987. Proteins as molecular chaperones. *Nature* **328:** 378–379.

Ewbank, J. and T.E. Creighton. 1993. Structural characterization of the disulfide folding intermediates of bovine a-lactalbumin. *Biochemistry* **32:** 3694–3707.

Feldheim, D., J. Rothblatt, and R. Schekman. 1992. Topology and functional domains of Sec63p, an endoplasmic reticulum membrane protein required for secretory protein translocation. *Mol. Cell. Biol.* **12:** 3288–3296.

Flaherty, K.M., C. DeLuca-Flaherty, and D.B. McKay. 1990. Three-dimensional structure of the ATPase fragment of a 70K heat-shock cognate protein. *Nature* **346:** 623–628.

Flajnik, M.F., C. Canel, J. Kramer, and M. Kasahara. 1991. Evolution of the major histocompatibility complex: Molecular cloning of major histocompatibility complex class I from the amphibian *Xenopus. Proc. Natl. Acad. Sci.* **88:** 537–541.

Flanagan, J.M., M. Kataoka, D. Sortie, and D.M. Engelman. 1991. Truncated staphylococcal nuclease is compact but disordered. *Proc. Natl. Acad. Sci.* **89:** 748–752.

Flynn, G.C., J. Rohl, M.T. Flocco, and J.E. Rothman. 1991. Peptide-binding specificity of the molecular chaperone BiP. *Nature* **353:** 726–730.

Frydman, J., E. Nimmesgern, H. Erdjument-Bromage, J.S. Wall, P. Tempst, and F.-U. Hartl. 1992. Function in protein folding of TRiC, a cytosolic ring-complex containing TCP1 and structurally related subunits. *EMBO J.* **11:** 4767–4778.

Gamer, J., H. Bujard, and B. Bukau. 1992. Physical interaction between heat shock proteins DnaK, DnaJ, and GrpE and the bacterial heat shock transcription factor sigma 32. *Cell* **69:** 833–842.

Gao, Y., I.E. Vainberg, R.L. Chow, and N.J. Cowan. 1993. Two cofactors and cytoplasmic chaperonin are required for the folding of alpha- and beta-tubulin. *Mol. Cell. Biol.* **13:** 2478–2485.

Gao, Y., J.O. Thomas, R.L. Chow, G.H. Lee, and N.J. Cowan. 1992. A cytoplasmic chaperonin that catalyzes beta-actin folding. *Cell* **69:** 1043–1050.

Georgopolous, C. 1992. The emergence of chaperone machines. *Trends Biol. Sci.* **17:**

295–299.

Georgopoulos, C., R.W. Hendrix, S.R. Casjens, and A.D. Kaiser. 1973. Host participation in bacteriophage lambda head assembly. *J. Mol. Biol.* **76:** 45–60.

Gething, M.-J. and J. Sambrook. 1992. Protein folding in the cell. *Nature* **355:** 33–45.

Goloubinoff, P., J.T. Christeller, A.A. Gatenby, and G.H. Lorimer. 1989. Reconstitution of active dimeric ribulose bisphosphate carboxylase from an unfolded state depends on two chaperonin proteins and MgATP. *Nature* **342:** 884–889.

Gragerov, A., E. Nudler, N. Komissarova, G.A. Gaitanaris, M.E. Gottesman, and V. Nikiforov. 1992. Cooperation of GroEL/GroES and DnaK/DnaJ heat shock proteins in preventing protein misfolding in *Escherichia coli. Proc. Natl. Acad. Sci.* **89:** 10341–10344.

Gray, T.E. and A.R. Fersht. 1991. Cooperativity in ATP hydrolysis by GroEL is increased by GroES. *FEBS Lett.* **292:** 254–258.

Gupta, R.S. 1990. Sequence and structural homology between a mouse T-complex protein TCP-1 and the "chaperonin" family of bacterial (GroEL, 65 kDa heat shock antigen) and eukaryotic proteins. *Biochem. Int.* **4:** 833–839.

Hartman, D.J., N.J. Hoogenraad, R. Condron, and P.B. Hoj. 1992. Identification of a mammalian 10-kDa heat shock protein, a mitochondrial chaperonin 10 homologue essential for assisted folding of trimeric ornithine transcarbamoylase *in vitro. Proc. Natl. Acad. Sci.* **89:** 3394–3398.

Hattori, H., Y.C. Liu, I. Tohnai, M. Ueda, T. Kaneda, T. Kobayashi, K. Tanabe, and K. Ohtsuka. 1992. Intracellular localization and partial amino acid sequence of a stress-inducuble 40-kDa protein in HeLa cells. *Cell. Struct. Funct.* **17:** 77–86.

Hemmingsen, S.M. and R.J. Ellis. 1986. Purification and properties of ribulose bisphosphate carboxylase large subunit binding protein. *Plant Physiol.* **80:** 269–276.

Hemmingsen, S.M., C. Woolford, d.V.S.M. van, K. Tilly, D.T. Dennis, C.P. Georgopoulos, R.W. Hendrix, and R.J. Ellis. 1988. Homologous plant and bacterial proteins chaperone oligomeric protein assembly. *Nature* **333:** 330–334.

Hendrick, J.P. and F.-U. Hartl. 1993. Molecular chaperone functions of heat-shock proteins. *Annu. Rev. Biochem.* **62:** 349–384.

Hendrick, J.P., T. Langer, T.A. Davis, F.-U. Hartl, and M. Wiedmann. 1993. Control of folding and membrane translocation by binding of the chaperone DnaJ to nascent polypeptides. *Proc. Natl. Acad. Sci.* **90:** 10216–10220.

Hendrix, R.W. 1979. Purification and properties of GroE, a host protein involved in bacteriophage assembly. *J. Mol. Biol.* **129:** 375–392.

Hightower, L.E. 1991. Heat shock, stress proteins, chaperones, and proteotoxicity. *Cell* **66:** 191–197.

Hohn, T., B. Hohn, A. Engel, M. Wortz, and P.R. Smith. 1979. Isolation and characterisation of the host protein GroE involved in bacteriophage lambda assembly. *J. Mol. Biol.* **129:** 359–373.

Holl, N.B., R. Rudolph, M. Schmidt, and J. Buchner. 1991. Reconstitution of a heat shock effect in vitro: Influence of GroE on the thermal aggregation of alpha-glucosidase from yeast. *Biochemistry* **30:** 11609–11614.

Horwich, A.L., K. Brooks Low, W.A. Fenton, I.N. Hirshfield, and K. Furtak. 1993. Folding *in vivo* of bacterial cytoplasmic proteins: Role of GroEL. *Cell* **74:** 909–917.

Jackson, G.S., R.A. Staniforth, D.J. Halsall, T. Atkinson, J.J. Holbrook, A.R. Clarke, and S.G. Burston. 1993. Binding and hydrolysis of nucleotides in the chaperonin catalytic cycle: Implications for the mechanism of assisted protein folding. *Biochemistry* **32:** 2554–2563.

Jaenicke, R. 1987. Folding and association of proteins. *Prog. Biophys. Mol. Biol.* **49:**

117–237.

Kang, P.J., J. Ostermann, J. Shilling, W. Neupert, E.A. Craig, and N. Pfanner. 1990. Requirement for hsp70 in the mitochondrial matrix for translocation and folding of precursor proteins. *Nature* **348:** 137–143.

Laminet, A.A., T. Ziegelhoffer, C. Georgopoulos, and A. Pluckthun. 1990. The *Escherichia coli* heat shock proteins GroEL and GroES modulate the folding of the beta-lactamase precursor. *EMBO J.* **9:** 2315–2319.

Landry, S.J. and L.M. Gierasch. 1991. The chaperonin GroEL binds a polypeptide in an alpha-helical conformation. *Biochemistry* **30:** 7359–7362.

Landry, S.J., R. Jordan, R. McMacken, and L.M. Gierasch. 1992. Different conformations for the same polypeptide bound to chaperones DnaK and GroEL. *Nature* **355:** 455–457.

Landry, S.J., J. Zeilstra-Ryalls, O. Fayet, C. Georgopoulos, and L.M. Gierasch. 1993. Characterization of a functionally important mobile domain of GroES. *Nature* **364:** 255–258.

Langer, T., G. Pfeifer, J. Martin, W. Baumeister, and F.-U. Hartl. 1992a. Chaperonin-mediated protein folding: GroES binds to one end of the GroEL cylinder which accommodates the protein substrate within its central cavity. *EMBO J.* **11:** 4657–4765.

Langer, T., C. Lu, H. Echols, J. Flanagan, M.K. Hayer, and F.-U. Hartl. 1992b. Successive action of molecular chaperones DnaK (Hsp70), DnaJ and GroEL (Hsp60) along the pathway of assisted protein folding. *Nature* **356:** 683–689.

Laskey, R.A., B.M. Honda, and J.T. Finch. 1978. Nucleosomes are assembled by an acidic protein which binds histones and transfers them to DNA. *Nature* **275:** 416–420.

Lewis, V.A., G.M. Heynes, D. Zheng, H. Saibil, and K. Willison. 1992. T-complex polypeptide-1 is a subunit of a heteromeric particle in the eucaryotic cytosol. *Nature* **358:** 249–252.

Liberek, K., T.P. Galitski, M. Zylicz, and C. Georgopoulos. 1992. The DnaK chaperone modulates the heat shock response of *Escherichia coli* by binding to the sigma 32 transcription factor. *Proc. Natl. Acad. Sci.* **89:** 3516–3520.

Liberek, K., J. Marszalek, D. Ang, C. Georgopoulos, and M. Zylicz. 1991a. *Escherichia coli* DnaJ and GrpE heat shock proteins jointly stimulate ATPase activity of DnaK. *Proc. Natl. Acad. Sci.* **88:** 2874–2878.

Liberek, K., D. Skowyra, M. Zylicz, C. Johnson, and C. Georgopoulos. 1991b. The *Escherichia coli* DnaK chaperone, the 70-kDa heat shock protein eukaryotic equivalent, changes conformation upon ATP hydrolysis, thus triggering its dissociation from a bound target protein. *J. Biol. Chem.* **266:** 14491–14496.

Lindquist, S. 1992. Won't you change partners and dance? *Curr. Biol.* **2:** 119–121.

Lubben, T.H., A.A. Gatenby, G.K. Donaldson, G.H. Lorimer, and P.V. Viitanen. 1990. Identification of a groES-like chaperonin in mitochondria that facilitates protein folding. *Proc. Natl. Acad. Sci.* **87:** 7683–7687.

Luke, M.M., A. Sutton, and K.T. Arndt. 1991. Characterization of SIS1, a *Saccharomyces cervisiae* homologue of bacterial dnaJ proteins. *J. Cell Biol.* **114:** 623–638.

Manning, K.U., P.E. Scherer, and G. Schatz. 1991. Sequential action of mitochondrial chaperones in protein import into the matrix. *EMBO J.* **10:** 3273–3280.

Martin, J., A.L. Horwich, and F.-U. Hartl. 1992. Role of chaperonin hsp60 in preventing protein denaturation under heat-stress. *Science* **258:** 995–998.

Martin, J., S. Geromanos, P. Tempst, and F.-U. Hartl. 1993a. Identification of nucleotide-binding regions in the chaperonin proteins GroEL and GroES. *Nature* **366:** 279–282.

Martin, J., M. Mayhew, T. Langer, and F.-U. Hartl. 1993b. The reaction cycle of GroEL and GroES in chaperonin-assisted refolding. *Nature* **366:** 228–233.

Martin, J., T. Langer, R. Boteva, A. Schramel, A.L. Horwich, and F.-U. Hartl. 1991. Chaperonin-mediated protein folding at the surface of groEL through a "molten globule"-like intermediate. *Nature* **352:** 36–42.

Melki, R., I.E. Vainberg, R.L. Chow, and N.J. Cowan. 1993. Chaperonin-mediated folding of vertebrate actin-related protein and gamma-tubulin. *J. Cell. Biol.* **122:** 1301–1310.

Mendoza, J.A., E. Rogers, G.H. Lorimer, and P.M. Horowitz. 1991. Chaperonins facilitate the *in vitro* folding of monomeric mitochondrial rhodanese. *J. Biol. Chem.* **266:** 13044–13049.

Mummers, E., R. Grimm, V. Speth, C. Eckerskorn, E. Schiltz, A.A. Gatenby, and E. Schafer. 1993. A TCP-1 related molecular chaperone from plants refolds phytochrome to its photoreversible form. *Nature* **363:** 644–648.

Murakami, H., D. Pain, and G. Blobel. 1988. 70-kD heat shock related protein is one of at least two distinct cytosolic factors stimulating protein import into mitochondria. *J. Cell Biol.* **107:** 2051–2057.

Nelson, R.J., T. Ziegelhoffer, C. Nicolet, M. Werner-Washburne, and E.A. Craig. 1992. The translation machinery and 70 kd heat shock protein cooperate in protein synthesis. *Cell* **71:** 97–105.

Nguyen, T.H., D.T. Law, and D.B. Williams. 1991. Binding protein BiP is required for translocation of secretory proteins into the endoplasmic reticulum in *Saccharomyces cerevisiae. Proc. Natl. Acad. Sci.* **88:** 1565–1569.

Ostermann, J., A.L. Horwich, W. Neupert, and F.-U. Hartl. 1989. Protein folding in mitochondria requires complex formation with hsp60 and ATP hydrolysis. *Nature* **341:** 125–130.

Palleros, D.R., W.J. Welch, and A.L. Fink. 1991. Interaction of hsp70 with unfolded proteins: Effects of temperature and nucleotides on the kinetics of binding. *Proc. Natl. Acad. Sci.* **88:** 5719–5723.

Palleros, D.R., K.L. Reid, L. Shi, W.J. Welch, and A.L. Fink. 1993. ATP-induced protein-HSP70 complex dissociation requires K^+ but not ATP hydrolysis. *Nature* **365:** 664–666.

Pelham, H.R.B. 1986. Speculations on the functions of the major heat shock and glucose-regulated proteins. *Cell* **46:** 959–961.

Phipps, B.M., A. Hoffmann, K.O. Stetter, and W. Baumeister. 1991. A novel ATPase complex selectively accumulated upon heat shock is a major cellular component of thermophilic archaebacteria. *EMBO J.* **10:** 1711–1722.

Phipps, B.M., D. Typke, R. Hegerl, S. Volker, A. Hoffmann, K.O. Stetter, and W. Baumeister. 1993. Structure of a molecular chaperone from a thermophilic archaebacterium. *Nature* **361:** 475–477.

Pratt, W.B. 1993. The role of heat shock proteins in regulating the function, folding, and trafficking of the glucocorticoid receptor. *J. Biol. Chem.* **268:** 21455–21458.

Rippmann, F., W.R. Taylor, J.B. Rothbard, and N.M. Green. 1991. A hypothetical model for the peptide binding domain of hsp70 based on the peptide binding domain of HLA. *EMBO J.* **10:** 1053–1059.

Sanders, S.L., K.M. Whitfield, J.P. Vogel, M.D. Rose, and R.W. Schekman. 1992. Sec61p and BiP directly facilitate polypeptide translocation into the ER. *Cell* **69:** 353–365.

Scherer, P.E., U.C. Krieg, S.T. Hwang, D. Vestweber, and G. Schatz. 1990. A precursor protein partly translocated into yeast mitochondria is bound to a 70 kd mitochondrial stress protein. *EMBO J.* **9:** 4315–4322.

Schmidt, M. and J. Buchner. 1992. Interaction of GroE with an all-β-protein. *J. Biol.*

Chem. **267:** 16829–16833.

Schroder, H., T. Langer, F.-U. Hartl, and B. Bukau. 1993. DnaK, DnaJ, GrpE form a cellular chaperone machinery capable of repairing heat-induced protein damage. *EMBO J.* **12:** 4137–4144.

Sheffield, W.P., G.C. Shore, and S.K. Randall. 1990. Mitochondrial precursor protein. Effects of 70-kilodalton heat shock protein on polypeptide folding, aggregation, and import competence. *J. Biol. Chem.* **265:** 11069–11076.

Skowyra, D., C. Georgopoulos, and M. Zylicz. 1990. The *E. coli dnaK* gene product, the hsp70 homologue, can reactivate heat-inactivated RNA polymerase in an ATP-hydrolysis-dependent manner. *Cell* **62:** 939–944.

Tilly, K., H. Murialdo, and C.P. Georgopoulos. 1981. Identification of a second *Escherichia coli groE* gene whose product is necessary for bacteriophage morphogenesis. *Proc. Natl. Acad. Sci.* **78:** 1629–1633.

Tissières, A., H.K. Mitchell, and V.M. Tracy. 1974. Protein synthesis in salivary glands of *Drosophila melanogaster*: Relation to chromosome puffs. *J. Mol. Biol.* **84:** 389–398.

Todd, M.J., P.V. Viitanen, and G.H. Lorimer. 1993. Hydrolysis of adenosine 5′-triphosphate by *Escherichia coli* GroEL: Effects of GroES and potassium ion. *Biochemistry* **32:** 8560–8567.

Trent, J.D., E. Nimmesgern, J.S. Wall, F.-U. Hartl, and A.L. Horwich. 1991. A molecular chaperone from a thermophilic archaebacterium is related to the eukaryotic protein t-complex polypeptide-1 (see comments). *Nature* **354:** 490–493.

Tsalkova, T., G. Zardeneta, W. Kudlicki, G. Kramer, P.M. Horowitz, and B. Hardesty. 1993. GroEL and GroES increase the specific enzymatic activity of newly-synthesized rhodanese if present during in vitro transcription/translation. *Biochemistry* **32:** 3377–3380.

Ursic, D. and M.R. Culbertson. 1991. The yeast homolog to mouse Tcp-1 affects microtubule-mediated processes. *Mol. Cell. Biol.* **11:** 2629–2640.

van derVies, S., P. Viitanen, A.A. Gatenby, G.H. Lorimer, and R. Jaenicke. 1992. Conformational states of ribulosebisphosphate carboxylase and their interaction with chaperonin 60. *Biochemistry* **31:** 3635–3644.

Viitanen, P., A.A. Gatenby, and G.H. Lorimer. 1992. Purified chaperonin 60 (groEL) interacts with the nonnative states of a multitude of *Escherichia coli* proteins. *Protein Sci.* **1:** 363–369.

Viitanen, P.V., G.K. Donaldson, G.H. Lorimer, T.H. Lubben, and A.A. Gatenby. 1991. Complex interactions between the chaperonin 60 molecular chaperone and dihydrofolate reductase. *Biochemistry* **30:** 9716–9723.

Viitanen, P.V., T.H. Lubben, J. Reed, P. Goloubinoff, D.P. O'Keefe, and G.H. Lorimer. 1990. Chaperonin-facilitated refolding of ribulosebisphosphate carboxylase and ATP hydrolysis by chaperonin 60 (groEL) are K^+ dependent. *Biochemistry* **29:** 5665–5671.

Vogel, J.P., L.M. Misra, and M.D. Rose. 1990. Loss of BiP/GRP78 function blocks translocation of secretory proteins in yeast. *J. Cell. Biol.* **110:** 1885–1895.

Wickner, S., J. Hoskins, and K. McKenney. 1991. Function of DnaJ and DnaK as chaperones in origin-specific DNA binding by RepA. *Nature* **350:** 165–167.

Yaffe, M.B., G.W. Farr, D. Miklos, A.L. Horwich, M.L. Sternlicht, and H. Sternlicht. 1992. TCP1 complex is a molecular chaperone in tubulin biogenesis. *Nature* **358:** 245–248.

Zheng, X., L.E. Rosenberg, F. Kalousek, and W.A. Fenton. 1993. GroEL, GroES, and ATP-dependent folding and spontaneous assembly of ornithine transcarbamylase. *J. Biol. Chem.* **268:** 7489–7493.

Zhong, T. and K.T. Arndt. 1993. The yeast SIS1 protein, a DnaJ homolog, is required for the initiation of translation. *Cell* **73:** 1175–1186.

Zylicz, M., D. Ang, K. Liberek, and C. Georgopoulos. 1989. Initiation of lambda DNA replication with purified host- and bacteriophage-encoded proteins: The role of the dnaK, dnaJ and grpE heat shock proteins. *EMBO J.* **8:** 1601–1608.

11

The Basis of Recognition of Nonnative Structure by the Chaperone SecB

Linda L. Randall and Traci B. Topping
Department of Biochemistry and Biophysics
Washington State University
Pullman, Washington 99164-4660

Simon J.S. Hardy
Department of Biology
University of York
York YO1 5DD United Kingdom

I. INTRODUCTION

Chaperones selectively bind polypeptide ligands that have no feature in common except that of being nonnative. In this chapter, we discuss this amazing ability with respect to SecB, a chaperone facilitating protein export in *Escherichia coli*. Intensive study of this protein has led to a rudimentary understanding of the mechanism of ligand binding and of how that binding relates to function. The principles emerging are, we believe, pertinent to the function of chaperones in other pathways.

II. THE FUNCTION OF SECB DURING PROTEIN EXPORT

SecB is involved in the export of a subset of the proteins that are located in either the periplasmic space or the outer membrane of *E. coli* (Kumamoto and Beckwith 1983, 1985). It interacts with them either

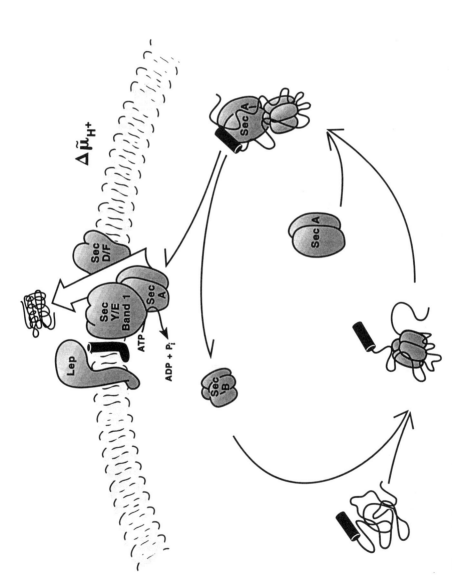

Figure 1 A model for the export of protein to the periplasmic space of *E. coli*. See the text for discussion.

while they are nascent growing polypeptide chains or after completion of synthesis but before they acquire their native structure (Kumamoto and Gannon 1988; Kumamoto 1989; Kumamoto and Francetic 1993). The primary functions of SecB are to maintain these polypeptide ligands in a state competent for translocation across the cytoplasmic membrane (Randall and Hardy 1986; Kumamoto and Gannon 1988; Kusters et al. 1989; Lecker et al. 1989; Weiss et al. 1988) and to deliver them to SecA, which in turn interacts with the integral membrane proteins SecY, SecE, and band 1 to form a translocase (see Fig. 1) (Hartl et al. 1990). The passage of the polypeptide through the cytoplasmic membrane mediated by the translocase complex requires both hydrolysis of ATP by SecA and proton motive force (Schiebel et al. 1991). On the periplasmic side of the membrane, leader peptidase removes the leader peptide to generate the matured amino terminus (Zwizinski and Wickner 1980; Wolfe et al. 1983). The membrane proteins SecD and SecF have a role late in the process (Gardel et al. 1990), possibly related to folding and release of the matured protein.

For a discussion of the role of other chaperones in protein localization, see Brodsky and Schekman (this volume), and for a detailed account of the biochemical and genetic studies of SecB, see Bassford (1990) and Collier (1993). Here, we examine the mechanism of binding of SecB to its ligands and discuss the implications for the role of molecular chaperones in general.

III. KINETIC PARTITIONING BETWEEN FOLDING AND INTERACTION WITH SECB

An important question about any chaperone is how it recognizes the nonnative polypeptides that are its ligands. All of the natural ligands of SecB are precursors of exported proteins, distinguished by the presence of an amino-terminal leader sequence, and it might be expected that SecB would specifically recognize that leader sequence. However, binding to the leader sequence alone would not provide any selectivity for nonnative structure. Furthermore, such binding would not maintain the precursor in an export-competent state by preventing the remainder of the polypeptide from folding, since the leader is not part of the native, folded structure. It is therefore not surprising that there is abundant evidence for association of SecB with the internal regions of nonnative precursors (Collier et al. 1988; Gannon et al. 1989; Lecker et al. 1989; Liu et al. 1989; Altman et al. 1990; Randall et al. 1990; Weiss and Bassford 1990; De Cock et al. 1992). The leader sequence plays a crucial, albeit indirect, role by retarding the rate of folding of the precursor polypeptide to such an extent that the rate of folding is lower than the rate of binding to SecB (Liu et al. 1989). Thus, most molecules of the precursor can bind SecB before they fold into a stable structure. Once folded, they are not capable of interaction with SecB. Binding of ligand then is mediated by a kinetic partitioning between folding and interaction with SecB, and the earliest function of the leader is to ensure that the partitioning is poised to favor binding to SecB. The most compelling evidence for this role of the leader comes from a study (Liu et al. 1989) in which it was shown that a species of maltose-binding protein completely lacking a leader sequence could efficiently bind SecB in vivo, provided it had a single amino-acyl substitution (the aspartyl residue at position 283 was replaced by a tyrosinyl residue), which drastically reduced the rate of folding of the mature polypeptide (Liu et al. 1988). Thus, it was concluded that in the binding to SecB, the function of the leader was to retard folding and thereby make regions of the polypeptide that would otherwise be inaccessible available to the chaperone.

The idea that there is a kinetic partitioning between folding and binding was further supported by studies of interaction in vitro of SecB with ligands (Hardy and Randall 1991). SecB and denatured, mature maltose-binding protein formed a complex with a dissociation constant of 10^{-9} M, whereas SecB showed no affinity for the fully folded native maltose-binding protein. The relative affinity for other polypeptide ligands could be determined because competition for binding to SecB resulted in relief of the blockage of folding of maltose-binding protein imposed by its interaction with the chaperone. The four proteins tested (ribose-binding protein, bovine pancreatic trypsin inhibitor, ribonuclease, and the α subunit of tryptophan synthetase) showed no affinity for SecB in their native states, but all bound when they were nonnative. Since there was no similarity among the sequences of the proteins that showed competition for binding to SecB, it was concluded that it is the nonnative state per se that is recognized by SecB rather than a specific sequence of amino acids.

The studies summarized to this point show that SecB binds only polypeptides that are nonnative and that for physiological ligands, the leader peptide enhances binding by increasing the probability that a polypeptide is in that state. But exactly what is it that allows a polypeptide to be recognized as nonnative? What type of interactions provide the energy of binding? Some insight into the mechanism was gained by a study of SecB and small peptide ligands (Randall 1992). An assay was developed based on the sensitivity to proteolysis of a complex, comprising SecB and a ligand, as compared to the sensitivity to proteolysis of the free components. SecB, when uncomplexed, is quantitatively cleaved by low concentrations of proteinase K to a form lacking the carboxy-terminal 50 amino-acyl residues. In contrast, the free ligand, carbox-amidomethylated bovine pancreatic trypsin inhibitor (R-BPTI) is resistant to cleavage under the same conditions. Formation of a complex renders the SecB resistant to cleavage and the R-BPTI sensitive to cleavage. As expected, native BPTI, which is not a ligand, offers no protection of SecB from proteolysis. The concentration of SecB used in this assay (0.6 μM) was well above the estimated dissociation constant for the complex with R-BPTI (5 nM) allowing one to determine the stoichiometry of the complex as 1 mole of ligand bound per 1 mole of monomeric SecB (monomeric molecular weight is 16,600). Since SecB functions as a tetramer (Watanabe and Blobel 1989), there are clearly multiple binding sites for peptides. A survey of a large number of peptides using this assay to assess binding revealed that the only feature that distinguished those that bound from those that did not was the presence of a net positive charge and a minimal length of 14 residues. Flexibility of the peptide also appeared to be important since those peptides that contained disulfide bonds, which would restrict their conformations, were more effective in protecting SecB from proteolysis when the disulfide bonds were reduced. A zinc finger peptide also showed an increased ability to bind SecB when Zn^{++}, which stabilizes its structure, was absent.

When multiple sites on the tetramer of SecB were occupied by the short hydrophilic peptides, a hydrophobic patch on SecB was exposed and could be detected by the binding of the fluorescent compound 1-anilino-naphthalene-8-sulfonate. It was proposed that this hydrophobic patch provides a binding site for hydrophobic regions that would be accessible in the physiological ligands, which are not short peptides, but nonnative polypeptides.

Characterization of a complex between SecB and one of its physiological ligands, maltose-binding protein, identified a binding frame within the polypeptide that spanned half of the primary sequence of maltose-binding protein and was poised around the center (T.B. Topping and L.L. Randall, unpubl.). Proteolysis of complexes allowed recovery of fragments of the ligand bound to SecB. Since nonoverlapping peptides from within the region defined as the binding frame were recovered, it is likely that one polypeptide is held independently at several sites by SecB, consistent with the demonstration of multiple sites in the peptide-binding study. There are candidates within the binding frame for stretches of ligand that could interact with the hydrophilic as well as the hydrophobic sites on SecB. The binding frame was identical whether the ligand was the mature or precursor form of either the wild-type or two slow-folding variants of maltose-binding protein.

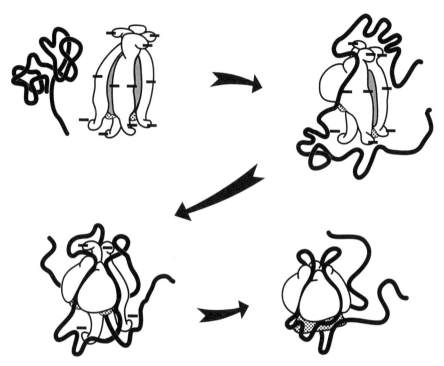

Figure 2 A model for the interaction of SecB with a nonnative polypeptide. See the text for discussion.

IV. A MODEL

A model that can account for the selective binding of SecB to nonnative ligands is depicted in Figure 2. The initial interaction is via an extended, flexible, hydrophilic stretch of the polypeptide. Such interaction might occur between SecB and a loop in a native, folded protein. However, the energy provided by one such interaction would result only in a weak binding. Thus, the affinity for native proteins would be low. Nonnative proteins would have several flexible regions that could simultaneously occupy the hydrophilic peptide-binding sites. Tethering the ligand at multiple points would drastically decrease the probability of dissociation of the complex. In addition, occupation of a sufficient number of the sites would induce a conformational change to create a hydrophobic site that could interact with hydrophobic regions, which would be accessible in nonnative proteins. Exposure of the hydrophobic binding site on SecB only after an initial interaction with a ligand would avoid problems of aggregation of uncomplexed SecB.

V. NATURE OF THE INTERACTIONS THAT PROVIDE THE BINDING ENERGY

Since SecB binds promiscuously and yet tightly, the mode of binding cannot be readily rationalized within the standard conceptual framework of protein-ligand interaction. The conventional view is that high selectivity in binding requires a degree of rigidity in proteins to provide the complementary stereospecific fit between the binding site and ligand (Creighton 1993). A generalization that seems to be true is that proteins do not undergo any substantial change in domain structure upon binding ligands. Changes that do occur usually involve the movement of domains relative to one another. When the ligand and the binding site are not complementary, the binding energy is usually low. How then does SecB interact with so many different stretches of polypeptide to form complexes with dissociation constants in the nanomolar range? Perhaps SecB in its noncomplexed state is unusually malleable and itself resembles a folding intermediate. The interaction between the loosely folded SecB and its nonnative polypeptide ligand might more closely resemble a folding interaction than one of the many ligand-protein interactions studied to date. It has been proposed by several groups that interaction between the backbones of extended strands in SecB and in the ligand might form β structures that would provide binding energy (Randall and Hardy 1989; MacIntyre et al. 1991; Breukink et al. 1992). Several observations make feasible a significant change in structure. First, a portion of SecB is loosely structured as indicated by its extreme sensitivity to proteolysis at low ionic strength. Second, this unstructured region is easily converted to a protease-resistant state by increased ionic strength, presence of divalent cations, or binding of ligand (Randall 1992). Third, circular dichroic spectra obtained under these different conditions provide evidence for significant changes in content of secondary structure (G. Fasman and L.L. Randall, unpubl.). Thus, it is possible that SecB contains an unstructured region, perhaps a loop, that provides a site for interaction with a wide variety of amino-acyl sequences, since it would be flexible enough to accommodate them. The binding energy contributed from each individual site would be low since one would not expect a perfect stereospecific fit, but simultaneous binding at the multiple sites on SecB would result in tight interaction with the ligand. The complex might gain further stability from association between different regions of SecB that can only closely approach each other after neutralization of charge by the ligand.

VI. IMPLICATIONS FOR FUNCTION OF OTHER CHAPERONES
DERIVED FROM STUDIES OF SECB

The primary role of SecB is to block folding and aggregation of precursors, thereby facilitating their export. Other chaperones in *E. coli* such as

DnaK/J and GroEL/S seem to have multiple roles (Georgopoulos 1992). Each of these chaperones is capable of both blocking the folding of some polypeptides to promote export (Kusukawa et al. 1989; Altman et al. 1991; Wild et al. 1992) and facilitating the folding of others (Jaenicke 1993). In eukaryotic cells, cytosolic hsp70 blocks folding during import of precursors into mitochondria, and once inside, the folding of the same polypeptides is facilitated by the closely related chaperones, mitochondrial hsp70 and mitochondrial hsp60 (Craig et al. 1993). Can the same principles be invoked to account for two apparently opposite functions: the blockage of folding and the facilitation of folding?

Before proceeding in our discussion, we must define terms because the words we use will direct the way in which we think about the molecular mechanism of chaperone function. First, we must distinguish between facilitation of a process and catalysis. Take as an example the generalized folding reaction of a polypeptide from unfolded (U) to native (N) state:

$$U \underset{k_{-1}}{\overset{k_1}{\rightleftarrows}} I \overset{k_f}{\rightarrow} N$$

where the conversion of a folding intermediate I to the native state N is rate limiting and essentially irreversible. The rate of flux through this pathway, or in other words the rate of appearance of the final product, the folded protein, is a function of both the concentration of the intermediate, I, and the rate constant k_f

$$\frac{d[N]}{dt} = [I]\, k_f$$

The rate constant is related to the energy barrier between the intermediate and the final state (the activation energy). Enzymatic catalysis increases the rate constant by lowering that barrier, often by stabilizing the transition state relative to the reactants and products. The rate of appearance of the folded protein can also be modulated by changing the concentration of the intermediate. For example, if the intermediate is prone to aggregation, not all of U will be converted to N. Some molecules of I will be diverted to form aggregates, I_n, and the decrease in concentration of free I will then be reflected in a decreased rate of appearance of N. Thus, chaperones might increase the rate of a reaction in two ways: They might act as catalysts to change the rate constants along the pathway or, as seems more likely, they might act by blocking aggregation to increase the amount of the free intermediate (Gething and Sambrook 1992; Jaenicke 1993). Now consider the process of protein export.

During export, the concentration of competent precursor is decreased if the polypeptide folds. In this example, chaperones would facilitate translocation by blocking folding, thereby increasing the effective concentration of the intermediate along the export pathway. In both examples, facilitation of folding and facilitation of export, the underlying principle is the same: The rate of the desirable process is enhanced by blocking other pathways that would deplete the intermediate.

Although it is possible that chaperones are true catalysts, we shall not consider this further. We discuss the alternative mechanisms for enhancing the rate of one pathway relative to another that have their action through affecting the levels of the reactants. If a reactant has access to two pathways, there will be a kinetic partitioning between the two pathways that is determined by the ratio of the appropriate rate constants. Consider aggregation and proper folding. Folding is a first-order reaction; thus, the units of rate constants for folding (k_f) are s^{-1}. Aggregation is of higher order, n, where the lowest possible value for n is 2. Rate constants for aggregation (k_a) have units of $M^{-(n-1)} s^{-1}$. The rate of the first-order reaction will be proportional to the concentration of I, whereas the rate of aggregation will depend on $[I]^n$. Consider a situation in which the rate of aggregation is second-order ($[I^2] k_a$) and is 100-fold higher than the rate of folding ($[I] k_f$). If the concentration of I were decreased 100-fold, the rates would then be equal and a decrease of 1000-fold would result in folding being 10-fold faster than aggregation. If aggregation were of an order higher than 2, the differential effect of changes of concentration would be more extreme. Now it can be readily seen how the presence of a chaperone could favor folding. In the absence of the chaperone, aggregation would decrease the concentration of free I. If the chaperone could bind the intermediate I more rapidly than the intermediate could aggregate, the concentration could be set to favor folding. To favor folding by a factor of 10, in the example above, the chaperone would need to decrease the level of I by 1000-fold. This would require that the chaperone be present at a concentration 1000 times greater than the dissociation constant for the complex between the chaperone and I. The dissociation constants for chaperones that facilitate folding are not yet known. However, the dissociation constants for the complexes of SecB and its ligands are in the nanomolar range, whereas the cellular concentration of SecB is approximately micromolar (Watanabe and Blobel 1989). Thus, at least for SecB, which must block aggregation to maintain the export-competent state, such a mechanism is feasible.

If chaperones are to mediate kinetic partitioning, they must associate with ligands at rates at least as high as those of the alternative pathways. A high proportion of collisions of nonnative polypeptides is likely to result in aggregation; thus, this competing process is near to being encounter-limited. We must therefore consider whether the rate of binding to chaperones could also approach this limit.

VII. HIGH RATE OF ASSOCIATION OF SECB AND LIGANDS

The binding of nonnative maltose-binding protein to SecB exhibits a high affinity (K_d = 1 nM) and yet it has a high rate constant for dissociation (k_{off}) as shown by the release within seconds of maltose-binding protein from a preformed complex with SecB following addition of a tenfold molar excess of R-BPTI, another nonnative ligand (L.L. Randall and S.J.S. Hardy, unpubl.). An apparent discrepancy exists between the observed dissociation constant and the rapid exchange of maltose-binding protein with R-BPTI if the rate constant for association is in the range previously observed for protein-protein interaction, i.e., k_{on}, 10^5 to 10^7 $M^{-1}s^{-1}$ (for a review of protein-protein interaction, see Janin and Chothia 1990). Since $K_d = k_{off}/k_{on}$, the k_{off} would be in the range 10^{-4} s^{-1} to 10^{-2} s^{-1}. Dissociation once a minute could not result in the rapid exchange with R-BPTI that was observed. This discrepancy would disappear if SecB could form a complex with a k_{on} of 10^8 $M^{-1}s^{-1}$ or greater. Consideration of the factors influencing the association rate constant leads us to believe that this is possible. Macromolecules of the size of an average protein collide with a rate constant of about 10^9 $M^{-1}s^{-1}$. Proteins that interact through specific binding sites have rate constants of association smaller than the rate constant for collision because not every encounter results in stable association. The binder and ligand must be oriented properly to allow stereospecific fit. Since the binding interface usually covers only 5–20% of each of two interacting proteins, approximately 1% of the collisions would involve the reactive surfaces (Janin and Chothia 1990). The higher the percentage of the surface that is reactive, the closer the rate of binding comes to being limited by the rate of collision. Since we know that 50% of the ligand maltose-binding protein is in contact with SecB when it binds, it is possible that most encounters are productive. Thus, the rate constant might easily be 10^8 $M^{-1}s^{-1}$ or even higher, and if so, the rate constant for dissociation would be in the range of 10^{-1} s^{-1} to 1 s^{-1}. Such a rate constant would be consistent with the observed data.

A further enhancement of the rate constant for association might come from an increase in the lifetime of the encounter complex if a collision between SecB and the regions outside of the binding site resulted in weak interaction. It is clear that relatively nonspecific binding energy can increase association rates as seen in the case of DNA-binding proteins that initially interact with DNA in a nonspecific way and subsequently associate with the specific sequence by a series of transfers (Von Hippel and Berg 1989).

With a rate constant for association of 10^8 $M^{-1}s^{-1}$, SecB could bind polypeptides before they aggregate, but since SecB has its primary role in maintaining precursors in a state competent for translocation across the membrane, it must also bind polypeptides before they fold. Does a rate constant for association of 10^8 $M^{-1}s^{-1}$ render binding to SecB capable of competing with folding? The early steps in folding such as collapse to a compact state and formation of secondary structure occur on a millisecond time scale (Matthews 1991). However, these states are in rapid equilibrium and have very little energy of stabilization. Even if SecB prefers one intermediate over another, all intermediate states would be populated so that SecB could bind and pull the equilibrium to block folding. The polypeptide is no longer a ligand only after passing through the rate-limiting step in folding, which for maltose-binding protein, as is the case for many proteins (Matthews 1991), is a late step and involves formation of the native structure (Chun et al. 1993). The presence of the leader peptide in the precursor decreases the rate of this step, increasing the probability that SecB will bind. The best estimate for the rate constant for folding of the precursor at 30°C is 0.25 s^{-1} (Hardy and Randall 1991). The concentration of SecB in the cell is estimated as 4 μM (Watanabe and Blobel 1989). If we assume that 10% of the SecB is free and that the rate constant for association is 10^8 $M^{-1}s^{-1}$ as discussed above, then we see that the pseudo-first-order rate constant for binding would be 40 s^{-1} and binding would be strongly favored over folding.

If the rate of binding to SecB approaches the encounter limit, binding would be competitive both with aggregation and with folding. All nonnative proteins that bind chaperones are likely to have large reactive surface areas and the arguments presented above would apply. A very high rate constant for binding may be a necessary feature if chaperones are to mediate kinetic partitioning.

VIII. CONCLUDING REMARKS

Chaperones are united by their remarkable ability to recognize nonnative structure. However, this common underlying principle is overlaid by an enormous diversity among chaperones as can be readily appreciated by perusal of the other chapters in this volume. Chaperones differ in the functions they mediate, in the protein partners required to accomplish those functions, in their oligomeric structures, in the role of ATP in the cycle of binding and release of ligands, and in the mechanism of binding of their ligands. Understanding of any one of the chaperones will provide a basis for further probing the similarities and differences that exist among the members of this intriguing family of proteins. The distinctive characteristics of the interaction of SecB with its ligands are (1) the ability to distinguish nonnative from native structure, (2) the high affinity of binding that is mediated through multiple relatively nonspecific binding sites for flexible lengths of polypeptide, and (3) a rate of binding that approaches the encounter limit. The action of other chaperones albeit with molecular interactions that differ from those of SecB may also depend on these same fundamental principles.

IX. EPILOGUE

As a soft silvery moon may appear from behind dense clouds on the darkest night, dimly to illuminate the tortuous path ahead, so knowledge of SecB offers fragmentary enlightenment to our bewildered minds as we clutch at understanding... .

> *Science! thou fair effusive ray*
> *From the great source of mental day,*
> *Free, generous and refined!*
> *Descend with all thy treasures fraught,*
> *Illumine each bewilder'd thought,*
> *And bless my laboring mind.*

Mark Akenside (1721–1770)

REFERENCES

Altman, E., S. Emr, and C. Kumamoto. 1990. The presence of both the signal sequence and a region of mature LamB protein is required for the interaction of LamB with the export factor SecB. *J. Biol. Chem.* **265:** 18154–18160.

Altman, E., C.A. Kumamoto, and S.D. Emr. 1991. Heat-shock proteins can substitute for SecB function during protein export in *Escherichia coli. EMBO J.* **10:** 239–245.

Bassford, P.J., Jr. 1990. Export of the periplasmic maltose-binding protein of *Escherichia coli. J. Bioenerg. Biomemb.* **22:** 401–439.

Breukink, E., R. Kusters, and B. de Kruijff. 1992. *In vitro* studies on the folding characteristics of the *Escherichia coli* precursor protein prePhoE. *Eur. J. Biochem.* **208:** 419–425.

Chun, S.-Y., S. Strobel, P.J. Bassford, Jr., and L.L. Randall. 1993. Folding of maltose-binding protein: Evidence for the identity of the rate-determining step *in vivo* and *in vitro. J. Biol. Chem.* **268:** 20855–20862.

Collier, D.N. 1993. SecB: A molecular chaperone of *Escherichia coli* protein secretion pathway. *Adv. Protein Chem.* **44:** 151–193.

Collier, D.N., V.A. Bankaitis, J.B. Weiss, and P.J. Bassford, Jr. 1988. The antifolding activity of SecB promotes the export of the *E. coli* maltose-binding protein. *Cell* **53:** 273–283.

Craig, E.A., B.D. Gambill, and R.J. Nelson. 1993. Heat shock proteins: Molecular chaperones of protein biogenesis. *Microbiol. Rev.* **57:** 402–414.

Creighton, T.E. 1993. *Proteins,* 2nd ed., pp. 334–346. W.H. Freeman, New York.

De Cock, H., W. Overeem, and J. Tommassen. 1992. Biogenesis of outer membrane protein PhoE of *Escherichia coli*: Evidence for multiple SecB-binding sites in the mature portion of the PhoE protein. *J. Mol. Biol.* **224:** 369–379.

Gannon, P.M., P. Li, and C.A. Kumamoto. 1989. The mature portion of *Escherichia coli* maltose-binding protein (MBP) determines the dependence of MBP on SecB for export. *J. Bacteriol.* **169:** 1286–1290.

Gardel, C., K. Johnson, A. Jacq, and J. Beckwith. 1990. The *secD* locus of *E. coli* codes for two membrane proteins required for protein export. *EMBO J.* **9:** 3209–3216.

Georgopoulos, C. 1992. The emergence of the chaperone machines. *Trends Biol. Sci.* **17:** 295–299.

Gething, M.-J. and J. Sambrook. 1992. Protein folding in the cell. *Nature* **355:** 33–45.

Hardy, S.J.S. and L.L. Randall. 1991. A kinetic partitioning model of selective binding of nonnative proteins by the bacterial chaperone SecB. *Science* **251:** 439–443.

Hartl, F.-U., S. Lecker, E. Schiebel, J.P. Hendrick, and W. Wickner. 1990. The binding cascade of SecB to SecA to SecY/E mediates preprotein targeting to the *E. coli* plasma membrane. *Cell* **63:** 269–279.

Jaenicke, R. 1993. Role of accessory proteins in protein folding. *Curr. Opin. Struct. Biol.* **3:** 104–112.

Janin, J. and C. Chothia. 1990. The structure of protein-protein recognition sites. *J. Biol. Chem.* **256:** 16027–16030.

Kumamoto, C.A. 1989. *Escherichia coli* SecB protein associates with exported protein precursors *in vivo. Proc. Natl. Acad. Sci.* **86:** 5320–5324.

Kumamoto, C.A. and J. Beckwith. 1983. Mutations in a new gene, *secB,* cause defective protein localization in *Escherichia coli. J. Bacteriol.* **154:** 253–260.

———. 1985. Evidence for specificity at an early step in protein export in *Escherichia coli. J. Bacteriol.* **163:** 267–274.

Kumamoto, C.A. and O. Francetic. 1993. Highly selective binding of nascent polypeptides by an *Escherichia coli* chaperone protein *in vivo. J. Bacteriol.* **175:** 2184–2188.

Kumamoto, C.A. and P.M. Gannon. 1988. Effects of *Escherichia coli secB* mutations on pre-maltose-binding protein conformation and export kinetics. *J. Biol. Chem.* **263:** 11554–11558.

Kusters, R., T. de Vrije, E. Breukink, and B. de Kruijff. 1989. SecB protein stabilizes a translocation-competent state of purified prePhoE protein. *J. Biol. Chem.* **264:** 20827–20830.

Kusukawa, N., T. Yura, C. Ueguchi, Y. Akiyama, and K. Ito. 1989. Effects of mutations in heat-shock genes *groES* and *groEL* on protein export in *Escherichia coli. EMBO J.* **8:** 3517–3521.

Lecker, S., R. Lill, T. Ziegelhoffer, C. Georgopoulos, P.J. Bassford, Jr., C.A. Kumamoto, and W. Wickner. 1989. Three pure chaperone proteins of *Escherichia coli,* SecB, trigger factor, and GroEL, form soluble complexes with precursor proteins *in vitro. EMBO J.* **9:** 2703–2709.

Liu, G., T.B. Topping, and L.L. Randall. 1989. Physiological role during export for the retardation of folding by the leader peptide of maltose-binding protein. *Proc. Natl. Acad. Sci.* **86:** 9213–9217.

Liu, G., T.B. Topping, W.H. Cover, and L.L. Randall. 1988. Retardation of folding as a possible means of suppression of a mutation in the leader sequence of an exported protein. *J. Biol. Chem.* **263:** 14790–14793.

MacIntyre, S., B. Mutschler, and U. Henning. 1991. Requirement of the SecB chaperone for export of a non-secretory polypeptide in *Escherichia coli. Mol. Gen. Genet.* **227:** 224–228.

Matthews, C.R. 1991. The mechanism of protein folding. *Curr. Opin. Struct. Biol.* **1:** 28–35.

Randall, L.L. 1992. Peptide binding by chaperone SecB: Implications for recognition of nonnative structure. *Science* **257:** 241–245.

Randall, L.L. and S.J.S. Hardy. 1986. Correlation of competence for export with lack of tertiary structure of the mature species: A study *in vivo* of maltose-binding protein in *E. coli. Cell* **46:** 921–928.

———. 1989. Unity in function in the absence of consensus in sequence: Role of leader peptides in export. *Science* **243:** 1156–1159.

Randall, L.L., T.B. Topping, and S.J.S. Hardy. 1990. No specific recognition of leader peptide by SecB, a chaperone involved in protein export. *Science* **248:** 860–863.

Schiebel, E., A.J.M. Driessen, F.-U. Hartl, and W. Wickner. 1991. $\Delta\mu H^+$ and ATP function at different steps of the catalytic cycle of preprotein translocase. *Cell* **64:** 927–939.

Von Hippel, P.H. and O.G. Berg. 1989. Facilitated target location in biological systems. *J. Biol. Chem.* **264:** 675–678.

Watanabe, M. and G. Blobel. 1989. Cytosolic factor purified from *E. coli* is necessary and sufficient for export of a preprotein and is a homotetramer of SecB. *Proc. Natl. Acad. Sci.* **86:** 2728–2732.

Weiss, J.B. and P.J. Bassford, Jr. 1990. The folding properties of the *Escherichia coli* maltose-binding protein influence its interaction with SecB *in vitro. J. Bacteriol.* **172:** 3023–3029.

Weiss, J.B., P.H. Ray, and P.J. Bassford, Jr. 1988. Purified SecB protein of *E. coli* retards folding and promotes membrane translocation of maltose-binding protein *in vitro. Proc. Natl. Acad. Sci.* **85:** 8978–8982.

Wild, J., E. Altman, T. Yura, and C.A. Gross. 1992. DnaK and DnaJ heat shock proteins participate in protein export in *Escherichia coli. Genes Dev.* **6:** 1165–1172.

Wolfe, P.B., W. Wickner, and J.M. Goodman. 1983. Sequence of the leader peptidase gene of *Escherichia coli* and the orientation of leader peptidase in the bacterial envelope. *J. Biol. Chem.* **258:** 12073–12080.

Zwizinski, C. and W. Wickner. 1980. Purification and characterization of leader (signal) peptidase from *Escherichia coli. J. Biol. Chem.* **255:** 7973–7977.

12

The Structure, Function, and Genetics of the Chaperonin Containing TCP-1 (CCT) in Eukaryotic Cytosol

Keith R. Willison and Hiroshi Kubota
Chester Beatty Laboratories
Institute of Cancer Research
London SW3 6JB, United Kingdom

I. INTRODUCTION

The double-torus chaperonins are one class of molecular chaperones found in eubacteria, mitochondria, and plastids that bind protein substrates at some intermediate stage in their protein folding pathways and use ATP hydrolysis to fold and release them as biologically active, correctly folded products (Horwich and Willison 1993; Saibil and Wood 1993; see Frydman and Hartl, this volume).

Recently, it has become clear that eukaryotic cytosol contains an abundant ring-shaped chaperonin that is composed of many kinds of subunits encoded by a family of related genes. The discovery of the cytosolic chaperonin containing TCP-1 (CCT) arose from the convergence of three independent lines of investigation. First, our laboratory had been investigating the mouse t-complex polypeptide 1 (*Tcp-1*) gene that maps to the t-complex region of mouse chromosome 17 (Silver et al. 1979) and is up-regulated during spermatogenesis (Dudley et al. 1984; Silver et al. 1987; Willison et al. 1990). We cloned mouse *Tcp-1* (Willison et al. 1986; Kubota et al. 1992b) and human *TCP1* (Willison et al. 1987; Kirchoff and Willison 1990) cDNAs and genes, made a set of seven monoclonal antibodies to the protein (Willison et al. 1989), and biochemically purified the chaperonin containing TCP-1 (Lewis et al. 1992). Second, investigation of the heat shock response in two species of thermophilic archaebacteria uncovered a novel, heat-inducible, double-toroid ATPase with in vitro chaperonin activity (Phipps et al. 1991, 1993; Trent et al. 1991). The subunits of these archaebacterial chaperonins are highly related to eukaryotic TCP-1, 40% identical in the case of TF55 from *Sulfolubus shibataea* (Trent et al. 1991). Third, examination of the protein folding activities of rabbit reticulocyte lysates programmed with actin and tubulin mRNAs showed that newly synthesized actin and tubulin form complexes with a chaperonin activity found endogenously in the lysate (Gao et al. 1992; Yaffe et al. 1992), and this chaperonin has turned out to be the chaperonin containing TCP-1.

The existence of chaperonins in eukaryotic cytosol raises many important questions and problems in areas as diverse as the biophysics of protein folding and the origin and evolution of the cytosol and the cytoskeleton. Here, we summarize what is known of the biochemistry, function, and structure of CCT and the gene family that encodes its subunits, and we discuss some ideas about the evolution of chaperonins and their substrates.

II.   SUBUNITS OF THE CHAPERONIN CONTAINING TCP-1 (CCT)

The first chaperonins to be characterized were the GroEL family found in eubacteria and endosymbiotic organelles of eukaryotes. These double-torus structures have sevenfold rotational symmetry and are generally composed of 14 identical 60-kD subunits (group I in Table 1). All members of the TCP-1 subfamily (group II in Table 1) of chaperonins are weakly but identifiably related in primary sequence to GroEL (Gupta 1990; Hemmingsen 1992; Lewis et al. 1992) and, like GroEL, are found as double-torus complexes. However, the rotational symmetry of the toruses is generally eightfold (Lewis et al. 1992; Phipps et al. 1993), although they may be able to assume ninefold (Trent et al. 1991; Lewis et al. 1992) or even sixfold symmetrical arrangements (Table 1) (Mummert et al. 1993). The archaebacterial TCP-1-like chaperonins are composed of one or two subunit species, but the eukaryotic chaperonins are composed of up to nine subunit species, each encoded by separate genes (Table 2). The subunits are numbered (S1–S9) according to ascending pI, measured by isoelectric focusing to equilibrium, and where we have isolated the gene for a particular subunit, the proteins are named CCT for chaperonin containing TCP-1 (Kubota et al. 1994). Thus, TCP-1 is spot S3, the protein is CCTα, and the gene is *Ccta*. The multiplicity of subunit species raises the possibility that toruses of different subunit com-

Table 1 Characteristics of Chaperonins

	Organism	Localization	Known substrate	Homology with		Rotational symmetry	Subunit species	References
				GroEL	TF55			
Subfamily of eubacterial chaperonins and eukaryotic intraorganellar chaperonins (group I chaperonins)								
GroEL	eubacteria	soluble	phage proteins	100%	weak	7	1	1–4
hsp60	eukaryotes	mitochondria	dihydrofolate reductase	60%	weak	7	1	5–6
Rubisco subunit-binding protein	plants	chloroplasts/plastids	Rubisco subunits	60%	weak	7	2	2,7
Subfamily of archaebacterial chaperonins and eukaryotic cytosolic chaperonins (group II chaperonins)								
TF55	archaebacteria (*S. shibatae*)	soluble	?	weak	100%	8 or 9	1	8
Thermosome	archaebacteria (*P. occultum*)	soluble	?	?	(70%)[a]	8	2	9–10
CCT (TCP-1 complex)	eukaryotes	cytosol	actin, tubulin	weak	40%	8 or 9	7–9	11–15
TCP-1-related chaperone	plants (oat)	cytosol	phytochrome	? (TCP-1 related)	? (TCP-1 related)	6	?	16

References: (1) Georgopoulos et al. 1973; (2) Hemmingsen et al. 1988; (3) Martin et al. 1991; (4) Langer et al. 1992; (5) Cheng et al. 1989; (6) Koll et al. 1992; (7) Martel et al. 1990; (8) Trent et al. 1991; (9) Phipps et al. 1991; (10) Phipps et al. 1993; (11) Lewis et al. 1992; (12) Yaffe et al. 1992; (13) Gao et al. 1992; (14) Frydman et al. 1992; (15) Kubota et al. 1994; (16) Mummert et al. 1993.
[a]Partial sequences of tryptic peptides.

Table 2 Correspondence between Subunit Proteins (Spots on Two-dimensional Gels) and Genes (Clones) of the Eukaryotic Hetero-oligomeric Chaperonin Containing TCP-1 (CCT)

Protein mouse spot	CCTα S3	CCTβ S4	CCTγ S5	CCTδ S8 or S9	CCTε S2	CCTζ S7	CCTη S8 or S9
Gene	*Ccta*	*Cctb*	*Cctg*	*Cctd*	*Ccte*	*Cctz*	*Ccth*
mouse	*Tcp-1*[1]	pTβ2[6]	pTγ7[6]	pTδ2[6]	pTε5[6]	pTζ12[6]	pCBL80[6]
human	*TCP1*[2]	p383[6]*	pAP3[8]*	IB713[9]*	p384[6]*	HTR3[10]*	–
C. elegans	pG13[3]*,[6]*	–	–	pG3[3]*,[6]*	pG4[3]*,[6]*	pG13[3]*,[6]*	–
S. cerevisiae	*TCP1*[4]	4950[7]	–	–	–	–	–
plants	*Tcp-1*[5]	–	–	–	–	–	–

Correspondence of mouse CCT subunit proteins (spots on 2D gels) to *Cct* genes is derived from Kubota et al. (1994). Clone names are shown for *Cct* genes of each species except for *Tcp-1* genes of mouse, human, *S. cerevisiae*, and plants. Asterisks indicate partial clones or partially sequenced clones.

References of genes and clones as follows: (1) Willison et al. 1986; Kubota et al. 1991, 1992b. (2) Willison et al. 1987; Kirchhoff and Willison 1990. (3) Waterston et al. 1992. (4) Ursic and Culbertson 1991. (5) Mori et al. 1992. (6) Kubota et al. 1994. (7) Miklos et al. 1994. (8) N.A. Walkley and A.N. Malik, unpubl. (9) Khan et al. 1992. (10) Segel et al. 1992.

positions may exist because some subunit species may be interchangeable (Lewis et al. 1992; Roobol and Carden 1993; Kubota et al. 1994). This idea is supported by the finding of a testis-specific subunit (S6) (G. Hynes et al., unpubl.).

III. *TCP-1*-RELATED GENES ENCODING CCT SUBUNITS

So far, we have isolated six mouse genes that are related to *Tcp-1* (Kubota et al. 1994), and we have shown the correspondence between six CCT subunits and their genes (Table 2). There are numerous partial cDNA sequences encoding TCP-1 or TCP-1-related polypeptides that have been discovered by single-pass sequencing of cDNAs from nematode (Waterston et al. 1992) and human brain (Khan et al. 1992) and by other work (Table 2). A genome-sequencing project in yeast discovered *TCP1β* (Miklos et al. 1994), the orthologue of the *Cctb* gene. Analysis of the mouse *Cct* genes showed that they encode proteins that are all approximately 30% related to one another, and phylogenetic analysis suggests that all of the *Cct* genes diverged from one another more than 1000 million years ago (Kubota et al. 1994), which predates the divergence of animals, plants, and yeasts. This is unusual behavior for a gene family because most other large eukaryotic families are much younger and suggests that the CCT chaperonin performs ancient and essential functions. We have suggested that each CCT subunit evolved an independent function early in the eukaryotic lineage. These independent functions and a common ATPase function seem to have been maintained in each subunit after divergence because it is possible to detect the orthologue of each mouse gene in *Saccharomyces cerevisiae* by nucleic acid hybridization with murine DNA probes (Kubota et al. 1994). Comparison of the complete polypeptide sequences of mouse and yeast *Ccta* and *Cctb* shows that they are 61% (332/544) and 66% (345/525) identical between the two species, respectively. The nucleotide sequences encoding the very conserved region C1 (see Fig. 1) are 70% identical between the two pairs of mouse and yeast *Ccta* and *Cctb* genes, and this homology probably accounts for the cross-hybridization between the mouse and yeast orthologues of each *Cct* gene.

IV. INTERACTIONS BETWEEN CCT AND SUBSTRATES

In vitro translation experiments in rabbit reticulocyte lysates programmed with synthetic mRNAs were the first to discover substrates of CCT. Gao et al. (1992) found that newly synthesized β-actin bound to CCT and could be released in its native conformation upon incubation

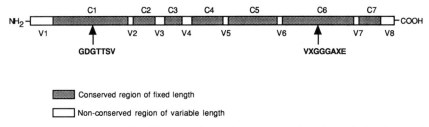

Figure 1   Structure of CCT subunits. The conserved region of fixed length (*stippled boxes*, C1–C7) and nonconserved region of variable length (*open boxes*, V1–V8) in seven CCT subunits are shown in a model structure of CCT. The large conserved regions C1 and C6 contain the very highly conserved motifs GDGTTSV and VXGGGAXE, respectively, postulated to be involved in ATP binding and ATPase activity (Lewis et al. 1992; Kubota et al. 1994).

with Mg-ATP. Yaffe et al. (1992) found that newly synthesized α and β tubulins both bound to CCT. Both these groups found that newly synthesized tubulin subunits bound to CCT could not be released effectively in assembly-competent form, and this suggested the requirement for cofactors, possibly the CCT equivalents of GroES (Gao et al. 1992; Yaffe et al. 1992). Gao et al. (1993) have purified two activities that are required to release assembly-competent tubulins, but the identity of the active molecules in these fractions is not known. However, it seems that like β-actin, neither γ-tubulin nor actin-related proteins (actin-RPV) require release factors (Melki et al. 1993). Frydman et al. (1992) found no evidence for cofactors required to release β-tubulin or luciferase from bovine testis CCT, and therefore this aspect of CCT function remains obscure. It is clear that during the heat shock response in two species of archaebacteria, the only major proteins induced are the 55–57-kD chaperonin subunits (Trent et al. 1990; Phipps et al. 1991), but possibly cofactors, required during normal growth conditions, might not be used under heat shock conditions. Possible cofactors could be other chaperones such as DnaK/hsp70 and DnaJ family members. Lewis et al. (1992) found that hsp70 proteins copurify with CCT, and Kubota et al. (1994) found a 45-kD polypeptide coimmunoprecipitating with mouse F9 cell CCT.

   A remarkable aspect of CCT substrate interactions is the length of time that substrates occupy the CCT. In vitro actin and tubulin bind with a $t_{1/2}$ of about 5 minutes (Gao et al. 1992; Yaffe et al. 1992) and the $t_{1/2}$ of luciferase folding by bovine CCT is 10 minutes (Frydman et al. 1992). These data have recently been confirmed in vivo in Chinese hamster ovary (CHO) cells, where the $t_{1/2}$ for binding to CCT is 5 minutes in the case of tubulin and 3 minutes with actin (Sternlicht et al. 1993). It may reflect conventional behavior of chaperonins, however, since we know that the $t_{1/2}$ of rhodanese folding by GroEL/GroES is several minutes (Martin et al. 1991; Langer et al. 1992) and that GroEL slows down the rate of folding of barnase 400-fold, compared to spontaneous folding, without substantially altering the folding pathway (Gray et al. 1993). Perhaps, to ensure accuracy of folding, the interaction of sites on the substrates with sites on the CCT folding surfaces is intricate and slow. It is also possible that the occupancy time on CCT depends on the characteristics of substrates, e.g., size, shape, pI, hydrophobicity, flexibility, number of domains or core structures, and interaction between domains. In addition, interaction with other chaperones or cofactors may slow down the folding process. It is probable that mechanisms exist in cells to regulate throughput of substrate on CCT.

Another important question about CCT is the spectrum of proteins that can be folded by CCT and cofactors. So far, actin and α and β tubulins have been identified as the main substrates of CCT in vivo and in vitro. However, actins and tubulins are extremely abundant proteins in cells and thus may be the easiest proteins to detect as CCT substrates in experiments in vivo (Sternlicht et al. 1993). It is very difficult to believe that CCT acts as a chaperone only for actins and tubulins, since they have no amino acid sequence homology. Melki et al. (1993) have shown that β-actin, actin-RPV, and α and β tubulins all cross-compete with one another for binary complex formation with CCT, suggesting that all four substrates can occupy CCTs of identical compositions. Frydman et al. (1992) reported that CCT refolded luciferase in vitro, and brain CCT was found to bind a neurofilament polypeptide fragment and release it by adding ATP (Roobol and Carden 1993). Sternlicht et al. (1993) found that CCT bound to several newly synthesized proteins in vivo in addition to actin and tubulin. These observations suggest a wider range of substrates for CCT and cofactors than just cytoskeletal proteins. It may be possible that different substrates bind to different sites on CCT at particular stages in their folding pathways.

V. *TCP1* GENETICS IN YEAST

As mentioned above, the *Cct* genes are highly conserved from *Saccharomyces cerevisiae* to mammals, and genetic analysis of CCT function is probably best performed in this unicellular organism. Before it was known that TCP1 (CCTα) was a subunit of CCT, it was shown by Ursic and Culbertson (1991) that *TCP1* is an essential gene and that a cold-sensitive mutation in *TCP1* affects microtubule-mediated processes and makes the mutant strain (*tcp-1*) particularly sensitive to inhibitors of microtubule function. More recently, Miklos et al. (1994) have purified CCT from yeast and shown that it is a heteromeric complex as found in mammalian cells (Lewis et al. 1992). Miklos et al. (1994) have also characterized the *TCP-1β/Cctb* gene and shown that it too is essential to yeast. New temperature-sensitive mutations in the *TCP1/Ccta* and *TCP1β/Cctb* genes both affect the function of microtubules (Miklos et al. 1994). It will be interesting to see if mutations in other subunit genes of yeast CCT affect tubulin or other substrates.

VI.   GENETIC MAPPING OF *Tcp-1* AND *Cct* GENES

More than a decade ago, we set out to isolate the gene encoding TCP-1, and we cloned a mouse cDNA fragment (pB1.4) of a gene highly expressed in testis by differential cDNA cloning (Dudley et al. 1984). Genetic mapping and mRNA selection/translation analysis of the full-length cDNA for pB1.4 (Willison et al. 1986) revealed that the cDNA encoded a protein named t-complex polypeptide 1 (Silver et al. 1979). The *Tcp-1* gene maps in the t-complex region of mouse chromosome 17 and this is the reason for its name. Thus, its name does not indicate its function, and this is why we thought that *Tcp-1* and all the related genes should be named *Cct* to indicate function. In addition, at least three of the novel mouse *Cct* genes do not map to the t-complex (A. Ashworth et al., unpubl.). Human *TCP1* (the human orthologue of *Ccta*) maps to the long arm of chromosome 6 unlinked to HLA on the short arm of chromosome 6 (Willison et al. 1987). Bovine *TCP1* has remained linked to BoLA (Andersson 1988), but rat *TCP1* (chromosome 1) is again unlinked to the MHC (chromosome 20) (Yasue et al. 1992) as in humans. Khan et al. (1992) mapped IB713 (Table 2) (the human orthologue of *Cctd*) to human chromosome 2, and A. Ashworth et al. (unpubl.) have mapped *Cctg* to mouse chromosome 3. Thus, the *Cct* family is probably dispersed over many chromosomal segments in mammals, and this feature is consistent with the ancient divergence time of the family.

An intriguing observation relevant to the possible involvement of *Tcp-1* in transmission ratio distortion of the mouse t-complex (Silver and Remis 1987) is that *Tcp-1*[a] (t-haplotype allele) has accumulated many nonsynonymous nucleotide changes compared to the *Tcp-1*[b] allele carried by standard laboratory strain chromosomes 17 (Kubota et al. 1991, 1992a; Morita et al. 1991, 1992). The evolution of TCP-1 on both t-haplotype and wild-type chromosomes 17 may have been accelerated during the combatitive evolution of the chromosomal types. The high number of nonsynonymous versus synonymous changes suggests that *Tcp-1*[a] evolution may be constrained because the t-complex region may also carry a gene that interacts with *Tcp-1*[a], possibly one that encodes another CCT subunit, a cofactor, or a substrate.

VII.   EXPRESSION OF THE *Tcp-1 (Ccta)* GENE

The TCP-1 protein is highly expressed in testis (Silvers et al. 1987; Willison et al. 1990), with *Tcp-1* mRNA being 20-fold more abundant in testis than in liver (Dudley et al. 1984). We have estimated the number of CCT particles present per cell to be between 10^5 (HEp2 cells) and 3 x 10^5 (spermatids) (Lewis et al. 1992). The slight disparity between *Tcp-1* mRNA levels and total CCT protein content per cell may reflect different turnover rates of CCT subunit mRNAs (our unpublished results). However, *Tcp-1* mRNA is not only extremely abundant in testis, but also highly expressed in embryos of early stages and rapidly growing cells in tissue culture (Kubota et al. 1992b). This may suggest that CCT is involved in cell-cycle-regulated events such as protein synthesis and cell division. The nucleotide sequence of the mouse *Tcp-1* structural gene shows 12 exons and a CpG island around the transcription start points of the TATA-less promoter (Kubota et al. 1992b). In this CpG island, several possible binding sites of transcription factors were found, but heat shock elements were not identified. This is consistent with the fact that TCP-1 protein is not heat inducible (our unpublished results) unlike many other chaperonins (GroEL, hsp60, TF55, thermosome), which are heat inducible. Among transcription factors that might bind the *Tcp-1* promoter sequence based on motif analysis, Sp1 and E2F are interesting. These may be important factors in the regulation of *Tcp-1* gene expression since Sp1 is up-regulated during spermatogenesis (Saffer et al. 1991) and E2F is involved in cell cycle control (Mudryj et al. 1991).

VIII.   EVOLUTION OF CCT

As first pointed out by Gupta (1990), GroEL and TCP-1 (*CCT*α) are weakly, although significantly, related throughout their length, and others support this view (Ellis 1990; Hemminsgen 1992; Lewis et al. 1992). The sequencing of the gene encoding TF55 from the archaebacteria, *Solfolobus shibataea*, showed that the predicted amino acid sequence of TF55 is 40% identical to mammalian TCP-1 (Trent et al. 1991), suggesting that all of the eukaryotic *Tcp-1* and *Cct* genes derive from a lineage giving rise to the *Archaea* (archaebacteria) and *Eukarya* (eukaryotes) (Fig. 2). Thus, there are two major subfamilies of chaperonins: the GroE and TCP-1 subfamilies (Ellis 1992; Lewis et al. 1992). All chaperonins are thought to have evolved from a common an-

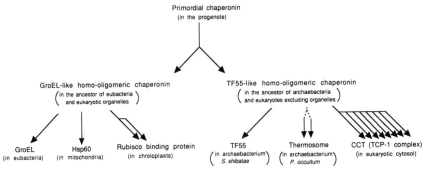

Figure 2 Evolution of chaperonins. This evolutionary diagram was constructed by consideration of the amino acid sequence homology between chaperonin subunits (see Table 1) and the endosymbiont hypothesis. An arrow indicates a subunit species.

cestral chaperonin in the progenote (Gupta 1990) and diverged in the eubacterial lineage and archaebacterial-eukaryotic lineage. After this first divergence, the latter lineage diverged to give rise to the archaebacterial and eukaryotic lineages (Fig. 2). We have speculated that the sequence conservation between the two subfamilies is likely to be due to the Mg-ATP binding and ATPase activity of the chaperonins, rather than the ability to form stacked toruses or bind substrates (Kubota et al. 1994). The conserved domains encompass the two glycine-rich motifs highlighted in Figure 1. The amino-terminal glycine-rich domain shows homology with the β-phosphate-binding loop of the cAMP-dependent protein kinase family (Lewis et al. 1992; Kubota et al. 1994). Interestingly, the more carboxy-terminal triple glycine region is related in sequence to members of the valosin-containing protein (VCP)/Cdc48p family (K.R. Willison and H. Kubota, unpubl.). VCP is a sixfold symmetrical stack-torus ATPase that binds clathrin heavy chains and may be a chaperone in membrane transport events.

To us, there seem to be a number of critical questions to be answered about the function of chaperonins containing TCP-1, and some of these are rather distinct from those being asked about GroEL function. In general, of course, one is interested in the nature of the regions on chaperonin subunits that can bind substrate polypeptides and how this binding is coupled to substrate folding, ATP hydrolysis, and release of folded substrates. From studies on GroEL, the prevailing view is that hundreds, at least, of proteins can be bound by this chaperonin and therefore that it performs some general action upon substrates. This general action could be something like, for example, burying or temporarily stabilizing exposed core-hydrophobic regions in a partially folded protein

(Martin et al. 1991). Thus, GroEL may not be an active folder of proteins but may merely function to prevent unproductive or dead-end intermediates occurring during folding.

If CCT functions like GroEL, why has it evolved so many subunits, each type of which having been very conserved during eukaryotic evolution? We have speculated that subunits of CCT evolved to cope with the folding of newly evolving proteins (Kubota et al. 1994), and if this was the case, it seems inescapable but to conclude that CCT subunits have specific binding sites for particular polypeptide structures or motifs found in proteins. It might be possible that CCT evolved solely to cope with the folding of a few cytoskeletal proteins like actin and tubulin and that it represents a special unusual type of chaperonin able to do useful tasks for the folding problems encountered by this particular class of proteins. It remains to be shown that CCT is a broad-spectrum folding machine, but it is hard to envisage how other proteins (other heat shock families?) in evolving eukaryotic cells took up the general functions performed by chaperonins, whereas the CCT evolved from having a general function to having a restricted one. We feel that CCT may work through a combination of general affinity for unfolded proteins and specific affinities for particular domains of partially folded proteins. Functional analysis of each subunit species through manipulation of the genes that encode them (Kubota et al. 1994) is now required to solve the mechanisms by which CCT facilitates protein folding in eukaryotic cytosol.

ACKNOWLEDGMENTS

This work was funded by the Cancer Research Campaign/Medical Research Council grant to the Institute for Cancer Research.

REFERENCES

Andersson, L. 1988. Genetic polymorphism of a bovine t-complex gene (TCP1) linkage to major histocompatibility genes. *J. Hered.* **79:** 1–5.

Cheng, M.Y., F.-U. Hartl, J. Martin, R.A. Pollock, F. Kalousek, W. Neupert, E.M. Hallberg, R.L. Hallberg, and A.L. Horwich. 1989. Mitochondrial heat-shock protein hsp60 is essential for assembly of proteins imported into yeast mitochondria. *Nature* **337:** 620–625.

Dudley, K., J. Potter, and K.R. Willison. 1984. Analysis of male sterile mutations in the mouse using haploid stage expressed cDNA probes. *Nucleic Acids Res.* **12:** 4281–4293.

Ellis, R.J. 1990. Molecular chaperones: The plant connection. *Science* **250:** 954–959.

———. 1992. Cytosolic chaperonin confirmed. *Nature* **358:** 191–192.

Frydman, J., E. Nimmesgern, H. Erdjument-Bromage, J.S. Wall, P. Tempst, and F.-U. Hartl. 1992. Function in protein folding of TRiC, a cytosolic ring complex containing TCP-1 and structurally related subunits. *EMBO J.* **11:** 4767–4778.

Gao, Y., I.E. Vainberg, R.I. Chow, and N.J. Cowan. 1993. Two cofactors and cytoplasmic chaperonin are required for the folding of α- and β-tubulin. *Mol. Cell Biol.* **13:** 2478–2485.

Gao, Y., J.O. Thomas, R.L. Chow, G.-H. Lee, and N.J. Cowan. 1992. A cytoplasmic chaperonin that catalyze β-actin folding. *Cell* **69:** 1043–1050.

Georgopoulos, C.P., R.W. Hendrix, S.R. Casjens, and A.D. Kaiser. 1973. Host participation in bacteriophage lambda head assembly. *J. Mol. Biol.* **76:** 45–60.

Gray, T.E., J. Eder, M. Bycroft, A.G. Day, and A.R. Fersht. 1993. Refolding of barnase mutants and pro-barnase in the presence and absence of GroEL. *EMBO J.* **12:** 4145–4150.

Gupta, R.S. 1990. Sequence and structural homology between a mouse t-complex protein TCP-1 and the "chaperonin" family of bacterial (GroEL, 60-65 kDa heat shock antigen) and eukaryotic proteins. *Biochem. Int.* **20:** 833–841.

Hemmingsen, S.M. 1992. What is a chaperonin? *Nature* **357:** 650.

Hemmingsen, S.M., C. Woolford, S.M. vanderVies, K. Tilly, D.T. Dennis, C.P. Georgopoulos, R.W. Hendrix, and R.J. Ellis. 1988. Homologous plant and bacterial proteins chaperone oligomeric protein assembly. *Nature* **333:** 330–334.

Horwich, A.L. and K.R. Willison. 1993. Protein folding in the cell: Function of two families of molecular chaperone, hsp60 and TF55-TCP1. *Philos. Trans. R. Soc. Lond.* **339:** 313–326.

Khan, A.S., A.S. Wilcox, M.H. Polymeropoulos, J.A. Hopkins, T.J. Stevens, M. Robinson, A.K. Orpana, and J.M. Sikela. 1992. Single pass sequencing and physical and genetic mapping of human brain cDNAs. *Nature Genet.* **2:** 180–185.

Kirchhoff, C. and K.R. Willison. 1990. Nucleotide and amino-acid sequence of human testis-derived *TCP1. Nucleic Acids Res.* **18:** 4247.

Koll, H., B. Guiard, J. Rassow, J. Osterman, A.L. Horwich, W. Neupert, and F.-U. Hartl. 1992. Antifolding activity of hsp 60 couples protein import into the mitochondrial matrix with export to the intermembrane space. *Cell* **68:** 1163–1175.

Kubota, H., T. Morita, Y. Satta, M. Nozaki, and A. Matsushiro. 1992a. Nucleotide sequence of a mouse *Tcp-1* pseudogene: A nucleotide record for a *t* complex gene carried by an ancestor of the mouse. *Mammal. Genome* **2:** 246–251.

Kubota, H., G. Hynes, A. Carne, A. Ashworth, and K. Willison. 1994. Identification of six *Tcp-1* related genes encoding divergent subunits of the TCP-1-containing chaperonin. *Curr. Biol.* **4:** (in press).

Kubota, H., T. Morita, T. Nagata, Y. Takemoto, M. Nozaki, G. Gachelin, and A. Matsushiro. 1991. Nucleotide sequence of mouse *Tcp-1ᵃ* cDNA. *Gene* **105:** 269–273.

Kubota, H., K. Willison, A. Ashworth, M. Nozaki, H. Miyamoto, A. Yamamoto, A. Matsushiro, and T. Morita. 1992b. Structure and expression of the gene encoding *t*-complex polypeptide 1 (*Tcp-1*). *Gene* **120:** 207–215.

Langer, T., G. Pfeifer, J. Martin, W. Baumeister, and F.-U. Hartl. 1992. Chaperonin-mediated protein folding: GroES binds to one end of the GroEL cylinder, which accomodates the protein substrate within its central cavity. *EMBO J.* **11:** 4757–4765.

Lewis, V.A., G.M. Hynes, D. Zheng, H. Saibil, and K. Willison. 1992. T-complex polypeptide-1 is a subunit of a heteromeric particle in the eukaryotic cytosol. *Nature* **358:** 249–252.

Martel, R., L.P. Cloney, L.E. Pelcher, and S.M. Hemmingsen. 1990. Unique composition of plastid chaperonin-60: α and β polypeptide-encoding genes are highly divergent. *Gene* **94:** 181–187.

Martin, J., T. Langer, R. Boteva, A. Schramel, A.L. Horwich, and F.-U. Hartl. 1991. Chaperonin-mediated protein folding at the surface of groEL through a "molten globule"-like intermediate. *Nature* **352:** 36–42.

Melki, R., I.E. Vainberg, R.L. Chow, and N.J. Cowan. 1993. Chaperonin-mediated folding of vertebrate actin-related protein and γ-tubulin. *J. Cell Biol.* **122:** 1301–1310.

Micklos, D., S. Caplan, D. Mertens, G. Hynes, Z. Pitluk, Y. Kashi, K. Harrison-Lavoie, C. Brown, B. Barrell, A.L. Horwich, and K. Willison. 1994. A second essential member of the heterooligomeric TCP1 chaperonin complex of yeast, TCP1β. *Proc. Natl. Acad. Sci.* (in press).

Mori, M., K. Murata, H. Kubota, A. Yamamoto, A. Matsushiro, and T. Morita. 1992. Cloning of a cDNA encoding the *Tcp-1* (*t* complex polypeptide 1) homologue of *Arabidopsis thaliana. Gene* **122:** 381–382.

Morita, T., H. Kubota, G. Gachelin, M. Nozaki, and A. Matsushiro. 1991. Cloning of cDNA encoding rat TCP-1. *Biochim. Biophys. Acta* **1129:** 96–99.

Morita, T., H. Kubota, K. Murata, M. Nozaki, C. Delarbre, K. Willison, Y. Satta, M. Sakaizumi, N. Takahata, G. Gachelin, and A. Matsushiro. 1992. Evolution of the mouse *t* haplotype: Recent and worldwide introgression to *Mus musculus. Proc. Natl. Acad. Sci.* **89:** 6851–6855.

Mudryj, M., S.H. Devoto, S.W. Heibert, T. Hunter, T. Pines, and J.R. Nevins. 1991. Cell cycle regulation of E2F transcription factor involves an interaction with cyclin A. *Cell* **65:** 1243–1253.

Mummert, E., R. Grimm, V. Speth, C. Eckerskorn, E. Schiltz, A.A. Gatenby, and E. Schafer. 1993. A TCP-1-related molecular chaperone from plants refolds phytochrome to its photoreversible form. *Nature* **363:** 644–648.

Phipps, B.M., A. Hoffmann, K.O. Stetter, and W. Baumeister. 1991. A novel ATPase complex selectively accumulated upon heat shock is a major cellular component of thermophilic archaebacteria. *EMBO J.* **10:** 1711–1722.

Phipps, B.M., D. Typke, R. Hegerl, S. Volker, A. Hoffmann, K.O. Stetter, and W. Baumeister. 1993. Structure of a molecular chaperone from a thermophilic archaebacterium. *Nature* **361:** 475–477.

Roobol, A. and M.J. Carden. 1993. Identification of chaperonin particles in mammalian brain cytosol and *t*-complex polypeptide 1 as one of their components. *J. Neurochem.* **60:** 2327–2330.

Saffer, J.D., S.P. Jackson, and M.B. Annarella. 1991. Developmental expression of Sp1 in the mouse. *Mol. Cell Biol.* **11:** 2189–2199.

Saibil, H. and S. Wood. 1993. Chaperonins. *Curr. Opin. Struct. Biol.* **3:** 207–213.

Segel, G.B., T.R. Boal, T.S. Cardillo, F.G. Murant, M.A. Lichtman, and F. Sherman. 1992. Isolation of a gene encoding a chaperonin-like protein by complementation of yeast amino acid transport mutants with human cDNA. *Proc. Natl. Acad. Sci.* **89:** 6060–6064.

Silver, L.M. and D. Remis. 1987. Five of the nine genetically defined regions of mouse t haplotype are involved in transmission ratio distortion. *Genet. Res.* **49:** 51–56.

Silver, L.M., K. Artzt, and D. Bennett. 1979. A major testicular cell protein specified by a mouse *T/t* complex gene. *Cell* **17:** 275–284.

Silver, L.M., K.C. Kleen, R.J. Distel, and N.B. Hecht. 1987. Synthesis of mouse *t* complex proteins during haploid stages of spermatogenesis. *Dev. Biol.* **119:** 605–608.

Sternlicht, H., G.W. Farr, M.L. Sternlicht, J.K. Driscoll, K. Willison, and M.B. Yaffe. 1993. TCP1-complex is a molecular chaperonin for tubulin and actin in vivo. *Proc. Natl. Acad. Sci.* **90:** 9422–9426.

Trent, J.D., J. Osipiuk, and T. Pinkau. 1990. Acquired thermotolerance and heat shock in the extremely thermophilic archaebacterium *Sulfolobus* spl. strain B12. *J. Bacteriol.* **172:** 1478–1484.

Trent, J.D., E. Nimmesgern, J.S. Wall, F.-U. Hartl, and A.L. Horwich. 1991. A molecular chaperone from a thermophilic archaebacterium is related to the eukaryotic protein t-complex polypeptide-1. *Nature* **354:** 490–493.

Ursic, D. and M.R. Culbertson. 1991. The yeast homolog to mouse *Tcp-1* affects microtubule-mediated processes. *Mol. Cell Biol.* **11:** 2629–2640.

Waterston, R., C. Martin, M. Craxton, A. Coulson, L. Hillier, R. Durbin, P. Green, R. Shownkeen, N. Halloran, M. Metzstein, T. Hawkins, R. Wilson, M. Berks, Z. Du, K. Tomas, J. Thierry-Meig, and J. Sulston. 1992. A survey of expressed genes in *Caenorhabditis elegans*. *Nature Genet.* **1:** 114–123.

Willison, K.R., K. Dudley, and J. Potter. 1986. Molecular cloning and sequence analysis of a haploid expressed gene encoding *t* complex polypeptide 1. *Cell* **44:** 727–738.

Willison, K.R., G. Hynes, P. Davies, A. Goldsborough, and V.A. Lewis. 1990. Expression of three *t*-complex genes, *Tcp-1*, *D17Leh117c3* and *D17Leh66*, in purified murine spermatogenic cell populations. *Gent. Res. Camb.* **56:** 193–201.

Willison, K., K. Dudley, P. Goodfellow, N. Spurr, V. Groves, P. Gorman, D. Sheer, and J. Trowsdale. 1987. The human homologue of the mouse *t*-complex gene, *TCP1* is located on chromosome 6 but is not near the *HLA* region. *EMBO J.* **6:** 1967–1974.

Willison, K., V. Lewis, K.S. Zuckerman, J. Cordell, C. Dean, K. Miller, M.F. Lyon, and M. Marsh. 1989. The *t* complex polypeptide 1 (TCP-1) is associated with the cytoplasmic aspect of Golgi membranes. *Cell* **57:** 621–632.

Yaffe, M.B., G.W. Farr, D. Mikolos, A.L. Horwich, M.L. Sternlicht, and H. Sternlicht. 1992. TCP1 complex is a molecular chaperone in tubulin biogenesis. *Nature* **358:** 245–248.

Yasue, M., T. Serikawa, T. Kuramoto, M. Mori, T. Higashiguchi, K. Ishizaki, and J. Yamada. 1992. Chromosomal assignments of 17 structural genes and 11 related DNA fragments in rats (*Rattus norvegicus*) by Southern blot analysis of rat x mouse somatic cell hybrid clones. *Genomics* **12:** 659–664.

13

Modulation of Steroid Receptor Signal Transduction by Heat Shock Proteins

Sean P. Bohen and Keith R. Yamamoto
Department of Biochemistry and Biophysics
Program in Biological Sciences
Biochemistry and Molecular Biology Program
University of California, San Francisco
San Francisco, California 94143-0448

I. INTRODUCTION

The 90-kD heat shock protein (hsp90) associates with numerous cytosolic and nuclear proteins involved in cell signaling, including certain steroid hormone receptors (Joab et al. 1984; Catelli et al. 1985; Sanchez et al. 1985; Wilhelmsson et al. 1990), tyrosine kinases (Oppermann et al. 1981; Lindquist and Craig 1988), serine-threonine kinases (Rose et al. 1987; Matts and Hurst 1989; Miyata and Yahara 1992; Stancato et al. 1993), and actin (Koyasu et al. 1986) and tubulin (Sanchez et al. 1988); in addition, hsp90 interacts with hsp56 and hsp70 (Sanchez et al. 1990a; Perdew and Whitelaw 1991). The association of hsp90 with

The Biology of Heat Shock Proteins and Molecular Chaperones
©1994 Cold Spring Harbor Laboratory Press 0-87969-427-0/94 $5 + .00

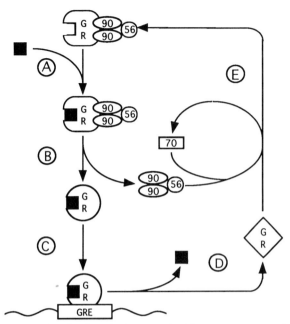

Figure 1 Cycle of glucocorticoid aporeceptor activation, deactivation, and reconstitution of the aporeceptor complex. (*A*) Ligand binds to the aporeceptor complex forming the transient *liganded aporeceptor complex.* (*B*) hsp90 and hsp56 dissociate from the complex, and the resultant *activated receptor* undergoes a conformational change. (*C*) The activated receptor binds to glucocorticoid response elements (GREs) and modulates transcription from nearby promoters. (*D*) Ligand dissociates from the receptor and this *deactivated aporeceptor* dissociates from the GREs. (*E*) The deactivated aporeceptor reassociates with hsp90 and hsp56 in the cytoplasm to reconstitute the activable *aporeceptor complex.* hsp70 appears to be transiently associated with the aporeceptor during assembly of the aporeceptor complex.

steroid hormone receptors has been most thoroughly studied and is the central focus of this chapter. We suggest that hsp90 may bind and alter the activity of a large class of cellular signaling proteins, thereby playing a common role in their function and regulation.

Members of the "nuclear receptor" superfamily respond to their cognate ligands, such as steroids, by binding to specific DNA sequences and altering the transcriptional activity of adjacent promoters (Yamamoto 1985; Evans 1988). Aspects of the signal transduction pathway that triggers these alterations are schematized in Figure 1. In their unliganded, inactive state, certain of these receptors are components of hetero-oligomers, which we term aporeceptor complexes. An *aporeceptor complex*

Table 1 Components and Activities of the Different Receptor Complexes

Receptor form	Components	Activities
Aporeceptor complex	aporeceptor, hsp90, hsp56, hsp70 (?)	binds ligand; transcriptionally inactive
Liganded aporeceptor complex	aporeceptor, ligand, hsp90, hsp56, hsp70 (?)	transient intermediate; transcriptionally inactive
Activated receptor	aporeceptor, ligand	binds DNA; modulates transcription
Free aporeceptor	aporeceptor only	insensitive to ligand; transcriptionally inactive

consists minimally of a receptor monomer, an hsp90 dimer, and an hsp56 monomer (Table 1) (Rexin et al. 1991; Rehberger et al. 1992). One function of the complex is to recognize and bind the cognate ligand. Concomitant with ligand binding, hsp90 and hsp56 dissociate from the complex, producing an *activated hormone-receptor complex* that is competent to bind DNA (Table 1); this process is called "receptor activation" and is a key step in signal transduction (for review, see Denis and Gustafsson 1989).

Aporeceptor complexes reside either in the cytoplasm or in the nuclei of target cells, depending on the type of receptor. Glucocorticoid, mineralocorticoid, and dioxin aporeceptor complexes (GR, MR, and DR, respectively) are primarily cytoplasmic (Picard and Yamamoto 1987; Hoffman et al. 1991; Robertson et al. 1993), whereas estrogen, progesterone, and androgen aporeceptor complexes (ER, PR, and AR, respectively) are predominantly nuclear as detected by immunofluorescence (King and Greene 1984; Welshons et al. 1984; Gasc et al. 1984; Sar et al. 1990). This difference in subcellular localization does not appear to reflect significant differences in receptor function. Receptors that are nuclear in the absence of ligand display low nuclear affinity and, as a result of their high dissociation rate, appear in the cytosolic fraction of cell extracts (Joab et al. 1984; Catelli et al. 1985). Moreover, molecular manipulations that convert apoGR to a constitutively nuclear form produce no apparent change in GR signal transduction or activity (Picard et al. 1988). Thus, although the equilibrium localization of different aporeceptor complexes may differ, their mechanisms of action are indistinguishable. We focus on findings that pertain to the glucocorticoid receptor, noting distinctions among the receptor types where relevant.

Extracellular hormone enters cells by an unknown mechanism and binds to the apoGR subunit of the aporeceptor complex (Fig. 1A). Glucocorticoid agonists and certain antagonists result in release of hsp90 and hsp56 from the complex (Fig. 1B). Agonist-GR complexes then associate tightly with the nucleus, bind to specific DNA sequences termed glucocorticoid response elements (GREs), and modulate transcriptional initiation from nearby promoters (Fig. 1C) (Yamamoto 1985; Evans 1988).

Upon hormone withdrawal, receptors release their hormonal ligands, becoming *deactivated* (Fig. 1D), and reassociate with hsp90 and hsp70 to form cytoplasmic aporeceptor complexes in a poorly understood process called "recycling" (Fig. 1E) (Rousseau et al. 1973; Raaka and Samuels 1983). Presumably, recycling recapitulates the assembly pathway traversed by newly translated aporeceptor protein (Dalman et al. 1989); in principle, receptors may undergo multiple cycles of activation and recycling. In this chapter, we consider three general issues:

1. What are the protein components of the aporeceptor complex inside the cell, how do these proteins interact, and how are complexes assembled? It is commonly held that hsp90 is specifically complexed with steroid receptors; in addition, at least six other proteins, including hsp70 and hsp56, have been reported to associate with various receptors.
2. What roles do the components of the aporeceptor complex have in maintaining the aporeceptor in an inactive state, and in subsequent signal transduction? Truncation mutants of GR that lack the region involved in hsp90 and ligand binding are constitutively active. Conversely, apoGR that is not associated with hsp90 is severely defective in signal transduction. Thus, aporeceptor inactivation and competence to respond to ligand correlate with hsp90 binding.
3. Are heat shock proteins regulators of the signaling activities of receptors? hsp90 interacts with components of a broad range of cell signaling machineries, and heat shock proteins themselves are targets of various kinases. Conceivably, cross-talk between signaling systems could occur through heat shock proteins; a selective alteration in the interaction of hsp90 with GR, for example, might modulate GR activity without affecting signaling by other systems.

II. COMPONENTS OF THE APORECEPTOR COMPLEX

A. Isolation of Aporeceptor Complexes

apoGR complexes recovered from the cytosol of untreated cells sediment at approximately 9S in sucrose gradients, whereas GR in extracts from ligand-treated cells sediments at approximately 4S (Vedeckis 1983). A ligand-bound approximately 9S complex is detected in cells held at 0–4°C during hormone treatment; presumably, this *liganded aporeceptor complex* represents an intermediate in the normal receptor transformation process that can be trapped at low temperature (Table 1). Chemical cross-linking of aporeceptor complexes indicates that they contain receptor, hsp90, and hsp56 at a stoichiometry of 1:2:1 (Rexin et al. 1991; Rehberger et al. 1992). hsp90 is an extremely abundant (1–2% of total soluble protein under nonstress conditions; Lai et al. 1984) and highly conserved dimeric protein found in all organisms that have been examined (Koyasu et al. 1986); it is essential for viability in yeast (see Borkovich et al. 1989) and likely other organisms. hsp90 possesses ATPase activity (Nadeau et al. 1993) and has been reported to act as a molecular chaperone (Miyata and Yahara 1992; Wiech et al. 1992). Metabolic labeling (Howard and Distelhorst 1988) and chemical cross-linking studies (Rexin et al. 1988) indicate that aporeceptor-hsp90 complexes form in vivo and are not simply artifacts generated in cell extracts.

In vitro and in vivo studies support the possibility that hsp56 is also a component of the aporeceptor complex. hsp56 copurifies with GR, PR, ER, and AR (Tai et al. 1986) and has been chemically cross-linked with apoGR and apoPR complexes (Rexin et al. 1991; Rehberger et al. 1992). hsp56 is a peptidyl prolyl isomerase (Chambraud et al. 1993) and an immunophilin, binding the immunosuppressant FK506 (Tai et al. 1992; hsp56 is also referred to as FKBP59). In addition to hsp90 and hsp56, hsp70 copurifies with PR (Kost et al. 1989) and with GR when GR is overexpressed in cultured cells (Sanchez et al. 1990b). However, attempts to cross-link hsp70 with aporeceptor complexes have failed (Rexin et al. 1991). hsp70 is an ATPase that binds reversibly to unfolded proteins in the absence of ATP and dissociates upon ATP hydrolysis (Pelham 1990). Receptor-free complexes containing hsp90, hsp70, and hsp56 have been recovered from cell extracts by coimmunoprecipitation (Perdew and Whitelaw 1991). In addition to hsp90, hsp70, and hsp56, at least four other proteins have been detected in various aporeceptor complexes (for review, see Smith and Toft 1993); of these, one protein, p60, is a stress-induced protein homologous to yeast *STI1* and is another component of the cytoplasmic heat shock protein complexes (Smith et al. 1993). Experiments to assess roles for these proteins in receptor signal transduction have not been reported; therefore, we do not consider these proteins further.

B. hsp90 Interacts Directly with Aporeceptors

Steroid hormone receptors can be divided into three segments: the highly conserved zinc-binding region that resides near the middle of the GR primary sequence and the polypeptide segments upstream (amino-termi-

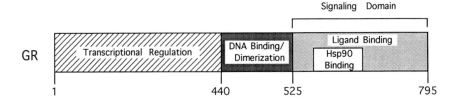

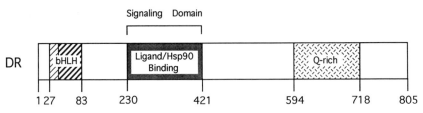

Figure 2 Functional domains of the glucocorticoid receptor (GR) and the dioxin receptor (DR). Ligand-binding activity of GR is distributed throughout the signaling domain (amino acids 525–795); the hsp90-binding region overlaps the ligand-binding region and represents a portion of the signaling domain. The ligand- and hsp90-binding activities of DR colocalize to the signaling domain (amino acids 230–421). DR and GR do not display any amino acid sequence similarity. The amino acid numbers cited are those of rat GR and mouse DR.

nal) and downstream (carboxy-terminal) from that motif. Each segment contains multiple discrete functional activities, some of which are identified in Figure 2; the zinc-binding region, for example, includes both DNA-binding and dimerization activities. The hsp90- and ligand-binding activities map to the carboxy-terminal segment, denoted as the *signaling domain*. hsp90 binding has been mapped to a proximal portion of the GR signaling domain (Pratt et al. 1988; Howard et al. 1990; Dalman et al. 1991), whereas mutations throughout this segment produce defects in ligand binding (Garabedian and Yamamoto 1992). The amino acid sequence of the signaling domain is conserved in GR, MR, PR, and AR, implying that these receptors may interact with hsp90 in a similar manner. However, the signaling domains of ER and DR are not similar to other nuclear receptors or to each other, and attempts to localize hsp90 binding to a discrete region of ER have failed (Chambraud et al. 1990). Experiments with DR indicate that hsp90 and ligand binding both reside within an approximately 200-amino-acid region of the receptor (Fig. 2) (Whitelaw et al. 1993). Taken together, the results suggest that hsp90 could conceivably bind to regions with similar structure, but binding does not appear to depend on a particular amino acid recognition sequence.

Ligand-mediated release of hsp90 from all receptors tested is temperature-dependent. Ligands can bind to the aporeceptor complexes with high affinity at 0–4°C, but liganded aporeceptor complexes must be heated to 25–30°C to induce the dissociation of hsp90 in a process that presumably mimics in vivo activation (Sanchez et al. 1987; Denis et al. 1988). Aporeceptor complexes can also be dissociated with high salt (Nemoto et al. 1990; Pongratz et al. 1992), although the different receptors display distinct salt sensitivities (Nemoto et al. 1990; Schlatter et al. 1992). Metal oxyanions, such as molybdate, and metals endogenous to the cytosol stabilize aporeceptor complexes against dissociation by heat and ligand, and by salt (Dahmer et al. 1984; Renoir et al. 1984). This stabilization is not well understood, but it may be mediated through the interaction of metals with hsp90, as complexes with various different proteins, such as GR, ER, and pp60^{v-src} (Hutchison et al. 1992c), are similarly affected. Metal oxyanions are known to inhibit various kinases and ATPases, but other kinase inhibitors fail to stabilize aporeceptor complexes (Denis and Gustafsson 1989). Conceivably, the oxyanions may affect aporeceptor complexes by inhibiting the ATPase activity of hsp90, hsp70, or some other protein.

hsp90 binding to GR does not appear to represent a simple equilibrium because complexes do not form spontaneously when purified receptor and purified hsp90 or hsp90 and hsp70 are mixed (Scherrer et al. 1992). In fact, complex assembly requires ATP, Mg^{++}, certain monovalent cations (notably K$^+$ but not Na$^+$), elevated temperature (30°C), and additional unidentified factors (see below; Smith et al. 1992). apoGR complexes reconstituted in vitro mimic the high-affinity ligand binding and low DNA-binding activity of their in vivo counterparts (Scherrer et al. 1990). The energy requirement and the finding that hsp90 and hsp70 are ATPases hypothesized to participate in protein folding are consistent with the model that some active conformational change in the aporeceptor and/or other complex components is required for complex formation.

Reconstitution of aporeceptor complexes in vitro provides an assay for fractionation of components that participate in this process (Scherrer et al. 1992). hsp90, hsp70, and hsp56 are abundant in a crude 30–50% ammonium sulfate precipitate fraction from reticulocyte lysates that displays weak reconstitution activity; addition of the 50% ammonium sulfate supernatant to the 30–50% fraction allows efficient reconstitution. Notably, reconstitution of functional heat-shock-protein-containing complexes in reticulocyte lysates has been observed for factors in addition to the steroid aporeceptors; the pp60$^{v\text{-}src}$ oncogene tyrosine kinase and c-Raf-1 serine/threonine kinase bind hsp90 in these extracts (Hutchison et al. 1992d; Stancato et al. 1993). However, some but not all cell extracts display this activity; reconstitution has failed, for example, in chick oviduct cytosol (Smith et al. 1990).

C. hsp56

The physical arrangement of the protein subunits in the aporeceptor complex is unknown, but present evidence suggests that hsp56 may not contact the aporeceptor subunit directly. Thus, hsp56 is a component of the affinity-purified or chemically cross-linked heterotetrameric aporeceptor complex, but unlike hsp90, hsp56 has not been efficiently cross-linked to apoGR (Rexin et al. 1991). Furthermore, salt treatment of the molybdate-stabilized complex selectively releases hsp56, leaving intact a partial complex composed of an aporeceptor monomer and an hsp90 dimer. In contrast, conditions that release hsp90 from the aporeceptor complex release hsp56 as well. Finally, as noted above, hsp56 and hsp90 associate, at least in vitro, in complexes in the absence of receptor. A simple interpretation of these findings is that hsp56 makes primary protein-protein contacts with hsp90 in the aporeceptor complex, rather than with the aporeceptor subunit itself.

D. hsp70

Compared with hsp90 and hsp56, which are found stoichiometrically in aporeceptor complexes, the presence or participation of hsp70 in these particles is much less certain. hsp70 has been detected in apoPR complexes (Kost et al. 1989) and in apoGR complexes under conditions of GR overexpression in cultured cells (Sanchez et al. 1990b), but it has not been shown to be associated with other aporeceptors that contain hsp90 and hsp56. In chemically cross-linked apoGR complexes isolated under nondenaturing conditions, hsp70 is selectively released by ATP treatment, showing that hsp70 itself is not cross-linked in those complexes (Rexin et al. 1991). On the other hand, hsp70 can associate with the aporeceptor subunit independent of hsp90 and hsp56 and remains associated with the aporeceptor subunit under conditions that release hsp90 and hsp56 (Kost et al. 1989). In fact, hsp70 dissociates from PR upon ligand binding, but it is released more slowly than hsp90 and hsp56 (Smith and Toft 1993). A potential complication is that hsp70 binds to hydrophobic regions of denatured proteins; indeed, such binding is reversed by ATP (Pelham 1990; Palleros et al. 1993). Thus, hsp70 may bind to partially unfolded surfaces of the aporeceptor complex, either artifactually in vitro or as part of its biological action in vivo. As discussed below, there is some evidence that hsp70 may participate in the recycling of PR and GR aporeceptors into functional complexes.

III. ROLE OF HEAT SHOCK PROTEINS IN RECEPTOR SIGNAL TRANSDUCTION

A. The Signaling Domain Possesses a "Protein Inactivation" Function

Truncated derivatives of GR that lack the signaling domain are constitutive transcriptional regulators (Godowski et al. 1987). A simple interpretation of this finding is that the signaling domain (perhaps together with other proteins of the aporeceptor complex) comprises a conformational switch: In the absence of hormone, the signaling region would fold in a way that precludes one or more receptor functions essential for its effects on transcription. According to this view, hormone binding would alter the conformation of the signaling domain, thus relieving the inhibition. Surprisingly, however, several types of experiments have established that functional inactivation by the unliganded signaling domain is an active, structure-independent process, rather than allostery. For example, GR can be rearranged, transferring the signaling domain from the carboxyl terminus to the amino terminus, without compromising its hormone dependence (Picard et al. 1988); i.e., the rearranged receptor is inactive in the absence of hormone and active after hormone binding. Alternatively, the signaling domain can be produced in *trans*, separate from the remainder of the receptor, and the two parts can be associated within the cell by a linked leucine zipper dimerization motif (Spanjaard and Chin 1993); this noncovalent association is sufficient for inactivation, which is relieved upon hormone binding. Finally, fusion of GR or ER signaling domains to heterologous proteins unrelated to the nuclear receptor family confers hormone responsiveness upon those proteins (Yamamoto et al. 1988; Eilers et al. 1989); i.e., they are inactive in the absence of hormone and reactivated upon hormone binding.

These findings imply that the unliganded signaling domain actively alters the structure of linked protein regions and that hormone reverses that effect. Clearly, this inactivation function can operate on various proteins and requires neither a specific organization of functional domains nor linkage through peptide bonds. Notably, however, inactivation is not fully promiscuous; i.e., reversible inactivation has been observed in fusions to various transcriptional regulators (Picard et al. 1988; Eilers et al. 1989), tyrosine kinases (Jackson et al. 1993), and luciferase (S.J. Holley and K.R. Yamamoto, unpubl.). In contrast, the intracellular localization but not the enzymatic activity of fusions to β-galactosidase is hormone-regulated (Picard et al. 1988), and a fusion to dihydrofolate reductase (DHFR) displayed about twofold higher enzymatic activity in the absence of hormone than in its presence (Israel and Kaufman 1993). Thus, the signaling domain may facilitate conformational alterations of linked protein segments, but some proteins appear to be relatively resistant to this effect; moreover, the DHFR fusion may indicate that imposed conformational change is not by necessity detrimental to the activity of the linked protein, suggesting that the signaling domain is not simply a reversible protein denaturant. (Because all known strong effects of the signaling domain are negative, we use the term "protein inactivation" in reference to this activity.)

Picard et al. (1990) found that GR, ER, and MR aporeceptors lose their signaling activity in yeast cells expressing 5% of the normal level of hsp82, a yeast homolog of hsp90. Importantly, the receptors were inactive, not constitutive, suggesting that signaling domains can inactivate receptors independent of hsp90. In addition, after dissociation of hsp90 from apoGR complexes in vitro by steroid, salt, or heat treatment, a significant fraction of free aporeceptors (40–65%) are unable to bind to DNA-cellulose (Hutchison et al. 1992a). Cleavage of these non-DNA-binding aporeceptors with specific proteases revealed that a fragment containing the DNA and signaling domains does not bind DNA, whereas a smaller fragment containing the DNA-binding domain but not the signaling domain is competent to bind DNA. Thus, although these experiments do not rule out participation of hsp90 in protein inactivation by receptor signaling domains, it appears that the apoGR signaling domain alone can inactivate DNA binding by apoGR in the absence of hsp90.

B. hsp90 Facilitates Relief of Inactivation

Hormonal activation of GR is compromised in cells expressing low levels of hsp82, implying that hsp90 may be important for hormone-mediated reversal of the protein inactivation conferred by the signaling domain. By this view, then, at least the hsp90 component of the aporeceptor complex is required for normal signal transduction. A class of hsp82 mutants has been isolated that forms aporeceptor complexes that respond abnormally to the hormone; this indicates that the conformation or activity of hsp90 in the aporeceptor complex, not merely its presence, is important for the signal response (Bohen and Yamamoto 1993). In cells expressing mutant hsp82 or low levels of the wild-type protein, different receptor types are differentially affected: In general, GR and MR responses are more sensitive than ER and PR to altered hsp82 dose or structure. Importantly, it is signal transduction and not transcriptional regulation that is abrogated in these cells, as a constitutive

GR derivative is fully active in the hsp82 mutant strains, and the full transcriptional response can be achieved at very high concentrations of hormone. In strains expressing low levels of wild-type hsp82, this activity may result from the activation of the few aporeceptors that are complexed with hsp90 or from uncomplexed aporeceptors that can be activated at high ligand levels. If the latter is true, this would indicate that hsp90 facilitates signal transduction but is not absolutely essential for the response.

Dissociation of hsp90 from apoGR, apoMR, or apoDR by various manipulations in vitro results in the loss of high-affinity ligand binding (Bresnick et al. 1989; Schulman et al. 1992; Pongratz et al. 1992). In fact, the rate of loss of hormone binding by GR correlates with the rate of dissociation of hsp90, suggesting that an apoGR-hsp90 complex is necessary for high-affinity hormone binding. Interestingly, rodent epididymal sperm, which lack hsp90, express apoGR that fails to bind glucocorticoids (Kaufmann et al. 1992), further supporting the notion that the apoGR-hsp90 complex is required in vivo. In contrast, ER, PR, and AR aporeceptors bind their ligands with high affinity at 0–4°C, even after dissociation of hsp90 (Eul et al. 1989; Chambraud et al. 1990; Nemoto et al. 1992). However, PR requires association with hsp90 to retain high ligand affinity at 37°C (Smith 1993). One interpretation of these findings is that the interaction of hsp90 with receptor signaling domains may induce a ligand-binding conformation whose stability in some cases requires bound hsp90, whereas other receptors may retain the functional conformation even after hsp90 release, at least at low temperature. Thus, different aporeceptors may differ in their requirements for continuous binding by hsp90.

Consistent with this view, three hsp82 mutants have been shown to affect multiple receptors, and GR and MR are in every case more strongly affected than are PR and ER (Bohen and Yamamoto 1993). Interestingly, these mutants link hsp90 function in receptor signal transduction with hsp90 activities required for cell growth; the mutants are viable, but display increased doubling time and temperature sensitivity. A fourth hsp82 mutant selectively compromises GR signaling, affecting neither signaling by other receptors nor cell growth. This indicates that certain aspects of the interaction of hsp82 with its targets are likely unique. These findings suggest that hsp90 may facilitate a spectrum of conformational changes during receptor activation, such as inducing a structure for high-affinity ligand binding, as well as formation of various functional surfaces in the activated receptor after dissociation of the aporeceptor complex. By this view, different hsp90 targets may require different subsets of hsp90 functions.

C. Is hsp90 Involved in Aporeceptor Inactivation?

Although the predominant role for hsp90 in receptor function appears to be facilitation of the signal response, it may be that hsp90 also participates in protein inactivation conferred by receptor signaling domains. Evidence supporting an effect of hsp90 on target protein conformation is consistent with this possibility, but a bona fide test of this notion will require careful assessment of the characteristics of protein inactivation in the presence and absence of hsp90. For such experiments, we suggest that it might be particularly useful to monitor the inactivation of heterologous proteins to which the signaling domain is fused and to choose fusion targets that are simple and well-characterized structurally and genetically.

D. hsp56 (FKBP59) and Ligand Sensitivity of the Aporeceptor Complex

The 56-kD receptor-associated protein, hsp56, is a member of the family of immunophilins that binds FK506, the FKBPs. This protein also displays peptidyl prolyl *cis-trans* isomerase activity (Chambraud et al. 1993). That both FK506 and glucocorticoids are immunosuppressants has led to speculation that the signal transduction pathways of these molecules may be linked, perhaps through hsp56. Indeed, cells treated with FK506 or rapamycin, a related immunosuppressant, respond to tenfold lower glucocorticoid concentrations than do untreated cells (Ning and Sánchez 1993); the level of maximal transcriptional activation is unchanged. Hutchison et al. (1993) detected no effects of FK506 on hormone-mediated dissociation of the apoGR complex, on reconstitution of the complex in vitro, or on nuclear localization or maximal transcriptional activation by the receptor. Unfortunately, however, that study was conducted at saturating hormone concentrations. Therefore, the possibility remains that FK506 binds to hsp56 and modifies interactions within the aporeceptor complex and, in turn, alters hormone-binding affinity or the probability of hsp90 and hsp56 release. By extension, this scheme would suggest that hsp56 itself may affect the sensitivity of the aporeceptor complex to activation by hormonal ligand.

An alternative or additional effect of FK506 may reside at the level of hormone transport across the cell membrane. The *mdr1* P-glycoprotein has been shown to export dexamethasone (Ueda et al. 1992), and the gene encoding P-glycoprotein is amplified in cultured thymoma cells selected for glucocorticoid insensitivity (Bourgeois et al. 1993). FK506 inhibits the activity of some transmembrane transporters, including P-glycoprotein (Takeguchi et al. 1993). Thus, FK506 might potentiate the dexamethasone response of GR by decreasing ligand export. Therefore, studies with FK506 may identify a step prior to ligand binding by receptor that modulates cell sensitivity to steroid ligands.

E. hsp70 and Aporeceptor Complex Cycling

Upon hormone withdrawal, apoGR complexes displaying high ligand affinity reappear in the cytoplasm at a rate that parallels the decline in nuclear GR levels (Rousseau et al. 1973; Raaka and Samuels 1983). Experiments in cells treated with inhibitors of protein synthesis or with dense amino acids indicate that the accumulation of cytoplasmic ligand-binding activity largely reflects receptor recycling, as opposed to de novo synthesis of receptors. These findings demonstrate that deactivated receptors can reassociate with components of the aporeceptor complex and suggest that receptors can undergo multiple cycles of activation.

Studies of aporeceptor reconstitution in vitro provide evidence that hsp70 is involved in complex assembly; the reconstitution activity of reticulocyte lysates is abrogated by depletion of ATP-binding proteins and restored by addition of purified hsp70 (Hutchison et al. 1994). In addition, the aporeceptor reconstitution activity and the ATP-driven release of hsp70 from its targets have similar monovalent cation requirements: For both reactions, K^+ is acceptable but Na^+ is not (Hutchison et al. 1992b; Palleros et al. 1993). Finally, a kinetic analysis revealed that hsp70 rapidly and transiently associates with apoPR (Smith 1993) and in particular that hsp90 binding to the aporeceptor lags slightly behind hsp70 binding. One interpretation is that hsp70 binds to the receptor and mediates the interaction of the signaling domain with hsp90. Assuming that in vitro reconstitution mimics receptor recycling, hsp70 may be vital to maintaining aporeceptors in ligand-responsive complexes.

Even during continuous hormone treatment, heterokaryon experiments imply that GR and PR shuttle between the nucleus and cytoplasm in vivo (Chandran and DeFranco 1992; Madan and DeFranco 1993). Although no attempt has been made to demonstrate that aporeceptor complexes are intermediates in this process, a logical view is that ligands dissociate from active receptors and that the resulting deactivated aporeceptors associate with cytoplasmic hsp90, hsp70, and hsp56, thus reconstituting functional aporeceptor complexes; in the heterokaryon experiments, the reactivated receptors would then enter either of the nuclei in the cell. The finding that ligand binding by GR is dramatically and reversibly reduced when ATP synthesis is inhibited is consistent with the ATP requirement for complex assembly (Mendel et al. 1986).

IV. POSSIBLE ROLES FOR HEAT SHOCK PROTEINS IN REGULATION OF STEROID RECEPTORS

A. Inducing a High-affinity Conformation in the Receptor Signaling Domain

The interaction of hsp90 with GR, MR, DR, and PR appears to be required for high-affinity ligand binding by these receptors. Thus, hsp90 induces a particular conformation in the ligand-binding domain of these receptors that sensitizes them to their cognate ligands. At least for GR and MR, hsp90 does not act as a "chaperone" as defined by Ellis and van der Vies (1991), because it must remain bound to the aporeceptor and thus participates actively in its function. In any case, it is interesting to consider that if the efficiency of the hsp90-receptor interaction could be regulated, for example, by modulation of hsp90 phosphorylation, the sensitivity of the hormone response could be regulated. In fact, regulation of hsp90 competence for receptor interaction would likely produce distinct effects on different receptors, as it is already apparent that different receptors are differentially sensitive to hsp82 mutation (Bohen and Yamamoto 1993); thus, even within a single cell, responsiveness to different signals could be individually "tuned" by this device.

B. Inducing Conformational Changes Necessary for Receptor Activity

hsp90, like some other heat shock proteins, may affect protein folding (Miyata and Yahara 1992; Wiech et al. 1992). We have suggested that hsp90, perhaps together with hsp56 in the aporeceptor complex, induces specific conformational changes in the aporeceptor that facilitate ligand binding and may also influence the extent or nature of inactivation by aporeceptor signaling domains, as well as subsequent "downstream" receptor activities such as transcriptional regulation after receptor activation. Such diverse effects of hsp90 might explain the spectrum of requirements for hsp90 displayed by different aporeceptors and explain the effects of hsp82 mutants on signaling by PR and ER. For example, MR, GR, and DR aporeceptors appear to depend strongly on conformational changes conferred by hsp90 for both ligand binding and transcriptional activation; in contrast, ER and AR might bind hormone relatively well without continuous contact with hsp90 but may require hsp90 for other aspects of signal transduction or for downstream events.

C. Cross-talk with Other Signaling Networks

It has been reported that PR and ER can be activated through the activation of certain transmembrane receptors, namely, those for dopamine

(Power et al. 1991; Smith et al. 1993) and epidermal growth factor (EGF) (Ignar-Trowbridge et al. 1992). Thus, it appears that alterations in aporeceptor complexes other than steroid binding can relieve inactivation by the signaling domains of these receptors, and transcription of target genes can be altered in response to signals other than steroids. Clearly, many potential mechanisms could account for these effects. However, as heat shock proteins are common components of the different receptor types, it is intriguing to speculate that certain second messenger systems might alter the properties of heat shock proteins and alter their interactions with signaling proteins, thereby modulating the activity of steroid receptors and, perhaps, other cellular signaling proteins. For example, such regulation may be mediated through changes in the phosphorylation of heat shock proteins or their targets. Several examples of cross-talk between different signaling systems have been identified (Power et al. 1991; Ignar-Trowbridge et al. 1992; Smith et al. 1993). Heat shock proteins interact with the components of a number of systems; it is intriguing to consider that heat shock proteins may serve as conduits for cross-talk between these diverse signals.

V. WHY DO STEROID RECEPTORS INTERACT WITH HSP90?

In contrast to the receptors discussed here, other members of the nuclear receptor family, such as the thyroid hormone receptors (TR) and retinoic acid receptors (RAR) are thought not to interact with hsp90 (Evans 1988). In vitro, apoTR and apoRAR bind DNA in the absence of ligand and hsp90 and are thought to occupy their cognate response elements in the absence of hormone in vivo. Thus, apoTR and apoRAR can alter the transcriptional activity of response-element-linked promoters, and hormonal activation modulates the regulatory effects of these receptors. In contrast, the aporeceptor complexes of GR, MR, PR, AR, ER, and DR are inactive as transcriptional regulators. In this sense, association of hsp90 with these aporeceptors can be considered an intermediate step in the folding process.

The aporeceptor complexes of GR and MR and, to a lesser extent, PR, ER, and AR may represent the recruitment of ubiquitous and abundant cellular protein folding machinery to preserve receptors in an inactive and perhaps structurally unstable state; i.e., the association of heat shock proteins with aporeceptors converts them from an inactive and relatively unresponsive form to an inactive but activable aporeceptor complex. Thus, the heat shock proteins appear to induce reversible changes in conformation that potentiate signaling activity. This general strategy appears to have been exploited by numerous signal transducing factors, including protein kinases, DR, and the sex pheromone receptor of the water mold, *Achlya ambisexualis* (Brunt et al. 1990). Thus, it would appear that various unrelated signaling proteins have independently acquired interactions with heat shock proteins during eukaryotic evolution. Given this observation, it is intriguing to postulate that heat shock proteins may exert regulatory effects on a broad range of intracellular signaling pathways and thus may serve as a common link between these systems.

Whatever the fate of these speculations, it is clear that investigations of the role of heat shock proteins in signal transduction by steroid hormone aporeceptor complexes have been particularly useful for advancing our understanding of hsp90 function. The biochemical and genetic approaches for such analyses are rapidly approaching a level of maturity that will allow investigators to answer specific questions about the roles of heat shock proteins in signal transduction and throughout the cell.

ACKNOWLEDGMENTS

We are grateful to S.J. Holley for helpful comments on the manuscript and to W.B. Pratt for communicating unpublished results. Studies of heat shock proteins and receptor signaling in our laboratory are supported by the National Science Foundation. S.P.B. is supported by a predoctoral fellowship from the American Heart Association, California affiliate.

REFERENCES

Bohen, S.P. and K.R. Yamamoto. 1993. Isolation of Hsp90 mutants by screening for decreased steroid receptor function. *Proc. Natl. Acad. Sci.* **90:** 11424–11428.

Borkovich, K.A., F.W. Farrelly, D.B. Finkelstein, J. Taulien, and S. Lindquist. 1989. Hsp82 is an essential protein that is required in higher concentrations for growth of cells at higher temperatures. *Mol. Cell. Biol.* **9:** 3919–3930.

Bourgeois, S., D.J. Gruol, R.F. Newby, and F.M. Rajah. 1993. Expression of an *mdr* gene is associated with a new form of resistance to dexamethasone-induced apoptosis. *Mol. Endocrinol.* **7:** 840–851.

Bresnick, E.H., F.C. Dalman, E.R. Sanchez, and W.B. Pratt. 1989. Evidence that the 90-kDa heat shock protein is necessary for the steroid binding conformation of the L cell glucocorticoid receptor. *J. Biol. Chem.* **264:** 4992–4997.

Brunt, S.A., R. Riehl, and J.C. Silver. 1990. Steroid hormone regulation of the *Achlya ambisexualis* 85-kilodalton heat shock protein, a component of the *Achlya* steroid receptor complex. *Mol. Cell. Biol.* **10:** 273–281.

Catelli, M.-G., N. Binart, I. Jung Testas, J.M. Renoir, E.-E. Baulieu, and J.R. Feramisco. 1985. The common 90-kd protein component of non-transformed "8S" steroid receptors is a heat-shock protein. *EMBO J.* **4:** 3131–3135.

Chambraud, B., M. Berry, G. Redeuilh, P. Chambon, and E.-E. Baulieu. 1990. Several regions of human estrogen receptor are involved in the formation of receptor-heat shock protein 90 complexes. *J. Biol. Chem.* **265:** 20686–20691.

Chambraud, B., N. Rouvière-Fourmy, C. Radanyi, K. Hsiao, D.A. Peattie, D.J. Livingston, and E.-E. Baulieu. 1993. Overexpression of p59-HBI (FKBP59), full length and domains, and characterization of PPIase activity. *Biochem. Biophys. Res. Commun.* **196:** 160–166.

Chandran, U.R. and D.B. DeFranco. 1992. Internuclear migration of chicken progesterone receptor, but not simian virus-40 large tumor antigen, in transient heterokaryons. *Mol. Endocrinol.* **6:** 837–844.

Dahmer, M.K., P.R. Housley, and W.B. Pratt. 1984. Effects of molybdate and endogenous inhibitors on steroid-receptor inactivation, transformation, and translocation. *Annu. Rev. Genet.* **46:** 67–81.

Dalman, F.C., L.C. Scherrer, L.P. Taylor, H. Akil, and W.B. Pratt. 1991. Localization of the 90-kDa heat shock protein-binding site within the hormone-binding domain of the glucocorticoid receptor by peptide competition. *J. Biol. Chem.* **266:** 3482–3490.

Dalman, F.C., E.H. Bresnick, P.D. Patel, G.H. Perdew, S.J. Watson, Jr., and W.B. Pratt. 1989. Direct evidence that the glucocorticoid receptor binds to hsp90 at or near the termination of receptor translation *in vitro. J. Biol. Chem.* **264:** 19815–19821.

Denis, M. and J.-Å. Gustafsson. 1989. The M_r ~90,000 heat shock protein: An important modulator of ligand and DNA-binding properties of the glucocorticoid receptor. *Cancer Res.* **49:** 2275s–2281s.

Denis, M., L. Poellinger, A.C. Wikström, and J.A. Gustafsson. 1988. Requirement of hormone for thermal conversion of the glucocorticoid receptor to a DNA-binding state. *Nature* **333:** 686–688.

Eilers, M., D. Picard, K.R. Yamamoto, and J.M. Bishop. 1989. Chimeras of *MYC* oncoprotein and steroid receptors cause hormone-dependent transformation of cells. *Nature* **340:** 66–68.

Ellis, R.J. and S.M. van der Vies. 1991. Molecular chaperones. *Annu. Rev. Biochem.* **60:** 321–347.

Eul, J., M.E. Meyer, L. Tora, M.T. Bocquel, C. Quirin-Stricker, P. Chambon, and H. Gronemeyer. 1989. Expression of active hormone and DNA-binding domains of the chicken progesterone receptor in *E. coli. EMBO J.* **8:** 83–90.

Evans, R.M. 1988. The steroid and thyroid hormone receptor superfamily. *Science* **240:** 889–895.

Garabedian, M.J. and K.R. Yamamoto. 1992. Genetic dissection of the signaling domain of a mammalian steroid receptor in yeast. *Mol. Biol. Cell* **3:** 1245–1257.

Gasc, J.-M., J.-M. Renoir, C. Radanyi, I. Joab, P. Tuohimaa, and E.-E. Baulieu. 1984. Progesterone receptor in chick oviduct: An immunohistochemical study with antibodies to distinct receptor components. *J. Cell Biol.* **99:** 1193–1201.

Godowski, P.J., S. Rusconi, R. Miesfeld, and K.R. Yamamoto. 1987. Glucocorticoid receptor mutants that are constitutive activators of transcriptional enhancement. (Published erratum appears in *Nature* **326:** 105.) *Nature* **325:** 365–368.

Hoffman, E.C., H. Reyes, F.-F. Chu, F. Sander, L.H. Conley, B.A. Brooks, and O. Hankinson. 1991. Cloning of a factor required for activity of the Ah (dioxin) receptor. *Science* **252:** 954–958.

Howard, K.J. and C.W. Distelhorst. 1988. Evidence for intracellular association of the glucocorticoid receptor with the 90-kDa heat shock protein. *J. Biol. Chem.* **263:** 3474–3481.

Howard, K.J., S.J. Holley, K.R. Yamamoto, and C.W. Distelhorst. 1990. Mapping the hsp90 binding region of the glucocorticoid receptor. *J. Biol. Chem.* **265:** 11928–11935.

Hutchison, K.A., M.J. Czar, and W.B. Pratt. 1992a. Evidence that the hormone-binding domain of the mouse glucocorticoid receptor directly represses DNA binding activity in a major portion of receptors that are "misfolded" after removal of hsp90. *J. Biol. Chem.* **267:** 3190–3195.

Hutchison, K.A., M.J. Czar, L.C. Scherrer, and W.B. Pratt. 1992b. Monovalent cation selectivity for ATP-dependent association of the glucocorticoid receptor with hsp70 and hsp90. *J. Biol. Chem.* **267:** 14047–14053.

Hutchison, K.A., K.D. Dittmar, M.J. Czar, and W.B. Pratt. 1994. Proof that hsp70 is required for assembly of the glucocorticoid receptor into a heterocomplex with hsp90. *J. Biol. Chem.* (in press).

Hutchison, K.A., L.F. Stancato, R. Jove, and W.B. Pratt. 1992c. The protein-protein complex between $pp60^{v-src}$ and hsp90 is stabilized by molybdate, vanadate, tungstate, and an endogenous cytosolic metal. *J. Biol. Chem.* **267:** 13952–13957.

Hutchison, K.A., B.K. Brott, J.H. De Leon, G.H. Perdew, R. Jove, and W.B. Pratt. 1992d. Reconstitution of the multiprotein complex of $pp60^{src}$, hsp90, and p50 in a cell-free system. *J. Biol. Chem.* **267:** 2902–2908.

Hutchison, K.A., L.C. Scherrer, M.J. Czar, Y. Ning, E.R. Sanchez, K.L. Leach, Jr., M.R. Deibel, and W.B. Pratt. 1993. FK506 binding to the 56-kilodalton immunophilin (hsp56) in the glucocorticoid receptor heterocomplex has no effect on receptor folding or function. *Biochemistry* **32:** 3953–3957.

Ignar-Trowbridge, D.M., K.G. Nelson, M.C. Bidwell, S.W. Curtis, T.F. Washburn, J.A. McLachlan, and K.S. Korach. 1992. Coupling of dual signaling pathways: Epidermal growth factor action involves the estrogen receptor. *Proc. Natl. Acad. Sci.* **89:** 4658–4662.

Israel, D.I. and R.J. Kaufman. 1993. Dexamethasone negatively regulates the activity of a chimeric dihydrofolate reductase/glucocorticoid receptor protein. *Proc. Natl. Acad. Sci.* **90:** 4290–4294.

Jackson, P., D. Baltimore, and D. Picard. 1993. Hormone-conditional tranformation by fusion proteins of c-Abl and its transforming variants. *EMBO J.* **12:** 2809–2820.

Joab, I., C. Radanyi, M. Renoir, T. Buchou, M.-G. Catelli, N. Binart, and J. Mester. 1984. Common non-hormone binding component in non-transformed chick oviduct. *Nature* **308:** 850–853.

Kaufmann, S.H., W.W. Wright, S. Okret, A.-C. Wikström, J.-Å. Gustafsson, N.L. Shaper, and J.H. Shaper. 1992. Evidence that rodent epididymal sperm contain the M_r 94,000 glucocorticoid receptor but lack the M_r 90,000 heat shock protein. *Endocrinology* **130:** 3074–3084.

King, W.J. and G.L. Greene. 1984. Monoclonal antibodies localize oestrogen receptor in the nuclei of target cells. *Nature* **307:** 745–747.

Kost, S.L., D.F. Smith, W.P. Sullivan, W.J. Welch, and D.O. Toft. 1989. Binding of heat shock proteins to the avian progesterone receptor. *Mol. Cell. Biol.* **9:** 3829–3838.

Koyasu, S., E. Nishida, T. Kadowaki, F. Matsuzaki, K. Iida, F. Harada, M. Kasuga, H. Sakai, and I. Yahara. 1986. Two mammalian heat shock proteins, HSP90 and HSP100, are actin-binding proteins. *Proc. Natl. Acad. Sci.* **83:** 8054–8058.

Lai, B.-T., N.W. Chin, A.E. Stanek, W. Keh, and K.W. Lanks. 1984. Quantitation and intracellular localization of the 85K heat shock protein by using monoclonal and polyclonal antibodies. *Mol. Cell. Biol.* **4:** 2802–2810.

Lindquist, S. and E.A. Craig. 1988. The heat-shock proteins. *Annu. Rev. Genet.* **22:** 631–677.

Madan, A.P. and D.B. DeFranco. 1993. Bidirectional transport of glucocorticoid receptors across the nuclear envelope. *Proc. Natl. Acad. Sci.* **90:** 3588–3592.

Matts, R.L. and R. Hurst. 1989. Evidence for the association of the heme-regulated eIF-

2a kinase with the 90-kDa heat shock protein in rabbit reticulocyte lysate *in situ. J. Biol. Chem.* **264:** 15542–15547.

Mendel, D.B., J.E. Bodwell, and A. Munck. 1986. Glucocorticoid receptors lacking hormone-binding activity are bound in nuclei of ATP-depleted cells. *Nature* **324:** 478–480.

Miyata, Y. and I. Yahara. 1992. The 90-kDa heat shock protein, HSP90, binds and protects casein kinase II from self-aggregation and enhances its kinase activity. *J. Biol. Chem.* **267:** 7042–7047.

Nadeau, K., A. Das, and C.T. Walsh. 1993. Hsp90 chaperonins possess ATPase activity and bind heat shock transcription factors and peptidyl prolyl isomerases. *J. Biol. Chem.* **268:** 1479–1487.

Nemoto, T., Y. Ohara-Nemoto, and O. Minoru. 1992. Association of the 90-kDa heat shock protein does not affect the ligand-binding ability of androgen receptor. *J. Steroid Biochem. Mol. Biol.* **42:** 803–812.

Nemoto, T., G.G.F. Mason, A. Wilhelmsson, S. Cuthill, J. Hapgood, J.-Å. Gustafsson, and L. Poellinger. 1990. Activation of the dioxin and glucocorticoid receptors to a DNA binding state under cell-free conditions. *J. Biol. Chem.* **265:** 2269–2277.

Ning, Y.-M. and E.R. Sánchez. 1993. Potentiation of glucocorticoid receptor-mediated gene expression by the immunophilin ligands FK506 and rapamycin. *J. Biol. Chem.* **268:** 6073–6076.

Oppermann, H., W. Levinson, and J.M. Bishop. 1981. A cellular protein that associates with the transforming protein of Rous sarcoma virus is also a heat-shock protein. *Proc. Natl. Acad. Sci.* **78:** 1067–1071.

Palleros, D.R., K.L. Reid, L. Shi, W.J. Welch, and A.L. Fink. 1993. ATP-induced protein-Hsp70 complex dissociation requires K^+ but not ATP hydrolysis. *Nature* **365:** 664–666.

Pelham, H.R.B. 1990. Functions of the hsp70 protein family: An overview. In *Stress proteins in biology and medicine* (ed. R.I. Morimoto et al.), pp. 287–299. Cold Spring Harbor Laboratory Press, Cold Spring Harbor, New York.

Perdew, G.H. and M.L. Whitelaw. 1991. Evidence that the 90-kDa heat shock protein (HSP90) exists in cytosol in heteromeric complexes containing HSP70 and three other proteins with M_r of 63,000, 56,000, and 50,000. *J. Biol. Chem.* **266:** 6708–6713.

Picard, D. and K.R. Yamamoto. 1987. Two signals mediate hormone-dependent nuclear localization of the glucocorticoid receptor. *EMBO J.* **6:** 3333–3340.

Picard, D., S.J. Salser, and K.R. Yamamoto. 1988. A movable and regulable inactivation function within the steroid binding domain of the glucocorticoid receptor. *Cell* **54:** 1073–1080.

Picard, D., B. Khursheed, M.J. Garabedian, M.G. Fortin, S. Lindquist, and K.R. Yamamoto. 1990. Reduced levels of hsp90 compromise steroid receptor action *in vivo*. *Nature* **348:** 166–168.

Pongratz, I., G.G.F. Mason, and L. Poellinger. 1992. Dual roles of the 90-kDa heat shock protein hsp90 in modulating functional activities of the dioxin receptor. *J. Biol. Chem.* **267:** 13728–13734.

Power, R.F., S.K. Mani, J. Codina, O.M. Conneely, and B.W. O'Malley. 1991. Dopaminergic and ligand-independent activation of steroid hormone receptors. *Science* **254:** 1636–1639.

Pratt, W.B., D.J. Jolly, D.V. Pratt, S.M. Hollenberg, V. Giguere, F.M. Cadepond, G. Schweizer-Groyer, M.-G. Catelli, R.M. Evans, and E.-E. Baulieu. 1988. A region in the steroid binding domain determines formation of the non-DNA-binding, 9S glucocorticoid receptor complex. *J. Biol. Chem.* **263:** 267–273.

Raaka, B.M. and H.H. Samuels. 1983. The glucocorticoid receptor in GH$_1$ cells: Evidence from dense amino acid labeling and whole cell studies for an equilibrium model explaining the influence of hormone on the intracellular distribution of receptor. *J. Biol. Chem.* **258:** 417–425.

Rehberger, P., M. Rexin, and U. Gehring. 1992. Heterotetrameric structure of the human progesterone receptor. *Proc. Natl. Acad. Sci.* **89:** 8001–8005.

Renoir, J.-M., T. Buchou, J. Mester, C. Radanyi, and E.-E. Baulieu. 1984. Oligomeric structure of the molybdate-stabilized, nontransformed 8S progesterone receptor from chicken oviduct cytosol. *Biochemistry* **23:** 6016–6023.

Rexin, M., W. Busch, and U. Gehring. 1988. Chemical cross-linking of heteromeric glucocorticoid receptors. *Biochemistry* **27:** 5593–5601.

———. 1991. Protein components of the nonactivated glucocorticoid receptor. *J. Biol. Chem.* **266:** 24601–24605.

Robertson, N.M., G. Schulman, S. Karnik, E. Alnemri, and G. Litwack. 1993. Demonstration of nuclear translocation of the mineralocorticoid receptor (MR) using an anti-MR antibody and confocal laser scanning microscopy. *Mol. Endocrinol.* **7:** 1226–1239.

Rose, D.W., R.E.H. Wettenhall, W. Kudlicki, G. Kramer, and B. Hardesty. 1987. The 90-kilodalton peptide of the heme-regulated eIF-2α kinase has sequence similarity with the 90-kilodalton heat shock protein. *Biochemistry* **26:** 6583–6587.

Rousseau, G.G., J.D. Baxter, S.J. Higgins, and G.M. Tomkins. 1973. Steroid-induced nuclear binding of glucocorticoid receptors in intact hepatoma cells. *J. Mol. Biol.* **79:** 539–554.

Sanchez, E.R., D.O. Toft, M.J. Schlesinger, and W.B. Pratt. 1985. Evidence that the 90-kDa phosphoprotein associated with the untransformed L-cell glucocorticoid receptor is a murine heat shock protein. *J. Biol. Chem.* **260:** 12398–12401.

Sanchez, E.R., S. Meshinchi, W. Tienrungroj, M.J. Schlesinger, D.O. Toft, and W.B. Pratt. 1987. Relationship of the 90-kDa murine heat shock protein to the untransformed and transformed states of the L-cell glucocorticoid receptor. *J. Biol. Chem.* **262:** 6986–6991.

Sanchez, E.R., T. Redmond, L.C. Scherrer, E.H. Bresnick, M.J. Welsh, and W.B. Pratt. 1988. Evidence that the 90-kilodalton heat shock protein is associated with tubulin-containing complexes in L cell cytosol and in intact PtK cells. *Mol. Endocrinol.* **2:** 756–760.

Sanchez, E.R., L.E. Faber, W.J. Henzel, and W.B. Pratt. 1990a. The 56-59-kilodalton protein identified in untransformed steroid receptor complexes is a unique protein that exists in cytosol in a complex with both the 70- and 90-kilodalton heat shock proteins. *Biochemistry* **29:** 5145–5152.

Sanchez, E.R., M. Hirst, L.C. Scherrer, H.-Y. Tang, M.J. Welsh, J.M. Harmon, S.S. Simons, Jr., G.M. Ringold, and W.B. Pratt. 1990b. Hormone-free mouse glucocorticoid receptors overexpressed in Chinese hamster ovary cells are localized to the nucleus and are associated with both hsp70 and hsp90. *J. Biol. Chem.* **265:** 20123–20130.

Sar, M., D.B. Lubahn, F.S. French, and E.M. Wilson. 1990. Immunohistochemical localization of the androgen receptor in rat and human tissues. *Endocrinology* **127:** 3180–3186.

Scherrer, L.C., F.C. Dalman, E. Massa, S. Meshinchi, and W.B. Pratt. 1990. Structural and functional reconstitution of the glucocorticoid receptor-Hsp90 complex. *J. Biol. Chem.* **265:** 21397–21400.

Scherrer, L.C., K.A. Hutchison, E.R. Sanchez, S.K. Randall, and W.B. Pratt. 1992. A heat shock protein complex isolated from rabbit reticulocyte lysate can reconstitute a functional glucocorticoid receptor-Hsp90 complex. *Biochemistry* **31:** 7325–7329.

Schlatter, L.K., K.J. Howard, M.G. Parker, and C.W. Distelhorst. 1992. Comparison of the 90-kilodalton heat shock protein interaction with *in vitro* translated glucocorticoid and estrogen receptors. *Mol. Endocrinol.* **6:** 132–140.

Schulman, G., P.V. Bodine, and G. Litwack. 1992. Modulators of the glucocorticoid receptor also regulate mineralocorticoid receptor function. *Biochemistry* **31:** 1734–1741.

Smith, C.L., O.M. Conneely, and B.W. O'Malley. 1993. Modulation of the ligand-independent activation of the human estrogen receptor by hormone and antihormone. *Proc. Natl. Acad. Sci.* **90:** 6120–6124.

Smith, D.F. 1993. Dynamics of heat shock protein 90-progesterone receptor binding and the disactivation loop model for steroid receptor complexes. *Mol. Endocrinol.* **7:** 1418–1429.

Smith, D.F. and D.O. Toft. 1993. Minireview: Steroid receptors and their associated proteins. *Mol. Endocrinol.* **7:** 4–11.

Smith, D.F., D.B. Schowalter, S.L. Kost, and D.O. Toft. 1990. Reconstitution of progesterone receptor with heat shock proteins. *Mol. Endocrinol.* **4:** 1704–1711.

Smith, D.F., B.A. Stensgard, W.J. Welch, and D.O. Toft. 1992. Assembly of progesterone receptor with heat shock proteins and receptor activation are ATP mediated events. *J. Biol. Chem.* **267:** 1350–1356.

Smith, D.F., W.P. Sullivan, T.N. Marion, K. Zaitsu, B. Madden, D.J. McCormick, and D.O. Toft. 1993. Identification of a 60-kilodalton stress-related protein, p60, which interacts with hsp90 and hsp70. *Mol. Cell. Biol.* **13:** 869–876.

Spanjaard, R.A. and W.W. Chin. 1993. Reconstitution of ligand-mediated glucocorticoid receptor activity by *trans*-acting functional domains. *Mol. Endocrinol.* **7:** 12–16.

Stancato, L.F., Y.-H. Chow, K.A. Hutchison, G.H. Perdew, R. Jove, and W.B. Pratt. 1993. Raf exists in a native heterocomplex with hsp90 and p50 that can be reconstituted in a cell-free system. *J. Biol. Chem.* **268:** 21711–21716.

Tai, P.-K.K., M.W. Albers, H. Chang, L.E. Faber, and S.L. Schreiber. 1992. Association of a 59-kilodalton immunophilin with the glucocorticoid receptor complex. *Science* **256:** 1315–1318.

Tai, P.-K.K., Y. Maeda, K. Nakao, N.G. Wakim, J.L. Duhring, and L.E. Faber. 1986. A 59-kilodalton protein associated with progestin, estrogen, androgen, and glucocorticoid receptors. *Biochemistry* **25:** 5269–5275.

Takeguchi, N., K. Ichimura, M. Koike, W. Matsui, T. Kashiwagura, and K. Kawahara. 1993. Inhibition of the multidrug efflux pump in isolated hepatocyte couplets by immunosuppressants FK506 and cyclosporine. *Transplantation* **55:** 646–650.

Ueda, K., N. Okamura, M. Hirai, Y. Tanigawara, T. Saeki, N. Kioka, T. Komano, and R. Hori. 1992. Human P-glycoprotein transports cortisol, aldosterone, and dexamethasone, but not progesterone. *J. Biol. Chem.* **267:** 24248–24252.

Vedeckis, W.V. 1983. Subunit dissociation as a possible mechanism of glucocorticoid receptor activation. *Biochemistry* **22:** 1983–1989.

Welshons, W.V., M.E. Lieberman, and J. Gorski. 1984. Nuclear localization of unoccupied oestrogen receptors. *Nature* **307:** 747–749.

Whitelaw, M.L., M. Göttlicher, J.-Å. Gustafsson, and L. Poellinger. 1993. Definition of a novel ligand binding domain of a nuclear bHLH receptor: Co-localization of ligand and hsp90 binding activities within the regulable inactivation domain of the dioxin receptor. *EMBO J.* **12:** 4169–4179.

Wiech, H., J. Buchner, R. Zimmermann, and U. Jakob. 1992. Hsp90 chaperones protein folding *in vitro. Nature* **358:** 169–170.

Wilhelmsson, A., S. Cuthill, M. Denis, A.-C. Wikström, J.-Å. Gustafsson, and L. Poellinger. 1990. The specific DNA binding activity of the dioxin receptor is modulated by the 90 kd heat shock protein. *EMBO J.* **9:** 69–76.

Yamamoto, K.R. 1985. Steroid receptor regulated transcription of specific genes and gene networks. *Annu. Rev. Genet.* **19:** 209–252.

Yamamoto, K.R., P.J. Godowski, and D. Picard. 1988. Ligand-regulated nonspecific inactivation of receptor function: A versatile mechanism for signal transduction. *Cold Spring Harbor Symp. Quant. Biol.* **53:** 803–811.

14

Expression and Function of the Low-molecular-weight Heat Shock Proteins

Andre-Patrick Arrigo
Laboratoire du Stress Cellulaire
Centre de Génétique Moléculaire et Cellulaire
CNRS UMR-106, Claude Bernard University
Lyon-I, 69622 Villeurbanne, France

Jacques Landry
Centre de recherche en cancérologie
de l'Université Laval, l'Hôtel-Dieu de Québec
Québec (Qué), Canada G1R 2J6

In this chapter, the name shSP includes all small heat shock proteins, cognate or heat inducible, defined as those proteins possessing the so-called α-crystallin protein domain. In general, the name used for individual shSP will be hsp*xx*, where *xx* corresponds to the two most significant digits of the apparent molecular weight. However, all mammalian shSP, excluding αA and αB crystallins, are called hsp27 irrespectively of slight interspecies variation in molecular weight, since they represent equivalent proteins. hsp*xx* from *Drosophila melanogaster* are called Dm-hsp*xx* to differentiate these proteins from the mammalian shSP.

The Biology of Heat Shock Proteins and Molecular Chaperones
©1994 Cold Spring Harbor Laboratory Press 0-87969-427-0/94 $5 + .00

I.   INTRODUCTION

Studies on the cellular response to heat shock and other physiological stresses have identified important families of proteins that are involved not only in cellular protection against these aggressions, but also in essential biochemical processes in unstressed cells. Among the protein families induced by heat shock, much has been learned about the hsp90, hsp70, and hsp60 families; these families accomplish different kinds of chaperonin function(s). This chapter deals with the family of small heat shock proteins (sHSP) which encompasses a large number of related protein species that share some structural features common to the lens protein α-crystallin and are represented in virtually all organisms, excluding perhaps prokaryotes. Neglected for a long time for several reasons including the fact that they did not appear to be as universally conserved as other heat shock proteins and that they were initially not observed in most mammalian cells, this group of heat shock proteins now generates renewed interest. sHSP are expressed differentially during development and growth cycle, and their expression correlates with differentiation and oncogenic status (for review, see Bond and Schlesinger 1987; de Jong et al. 1989, 1993; Pauli and Tissières 1990; Arrigo and Tanguay 1991; Pauli et al. 1992; Arrigo and Mehlen 1994). Elevated expression of sHSP efficiently confers protection against heat shock and a variety of toxic chemicals used in chemotherapy of cancer. Recent data suggest a homeostatic function at the level of signal transduction and a role in the growth, differentiation, and transformation process. We review here studies on the heat-induced and constitutive expression of the sHSP and on the mechanisms that regulate their phosphorylation. We then conclude by discussing the role(s) played by the sHSP in normal and stressed cells with a special emphasis on a possible role of some of the proteins at the level of actin microfilament assembly.

II.   THE FAMILY OF SMALL HEAT SHOCK PROTEINS

Among species, the sHSP are less conserved than the high-molecular-weight heat shock proteins (e.g., hsp70). Their number is variable (at least 4 major proteins in *Drosophila*, 3 in mammals including αA- and αB-crystallin, 1 in yeast, and >20 in plants) and their molecular masses are between 15 and 30 kD. The major characteristic of all the sHSP is a conserved domain, often referred to as the α-crystallin domain, which for *Drosophila* and mammalian sHSP consist of some 80 residues in the second half of the protein (Ingolia and Craig 1982; Southgate et al. 1983; Wistow 1985). From analyzing the sequence of several sHSP including α-crystallins, Wistow (1985) concluded that these proteins derived at least in part from the duplication of an ancestral gene. In *Drosophila*, the four sHSP (22, 23, 26, and 27 kD; indicated here as Dm-hsp22, 23, 26, and 27) (McKenzie et al. 1975; Mirault et al. 1978) are encoded by genes that are clustered within 12 kb of DNA at the 67B locus of chromosome 3L (Petersen et al. 1979; Corces et al. 1980; Craig and McCarthy 1980; Wadsworth et al. 1980; Voellmy et al. 1981; Ayme and Tissières 1985; for review, see Southgate et al. 1985; Pauli and Tissières 1990; Arrigo and Tanguay 1991; Pauli et al. 1992; Arrigo and Mehlen 1994). In plants, the sHSP belong to at least two multigene families (Raschke et al. 1988). In humans, hsp27 (also denoted hsp28) is encoded by a single active gene located on chromosome 7; two additional pseudogenes located on chromosomes 3 and X have been detected (Hickey et al. 1986; McGuire et al. 1989; L.A. Weber and E. Hickey, pers. comm.). In the mouse, hsp27 (also denoted hsp25) is also encoded by a single active gene (Frohli et al. 1993; Gaestel et al. 1993). In humans, both αA and αB crystallins are encoded by single-copy genes, located on chromosomes 21 and 11, respectively (Quax-Jeuken et al. 1985; Ngo et al. 1989). α-crystallin proteins should be considered as true heat shock proteins; the divergence in sequences between hsp27 and α crystallins is comparable to that found between the individual sHSP in *Drosophila*. A complete phylogenic tree of members of sHSP has been constructed by de Jong et al. (1988, 1993).

After examining the sequences of sHSP and α-crystallin molecules, Wistow (1985) and de Jong et al. (1988) suggested that the structure of the proteins of the sHSP family is organized into two major domains, each consisting of two structurally related motifs. Domain I of sHSP corresponds to exon I of α-crystallin (residues 1 to ~65) or to residues 1–87 of human hsp27. This domain is not very well conserved between sHSP of different species and is only moderately conserved as a whole even between sHSP of the same species. Nevertheless, small portions are very well conserved, most likely pointing to the most important regions. For example, three of the *Drosophila* sHSP share a region comprising the first 15 amino-terminal amino acids that resembles signal peptides (Southgate et al. 1983; Arrigo and Pauli 1988). In addition, the end of domain I in chick, mammalian, and *Drosophila* sHSP includes conserved phosphorylation sequences (see Section V). Domain II (the so-called α-crystallin domain) corresponds to residues 88–167 of human hsp27 and is highly conserved between the individual sHSP of a given species. Even between distant species, portions of this domain are surprisingly well conserved. For example, residues 122–143 of human hsp27 are 70% identical to corresponding sequences of Dm-hsp27 and Dm-hsp26. The sHSP from plant, yeast, and *Caenorhabditis elegans*, which appear to have diverged considerably from the putative ancestral gene, have homology with other sHSP only in this region. For example, 50% of the residues corresponding to the human sequence 161–171 are conserved among yeast, soybean, and humans. The yeast sHSP has only one additional region of significant homology with human hsp27 and is located at residues 138–149 (9 of 12 residues conserved). Regions of homology in this domain are also found in surface antigens from parasitic eukaryotes and bacteria. These include the egg antigen p40 of the blood parasite *Schistosoma mansoni* (Nene et al. 1986; de Jong et al. 1988), the 18-kD immunodominant antigen of *Mycobacterium leprae* (Nerland et al. 1988), and a 14-kD protein of *Mycobacterium tuberculosis* (Verbon et al. 1992). The remaining carboxy-terminal sequences are usually not well conserved, except for closely related sHSP; e.g., the *Drosophila* proteins display significant homology in the 25 residues located immediately after the crystallin domain (Southgate et al. 1983). Very little is known concerning the secondary structure of the sHSP. Secondary-structure predictions suggest the predominance of β-sheet conformation, with less than 5% α-helix structures (Walsh et al. 1991). Amphiphilic α helices with high hydrophobic moment are also observed at the amino and carboxyl termini of some sHSP, which may promote their interaction with membranes or other proteins (Plesofsky-Vig and Brambl 1990).

III. EXPRESSION OF SMALL HEAT SHOCK PROTEINS

Early experiments performed in *Drosophila* tissue-culture cells and more recent studies done with mammalian cell lines have shown that the sHSP are among the most strongly induced heat shock proteins when cells are exposed to heat shock or other stresses. In mammalian cells, their basal levels of expression are highly variable in the different cell lines examined, and in some instances, they are very low, rendering the detection problematic (Klemenz et al. 1991a, 1993; Landry et al. 1991; Inaguma et al. 1992; J. St-Amand and J. Landry, unpubl.). Typically, induction values on the order of 10–20-fold are obtained. For example, hsp27 concentration is increased by more than 10-fold, amounting to up to 1% of all proteins after heat shock of mouse Swiss-3T3 and Chinese hamster CCL39 cells. In contrast, in mouse NIH-3T3 cells, hsp27 is also induced severalfold by heat shock but does not attain levels exceeding 0.01% of total proteins. αB-crystallin represents less than 0.002% of the proteins in the three cell lines; after heat shock, its concentration increases up to about 0.05% in Swiss-3T3 and NIH-3T3 cells but still remains undetectable (<0.002%) in CCL39 cells. A level of 0.05% was also attained for αB-crystallin after heat shock of rat glioma cells. For comparison, in the lens, the tissue where both proteins are most expressed, hsp27 and αB-crystallin represent about 0.2–1% and 2–10% of total proteins, respectively.

Transcriptional activation of sHSP genes is thought to be the major regulatory mechanism explaining sHSP accumulation after stress in *Drosophila*, *C. elegans*, yeast, soybean, and *Xenopus laevis* (Klemenz and Gehring 1986; Hoffman et al. 1987; Czarnecka et al. 1989; Jones et al. 1989; Krone and Heikkila 1989; Susek and Lindquist 1990). Although not thoroughly investigated in the case of the sHSP, the stress induction of sHSP genes is probably mediated similarly to the other heat shock genes by the binding of a heat shock transcription factor (HSF) to repeats of the regulatory sequence HSE (heat shock element) (Morimoto et al. 1990). Such repeats have been localized upstream of all sHSP genes investigated so far, including the four *Drosophila* sHSP, the human and mouse hsp27, and αB-crystallin (Hickey et al. 1986; Klemenz et al. 1991a; DasGupta et al. 1992; Frohli et al. 1993; Gaestel et al. 1993). However, notable differences exist in the stress-induced expression of the sHSP compared to that of the other heat shock proteins. For example, sHSP of *Drosophila* are usually expressed at temperatures that are lower than those needed for the induction of hsp70 (Yost et al. 1990). In mammalian cells, the sHSP accumulate with slower kinetics and are synthesized for a longer time after stress (Arrigo and Welch 1987; Landry et al. 1991; Klemenz et al. 1993). In chicken embryo cells, the more than tenfold increase in the concentration of hsp27 after heat shock, in contrast to accumulation of hsp71 and hsp88, is not the result of transcriptional activation (Edington and Hightower 1990).

The genes encoding the sHSP are also expressed in the absence of stress. This was first observed following exposure of *Drosophila* embryonic cells to drugs that induce their differentiation (Buzin and Bournias-Vardiabasis 1982) and during specific stages of the development of *Drosophila* (Sirotkin and Davidson 1982; Cheney and Shearn 1983), which correlated with peaks of accumulation of the molting hormone β-ecdysterone (Handler 1982; Mason et al. 1984; Thomas and Lengyel 1986; Dubrovsky and Zhimulev 1988). This phenomenon was further analyzed at the protein level by using specific antibodies (Arrigo 1987; Arrigo and Pauli 1988; Pauli et al. 1989; Arrigo and Tanguay 1991; Marin et al. 1993). Another hint demonstrating the importance of β-ecdysterone in the induction of the sHSP was the observation of the strong synthesis of these proteins in tissue-culture cells or imaginal discs treated with this hormone (Ireland and Berger 1982; Ireland et al. 1982; Vitek and Berger 1984). It was then shown that this hormonal induction was regulated by steroid-receptor-binding sequences found far upstream of the beginning of the sHSP genes (Mestril et al. 1986; Riddihough and Pelham 1986, 1987). Interestingly, similar results were obtained in the mammalian system. Indeed, hsp27 is one of the major polypeptides synthesized by certain mammalian cells following treatment with steroid hormones, including estrogens (Fuqua et al. 1989), and is expressed in several estrogen-sensitive human tissues and breast tumors (Ciocca et al. 1983; Seymour et al. 1990b). Moreover, the mouse *hsp27* gene contains an estrogen-responsive element in direct proximity to the TATA box (Gaestel et al. 1993).

In addition to hormonal induction, the *Drosophila* sHSP genes appear to be under the control of other complex mechanisms since studies using *P*-element transformation demonstrated the presence of additional *cis*- and *trans*-regulatory elements upstream of these genes (Cohen and Meselson 1985; Glaser et al. 1986; Klemenz and Gehring 1986; Hoffman et al. 1987; Glaser and Lis 1990). However, the identification of specific sequences appear to be complex since, for the same sHSP gene, they differ from one tissue, or one developmental stage, to another (Cohen and Meselson 1985; Glaser et al. 1986). Other mechanisms also must exist to explain the complex tissue-specific expression of the mammalian sHSP.

A. Analysis of the Constitutive and Tissue-specific Expression of the *Drosophila* sHSP during Development

In contrast to their coordinated synthesis following heat shock, the four *Drosophila* sHSP display different patterns of expression during development. This is illustrated in Figure 1 where a quantitative analysis of Dm-hsp23 and Dm-hsp27 mRNAs and proteins during development is presented. As seen in this figure, only Dm-hsp27 is strongly expressed during embryogenesis. Later, the level of the transcripts from both genes is maximally abundant in white prepupae, but low levels of their corresponding polypeptides are detected. The maximal accumulation of the proteins is observed later in the middle of the pupal stage; at that stage, their corresponding mRNAs have almost completely disappeared. This implies that the sHSP are probably more stable than their corresponding mRNAs, resulting in a 30-fold increase in the level of the sHSP during the pupal phase of the insect (Arrigo 1987).

An additional characteristic of the sHSP concerns their tissue-specific expression during development (Glaser et al. 1986; Glaser and Lis 1990; Hass et al. 1990; Pauli et al. 1990; Arrigo and Tanguay 1991; Marin et al. 1993; Arrigo and Mehlen 1994). Again, this contrasts with the coordinated synthesis of these proteins in almost all tissues following heat shock (for review, see Arrigo and Tanguay 1991; Pauli and Tissières

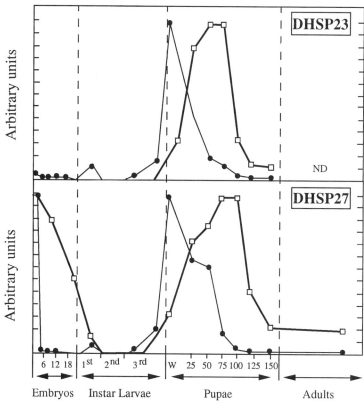

Figure 1 Quantification of Dm-hsp23 (DHSP23) and Dm-hsp27 (DHSP27) transcripts and proteins during *Drosophila* development. Protein and mRNA levels were analyzed from selected stages of the developing insect. Northern blot analysis was performed with specific probes that do not cross-hybridize with other sHSP mRNAs. Specific antisera were used to probe the corresponding immunoblots. (W) White prepupae; (*thin line*) mRNA; (*thick line*) proteins. (Reprinted, by copyright permission of the National Research Council of Canada, from Pauli et al. 1989.)

1990). The first evidence of a tissue-specific expression of some sHSP came from the work of Zimmerman et al. (1983), who detected the transcription of Dm-hsp26 and Dm-hsp27 genes in the ovarian nurse cells and in the developing oocyte. Glaser et al. (1986) were the first to detect the presence of a fusion gene Dm-hsp26-*lacZ* in neurocytes. Other studies performed with affinity-purified antisera have described the spatial expression of Dm-hsp27, Dm-hsp26, and Dm-hsp23 in the body of the developing fly (Arrigo 1987; Arrigo and Pauli 1988; Hass et al. 1990; Pauli et al. 1990; Marin et al. 1993). As described in more detail below, the developmental expression of these proteins is tissue-specific. In contrast to the other *Drosophila* shSP, Dm-hsp22 is not expressed during development (Arrigo and Tanguay 1991) but accumulates in embryonic cells treated with drugs that are teratogenic to humans (Buzin and Bounias-Vardiabasis 1982).

1. Dm-hsp27

Dm-hsp27 is probably the *Drosophila* shSP whose expression during development has been the most intensively studied (Arrigo and Pauli 1988; Hass et al. 1990; Pauli et al. 1990). Using immunological detection on thin sections of the developing fly, this protein was localized at the level of the neural cord of late embryos. During the first, second, and early third instar larval stages, Dm-hsp27 is present in the central nervous system (CNS) (Fig. 2A) and the gonads. These are the only larval tissues, together with imaginal discs, that contain high levels of mitotic cells. Interestingly, a strong accumulation of Dm-hsp27 was observed in the imaginal discs of late third instar larvae (Fig. 2B). This phenomenon occurs when the cells of this organ stop to divide (in late S/G$_2$ phase) and begin to differentiate. An analysis of this phenomenon at the level of the eye disc shows that Dm-hsp27 is present during the complete differentiation of this organ. In late pupae, this protein is still present at the top of the omatidia (Fig. 2C) but is not present in the eye of the newborn fly. A similar conclusion is made concerning the expression of Dm-hsp27 in the other disc-derived adult tissues. During larva-pupa development, Dm-hsp27 is still present in the cortex of the brain and thoracic ganglion (Fig. 2C,D), but, in late pupae and newborn flies, large portions of these organs are devoid of this protein. These observations suggest important roles for Dm-hsp27 in cells that are actively dividing as well as in cells that have the capability to differentiate.

2. Dm-hsp26 and Dm-hsp23

Earlier work by Glaser et al. (1986) reveals a tissue-specific expression of Dm-hsp26-LacZ fusion protein during *Drosophila* development that resembles that of Dm-hsp27. However, recent analysis using specific antisera has shown that the pattern of accumulation of these two proteins differed at the quantitative level. In contrast to Dm-hsp27, the level of Dm-hsp26 in the brain is low, but this protein is abundant in the gonads throughout larval and pupal development (Marin et al. 1993), suggesting a role of this protein in this organ. In contrast to Dm-hsp27 and Dm-hsp26, only low levels of Dm-hsp23 are observed during the embryogenesis. Recent results suggest that this is due to the expression of Dm-hsp23 in only a few embryonic cells located at regular intervals along the

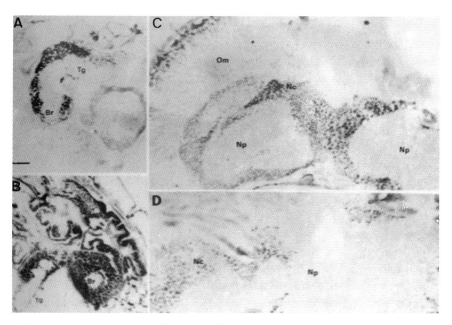

Figure 2 Accumulation of Dm-hsp27 during *Drosophila* development. Dm-hsp27 was localized in thin section of the developing insect by immunohisto-logical detection. The presence of this protein is visualized by a dark coloration. (*A*) Sagittal section of second instar larva. Dm-hsp27 is present in the central nervous system (CNS). Bar, 50 μm. (*B*) Sagittal section of late third instar larva; note the presence of Dm-hsp27 in the CNS and imaginal discs, particularly at the level of the eye disc which is localized on the external side of the larval body close to the top right side of the picture. (*C*) Partial view of a longitudinal section of the head of a late pupa; note the accumulation of Dm-hsp27 at the top of the eye omatidies and at the level of neurocytes which are localized in the cortex of several brain lobes. (*D*) Longitudinal section of the thoracic ganglion of a late pupa; Dm-hsp27 accumulates in neurocytes present in the cortex. (Br) Brain; (Tg) thoracic ganglion; (Nc) neurocyte; (Np) neuropile; (Om) ommatidia. (Reprinted, by copyright permission of the Rockefeller University Press, from Pauli et al. 1990.)

CNS, which have been identified as midline precursor cells (MPC) (Tanguay 1989; Hass et al. 1990; Arrigo and Tanguay 1991). Expression of Dm-hsp27 or Dm-hsp26 has not been observed in these cells. In young adults, Dm-hsp23 is still present in the gonads, in neurocytes and glia cells of the CNS, and in the leg nerves (Hass et al. 1990; Marin et al. 1993). The pattern of expression of Dm-hsp23 and Dm-hsp26 in differentiating imaginal discs is unknown.

These observations demonstrate that in contrast to their coordinated synthesis following heat shock, at least three *Drosophila* shSP display their own characteristic pattern of expression during development. This favors the hypothesis that each individual shSP may have similar or related function(s) in different tissues of the developing fly.

B. Analysis of the Constitutive and Tissue-specific Expression of the shSP in Other Species Including Mammals

The constitutive expression of the shSP in species other than *Drosophila* is less documented. However, there are some reports which indicate that shSP are also expressed in the absence of stress in other species. For example, Kurtz et al. (1986) observed that the unique shSP of yeast is constitutively expressed during meiosis and ascospore formation. Similarly, low levels of hsp30 have been detected in unstressed *Xenopus* embryos (Heikkila et al. 1991). Sensory ganglia from tadpoles (but not from adult bullfrog) synthesize small heat-shock-like proteins in vitro at normal temperature (Hammerschlag et al. 1989). This suggests that some amphibian shSP are expressed at the level of the developing CNS.

In mammalian cells, hsp27 is constitutively expressed at low level in a number of different tissue-culture cells (including human HeLa and HL60, monkey CV-1 and COS, rat REF52, mouse Swiss-3T3, and Chinese hamster CCL39). However, it is not, or is barely, detectable in several mouse cell lines such as NIH-3T3 and L929 fibroblasts. In MCF-7 and T47D human mammary tumor cells, the expression of the shSP has been found to be regulated by steroid hormone (Fuqua et al. 1989), a situation reminiscent of the induction of the *Drosophila* shSP by β-ecdysterone. Another example of the expression of hsp27 in the absence of stress is illustrated by the change in expression of this protein associated with change in the state of differentiation or proliferation of a number of different cell types such as embryonal carcinoma cells, embryonic stem cells (Stahl et al. 1992), mouse Ehrlich ascite tumor cells (Benndorf et al. 1988), normal B cells, B lymphoma cells (Spector et al. 1992), HL60 cells (Spector et al. 1993), osteoblasts, promyelocytic leukemia cells (Shakoori et al. 1992), and normal T cells (Hanash et al. 1993). Overexpression of hsp27 appears to reduce growth rate in some cell lines (Knauf et al. 1992) but not in others (Spector et al. 1992; J.N. Lavoie and J. Landry, unpubl.).

During mouse development, hsp27 has been observed to accumulate, in absence of stress, in several tissues, including neurons of the spinal cord and Purkinje cells (Gernold et al. 1993). It is also expressed in tissues containing muscle cells. The presence of hsp27 was also observed at the level of neural tube closure during early development of the rat (Walsh et al. 1991). During mouse development, αB-crystallin has been observed in the CNS, particularly at the level of glial cells (Iwaki et al. 1990; Gernold et al. 1993).

In adult mice and rats, hsp27 and αA and αB crystallins appear to each have specific patterns of expression. αA-crystallin is restricted to lens and spleen cells, whereas αB-crystallin and hsp27 are detected in several different tissues but are absent in the spleen (Bhat and Nagineni 1989; Dubin et al. 1989; Srinivasan et al. 1992; Klemenz et al. 1993). hsp27 and αB-crystallin are both observed in tissues that contain smooth or skeletal muscles; however, αB-crystallin but not hsp27 is strongly expressed in the CNS (Iwaki et al. 1990; Gernold et al. 1993; Klemenz et al. 1993; Tanguay et al. 1993).

A modulation in the levels of expression of mammalian sHSP has also been associated with pathological conditions. αB-crystallin is particularly abundant in the human brain, particularly in Rosenthal fibers of astrocytes from patients suffering from Alexander's disease (infantil neuronal degenerescence) (Iwaki et al. 1989). This protein is also observed at the level of intracellular inclusions in degenerative disorders of brain and liver, associated with extensive cytoskeletal and organellar rearrangements (Mayer et al. 1991). The presence of hsp27 complexed with a 22-kD protein, later identified as αB-crystallin, has been correlated with tumorigenic potential of adenovirus-transformed cells, and hsp27 was found to be repressed upon transformation of baby rat kidney (BRK) cells with E1A and c-Ha-*ras* oncogenes (Zantema et al. 1989). In addition, αB-crystallin shows elevated levels of expression after v-*mos* and activated c-Ha-*ras* are induced in NIH-3T3 cells (Klemenz et al. 1991b). A link between tumorigenicity and sHSP expression was thus suggested. In humans, the level of expression of hsp27 in individual tumors was found to be highly variable and was investigated for prognostic value (Puy et al. 1989; Ciocca et al. 1989, 1992; Navarro et al. 1989; Seymour et al. 1990a; Thor et al. 1991; Têtu et al. 1992). In human breast tumors, significant correlations were observed between hsp27 expression and estrogen receptor content, pS2 expression, nodal metastases, advanced T-stage lymphatic/vascular invasion, and disease-free survival (Thor et al. 1991). In malignant fibrous histiocytoma, hsp27 overexpression was associated with a more favorable prognosis, and a significant correlation was observed with overall survival and metastases-free survival (Têtu et al. 1992).

It can thus be concluded that in addition to their role during stress, the sHSP may have an additional role(s) in the normal unstressed cell and under pathological conditions. Several examples, from yeast, *Drosoph-*

ila, and mammals, are consistent with the idea that the sHSP accumulate when cells stop dividing (in late S/G$_2$) and enter the differentiation pathway. In other cells, it can be speculated that the high levels of sHSP may be required to regulate the rate of division of cells with a high mitotic index (such as those of the developing nervous system and the gonads). The presence of the mammalian hsp27 (and αB-crystallin), but not the *Drosophila* sHSP, at the level of muscle cells is consistent with a function of this mammalian protein at the level of the cytoskeleton, as described in more detail below.

IV. CELLULAR LOCALIZATION AND BIOCHEMICAL PROPERTIES OF THE SMALL HEAT SHOCK PROTEINS

A. sHSP in Unstressed Cells or in Cells That Have Recovered from a Heat Stress

In human and monkey cells, the constitutively expressed hsp27 is cytoplasmic and often observed concentrated in a polarized perinuclear zone; a similar locale of this protein is observed in cells that have recovered from stress (Arrigo et al. 1988; Arrigo 1990a). αB-crystallin appears to share this particular cellular locale (P. Mehlen and A.-P. Arrigo, unpubl.), whereas hsp25 from chicken is more diffusely distributed in the cytoplasm (Collier and Schlesinger 1986). Other studies have shown that Dm-hsp23 is concentrated at the level of cytoplasmic granules (Tanguay et al. 1985; Duband et al. 1986), a situation reminiscent of the cellular locale of the sHSP of tomato cells (Nover et al. 1989). sHSP are also found in organelles: Dm-hsp22 (R.M. Tanguay et al., in prep.) and hsp30 of *Neurospora crassa* (Plesofsky-Vig and Brambl 1990) are concentrated in mitochondria; some sHSP of higher plants and algae have also been observed in chloroplasts (Vierling et al. 1988; Chen and Vierling 1991). In contrast, Dm-hsp27 is a nuclear protein. A nuclear localization of Dm-hsp27 has been observed following β-ecdysterone stimulation of embryonic tissue-culture cells (Beaulieu et al. 1989) during development (Pauli et al. 1990) or after heat shock recovery (Arrigo and Pauli 1988; Beaulieu et al. 1989; Pauli et al. 1990). Moreover, as shown in Figure 3, Dm-hsp27 accumulates in the nucleus when expressed in monkey COS cells through an expression vector containing the coding sequence of this protein linked to the constitutive SV40 late promoter. In these cells, the endogenous hsp27 protein remains cytoplasmic.

Cell fractionation analyses have shown that the constitutively expressed sHSP or those that are expressed during development or following hormonal induction are generally recovered in the soluble fraction

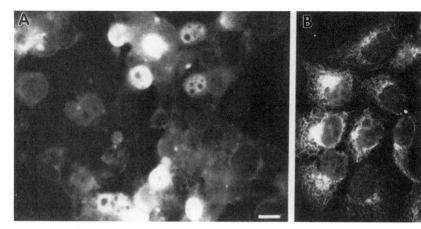

Figure 3 Immunofluorescence analysis of COS cells expressing Dm-hsp27. COS cells transiently expressing Dm-hsp27 were fixed and processed for indirect immunofluorescence analysis using anti-Dm-hsp27 (*A*) or anti-hsp27 (*B*) sera. Note the nuclear staining in every cell that expresses the *Drosophila* antigen and the cytoplasmic staining of the endogenous hsp27 of COS cells. Bar, 10 μm. (Reprinted, with permission, from Mehlen et al. 1993.)

following cell lysis in the presence of nonionic detergents. Even a fraction of the nuclear Dm-hsp27, expressed in *Drosophila* or COS cells, is extractable from the nuclear pellet and is recovered in the soluble fraction following cell lysis (Arrigo and Pauli 1988; Beaulieu et al. 1989; Mehlen et al. 1993). All of the sHSP so far examined were found to form oligomeric structures that sediment at 15S–20S in sucrose gradient (Arrigo and Ahmad-Zadeh 1981; Arrigo 1987; Arrigo and Pauli 1988) and display native molecular masses between 300 and 800 kD (Arrigo and Welch 1987; Arrigo et al. 1988; Bentley et al. 1992). These properties are shared by α-crystallin molecules (Seizen et al. 1978). Moreover, electron microscopy analyses have shown that α-crystallin and sHSP molecules form similar granule-like structures of 10–15 nm in diameter (Seizen et al. 1978; Arrigo et al. 1988). For mouse hsp27, Behlke et al. (1991) proposed a spherical structure composed of 32 monomers, arranged in hexagonal packing. Two different theories have been proposed to explain α-crystallin structure: the three-layered spherical (Bindels et al. 1979; Tardieu et al. 1986) and the micellar (Augusteyn and Koretz 1987) models. In the micellar model, the amino-terminal part of α-crystallin appears to be responsible for interactions with other molecules, suggesting that the conserved carboxy-terminal sequence may not be crucial for the overall structure of the protein. Alternatively, it has been suggested that this carboxy-terminal domain, which is shared by α-crystallin and shSP, may be responsible for the thermodynamic stability of these proteins (Ingolia and Craig 1982; Wistow, 1985; de Jong et al. 1988). α-crystallin molecules are particularly thermostable, with exceptionally long half-lives in the lens. Deletions in these particular domains (Mehlen et al. 1993), as well as X-ray structural analysis, will be needed to unravel the molecular structure of these proteins.

B. shSP in Stressed Cells

Studies on the cellular localization of the shSP using immuno-fluorescence or electron microscopy analyses have revealed a common feature to these proteins: their redistribution inside or around the nucleus during heat shock or other forms of stress. This was first observed in *Chironomus* (Vincent and Tanguay 1979) and *Drosophila* cells (Arrigo 1980; Arrigo and Ahmad-Zadeh 1981; Vincent and Tanguay 1982; Arrigo et al. 1988; Beaulieu et al. 1989; Mehlen et al. 1993) and occurs in every cell tested so far (illustrated in Fig. 4) (Collier and Schlesinger 1986; Arrigo and Welch 1987; Arrigo et al. 1988; Collier et al. 1988; Nover et al. 1989; Rossi and Lindquist 1989; Lavoie et al. 1993a). This phenomenon appears to be modulated by the metabolic state (Rossi and Lindquist 1989) and the degree of thermoresistance (Arrigo et al. 1988; Lavoie et al. 1993a) of the cells. Immunoelectron microscopy analyses of cell sections have confirmed the presence of hsp27 inside the nucleus during heat shock and its absence from nucleolar structures (Arrigo et al. 1988). In a recent report, Klemenz et al. (1991a) have shown that the nuclear redistribution of the shSP is also shared by αB-crystallin. Another feature of the shSP concerns their gradual redistribution in the cytoplasm during heat shock recovery (Arrigo et al. 1980, 1988), a phenomenon inhibited by agents that affect intracellular pH, such as the monovalent ionophore monensin (Arrigo 1990a).

During heat shock, the size of hsp27 oligomers increases, leading to the formation of super-aggregated structures ($>10^6$ daltons) that redistribute inside the nucleus (Arrigo et al. 1988). Consequently, the shSP are recovered in the insoluble fraction upon cell lysis. None of these phenomena are observed in thermotolerant cells exposed to heat shock (Arrigo 1987; Arrigo et al. 1988). In heat-shocked tomato cells, the granules formed by the shSP concentrate at the periphery of the nucleus and appear to be associated with untranslated mRNAs (Nover et al. 1989). However, no such association has been described in other cells (Collier et al. 1988). Interestingly, α-crystallins also have the tendency to form super-aggregated structures (Klemenz et al. 1991a). Such a phenomenon occurs in lens cells during aging as well as in individuals

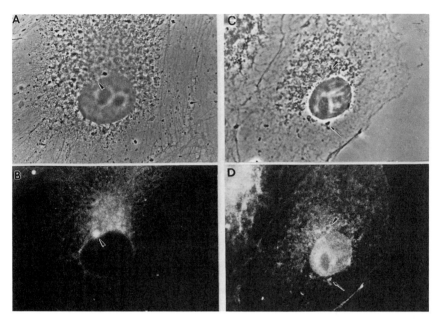

Figure 4 Cellular localization of hsp27 in unstressed and heat-shock-treated CV-1 cells. Monkey CV-1 cells growing on glass coverslips at 37°C were fixed with cold methanol and processed for indirect immunofluorescence using an antibody recognizing hsp27. (*A,B*) Cells kept at 37°C; (*C,D*) cells exposed to a 30-min heat shock treatment at 45°C before fixation; (*A,C*) phase contrast; (*B,D*) immunofluorescence. Note in unstressed cells the high concentration of hsp27 in the perinuclear region and the redistribution of this protein in the nucleus, but not in the nucleolus, during heat shock. The arrowheads and arrows indicate the position of dense cytoplasmic structure stained with anti-hsp27 serum. Bar, 8 μm. (Reprinted, with permission, from Arrigo et al. 1988.)

developing cataract, a pathology resulting in the opacification of the lens (Seizen et al. 1978). In addition, the α-crystallin structure is influenced by factors such as pH, temperature, calcium ions, and ionic strength (Seizen et al. 1980). The functional significance of the transient super-aggregation and nuclear redistribution of the sHSP and α-crystallins during heat shock is unclear. It is not known whether this phenomenon reflects a transient inactivation of these proteins or a mechanism leading to the protection of nuclear structure.

In addition to having their own tissue-specific expression in unstressed cells, the sHSP appear to have their own particular intracellular localization. This suggests that they may have similar roles but probably in different cells and also in different parts of these cells.

V. PHOSPHORYLATION OF THE SMALL HEAT SHOCK PROTEINS

One intriguing aspect of shSP biochemistry is the ability of many of these proteins to become phosphorylated in response to a large variety of stimuli, suggesting that in addition to the level of expression, the level of phosphorylation can modulate their function in cellular physiology. Several studies predating the characterization of hsp27 described a 27-kD protein as being a major early substrate of phosphorylation in response to growth factors (Chambard et al. 1983; Feuerstein and Cooper 1984). Kim et al. (1984) first observed in rat cells that hsp27 exists as multiple isoforms and that the most acidic species incorporated ^{32}P upon incubation of the cells with radioactively labeled orthophosphate. They suggested that these phosphopeptides represented phosphorylated iso-forms of the same protein. It was later demonstrated that the different isoforms of human hsp27, generally denoted a, b, c, and d, correspond to the same gene product phosphorylated at 0, 1, 2, and 3 sites, respectively (Landry et al. 1989, 1992). The identity of hsp27 with the 27-kD protein described in the early studies was first suggested by Welch (1985), who reported that hsp27 was phosphorylated when signal transduction pathways were stimulated by serum, phorbol esters, or calcium-active agents. Numerous agents are now known to induce the phosphorylation of mammalian hsp27. This includes, in addition to toxic agents such as heat shock (Landry et al. 1991, 1992), arsenite (Crête and Landry 1990; Landry et al. 1992) and hydrogen peroxide (Shibanuma et al. 1992), mitogens and differentiation factors such as thrombin (Mendelsohn et al. 1991; Landry et al. 1992), bombesin (Bitar et al. 1991), bradykinin (Grose et al. 1990; Saklatvala et al. 1991), fibroblast growth factor (FGF) (Saklatvala et al. 1991), serum (Welch 1985; Landry et al. 1992), platelet-derived growth factor (PDGF) (Saklatvala et al. 1991), trans-forming growth factor-β1 (TGF-β1) (Shibanuma et al. 1992), phorbol ester (Welch 1985; Regazzi et al. 1988; Darbon et al. 1990; Arrigo 1990b; Saklatvala et al. 1991), A23187 (Welch 1985; Crête and Landry 1990; Saklatvala et al. 1991), retinoic acid (Spector et al. 1994), leuke-mia inhibitory factor (LIF) (Michishita et al. 1991), and inflammatory cytokines such as tumor necrosis factor α (TNFα) (Hepburn et al. 1988; Robaye et al. 1989; Schutze et al. 1989; Arrigo 1990b; Arrigo and Michel 1991; Saklatvala et al. 1991; Landry et al. 1992; Guesdon et al. 1993), and interleukin 1α (IL-1α) (Saklatvala et al. 1991; Guesdon and Saklatvala 1991; Guesdon et al. 1993). Two of the *Drosophila* shSP (Dm-hsp27 and Dm-hsp26) are also phosphorylated in response to ec-dysterone (Rollet and Best-Belpomme 1986) and probably also during development (Arrigo and Pauli 1988). In some cases, circumstantial evi-dence has accumulated suggesting a role of hsp27 phosphorylation in the

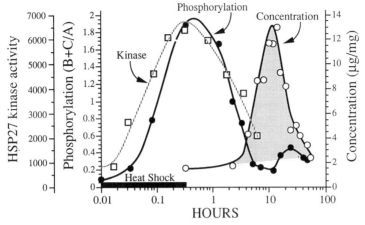

Figure 5 Heat-shock-induced hsp27 accumulation and phosphorylation and hsp27 kinase activity. Chinese hamster O23 cells were heat shocked for 20 min at 44°C and then returned to 37°C. At various times during heat shock or recovery, the cells were lysed in appropriate buffer to determine hsp27 kinase activity using recombinant hsp27 as substrate, hsp27 phosphorylation levels (ratio of phosphorylated to unphosphorylated isoforms), and hsp27 concentration relative to total proteins. (Adapted from Zhou et al. 1993 and Landry et al. 1991.)

physiological events associated with growth and differentiation in both normal and cancerous cells. In normal esteoblastic cells, phosphorylation of hsp27 induced in the late G_1 phase of the cell cycle by TGF-β1, phorbol esters, or H_2O_2 correlates with an inhibition of DNA synthesis induced by these agents (Shibanuma et al. 1991, 1992). hsp27 phosphorylation by thrombin may be involved in activation of platelets (Mendelsohn et al. 1991). Differentiation stage-specific phosphorylation of hsp27 was observed in cells of acute lymphoblastic leukemia (Strahler et al. 1991).

Most of these agents produce a very rapid increase in phosphorylation detectable within a few minutes, but they do not activate sHSP gene transcription and the accumulation of the protein. In the case of heat shock, which induces both processes, phosphorylation precedes accumulation of the protein and affects mostly pre-existing proteins (Fig. 5). In Chinese hamster cells exposed to 44°C heat shock, phosphorylation is maximal within 20 minutes and is back to normal basal levels within 2–3 hours, whether or not the high temperature is maintained. Under the same condition, accumulation of hsp27 is not detectable before 3 hours (Landry et al. 1991).

Human hsp27 is phosphorylated at three sites identified as Ser-82, Ser-78, and Ser-15. In rodent hsp27, Ser-78 (human sequence) is replaced by asparagine (Gaestel et al. 1989; Lavoie et al. 1990), but the two other sites, corresponding to human hsp27 Ser-82 and Ser-15, are phosphorylated (Gaestel et al. 1991). Analyses of the in vivo phosphorylation of all possible single, double, or triple phosphorylation mutants of the human protein indicated that phosphorylation does not occur in a defined ordered sequence and that Ser-82 and Ser-15 are the preferred sites phosphorylated after heat shock, arsenite, thrombin, or serum stimulation (Landry et al. 1992; J. Landry et al., unpubl.). It may be significant that Ser-78 and Ser-82 are located just outside the amino-terminal end of the α-crystallin domain. Moreover, all phosphorylation sites of hsp27 are located within the common sequence motif, Arg-X-X-Ser, motifs that are also found in two copies at similar locations in Dm-hsp27 and αB-crystallin. The possibility that phosphorylation at these sites could regulate the proposed functions of stabilization or multimerization ascribed to the α-crystallin domain is just beginning to be investigated. Interestingly, although phosphorylation does not appear to modify the apparent half-life of the protein (H. Lambert and J. Landry, unpubl.), it does affect the size of the sHSP multimers (P. Mehlen and A.-P. Arrigo; J.N. Lavoie and J. Landry; both unpubl.).

VI. REGULATION OF HSP27 PHOSPHORYLATION

The kinetics of phosphorylation/dephosphorylation of hsp27 in vivo can be closely correlated with the kinetics of increase and decline in protein kinase activities measured in cell extracts using recombinant hsp27 as a substrate (Fig. 5). Such a correlation is obtained after stimulation of cells with serum, FGF, thrombin, arsenite, H_2O_2, or TNFα, suggesting that activation of hsp27 kinase activity is the major mechanism regulating the phosphorylation of hsp27 (Zhou et al. 1993; J. Landry and J. Huot, unpubl.). A heat-shock-induced hsp27 kinase was recently purified from extracts of Chinese hamster cells. The enzyme has a Stokes' radius equivalent to that of a globular protein of about 50 kD. After electrophoresis and in-gel renaturation, two polypeptides with relative molecular weights of 45,000 and 54,000 and hsp27 kinase activity were identified, suggesting that the two polypeptides were active as monomers under native conditions. The same enzymes were also induced after treatment with sodium arsenite, TNFα, or H_2O_2, or stimulation of quiescent cells with serum, thrombin, or FGF, indicating that activation of the pp45-54 hsp27 kinase is the major converging point of several distinct signal transduction pathways responsible for phosphorylation of hsp27 (J. Landry and J. Huot, unpubl.). pp45-54 kinase may be homologous to an IL-1-induced hsp27 kinase that on gel filtration behaves as a 45-kD protein (Guesdon et al. 1993) and to the pp53-60 MAPKAP kinase II, an insulin-stimulated glycogen synthase kinase that was purified to homogeneity from rabbit skeletal muscle (Stokoe et al. 1992b). The latter recognizes the motif Leu-X-Arg-X-X-Ser and does efficiently phosphorylate hsp27 in vitro at the same sites that are phosphorylated in vivo (Stokoe et al. 1992a).

Levels of hsp27 phosphorylation may also be regulated by phosphatases, since phosphorylated hsp27 is a substrate for calcium/calmodulin (2B type) protein phosphatases (Gaestel et al. 1992). Moreover, Guy et al. (1993) showed that an okadaic- and oxyradical-sensitive phosphatase may also regulate phosphorylation of hsp27. These authors suggested that, like the known tyrosine phosphatase, the hsp27 phosphatase may contain a critical cysteine residue in its active domain, which would be sensitive to redox changes. Oxyradical-generating agents are also among the most potent inducers of pp45-54 kinase (J. Huot and J. Landry, unpubl.). Both activation of kinases and inactivation of phosphatases may thus contribute to the extremely high degree of hsp27 phosphorylation induced by cytokine and oxyradical-generating agents.

Several other observations suggest that oxyradicals are involved in the signal transduction pathways linking agonist stimulation to hsp27 phosphorylation. For example, reducing the oxyradical level in cells by pre-incubating with N-acetyl-cysteine, an intracellular free-radical scavenger (J. Huot and J. Landry, in prep.), or by transfection with the gene encoding the antioxidant glutathione peroxidase (P. Mehlen and A.-P. Arrigo, in prep.) reduces or inhibits phosphorylation of hsp27 following treatments with either H_2O_2, TNFα, or serum. In the case of N-acetyl-cysteine, it was shown that the inhibition results at least in part from a block in the stimulation of hsp27 kinase activity (J. Huot and J. Landry, in prep.). None of these conditions, however, significantly inhibited heat shock induction of hsp27 phosphorylation or hsp27 kinase, suggesting that pathways independent of oxyradicals also exist. Moreover, thermotolerant cells that had received a prior mild heat shock display only a weak stimulation of hsp27 phosphorylation and hsp27 kinase by heat shock or TNFα, indicating that complex mechanisms of regulations operate (Arrigo and Michel 1991; Landry et al. 1991; Zhou et al. 1993).

pp45-54 hsp27 kinase loses its activities upon treatment in vitro with protein phosphatase, suggesting that it is itself activated by phosphorylation (Zhou et al. 1993; H. Lambert and J. Landry, in prep.). Although not directly involved in the phosphorylation of hsp27 (Zhou et al. 1993), kinase C and kinase A are possible upstream activators of hsp27 kinase. Indeed, treatment of cells with phorbol esters (Welch 1985; Regazzi et al. 1988; Darbon et al. 1990; Arrigo 1990b; Saklatvala et al. 1991) or microinjection of cAMP into live cells (Lamb et al. 1989) leads to phosphorylation of hsp27. However, down-regulating kinase C with phorbol ester (PMA) pretreatment has little effect on subsequent stimulation of hsp27 phosphorylation by TNFα or heat shock but inhibits the phosphorylation of this protein by PMA (Arrigo 1990b). These observations indicate that kinase C is only an optional upstream activator located distally in the pathways of hsp27 phosphorylation.

A more proximal upstream activator of hsp27 kinase may be the mitogen-activated protein (MAP) kinases. Virtually all known inducers of hsp27 phosphorylation, including heat shock and H_2O_2, also induce MAP kinase activities (Dubois and Bensaude 1993; H. Lambert and J. Landry, unpubl.). Inactive pp45-54 hsp27 kinase purified from unstimulated cells can be activated by MAP kinase purified from heat-shocked cells or by activated recombinant MAP kinase (H. Lambert and J. Landry, unpubl.). MAP kinase is recognized as a molecule of central importance in mediating cell response to most external stimuli including growth factors, mitogens, and differentiating agents (for review, see Davis 1993); it has several known substrates involved in important physiological processes. MAP kinase phosphorylates and activates MAP-KAP kinase I (pp90^{rsk}), which phosphorylates the ribosomal protein S6 and the glycogen-binding subunit of protein phosphatase 1. It also activates MAPKAP kinase 2, which phosphorylates in vitro glycogen synthetase and hsp27. MAP kinase can also phosphorylate and modulate the activity of several transcription factors. The identification of the pp45-54 hsp27 kinase as a substrate of MAP kinase suggests that sHSP accomplish important functions linked to cell response to environmental stimuli of diverse natures.

VII.   ROLE IN UNSTRESSED CELLS: MAMMALIAN HSP27 IS ESSENTIAL IN SIGNAL TRANSDUCTION TO MICROFILAMENTS

As described above, circumstantial evidence has been accumulated, suggesting a role of sHSP in events associated with growth and differentiation in both normal and cancerous cells as well as during the development of organisms such as *Drosophila* and mouse. It is only recently that a precise biochemical function could be proposed.

A first evidence came from the work of Miron et al. (1988, 1991) on IAP, an inhibitor of actin polymerization purified as a contaminant of the vinculin-rich fraction of turkey gizzard. IAP behaves in vitro as an actin cap-binding protein, inhibits actin polymerization, and can depolymerize actin. Sequence analysis revealed that IAP was highly homologous to mouse and human sHSP and was undoubtly the chicken sHSP. Additional evidence suggested that mammalian hsp27 may similarly interact with actin filaments in vivo. Immunocytofluorescence studies in Chinese hamster fibroblasts permeabilized with saponin, an agent that better preserves plasma membrane protrusions and cortical actin structures, revealed that hsp27, in addition to being concentrated in a perinuclear region, was enriched in a region of highly motile cytoplasm such as the lamellipodia and ruffles, which are active sites of actin polymerization in fibroblasts (Fig. 6) (Lavoie et al. 1993b). Moreover, overexpression of hsp27 in Chinese hamster fibroblasts partially prevented cyctochalasin-D-induced microfilament disruption, suggesting that hsp27 might affect some aspects of actin filament dynamics in vivo (Lavoie et al. 1993a). Finally, Bitar et al. (1991) demonstrated that microinjection of antibodies against hsp27 in mouse cells blocked bombesin- and kinase-C-induced sustained contraction. Because hsp27 is a ubiquitous target of phosphorylation upon cell stimulation by agonists, these findings raised the possibility that hsp27 could be involved in regulating the dynamics of actin microfilaments in response to growth factor stimulation.

To test this possibility, Chinese hamster cell lines that overexpressed the human hsp27 protein or a nonphosphorylatable form of the protein were developed by transfection of the wild-type or a mutant form of the human gene in which codons for Ser-15, Ser-78, and Ser-82 were replaced by glycine codons (Lavoie et al. 1993b). Overexpression of hsp27 caused an increased concentration of filamentous actin at the cell cortex and elevated pinocytotic activity, a process that is dependent on actin dynamics. In contrast, overexpression of the nonphosphorylatable hsp27 reduced cortical F-actin concentration and decreased pinocytotic activity (Fig. 7). In fibroblastic cells, addition of mitogens to serum-starved cells causes a rapid accumulation of F-actin under membrane ruffles, cell rounding, and formation of surface blebbing. Interestingly, total F-actin in hsp27-overexpressing cells increased to approximately double the amount induced in control cells in response to thrombin or FGF, two factors that induce the phosphorylation of hsp27. The mutant hsp27 had an opposite effect. In cells overexpressing the nonphosphorylatable hsp27, little increase in F-actin content was detected following addition of the factors (Fig. 8); the nonphosphorylatable form of hsp27 appeared to act as a dominant negative mutant, inhibiting both thrombin and FGF induction of F-actin accumulation and formation of ruffles.

It thus appears that hsp27 is a necessary component of a signaling pathway between mitogens and actin polymerization at the membrane. The phenotype of the cells expressing the mutant hsp27 is intriguingly similar to that described for cells expressing the dominant negative Rac1 GTP-binding protein (Ridley and Hall 1992; Ridley et al. 1992). As

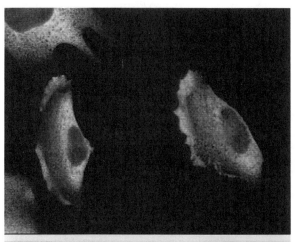

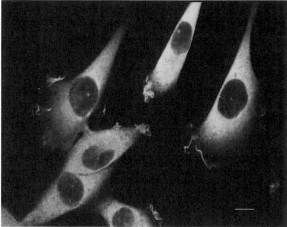

Figure 6 Localization of hsp27 in lamellipodia and membrane ruffles of un-stressed cells. Chinese hamster CCL39 cells were fixed with formaldehyde, permeabilized with saponin to better preserve plasma membrane protrusions, and processed for indirect immunofluorescence using an antibody recognizing hamster hsp27. Cells were examined by laser confocal microscopy. (*Top*) In cells growing on a highly adherent fibronectin-coated surface, hsp27 is abundant in well-extended lamellipodia. (*Bottom*) On less adherent plastic substratum, the lamellipodia retract, forming membrane ruffles containing enhanced concentration of hsp27. Bar, 10 μm.

found for mutant hsp27, accumulation of a dominant negative mutant Rac1 protein was reported to prevent cortical microfilament assembly and membrane ruffling in stimulated cells. Rac1 and hsp27 may thus function in the same pathway that transduces signals to cortical actin.

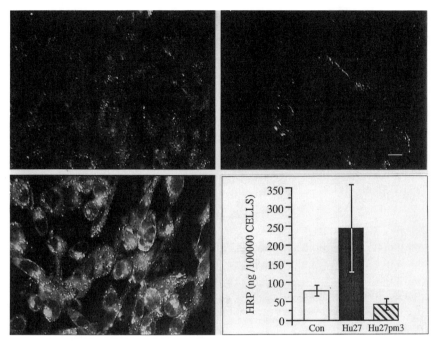

Figure 7   Effect of overexpressing human hsp27 or a nonphosphorylatable hsp27 on pinocytotic activities of Chinese hamster cells. Pinocytotic activity was visualized by confocal microscopy after incubating control cells (CON, *top left*), cells expressing hsp27 (Hu27, *bottom left*), or cells expressing the phosphorylation mutant hsp27 (Hu27pm3, *top right*) for 20 min in medium containing Texas-red-labeled dextran-lysine. For quantification (*bottom right*), the cells were incubated for 20 min in the presence of horseradish peroxidase, and the amount of peroxidase inside the cells was determined enzymatically. Note that, by comparison to control cells, Hu27 cells have more numerous and more heavily stained vacuoles. Pinocytosis is inhibited in Hu27pm3 cells. The histogram is adapted from Lavoie et al. (1993b).

The opposite effects of wild-type and mutant hsp27 on the dynamics of actin microfilaments are consistent with the possibility that the actin capping activity of hsp27, described by Miron et al. (1988, 1991), is regulated by phosphorylation. Phosphorylation of hsp27 during stimulation by growth factors might cause a change in the conformation of the protein, resulting in the dissociation of hsp27 from the barbed end of actin filaments. Local phosphorylation of cortical hsp27 could thus regulate the spatial organization of F-actin by freeing barbed ends of microfilaments for addition of monomers. A higher concentration of hsp27 in cells would increase the proportion of hsp27 relative to other actin capping proteins, thus enhancing polymerization in response to growth factors.

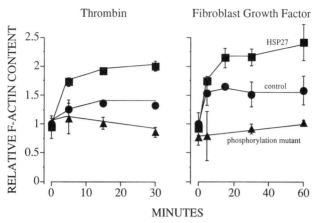

Figure 8 hsp27 enhances, whereas the nonphosphorylatable hsp27 inhibits, FGF and thrombin-induced actin polymerization in serum-starved cells. F-actin content was measured in control cells (*circles*), cells expressing hsp27 (*squares*), or cells expressing the phosphorylation mutant hsp27 (*triangles*), at various times after addition of FGF or thrombin. The results imply that hsp27 is a component of the signaling pathway between mitogens and actin polymerization. (Adapted from Lavoie et al. 1993b.)

The presence of the nonphosphorylatable form of hsp27 in high amount would then increase the lifetime of hsp27 at the barbed end, therefore reducing the microfilament response to growth factors. This model is purely hypothetical. Additional information on the biochemistry of the interaction of hsp27 with actin is required.

VIII.   ROLE OF SMALL HEAT SHOCK PROTEINS IN THE CELLULAR RESPONSE TO TOXIC STRESS

In virtually all cell lines and organisms investigated, accumulation of heat shock proteins following heat shock or exposure to chemical stresses is accompanied by the acquisition of a dramatic increased capacity of the cells to survive subsequent severe hyperthermic treatments. These newly acquired thermal properties are transient, lasting the time required for the cells to restore normal levels of heat shock proteins (Landry et al. 1982, 1991). It was thus suggested that heat shock proteins have thermoprotective functions.

The first evidence for a thermoprotective function of sHSP was obtained in *Dictyostelium*, where mutants defective for synthesis of sHSP were found to be unable to develop thermotolerance despite induction of the other heat shock proteins (Loomis and Wheeler 1982). In *Drosophila*, selective induction of the sHSP by ecdysterone in a hormone-sensitive cell line was found to bring about the thermotolerant phenotype in the absence of heat shock (Berger and Woodward 1983). Similarly, induction by dexamethasone of αB-crystallin, in the absence of accumulation of other heat shock proteins, was accompanied by increased resistance to lethal heat shock in NIH-3T3 cells (Aoyama et al. 1993). In addition, stable thermoresistant mutants developed from Chinese hamster cells were found to overexpress hsp27 constitutively (Chrétien and Landry 1988).

Recently, a number of gene transfection studies clearly demonstrated a protective function of the sHSP during hyperthermia. Stable Chinese hamster or mouse cell lines expressing elevated levels of hsp27 were obtained after transfecting recombinant plasmids containing the human *hsp27* gene (Landry et al. 1989; Huot et al. 1991). Transfected cell lines that expressed constitutively high levels of human hsp27 were extremely thermoresistant compared to the parental cell line. Analysis of individual clonal isolates revealed a good correlation between the amount of hsp27 present in each cell line and intrinsic thermal resistance: The more human hsp27 was expressed constitutively by a transfectant clone, the more resistant it was to thermal killing. Thermoprotection was also obtained after transfecting Chinese hamster cells (Rollet et al. 1992) or COS cells (Mehlen et al. 1993) with Dm-hsp27, or mouse cells with αB-crystallin (Aoyama et al. 1993). The effect of a transient overexpression of sHSP, as it occurs during induction of thermotolerance by heat shock, was investigated in mouse cells conditionally expressing the Chinese hamster hsp27. Cell lines were developed by transfecting NIH-3T3 cells with the Chinese hamster hsp27 sequences under the control of the metallothionein MT-1 promoter. Studies performed with stably transfected clonal cell lines demonstrated that exposure of the cells to low concentrations of $CdCl_2$ induced accumulation of Chinese hamster hsp27 and a progressive development of thermoresistance that attained a level approaching heat-shock-induced thermotolerance. After $CdCl_2$ removal, thermal resistance and hsp27 decayed in a coordinated manner (Lavoie et al. 1993a). These results indicated that there is an immediate hsp27 mass effect on cellular capacity to survive thermal injuries, suggesting a direct effect of hsp27 on some heat-sensitive targets.

It is interesting to note that the thermal resistance conferred by sHSP is often accompanied by an increased cellular resistance to other toxic agents. For example, protection against oxidative stress was observed in COS cells expressing Dm-hsp27 (Mehlen et al. 1993). Stable L929 mouse cell lines expressing either Dm-hsp27 or αB-crystallin display increased resistance to thermal and oxidative stresses as well as to the cytotoxic effects of the inflammatory cytokine TNFα (P. Melhen et al., in prep.). However, overexpression of the human hsp27 in a murine fibrosarcoma cell line had no effect on cellular susceptibility to TNFα (Jaattela et al. 1992). Finally, some Chinese hamster cell lines over-expressing the human hsp27 protein are also resistant to a variety of cytotoxic drugs (Huot et al. 1991).

Susek and Lindquist (1989) and Bentley et al. (1992) observed only a slight increase in thermoresistance in yeast cells expressing high levels of hsp26. Moreover, Petko and Lindquist (1986) found no detectable phenotypic effect, and, in particular, no increase in the thermosensitivity, and no effect on the capacity to develop thermotolerance, in yeast after inactivation of the gene encoding hsp26, which appears to be the unique protein member of the sHSP family in these cells. Clearly, sHSP from different organisms have evolved to develop different biological ac-tivities. Considering that other heat shock proteins also have protective functions against stress (see Parsell and Lindquist, this volume), the specific protective function that sHSP have in mammalian cells may have been lost in some organisms during evolution and, perhaps, gained by another heat shock protein.

IX. MECHANISMS OF SMALL HEAT SHOCK PROTEIN PROTECTION AGAINST STRESSES

What is the target that is protected by hsp27 during heat shock and responsible for the generalized state of resistance observed? In cells made thermotolerant and that express the full spectrum of heat shock proteins, virtually all cellular structures and activities are protected from the immediate effects of heat shock, and moreover, the damages that are still induced at elevated temperature recover faster upon return to normal temperature (for review, see Laszlo 1992). In contrast to thermotolerant cells that express the full spectrum of heat shock proteins, cells over-expressing hsp27 alone were found not to be protected against heat-induced inhibition of protein synthesis, transcription of ribosomal genes, processing of ribosomal RNA, transcription of specific RNA, and trans-location of hsp70 to the nucleus. For all these parameters, the immediate deleterious effect of heat shock was the same in control versus hsp27-overexpressing cells. However, in all cases, the recovery from the heat-induced perturbations was faster in the transfected cells. The only end-point for which an immediate protection was observed was the microfila-ment. In cells that expressed hsp27 either conditionally or constitutively, the microfilament network was protected from immediate heat-induced disruptions (Laszlo et al. 1993; Lavoie et al. 1993a; L.A. Weber et al., in prep.). All of these results are consistent with the hypothesis that hsp27 confers thermal protection by stabilizing the microfilament network, thereby facilitating the repairs (possibly accomplished by other heat shock proteins) at other cellular levels after heat shock. As discussed above, this could result from higher actin-polymerizing activities due to the higher concentration of hsp27 and its activation by phosphorylation early during heat shock. Experiments performed with cells overexpressing a nonphosphorylatable mutant of hsp27 supported this idea. As discussed, these cells are nonresponsive to growth factor induction of actin polymerization. In contrast to cells overexpressing a wild-type hsp27 protein, which were found to be resistant to cytotoxic and growth inhibitory actions of cytochalasin D (Lavoie et al. 1993a), cells overexpressing the nonphosphorylatable mutant were sensitized to the action of cytochalasin D. Moreover, when heat resistance was investigated, it was found that the mutant hsp27 was much less efficient in protection than the wild-type protein (J.N. Lavoie and J. Landry, in prep.). It was thus suggested that most of the protective function of hsp27 is dependent on its phosphorylation and action at the level of actin filament dynamics.

A number of other possible functions or characteristics ascribed to the diverse sHSP, however, might also contribute to a protective function, including protease inhibitory (Orthwerth and Olesen 1992) and chaperon-like (Horwitz 1992; Jakob et al. 1993; Merck et al. 1993) activities, long-term stability (Tardieu and Delaye 1988), and the ability to be cross-linked to other proteins via transglutamination (Groenen et al. 1992). Such transaminase activation may be mediated by the high levels of intracellular calcium observed during stress. It is possible that the distinct sHSP species in a given organism have specialized functions in protection and that the expression of several distinct sHSP allows a wider spectrum of protection.

This idea is supported by recent data obtained on the mechanism of protection mediated by Dm-hsp27. Analysis of stable transfectants of L929 cells expressing Dm-hsp27 revealed a striking difference in the protective activity of this *Drosophila* protein compared to that of hsp27, namely, that Dm-hsp27 provides a strong protection against the inhibition of protein synthesis observed during heat shock (P. Chareyron et al., in prep.). Such differences in the mechanisms of protection are not surprising considering the distinct cellular localization of the two proteins. Indeed, as described above, Dm-hsp27 is mainly localized in the nucleus, excluding an interaction with cytoplasmic actin as described for hsp27.

Structure-function relationship studies constitute a powerful approach to unravel the specific functions that individual sHSP may have. As mentioned above, mutation analyses of the phosphorylation sites of hsp27 clearly demonstrated the importance of phosphorylation on the functions of the proteins. Analysis of deletion mutants of Dm-hsp27 revealed that the first (amino-terminal) part of the α-crystallin domain is probably not of major importance in the protective activity of the protein. In contrast, the deletion of the last 42 amino acids of the α-crystallin domain of this protein abolished its activity (Mehlen et al. 1993). This suggests that the carboxy-terminal part of the α-crystallin domain may be directly or indirectly implicated in the protective function of Dm-hsp27. Ongoing studies aim at evaluating in a comparative manner the effects of more subtle mutations on both Dm-hsp27 and hsp27.

X. CONCLUDING REMARKS

Recently, several studies have provided decisive hints in the elucidation of the mechanisms regulating the expression and functions of the sHSP. Analyses of several members of this family of proteins have lead to the conclusion that in unstressed cells, each sHSP has its own tissue-specific expression and particular intracellular localization. Hence, the sHSP have probably similar roles but in different cells and also in a different part of these cells. Some of these proteins appear to be involved in the maintenance of microfilament integrity, and others may have roles in the control of cell division/differentiation or act as homeostatic elements in different parts of the cell, such as the cytoplasm, nuclei, mitochondria, and chloroplasts. They are also structural elements in lens cells. In contrast, during stress, most of the sHSP so far analyzed are expressed in every cell and at least three of them have demonstrated protective functions. This suggests that they belong to the stress-induced machinery that protects against and/or repairs cellular damages. Another fascinating aspect of the sHSP concerns their constitutive expression during development. This may represent a molecular mechanism that controls the growth and differentiation of specific tissues and protects them against environmental or physiological stresses. Future studies should bring us further understanding of the mechanisms regulating the tissue-specific expression and the role of these sHSP during development. At the biochemical level, more work is necessary for a better understanding of their complex structure-function relationships. It will be of prime importance to define the active sHSP oligomers, the role of their phosphorylation status, and the transduction mechanisms regulating this modification. The role of the sHSP as molecular chaperones and the role of the mammalian sHSP at the level of actin microfilament assembly are among the more promising studies. They may unravel major roles of these heat shock proteins during cell division and differentiation. In addition, the study of the molecular mechanisms regulating the aggregation of these proteins could be of interest for a better understanding of the biochemistry of the lens α crystallins.

ACKNOWLEDGMENTS

We thank Josée Lavoie for providing the unpublished confocal images shown in Figures 6 and 7. This work was supported by grants 6011 from the Association pour la Recherche sur le Cancer and 91.C.388 from the Ministère de la Recherche et de la Technologie (to A.-P.A.), and grant MT-1088 from the Medical Research Council of Canada (to J.L.).

REFERENCES

Aoyama, A., E. Fröhli, R. Schafer, and R. Klemenz. 1993. αB-crystallin expression in mouse NIH 3T3 fibroblasts: Glucocorticoid responsiveness and involvement in thermal protection. *Mol. Cell. Biol.* **13:** 1824–1835.

Arrigo, A.-P. 1980. Investigation of the function of the heat shock proteins in *Drosophila melanogaster* tissue culture cells. *Mol. Gen. Genet.* **178:** 517–524.

———. 1987. Cellular localization of hsp23 during *Drosophila* development and subsequent heat shock. *Dev. Biol.* **122:** 39–48.

———. 1990a. The monovalent ionophore monensin maintains the nuclear localization of the human stress protein hsp28 during heat shock recovery. *J. Cell Sci.* **96:** 419–427.

———. 1990b. Tumor necrosis factor alpha induces the rapid phosphorylation of the mammalian low molecular weight heat shock protein hsp28. *Mol. Cell. Biol.* **10:** 1276–1280.

Arrigo, A.-P. and C. Ahmad-Zadeh. 1981. Immunofluorescence localization of a small heat shock protein (hsp23) in *Drosophila melanogaster. Mol. Gen. Genet.* **184:** 73–79.

Arrigo, A.-P. and P. Mehlen. 1994. Expression, cellular location and function of low molecular weight heat shock proteins (hsp20s) during development of the nervous system. In *Heat shock or cell stress proteins and the nervous system* (ed. J. Mayer and I. Brown). Academic Press, New York. (In press.)

Arrigo, A.P. and M.R. Michel. 1991. Decreased heat- and tumor necrosis factor-mediated hsp28 phosphorylation in thermotolerant HeLa cells. *FEBS Lett.* **282:** 152–156.

Arrigo, A.-P. and D. Pauli. 1988. Characterization of hsp27 and of three immunologically related polypeptides during *Drosophila* development. *Exp. Cell. Res.* **175:** 169–183.

Arrigo, A.-P. and R.M. Tanguay. 1991. Expression of heat shock proteins during development in *Drosophila*. In *Heat shock and development* (ed. L. Hightower and L. Nover), pp. 106–119. Springer-Verlag, Berlin.

Arrigo, A.-P. and W.J. Welch. 1987. Characterization and purification of the mammalian 28,000 dalton heat shock protein. *J. Biol. Chem.* **262:** 15359–15369.

Arrigo, A.-P., S. Fakan, and A. Tissières. 1980. Localization of the heat shock induced proteins in *Drosophila melanogaster* tissue culture cells. *Dev. Biol.* **78:** 86–103.

Arrigo, A.-P., J. Suhan, and W.J. Welch. 1988. Dynamic changes in the structure and locale of the mammalian low molecular weight heat shock protein. *Mol. Cell. Biol.* **8:** 5059–5071.

Augusteyn, R. and J.F. Koretz. 1987. A possible structure for α-crystallin. *FEBS Lett.*

222: 1–5.

Ayme, A. and A. Tissières. 1985. Locus 67B of *Drosophila melanogaster* contains seven, not four, closely related heat shock genes. *EMBO J.* **4:** 2949–2954.

Beaulieu, J.F., A.-P. Arrigo, and R.M. Tanguay. 1989. Interaction of *Drosophila* 27Kd heat shock protein with the nucleus of heat-shocked and ecdysterone-stimulated cultured cells. *J. Cell. Sci.* **92:** 29–36.

Behlke, J., G. Lutsch, M. Gaestel, and H. Bielka. 1991. Supramolecular structure of the recombinant murine small heat shock protein hsp25. *FEBS Lett.* **288:** 119–122.

Benndorf, R., R. Kraft, A. Otto, J. Stahl, H. Böhm, and H. Bielka. 1988. Purification of the growth-related protein p25 of the Ehrlich ascites tumor and analysis of its isoforms. *Biochem. Int.* **17:** 225–234.

Bentley, N.J., I.T. Fitch, and M.F. Tuite. 1992. The small heat-shock protein hsp26 of *Saccharomyces cerevisiae* assembles into a high molecular weight aggregate. *Yeast* **8:** 95–106.

Berger. E.M. and M.P. Woodward. 1983. Small heat shock proteins of *Drosophila* may confer thermal tolerance. *Exp. Cell. Res.* **147:** 437–442.

Bhat, S.P. and C.N. Nagineni. 1988. AlphaB-subunit of lens-specific alphaB-crystallin is present in other ocular and non-ocular tissues. *Biochem. Biophys. Res. Commun.* **158:** 319–325.

Bitar, K.N., M.S. Kaminski, N. Hailat, K.B. Cease, and J.R. Strahler. 1991. Hsp27 is a mediator of sustained smooth muscle contraction in response to bombesin. *Biochem. Biophys. Res. Commun.* **181:** 1192–1200.

Bindels, J.G., R.J. Siezen, and H.J. Hoenders. 1979. A model for the architecture of α-crystallin. *Ophthalmic Res.* **11:** 441–452.

Bond, U. and M.J. Schlesinger. 1987. Heat shock proteins and development. *Adv. Genet.* **24:** 1–29.

Buzin, C.H. and N. Bournias-Vardiabasis. 1982. The induction of a subset of heat shock proteins by drugs that inhibit differentiation in *Drosophila* embryonic cell cultures. In *Heat shock from bacteria to man* (ed. M.J. Schlesinger et al.), pp. 387–394. Cold Spring Harbor Laboratory, Cold Spring Harbor, New York.

Chambard, J.-C., A. Franchi, A. Le Cam, and J. Pouysségur. 1983. Growth factor-stimulated protein phosphorylation in G0/G1-arrested fibroblasts. Two distinct classes of growth factors with potentiating effects. *J. Biol. Chem.* **258:** 1706–1713.

Cheney, C.M. and A. Shearn. 1983. Developmental regulation of *Drosophila* imaginal discs proteins: Synthesis of a heat-shock protein under non-heat shock conditions. *Dev. Biol.* **95:** 325–330.

Chen, Q. and E. Vierling. 1991. Analysis of conserved domains identifies a unique structural feature of a chloroplast heat shock protein. *Mol. Gen. Genet.* **226:** 425–431.

Chrétien, P. and J. Landry. 1988. Enhanced constitutive expression of the 27-kDa heat shock proteins in heat-resistant variants from Chinese hamster cells. *J. Cell Physiol.* **137:** 157–166.

Ciocca, D.R., L.A. Puy, and L.C. Fasoli. 1989. Study of estrogen receptor, progesterone receptor, and the estrogen-regulated Mr 24,000 protein in patient with carcinomas of the endometrium and cervix. *Cancer Res.* **49:** 4298–4304.

Ciocca, D.R., D.J. Adams, D.P. Edwards, R.J. Bjercke, and W.L. McGuire. 1983. Distribution of an estrogen-induced protein with a molecular weight of 24,000 in normal and malignant human tissues and cells. *Cancer Res.* **43:** 1204–1210.

Ciocca, D.R., G. Lo Castro, L.V. Alonio, M.F. Cobo, H. Lotfi, and A. Teyssie. 1992. Effect of human papillomavirus infection on estrogen receptor and heat shock protein hsp27 phenotype in human cervix and vagina. *Int. J. Gynecol. Pathol.* **11:** 113–121.

Cohen, R.S and M. Meselson. 1985. Separate regulatory element for the heat-inducible and ovarian expression of the *Drosophila* hsp26 gene. *Cell* **43**: 737–746.

Collier, N.C. and M. Schlesinger. 1986. The dynamic state of heat shock proteins in chicken embryo fibroblasts. *J. Cell. Biol.* **103**: 1495–1507.

Collier, N.C., J. Heuser, M. Aach-Levy, and M. Schlesinger. 1988. Ultrastructural and biochemical analysis of the stress granules in chicken embryo fibroblasts. *J. Cell Biol.* **106**: 1131–1139.

Corces, V., R. Holmgren, R. Freund, R. Morimoto, and M. Meselson. 1980. Four heat shock proteins of *Drosophila melanogaster* coded within a 12-kilobase region in chromosome subdivision 67B. *Proc. Natl. Acad. Sci.* **77**: 5390–5393.

Craig, E.A. and B.J. McCarthy. 1980. Four *Drosophila* heat shock genes at 67B: Characterization of recombinant plasmids. *Nucleic Acids Res.* **8**: 4441–4457.

Crête, P. and J. Landry. 1990. Induction of HSP27 phosphorylation and thermoresistance in Chinese hamster cells by arsenite, cycloheximide, A23187, and EGTA. *Radiat. Res.* **121**: 320–327.

Czarnecka, E., J.L. Key, and W.B. Gurley. 1989. Regulatory domains of the Gmhsp17.5-E heat shock promoter of soybean. *Mol. Cell Biol.* **9**: 3457–3463.

Darbon, J.M., M. Issandou, J.F. Tournier, and F. Bayard. 1990. The respective 27 kDa and 28 kDa protein kinase C substrates in vascular endothelial and MCF-7 cells are most probably heat shock proteins. *Biochem. Biophys. Res. Commun.* **168**: 527–536.

Davis, R.J. 1993. The mitogen-activated protein kinase signal transduction pathway. *J. Biol. Chem.* **268**: 14553–14556.

DasGupta, S., T.C. Hohman, and D. Carper. 1992. Hypertonic stress induces αB-crystallin expression. *Exp. Cell. Res.* **54**: 461–470.

de Jong, W.W., J.A.M. Leeunissen, and C.E.M. Voorter. 1993. Evolution of the α-crystallin/small heat-shock protein family. *Mol. Biol. Evol.* **10**: 103–126.

de Jong, W.W., W. Hendricks, J.W.M. Mulders, and H. Bloemendal. 1989. Evolution of eye lens crystallin: The stress connection. *Trends Biochem. Sci.* **14**: 365–368.

de Jong, W.W., J.A.M. Leeunissen, P.J.M. Leenen, A. Zweers, and M. Versteeg. 1988. Dogfish alpha-crystallin sequences: Comparison with small heat shock proteins and Schistosoma egg antigen, *J. Biol. Chem.* **263**: 5141–5149.

Duband, J.-L., F. Lettre, A.-P. Arrigo, and R.M. Tanguay. 1986. Expression and cellular localization of HSP23 in unstressed and heat shocked *Drosophila* culture cells. *Canada. J. Genet. Cytol.* **28**: 1088–1092.

Dubin, R.A., E.F. Wawrousek, and B. Piatigorsky. 1989. Expression of the murine alphaB-crystallin gene is not restricted to the lens. *Mol. Cell Biol.* **9**: 1083–1091.

Dubois, M.F. and O. Bensaude. 1993. MAP kinase activation during heat shock in quiescent and exponentially growing mammalian cells. *FEBS Lett.* **324**: 191–195.

Dubrovsky, E.B. and I.F. Zhimulev. 1988. Trans-regulation of ecdysterone-induced protein synthesis in *Drosophila melanogaster* salivary glands. *Dev. Biol.* **127**: 33–44.

Edington, B.V. and L.E. Hightower. 1990. Induction of a chicken small heat shock (stress) protein: Evidence of multilevel posttranscriptional regulation. *Mol. Cell Biol.* **10**: 4886–4898.

Feuerstein, N. and H.L. Cooper. 1984. Rapid phosphorylation-dephosphorylation of specific proteins induced by phorbol ester in HL-60 cells. Further characterization of the phosphorylation of 17-kilodalton and 27-kilodalton proteins in myeloid leukemic cells and human monocytes. *J. Biol. Chem.* **259**: 2782–2788.

Frohli, E., A. Aoyama, and R. Klemenz. 1993. Cloning of the mouse hsp25 gene and an extremely conserved hsp25 pseudogene. *Gene* **128**: 273–227.

Fuqua, S.A.W., M. Blum-Salingaros, and W.L. McGuire. 1989. Induction of the estrogen-regulated "24K" protein by heat shock. *Cancer Res.* **49**: 4126–4129.

Gaestel, M., R. Gotthardt, and T. Muller. 1993. Structure and organisation of a murine gene encoding small heat-shock protein Hsp25. *Gene* **128**: 279–283.

Gaestel, M., R. Benndorf, K. Hayess, E. Priemer, and K. Engel. 1992. Dephosphorylation of the small heat shock protein hsp25 by calcium/calmodulin-dependent (type 2B) protein phosphatase. *J. Biol. Chem.* **267**: 21607–21611.

Gaestel, M., W. Schröder, R. Benndorf, C. Lippman, K. Buchner, F. Huchot, V.A. Ermann, and H. Bielka. 1991. Identification of the phosphorylation sites of the murine small heat shock protein 25. *J. Biol. Chem.* **266**: 14721–14724.

Gaestel, M., B. Gross, R. Benndorf, M. Strauss, W.-H. Schunk, R. Kraft, A. Otto, H. Böhm, J. Stahl, H. Drabsch, and H. Bielka. 1989. Molecular cloning, sequencing and expression in *Escherichia coli* of the 25-kDa growth-related protein of Ehrlich ascites tumor and its homology to mammalian stress proteins. *Eur. J. Biochem.* **179**: 209–213.

Gernold, M., U. Knauf, M. Gaestel, J. Stahl, and P.M. Kloetzel. 1993. Development and tissue-specific distribution of mouse small heat shock protein hsp25. *Dev. Genet.* **14**: 103–111.

Glaser, R.L. and J.T. Lis. 1990. Multiple, compensatory regulatory elements specify spermatocyte-specific expression of the *Drosophila melanogaster* hsp26 gene. *Mol. Cell. Biol.* **10**: 131–137.

Glaser, R.L., M.F. Wolfner, and J.T. Lis. 1986. Spatial and temporal pattern of HSP26 expression during normal development. *EMBO J.* **5**: 747–754.

Groenen, P.J.T.A., H. Bloemendal, and W.W. de Jong. 1992. The carboxy-terminal lysine of α-B-crystallin is an amine-donor substrate for tissue transglutaminase. *Eur. J. Biochem.* **205**: 671–674.

Grose, J.H., L. Caron, M. Lebel, and J. Landry. 1990. Prostacyclin secretion and specific intracellular protein phosphorylation. *Adv. Prostaglandin Thromboxane Leukotriene Res.* **21A**: 145–148.

Guesdon, F. and J. Saklatvala. 1991. Identification of a cytoplasmic protein kinase regulated by IL-1 that phosphorylates the small heat shock protein, hsp27. *J. Immunol.* **147**: 3402–3407.

Guesdon, F., N. Freshney, R.J. Waller, L. Rawlinson, and J. Saklatvala. 1993. Interleukin 1 and tumor necrosis factor stimulate two novel protein kinases that phosphorylate the heat shock protein hsp27 and beta-casein. *J. Biol. Chem.* **268**: 4236–4243.

Guy, G.R., J. Cairns, S.B. Ng, and Y.H. Tan. 1993. Inactivation of a redox-sensitive protein phosphatase during the early events of tumor necrosis factor/interleukin-1 signal transduction. *J. Biol. Chem.* **268**: 2141–2148.

Haas, C., U. Klein, and P.M. Kloetzel. 1990. Developmental expression of *Drosophila melanogaster* small heat shock proteins. *J. Cell Sci.* **96**: 413–418.

Hanash, S.M., J.R. Strahler, Y. Chan, R. Kuick, D. Teichroew, J.V. Neel, N. Hailat, D.R. Keim, J. Gratiot-Deans, D. Ungar, R. Melhem, X.X. Zhu, P. Andrews, F. Loottspeich, C. Eckerskorn, E. Chu, I. Ali, D.A. Fox, and B.C. Richardson. 1993. Data base analysis of protein expression pattern during T-cell ontogeny and activation. *Proc. Natl. Acad. Sci.* **90**: 3314–3318.

Hammerschlag, R., S. Maines, and M. Ando. 1989. Sensory glia from tadpoles but not adult bullfrogs synthesize heat shock-like proteins in vitro at non-heat shock temperature. *J. Neurosci. Res.* **23**: 416–424.

Handler, A.M. 1982. Ecdysteroid titers during pupal and adult development in *Drosophila melanogaster*. *Dev. Biol.* **93**: 73–82.

Heikkila, J.J., P.H. Krone, and N. Ovseneck. 1991. Regulation of heat shock gene expression during *Xenopus* development. In *Heat shock and development* (ed. L. Hightower and L. Nover), pp. 120–137. Springer-Verlag, Berlin.

Hepburn, A., D. Demolle, J.M. Boeynaems, W. Fiers, and J.E. Dumont. 1988. Rapid phosphorylation of a 27 kDa protein induced by tumor necrosis factor. *FEBS Lett.* **227:** 175–178.

Hoffman, E.P., S.L. Gerring, and K.J. Corces. 1987. The ovarian, ecdysterone and heat-shock-responsive promoters of *Drosophila melanogaster* hsp27 gene react differently to perturbation of DNA sequence. *Mol. Cell. Biol.* **7:** 973–981.

Hickey, E., S.E. Brandon, R. Potter, G. Stein, J. Stein, and L.A. Weber. 1986. Sequence and organization of genes encoding the human 27KDa heat shock protein. *Nucleic Acids Res.* **14:** 4127–4144.

Horwitz, J. 1992. Alpha-crystallin can function as a molecular chaperone. *Proc. Natl. Acad. Sci.* **89:** 10449–10453.

Huot, J., G. Roy, H. Lambert, P. Chrétien, and J. Landry. 1991. Increased survival after treatments with anticancer agents of Chinese hamster cells expressing the human M_r 27,000 heat shock protein. *Cancer Res.* **51:** 5245–5252.

Inaguma, Y., H. Shinohara, S. Goto, and K. Kato. 1992. Translocation and induction of alpha-B crystallin by heat shock in rat glioma (GA-1) cells. *Biochem. Biophys. Res. Commun.* **182:** 844–850.

Ingolia, T.D. and E. Craig. 1982. Four small *Drosophila* heat shock proteins are related to each other and to mammalian alpha-crystallin. *Proc. Natl. Acad. Sci.* **79:** 2360–2364.

Ireland, R.C. and E.M. Berger. 1982. Synthesis of the low molecular weight heat shock proteins stimulated by ecdysterone in a cultured *Drosophila* cell line. *Proc. Natl. Acad. Sci.* **79:** 855–859.

Ireland, R.C., E.M. Berger, K. Sirotkin, M.A. Yund, D. Osterburg, and J. Fristom. 1982. Ecdysterone induces the transcription of four heat shock genes in *Drosophila* S3 cells and imaginal discs. *Dev. Biol.* **93:** 498–507.

Iwaki, T., A. Kume-Iwaki, R.K.H. Liem, and J.E. Golman. 1989. αB-crystallin is expressed in non lenticular tissues and accumulates in Alexander's disease brain cells. *Cell* **57:** 71–78.

Jaattela, M., D. Wissing, P.A. Bauer, and G.C. Li. 1992. Major heat shock protein hsp70 protects tumor cells from tumor necrosis factor cytotoxicity. *EMBO J.* **11:** 3507–3512.

Jakob, U., M. Gaestel, K. Engel, and J. Buchner. 1993. Small heat shock proteins are molecular chaperones. *J. Biol. Chem.* **268:** 1517–1520.

Jones, D., D.K. Dixon, R.W. Graham, and E.P. Candido. 1989. Differential regulation of closely related members of the hsp16 gene family in *Caenorhabditis elegans*. *DNA* **8:** 481–490.

Kim, Y.J., J. Shuman, M. Sette, and A. Przybyla. 1984. Nuclear localization and phosphorylation of three 25-kilodalton rat stress proteins. *Mol. Cell. Biol.* **4:** 468–474.

Klemenz, R. and W.J. Gehring. 1986. Sequence requirement for expression of the *Drosophila melanogaster* heat shock protein hsp22 gene during heat shock and normal development. *Mol. Cell. Biol.* **6:** 2011–2019.

Klemenz, R., A.-C. Andres, E. Froehli, R. Schaefer, and A. Aoyama. 1993. Expression of the murine small heat shock protein hsp25 and αB-crystallin in the absence of stress, *J. Cell. Biol.* **120:** 639–645.

Klemenz, R., E. Froehli, R.H. Steiger, R. Schaefer, and A. Aoyama. 1991a. Alpha-crystallin is a small heat shock protein. *Proc. Natl. Acad. Sci.* **88:** 3652–3656.

Klemenz, R., E. Frohli, A. Aoyama, S. Hoffmann, R.J. Simpson, R.L. Moritz, and R. Schafer. 1991b. αB crystallin accumulation is a specific response to Ha-ras and v-mos oncogene expression in mouse NIH 3T3 fibroblasts. *Mol. Cell. Biol.* **11:** 803–812.

Knauf, U., H. Bielka, and M. Gaestel. 1992. Over-expression of the small heat shock protein, hsp25, inhibits growth of Ehrlich ascites tumor cells. *FEBS Lett.* **308:** 297–302.

Krone, P.H. and J.J. Heikkila. 1989. Expression of microinjected hsp 70/CAT and hsp 30/CAT chimeric genes in developing *Xenopus laevis* embryos. *Development* **106:** 271–281.

Kurtz, S., J. Rossi, L. Petko, and S. Lindquist. 1986. An ancient developmental induction: Heat-shock protein induced in sporulation and oogenesis. *Science* **231:** 1154–1157.

Lamb, N.J.C., A. Fernandez, J.R. Feramisco, and W.J. Welch. 1989. Modulation of vimentin containing intermediate filament distribution and phosphorylation in living fibroblasts by the cAMP-dependent protein kinase. *J. Cell Biol.* **108:** 2409–2422.

Landry, J., P. Chrétien, A. Laszlo, and H. Lambert. 1991. Phosphorylation of HSP27 during development and decay of thermotolerance in Chinese hamster cells. *J. Cell Physiol.* **147:** 93–101.

Landry, J., P. Chrétien, H. Lambert, E. Hickey, and L.A. Weber. 1989. Heat shock resistance conferred by expression of the human HSP27 gene in rodent cells. *J. Cell. Biol.* **109:** 7–15.

Landry, J., D. Bernier, P. Chrétien, L.M. Nicole, R.M. Tanguay, and N. Marceau. 1982. Synthesis and degradation of heat shock proteins during development and decay of thermotolerance. *Cancer Res.* **42:** 2457–2461.

Landry, J., H. Lambert, M. Zhou, J.N. Lavoie, E. Hickey, L.A. Weber, and C.W. Anderson. 1992. Human HSP27 is phosphorylated at serines 78 and 82 by heat shock and mitogen-activated kinases that recognize the same amino acid motif as S6 kinase II. *J. Biol. Chem.* **267:** 794–803.

Laszlo, A. 1992. The effects of hyperthermia on mammalian cell structure and function. *Cell Prolif.* **25:** 59–87.

Laszlo, A., T. Davidson, A. Hu, J. Landry, and J. Bedford. 1993. Putative determinants of the cellular response to hyperthermia. *Int. J. Radiat. Biol.* **63:** 569–581.

Lavoie, J., P. Chrétien, and J. Landry. 1990. Sequence of the Chinese hamster small heat shock protein HSP27. *Nucleic Acids Res.* **18:** 1637.

Lavoie, J.N., G. Gingras-Breton, R.M. Tanguay, and J. Landry. 1993a. Induction of Chinese hamster HSP27 gene expression in mouse cells confers resistance to heat shock. HSP27 stabilization of the microfilament organization. *J. Biol. Chem.* **268:** 3420–3429.

Lavoie, J.N., E. Hickey, L.A. Weber, and J. Landry. 1993b. Modulation of actin microfilament dynamics and fluid phase pinocytosis by phosphorylation of heat shock protein 27. *J. Biol. Chem.* **268:** 24210–24214.

Loomis, W.F. and S. Wheeler. 1982. Chromatin-associated heat shock proteins in *Dictyostelium*. *Dev. Biol.* **90:** 412–418.

Marin, R., J.-P. Valet, and R.M. Tanguay. 1993. Hsp23 and hsp26 exhibit distinct spatial and temporal patterns of constitutive expression in *Drosophila* adults. *Dev. Genet.* **14:** 112–118.

Mason, P.J., L.M.C. Hall, and J. Gausz. 1984. The expression of heat shock genes during normal development in *Drosophila melanogaster*. *Mol. Gen. Genet.* **194:** 73–78.

Mayer, R.J., J. Arnold, L. Laslo, M. Landon, and J. Lowe. 1991. Ubiquitin in health and disease. *Biochim. Biophys. Acta* **1089:** 141–157.

McGuire, S.E., S.A.W. Fuqua, S.L. Naylor, D.A. Helin-Davis, and W.L. McGuire. 1989. Chromosomal assignment of the human 27-kDa heat shock protein gene family. *Somat. Cell. Mol. Genet.* **15:** 167–171.

McKenzie, S.L., S. Henikoff, and M. Meselson. 1975. Localization of RNA from heat-induced polysomes at puff sites in *Drosophila melanogaster*. *Proc. Natl. Acad. Sci.* **72:**

1117–1121.

Mehlen, P., J. Briolay, L. Smith, C. Diaz-Latoud, N. Fabre, D. Pauli, and A.-P. Arrigo. 1993. Analysis of the resistance to heat and hydrogen peroxide stresses in COS cells transiently expressing wild type or deletion mutants of the *Drosophila* 27-kDa heat shock protein. *Eur. J. Biochem.* **215:** 277–284.

Mendelsohn, M.E., Y. Zhu, and S. O'Neill. 1991. The 29-kDa proteins phosphorylated in thrombin-activated human platelets are forms of the estrogen receptor-related 27-kDa heat shock protein. *Proc. Natl. Acad. Sci.* **88:** 11212–11216.

Merck, K.B., P.J. Groenen, C.E. Voorter, W.A. de Haard-Hoekman, J. Horwitz, H. Bloemendal, and W.W. de Jong. 1993. Structural and functional similarities of bovine alpha-crystallin and mouse small heat-shock protein. A family of chaperones. *J. Biol. Chem.* **268:** 1046–1052.

Mestril, R., P. Shiller, J. Amin, H. Klapper, A. Jayakumar, and R. Voellmy. 1986. Heat shock and ecdysterone activation of *Drosophila melanogaster* hsp23 gene: A sequence element implied in developmental regulation. *EMBO J.* **5:** 1667–1673.

Michishita, M., M. Satoh, M. Yamaguchi, K. Hirayoshi, M. Okuma, and K. Nagata. 1991. Phosphorylation of the stress protein hsp27 is an early event in murine myelomonocytic leukemic cell differentiation induced by leukemia inhibitory factor/D-factor. *Biochem. Biophys. Res. Commun.* **176:** 979–984.

Mirault, M.E., M. Goldschmidt-Clermant, L. Moran, A.-P. Arrigo, and A. Tissières. 1978. The effect of heat shock on gene expression in *Drosophila melanogaster. Cold Spring Harbor Symp. Quant. Biol.* **42:** 819–827.

Miron, T., M. Wilchek, and B. Geiger. 1988. Characterization of an inhibitor of actin polymerization in vinculin-rich fraction of turkey gizzard smooth muscle. *Eur. J. Biochem.* **178:** 543–553.

Miron, T., K. Vancompernolle, J. Vanderkerckhove, M. Wilchek, and B. Geiger. 1991. A 25-kD inhibitor of actin polymerization is a low molecular mass heat shock protein. *J. Cell. Biol.* **114:** 255–261.

Morimoto, R.I., A. Tissières, and C. Georgopoulos, eds. 1990. *Stress proteins in biology and medicine.* Cold Spring Harbor Laboratory Press, Cold Spring Harbor, New York.

Navarro, D., J.J. Cabrera, O. Falcon, P. Jimenez, A. Ruiz, R. Chirino, A. Lopez, J.F. Rivero, J.C. Diaz-Chico, and B.N. Diaz-Chico. 1989. Monoclonal antibody characterization of progesterone receptors, estrogen receptors and the stress-responsive protein of 27 kDa (SRP27) in human uterine leiomyoma. *J. Steroid Biochem.* **34:** 491–498.

Nene, V., D.V. Dunne, K.S. Johnson, D.W. Taylor, and J.S. Cordingley. 1986. Sequence and expression of a major egg antigen from *Schistosoma mansoni*: Homologies to heat shock proteins and α-crystallins. *Mol. Biochem. Parasitol.* **21:** 179–188.

Nerland, A.H., A.S. Mustafa, D. Sweetser, T. Godal, and R.A. Young. 1988. A protein antigen of *Mycobaterium leprae* is related to a family of small heat shock proteins. *J. Bacteriol.* **170:** 5919–5921.

Ngo, J.T., I. Klisak, R.A. Dubin, J. Piatigorsky, T. Mohandas, R.S. Sparkes, and J.B. Bateman. 1989. Assignment of the alpha B-crystallin gene to human chromosome 11. *Genomics* **5:** 665–669.

Nover, L. K.D. Sharf, and D. Neumann. 1989. Cytoplasmic heat shock granules are formed from precursor particles and are associated with a specific set of mRNAs. *Mol. Cell. Biol.* **9:** 1298–1308.

Orthwerth, B.J. and P.R. Olesen. 1992. Characterization of the elastase inhibitor properties of α-crystallin and the water insoluble fraction from bovine lens. *Exp. Cell Res.* **54:** 103–111.

Pauli, D. and A. Tissières. 1990. Developmental expression of the heat shock genes in

Drosophila melanogaster. In *Stress proteins in biology and medicine* (ed. R. Morimoto et al.), pp. 361–378. Cold Spring Harbor Laboratory Press, New York.

Pauli, D., A.-P. Arrigo, and A. Tissières. 1992. Heat shock response in *Drosophila. Experientia* **48:** 623–629.

Pauli, D., C.-H. Tonka, A. Tissières, and A.-P. Arrigo. 1990. Tissue specific expression of the heat shock protein hsp27 during *Drosophila melanogaster* development. *J. Cell Biol.* **111:** 817–828.

Pauli, D., A.-P. Arrigo, J. Vasquez, C.-H. Tonka, and A. Tissières. 1989. Expression of the small heat shock genes during *Drosophila* development: Comparison of the accumulation of HSP23 and hsp27 mRNAs and polypeptides. *Genome* **31:** 671–676.

Petersen, N.S., G. Moeller, and H.K. Mitchell. 1979. Genetic mapping of the coding regions for three heat shock proteins in *Drosophila melanogaster. Genetics* **92:** 891–902.

Petko, L. and S. Lindquist. 1986. Hsp26 is not required for growth at high temperature, not for thermotolerance, spore development or germination. *Cell* **45:** 885–894.

Plesorfsky-Vig, N. and R. Brambl. 1990. Gene sequence analysis of hsp30, a small heat shock protein of *Neurospora crassa* which associate with mitochondria. *J. Biol. Chem.* **265:** 15432–15440.

Puy, L.A., G. Lo Castro, J.E. Olcese, H.O. Lotfi, H.R. Brandi, and D.R. Ciocca. 1989. Analysis of a 24-kilodalton (KD) protein in the human uterine cervix during abnormal growth. *Cancer* **64:** 1067–1073.

Quax-Jeuken, Y., W. Quax, G.L.N. van Rens, P. Meera Khan, and H. Bloemendal. 1985. Assignment of the human αA-crystallin gene (CRYA1) to chromosome 21. *Cytogenet. Cell Biol.* **40:** 727–728.

Raschke, E.G., G. Baumann, and F. Schöffl. 1988. Nucleotide sequence analysis of soybean small heat shock protein genes belonging to two different multigenic families. *J. Mol. Biol.* **199:** 549–557.

Regazzi, R., U. Eppenberger, and D. Fabbro. 1988. The 27,000 dalton stress proteins are phosphorylated by protein kinase C during the tumor promoter-mediated growth inhibition of human mammary carcinoma cells. *Biochem. Biophys. Res. Commun.* **152:** 62–68.

Riddihough, G. and H.R.B. Pelham. 1986. Activation of the *Drosophila* hsp27 promoter by heat shock and by ecdysome involves independent and remote regulatory sequences. *EMBO J.* **5:** 1653–1658.

———. 1987. An ecdysone response element in the *Drosophila* hsp27 promoter. *EMBO J.* **6:** 3729–3734.

Ridley, A.J. and A. Hall. 1992. The small GTP-binding protein rho regulates the assembly of focal adhesions and actin stress fibers in response to growth factors. *Cell* **70:** 389–399.

Ridley, A.J., H.F. Paterson, C.L. Johnston, D. Diekmann, and A. Hall. 1992. The small GTP-binding protein rac regulates growth factor-induced membrane ruffling. *Cell* **70:** 401–410.

Robaye, B., A. Hepburn, R. Lecocq, W. Fiers, J.M. Boeynaems, and J.E. Dumont. 1989. Tumor necrosis factor-alpha induces the phosphorylation of 28kDa stress proteins in endothelial cells: Possible role in protection against cytotoxicity. *Biochem. Biophys. Res. Commun.* **163:** 301–308.

Rollet, E. and M. Best-Belpomme. 1986. HSP26 and 27 are phosphorylated in response to heat and ecdysterone in *Drosophila melanogaster* cells. *Biochem. Biophys. Res. Commun.* **141:** 426–433.

Rollet, E., J.N. Lavoie, J. Landry, and R.M. Tanguay. 1992. Expression of *Drosophila*'s

27kDa heat shock protein into rodent cells confers thermal resistance. *Biochem. Biophys. Res. Commun.* **185:** 116–120.

Rossi, J. and S. Lindquist. 1989. The intracellular location of yeast heat shock protein 26 varies with metabolism. *J. Cell Biol.* **108:** 425–439.

Saklatvala, J., P. Kaur, and F. Guesdon. 1991. Phosphorylation of the small heat-shock protein is regulated by interleukin 1, tumour necrosis factor, growth factors, bradykinin and ATP. *Biochem. J.* **277:** 635–642.

Schutze, S., P. Scheurich, K. Pfizenmaier, and M. Kronke. 1989. Tumor necrosis factor signal transduction. Tissue-specific serine phosphorylation of a 26-kDa cytosolic protein. *J. Biol. Chem.* **264:** 3562–3567.

Seizen, R.J., J.G. Bindes, and H.J. Hoenders. 1978. The quaternary structure of bovine alpha-crystallin. *Eur. J. Biochem.* **91:** 387–396.

———. 1980. The quaternary structure of bovine α-crystallin. *Eur. J. Biochem.* **111:** 435–444.

Seymour, L., W.R. Bezwoda, and K. Meyer. 1990a. Tumor factors predicting for prognosis in metastatic breast cancer. The presence of P24 predicts for response to treatment and duration of survival. *Cancer* **66:** 2390–2394.

Seymour, L., W.R. Bezwoda, K. Meyer, and C. Behr. 1990b. Detection of P24 protein in human breast cancer: Influence of receptor status and oestrogen exposure. *Br. J. Cancer* **61:** 886–890.

Shakoori, A.R., A.M. Oberdorf, T.A. Owen, L.A. Weber, E. Hickey, J.L. Stein, J.B. Lian, and G.S. Stein. 1992. Expression of heat shock genes during differentiation of mammalian osteoblasts and promyelocytic leukemia cells. *J. Cell Biochem.* **48:** 277–287.

Shibanuma, M., T. Kuroki, and K. Nose. 1991. Release of H_2O_2 and phosphorylation of a 30 kilodalton protein as early responses of cell cycle-dependent inhibition of DNA synthesis by transforming growth factor β1. *Cell Growth Differ.* **2:** 583–591.

———. 1992. Cell cycle dependent phosphorylation of HSP28 by TGFβ1 and H_2O_2 in normal mouse osteoblastic cells (MC3T3-E1), but not in their Ras-transformants. *Biochem. Biophys. Res. Commun.* **187:** 1418–1425.

Sirotkin, K. and N. Davidson. 1982. Developmentally regulated transcription from *Drosophila melanogaster* site 67B. *Dev. Biol.* **89:** 196–210.

Southgate, R., A. Ayme, and R. Voellmy. 1983. Nucleotide sequence analysis of the *Drosophila* small heat heat shock gene cluster at locus 67B. *J. Mol. Biol.* **165:** 35–57.

Southgate, R., M.E. Mirault, A. Ayme, and A. Tissières. 1985. Organization, sequences, and induction of heat shock genes. In *Changes in eukaryotic gene expression in response to environmental stress* (ed. B.G. Atkinson and D.B. Walden), pp. 3–30. Academic Press, New York.

Spector, N.L., C. Ryan, W. Samson, L.M. Nadler, and A.-P. Arrigo. 1993. Hsp28 is a unique marker of growth arrest during macrophage differentiation of HL-60 cells. *J. Cell. Physiol.* **156:** 619–625.

Spector, N.L., W. Samson, C. Ryan, J. Gribben, W. Urba, W.J. Welch, and L.M. Nadler. 1992. Growth arrest of human B lymphocytes is accompanied by induction of the low molecular weight mammalian heat shock protein (Hsp28). *J. Immunol.* **148:** 1668–1673.

Spector, N.L., P. Mehlen, C. Ryan, L. Lys, W. Samson, H. Levine, L.M. Nadler, N. Fabre, and A.-P. Arrigo. 1994. Regulation of the 28 KDa heat shock protein by retinoic acid during differentiation of human leukemic HL-60. *FEBS Lett.* (in press).

Srinivasan, A.N., C.N. Nagineni, and S.P. Bhat. 1992. αA-crystallin is expressed in nonocular tissues. *J. Biol. Chem.* **267:** 23337–23341.

Stahl, J., A.M. Wobus, S. Ihrig, G. Lutsch, and H. Bielka. 1992. The small heat shock protein hsp25 is accumulated in P19 embryonal carcinoma cells and embryonic stem cells of line BLC6 during differentiation. *Differentiation* **51**: 33–37.

Stokoe, D., K. Engel, D.G. Campbell, P. Cohen, and M. Gaestel. 1992a. Identification of MAPKAP kinase 2 as a major enzyme responsible for the phosphorylation of the small mammalian heat shock proteins. *FEBS Lett.* **313**: 307–313.

Stokoe, D., D.G. Campbell, S. Nakielny, H. Hidaka, S.J. Leevers, C. Marshall, and P. Cohen. 1992b. MAPKAP kinase-2: A novel protein kinase activated by mitogen-activated protein kinase. *EMBO J.* **11**: 3985–3994.

Strahler, J.R., R. Kuick, and S.M. Hanash. 1991. Diminished phosphorylation of a heat shock protein (HSP 27) in infant acute lymphoblastic leukemia. *Biochem. Biophys. Res. Commun.* **175**: 134–142.

Susek, R.E. and S.L. Lindquist. 1989. hsp26 of *Saccharomyces cerevisiae* is related to the superfamily of small heat shock proteins but is without a demonstrable function. *Mol. Cell Biol.* **9**: 5265–5271.

———. 1990. Transcriptional derepression of the *Saccharomyces cerevisiae* HSP26 gene during heat shock. *Mol. Cell Biol.* **10**: 6362–6373.

Tanguay, R.M. 1989. Localized expression of a small heat shock protein, hsp23, in specific cells of the central nervous system during early embryogenesis in *Drosophila*. *J. Cell. Biol.* **109**: 155a.

Tanguay, R.M., Y. Wu, and E.W. Khandjian. 1993. Tissue-specific expression of heat shock proteins of the mouse in the absence of stress. *Dev. Genet.* **14**: 112–118.

Tanguay, R.M., J.-L. Duband, F. Lettre, J.-P. Valet, A.-P. Arrigo, and L. Nicole. 1985. Biochemical and immunocytochemical localization of heat shock proteins *Drosophila* culture cells. *Ann. N.Y. Acad. Sci.* **455**: 711–714.

Tardieu, A. and M. Delaye. 1988. Eye lens proteins and transparency: From light transmission theory to solution X-ray structural analysis. *Annu. Rev. Byophys. Chem.* **17**: 47–70.

Tardieu, A.D., D. Laporte, P. Licino, B. Krop, and M. Delaye. 1986. Calf lens α-crystallin quaternary structure: A three layer tetrahedral model. *J. Mol. Biol.* **192**: 711–724.

Têtu, B., B. Lacasse, H.L. Bouchard, R. Lagacé, J. Huot, and J. Landry. 1992. Prognostic influence of HSP-27 expression in malignant fibrous histiocytoma: A clinicopathological and immunohistochemical study. *Cancer Res.* **52**: 2325–2328.

Thomas, S.R. and J.A. Lengyel. 1986. Ecdysteroid-regulated heat shock gene expression during *Drosophila melanogaster* development. *Dev. Biol.* **115**: 434–438.

Thor, A., C. Benz, D. Moore, E. Goldman, S. Edgerton, J. Landry, L. Schwartz, B. Mayall, E. Hickey, and L.A. Weber. 1991. Stress response protein (srp-27) determination in primary human breast carcinomas: Clinical, histologic, and prognostic correlations. *J. Natl. Cancer Inst.* **83**: 170–178.

Verbon, A., A. Hartskeerl, A. Schuitma, A.H.J. Kolk, D.B. Young, and R. Lathigra. 1992. The 14,000-molecular-weight antigen of *Mycobacterium tuberculosis* is related to the alpha-crystallin family of low molecular weight heat shock proteins. *J. Bacteriol.* **174**: 1352–1359.

Vierling, E., R.T. Nagao, E. Derocher, and L.M. Harris. 1988. A heat shock protein localized to chloroplasts is a member of an eukaryotic superfamilly of heat shock proteins. *EMBO J.* **7**: 575–581.

Voellmy, R., M. Goldschmidt-Clermont, R. Southgate, A. Tissieres, R. Levis, and W.J. Gehring. 1981. A DNA segment isolated from chromosomal site 67B in *D. melanogaster* contains four closely linked heat shock genes. *Cell* **45**: 185–193.

Vincent, M. and R.M. Tanguay. 1979. Heat shock induced proteins present in the cell nucleus of *Chironomus tentans* salivary gland. *Nature* **281:** 501–503.

————. 1982. Different intracellular distribution of heat shock and arsenite induced proteins in *Drosophila* KC cells. *J. Mol. Biol.* **162:** 365–378.

Vitek, M. and E. Berger. 1984. Steroid and high temperature induction of the small protein genes in *Drosophila. J. Mol. Biol.* **178:** 173–189.

Voorter, C.E.M., W.A. de Haard-Hoekman, E.S. Roersma, H.E. Meyer, H. Bloemendal, and W.E. de Jong. 1989. The *in vivo* phosphorylation sites of bovine αB-crystallin. *FEBS Lett.* **259:** 50–52.

Wadsworth, S., E.A. Craig, and B.J. McCarthy. 1980. Genes for three *Drosophila* heat shock induced proteins at a single locus. *Proc. Natl. Acad. Sci.* **77:** 2134–2137.

Walsh, D., K. Li, C. Crowther, D. Marsh, and M. Edwards. 1991. Thermotolerance and heat shock response during early development of the mammalian embryo. In *Heat shock and development* (ed. L. Hightower and L. Nover), pp 58–70. Springer-Verlag, Berlin.

Walsh, M.T., A.C. Sen, and B. Chakrabarti. 1991. Micellar subunit assembly in a three-layer model of oligomeric α-crystallin. *J. Biol. Chem.* **266:** 20079–20084.

Welch, W.J. 1985. Phorbol ester, calcium ionophore, or serum added to quiescent rat embryo fibroblast cells result in the elevated phosphorylation of two 28,000 dalton mammalian stress proteins. *J. Biol. Chem.* **260:** 3058–3062.

Wistow, G. 1985. Domain structure and evolution in alpha-crystallins and small heat shock proteins. *FEBS Lett.* **181:** 1–6.

Yost, H.J., R.B. Petersen, and S. Lindquist. 1990. Posttranslational regulation of heat shock protein synthesis in *Drosophila*. In *Stress proteins in biology and medicine* (ed. R. Morimoto et al.), pp. 379–409. Cold Spring Harbor Laboratory Press, Cold Spring Harbor, New York.

Zantema, A., E. de Jong, R. Lardenoije, and A.J. van der Eb. 1989. The expression of heat shock protein hsp27 and a complexed 22-kilodalton protein is inversely correlated with oncogenicity of adenovirus-transformed cells. *J. Virol.* **63:** 3368–3375.

Zimmerman, J.L., W. Petri, and M. Meselson. 1983. Accumulation of a specific subset of *Drosophila melanogaster* mRNAs in normal development without heat shock. *Cell* **32:** 1161–1170.

Zhou, M., H. Lambert, and J. Landry. 1993. Transient activation of a distinct serine protein kinase is responsible for 27-kDa heat shock protein phosphorylation in mitogen-stimulated and heat-shocked cells. *J. Biol. Chem.* **268:** 35–43.

15

Structure and Regulation of Heat Shock Gene Promoters

Mary Fernandes, Thomas O'Brien,
and John T. Lis
Section of Biochemistry
Molecular and Cell Biology,
Cornell University, Ithaca, New York 14853

I. INTRODUCTION

The heat shock gene system of *Drosophila* has several features that make it an attractive model to investigate the transcriptional activation of genes. First, the major heat shock genes are induced approximately 200-fold by heat shock (Lis et al. 1981). This, coupled with the repression of most preexisting transcription, simplifies the detection of RNA and protein products from these genes. Second, the induction is mediated by proteins that are present in the uninduced cells (Zimarino and Wu 1987), and the immediacy of the response facilitates kinetic investigations of the mechanism of activation of these genes (O'Brien and Lis 1993). Third, investigations by many laboratories provide a strong foundation of information on the promoter sequences of these genes (Bienz and Pelham 1987), their chromatin structure (Eissenberg et al. 1985), and DNA sequence elements and protein factors that participate in the regulation of their transcription (Lis and Wu 1993). Finally, the compatibility of components of heat shock promoters with those of other unrelated genes indicates that at least some of the principles identified in the study of transcription of heat shock genes are likely to have general relevance to transcriptional regulation (Garabedian et al. 1986; Fischer et al. 1988; Kraus et al. 1988; Martin et al. 1989). In this chapter, we discuss the heat shock factor (HSF) that modulates transcription through its binding to heat shock elements (HSEs). This interaction and its interplay with other factors and RNA polymerase are considered in the context of heat shock promoters that are primed for rapid induction.

II. HEAT SHOCK FACTOR

HSF was identified in heat-shocked *Drosophila melanogaster* nuclear extracts as a protein component that could specifically bind to the regulatory site of the *hsp70* gene (Parker and Topol 1984; Wu 1984b) and as a purified fraction that could promote transcription in an in vitro cell-free system (Parker and Topol 1984). This transcription factor activity was subsequently purified from yeast (Sorger and Pelham 1987; Wiederrecht et al. 1988), *Drosophila* (Wu et al. 1987), and cultured human cells (Goldenberg et al. 1988). The genes encoding HSF from these and other species have been cloned and characterized (Sorger and Pelham 1988; Wiederrecht et al. 1988; Clos et al. 1990; Scharf et al. 1990; Jakobsen and Pelham 1991; Rabindran et al. 1991; Sarge et al. 1991; Schuetz et al. 1991; Gallo et al. 1993; Nakai and Morimoto 1993). In the yeasts *Saccharomyces cerevisiae*, *Kluyveromyces lactis*, and *Schizosaccharomyces pombe*, HSF is encoded by a unique, single-copy gene. Although only one gene encoding HSF has been isolated from *Drosophila*, all other higher eukaryotes possess multiple HSFs (two for mouse and humans and three for chicken and tomato).

DNA-binding studies with HSFs of *Drosophila* and *S. cerevisiae* (DmHSF and ScHSF, respectively) have revealed that HSF does not fall into the known categories of DNA-binding proteins. This conclusion is supported by several observations. First, a comparison of footprinting and gel shift assays on synthetic binding sites is consistent with DmHSF binding as a trimer (Perisic et al. 1989). Second, chemical cross-linking of ScHSF or DmHSF in solution produces trimer-sized products with high efficiency (Perisic et al. 1989; Sorger and Nelson 1989). Third, four oligomeric complexes with the ratio 1:3:3:1 are formed on cosynthesis of a long and a short derivative of ScHSF (Sorger and Nelson 1989).

The trimeric structure of DmHSF is further supported by gel filtration and glycerol gradient sedimentation analyses (Westwood and Wu 1993). Moreover, both the monomer and the trimer were found to have unusually high frictional ratios, indicating that they are asymmetrically shaped. This accounts for the large apparent size seen on native gels that led to early overestimates of the size of the native protein (Clos et al. 1990). Although the multimeric state is conserved, the monomeric size of HSF can vary between species. The monomeric sizes of HSFs from *S. cere-*

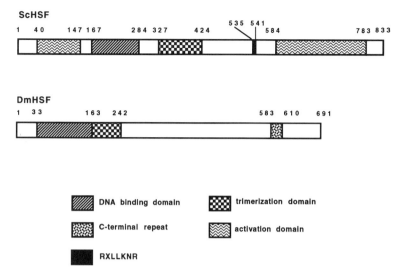

Figure 1 Schematic representation of ScHSF and DmHSF.

visiae, Drosophila, and humans are 93, 77, and 57 kD, respectively. HSFs isolated from all organisms show strong sequence similarity only in their DNA-binding and multimerization domains. A comparison of the primary amino acid sequences of DmHSF and ScHSF reveals that the two proteins are extremely divergent, except for two major regions of conservation at the amino terminus and two minor regions of conservation further interior (Clos et al. 1990). The minimal DNA-binding domains of DmHSF and ScHSF contain the region most conserved between the two species. Although this suggests the presence of a common motif for DNA binding, the only discernible feature is a short but good match to the putative recognition helix (helix 4.2) of bacterial σ factors (Clos et al. 1990). The finding that the Sfl1 (suppressor of flocculation) protein in *S. cerevisiae* possesses a domain highly homologous to the DNA-binding domain of ScHSF suggests that HSF may be a member of a unique family of DNA-binding domains (Fujita et al. 1989). Alternatively, it may be structurally related to a known DNA-recognition motif(s) without having any apparent sequence similarity.

The trimerization domain of HSF lies carboxy-terminal to the DNA-binding domain (Fig. 1) and includes the second-most conserved region. The conserved residues in the trimerization domain define arrays of hydrophobic heptad repeats or "leucine zipper" coiled-coil motifs similar to those found in the dimerization domains of bZIP transcription factors (Sorger and Nelson 1989; Clos et al. 1990). A helical wheel diagram of the heptad repeats shows hydrophobic residues juxtaposed at four positions on one helical face. These have the potential to form strong hydrophobic associations with the hydrophobic faces of other α helices. Physicochemical studies of the trimerization domains of HSFs from *S. cerevisiae* and *K. lactis* (ScHSF and KlHSF, respectively) indicate that they form stable and predominantly α-helical homotrimers in solution, consistent with a triple-stranded, α-helical coiled-coil structure (Peteranderl and Nelson 1992). Such a multimeric structure is analogous to that of the hemagglutinin protein of influenza virus, which has been defined at 3 Å resolution (Wilson et al. 1981), but is unique for a DNA-binding protein. Structurally, this presents a challenging problem in that three DNA-binding domains with presumably threefold rotational symmetry must bind to the HSE sequence that has twofold rotational symmetry. In support of the functional importance of the leucine-zipper motifs is their remarkable degree of conservation, as evidenced by the presence of functionally similar hydrophobic amino acids at the appropriate positions in the trimerization domains of all HSFs. HSFs from *S. cerevisiae* and *K. lactis* are constitutively trimeric and bound to DNA under nonheat shock conditions, whereas the metazoan HSFs and HSF from *S. pombe* exist as monomers and trimerize only upon heat shock, thereby exhibiting heat-inducible DNA binding. Metazoans with multiple HSFs have both constitutive and inducible forms of HSFs.

In addition to the hydrophobic repeats found in the trimerization domain, another array of hydrophobic heptad repeats is present in the carboxy-terminal region of HSFs that exhibit heat-inducible DNA binding but not in the constitutively trimeric ScHSF and KlHSF. This repeat has been speculated to be involved in the formation of inactive HSF or in the suppression of trimer assembly under nonheat shock conditions. Nonconservative substitutions of the hydrophobic residues within the carboxy-terminal heptad repeat of human HSF1 led to constitutive trimer formation and high-affinity DNA-binding activity, suggesting that this region is involved in the suppression of trimer assembly, possibly by means of an intramolecular coiled-coil interaction between amino-terminal and carboxy-terminal motifs. Deletion mapping indicates that the carboxy-terminal motif may be part of a larger domain involved in negative regulation (Rabindran et al. 1993).

The two marginally conserved regions of ScHSF and DmHSF have been speculated to be involved in transcriptional activation. These regions are rich in serines and threonines that could function as potential phosphorylation sites during the process of transcriptional activation. However, activating domains have so far only been localized for ScHSF and KlHSF (Nieto-Sotelo et al. 1990; Sorger 1990). These domains were identified by their ability to function when fused to a heterologous DNA-binding domain. Interestingly, these activators are unrelated in sequence but functionally interchangeable. In addition, an amino-terminal *trans*-activator has been identified for ScHSF (Sorger 1990). A central region of ScHSF that includes the DNA-binding and trimerization domains and a short carboxy-terminal heptapeptide motif (RXLLKNR) has been shown to regulate the activating functions negatively (Nieto-Sotelo et al. 1990; Jakobsen and Pelham 1991). Point mutations in these "repressor" domains unmask the *trans*-activation functions (Bonner et al. 1992). The transcriptional activity of HSF is presumably modulated by controlling the interplay between these activation and repressing domains and the binding of the activated trimer to upstream binding sites. For further discussion of HSF and the regulation of its localization and multimeric state, see chapters by Wu et al. and Morimoto et al. (this volume).

THE HEAT SHOCK ELEMENT

The binding of activated HSF to *cis*-acting, upstream DNA sequence elements commonly termed heat shock elements (HSEs), specifies the transcriptional activation of heat shock genes. The HSE was first identified as a sequence required for heat inducibility of the *Drosophila hsp70* gene (Mirault et al. 1982; Pelham 1982). Further analyses led to the definition of an HSE as a repeating array of the 5-bp sequence 5'-nGAAn-3' ($\rightarrow$), where each repeat is inverted relative to the immediately adjacent repeat (Amin et al. 1988; Xiao and Lis 1988). All heat shock promoters, from different genes and diverse organisms, have the same 5-bp building block. The number of 5-bp units in a functional HSE can vary but usually ranges from three to six. The number of HSEs associated with different heat shock genes can also vary, as can the distance between these HSEs. For example, the promoter region of the *Drosophila hsp70* gene has four HSEs numbered I, II, III, and IV, each containing three or four 5-bp units (Fig. 2). HSE I and HSE II are three nucleotides apart and separated by 85 nucleotides from HSE III and HSE IV. The *hsp83* gene, on the other hand, has a single HSE containing eight 5-bp units. The location of HSEs within the promoter is another variable and may range from about 40 bases upstream of the transcriptional start site as for the *hsp70* gene to approximately 270 bases in case of the *hsp27* gene. HSEs can start with either a 5'-nGAAn repeat or with its complement nTTCn. In addition, an HSE can tolerate a 5-bp insertion between repeating units, provided the phase of the repeats is maintained (Amin et al. 1988).

The consensus sequence nGAAn was defined by sequence comparison of functional HSEs in *Drosophila* heat shock genes (Lis et al. 1990). The G at position 2 is absolutely conserved, with base substitu-

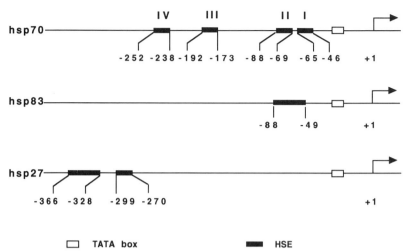

Figure 2 Heat shock gene promoter and regulatory regions of the *Drosophila hsp70*, *hsp83*, and *hsp27* genes.

tions here abolishing heat-induced expression (Xiao and Lis 1988), whereas the As at positions 3 and 4 are less conserved since base substitutions at these sites occur in some functional HSEs. Recent analysis of binding affinities has shown that the base at position 1 is nonrandom and that an A is preferred at this position (Fernandes et al. 1994). Substitution of this A with a T dramatically reduces the binding of both DmHSF and ScHSF. This substitution was also shown to greatly reduce heat-induced expression of the *Drosophila hsp70* gene. The base at position 5 is least conserved, and base changes here have minor effects on the binding affinity of HSF (Fernandes et al. 1994) and transcriptional activity (Xiao and Lis 1988). In general, however, the nucleotide sequence of the HSE has been evolutionarily conserved among eukaryotes. The heat shock genes of organisms as diverse as ciliated protozoa (*Tetrahymena*), nematodes (*Caenorhabditis elegans*), yeasts (*K. lactis, S. cerevisiae*), insects (*Drosophila*), amphibians (*Xenopus*), chickens, rodents (mice and rats), and humans all utilize the same basic unit in promoter elements controlling the heat shock response (Mirault et al. 1982; Pelham 1982; Craig and Jakobsen 1984; Bienz and Pelham 1987; Kingston et al. 1987; Amin et al. 1988; Neves et al. 1988). Most of the results obtained originally with *Drosophila* and *Saccharomyces* HSFs have been corroborated by experiments with HSFs from other organisms. Thus, analysis of the HSEs in the human *hsp70* gene has revealed the importance of the conserved central core of the 5-bp unit, in addition to the significance of the A at position 1 (Cunniff and Morgan 1993).

It is worth noting that HSEs mediate the response not only to heat shock, but also to some other forms of stress such as exposure to heavy metal ions or amino acid analogs. In addition, some metazoan HSEs may be involved in gene expression under nonstress conditions. One such example is the human *hsp70* gene, which is expressed during erythroid cell differentiation. This process can be induced by hemin, and the hemin-induced HSF has been shown to interact with HSEs of the *hsp70* gene in these cells (Theodorakis et al. 1989). In-vitro-binding experiments have shown that the binding of HSF to HSEs can be induced by agents that affect protein conformation, such as urea, nonionic detergents, hydrogen ions, and antibody to HSF (Mosser et al. 1990; Zimarino et al. 1990). Binding of HSF can also be activated by Ca^{++}, Mn^{++}, and La^{++} (Mosser et al. 1990) and by cadmium, azetidine, dinitrophenol, sodium salicylate, and sodium arsenite (Mosser et al. 1988).

IV. HSF-HSE INTERACTIONS

The native HSF trimer displays a remarkable flexibility in its ability to interact with HSEs containing different numbers and arrangements of 5-bp units. The smallest array that shows detectable binding of DmHSF in vitro contains two 5-bp units in either a head-to-head (nGAAnnTTCn; →←) or tail-to-tail (←→) configuration. Interestingly, HSF binds to both these sequences with similar affinity. High-resolution footprinting assays using the chemical cleavage agent methidiumpropyl-EDTA·Fe(II) have shown that the bound HSF covers and is centered similarly on both of these sequences. DNase I footprinting assays have shown that as the length of an HSE containing two 5-bp units is increased in 5-bp installments, there is a proportional (~5-bp) increase in the footprint produced by HSF (Perisic et al. 1989). Taken together, these results suggest that the 5-bp unit is a site of interaction with each DNA-binding domain of HSF and that detectable binding is provided by two adjacent binding sites in opposite orientation (Fig. 3) (Perisic et al. 1989). A complete, minimal binding site for trimeric HSF is provided by three 5-bp units, i.e., →←→ or ←→←.

The binding affinity of HSF has been shown to increase as the number of 5-bp units within an HSE increases. Thus, HSF binds cooperatively to 5-bp units (Topol et al. 1985; Xiao et al. 1991). This cooperativity leads to particularly tight binding when the number of 5-bp units in an array is six or more. Such HSF-HSE complexes dissociate with a half-life of greater than 48 hours (Xiao et al. 1991). Cooperative binding of HSF is even more pronounced at full heat shock temperatures, suggesting that longer arrays of 5-bp units have a particularly striking advantage

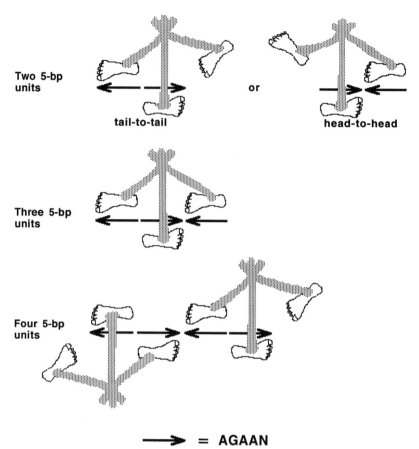

➤ = AGAAN

Figure 3 Model for interaction of HSF trimers with different arrays of 5-bp units. Arrows designate the 5-bp units and their relative orientation. The foot represents the DNA-binding domain of HSF, and the legs join the subunits of the trimer. (Adapted from Lis et al. 1990.)

in sequestering HSF at these temperatures. Since different members of the heat shock gene family have varying numbers of 5-bp units, the resulting range in affinities for HSF may account in part for their differential response to heat shock (Lindquist 1980). A truncated DmHSF containing only the DNA-binding and trimerization domains (from amino acid 33 to 252) has been shown to bind HSEs cooperatively (M. Fernandes and J.T. Lis, unpubl.), suggesting that these domains are sufficient for cooperative protein-protein interactions between trimers.

Methylation interference experiments have indicated that the contacts of both DmHSF and ScHSF with the DNA of the HSE are primarily in the major groove (Perisic et al. 1989; Fernandes et al. 1994). DmHSF has also been shown to cause a modest bend in the DNA of the HSE (Shuey and Parker 1986; M. Fernandes and J.T. Lis, unpubl.). The HSF-HSE interaction appears to be strikingly similar in all organisms. In support of this are in vitro interference and footprinting experiments and in vivo transfection and footprinting experiments that reveal a similar modular recognition of the 5-bp unit by HSFs from different species and a remarkable consistency in the topography of protein-DNA contacts within the motif (Kingston et al. 1987; Morgan et al. 1987; Wiederrecht et al. 1987; Mosser et al. 1988; Perisic et al. 1989; Williams et al. 1990; Abravaya et al. 1991; Cunniff et al. 1991; Gallo et al. 1991; Sarge et al. 1991).

V. UNINDUCED HEAT SHOCK GENES ARE PRIMED FOR A RAPID RESPONSE

The promoters of *Drosophila* heat shock genes are primed for a rapid response to a heat shock stimulus. The chromatin structures of the uninduced promoters are in an open configuration as evidenced by their nuclease hypersensitivity (Costlow and Lis 1984; Wu 1980). At least two different transcription factors are bound to the uninduced promoters, GAGA factor and TBP (TATA-binding protein) (Wu 1984a; Thomas and Elgin 1988; Giardina et al. 1992). GAGA factor is a transcription factor that has been shown to bind to alternating GA-CT sequences upstream of many *Drosophila* genes (Biggin and Tjian 1988; Soeller et al. 1988; Thomas and Elgin 1988; Gilmour et al. 1989). TBP is the DNA-sequence-specific binding component of TFIID, the general transcription factor complex that binds to the TATA region approximately 30 bp upstream of the site of transcript initiation in many RNA polymerase II promoters (Hoey et al. 1990).

RNA polymerase II is also associated with the uninduced heat shock promoter. In vivo UV cross-linking has demonstrated that the uninduced *hsp70* promoter has approximately one RNA polymerase II complex located in the region between nucleotides –12 and +65 relative to the transcriptional start site (+1) (Gilmour and Lis 1986). Nuclear run-on assays showed that this polymerase is transcriptionally engaged but paused after synthesizing about 25 nucleotides (Rougvie and Lis 1988). This paused polymerase was defined at higher resolution to be between +17 and +37 by using the DNA-modifying reagent $KMnO_4$ to map the transcription bubble (Giardina et al. 1992). A similar interval was derived by determining the length of the short RNAs associated with paused polymerase (Rasmussen and Lis 1993). Within this interval are two preferred positions, separated by approximately one turn of the DNA helix. The distribution of paused polymerase on the uninduced *hsp26* and *hsp27* genes is similar to that found for the *hsp70* gene but begins about 10 bp farther into the gene (Giardina et al. 1992; Rasmussen and Lis 1993).

A region upstream of the *hsp70* TATA box, which includes GAGA sequences and the HSEs, can program the formation of a paused polymerase on a nonheat shock gene promoter that normally displays no detectable pausing (Lee et al. 1992). Although mutations in the GAGA element reduce the level of paused polymerase severalfold, mutations in the HSE have little effect on generating the pause. In addition, sequences around the start site can also have a role in generating paused RNA polymerase. Alterations to the hybrid *hsp70* promoter that reduce pausing also reduce the heat inducibility of the promoter, indicating that the formation of paused polymerase is an important intermediate in the pathway to full transcriptional activation (Lee et al. 1992).

Mutations in upstream GAGA repeats also affect the nuclease hypersensitivity of the promoter region as seen in transgenic *Drosophila* fly lines that contain variants of the *hsp26* gene promoter (Lu et al. 1992, 1993). In contrast, mutations in the HSEs have little consequence on the nuclease hypersensitivity of heat shock promoters (Lu et al. 1993). The binding of GAGA factor to GAGA elements appears to be critical for the interruption of normal nucleosome packaging of at least some heat shock promoters, and it may act by directly displacing nucleosomes from the heat shock promoter. Since TBP is unable to bind to a TATA sequence covered by a nucleosome in vitro (Workman and Buchman 1993), the action of GAGA factor may be a necessary first step in opening the promoter. This "open" promoter may in turn allow entry and pausing of RNA polymerase II.

What features of RNA polymerase II may be involved in generating and maintaining this pause? The large subunit of RNA polymerase II contains a carboxy-terminal domain (CTD) that is essential in vivo (Nonet et al. 1987; Allison et al. 1988; Bartolomei et al. 1988; Zehring et al. 1988). The CTD consists of a heptapeptide repeat that is repeated 42 times in *Drosophila* (Allison et al. 1988; Zehring et al. 1988) and found in either an unphosphorylated or hyperphosphorylated form (Cadena and Dahmus 1987; Kolodziej et al. 1990). In vitro, the unphosphorylated form preferentially enters into the pre-initiation complex (Laybourn and Dahmus 1989; Payne et al. 1989; Lu et al. 1991), and its CTD then becomes phosphorylated upon initiation of transcription (Laybourn and Dahmus 1990). In support of this, Usheva et al. (1992) have shown that only the unphosphorylated form of the CTD is capable of interacting with TBP in vitro. However, since it has recently been shown that polymerase with an unphosphorylated CTD is able to elongate from a promoter (Serizawa et al. 1993), it is possible that phosphorylation of the CTD does not occur immediately. The CTD of an initiated polymerase is in close proximity to the promoter and may still be interacting with TBP. This interaction is flexible because each of the 42 repeats may be capable of binding to TBP and could account for polymerase pausing over a 20-bp region. The interaction may be maximal when both RNA polymerase and TBP are on the same side of the DNA helix, accounting for the observed bimodal distribution of paused polymerase (Rasmussen and Lis 1993). If this type of interaction is occurring, the CTD of the paused polymerase should be unphosphorylated. Using an in vivo UV cross-linking technique and antibodies specific for the unphosphorylated and the hyperphosphorylated forms of the *Drosophila* CTD (Weeks et al. 1993), the polymerase at the 5′end of the uninduced *hsp70* gene has been shown to possess an unphosphorylated CTD region (T. O'Brien et al., unpubl.). Although phosphorylation of the CTD does not appear to be essential for elongation of polymerase to the pause in vivo, it may have a role in stimulating a paused polymerase to elongate into the body of the gene.

VI. HEAT SHOCK INDUCTION

The presence of a paused polymerase at the 5′end of uninduced heat shock genes suggests that these genes can respond rapidly to a heat shock. Indeed, a kinetic analysis of the distribution of polymerase following heat shock as measured by nuclear run-on and in vivo UV cross-linking assays reveals a rapid induction of transcription (O'Brien and Lis 1993). The paused polymerase elongates from the *hsp70* promoter within 30–60 seconds, and the first wave of polymerase reaches the 3′end of the gene within 120 seconds (giving an elongation rate of ~1.2 kb/min). A similar rapid induction of transcription is seen for the small heat shock genes (*hsp22*, *hsp23*, *hsp26*, and *hsp27*). Consistent with this rapid induction, HSF, the transcriptional activator of heat shock genes, acquires the ability to bind to HSEs within 30 seconds after heat shock (Zimarino and Wu 1987).

The rapid binding of HSF to its target HSEs is the major change that accompanies heat shock. This binding is likely to be facilitated by a nucleosome-free promoter, as HSF (at least human HSF) is unable to bind to HSEs packaged in nucleosomes (Taylor et al. 1991). Although heat shock changes the architecture of a heat shock promoter, many features of the uninduced promoter persist (Fig. 4). Both the GAGA and the TATA elements remain occupied after heat shock induction. Although the transcription bubble associated with the paused polymerase also per-

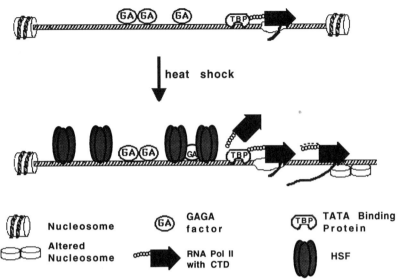

Figure 4 Architecture of the *hsp70* promoter before and after heat shock. The uninduced *hsp70* promoter contains a polymerase that has initiated transcription but has paused after synthesizing 20–40 bases. The CTD of this paused polymerase is unphosphorylated and may be directly interacting with TBP. GAGA factor is also bound to the uninduced promoter and may be required to "open" up the promoter, making it more accessible to TBP and RNA polymerase. Upon heat shock, HSF binds upstream as a trimer and activates transcription. Pausing of polymerase persists after induction, but the level of elongating polymerases is 100-fold higher than that in uninduced cells. Elongation of polymerase from the pause is coincident with phosphorylation of the CTD.

sists after full heat shock (Giardina et al. 1992), additional melting of DNA in the region of the start site is detected, presumably due to the entry of the next polymerase. Pausing of polymerase continues to occur even after the *hsp70* gene becomes transcriptionally active as assayed by nuclear run-ons and by measuring the short RNAs produced by the paused polymerase (O'Brien and Lis 1991; Giardina et al. 1992; Rasmussen and Lis 1993). Therefore, elongation of polymerase from the pause remains the rate-limiting step in transcription even in induced cells. Entry of new polymerases to this open promoter is faster than the rate at which paused polymerases escape into a productive elongation mode, i.e., once per 6 seconds.

Heat shock causes two major changes in HSF. First, HSF multimerizes to form a trimer and acquires the ability to bind HSEs (Westwood et al. 1991). Second, it undergoes a posttranslational modification, phosphorylation at multiple positions (Sorger 1990). Activated HSF may interact directly with the paused polymerase to stimulate elongation. This type of modification is postulated for the action of the bacteriophage λ Q protein on the paused polymerase at the 5' end of the late operon (Roberts 1988). Alternatively, HSF may act by increasing the rate of initiation of transcription of additional polymerases. The entry of a new polymerase may be essential to obtain release of the paused polymerase, resulting in the coupling of the two steps (initiation and elongation). HSF may also increase the rate of elongation by severing interactions between the paused polymerase and the basal transcription factors. It may do this by directly or indirectly stimulating phosphorylation of the CTD of the paused polymerase. Upon heat shock, the CTD of elongating polymerases are partially or hyperphosphorylated (T. O'Brien et al., unpubl.), and this phosphorylation may have a role in stimulating a paused polymerase to resume elongation. There is also the possibility that HSF may facilitate polymerase elongation by removing or altering nucleosomes or by antagonizing H1, which can act as a repressor of transcription (Croston et al. 1991; Laybourn and Kadonaga 1991).

VII. GENERALITY OF POLYMERASE PAUSING

A paused RNA polmerase is found not only on the 5' ends of *Drosophila* heat shock gene promoters, but also on those of several nonheat shock genes. These include the α and β1 tubulin, polyubiquitin, and glyceraldehyde-3-phosphate dehydrogenase 1 and 2 genes of *Drosophila*. These paused RNA polymerases have been detected by nuclear run-on assays (Rougvie and Lis 1990; Giardina et al. 1992) and in the case of β1 tubulin by high-resolution mapping of transcription bubbles with $KMnO_4$ (Giardina et al. 1992). Krumm et al. (1992) have also observed a paused RNA polymerase on the 5' end of the human c-*myc* gene around position +30, using both in vivo $KMnO_4$ mapping and nuclear run-on assays (Krumm et al. 1992). Remarkably, this pause is in a position that is very similar to that of *Drosophila hsp70*. Likewise, the transthyretin gene, one of two mammalian genes analyzed by Mirkovitch and Darnell (1992), shows this paused polymerase pattern. There is therefore an emerging class of eukaryotic genes transcribed by RNA polymerase II whose regulation is not confined to the level of preinitiation or initiation but is exercised at a postinitiation step. The heat shock system should serve as a simple model to explore the postinitiation control of this class of promoters by upstream factors.

ACKNOWLEDGMENT

This work was supported by a U.S. Public Health Service grant GM-25232 from the National Institutes of Health.

REFERENCES

Abravaya, K., B. Phillips, and R.I. Morimoto. 1991. Heat shock-induced interactions of heat shock transcription factor and the human hsp70 promoter examined by in vivo footprinting. *Mol. Cell Biol.* **11:** 586–592.

Allison, L.A., J.K.C. Wong, V.D. Fitzpatrick, M. Moyle, and C.J. Ingles. 1988. The carboxy-terminal domain of the largest subunit of RNA polymerase II of *Saccharomyces cerevisiae, Drosophila melanogaster,* and mammals: A conserved structure with an essential function. *Mol. Cell. Biol.* **8:** 321–329.

Amin, J., J. Ananthan, and R. Voellmy. 1988. Key features of heat shock regulatory elements. *Mol. Cell. Biol.* **8:** 3761–3769.

Bartolomei, M.S., F.F. Halden, C.R. Cullen, and J.L. Corden. 1988. Genetic analysis of the repetitive carboxyl-terminal domain of the largest subunit of mouse RNA polymerase II. *Mol. Cell. Biol.* **8:** 330–339.

Bienz, M. and H.R.B. Pelham. 1987. Mechanisms of heat-shock gene activation in higher eukaryotes. *Adv. Genet.* **24:** 31–73.

Biggin, M.D. and R. Tjian. 1988. Transcription factors that activate the Ultrabithorax promoter in developmentally staged extracts. *Cell* **53:** 699–711.

Bonner, J., S. Heyward, and D. Fackenthal. 1992. Temperature-dependent regulation of a heterologous transcriptional activation domain fused to yeast heat shock transcription factor. *Mol. Cell. Biol.* **12:** 1021–1030.

Cadena, D.L. and M.E. Dahmus. 1987. Messenger RNA synthesis in mammalian cells is catalyzed by the phosphorylated form of RNA polymerase II. *J. Biol. Chem.* **262:** 12468–12474.

Clos, J., T. Westwood, P.B. Becker, S. Wilson, K. Lambert, and C. Wu. 1990. Molecular cloning and expression of a hexameric *Drosophila* heat shock factor subject to negative regulation. *Cell* **63:** 1085–1097.

Costlow, N. and J.T. Lis. 1984. High-resolution mapping of DNase I-hypersensitive sites of *Drosophila* heat shock genes in *Drosophila melanogaster* and *Saccharomyces cerevisiae. Mol. Cell. Biol.* **4:** 1853–1863.

Craig, E.A. and K. Jakobsen. 1984. Mutations of the heat inducible 70 kilodalton genes of yeast confer temperature sensitive growth. *Cell* **38:** 841–849.

Croston, G.E., L.A. Kerrigan, L.M. Lira, D.R. Marshak, and J.T. Kadonaga. 1991. Sequence-specific antirepression of histone H1-mediated inhibition of basal RNA polymerase II transcription. *Science* **251:** 643–649.

Cunniff, N.F.A. and W.D. Morgan. 1993. Analysis of heat shock element recognition by saturation mutagenesis of the human hsp70.1 gene promoter. *Mol. Cell. Biol.* **268:** 8317–8324.

Cunniff, N.F.A., J. Wagner, and W.D. Morgan. 1991. Modular recognition of 5-base-pair DNA sequence motifs by human heat shock transcription factor. *Mol. Cell. Biol.* **11:** 3504–3514.

Eissenberg, J.C., I.L. Cartwright, G.H. Thomas, and S.C.R. Elgin. 1985. Selected topics in chromatin structure. *Annu. Rev. Genet.* **19:** 485–536.

Fernandes, M., H. Xiao, and J.T. Lis. 1994. Fine structure analysis of the *Drosophila* and

Saccharomyces heat shock factor—Heat shock element interactions. *Nucleic Acids Res.* (in press).

Fischer, J.A., E. Giniger, T. Maniatis, and M. Ptashne. 1988. GAL4 activates transcription in *Drosophila. Nature* 332: 853–856.

Fujita, A., Y. Kikuchi, S. Kuhara, Y. Misumi, S. Matsumoto, and H. Kobashi. 1989. Domains of the SFL1 protein of yeasts are homologous to Myc oncoproteins or yeast heat-shock transcription factor. *Gene* 85: 321–328.

Gallo, G.J., H. Prentice, and R.E. Kingston. 1993. Heat shock factor is required for growth at normal temperatures in the fission yeast *Schizosaccharomyces pombe. Mol. Cell. Biol.* 13: 749–761.

Gallo, G.J., T.J. Schuetz, and R.E. Kingston. 1991. Regulation of heat shock factor in *Schizosaccharomyces pombe* more closely resembles regulation in mammals than in *Saccharomyces cerevisiae. Mol. Cell. Biol.* 11: 281–288.

Garabedian, M.J., B.M. Shepherd, and P.C. Wensink. 1986. A tissue-specific transcription enhancer from the *Drosophila* yolk protein 1 gene. *Cell* 45: 859–867.

Giardina, C., M. Perez-Riba, and J.T. Lis. 1992. Promoter melting and TFIID complexes on *Drosophila* genes in vivo. *Genes Dev.* 6: 2190–2200.

Gilmour, D.S. and J.T. Lis. 1986. RNA polymerase II interacts with the promoter region of the noninduced *hsp70* gene in *Drosophila melanogaster* cells. *Mol. Cell. Biol.* 6: 3984–3989.

Gilmour, D.S., G.H. Thomas, and S.C. Elgin. 1989. *Drosophila* nuclear proteins bind to regions of alternating C and T residues in gene promoters. *Science* 245: 1487–1490.

Goldenberg, C.J., Y. Luo, M. Fenna, R. Baler, R. Weinmann, and R. Voellmy. 1988. Purified human factor activates heat shock promoter in a HeLa cell-free transcription system. *J. Biol. Chem.* 263: 19734–19739.

Hoey, T., B.D. Dynlacht, M.G. Peterson, B.F. Pugh, and R. Tjian. 1990. Isolation and characterization of the *Drosophila* gene encoding the TATA box binding protein, TFIID. *Nature* 61: 1179–1186.

Jakobsen, B.K. and H.R.B. Pelham. 1991. A conserved heptapeptide restrains the activity of the yeast heat shock transcription factor. *EMBO J.* 10: 369–375.

Kingston, R.E., T.J. Schuetz, and Z. Larin. 1987. Heat-inducible human factor that binds to a human *hsp70* promoter. *Mol. Cell. Biol.* 7: 372–375.

Kolodziej, P.A., N. Woychik, S.-M. Liao, and R.A. Young. 1990. RNA polymerase II subunit composition, stoichiometry, and phosphorylation. *Mol. Cell. Biol.* 10: 1915–1920.

Kraus, K.W., Y. Lee, J.T. Lis, and W.F. Wolfner. 1988. Sex-specific control of *Drosophila melanogaster* yolk protein 1 gene expression is limited to transcription. *Mol. Cell. Biol.* 8: 4756–4764.

Krumm, A., T. Meulia, M. Brunvand, and M. Groudine. 1992. The block to transcriptional elongation within the human c-*myc* gene is determined in the promoter-proximal region. *Genes Dev.* 6: 2201–2213.

Laybourn, P.J. and M.E. Dahmus. 1989. Transcription-dependent structural changes in the C-terminal domain of mammalian RNA polymerase subunit IIa/o. *J. Biol. Chem.* 264: 6693–6698.

———. 1990. Phosphorylation of RNA polymerase IIA occurs subsequent to interaction with the promoter and before the initiation of transcription. *J. Biol. Chem.* 265: 13165–13173.

Laybourn, P.J. and J.T. Kadonaga. 1991. Role of nucleosomal cores and histone H1 in regulation of transcription by RNA polymerase II. *Science* 254: 238–245.

Lee, H.-S., K.W. Kraus, M.F. Wolfner, and J.T. Lis. 1992. DNA sequence requirements for generating paused polymerase at the start of *hsp 70*. *Genes Dev.* **6:** 284–295.

Lindquist, S. 1980. Varying patterns of protein synthesis in *Drosophila* during heat shock: Implications for regulation. *Dev. Biol.* **77:** 463–479.

Lis, J.T. and C. Wu. 1993. Protein traffic on the heat shock promoter: Parking, stalling, and trucking along. *Cell* **74:** 1–4.

Lis, J.T., H. Xiao, and O. Perisic. 1990. Modular units of heat shock regulatory regions: Structure and function. In *Stress proteins in biology and medicine* (ed. R. Morimoto et al.), pp. 411–428. Cold Spring Harbor Laboratory Press, Cold Spring Harbor, New York.

Lis, J.T., W. Neckameyer, R. Dubensky, and N. Costlow. 1981. Cloning and characterization of nine heat-shock-induced mRNAs of *Drosophila melanogaster*. *Gene* **15:** 67–80.

Lu, H., O. Flores, R. Weinmann, and D. Reinberg. 1991. The nonphosphorylated form of RNA polymerase II preferentially associates with the preinitiation complex. *Proc. Natl. Acad. Sci.* **88:** 10004–10008.

Lu, Q., L.L. Wallrath, H. Granok, and S.C.R. Elgin. 1993. Distinct roles of $(CT)_n \cdot (GA)_n$ repeats and heat shock elements in chromatin structure and transcriptional activation of the *Drosophila hsp26* gene. *Mol. Cell. Biol.* **13:** 2802–2814.

Lu, Q., L.L. Wallrath, B.D. Allan, R.L. Glaser, J.T. Lis, and S.C.R. Elgin. 1992. A promoter sequence containing $(CT) \cdot (GA)_n$ repeats is critical for the formation of the DNase 1 hypersensitive sites in the *Drosophila hsp26* gene. *J. Mol. Biol.* **226:** 985–997.

Martin, M., A. Giangrande, C. Ruiz, and G. Richards. 1989. Induction and repression of the *Drosophila* Sgs-3 glue gene are mediated by distinct sequences in the proximal promoter. *EMBO J.* **8:** 561–568.

Mirault, M.-E., R. Southgate, and E. Delwart. 1982. Regulation of heat-shock genes: A DNA sequence upstream of *Drosophila hsp70* genes is essential for their induction in monkey cells. *EMBO J.* **1:** 1279–1285.

Mirkovitch, J. and J.E. Darnell. 1992. Mapping of RNA polymerse II on mammalian genes in cells and nuclei. *Mol. Biol. Cell.* **3:** 1085–1094.

Morgan, W.D., G.T. Williams, R.I. Morimoto, J. Greene, R.E. Kingston, and R. Tjian. 1987. Two transcriptional activators, CCAAT-box binding transcription factor and heat shock transcription factor, interact with human hsp70 gene promoter. *Mol. Cell. Biol.* **7:** 1129–1138.

Mosser, D.D., N.G. Theodorakis, and R.I. Morimoto. 1988. Coordinate changes in heat shock element-binding activity and *hsp70* gene transcription rates in human cells. *Mol. Cell. Biol.* **8:** 4736–4744.

Mosser, D.D., P.T. Kotzbauer, K.D. Sarge, and R.I. Morimoto. 1990. In vitro activation of heat shock transcription factor DNA-binding by calcium and biochemical conditions that affect protein conformation. *Proc. Natl. Acad. Sci.* **87:** 3748–3752.

Nakai, A. and R. I. Morimoto. 1993. Characterization of a novel chicken heat shock transcription factor, heat shock factor 3, suggests a new regulatory pathway. *Mol. Cell. Biol.* **13:** 1983–1997.

Neves, A.M., I. Barahona, L. Galego, and C. Rodrigues-Pousada. 1988. Ubiquitin genes in *Tetrahymena pyriformis* and their expression during heat shock. *Gene* **73:** 87–96.

Nieto-Sotelo, J., G. Wiederrecht, A. Okuda, and C.S. Parker. 1990. The yeast heat shock transcription factor contains a transcriptional activation domain whose activity is repressed under nonshock conditions. *Cell* **62:** 807–817.

Nonet, M., D. Sweetser, and R.A. Young. 1987. Functional redundancy and structural polymorphism in the large subunit of RNA polymerase II. *Cell* **50:** 909–915.

O'Brien, T. and J.T. Lis. 1991. RNA polymerase II pauses at the 5' end of the transcriptionally induced *Drosophila hsp70* gene. *Mol. Cell. Biol.* **11:** 5285–5290.

―――. 1993. Rapid changes in *Drosophila* transcription after an instantaneous heat shock. *Mol. Cell. Biol.* **13:** 3456–3463.

Parker, C.S. and J. Topol. 1984. A *Drosophila* RNA polymerase II transcription factor contains a promoter-region-specific DNA-binding activity. *Cell* **36:** 357–369.

Payne, J.M., P.J. Laybourn, and M.E. Dahmus. 1989. The transition of RNA polymerase II from initiation to elongation is associated with phosphorylation of the carboxyl-terminal domain of subunit IIa. *J. Biol. Chem.* **264:** 19621–19629.

Pelham, H.R.B. 1982. A regulatory upstream promoter element in the *Drosophila hsp70* heat-shock gene. *Cell* **30:** 517–528.

Perisic, O., H. Xiao, and J.T. Lis. 1989. Stable binding of *Drosophila* heat shock factor to head-to-head and tail-to-tail repeats of a conserved 5 bp recognition unit. *Cell* **59:** 797–806.

Peteranderl, R. and H.C.M. Nelson. 1992. Trimerization of the heat shock transcription factor by a triple-stranded a-helical coiled-coil. *Biochemistry* **31:** 12272–12276.

Rabindran, S.K., G. Giorgi, J. Clos, and C. Wu. 1991. Molecular cloning and expression of a human heat shock factor, HSF1. *Proc. Natl. Acad. Sci.* **88:** 6906–6910.

Rabindran, S.K., R.I. Haroun, J. Clos, J. Wisniewski, and C. Wu. 1993. Regulation of heat shock factor trimer formation: Role of a conserved leucine zipper. *Science* **259:** 230–234.

Rasmussen, E. and J.T. Lis. 1993. In vivo transcriptional pausing and cap formation on three *Drosophila* heat shock genes. *Proc. Natl. Acad. Sci.* **90:** 7923–7927.

Roberts, J.W. 1988. Phage lambda and the regulation of transcription termination. *Cell* **52:** 5–6.

Rougvie, A.E. and J.T. Lis. 1988. The RNA polymerase II molecule at the 5' end of the uninduced *hsp70* gene of *D. melanogaster* is transcriptionally engaged. *Cell* **54:** 795–804.

―――. 1990. Post-initiation transcriptional control in *Drosophila melanogaster*. *Mol. Cell. Biol.* **10:** 6041–6045.

Sarge, K.D., V. Zimarino, K. Holm, C. Wu, and R.I. Morimoto. 1991. Cloning and characterization of two mouse heat shock factors with distinct inducible and constitutive DNA-binding ability. *Genes Dev.* **5:** 1902–1911.

Scharf, K.-D., S. Rose, W. Zott, F. Schoff, and L. Nover. 1990. Three tomato genes code for heat stress transcription factors with a region of remarkable homology to the DNA-binding domain of the yeast HSF. *EMBO J.* **9:** 4495–4501.

Schuetz, T.J., G.J. Gallo, L. Sheldon, P. Tempst, and R.E. Kingston. 1991. Isolation of a cDNA for HSF2: Evidence for two heat shock factor genes in humans. *Proc. Natl. Acad. Sci.* **88:** 6911–6915.

Serizawa, H., J.W. Conaway, and R.C. Conaway. 1993. Phosphorylation of C-terminal domain of RNA polymerase II is not required in basal transcription. *Nature* **363:** 371–374.

Shuey, D.J. and C.S. Parker. 1986. Bending of promoter DNA on binding of heat shock transcription factor. *Nature* **323:** 459–461.

Soeller, W.C., S.J. Poole, and T. Kornberg. 1988. In vitro transcription of the *Drosophila* engrailed gene. *Genes Dev.* **2:** 68–81.

Sorger, P.K. 1990. Yeast heat shock factor contains separable transient and sustained response transcriptional activators. *Cell* **62:** 793–805.

Sorger, P.K. and H.C.M. Nelson. 1989. Trimerization of a yeast transcriptional activator via a coiled-coil motif. *Cell* **59:** 807–813.

Sorger, P.K. and H.R.B. Pelham. 1987. Purification and characterization of a heat-shock element binding protein from yeast. *EMBO J.* **6:** 3035–3041.

―――. 1988. Yeast heat shock factor is an essential DNA-binding protein that exhibits temperature-dependent phosphorylation. *Cell* **54:** 855–864.

Taylor, I.C.A., J.L. Workman, T.J. Schuetz, and R.E. Kingston. 1991. Facilitated binding of GAL4 and heat shock factor to nucleosomal templates: Differential function of DNA-binding domains. *Genes Dev.* **5:** 1285–1298.

Theodorakis, N.G., D.J. Zand, P.T. Kotzbauer, G.T. Williams, and R. Morimoto. 1989. Hemin-induced transcriptional activation of the *hsp70* gene during erythroid maturation in K562 cells is due to a heat shock factor-mediated stress response. *Mol. Cell. Biol.* **9:** 3166–3173.

Thomas, G.H. and S.C.R. Elgin. 1988. Protein/DNA architecture of the DNase I hypersensitive region of the *Drosophila hsp26* promoter. *EMBO J.* **7:** 2191–2201.

Topol, J., D.M. Ruden, and C.S. Parker. 1985. Sequences required for in vitro transcriptional activation of a *Drosophila hsp70* gene. *Cell* **42:** 527–537.

Usheva, A., E. Maldonado, A. Goldring, H. Lu, C. Houbavi, D. Reinberg, and Y. Aloni. 1992. Specific interaction between the nonphosphorylated form of RNA polymerase II and the TATA-binding protein. *Cell* **69:** 871–881.

Weeks, J.R., S.E. Hardin, S. Jianjun, and A.L. Greenleaf. 1993. Locus-specific variation in phosphorylation state of RNA polymerase II in vivo: Correlations with gene activity and transcript processing. *Genes Dev.* **7:** 2329–2344.

Westwood, J.T. and C. Wu. 1993. Activation of *Drosophila* heat shock factor: Conformational change associated with a monomer-to-trimer transition. *Mol. Cell. Biol.* **13:** 3481–3486.

Westwood, J.T., J. Clos, and C. Wu. 1991. Stress-induced oligomerization and chromosomal relocalization of heat-shock factor. *Nature* **353:** 822–827.

Wiederrecht, G., D. Seto, and C.S. Parker. 1988. Isolation of the gene encoding the *S. cerevisiae* heat shock transcription factor. *Cell* **54:** 841–853.

Wiederrecht, G., D.J. Shuey, W.A. Kibbe, and C.S. Parker. 1987. The *Saccharomyces* and *Drosophila* heat shock gene transcription factors are identical in size and DNA binding properties. *Cell* **48:** 507–515.

Williams, G. and R. Morimoto. 1990. Maximal stress-induced transcription from the human hsp70 promoter requires interactions with the basal promoter elements independent of rotational alignment. *Mol. Cell. Biol.* **10:** 3125–3136.

Wilson, I.A., J.J. Skehel, and D.C. Wiley. 1981. Structure of the haemagglutinin membrane glycoprotein of influenza virus of 3 Å resolution. *Nature* **289:** 366–378.

Workman, J.L. and A.R. Buchman. 1993. Multiple functions of nucleosomes and regulatory factors in transcription. *Trends Biochem Sci.* **18:** 90–95.

Wu, C. 1980. The 5′ ends of *Drosophila* heat shock genes in chromatin are hypersensitive to DNase I. *Nature* **286:** 854–860.

―――. 1984a. Two protein-binding sites in chromatin implicated in the activation of heat-shock genes. *Nature* **309:** 229–234.

―――. 1984b. Activating protein factor binds in vitro to upstream control sequences in heat shock gene chromatin. *Nature* **311:** 81–84.

Wu, C., S. Wilson, B. Walker, I. Dawid, T. Paisley, V. Zimarino, and H. Ueda. 1987. Purification and properties of *Drosophila* heat shock activator protein. *Science* **238:** 1247–1253.

Xiao, H. and J.T. Lis. 1988. Germline transformation used to define key features of heat-shock response elements. *Science* **239:** 1139–1142.

Xiao, H., O. Perisic, and J.T. Lis. 1991. Cooperative binding of *Drosophila* heat shock factor to arrays of a conserved 5 bp unit. *Cell* **64:** 585–593.

Zehring, W.A., J.M. Lee, J.R. Weeks, R.S. Jokerst, and A.L. Greenleaf. 1988. The C-terminal repeat domain of RNA polymerase II largest subunit is essential in vivo but is not required for accurate transcription initiation in vitro. *Proc. Natl. Acad. Sci.* **85:** 3698–3702.

Zimarino, V. and C. Wu. 1987. Induction of sequence-specific binding of *Drosophila* heat shock activator. *Nature* **327:** 727–730.

Zimarino, V., S. Wilson, and C. Wu. 1990. Antibody-mediated activation of *Drosophila* heat shock factor in vitro. *Science* **249:** 546–549.

16

Structure and Regulation of Heat Shock Transcription Factor

Carl Wu, Joachim Clos,[1] Gisele Giorgi,[2] Raymond I. Haroun,[3] Soon-Jong Kim, Sridhar K. Rabindran,[4] J. Timothy Westwood,[5] Jan Wisniewski, and Gloria Yim[6]
Laboratory of Biochemistry
National Cancer Institute
National Institutes of Health
Bethesda, Maryland 20892

I. INTRODUCTION

The transcriptional induction of heat shock genes by elevated temperatures and other forms of physiological stress in eukaryotes is mediated by the transcription factor heat shock factor (HSF). Considerable progress in our understanding of the structure of HSF and how its activity is regulated by heat shock has been achieved since the identification and purification of HSF proteins from a variety of metazoan species. For a review of the early studies, see Lis et al. (1990) and Wu et al. (1990). Here, we summarize recent advances in the structure and regulation of

Present addresses: [1]Bernhard Nocht Institute for Tropical Medicine, Bernard Nocht Strasse 74, 20359 Hamburg, Germany; [2]Department of Molecular Cell Biology, University of California, Berkeley, California 94720; [3]Columbia University College of Physicians & Surgeons, 630 West 168 Street, Box #333, New York, New York 10062; [4]American Cyanamid Co., Medical Research Division, 401 N. Middletown Rd., Pearl River, New York 10965; [5]Department of Zoology, Erindale College, University of Toronto, 3359 Mississauga Rd., Mississauga, Ontario L5L 1C6, Canada; [6]University of Chicago School of Medicine, Chicago, Illinois.

The Biology of Heat Shock Proteins and Molecular Chaperones
©1994 Cold Spring Harbor Laboratory Press 0-87969-427-0/94 $5 + .00

HSF, drawing primarily from studies in our laboratory on the *Drosophila* and human HSF proteins. Further elaborations on HSF, including the multiplicity of HSF genes and how heat shock promoters are poised to respond to HSF binding, are treated elsewhere in this volume (see Sarge et al.; Lis et al.).

II.   DNA BINDING BY A MONOMER-TRIMER TRANSITION

Like many inducible transcriptional regulators, the HSF protein is synthesized constitutively and stored in a latent form under normal conditions. This property of HSF, originally observed for human and *Drosophila* HSF proteins (Kingston et al. 1987; Zimarino and Wu 1987), appears to be common to all eukaryotic species studied. With the exception of budding yeasts, the latent HSF is activated in response to heat shock by the acquisition of high-affinity DNA-binding activity, which is accomplished by a conversion of HSF protein from a monomer to a homotrimer (Perisic et al. 1989; Westwood et al. 1991; Baler et al. 1993; Sarge et al. 1993; Westwood and Wu 1993). The HSF trimer binds to conserved, upstream heat shock elements (HSEs) that are composed of the 5-bp module [nGAAn] arranged as contiguous inverted repeats. Three 5-bp modules [nGAAnnTTCnnGAAn] constitute the minimal sequence for high-affinity binding to HSF (Perisic et al. 1989).

The control over HSF binding to DNA is bypassed in the budding yeasts *Saccharomyces cerevisiae* and *Kluyveromyces lactis*, which have constitutively trimeric HSF proteins that remain bound to HSEs under both normal and heat shock conditions (Sorger and Nelson 1989; Gross et al. 1990; Jakobsen and Pelham 1991). These yeast proteins stimulate transcription from heat shock promoters upon temperature elevation, which is correlated with increased phosphorylation of *S. cerevisiae* HSF (ScHSF) at serine and threonine residues (Sorger 1990). However, whether the hyperphosphorylation is causally related to the acquisition of transcriptional competence remains to be resolved. There are indications that a human HSF protein also undergoes heat-shock-induced phosphorylation (Larsen et al. 1988; S.R. Rabindran, unpubl.).

The trimeric state of the activated form of HSF was initially shown by chemical cross-linking experiments for the *Drosophila melanogaster* HSF (DmHSF) and ScHSF proteins (Perisic et al. 1989; Sorger and Nelson 1989). The monomer and trimer states of DmHSF are revealed when extracts prepared from nonshocked and heat-shocked cells are analyzed by chemical cross-linking and Western blotting (Fig. 1). The oligomerization states of DmHSF were also determined by gel filtration and

A.   B.

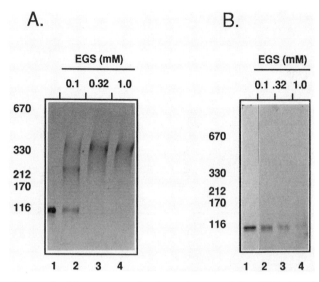

Figure 1 Monomer and trimer forms of DmHSF. Unshocked (*A*) and heat-shocked (*B*) SL2 cell extracts were fractionated by Superose 6 chromatography and cross-linked with EGS at the indicated concentrations. Reaction products were resolved by SDS-PAGE (3.5–7% polyacrylamide) and analyzed by Western blotting. The migration of protein markers (K) are indicated at the left of each panel. (Reprinted with permission, from Westwood and Wu 1993.)

glycerol gradient sedimentation analyses of cell extracts (Westwood and Wu 1993). Although gel-filtration chromatography and density gradient sedimentation employed as separate techniques do not give the molecular weight of a particle, both techniques used in conjunction allow calculation of the native molecular weight as well as the frictional ratio, an indicator of particle shape. The hydrodynamic studies reveal that the activation of DmHSF binding is best described as a transition from monomer to trimer (Table 1). The frictional ratios determined for both monomer and trimer forms of DmHSF indicate significant deviation from a globular shape (f/f_0 = 1.9 and 2.6, respectively; the frictional ratio of a sphere being 1.0). Interestingly, the degree of asymmetry of the DmHSF trimer is greater than the asymmetry of the monomer, suggesting that the inert, monomeric form of DmHSF may undergo a heat-shock-induced conformational change to a more extended structure in the transition from monomer to trimer. The asymmetrical shape provides an explanation for our previous overestimation of the oligomeric state of the HSF trimer, which was based on the native electrophoretic mobility relative to globular protein standards (see Clos et al. 1990).

Table 1 Native Molecular Weight of *Drosophila* HSF

Method	Heat-shocked HSF	Unshocked HSF
Gel filtration chromatography		
apparent M_r (x10^3)	900	270
Stokes' radius (Å)	109 ± 8	55 ± 3
Glycerol gradient centrifugation		
apparent M_r (x10^3)	115	60
$S_{20,w}$ (x10^{-13}) (s)	6.2 ± 0.3	4.1 ± 0.3
Corrected M_r (x10^3)	264 ± 32	88 ± 11
Predicted M_r (x10^3)	231	77
Frictional ratio (f/f_o)	2.6	1.9

III. CLONING HSF GENES

With the development of a protocol to purify HSF, it became possible to clone HSF genes by using HSF-specific antibodies or microsequence information generated from HSF peptides. Antibodies to HSF were used to isolate a cDNA for ScHSF from expression libraries by Sorger and Pelham (1987) and Weiderrecht et al. (1988). The DmHSF gene was obtained by synthesizing degenerate oligonucleotides based on DmHSF peptide sequences and probing a genomic library (Clos et al. 1990). Several cDNAs for tomato HSF were also isolated by screening an expression library with labeled HSE sequences (Scharf et al. 1990). Complementary DNA clones encoding HSFs from a wide variety of species have subsequently been isolated by means of these and other techniques. Whereas a single HSF gene appears to be present in *S. cerevisiae, K. lactis*, and *Drosophila*, two or three related HSF genes occur in tomatoes, humans, mice, and chickens. The presence of multiple HSF genes in some species appears to reflect functional diversification and is discussed in detail in this volume by Morimoto et al.

IV. FUNCTIONAL DOMAINS OF HSF

A phylogenetic comparison of the predicted amino acid sequences of DmHSF and ScHSF provided useful insights on functional domains (Clos et al. 1990). The predicted sequences of these two HSF proteins are strikingly divergent, except for two conserved regions in the amino-terminal part of HSF protein that represent the DNA-binding and trimerization domains (Fig. 2A). The minimal DNA-binding domain as defined by deletion mapping for ScHSF (Wiederrecht et al. 1988) is contained within a region of 118 amino acids that is essentially coincident

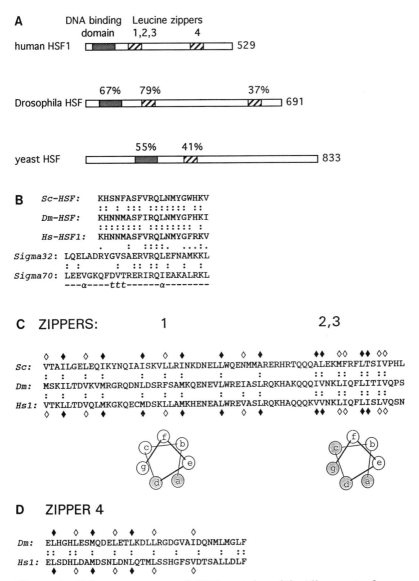

Figure 2   Primary structure of HSF proteins. (*A*) Alignment of conserved domains; (*B*) sequence comparison of the most conserved segment of the HSF DNA-binding domain with the putative helix-turn-helix motif of σ factors; (*C*) sequence alignment of amino-terminal leucine zipper motifs. (*Closed* and *open diamonds*) Positions of the hydrophobic residues in the 4-3 heptad repeat; (*shaded areas*) the same positions along an idealized α-helical wheel. (*D*) Sequence alignment of carboxy-terminal leucine zipper motif.

with the conserved region near the amino terminus. Except for a short but good match to the putative recognition helix of bacterial σ factors (Clos et al. 1990) (Fig. 2B), and a modest sequence similarity to the HNF3/*forkhead* DNA-binding region (Scharf et al. 1994), the overall sequence of this domain does not show extensive similarity to any known category of DNA-binding motifs. It was therefore anticipated that a high-resolution structure determination should reveal a new motif for specific DNA recognition. However, recent X-ray and nuclear magnetic resonance (NMR) studies indicate that the modest sequence similarities belie a greater structural conservation between the HSF DNA-binding domain and a superclass of DNA-binding motifs exemplified by the helix-turn-helix and HNF3 protein families (Harrison et al. 1994; Vuister et al. 1994). Titration of the DmHSF DNA-binding domain with a 13-mer DNA duplex containing a single nGAAn module indicates that the region with sequence similarity to the putative σ factor recognition helix is indeed α-helical and is involved in contacting DNA (Vuister et al. 1994).

The trimerization domain of HSF is situated carboxy-terminal to the DNA-binding domain, separated by a short linker. The conserved residues in this region essentially define several arrays of hydrophobic heptad repeats, or coiled-coil motifs (Sorger and Nelson 1989; Clos et al. 1990), arranged in a manner reminiscent of, but distinct from, the helix-loop-helix dimerization motif found in other transcription factors (Fig. 2A,C) (Murre et al. 1989). The first heptad repeat is long and occupies the amino-terminal half of the region. The other two short heptad repeats overlap and are positioned one residue out of phase. When plotted on an idealized α helix, the short overlapping heptad repeats reveal hydrophobic residues juxtaposed at four positions on one helical face, suggesting potential for hydrophobic associations with the hydrophobic faces of one or more α helices (Fig. 2C).

Physicochemical studies of the ScHSF or *K. lactis* HSF (KlHSF) trimerization domains synthesized as approximately 90-residue polypeptides indicate that they form stable homotrimers in solution with predominantly α-helical character, suggesting a triple-stranded α-helical coiled-coil structure (Peteranderl and Nelson 1993). Although the assembly of coiled-coils in bundles of three or more has been observed in protein structures, the assembly of a DNA-binding protein trimer is novel. This presents an interesting structural problem that must accommodate binding of three DNA-binding domains arranged with presumably threefold rotational symmetry to the HSE sequence containing multiple twofold symmetry.

The remaining carboxy-terminal half of the HSF protein sequences is highly divergent and variable in length, except for a short region near the carboxyl terminus that includes yet another conserved hydrophobic heptad repeat (Fig. 2A,D). This conserved region, initially identified by comparison of the DmHSF and human HSF1 sequences (Clos et al. 1990; Rabindran et al. 1991), is found among the studied HSFs that show heat-shock-inducible DNA binding and trimerization but is absent in the constitutively trimeric ScHSF and KlHSF polypeptides. Nonconservative substitutions of hydrophobic residues within this carboxy-terminal heptad repeat for human HSF1 led to constitutive trimer formation and high-affinity DNA-binding activity, indicating that this region participates in the suppression of trimer assembly (Fig. 3) (Rabindran et al. 1993). Deletion mapping studies reveal a second region carboxy-terminal to the zipper whose integrity is also important for monomer stability (Rabindran et al. 1993; Sarge et al. 1993). These studies suggest that the carboxy-terminal zipper motif may be a part of a larger domain involved in negative regulation of oligomerization, possibly by means of an intramolecular coiled-coil interaction between amino-terminal and carboxy-terminal zipper motifs (Fig. 5). Sequences in the carboxy-terminal portion of the tomato HSF including short peptide motifs with a central tryptophan residue are also involved in *trans*-activation (Treuter et al. 1993).

The regions of HSF important for *trans*-activation have been analyzed for two yeast HSF proteins. Both ScHSF and KlHSF have potent carboxy-terminal activators that function when fused to a heterologous DNA-binding domain, and an amino-terminal *trans*-activator has also been identified for ScHSF (Nieto-Sotelo et al. 1990; Sorger 1990). These activities are negatively regulated by the central region of HSF that includes the DNA-binding and trimerization domains and a heptapeptide motif (RXLLKNR) in the carboxy-terminal portion (Nieto-Sotelo et al. 1990; Jakobsen and Pelham 1991). Analysis of point mutations in ScHSF confirm the participation of the DNA-binding domain and the conserved heptapeptide motif in masking the *trans*-activation function of the yeast protein (Bonner et al. 1992). Hence, despite the apparent differences in the levels of control exercised by the budding yeast and metazoan HSF proteins, the activation of all HSFs is via the relief of repression.

V. REGULATION OF THE MONOMER-TRIMER TRANSITION

The binding of HSF to DNA can be induced in vitro by treating unshocked cell extracts with a variety of agents affecting protein structure, including heat, mildly acidic pH (6.5), detergents, and polyclonal antibodies raised against the activated (trimeric) form of HSF (Larson et al. 1988; Mosser et al. 1990; Zimarino et al. 1990). These studies initially suggested that the latent HSF could be activated in vitro through a simple

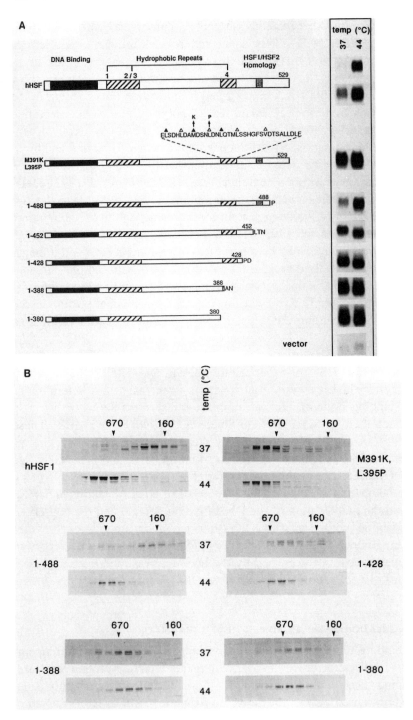

Figure 3 (See facing page for legend.)

and direct change in conformation or oligomeric state. However, the inability to convert the activated trimer back to the monomeric species after restoration of normal conditions in vitro indicates that interconversion between HSF monomers and trimers is not likely to be the reversible response of a single-component system to physical changes in the environment. The constitutive assembly of HSF trimers in *Escherichia coli* at nonheat shock temperatures, and the persistence of the trimeric state when HSF proteins are incubated at nonshock temperatures in vitro, further indicates that other intracellular components have a role in folding the nascent HSF polypeptide during synthesis and disaggregating the HSF trimer upon return to normal conditions (Clos et al. 1990; Rabindran et al. 1991). Indeed, HSF can be folded to the latent form when expressed by microinjection in frog oocytes, by DNA transfection in tissue-culture cells, or by translation in reticulocyte lysates (Clos et al. 1990; Sarge et al. 1991; Baler et al. 1993; Rabindran et al. 1993).

The importance of the cellular environment in modulating HSF activity is also demonstrated by the behavior of the protein when it is expressed in a heterologous host. When human HSF1 is expressed by DNA transfection in *Drosophila* cells, the temperature that induces DNA binding and trimerization of human HSF1 is decreased by about 10°C to the induction temperature for *Drosophila* cells (Fig. 4A,B). The human HSF1 protein is also activated below its normal temperature threshold when expressed in tobacco protoplasts or in frog oocytes (Baler et al. 1993; Treuter et al. 1993). In contrast, DmHSF is constitutively trimeric when expressed in human cells, even when the host cell temperature is

Figure 3 (*A*) Activity of wild-type and mutant human HSF1 proteins transiently expressed in 293 cells. (*Left*) Map of wild-type and mutant human HSF1 ORFs. Numbers at the right indicate the endpoint of the truncated fragments; amino acids appended by cloning are represented by single-letter code. The DNA-binding region and hydrophobic repeats as delimited by conserved sequence blocks are indicated by *closed* and *hatched* areas, respectively. (*Gray areas*) A 12-residue element conserved among vertebrate HSFs. (*Right*) Gel mobility shift analyses on extracts from unshocked (37°C) and heat-shocked (44°C) transfected cells were carried out as described by Rabindran et al. (1993). Only the complex of HSF bound to the HSE is shown. (*B*) Gel-filtration chromatography of extracts from nonshocked and heat-shocked 293 cells transfected with wild-type and mutant human HSF1 genes (hHSF1). Whole-cell extracts were fractionated on a Superose 6 PG3.2/30 column (Pharmacia). Fractions were analyzed by SDS-gel electrophoresis and Western blotting with anti-HSF1 serum. Arrows indicate positions of molecular-mass markers (kD). (Reprinted, with permission, from Rabindran et al. 1993, © American Association for the Advancement of Science.)

decreased to 25°C, the normal growth temperature for *Drosophila* (Fig. 4C,D). These results indicate that the activity of HSF in vivo cannot be a simple function of the absolute environmental temperature but is influenced by cellular factors whose ability to act on HSFs from a different species is not necessarily conserved.

What then is the primary sensor of the heat shock signal? Possible candidates for this function are molecular chaperones, including the heat shock proteins themselves, which have been proposed as homeostatic suppressors of HSF activity (Clos et al. 1990; Morimoto et al. 1990). Recent studies provide some evidence for an association between HSF and the major heat shock protein hsp70 (Abravaya et al. 1992; Baler et al. 1992; Mosser et al. 1993; Schlessinger and Ryan 1993). Post-translational modifications, including phosphorylation, might also affect the interconversion between monomer and trimer. Modifying enzyme(s)

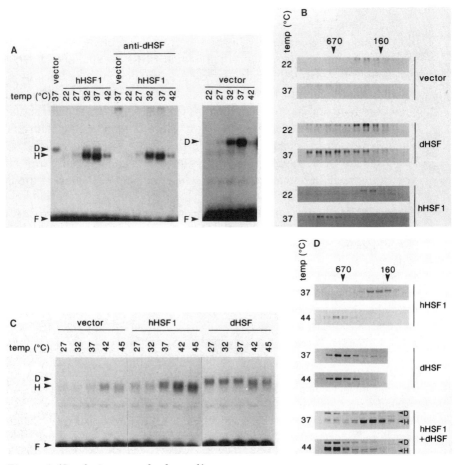

Figure 4 (See facing page for legend.)

could serve as the primary sensor of the heat shock signal or may act constitutively, subtly altering HSF to render the modified monomer directly sensitive to temperature changes. Given our present level of understanding, the list of potential candidates for the heat shock sensor should include, at a minimum, HSF itself, intracellular components such as the pool of available molecular chaperones, and HSF-modifying enzymes or enzyme pathways (Fig. 5). Independent of the sensory mechanism, the heat shock signal ultimately becomes transduced into a change in the overall protein conformation of HSF: either a disruption of interactions that normally suppress trimerization for the monomeric forms of HSF or a disruption of interactions that suppress the *trans*-activation domain(s), or both.

VI. LOCALIZATION OF HSF IN CELL NUCLEI PRIOR TO AND AFTER HEAT SHOCK

As expected for an active DNA-binding protein, the DmHSF trimer is localized by indirect immunofluorescence in the nuclei of fixed *Drosophila* cells after heat shock (Fig. 6A) (Westwood et al. 1991). Areas of punctate staining are observed, with a distinct lack of staining in the nucleolus. The DmHSF monomer present in unshocked *Drosophila* cells is also predominantly localized in the nucleus. This finding was unexpected in view of previous biochemical studies that placed the inactive

Figure 4 (*A*) Activity of human HSF1 (hHSF1) in *Drosophila* SL2 cells. Gel mobility shift analysis of extracts from cells transfected with a human HSF1 construct or with the expression vector. Cells were heat shocked for 15 min at the indicated temperatures. Positions of free DNA (F) and DNA bound to human HSF1 (H) and the endogenous DmHSF (D) are shown. Samples treated with antibodies to DmHSF (anti-dHSF) are indicated. The autoradiogram on the right was exposed eight times longer than the one on the left. (*B*) Gel-filtration chromatography and Western blotting of extracts prepared from control (22°C) or heat-shocked (37°C) cells transfected with the vector alone, and with constructs carrying DmHSF and human HSF1. Arrows indicate molecular-mass standards (kD). (*C*) Activity of human HSF1 and DmHSF in human 293 cells. Gel mobility shift analysis of extracts prepared from cells transfected with the expression vector or with plasmids encoding human HSF1 or DmHSF. Cells were incubated for 30 min at the indicated temperatures. (*D*) Gel-filtration chromatography and Western blotting of cell extracts prepared from control (37°C) or heat-shocked (44°C) cells. (*Upper panel*) Endogenous human HSF1, detectable by increasing the extract concentration by fivefold; (*lower panels*) DmHSF and human HSF1, as indicated. (Reprinted, with permission, from Clos et al. 1993, © Macmillan Magazines Limited.)

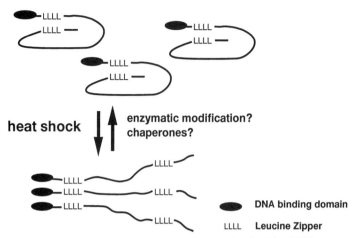

Figure 5 HSF trimerization during heat shock. The HSF monomer contains a DNA-binding domain (*closed oval*) near its amino terminus and an adjacent trimerization domain that is composed of hydrophobic heptad repeats (LLLL, not necessarily leucines). The maintenance of the monomeric state depends, in part, on the more carboxy-terminal hydrophobic repeats that may interact direct-ly with the amino-terminal repeats to suppress trimerization. Upon heat shock, the monomer's structure is disrupted and trimers form, probably through a triple-stranded coiled-coil. The mechanism of control of the monomer-trimer transition is unknown, but it could be dependent on modifications of HSF, on in-teractions of HSF with protein chaperones, and on physical changes in the en-vironment.

HSF protein in the cytoplasmic fraction of unshocked *Drosophila* or mammalian cells (Larson et al. 1988; Mosser et al. 1990; Zimarino et al. 1990). To reconcile this discrepancy, we examined by immunostaining the localization of DmHSF during biochemical fractionation. When cytoplasmic extracts were prepared by hypotonic lysis of unshocked *Drosophila* cells in 10 mM KCl buffer (for method, see Dignam et al. 1983), the resulting nuclei displayed little staining for DmHSF, indicat-ing a susceptibility of the DmHSF monomer to hypotonic extraction (Fig. 6B). In contrast, the same procedure applied to isolated nuclei from heat-shocked cells showed substantial retention of nuclear staining, and this staining was eliminated by extraction with high salt (0.4 M NaCl) (Fig. 6B). Gel mobility shift assays coupled with antibody-induced ac-tivation of HSF binding in vitro confirmed that the cytoplasmic extract from unshocked cells did contain the bulk of DmHSF, whereas it was the nuclear (0.4 M NaCl) extract from heat-shocked cells that had high levels of DmHSF (Fig. 6C). The results indicate that DmHSF monomers are mostly localized in the nucleus before heat shock, but they undergo leakage during cell lysis and subcellular fractionation. The apparent dichotomy of nuclear localization in fixed cells and cytoplasmic detection after subcellular fractionation are also shared by several other transcription factors, including the estrogen receptor (Welshons et al. 1984), the retinoblastoma protein (Mittnacht and Weinberg 1991), and the Ets-1 protein (Pognonec et al. 1989). Accordingly, the identification of HSF proteins in the cytoplasmic fraction after hypotonic lysis should not be construed as indicating a cytoplasmic location for HSF in intact cells.

Similar to the nuclear localization of DmHSF, the human HSF1 monomer in HeLa cells is predominantly nuclear, with a diffuse staining pattern and exclusion from the nucleolus under nonshock conditions (Fig. 6D). The diffuse nuclear staining becomes punctate after heat shock, suggesting an aggregation of human HSF1 trimers, possibly at chromosomal target sites. The predominance of the nuclear staining for mammalian HSF1 proteins is curiously dependent on the source of the polyclonal antisera used for indirect immunofluorescence. Sarge et al. (1993) have found both nuclear and cytoplasmic staining HSF in fixed, unshocked HeLa cells using an antiserum raised against bacterially expressed murine HSF1 protein, purified by SDS-gel electrophoresis. However, only nuclear staining was observed using antiserum from our laboratory, which was generated against bacterially expressed human HSF1, purified under native conditions (K. Sarge, S. Murphy, R. Morimoto, S.K. Rabindran, and C. Wu, unpubl.). These results indicate the interesting possibility that a subpopulation of HSF, present in the cytoplasmic compartment and recognizable only with antisera raised to the denatured protein, undergoes translocation to the nucleus after heat shock. We have also noticed a somewhat weaker recognition of the inactive form of HSF using the polyclonal sera prepared in our laboratory; this may be related to differences in the conformation or modification of the proteins.

VII.   DISTRIBUTION OF HSF ON POLYTENE CHROMOSOMES

The larval salivary gland polytene chromosomes of *D. melanogaster* provide unique opportunities for a cytogenetic visualization of HSF binding to the *Drosophila* genome. As revealed by indirect immunofluorescence, the DmHSF monomer is diffusely located over the entire chromatin in the absence of a heat shock (Westwood et al. 1991). After heat shock for 25 minutes at 37°C, the diffuse pattern is replaced by discrete staining at more than 200 distinct cytological loci. The loci include all nine major heat shock puff sites described by Ashburner (1970), as well as major de-

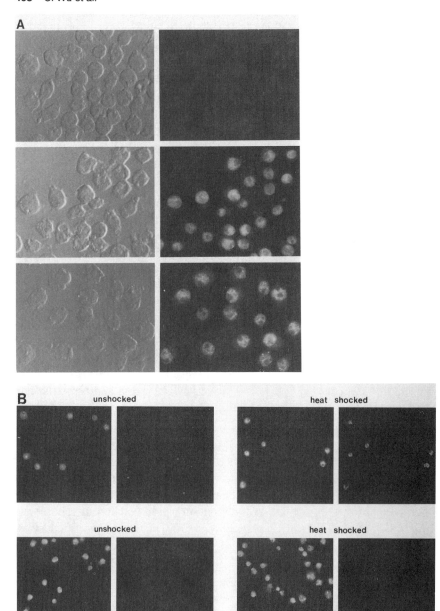

Figure 6 (*A*) Nuclear localization of DmHSF. (*Left*) *Drosophila* SL2 cells visualized by differential interference microscopy. (*Right*) SL2 cells stained with preimmune serum (*top*) and anti-DmHSF serum, followed by FITC-conjugated secondary antibody (*middle, bottom*). Cells were unshocked in the *top and middle panels,* and heat shocked for 15 min at 37°C in the *bottom panel* (Reprinted, with permission, from Westwood et al. 1991, © Macmillan Magazines Limited.) (*B*) Loss of nuclear localization during biochemical extraction. Isolated nuclei

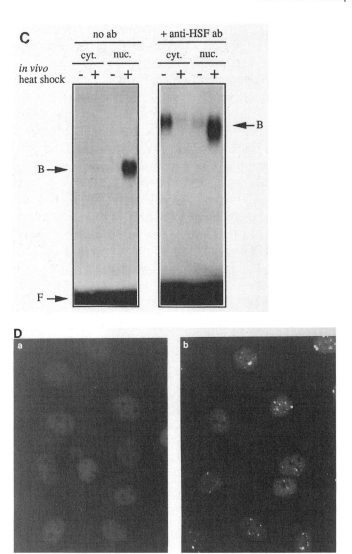

from unshocked and heat-shocked SL2 cells were stained for DNA using Hoescht 33342 (*left panel* of each pair) and for HSF (*right panel* of each pair) after incubation with hypotonic buffer (*upper panels*) or with buffer containing 0.4 M NaCl (*lower panels*). (*C*) Distribution of DmHSF in the cytoplasmic fraction (cyt.) of unshocked SL2 cells and in the nuclear fraction (nuc.) of heat-shocked cells. Gel mobility shift assay of DmHSF-binding activity in the absence (*left panel*) or presence (*right panel*) of polyclonal antibodies to DmHSF which are capable of activating the inert form of HSF (Zimarino et al. 1990). Cytoplasmic and nuclear extracts were prepared from unshocked and heat-shocked cells as described (Westwood and Wu 1993). (*D*) Nuclear localization of human HSF1 in unshocked HeLa cells (*left*) and aggregation of the protein in discrete spots after heat shock (*right*).

velopmental loci, which could reflect an alternate function for DmHSF trimers in the repression of normal gene expression during heat stress. To examine the in vivo kinetics of DmHSF trimer binding to its cognate sites at the earliest stages of induction, we analyzed the chromosomal distribution of DmHSF after heat shock of salivary glands in culture for 2, 10, and 30 seconds and 1 minute (Fig. 7). A 2-second heat shock results in an HSF-staining pattern that is virtually indistinguishable from the pattern obtained with unshocked glands (Westwood et al. 1991). DmHSF is found diffusely staining the chromosomes with a slight accumulation at some sites, including the 63BC (*hsp82*) locus which shows fairly bright staining (Fig. 7a). This preferential accumulation at 63BC may reflect the presence of a small amount of DmHSF trimers bound to the highest-affinity target known for DmHSF, the *hsp82* promoter carrying eight contiguous 5-bp modules (Xiao et al. 1991). The first discernible change in the distribution of DmHSF is evident after 10 seconds of heat shock (Fig. 7b).

More than 20 distinct loci show preferential DmHSF staining, including the heat shock loci at 63BC (*hsp82*), 64F, 67B (*hsp27, hsp26, hsp23, hsp22*), 87A/87C (*hsp70*), as well as the developmental puffs at 74E and 75AB. By 30 seconds of heat shock, DmHSF is localized at more than 50 distinct loci and the diffuse staining of HSF over the rest of the chromatin is decreased. DmHSF is now found at all major heat shock loci with bright staining seen at 63BC, 67B, 87A, and 87C, prior to the onset of puffing at these loci (Fig. 7c). After 1 minute of heat shock, DmHSF is found at more than 100 loci, including all of the major heat shock loci, now in the puffed state (Fig. 7d).

The distribution of DmHSF on polytene chromosomes after a more extended heat shock of whole larvae (5 and 15 min) revealed similar intense staining at the heat shock puffs at 63BC, 67B, 87A/87C, and the other major heat shock puffs. Only the heat shock puff at 93D, which lacks multiple HSEs and appears not to be under the primary control of HSF, fails to show strong staining for HSF (Fig. 8a,b) (Westwood et al. 1991). The staining pattern is essentially the same at 30 minutes of induction, although in many of the chromosome preparations, the heat shock puffs appear to be in the initial phase of regression (data not shown). After 60 minutes of induction, some chromosome spreads show considerable puff regression and little specific localization of DmHSF, whereas others from the same salivary gland retain the heat shock puffs and the specific DmHSF-staining pattern. However, after 120 and 180 minutes of continuous heat shock, the site-specific localization of DmHSF is almost absent for all chromosome spreads, with residual staining at a few loci, including 63BC (Fig. 8c). The return of the HSF-staining pattern to the normal state after prolonged incubation at the high

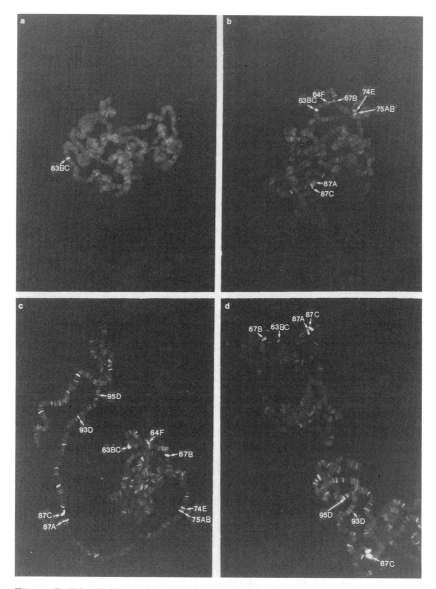

Figure 7 Distribution of DmHSF on polytene chromosomes after an in vitro heat shock for 2 sec (*a*), 10 sec (*b*), 30 sec (*c*), and 1 min (*d*). Locations of the heat shock puff sites (33B, 63BC, 64F, 67B, 70A, 87A, 87C, 93D, and 95D) and two developmental puff sites (74E and 75AB) are indicated. Late third instar larvae grown at room temperature (23–24°C) were quickly dissected in PBS and transferred to a depression slide containing TB1 buffer (Bonner 1981). The slide was kept in a humid chamber maintained at 36.0 ± 0.1°C prior to and during the heat treatment. After heat shock, glands were transferred to fixative (50% acetic acid, 3.7% formaldehyde), squashed, and analyzed by indirect immunofluorescence as described by Westwood et al. (1991).

412 C. Wu et al.

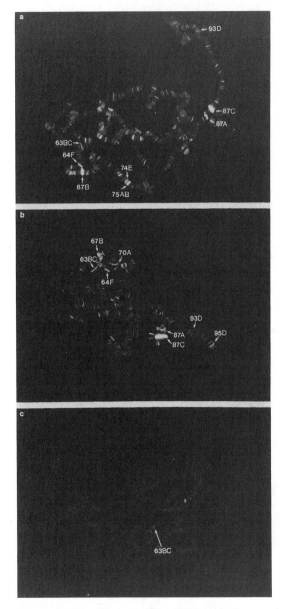

Figure 8 Distribution of DmHSF on polytene chromosomes after an in vivo heat shock of 5 (*a*), 15 (*b*), and 180 (*c*) min. The heat shock puffs and the 74E and 75AB puffs are indicated where possible. Late third instar larvae grown at room temperature were transferred to prewarmed vials containing moistened filter paper, which were then sealed and submerged in a 36.0 ± 0.1°C water bath. After heat shock, chromosome spreads were prepared and analyzed by indirect immunofluorescence as described above.

temperature underscores the involvement of cellular factors that must act to deactivate the protein even at the heat shock temperature.

VIII. MECHANISM OF *TRANS*-ACTIVATION BY HSF

The mechanism by which the binding of HSF trimers leads to a stimulation of transcription is unknown. In *Drosophila*, heat shock promoters are primed for rapid induction, with an accessible chromatin structure that is hypersensitive to nucleolytic probes (Wu 1980, 1984; Costlow and Lis 1984). At least two transcription factors are known to bind heat shock promoter sequences under uninduced conditions: the GAGA factor and the TATA-binding protein complex (Wu 1984; Thomas and Elgin 1988; Giardina et al. 1992). These constitutive factors are likely to be involved in generating a nucleosome-free region for the subsequent binding of DmHSF. In addition, the presence of a paused RNA polymerase at the 5′ end of uninduced heat shock genes indicates that RNA polymerase binding and initiation have been completed prior to the arrival of activated DmHSF trimers (Gilmour and Lis 1986; Rougvie and Lis 1988). Therefore, DmHSF must act directly or indirectly to accelerate the rate of escape of the paused polymerase. The generation of accessible heat shock promoters and establishment of the paused RNA polymerase are discussed in detail in the chapter by Fernandes et al. (this volume). The signaling mechanisms for HSF trimer assembly and disassembly, the generation of an accessible chromatin structure, and the activation of the paused RNA polymerase II present unique opportunities for future investigations on the induction of the heat shock response and on general aspects of eukaryotic transcriptional control.

ACKNOWLEDGMENTS

This study was supported by the Intramural Research Program of the National Cancer Institute. R.I.H. was supported by an HHMI-NIH Research Scholar Award.

REFERENCES

Abravaya, K., M.P. Myers, S.P. Murphy, and R.I. Morimoto. 1992. The human heat shock protein *hsp70* interacts with HSF, the transcription factor that regulates heat shock gene expression. *Genes & Dev.* **6:** 1153–1164.

Ashburner, M. 1970. Patterns of puffing activity in the salivary gland chromosomes of *Drosophila*. V. Responses to environmental treatments. *Chromosoma* **31:** 356–376.

Baler, R., G. Dahl, and R. Voellmy. 1993. Activation of heat shock genes is accompanied by oligomerization, modification, and rapid translocation of heat shock transcription factor HSF1. *Mol. Cell. Biol.* **13:** 2486–2496.

Baler, R., W.J. Welch, and R. Voellmy. 1992. Heat shock gene regulation by nascent polypeptides and denatured proteins: hsp70 as a potential autoregulatory factor. *J. Cell Biol.* **117:** 1151–1159.

Bonner, J.J. 1981. Induction of *Drosophila* heat-shock puffs in isolated polytene nuclei. *Dev. Biol.* **86:** 409–418.

Bonner, J., S. Heyward, and D. Fackenthal. 1992. Temperature-dependent regulation of a heterologous transcriptional activation domain fused to yeast heat shock transcription factor. *Mol. Cell. Biol.* **12:** 1021–1030.

Clos, J., T. Westwood, P.B. Becker, S. Wilson, K. Lambert, and C. Wu. 1990. Molecular cloning and expression of a hexameric *Drosophila* heat shock factor subject to negative regulation. *Cell* **63:** 1085–1097.

Costlow, N. and J.T. Lis. 1984. High-resolution mapping of DNaseI-hypersensitive sites of *Drosophila* heat shock genes in *Drosophila melanogaster* and *Saccharomyces cerevisiae. Mol. Cell. Biol.* **4:** 1853–1863.

Dignam, J.D., P.L. Martin, B.S. Shastry, and R.G. Roeder. 1983. Eukaryotic gene transcription with purified components. *Methods Enzymol.* **101:** 582–598.

Giardina, C., M. Perez-Riba, and J.T. Lis. 1992. Promoter melting and TFIID complexes on *Drosophila* genes in vivo. *Genes Dev.* **6:** 2190–2200.

Gilmour, D.S. and J.T. Lis. 1986. RNA polymerase II interacts with the promoter region of the noninduced hsp70 gene in *Drosophila melanogaster* cells. *Mol. Cell. Biol.* **6:** 3984–3989.

Gross, D.S., K.E. English, K.W. Collins, and S. Lee. 1990. Genomic footprinting of the yeast HSP82 promoter reveals marked distortion of the DNA helix and constitutive occupancy of heat shock and TATA elements. *J. Mol. Biol.* **216:** 611–631.

Harrison, C., A. Bohm, and H. Nelson. 1994. Crystal structure of the DNA-binding domain of the heat shock transcription factor. *Science* (in press).

Jakobsen, B.K. and H.R.B. Pelham. 1991. A conserved heptapeptide restrains the activity of the yeast heat shock transcription factor. *EMBO J.* **10:** 369–375.

Kingston, R.E., T.J. Schuetz, and Z. Larin. 1987. Heat-inducible human factor that binds to a human *hsp70* promoter. *Mol. Cell. Biol.* **7:** 372–375.

Larson, J.S., T.J. Schuetz, and R.E. Kingston. 1988. Activation in vitro of sequence-specific DNA binding by a human regulatory factor. *Nature* **335:** 372–375.

Lis, J.T., H. Xiao, and O. Perisic. 1990. Modular units of heat shock regulatory regions: Structure and function. In *Stress proteins in biology and medicine* (ed. R. Morimoto et al.), pp. 411–428. Cold Spring Harbor Laboratory Press, Cold Spring Harbor, New York.

Mittnacht, S. and R.A. Weinberg. 1991. G1/S phosphorylation of the retinoblastoma protein is associated with an altered affinity for the nuclear compartment. *Cell* **65:** 381–393.

Morimoto, R.I., A. Tissieres, and C. Georgopoulos. 1990. The stress response, function of the proteins, and perspectives. In *Stress proteins in biology and medicine* (ed. R. Morimoto et al.), pp. 1–36. Cold Spring Harbor Laboratory Press, Cold Spring Harbor, New York.

Mosser, D.D., J. Duchaine, and B. Massie. 1993. The DNA-binding activity of human heat shock transcription factor is regulated in vivo by hsp70. *Mol. Cell. Biol.* **13:** 5427–5438.

Mosser, D.D., P.T. Kotzbauer, K.D. Sarge, and R.I. Morimoto. 1990. In vitro activation of heat shock transcription factor DNA-binding by calcium and biochemical conditions that affect protein conformation. *Proc. Natl. Acad. Sci.* **87:** 3748–3752.

Murre, C., P. Schonleber-McCaw, and D. Baltimore. 1989. New DNA binding and dimerization motif in immunoglobulin enhancer binding, daugterless, MyoD, and myc proteins. *Cell* **56:** 777–783.

Nieto-Sotelo, J., G. Wiederrecht, A. Okuda, and C.S. Parker. 1990. The yeast heat shock transcription factor contains a transcriptional activation domain whose activity is repressed under nonshock conditions. *Cell* **62:** 807–817.

Perisic, O., H. Xiao, and J.T. Lis. 1989. Stable binding of *Drosophila* heat shock factor to head-to-head and tail-to-tail repeats of a conserved 5 bp recognition unit. *Cell* **59:** 797–806.

Peteranderl, R. and H.C.M. Nelson. 1993. Trimerization of the heat shock transcription factor by a triple-stranded a-helical coiled-coil. *Biochemistry* **13:** 12272–12276.

Pognonec, P., K.E. Boulukos, and J. Ghysdael. 1989. The c-ets-1 protein is chromatin associated and binds to DNA in vitro. *Oncogene* **4:** 691–697.

Rabindran, S.K., G. Giorgi, J. Clos, and C. Wu. 1991. Molecular cloning and expression of a human heat shock factor, HSF1. *Proc. Natl. Acad. Sci.* **88:** 6906–6910.

Rabindran, S.K., R.I. Haroun, J. Clos, J. Wisniewski, and C. Wu. 1993. Regulation of heat shock factor trimerization: Role of a conserved leucine zipper. *Science* **259:** 230–234.

Rougvie, A.E. and J.T. Lis. 1988. The RNA polymerase II molecule at the 5′ end of the uninduced *hsp70* gene of *D. melanogaster* is transcriptionally engaged. *Cell* **54:** 795–804.

Sarge, K., S.P. Murphy, and R.I. Morimoto. 1993. Activation of heat shock transcription by HSFI involves oligomerization, acquisition of DNA binding activity, and nuclear localization and can occur in the absence of stress. *Mol. Cell. Biol.* **13:** 1392–1407.

Sarge, K., V. Zimarino, K. Holm, C. Wu, and R.I. Morimoto. 1991. Cloning and characterization of two mouse heat shock factors with distinct inducible and constitutive DNA binding activity. *Genes Dev.* **5:** 1902–1911.

Scharf, K.-D., T. Materna, E. Treuter, and L. Nover. 1994. Heat stress promoters and transcription factors. In *Plant promoters and transcription factors* (ed. L. Nover), pp. 121–158. Springer-Verlag, Berlin.

Scharf, K.D., S. Rose, W. Zott, F. Schoffl, and L. Nover. 1990. Three tomato genes code for heat stress transcription factors with a region of remarkable homology to the DNA binding domain of yeast HSF. *EMBO J.* **9:** 4495–4501.

Schlesinger, M. and C. Ryan. 1993. An ATP- and hsc70-dependent oligomerization of nascent heat-shock factor (HSF) polypeptide suggests that HSF itself could be a "sensor" for the cellular stress response. *Protein Sci.* **2:** 1356–1360.

Sorger, P.K. 1990. Yeast heat shock factor contains separable transient and sustained response transcriptional activators. *Cell* **62:** 793–805.

Sorger, P.K. and H.C.M. Nelson. 1989. Trimerization of a yeast transcriptional activator via a coiled-coil motif. *Cell* **59:** 807–813.

Sorger, P.K. and H.R.B. Pelham. 1987. Purification and characterization of a heat shock element binding protein from yeast. *EMBO J.* **6:** 3035–3041.

Thomas, G.H. and S.C.R. Elgin. 1988. Protein/DNA architecture of the DNase I hypersensitive region of the *Drosophila hsp26* promoter. *EMBO J.* **7:** 2191–2201.

Treuter, E., L. Nover, K. Ohme, and K.-D. Scharf. 1993. Promoter specificity and deletion analysis of three tomato heat stress transcription factors. *Mol. Gen. Genet.* **240:** 113–125.

Vuister, G.W., S.-J. Kim, C. Wu, and A. Bax. 1994. NMR evidences for similarities between the DNA-binding regions of *Drosophila melanogaster* heat shock factor and the helix-turn-helix and HNF/*fork head* families of transcription factors. *Biochemistry* **33:**

10–16.

Welshons, W.V., M.E. Liberman, and J. Gorski. 1984. Nuclear localization of unoccupied oestrogen receptors. *Nature* **307:** 747–749.

Westwood, J.T. and C. Wu. 1993. Activation of *Drosophila* heat shock factor: Conformational change associated with monomer to trimer transition. *Mol. Cell. Biol.* **13:** 3481–3486.

Westwood, J.T., J. Clos, and C. Wu. 1991. Stress-induced oligomerization and chromosomal relocalization of heat-shock factor. *Nature* **353:** 822–827.

Wiederrecht, G., D. Seto, and C.S. Parker. 1988. Isolation of the gene encoding the *S. cerevisiae* heat shock transcription factor. *Cell* **54:** 841–853.

Wu, C. 1980. The 5′ ends of *Drosophila* heat shock genes in chromatin are hypersensitive to DNaseI. *Nature* **286:** 854–860.

———. 1984. Two protein-binding sites in chromatin implicated in the activation of heat-shock genes. *Nature* **309:** 229–234.

Wu, C., V. Zimarino, C. Tsai, B. Walker, and S. Wilson. 1990. Transcriptional regulation of heat shock genes. In *Stress proteins in biology and medicine* (ed. R. Morimoto et al.), pp. 429–442. Cold Spring Harbor Laboratory Press, Cold Spring Harbor, New York.

Xiao, H., O. Perisic, and J.T. Lis. 1991. Cooperative binding of *Drosophila* heat shock factor to arrays of a conserved 5 bp unit. *Cell* **64:** 585–593.

Zimarino, V. and C. Wu. 1987. Induction of sequence-specific binding of *Drosophila* heat shock activator. *Nature* **327:** 727–730.

Zimarino, V., S. Wilson, and C. Wu. 1990. Antibody-mediated activation of *Drosophila* heat shock factor in vitro. *Science* **249:** 546–549.

17

Regulation of Heat Shock Gene Transcription by a Family of Heat Shock Factors

Richard I. Morimoto, Donald A. Jurivich, Paul E. Kroeger, Sameer K. Mathur, Shawn P. Murphy, Akira Nakai, Kevin Sarge, Klara Abravaya, and Lea T. Sistonen
Department of Biochemistry,
Molecular Biology, and Cell Biology
Northwestern University
Evanston, Illinois 60208

I. INTRODUCTION

All bacterial, plant, and animal cells must cope with rapid changes in their environment, including exposure to elevated temperatures, heavy metals, toxins, oxidants, and bacterial and viral infection, by a rapid and often dramatic change in the patterns of gene expression, resulting in the elevated synthesis of a family of heat shock proteins and molecular chaperones (Lindquist and Craig 1988; Morimoto et al. 1990). Heat shock proteins ensure survival under stressful conditions, which if left unchecked, lead to irreversible cell damage and untimely cell death. They have essential roles in protein biosynthesis, specifically in the

The Biology of Heat Shock Proteins and Molecular Chaperones
©1994 Cold Spring Harbor Laboratory Press 0-87969-427-0/94 $5 + .00

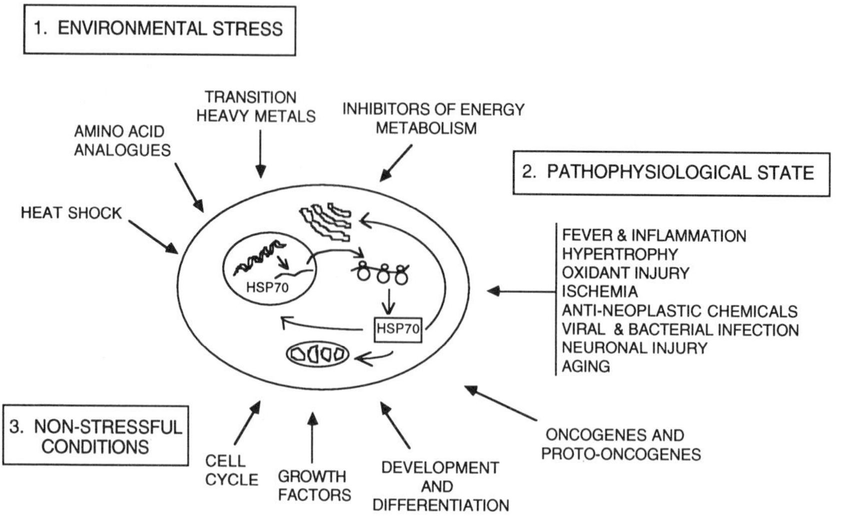

Figure 1 Conditions that result in the induction of heat shock gene expression in eukaryotes. Representation of three general classes of conditions known to result in the elevated expression of stress proteins, including (1) environmental and physiological stress, (2) pathophysiological states including conditions of disease, and (3) nonstressful conditions such as cell growth and development. Each condition acts on the cell as diagrammed in this figure and, in the case of environmental stress and certain pathophysiological states, leads to the activation of heat shock gene expression and the synthesis of heat shock proteins.

synthesis, transport, and translocation of proteins and in the regulation of protein conformation, and are also referred to as molecular chaperones (Craig et al.; Langer and Neupert; Brodsky and Schekman; Gething et al.; Dice et al.; Hightower et al.; Georgopoulos et al.; Frydman and Hartl; Randall et al.; Willison and Kuboda; Bohen and Yamamoto; all this volume). Heat shock proteins constitute a surprisingly large fraction of the protein within a cell, amounting to 5–10% of the total protein mass in cells growing under ambient conditions. Yet, despite their abundance, the genes encoding heat shock proteins are rapidly induced in response to stressful conditions. The consequence of this genetic switch, which is activated in response to a perturbation of the physiological state of the cell, is the elevated expression of constitutively expressed heat shock proteins and the de novo induction of heat shock genes that are expressed primarily in response to stress.

A representation of the variety of physiological and environmental stimuli known to induce heat shock gene expression is shown in Figure 1 and include (1) environmental stresses such as exposure to heat shock, amino acid analogs, and heavy metals; (2) pathophysiology and disease states including oxidative stress, fever, and inflammation; and (3) non-stress conditions including the cell cycle, growth factors, serum stimulation, development, differentiation, and activation by certain oncogenes. The diversity of these conditions has elicited numerous questions to understand how cells detect a change in their physiological state or more broadly stated, "stress." Does this occur through receptor-ligand interactions? If so, what is the receptor and what are the ligands? Such a mechanism will need to accommodate the simple observations that the temperature of activation of heat shock transcription varies among organisms proportional to the optimal ambient temperature and that, as indicated by recent data, the threshold of activation can be manipulated experimentally even within a particular cell type (Jurivich et al. 1994). Therefore, this putative receptor for physiological stress is likely to correspond to an ancient yet novel form of signal transduction, with origins from a time in evolution when exposure to adverse environmental conditions was commonplace. Furthermore, the nature of the receptor or receptor complex may reflect the multitude of conditions known to activate the heat shock transcriptional response. Given the diversity of these cellular responses, a worthy goal is to understand the mechanism(s) by which physiological stress is detected within the cell and to establish approaches to quantify the effects of stress and how this information is transduced to the transcriptional apparatus.

In eukaryotic cells, the heat shock response involves transcriptional activation mediated by heat shock factors (HSFs). In this chapter, we consider the role of a family of vertebrate HSFs in the transcriptional regulation of heat shock genes. We emphasize the differential activation of HSF1, HSF2, and HSF3 in the detection of cellular "stress" signals, concentrating on the regulation and biochemical properties of HSF1 and HSF2. Other topics include studies on the uncoupling of heat shock transcriptional responses through the modulation of HSF expression or through access to specific heat shock elements, the pharmacological manipulation of HSF activity, and the role of the heat shock transcriptional response in inflammation and disease.

II. STRESS-INDUCIBLE TRANSCRIPTION BY A FAMILY OF HEAT SHOCK FACTORS: GENERAL FEATURES AND DOMAIN ORGANIZATION

In yeast and *Drosophila*, only one heat shock transcription factor has been cloned (Wiederrecht et al. 1988; Clos et al. 1990; see Wu et al., this volume). However, in larger eukaryotes, there exists a family of HSFs, at least three in chicken and tomato, and two in the mouse and humans (Scharf et al. 1990; Rabindran et al. 1991; Sarge et al. 1991; Schuetz et al. 1991; Nakai and Morimoto 1993). Expression of the HSF genes in most species is constitutive and not stress-responsive, whereas some of the tomato HSF genes have been shown to be stress-induced (Scharf et al. 1990). Vertebrate HSF1 is the functional homolog to the general HSF found in smaller eukaryotes and has been shown to be activated in vivo by heat shock and numerous other forms of physiological stress (Baler et al. 1993; Sarge et al. 1993; Holbrook and Udelsman, this volume). In contrast to the stress-responsive role for HSF1, HSF2 acquires DNA-binding activity during hemin treatment of human K562 cells (Theodorakis et al. 1989; Sistonen et al. 1992, 1994) and during murine spermatocyte differentiation (K.D. Sarge et al., in prep.). Furthermore, HSF2 DNA-binding activity is constitutive in certain embryonal carcinoma cell lines and during the blastocyst stage of early mouse development (Mezger et al. 1989; S. Murphy, pers. comm.; V. Mezger, unpubl.). Less is known about the properties of HSF3, which has been cloned from chicken (cHSF3). HSF3 is expressed ubiquitously, and its DNA-binding properties are negatively regulated as is known for HSF1 and HSF2 (Nakai and Morimoto 1993). However, unlike HSF1, which is rapidly activated by stress in all cells, HSF3 exhibits delayed heat shock activation properties for DNA-binding activity in a cell-type-specific manner (A. Nakai et al., in prep.). In vertebrate cells, multiple HSFs are coexpressed, yet the DNA-binding activity of each factor is negatively regulated and becomes activated in response to different cellular signals, suggesting multiple regulatory mechanisms to control HSF activity.

The general structural features of mouse and chicken HSF domains are shown in Figure 2. The cloned HSFs vary in size: 301 amino acids for tomato HSF24, 512 for tomato HSF8, 503 and 529 for mouse and human HSF1, 517 and 536 for mouse and human HSF2, 691 for *Drosophila* HSF, and 833 for *Saccharomyces cerevisiae* HSF (Scharf et al. 1990; Rabindran et al. 1991; Sarge et al. 1991; Schuetz et al. 1991; Nakai and Morimoto 1993). Despite this wide variation in size, and limited overall homology of approximately 40% amino acid identity between different HSFs, all HSFs have two highly conserved features: the amino-terminal localized DNA-binding domain of approximately 100 amino acids and a motif of hydrophobic heptad repeats that mediates the oligomerization of HSF (Perisic et al. 1989; Sorger and Nelson 1989; Clos et al. 1990; Nieto-Sotelo et al. 1990; Rabindran et al. 1993). The level of conservation in the DNA-binding domain is not surprising as each HSF binds to the highly conserved heat shock element (HSE) DNA-binding motif. A feature found only among the animal HSFs is another leucine zipper located in the carboxy-terminal domain. This zipper has been demonstrated to have a role in the negative regulation of HSF

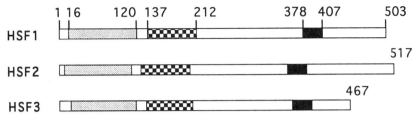

Figure 2   HSF gene family and domains of HSFs. Mouse HSF1, mHSF2, and chicken HSF3 are schematically diagrammed and provide a reference for structural motifs in each class of HSF. The approximately 100-amino-acid DNA-binding domain (*stippled box*) is designated at the amino terminus. Adjacent to the DNA-binding domain is the conserved hydrophobic heptad repeats (*checkered box*). At the carboxyl terminus, there is an additional leucine zipper motif (*closed box*) involved in negative regulation of HSF activity. For reference, the amino acids that demarcate the borders of each domain in mHSF1 have been designated above the schematic, and the length of each protein is noted at the carboxyl terminus.

DNA-binding activity through interactions with the coiled-coil domain adjacent to the DNA-binding domain (Rabindran et al. 1993). Deletion or mutation of the carboxy-terminal leucine zipper results in constitutive DNA-binding activity of the *Drosophila* HSF or cHSF3 (Nakai and Morimoto 1993; Rabindran et al. 1993).

The analysis of HSF structure has been extended to the characterization of domains responsible for activation of transcription. A detailed analysis of the activation domains of the HSFs from the yeasts *S. cerevisiae* and *Kluyveromyces lactis* has identified separate domains for full transcriptional activity (Nieto-Sotelo et al. 1990; Jakobsen and Pelham 1991; Chen et al. 1993). A comprehensive deletional analysis performed by Chen et al. (1993) has provided direct support for distinct amino acid residues required in *K. lactis* HSF for activation of HSF during the heat shock response that are distinct from the residues required for sustained response and growth at high temperature. A more restricted analysis of the activation domains of mouse HSF1 (mHSF1) and mHSF2 through the use of deletional analysis and Gal4 fusion proteins has revealed that the activation domains of mHSF1 (425–471 amino acids) and mHSF2 (473–517 amino acids) are carboxy-terminal to the fourth leucine zipper (P.E. Kroeger and R.I. Morimoto, unpubl.). These sequences are acidic in character and function as constitutive activators in this context. Heat shock does not further induce the level of transcription, suggesting that appending the activation domain to Gal4 relieves the negative regulation that normally occurs in HSF at control temperatures.

Thus, in contrast to yeast HSF, where multiple activation domains have been defined that regulate constitutive and inducible transcriptional activation, the mouse HSFs appear to have a single activation domain. Perhaps the appearance of additional HSFs in higher eukaryotes is the result of division of labor, with multiple HSFs filling the requirements once regulated by a single HSF in yeast. Characterization of the activation domain(s) from each HSF family member and their regulation will further substantiate these ideas. Additionally, it will also be important to perform analyses of different members of the mammalian HSF family in a null background generated through the use of recombination knock-out technology.

III. ACTIVATION OF HSF1 AND HSF2 DNA-BINDING ACTIVITY IN RESPONSE TO DISTINCT PHYSIOLOGICAL EVENTS

The discovery of multiple HSFs in larger eukaryotes immediately posed questions as to their role in the transcriptional regulation of heat shock genes. Foremost was to identify conditions that regulate the activities of each HSF. Do they provide redundant functions or does each HSF confer distinct biological roles in the differential response to the myriad of conditions that elicit inducible transcription? Among the cloned HSF genes, an interesting story has developed for the vertebrate HSF1 and HSF2 proteins (Fig. 3).

Comparison of the functional characteristics and biochemical properties of these two factors in unstressed and heat-shocked cells was facilitated by polyclonal antibodies specific to HSF1 and HSF2 (Fig. 4) (Sistonen et al. 1992; Sarge et al. 1993). This analysis revealed that HSF1 exhibited the properties expected for the transcription factor responsible for induction of heat shock gene transcription in response to stress, including oligomerization from a monomer to trimer, acquisition of DNA-binding ability, increase in phosphorylation, and relocalization to the nucleus (Sistonen et al. 1992, 1994; Baler et al. 1993; Sarge et al. 1993). These results were consistent with previous findings that the amino acid sequence of peptide fragments of HSF purified from heat-shocked human cells corresponded to the derived amino acid sequences of the cloned HSF1 cDNA (Rabindran et al. 1991; Schuetz et al. 1991). It has been possible to establish the order of events in HSF1 activation, in particular the role of phosphorylation, by examination of the kinetics of activation and through the use of anti-inflammatory drugs that activate HSF1 DNA-binding activity. Conversion of the control monomer to activated trimer correlates with DNA-binding activity measured by in vitro

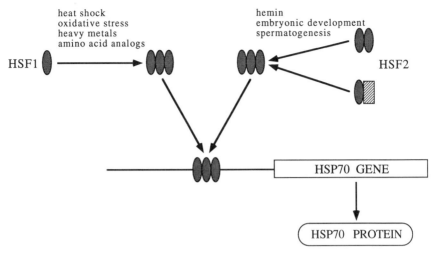

Figure 3 Differential regulation of HSF1 and HSF2 DNA-binding activity. Vertebrate cells express multiple HSFs including HSF1 and HSF2. HSF1 is maintained in the control non-DNA-binding state as a monomer that becomes stress-activated by heat shock, oxidative stress, heavy metals, and exposure to amino acid analogs to a homotrimer that binds to the promoter of the *hsp70* gene, leading to elevated hsp70 accumulation. HSF2 is maintained in the control state as a dimer (represented here as either a homodimer or heterodimer) and becomes activated to the trimeric state during hemin treatment and during spermatogenesis. HSF2 is constitutively active as a trimer during early embryonic development. Activation of either HSF1 or HSF2 leads to elevated transcription of the *hsp70* gene and other genes that contain HSEs in the promoter region.

and in vivo methods and can occur in the absence of inducible phosphorylation of HSF1. Sodium salicylate induces HSF1 DNA-binding activity and in vivo occupancy of the hsp70 HSEs, yet the salicylate-induced HSF1 is not transcriptionally competent (Jurivich et al. 1992). Whether this uncoupling of HSF activation properties is because salicylate-induced HSF1 is not inducibly phosphorylated remains unknown. The inducible phosphorylation of HSF1 is stress dependent with heat shock, heavy metals, and arachidonate treatment leading to the fully phosphorylated state, whereas amino acid analogs result in the activation of the unphosphorylated state of HSF1 (Sarge et al. 1993; Jurivich et al. 1994).

   In contrast to the inducible regulation of HSF1, the properties of mouse and human HSF2, including DNA-binding ability, oligomeric state, and covalent modification, are not altered by heat shock treatment, indicating that HSF2 does not have a primary role in stress-induced heat shock gene transcription (Sistonen et al. 1992, 1994; Sarge et al. 1993).

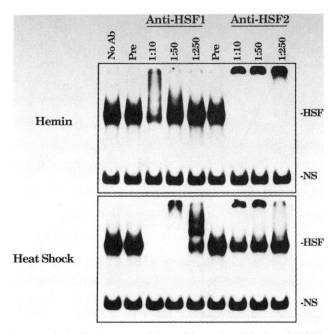

Figure 4 Antibody recognition of heat-shock-induced HSF1 and hemin-induced HSF2. Whole-cell extracts from hemin-treated (*upper panel*) and heat-shocked (*lower panel*) K562 cells were incubated either alone (no Ab), with a 1:10 dilution of preimmune serum (Pre), or with a 1:10, 1:50, or 1:250 dilution of HSF1 or HSF2 antiserum at room temperature for 20 min before being subjected to gel mobility shift analysis.

These results suggest that HSF1 and HSF2 have evolved to perform distinctive regulatory aspects of heat shock gene regulation. A key insight to the regulation of HSF2 followed from studies on the expression of the major heat shock genes during nonterminal erythroid differentiation. Exposure of human K562 erythroleukemia cells to hemin induced the synthesis of hsp70 and hsp90 and the activation of HSF (Singh and Yu 1984; Theodorakis et al. 1989). Through the use of polyclonal antibodies specific to HSF1 and HSF2, it was shown that the HSE DNA-binding activity in hemin-treated K562 cells corresponded to HSF2 (Fig. 4) (Sistonen et al. 1992). The acquisition of HSF2 DNA-binding activity exhibited slower activation kinetics, relative to the rapid induction of HSF1 DNA-binding activity following heat shock. Other distinctions are that HSF2 DNA-binding activity induced by hemin treatment can be maintained in the activated state for extended periods up to 60 hours, whereas HSF1 is rapidly and transiently activated by heat shock, and that equivalent levels of HSF1 and HSF2 DNA-binding activity do not stimulate *hsp70* gene transcription to the same extent. The relative rate of transcription of the *hsp70* gene is stimulated to a lesser extent by HSF2, consistent with other studies that recombinant mHSF1 stimulates transcription in vitro more efficiently than mHSF2 (Kroeger et al. 1993). The distinct transcriptional activities of HSF1 and HSF2 may be due to the difference in occupancy of the hsp70 promoter as observed by in vivo footprinting analysis. Genomic footprinting analysis of the proximal and distal HSEs of the hsp70 promoter reveals an overall weaker protection of DNA isolated from hemin-treated cells than of DNA isolated from heat-shocked cells (Sistonen et al. 1992). Furthermore, the site-1 consensus guanine residue of the proximal HSE is not protected in hemin-treated cells, a result that is consistent with the differences detected by in vitro DNase I footprinting analysis of the hsp70 promoter bound to either HSF1 or HSF2 (Kroeger et al. 1993).

Additional support for HSF2 as a developmental activator of non-stress-induced heat shock gene transcription is demonstrated by studies on the heat shock response in early mouse embryos. Mezger et al. (1989) have identified a constitutive HSF DNA-binding activity in embryonal carcinoma cells; this activity has been subsequently shown to correspond to HSF2 DNA-binding activity (V. Mezger et al.; S. Murphy et al.; both pers. comm.). Exposure of murine embryonal carcinoma cells to heat shock leads to the appearance of HSF1 as the predominant DNA-binding activity and a striking reduction in HSF2 DNA-binding activity accompanied by reduced levels of HSF2 protein. Although these studies establish that HSF2 DNA-binding activity is regulated during mouse embryogenesis, the role of HSF2 on heat shock gene expression or on the expression of other potential regulatory targets during early embryogenesis remains unknown. Another observation that supports a role for HSF2 during development followed from a survey of the expression of HSF1 and HSF2 in different mouse tissues. The expression of HSF2 message was found to be elevated in the testes (Sarge et al. 1991). HSF2 is expressed in primary spermatocytes as shown by in situ hybridization and immunohistochemical analysis, and HSF2 DNA-binding activity is detected by gel mobility shift assay in extracts from the adult testes (K.D. Sarge et al., in prep.). Although it is tempting to suggest that the appearance of HSF2 during spermatogenesis is related to the expression of a testes-specific *hsp70* gene, the data are only suggestive at this point. These observations together with the demonstration that HSF2 DNA-binding activity is regulated upon hemin treatment of K562 cells provide additional support for the suggestion that HSF2 could have a critical role as a developmental HSF.

Table 1 Comparisons of the Properties of Mammalian HSF1 and HSF2

	HSF1		HSF2		
	37°C	42°C	37°C	42°C	hemin
Protein size					
denatured (kD)	72	87	72	72	72
native (kD)	72	270	140	140	210
Oligomeric					
state	monomer	trimer	dimer	dimer	trimer
Inducible					
phosphorylation	–	++	–	–	–
Subcellular					
localization	cyto/nuc	nuc	cyto/nuc	cyto/nuc	nuc
Gene location		15		15	
Gene copy number		1		1	

Biochemical studies of human and mouse HSFs have demonstrated that HSF is a monomer in the control form, whereas HSF2 is a dimer (Sistonen et al. 1994). The conversion of the HSF1 monomer to a trimer has been observed in human and mouse cells (Baler et al. 1993; Rabindran et al. 1993; Sarge et al. 1993) and confirmed by analysis of the recombinant protein (Kroeger et al. 1993). Subcellular localization of mammalian HSFs has been examined in both human and mouse cell lines (Sarge et al. 1993; Sistonen et al. 1994). Both mouse HSF genes have been localized to chromosome 15 and are single copy (K. Sarge, pers. comm.).

IV. COMPARISON OF THE ACTIVATION PATHWAYS OF HSF1 AND HSF2: BIOPHYSICAL PROPERTIES OF HSF1 AND HSF2

How is each factor maintained in a non-DNA-binding conformation in vivo and yet able to respond to distinct extracellular signals by activation of either HSF? One approach has been to compare the biochemical properties of both factors in their control or activated states (Table 1). The oligomeric state and shape of inactive and active HSF2 from control and hemin-treated cells and HSF1 in control and heat-shocked K562 cells were established by gel filtration chromatography and glycerol gradient sedimentation analysis (Sistonen et al. 1994). The estimates of the native molecular mass for the control and activated forms of HSF1 are approximately 69 kD and 178 kD, respectively. These observations are consistent with studies showing that HSF1 undergoes a transition from a monomer to a trimer upon activation (Baler et al. 1993; Rabindran et al. 1993; Sarge et al. 1993). For the control and activated forms of HSF2, a calculated native molecular mass of approximately 127 kD and 202 kD was obtained (Sistonen et al. 1994). Taken together with the results from chemical cross-linking analysis that both inactive HSF2 and active HSF2 have a size of approximately 140 kD and 200 kD, respectively, the control non-DNA-binding form of HSF2 is primarily a dimer, whereas the activated DNA-binding form of HSF2 is a trimer. These results reveal that HSF1 and HSF2 share a common feature of adopting the trimeric form upon activation to the DNA-binding state, yet they differ in that the native state for the control non-DNA-binding form of HSF1 is a monomer, whereas HSF2 is as a dimer.

The dimeric form of HSF2 is localized primarily in the cytoplasmic fraction of control cells (Sistonen et al. 1994). During the time course of hemin treatment, the amount of cytoplasmic HSF2 remains constant, whereas the level of nucleus-localized trimeric HSF2 increases significantly, correlating with the kinetics of HSF2 DNA-binding activity and transcription. In contrast, HSF1 in control cells is in a monomeric state distributed in both the cytoplasmic and nuclear fraction, with the major proportion of HSF1 in the cytoplasm. Following heat shock and other stress, HSF1 undergoes trimerization and is localized to the nucleus. These biochemical studies have been corroborated by indirect immunofluorescence studies and biochemical fractionation which indicate that the phosphorylated, heat-activated form of HSF1 is localized in the nucleus (Sarge et al. 1993). The results on the properties of HSF1 and HSF2 begin to reveal that the mechanisms used by HSF1 and HSF2 to control the DNA-binding activities are likely to be distinct. The further understanding of these differences, specifically the identification of the nuclear localization sequences and an understanding of their regulation, may reveal how each factor has acquired the ability to respond to distinct signals for activation.

The biochemical properties of mHSF1 and mHSF2 proteins were also established using recombinant proteins overexpressed in *Escherichia coli* (Kroeger et al. 1993). Like other HSFs, when expressed in *E. coli*, mHSF1 and mHSF2 exhibit constitutive DNA-binding properties. The native size of mHSF1 and mHSF2 corresponds to a trimer with a cross-linked molecular mass of approximately 200–210 kD (Kroeger et al. 1993), which agreed with the previous in vivo analysis (Sarge et al. 1993; Sistonen et al. 1994). Given that the apparent molecular mass of mHSF1 and mHSF2 on SDS-polyacrylamide gels is approximately 70 kD, these data offer additional support that both mHSF1 and mHSF2 are trimers in solution (Perisic et al. 1989; Sorger and Nelson 1989; Sarge et al. 1993). mHSF1 and mHSF2 have apparent molecular masses of 350–400 kD corresponding to a Stokes' radius of approximately 65 Å and $s_{20,w}$ values of 6.6 and 5.5, respectively. From these experimental values, the frictional ratio for mHSF1 and mHSF2 was calculated to be about 1.7. This corresponded to an axial ratio of 14–16 to 1, which indicated that the mHSFs also have a very elongated shape consistent with the *Drosophila* studies (Westwood and Wu 1993; Wu et al., this volume). The analysis of control and activated forms of human HSF1 and HSF2 found that both factors were elongated in the activated state with frictional ratios of approximately 1.7, but in contrast to the *Drosophila* studies, there was not the apparent change in frictional ratio upon activation (Sistonen et al. 1994).

The observations on the vertebrate HSFs are consistent with the analysis of *Drosophila* HSF, indicating that HSF has an elongated shape in both the control and activated forms and that activation may involve an unfolding of the control form of HSF prior to trimerization. A rigorous biochemical analysis of the oligomerization domain of yeast HSF has revealed the presence of a three-stranded α-helical coiled-coil (Peteranderl and Nelson 1992) similar to that of influenza virus hemagglutinin (Wilson et al. 1983). These features seem to be unusual for a transcription factor but may be required for the increased affinity of DNA binding that occurs through multimerization (see below). Studies by Rabindran et al. (1993) have supported this hypothesis and suggest that HSF1 is maintained in the control form by association of the carboxy-terminal fourth leucine zipper with leucine zippers 1–3 of the oligomerization domain. Consistent with the observations on HSF1, deletion of the fourth leucine zipper in cHSF3 also leads to the acquisition of DNA-binding properties (Nakai and Morimoto 1993). Certainly, the activation of HSF1 from the latent state to its active trimeric form is still a question that requires analysis. Additionally, the discovery that HSF2 is a dimer in the control form (see above) suggests that there may be multiple ways in which the cell regulates the activity of HSFs.

V. AUTOREGULATION OF HSF1 DNA-BINDING ACTIVITY: ROLE FOR HEAT SHOCK PROTEINS

The heat shock transcriptional response, like many inducible responses of the cell, occurs rapidly upon heat shock. Within minutes of temperature elevation, the transcription of the two major heat shock genes, *hsp70* and *hsp90*, is induced. However, unlike other inducible responses in which the expression of target genes is often maintained at elevated levels, the transcription of heat shock genes and the levels of HSF DNA-binding activity attenuate in a temperature-dependent manner (Abravaya et al. 1991a). HSF levels and transcription rates are exquisitely sensitive to the precise temperature at which the cells were maintained prior to heat shock, attenuating rapidly upon return to control temperatures, whereas at extreme heat shock temperatures (≥43°C), HSF DNA-binding activity and transcription rates persist at elevated levels. Of particular interest, the attenuation response is gradual in cells exposed to intermediate heat shock temperatures. This latter point reveals that the heat shock transcriptional response is titrated and proportional to the magnitude and duration of the stress. In this section, we examine evidence that supports a role for heat shock proteins in the regulation of HSF1 activity.

One line of evidence to indicate that the DNA-binding properties of HSF are regulated is based on studies that examined the in vivo occupancy of HSF1 with the HSE of the hsp70 promoter in human cells maintained at control temperatures or exposed to different heat shock temperatures of increasing duration (Abravaya et al. 1991a,b). The rapid increase in transcription rate is accompanied by the simultaneous appearance of HSF1 DNA-binding activity measured in vitro by gel mobility shift analysis (Kingston et al. 1987; Zimarino and Wu 1987; Mosser et al. 1988; Zimarino et al. 1990). However, these in vitro assays do not reveal whether the levels of activated HSF1 reflect the amount of HSF1 that is bound in vivo to the HSE. To address this problem, Abravaya et al. (1991a,b) performed in vivo genomic footprinting analysis on the hsp70 promoter and demonstrated that in human cells, the HSE is unoccupied prior to heat shock. Upon heat shock, the HSE becomes associated with HSF1, with occupancy correlating closely with the levels of HSF1 DNA-binding activity measured using the gel shift assay. Likewise, during the process of gradual attenuation, the HSF levels decrease, and correspondingly, the HSEs are no longer occupied by HSF as measured by the in vivo footprint assay (Abravaya et al. 1991a). Quantitation of the occupancy of each HSE-binding site during activation and recovery revealed that all five pentamers of the HSE were equally bound by HSF1. Taking advantage of the rapid recovery from heat shock, which occurs when cells are shifted from 42°C to 37°C, it was possible to examine details of HSF interaction in a manner analogous to a pulse-chase experiment (Fig. 5). Within 15 minutes of downshift to 37°C, the level of HSF DNA-binding activity declined to 50%, and after 60 minutes, DNA-binding activity was no longer detected. Through the use of in vivo footprinting (Fig. 5C), it was possible to establish that the in vivo equilibrium dissociation rate for HSF has a half-life of 10 minutes, whereas the in vitro dissociation rate of the HSF-HSE complex is greater than 100 minutes (Fig. 5D). These results suggest that the activated form of HSF may be modified during the recovery phase so as to interfere with its DNA-binding activity and the conversion from the active trimer to the inert monomeric form.

A number of experimental observations are consistent with the suggestion that HSF is negatively regulated. HSF1 DNA-binding properties and oligomeric state are reversibly modulated as a transition from the non-DNA-binding state to the DNA-binding state within the cell. This is not an intrinsic property of HSFs as the recombinant *Drosophila*, chicken, mouse, and human HSFs are expressed in *E. coli* as native trimers

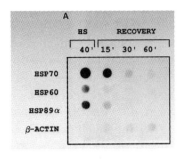

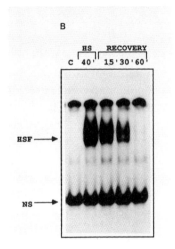

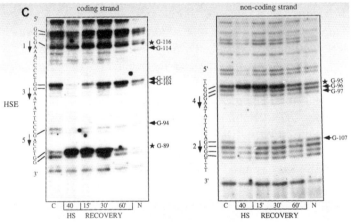

Figure 5 Equilibrium dissociation rate of HSF during heat shock and recovery. HeLa S3 cells, which had been heat-shocked for 40 min (HS), were allowed to recover at 37°C (Recovery). Transcription rates, HSF DNA-binding levels, in vivo footprinting, and dissociation rates of HSF complexes were measured. (*See facing page for part D and legend.*)

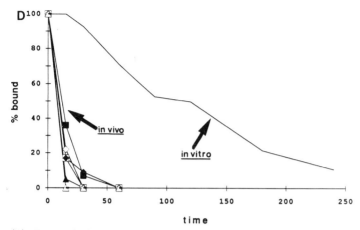

(*A*) Transcription rates of hsp70, hsp60, hsp89α, and β-actin genes were measured by run-on assays after 40 min of heat shock and at the indicated times during recovery. (*B*) Levels of activated HSF were measured at the indicated times by gel-shift assays. Complexes due to nonspecific (NS) DNA-binding proteins and heat shock factor (HSF) are shown by arrows. (*C*) Genomic footprinting of the hsp70 proximal HSE region (coding and noncoding strand) during recovery at 37°C. As part of the same experiment shown in panel *A*, HeLa S3 cells that had been heat shocked for 40 min (HS) were allowed to recover at 37°C (Recovery). Methylation patterns in genomic DNA isolated from cells at 40 min of heat shock and at the indicated times during recovery. (Lane N [Naked]) Deproteinized DNA, methylated in vitro; (lane C [control]) DNA isolated from nonheat-shocked cells. The HSE sequence, where the GAA sites are positioned are shown adjacent to the methylation patterns for the coding (*left*) and noncoding (*right*) strand. Arrows denote guanine residues protected from methylation; stars denote guanines hypersensitive to methylation. Band intensities were determined by densitometric scanning, and guanines were judged hypersensitive or protected from methylation on the basis of normalization to neighboring bands that did not show any altered reactivity. (*D*) Comparison of in vivo and in vitro rates of dissociation of HSF from the HSE. The decrease in protection from methylation of indicated guanines during recovery from a 42°C, 40-min heat shock is denoted by the arrow labeled in vivo. Densitometric scanning of the autoradiogram was used to determine the intensities of bands corresponding to the essential guanines (noted with arrows), normalizing to neighboring bands representing guanines whose reactivity to methylation is not affected by heat shock. The intensities of these bands in the 40-min, heat-shocked sample were arbitrarily taken to represent maximal binding, and the intensities of these bands in the nonheat-shocked sample were taken to represent the absence of HSF binding. The rate of dissociation of the HSF-HSE complex in vitro, denoted by the arrow labeled in vitro, is also plotted, with the amount of HSF-HSE complex prior to the addition of competitor being taken to represent 100% binding.

and exhibit constitutive DNA-binding activity (Clos et al. 1990; Kroeger et al. 1993; Nakai and Morimoto 1993; Sarge et al. 1993). To account for this discrepancy in HSF properties, it has been proposed that a regulatory protein expressed in eukaryotic cells may modulate the oligomeric state of HSF and its DNA-binding activity. The putative regulatory proteins for HSF1 are present in reticulocyte lysates as in-vitro-translated HSF1 is maintained as a monomer and does not exhibit DNA-binding activity. However, upon exposure to heat shock conditions, in-vitro-translated HSF1 trimerizes and acquires DNA-binding activity. As to the possible candidates for the negative regulation of the heat shock response, it has long been speculated from early studies in *Drosophila* and yeast (see Craig et al., this volume) that heat shock proteins were potential candidates. An autoregulatory loop has been proposed in which the increased levels of malfolded proteins induced during heat shock and other forms of stress sequester hsp70, resulting in the activation of HSF (DiDomenico et al. 1982; Morimoto et al. 1990; Abravaya et al. 1992; Baler et al. 1992; Morimoto 1993). Additional support for the autoregulatory hypothesis comes from experimental evidence that links the activation of the heat shock response to increased levels of denatured and misfolded proteins (Baler et al. 1992). Pretreatment with protein synthesis inhibitors blocks the induction of HSF and heat shock gene transcription, a result which suggests that nascent protein chains represent a critical target for protein damage (Mosser et al. 1988; Amici et al. 1992). Taken together, these results provide strong evidence for the hypothesis that the appearance and accumulation of misfolded nascent polypeptides are directly involved as a component of the sensor or trigger of the heat shock response. Molecular chaperones such as members of the hsp70 family therefore represent attractive candidates in the autoregulation of the heat shock response as they have been shown to facilitate protein folding by stabilizing intermediate folded states of nascent proteins, thus preventing them from engaging in inappropriate interactions that may lead to irreversible, nonspecific aggregation (Gething and Sambrook 1992; Craig et al; Frydman and Hartl et al; both this volume).

What is the experimental evidence to support a regulatory role for hsp70 in HSF1 activation? The initial biochemical evidence was provided by in vitro studies in which the inert monomer was converted to the active trimer using cytoplasmic S-100 extracts from non-heat-shocked HeLa cells by exposure to heat, nonionic detergents, or low pH (Larson et al. 1988; Mosser et al. 1990). Addition of exogenous hsp70 blocks the in vitro activation of HSF; furthermore, the inhibitory effect of hsp70 on HSF1 activation is relieved by addition of ATP, a biochemical feature indicative of heat shock protein function (Abravaya et al. 1992).

This inhibitory effect of hsp70 is dependent on the entire heat shock protein, suggesting that the inhibition is achieved through the effects of the ligand-free state of hsp70 on HSF1 (Abravaya et al. 1992). Although these observations support a role for hsp70 in regulating the process of HSF1 activation, they do not reveal whether the effect is through direct interaction. Indeed, it seems unlikely that the control form of HSF1 is stably associated with hsp70 or any other protein as the analysis of the physical properties of HSF1 in humans and *Drosophila* reveals that the protein is primarily a monomer. Additional support for the regulatory role of hsp70 was provided by in vivo studies using cell lines that constitutively express higher levels of the cytoplasmic human hsp70 and, as a consequence, are refractory to the activation of HSF1 (Mosser et al. 1993). Taken together, both studies lend support to the hypothesis that the relative levels of a heat shock protein such as hsp70 could influence HSF activation. If indeed it is the balance or ratio of hsp70 that is either directly or indirectly involved in maintaining the non-DNA-binding form of HSF, then it might be expected that overexpression of HSF1 would be either constitutively active or more readily activated. To test this possibility, Sarge et al. (1993) placed the HSF1 gene under the control of the β-actin promoter. Following transient transfection, HSF1 was found solely in the nucleus, primarily in an unphosphorylated, trimeric, DNA-binding competent state. These results reveal that HSF1 is not directly responding to biochemical events associated with physiological stress, as the constitutive trimerization and nuclear localization of HSF1 occur at control cell temperatures. Furthermore, the results establish that heat shock is not an obligatory step in HSF activation and are consistent with the hypothesis that there is a critical balance between HSF1 and its negative regulatory molecules to maintain HSF1 in either the non-DNA-binding state or the DNA-binding state.

Many of the current observations are consistent with a model for regulation of HSF1 DNA-binding activity that is schematically represented in Figure 6. Under nonstressful conditions, HSF1 is maintained in a non-DNA-binding form through transient interactions with hsp70, possibly by influencing or stabilizing a specific conformation of HSF1 void of DNA-binding activity. During heat shock, the appearance of misfolded or aggregated proteins creates a large pool of new protein substrates that compete with HSF1 for association with hsp70. Thus, heat shock and other stresses initiate the events that remove the negative regulatory influence on HSF1 DNA-binding activity. HSF1 oligomerizes, binds DNA, undergoes a stress-dependent inducible serine phosphorylation (M. Kline and R.I. Morimoto, unpubl.), and acquires transcriptional activity. Thus, the activation of HSF1 DNA binding leads to

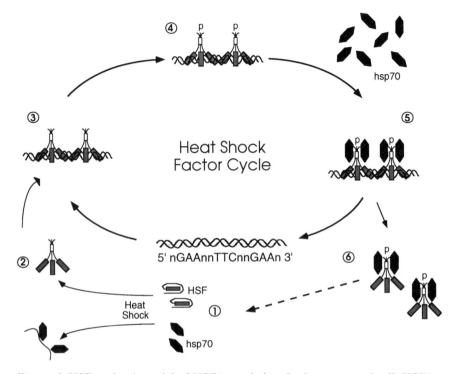

Figure 6 HSF cycle: A model of HSF1 regulation. In the unstressed cell, HSF1 is maintained in a monomeric, non-DNA-binding form through its interactions with hsp70 (*1*). Upon heat shock or other forms of stress, HSF1 assembles into a trimer (*2*), HSF1 trimers bind to specific sequence element, the heat shock element, in heat shock gene promoters (*3*), and becomes phosphorylated (*4*). Transcriptional activation of the heat shock genes leads to increased levels of hsp70 and to formation of an HSF1-hsp70 complex (*5*). Finally, HSF1 dissociates from the DNA and is eventually converted to non-DNA-binding monomers (*6*).

the elevated transcription, synthesis, and accumulation of heat shock proteins, in particular hsp70, which then associates with HSF1. The interaction of HSF1 with hsp70, and other heat shock proteins, interferes with HSF1 activity. These events may be important in the regulation of its transcriptional activity and/or conversion back to the control form. Although experimental evidence supports many aspects of this model, these ideas should only be considered as a framework for further studies.

VI. MULTIPLE HSFS CAN BE SIMULTANEOUSLY ACTIVATED IN VIVO: IMPLICATIONS FOR COMPLEX FORMS OF HEAT SHOCK GENE REGULATION

Another question prompted by the discovery of multiple HSFs is whether these factors can be activated simultaneously, and if so to examine their effects as agonists or antagonists of heat shock gene transcription. This problem has been examined in the human erythroleukemia cell line K562, as these cells have become invaluable in the dissection of the differential regulation of HSF1 and HSF2.

The active forms of HSF1 and HSF2 can co-exist in K562 cells treated with hemin for 20 hours and subjected to heat shock (Sistonen et al. 1994). The combined treatment of hemin and heat shock results in a twofold increase in HSE-binding activity as compared to the levels attained in cells treated with hemin alone. As in cells exposed to heat shock alone, HSE-binding activity attenuates during continuous heat shock at 42°C and returns to the preheat shock level by 4 hours. Thus, despite the high constitutive levels of HSF2 DNA-binding activity in hemin-treated cells, heat shock further induces the levels of HSE-binding activity by activation of HSF1. To establish whether the additive increase in DNA-binding activity was due to the simultaneous activation of both HSF1 and HSF2 or due to the effects of heat shock on HSF2 alone, the properties of subcellular localization, oligomeric state, and DNA-binding activity of HSF1 and HSF2 in K562 cells were examined in cells exposed to the combined treatment of hemin and heat shock. Both factors are in a trimeric state in the nucleus of hemin-pretreated heat-shocked cells. HSF1 is phosphorylated in hemin-pretreated heat-shocked cells as was observed for cells subjected to heat shock alone. In addition, analysis of HSF1 and HSF2 DNA-binding activity by antibody perturbation assay indicates that both factors have acquired DNA-binding activity. These results reveal that both HSF1 and HSF2 can be simultaneously activated, however, in response to apparently different signals and that both factors display properties identical to those observed for HSF1 and HSF2 in cells exposed to either heat shock or hemin treatment, respectively. The activation and attenuation of HSF1, even in the presence of HSF2, suggest that HSF2 is not regulated in a manner similar to that described for HSF1.

Does the simultaneous co-activation of HSF1 and HSF2 affect transcription of heat shock genes relative to the relative rates of transcription achieved by either factor alone? Surprisingly, when hemin-treated K562 cells are subsequently subjected to heat shock, the rate of hsp70 gene transcription is stimulated to an even higher level than was expected based on the additive effect on DNA-binding activity (Sistonen et al. 1994). Therefore, co-activation of HSF1 and HSF2 in hemin-pretreated heat-shocked cells has a synergistic effect on the transcription of the *hsp70* gene. Other characteristic features of the heat shock response are maintained in the hemin-treated and heat-shocked cells; for example, the attenuation of hsp70 transcription during later time points of heat shock can still be observed, although attenuation is slightly delayed relative to that in non-hemin-treated cells (Sistonen et al. 1994).

One explanation for the relatively high transcriptional induction of hsp70 in hemin-pretreated heat-shocked cells is that binding of HSF1 and/or HSF2 to the HSE of the endogenous hsp70 promoter is altered relative to the binding of HSF1 in heat-shocked cells and the binding of HSF2 in hemin-treated cells. To examine this, the HSF-HSE interactions in hemin-treated heat-shocked cells as well as in both hemin-treated and heat-shocked cells using in vivo genomic footprinting were compared. The pattern of the in vivo footprint in hemin-treated heat-shocked cells mimics that of HSF1 (also the consensus guanine residue at site 1 of the proximal HSE is protected) from cells exposed to heat shock alone, suggesting that the HSF2 bound to the HSEs of hemin-treated cells may be replaced by HSF1 (Sistonen et al. 1994). One possibility that might explain the synergistic induction of hsp70 gene transcription is that HSF1 and HSF2 trimers could simultaneously occupy the hsp70 promoter, in contrast to the situation in heat-shocked or hemin-treated cells, where the hsp70 promoter is occupied by two trimers of either HSF1 or HSF2, respectively (Abravaya et al. 1991a; Sistonen et al. 1992, 1994; Kroeger et al. 1993). Alternatively, HSF1 and HSF2 subunits may associate to form mixed trimers with enhanced transcriptional properties relative to pure HSF1 or HSF2 trimers. However, to date, there is no evidence that mixed heterotrimers of HSF1 and HSF2 can form either in vitro or in vivo.

VII.   INTERACTION OF HSF WITH THE HEAT SHOCK ELEMENT: COMPARISON OF HSFs AND IDENTIFICATION OF NEW BINDING SITES

Since HSF1 and HSF2 are activated in response to distinct stimuli and interact differently with the hsp70 HSE, we examined the nucleotide sequence recognition properties of both factors using recombinant mHSF1 and mHSF2. This approach, through the selection of new binding sites, establishes in an unbiased manner the consensus sequence of the HSE and further characterizes the interaction of mHSF1 and mHSF2 with natural binding sites from heat shock gene promoters and the selected sequences. This type of selection analysis identified new HSF-binding sites; the analysis of these binding sites corroborated existing views of the HSE consensus sequence and revealed information on the properties of the HSE (Amin et al. 1988; Xiao and Lis 1988; Perisic et al. 1989). The preferred mHSF1- and mHSF2-binding sites were composed of inverted adjacent pentamers that contained the primary sequence 5'-nGAAn-3'. The affinity of interaction is dictated by the overall composition of the binding site, primarily the number and arrangement of consensus pentamers (Table 2). The identities of the nucleotides in the "n" positions of the consensus, nGAAn, also contribute to the affinity of HSF binding. mHSF1 exhibited a preference for nucleotides in the first and fifth positions of the pentamer, notably an alternating preference for A or C in the first position (1 and 1′) of adjacent pentamers. As predicted from previous studies, the spacing of the pentamers was also shown to be critical for high-affinity interactions. These data reinforced the model in which all pentameric sites must be on the same face of the DNA helix for efficient HSF interaction to occur (Amin et al. 1988; Xiao and Lis 1988; Perisic et al. 1989). The diversity of sites isolated indicates that HSF is a flexible protein tolerant to sequence changes. Thus, multiple factors should be taken into account when trying to establish the functional properties of a potential HSE sequence, including the total number of adjacent inverted arrays of pentamers, the number of consensus sites, the presence of dimeric consensus sites, and the identity of nucleotides flanking the pentamers.

The affinity of protein binding to certain sites can be modulated by subtle changes in the consensus sequence. Thus, one way for the cell to modulate transcriptional activity of a gene regulated by a family of factors is to utilize binding sites that have a specificity or preference for one or another of the factors. For example, the mHSF1 selected sites, 1B5-2, 1B5-30, and 1B5-12, were bound with higher affinity by mHSF1 than by mHSF2 (Table 2). Conversely, the mHSF2 selected site, 2U5-19, was bound more avidly by mHSF2 than by mHSF1. Binding sites preferred by mHSF1 are often composed of five to six pentamers, whereas mHSF2 preferred sites contain three pentamers. These results suggest that along with the other mechanisms involved in the regulation of HSF activity (oligomerization, nuclear translocation, and phosphorylation), the composition of the HSE might dictate whether a particular HSF can bind.

Comparison of the results obtained by the analysis of random selection of mHSF1- and mHSF2-binding sites with in vitro and in vivo footprinting studies on the hsp70 promoter HSE reveals that mHSF2 differs from mHSF1 in its overall DNA-binding properties. A key observation is that mHSF2 was unable to bind to the first pentamer in the HSE, whereas mHSF1 binds with equal avidity to all five sites (Fig. 7). The reason for the difference in binding was suggested to be due to the lack of cooperativity between mHSF2 trimers. Two cloned randomly selected sequences (1B5-40 and 1B5-10) mimic the hsp70 footprints by mHSF1 and mHSF2 and offer additional support that differences in cooperativity between adjacent trimers can account for these observations. Trimer-trimer cooperativity appears to have a greater role in the binding of mHSF1 to

Table 2 Compilation of the Equilibrium Dissociation Constants for Representative mHSF1 and mHSF2 Selected Binding Sites

		K_d $(\times 10^{-9})$	
		mHSF1	mHSF2
Genomic sites			
hsp70	cGAAacCCCtgGAAtaTTCccGACc	0.83	2.9
hsp90	cTTCcgGAAggTTCggGAGgcTTCtgGAAa	0.24	0.31
Angiotensinogen	gTGCtaGAAgtTTCccaaagTGCggGAAggGACtgGAAgcCCCt	0.49	1.21
MDR1	aAGCcaGAAcaTTCc......cTTCagGAAgcAACc	>10.0	>15.0
Selected Sites			
mHSF1			
1B5-2	cgagattGAAagTTCctGAAagATCacGAGgcg	1.14	>15.0
1B5-10	cgTACtcGAAtaGCCtaGAAcgTGCttGAGg	1.42	2.93
1B5-12	tGTCgaGAAtgTTCggGACcgATCagGGTgaGGCg	0.60	2.79
1B5-30	cGCTtgGACagGTTatTGCtaGAActGTCgaGGCg	3.91	>10.0
1B5-13	cGAGtgTTCtaGATcaGCCgaGGAtattgaggcg	1.55	5.16
1B5-34	cgttcTTCtgGATggTTCtgGAAggTCCcgaggcg	0.17	0.22
1B5-40	cGATgcTACtaGAAggTTCacGAGtagaggcg	0.42	1.0
mHSF2			
2U5-19	cgacaGAAtcTTCacGTAagttagggcgaggcg	7.0	3.5
2U5-22	cggGCAggTACaaGGAagATCtgGAAgaggccg	>12.0	>15.0

A total of 41 mHSF1 and 34 mHSF2 selected sites were analyzed (P.E. Kroeger and R.I. Morimoto, in prep.) and those discussed in the text are presented. The mHSF-binding sites were selected by repeated binding and amplification of an oligonucleotide randomized in the middle 27 bases. The selected sites were cloned, sequenced, and used as probes for quantitative DNase I footprinting to determine the K_d values for mHSF1 and mHSF2 binding as shown in Fig. 6. For reference, the HSEs from the hsp70 (−91 bp), hsp90 (−57 bp), angiotensinogen (−531 bp), and MDR1 (−200/−139 bp) promoters are also shown with their respective K_d values for mHSF1 and mHSF2 binding.

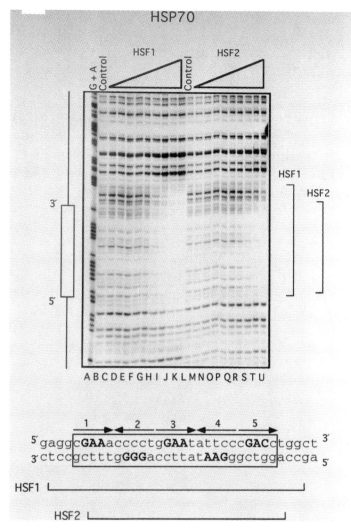

Figure 7 Differential binding of HSF1 and HSF2 on the hsp70 promoter. Equilibrium DNase I footprinting of mHSF1 (lanes *C–K*) and mHSF2 *(M–U)* binding to the hsp70 HSE. The concentration of hsp70 HSE probe in all reactions was 0.1 nM, and the probe was labeled at the 5′ end on the bottom strand. The concentration of mHSF1 in lanes *C–K* was 0.026, 0.052, 0.103, 0.206, 0.413, 0.826, 0.65, 3.3, and 6.6 nM. (Lanes *A,B,L*) G + A sequencing ladder and control DNase I footprinting reactions, respectively. The extent of mHSF1 and mHSF2 binding is indicated at the right with brackets. Below, the sequence of the hsp70 HSE is shown schematically, and the boundaries of mHSF1 and mHSF2 interaction are denoted with brackets. The box at the left is marked with the orientation of the sequence and is also overlaid on the sequence below for additional assistance; orientation of sites 1–5 are indicated with arrows, and the core motifs are boldfaced.

DNA, rather than for mHSF2. The region of the mHSF1 protein responsible for increased cooperativity was identified by the analysis of chimeric HSF proteins. The increased trimer-trimer cooperativity of mHSF1 could be transferred to mHSF2 when the DNA-binding domain and part of the oligomerization domain of the two proteins were exchanged. Therefore, the differences in mHSF1 and mHSF2 binding on most templates are due to cooperative interactions between adjacent trimers, and the DNA-binding domain or adjacent sequences of mHSF1 contain a domain responsible for this increased cooperativity. Cooperativity was known to be important for *Drosophila* HSF binding since a previous analysis demonstrated that the number of pentamers in a binding site correlated well with the stability of the interaction (Xiao et al. 1991).

What are the implications of these diverse HSF-binding sites with respect to the regulation of known heat shock genes or other genes that contain HSEs within the promoter? The representation of a particular pentameric sequence within the genome can be calculated. For example, the human genome should contain approximately 10,000–12,000 copies of a perfect consensus trimeric array (5′-nGAAnnTTCnnGAAn-3′). Comparably, there would then be only three copies of a perfect five-unit array (5′-nGAAnnTTCnnGAAnnTTCnnGAAn-3′) per genome, yet we know that there are more than three genes that are highly inducible upon heat shock. Therefore, sequence divergence in the HSEs, while maintaining the essential guanine residue, is essential to attain a sufficient number of high-affinity HSF1-binding sites. For example, a putative HSE as represented by 1B5-34, which contains three consensus sites and two nonconsensus sites, would be found at 150–200 sites per genome. A dyad of consensus pentamer sites within a larger array provides the necessary DNA sequences to maintain mHSF1 interaction. Likewise, the five-unit arrays that conserve the guanine at the first position, and some of the adenine residues in the third and fourth positions, and have only two consensus sites would be represented on the order of 200–800 sites per genome. Are these estimations for the number of sites correct? Immunofluorescence studies of *Drosophila* HSF binding during heat shock have suggested that there are more than 150 additional sites in addition to the heat shock genes with which HSF interacts; therefore, this estimation seems reasonable given the size of the human genome (Westwood et al. 1991).

From a knowledge of the sites and the requirements for binding, it is possible to make predictions on HSF occupancy at an HSF1 or HSF2 target site. Since the number of molecules of mHSF1 per mammalian cell is known, the molarity of mHSF1 trimers in the nucleus during heat shock can be calculated, which for HeLa cells would be approximately 2.5×10^{-7} M (Letovsky and Dynan 1989; Sarge et al. 1993). For mHSF2 during hemin treatment, the concentration of trimeric protein in the nucleus would be approximately 1.3×10^{-7} M. Given this concentration of both HSFs, most high-affinity sites would be occupied for a significant amount of the time as the local concentration of HSF in the nucleus during activation exceeds the equilibrium dissociation constants for most binding sites. Lower-affinity sites, of which there would be upwards of 10^4 to 10^5, would be only partially occupied as the number of sites exceeds the number of HSF molecules. Of course, this analysis is in the absence of other factors such as the effects of proteins that might interfere with HSF binding, the affinity of HSF for nonspecific sites within the genome, and the location and accessibility of these sites. With regard to the latter point, previous studies have demonstrated that HSF binds poorly in vitro to a chromatin template unless the chromatin has been disrupted by the binding of TFIID or the GAGA factor (Taylor et al. 1991; Lu et al. 1992; Lis and Wu 1993). Therefore, although there may be many potential binding sites for HSF in the genome, if they are masked by chromosomal proteins, then HSF would not be able to bind and would require an additional factor to gain access to the HSE. One additional observation that may have an impact on our calculations is that, at least in HeLa cells, HSF1 is localized within the nucleus of heat-shocked cells to punctate nonnucleolar structures (Fig. 8) (Sarge et al. 1993). This suggests that the local concentration of HSF1 in the nucleus may be significantly higher than estimated from the number of molecules present in the cell.

Consistent with the proposition that there should be other HSEs in the genome, it has been possible to identify a high-affinity site for mHSF1 and mHSF2 in the promoter of the angiotensinogen gene and a low-affinity site in the multidrug resistance (MDR1) promoter. The MDR1 promoter, which was known to be weakly heat- and arsenite-responsive (Chin et al. 1990; Miyazaki et al. 1992), was bound in several sites, with very low affinity by both HSFs (K_d ~10–20 nM). The binding sites for HSF in the MDR1 promoter were out of frame with respect to each other so as to place binding sites on opposite sides of the DNA helix. The result was low-affinity binding since there was no opportunity for cooperative interactions between trimers. The rat angiotensinogen HSE was bound with an affinity comparable to that of the *hsp70* gene by both mHSF1 and mHSF2. The interaction of HSF with the angiotensinogen promoter was also examined by transient transfection analysis and shown to be heat-shock-inducible in HeLa cells. The HSEs are located in the angiotensinogen promoter between an acute-phase response element and

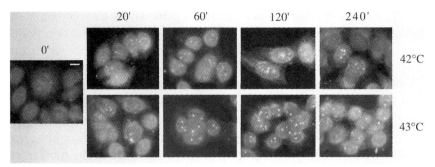

Figure 8 Heat shock induces the formation of nucleus-localized granules of HSF1. Immunofluorescence analysis was performed on HeLa cells prior to heat shock (0′) or after incubation at 42°C or 43°C for periods of 20, 40, 60, 120, and 240 min using the HSF1 antiserum; 42°C heat shock leads to the transient appearance of HSF1 granules, whereas 43°C heat shock results in the continuous appearance of granules. Bar, 10 μm.

a glucocorticoid regulatory element (GRE) that activates the gene in response to physiological stress and cytokine signals including glucocorticoids (Brasier et al. 1990a,b; Ron et al. 1990, 1991). Glucocorticoid production is known to be stimulated by adrenocorticotropic hormone (ACTH), and ACTH has been shown to activate HSF in vivo (Besedovsky et al. 1986; Blake et al. 1991; Holbrook and Udelsman, this volume). The location of the HSE in the angiotensinogen promoter makes the link between glucocorticoids, ACTH, and HSF even more provocative. It is tempting to speculate that HSF is involved in the regulation of angiotensinogen gene expression either independently or in concert with other factors implicated in the regulation of this promoter. There are undoubtedly numerous other genes that can be activated or repressed by the interaction of HSF, and further analysis of the HSF-binding sites will aid in understanding the functionality of new HSEs as they are discovered.

VIII. DOES ACTIVATION OF HSF ALWAYS LEAD TO A COORDINATE TRANSCRIPTIONAL RESPONSE?

Many of the general features of vertebrate heat shock gene transcription have been established from studies on tissue-culture cell lines derived from human, murine, or avian tissues. Although these studies together with the observation from yeasts, *Drosophila*, and plants have collectively established the conserved features of the heat shock response, there have been a number of observations which indicate that certain tissues or cell lines exhibit either a diminished or restricted response to physiological stress (Morange et al. 1984; Aujame and Morgan 1985; Mezger et al. 1987; Aujame 1988; Hensold et al. 1990; Marini et al. 1990; Nishimura et al. 1991; Pirity et al. 1992). For example, studies in which the entire organism (rodent) was subjected to hyperthermia have indicated that the magnitude of the heat shock response can vary substantially among tissues (Blake et al. 1990) and that this variation is enhanced as a consequence of aging (Fargnoli et al. 1987; Liu et al. 1991; Heydari et al. 1993). Although the tissues and cell lines used in these studies vary, the consistent observation has been the diminished induction of hsp70 message and absence of elevated hsp70 protein synthesis. Whether the apparent defect is specific to hsp70 or extends to the expression of other heat shock genes has not been generally addressed. Studies on plasmacytoma and erythroleukemic (MEL) cell lines have shown that the inability of the *hsp70* gene to be induced is not due to mutations in the hsp70 promoter region (Aujame and Morgan 1985; Aujame 1988; Hensold et al. 1990). Of particular interest have been studies on primary cultures of specific neuronal cells in which hsp70 expression is notably deficient (Marini et al. 1990; Nishimura et al. 1991). Considering the well-studied cytoprotective role of hsp70, the observations that neuronal cells lack hsp70 expression may be relevant to the unusual sensitivity of these cells, in vivo, to physiological stresses such as tissue injury, ischemia, neurotransmitter toxicity, and elevated temperatures caused by inflammation and fever (Brown 1990).

Comparison of the heat shock response in murine and human neuronal and glial cell lines by [^{35}S]methionine incorporation and SDS-PAGE has revealed that neuronal cells exhibit a severe reduction in the expression of hsp70 (I.R. Brown et al., unpubl.). This apparent deficiency in neuronal cells (Y79 retinoblastoma cells) was not due to the lack of HSF1 or the inability of HSF1 to function in vivo as the hsp90α promoter in neuronal and glial cells exhibited equivalent levels of HSF1 as measured by in vivo occupancy and hsp90α transcription (S.K. Mathur et al., in prep.). In contrast, despite the presence of functional HSF1, the hsp70 HSE was not occupied by HSF following heat shock. However, the inability of HSF1 to bind in vivo to the HSEs in the hsp70 promoter may be related to the general absence of transcription factors bound to the basal elements in the hsp70 promoter. These basal factors are critical for expression of the human *hsp70* gene at the G_1/S boundary of the cell cycle, in response to certain growth factors, hormones, mitogens, and serum stimulation (Williams and Morimoto 1990; Abravaya et al. 1991a). These multiple *cis*-regulatory elements are necessary and confer a range of constitutive and inducible transcriptional responses to the *hsp70* gene. The inability of both basal and inducible factors to associate with the hsp70 promoter reveals a locus-specific restriction that may represent an important regulatory step since the hsp90α promoter in neuronal cells is associated with HSF1, thus leading to normal levels of hsp90α gene transcription. Furthermore, the heat responsiveness of the transfected hsp70 promoter constructs in Y79 cells indicates that the in vivo deficiency of hsp70 is likely to be the consequence of a restriction at the level of chromatin structure. These observations suggest that in addition to the requirement for HSF1, the inducible transcription of hsp70 in addition to HSF1 may require other factors, perhaps involved in regulating the access to the hsp70 locus in the major histocompatibility complex (MHC) class III region of chromosome 6 (Harrison et al. 1987; Sargent et al. 1989). What is the basis for the gene-specific inaccessibility to transcription factors and what is the extent of inaccessibility in the vicinity of the *hsp70* gene? Given the location of the *hsp70* gene on chromosome 6 in a genetically important region, it will be of some interest to examine the expression and regulation of the chromosomal genes (transforming nerve factor α and β and complement) that flank the *hsp70* gene.

A critical feature of HSF1 regulation in vivo relates to its negative control; the heat shock promoter element in animal cells becomes factor-bound only upon HSF activation (Abravaya et al. 1991a). Therefore, it has been typically considered that the elevated stress-dependent transcription of a heat shock gene is dependent solely on the presence of the HSE. Furthermore, studies on the heat shock response during development have not indicated any unusual features in the inducible response. These and other studies have led to the general concepts that heat shock genes exhibit a highly conserved and ubiquitous regulation that is independent of cell type or state of differentiation. However, among these observations have been those which indicate that the heat shock response can vary in a cell-type-specific manner. The observations presented here offer a molecular basis for the differential stress responsiveness of heat shock gene transcription. The lack of HSF1 binding to the hsp70 promoter in Y79 cells may be directly related to the lack of basal factor interactions in sequences flanking the HSE and suggests that the inability of HSF1 to bind to the chromatin template reveals a requirement for other *trans*-factor interactions that must precede HSF1 binding. Two lines of evidence from in vitro and in vivo studies support this suggestion. In vitro studies on the assembly of chromatin templates reveals that HSF1 is itself incapable of binding to nucleosome-bound DNA unless a basal factor (TFIID) is also associated (Taylor et al. 1991). A related observation is in *Drosophila* in which the magnitude of the heat shock transcriptional response is affected by the GAGA factor, which has been suggested to have an important role in regulating chromatin structure and the accessibility of factors such as HSF to the DNA molecule (Lis and Wu 1993; Lu et al. 1993). Whether there is a similar factor that has a critical role in regulating access of HSF to the DNA of mammalian heat shock promoters remains to be determined.

The implication of cell- or tissue-specific variation in the heat shock response is likely to have significant biological import given the increasingly important role of heat shock proteins and molecular chaperones that are transcriptionally regulated by HSF. As heat shock proteins have been implicated in inflammation and ischemia, their involvement in neuronal cell function may become increasingly relevant. A significant body of experimental data on the stress response of neuronal and nonneuronal cells of the brain has revealed that neuronal cells exhibit a diminished response. In part, these differences may be linked to the sensitivity of neuronal cells to ischemic-, chemical-, and neurotransmitter-induced cell damage and the consequences of prolonged or extreme fever.

IX. ROLE FOR THE HEAT SHOCK TRANSCRIPTIONAL RESPONSE IN INFLAMMATION AND DISEASE

Increasing evidence supports a role for heat shock proteins in a number of disease states. The chapters in this volume on ischemia in the heart (Benjamin and Sanders Williams), the brain (Nowak and Abe), and during aging (Holbrook and Udelsman) offer evidence to support a role for heat shock proteins in disease. Whether the appearance of heat shock proteins indicates a disease state or a response to suboptimal physiological conditions remains unclear. A common component of tissue injury and infection is the inflammatory response; therefore, it is relevant to address whether the heat shock response is involved directly or indirectly with the inflammatory response. In this section, we also examine the role of fatty acids and nonsteroidal anti-inflammatory drugs in the regulation of heat shock gene transcription.

Arachidonate has a role of central importance as an essential fatty acid that is a precursor molecule in the biosynthesis of two important pathways for the mediation of cellular inflammatory responses (Jurivich et al. 1992). The initial observation to link heat shock gene expression with the metabolic pathway of arachidonic acid was based on studies which demonstrated that treatment of human cells with prostaglandins A1, A2, and J2 led to the induction of heat shock protein synthesis (Ohno et al. 1988; Santoro et al. 1989). Subsequent characterization of prostaglandin A (PGA) A1 and A2 effects on human K562 cells revealed that both fatty acid metabolites activated HSF and heat shock gene transcription. Additional support for this hypothesis comes from observations on the effects of sodium salicylate (Jurivich et al. 1992) and indomethacin (B. Lee et al., in prep.) as potent inducers of heat shock response. Treatment of human tissue-culture cells with either sodium salicylate or indomethacin leads to the rapid activation of HSF1 DNA-binding activity (Fig. 9). Treatment with indomethacin leads to a prolonged activation of HSF1, whereas pretreatment with indomethacin lowers the temperature threshold of HSF activation. A common feature of salicylate and indomethacin is their role as inhibitors of arachidonate metabolism. On the basis of these observations, the effects of arachidonate were tested on human cells. Arachidonate itself was found to be a potent activator of HSF1 and heat shock gene transcription (Jurivich et al. 1994). Furthermore, pretreatment with arachidonate lowers the temperature threshold for induction of heat shock gene transcription (Fig. 10). The discovery of arachidonate as an inducer of the heat shock response is of interest for a number of reasons. First, arachidonate represents a new class of signaling molecules that can influence the heat shock response. Second, the observations presented here imply that conditions which affect arachidonate metabolism, such as inflammation, may intersect with the effects of physiologic stress on gene expression. Third, pretreatment with fatty acids or inhibitors of fatty acid metabolism reduces the threshold of HSF activation from extreme temperatures (>42°C) to temperatures achieved during inflammation and fever (38–40°C). Taken together, these observations reveal a critical link between arachidonate, its metabolites, and the heat shock response.

The ability of the fatty acid arachidonate to lower the temperature threshold of heat shock gene expression offers an interesting insight on the regulatory mechanisms for the heat shock response. Previous studies have indicated that cells derived from related species exhibit a narrow and constant temperature profile for induction of the heat shock response (Lindquist and Craig 1988; Abravaya et al. 1991b). For example, among rodent and primate cells in culture, the heat shock response is typically induced upon exposure to temperatures above 41°C, with most studies employing extreme temperatures of 42–45°C. Although the temperature range for activation of HSF1 DNA binding is restricted to a narrow window in cultured cells, it is worth noting that studies performed in whole animals exposed to hyperthermia have shown that different tissues exhibit a wide variation in the expression of hsp70 message or protein synthesis (Blake et al. 1990). One interpretation of these observations is that there may be additional factors that participate in the signaling events that lead to the induction of heat shock gene transcription. One of

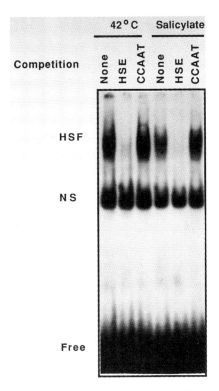

Figure 9 Salicylate treatment induces HSF DNA-binding activity without inducing hsp70 transcription. Gel mobility shift analysis (*top*) of whole-cell extracts treated with either 42°C heat shock or 20 mM sodium salicylate at 37°C for 10 min. Analysis was performed with a ^{32}P-labeled oligonucleotide containing the heat shock element (HSE) from the human heat shock protein promoter. Competition with a 100-fold excess of unlabeled HSE oligonucleotide or the CCAAT element oligonucleotide is marked above. (HSF) Heat shock transcription factor; (NS) nonspecific protein-DNA binding; (Free) unbound HSE oligonucleotide. Transcription rates (*bottom*) were measured by nuclear run-on assay of human heat shock genes hsp70, hsc70, hsp90, and β-actin. HeLa cells were heat-shocked at 42°C or exposed to 20 mM or 30 mM sodium salicylate at 37°C as indicated.

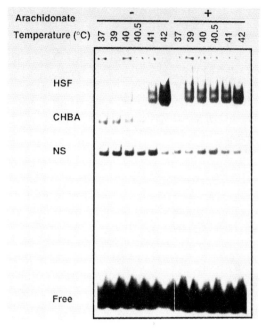

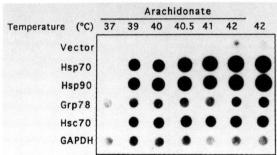

Figure 10   Arachidonate treatment lowers the threshold of the heat shock response. Cells pretreated with arachidonate exhibit a reduced temperature threshold for HSF1 DNA binding and heat shock gene transcription. (*Top*) HeLa S3 cells exposed to increasing temperatures (39–42°C) for 30 min were compared to cells simultaneously treated with 10 μM arachidonate and heat. Gel mobility shift analysis shows that HSF DNA binding can be induced in arachidonate-treated cells at 1.5–2.0°C lower than cells not exposed to arachidonate. (*Bottom*) Transcription rates of heat shock genes, determined by a nuclear run-on assay, show an increase from cells simultaneously treated with 10 μM arachidonate and a mild heat shock (39–40.5°C) that is comparable to that of a vigorous heat shock alone (42°C). Cells that are not exposed to arachidonate do not increase heat shock gene expression until 41–42°C.

these factors could be arachidonate, which, as we have shown in this study, could act either by itself to fully induce heat shock gene transcription or in concert with temperature to promote synergistically expression of the heat shock response. Other factors may contribute to gradations of heat shock gene transcription among different tissues, but the observations with arachidonate offer a potential mechanism for in vivo modulation of heat shock gene expression. These results also establish that at least arachidonate and perhaps other molecules elicited by inflammation are likely to stimulate the heat shock response at temperatures attained during inflammatory and infectious diseases. The data also raise an interesting possibility that multiple factors associated with the inflammatory response converge upon the triggering mechanism for HSF oligomerization and phosphorylation. In this regard, HSF1 phosphorylation during arachidonate treatment may be related to a recently identified arachidonate-inducible ε protein kinase C (PKC) (Koide et al. 1992). Several potential PKC substrate sites are available on HSF1; however, it is not known whether these sites are utilized by this or other protein kinases that might be regulated by arachidonate.

It is important to note that the effects of arachidonate treatment parallel many of the effects of heat shock on HSF1. Both treatments induce HSF1 translocation into the nucleus, resulting in the formation of the phosphorylated trimeric DNA-binding form of HSF1. A commonly held hypothesis is that heat shock and other forms of physiologic stress transiently increase the level of malfolded proteins, which results in the activation of HSF1 (Hightower 1980, 1991; Ananthan et al. 1986). Since both heat shock and arachidonate induce HSF1, our data could suggest that arachidonate, directly or indirectly, affects protein conformation. Indeed, there is experimental evidence to support a role for arachidonate as an inducer of oxidative stress (Clark et al. 1987; Chan et al. 1988; Rotman et al. 1992), thus leading to damage of pre-existing or newly synthesized proteins. Whether arachidonate or heat shock have in common their effects on the accumulation of malfolded proteins as the mediating signal for HSF DNA binding remains unclear. An alternative explanation for arachidonate's action is that it affects HSF1 directly or influences HSF1 DNA binding indirectly through its effects on heat shock cognate (hsc)/hsp70 proteins (Abravaya et al. 1992; Mosser et al. 1993). Other free fatty acids including palmitic and stearic acid have been shown to associate with hsc70 (Guidon and Hightower 1986a,b), thus offering an additional link between fatty acids and their potential regulatory role during the heat shock response.

The observation that effectors of inflammation such as heat and arachidonate can collectively influence HSF1 activation, leading to elevated heat shock gene expression, raises questions on the role of the heat shock response or heat shock proteins in inflammation. This transcriptional response may be necessary to shift rapidly the cellular state in which there are increased needs for molecular chaperones during inflammation. For example, certain cytokine responses, in particular those that increase arachidonate levels (Hannigan and Williams 1991), may require higher levels of heat shock proteins for the synthesis, folding, translocation, and secretion of proteins synthesized during inflammation. Separately, heat shock proteins induced during inflammation have a critical role in cytoprotection. Phagocytosis and bacterial infection induce heat shock protein synthesis, and these events have been shown to protect cells against subsequent exposure to cytotoxic forms of stress (Clerget and Polla 1990; Kaufmann et al. 1991; Landry et al. 1991; Li et al. 1992; Mosser and Martin 1992; Sanchez et al. 1992). Thus, induction of the heat shock response during inflammation could provide both regulatory and cytoprotective functions. The molecular basis for inflammatory responses is undoubtedly complex, but further analysis of arachidonate-mediated HSF1 DNA binding will provide new insights to understanding how inflammatory mediators regulate gene expression.

ACKNOWLEDGMENTS

The studies from our laboratory were supported by grants from the National Institutes of General Medicine, the March of Dimes Foundation, individual NRSA (S.M., K.S., P.K.), the American Cancer Society (K.A.), the Japanese Cancer Foundation (A.N.), the Academy of Finland and the Fogarty International Foundation (L.S.), and a National Institutes of Health Physician Scientist award (D.J.).

REFERENCES

Abravaya, K., B. Phillips, and R.I. Morimoto. 1991a. Attenuation of the heat shock response in HeLa cells is mediated by the release of bound heat shock transcription factor and is modulated by changes in growth and in heat shock temperatures. *Genes Dev.* **5:** 2117–2127.

——. 1991b. Heat shock-induced interactions of heat shock transcription factor and the human hsp70 promoter examined by in vivo footprinting. *Mol. Cell. Biol.* **11:** 586–592.

Abravaya, K., M.P. Myers, S.P. Murphy, and R.I. Morimoto. 1992. The human heat shock protein hsp70 interacts with HSF, the transcription factor that regulates heat shock gene transcription. *Genes Dev.* **6:** 1153–1164.

Amici, C., L. Sistonen, M.G. Santoro, and R.I. Morimoto. 1992. Antiproliferative prostaglandins activate heat shock transcription factor. *Proc. Natl. Acad. Sci.* **89:** 6227–6231.

Amin, J., J. Ananthan, and R. Voellmy. 1988. Key features of heat shock regulatory elements. *Mol. Cell. Biol.* **8:** 3761–3769.

Ananthan, J., A.L. Goldberg, and R. Voellmy. 1986. Abnormal proteins serve as eukaryotic stress signals and trigger the activation of heat shock genes. *Science* **232:** 522–524.

Aujame, L. 1988. The major heat-shock protein hsp68 is not induced by stress in mouse erythroleukemia cell lines. *Biochem. Cell Biol.* **66:** 691–701.

Aujame, L. and C. Morgan. 1985. Non expression of a major heat shock gene in mouse plasmacytoma MPC-11. *Mol. Cell. Biol.* **5:** 1780–1783.

Baler, R., G. Dahl, and R. Voellmy. 1993. Activation of human heat shock genes is accompanied by oligomerization, modification, and rapid translocation of heat shock transcription factor HSF1. *Mol. Cell. Biol.* **13:** 2486–2496.

Baler, R., W.J. Welch, and R. Voellmy. 1992. Heat shock gene regulation by nascent polypeptides and denatured proteins: hsp70 as a potential autoregulatory factor. *J. Cell Biol.* **117:** 1151–1159.

Besedovsky, H., A. Del Rey, E. Sorkin, and C. A. Dinarello. 1986. Immunoregulatory feedback between interleukin-1 and glucocorticoid hormones. *Science* **233:** 652–654.

Blake, M.J., T.S. Nowak, and N.J. Holbrook. 1990. In vivo hyperthermia induces expression of hsp70 mRNA in brain regions controlling the neuroendocrine response to stress. *Mol. Brain Res.* **8:** 89–92.

Blake, M.J., R. Udelsman, G.J. Feulner, D.D. Norton, and N.J. Holbrook. 1991. Stress-induced heat shock protein 70 expression in adrenal cortex: An adrenocorticotropic hormone-sensitive, age-dependent response. *Proc. Natl. Acad. Sci.* **88:** 9873–9877.

Braiser, A.R., D. Ron, J.E. Tate, and J.F. Habener. 1990a. A family of constitutive C/EBP-like DNA binding proteins attenuate the IL-1 alpha induced, NF kappa B mediated trans-activation of the angiotensinogen gene acute-phase response element. *EMBO J.* **9:** 3933–3944.

———. 1990b. Synergistic enhancons located within an acute phase responsive enhancer modulate glucocorticoid induction of angiotensinogen gene transcription. *Mol. Endocrinol.* **4:** 1921–1933.

Brown, I.R. 1990. Induction of heat shock (stress) genes in the mammalian brain by hyperthermia and other traumatic events: A current perspective. *J. Neurosci. Res.* **27:** 247–255.

Chan, P.H., S.F. Chen, and A.C. Yu. 1988. Induction of intracellular superoxide radical formation by arachidonic acid and by polyunsaturated fatty acids in primary astrocytic cultures. *J. Neurochem.* **50:** 1185–1193.

Chen, Y., N.A. Barlev, O. Westergaard, and B.K. Jakobsen. 1993. Identification of the C-terminal activator domain in yeast heat shock factor: Independent control of transient and sustained transcriptional activity. *EMBO J.* **12:** 5007–5018.

Chin, K.-V., S. Tanaka, G. Darlington, I. Pastan, and M. Gottesman. 1990. Heat shock and arsenite increase expression of the multidrug resistance (MDR1) gene in human renal carcinoma cells. *J. Biol. Chem.* **265:** 221–226.

Clark, R.A., K.G. Leidal, D.W. Pearson, and W.M. Nauseef. 1987. NADPH oxidase of human neutrophils. Subcellular localization and characterization of an arachidonate-activatable superoxide-generating system. *J. Biol. Chem.* **262:** 4065–4074.

Clerget, M. and B.S. Polla. 1990. Erythrophagocytosis induces heat shock protein synthesis by human monocytes-macrophages. *Proc. Natl. Acad. Sci.* **87:** 1081–1085.

Clos, J., J.T. Westwood, P.B. Becker, S. Wilson, K. Lambert, and C. Wu. 1990. Molecular cloning and expression of a hexameric *Drosophila* heat shock factor subject to negative regulation. *Cell* **63:** 1085–1097.

DiDomenico, B.J., G.E. Bugaisky, and S. Lindquist. 1982. The heat shock response is regulated at both the transcriptional and posttranscriptional levels. *Cell* **31**: 593–603.

Fargnoli, J., T. Kunisada, J.A.J. Fornace, E.L. Schneider, and N.J. Holbrook. 1987. Decreased expression of heat shock protein 70 mRNA and protein after heat treatment in cells of aged rats. *Proc. Natl. Acad. Sci.* **87**: 846–850.

Gething, M.-J. and J. Sambrook. 1992. Protein folding in the cell. *Nature* **355**: 33–45.

Guidon, P.T.J. and L.E. Hightower. 1986a. The 73 kilodalton heat shock cognate protein purified from rat brain contains nonesterified palmitic and stearic acids. *J. Cell. Physiol.* **128**: 239–245.

———. 1986b. Purification and initial characterization of the 71-kilodalton rat heat-shock protein and its cognate as fatty acid binding proteins. *Biochemistry* **25**: 3231–3239.

Hannigan, G.E. and B.R.G. Williams. 1991. Signal transduction by interferon-alpha through arachidonic acid metabolism. *Science* **251**: 204–207.

Harrison, G.S., H.A. Drabkin, F.T. Kao, J. Hartz, I.M. Hart, E.H. Chu, B.J. Wu, and R.I. Morimoto. 1987. Chromosomal location of human genes encoding major heat-shock protein HSP70. *Somat. Cell. Mol. Genet.* **13**: 119–130.

Hensold, J.O., C.R. Hunt, S.K. Calderwood, D.E. Housman, and R.E. Kingston. 1990. DNA binding of heat shock factor to the heat shock element is insufficient for transcriptional activation in murine erythroleukemia cells. *Mol. Cell. Biol.* **10**: 1600–1608.

Heydari, A.H., B. Wu, R. Takahashi, R. Strong, and A. Richardson. 1993. Expression of heat shock protein 70 is altered by age and diet at the level of transcription. *Mol. Cell. Biol.* **13**: 2909–2918.

Hightower, L.E. 1980. Cultured animal cells exposed to amino acid analogues or puromycin rapidly synthesize several polypeptides. *J. Cell. Physiol.* **102**: 407–427.

———. 1991. Heat shock, stress proteins, chaperones, and proteotoxicity. *Cell* **66**: 191–197.

Jakobsen, B.K. and H.R. Pelham. 1991. A conserved heptapeptide restrains the activity of the yeast heat shock transcription factor. *EMBO J.* **10**: 369–375.

Jurivich, D.A., L. Sistonen, R.A. Kroes, and R.I. Morimoto. 1992. Effect of sodium salicylate on the human heat shock response. *Science* **255**: 1243–1245.

Jurivich, D.A., L. Sistonen, K.D. Sarge and R.I. Morimoto. 1994. Arachidonate is a potent modulator of human heat shock gene transcription. *Proc. Natl. Acad. Sci.* (in press).

Kaufmann, S.H., B. Schoel, J.D. van Embden, T. Koga, A. Wand-Wurttenberger, M.E. Munk, and U. Steinhoff. 1991. Heat-shock protein 60: Implications for pathogenesis of and protection against bacterial infections. *Immunol. Rev.* **121**: 67–90.

Kingston, R.E., T.J. Schuetz, and Z. Larin. 1987. Heat-inducible human factor that binds to a human hsp70 promoter. *Mol. Cell. Biol.* **7**: 1530–1534.

Koide, H., K. Ogita, U. Kikkawa, and Y. Nishizuka. 1992. Isolation and characterization of the epsilon subspecies of protein kinase C from rat brain. *Proc. Natl. Acad. Sci.* **89**: 1149–1153.

Kroeger, P.E., K.D. Sarge, and R.I. Morimoto. 1993. Mouse heat shock transcription factors 1 and 2 prefer a trimeric binding site but interact differently with the HSP70 heat shock element. *Mol. Cell. Biol.* **13**: 3370–3383.

Landry, J., P. Chretien, A. Laszlo, and H. Lambert. 1991. Phosphorylation of HSP27 during development and decay of thermotolerance in Chinese hamster cells. *J. Cell. Physiol.* **147**: 93–101.

Larson, J.S., T.J. Schuetz, and R.E. Kingston. 1988. Activation in vitro of sequence-specific DNA binding by a human regulatory factor. *Nature* **335**: 372–375.

Letovsky, J. and W.S. Dynan. 1989. Measurement of the binding of transcription factor Sp1 to a single GC box. *Nucleic Acids Res.* **17:** 2639–2653.

Li, G.C., L. Li, R.Y. Liu, M. Rehman, and W.M. Lee. 1992. Heat shock protein hsp70 protects cells from thermal stress even after deletion of its ATP-binding domain. *Proc. Natl. Acad. Sci.* **89:** 2036–2040.

Lindquist, S. and E.A. Craig. 1988. The heat shock proteins. *Annu. Rev. Genet.* **22:** 631–677.

Lis, J. and C. Wu. 1993. Protein traffic on the heat shock promoter: Parking, stalling and trucking along. *Cell.* **74:** 1–4.

Liu, A.Y.-C., H.-S. Choi, Y.-K. Lee, and K.Y. Chen. 1991. Molecular events involved in transcriptional activation of heat shock genes become progressively refractory to heat stimulation during aging of human diploid fibroblasts. *J. Cell. Physiol.* **149:** 560–566.

Lu, Q., L.L. Wallrath, H. Granok, and S.C.R. Elgin. 1993. (CT)$_n$•(GA)$_n$ repeats and heat shock elements have distinct roles in chromatin structure and transcriptional activation of the *Drosophila* hsp26 gene. *Mol. Cell. Biol.* **13:** 2802–2814.

Marini, A.M., M. Kozuka, R.H. Lipsky, and T.S. Nowak. 1990. 70-kilodalton heat shock protein induction in cerebellar astrocytes and cerebellar granule cells in vitro: Comparison with immunocytochemical localization after hyperthermia in vivo. *J. Neurochem.* **54:** 1509–1516.

Mezger, V., O. Bensaude, and M. Morange. 1987. Deficient activation of heat shock gene transcription in embryonal carcinoma cells. *Dev. Biol.* **124:** 544–550.

———. 1989. Unusual levels of heat shock element-binding activity in embryonal carcinoma cells. *Mol. Cell. Biol.* **9:** 3888–3896.

Miyazaki, M., K. Kohno, T. Uchiumi, H. Tanimura, K. Matsuo, M. Nasu, and M. Kuwano. 1992. Activation of human multidrug resistance-1 gene promoter in response to heat shock stress. *Biochem. Biophys. Res. Commun.* **187:** 677–684.

Morange, M., A. Diu, O. Bensaude, and C. Babinet. 1984. Altered expression of heat shock proteins in embryonal carcinoma and mouse early embryonic cells. *Mol. Cell. Biol.* **4:** 730–735.

Morimoto, R.I. 1993. Cells in stress: Transcriptional activation of heat shock genes. *Science* **259:** 1409–1410.

Morimoto, R.I., A. Tissières, and C. Georgopoulos. 1990. The stress response, function of the proteins, and perspectives. In *Stress proteins in biology and medicine.* (ed. R.I. Morimoto et al.), p. 1–36. Cold Spring Harbor Laboratory Press, Cold Spring Harbor, New York.

Mosser, D.D. and L.H. Martin. 1992. Induced thermotolerance to apoptosis in a human T lymphocyte cell line. *J. Cell. Physiol.* **151:** 561–570.

Mosser, D.D., J. Duchaine, and B. Massie. 1993. The DNA-binding activity of the human heat shock transcription factor is regulated in vivo by hsp70. *Mol. Cell. Biol.* **13:** 5427–5438.

Mosser, D.D., N.G. Theodorakis, and R.I. Morimoto. 1988. Coordinate changes in heat shock element-binding activity and hsp70 gene transcription rates in human cells. *Mol. Cell. Biol.* **8:** 4736–4744.

Mosser, D.D., P.T. Kotzbauer, K.D. Sarge, and R.I. Morimoto. 1990. In vitro activation of heat shock transcription factor DNA-binding by calcium and biochemical conditions that affect protein conformation. *Proc. Natl. Acad. Sci.* **87:** 3748–3752.

Nakai, A. and R.I. Morimoto. 1993. Characterization of a novel chicken heat shock transcription factor, HSF3, suggests a new regulatory pathway. *Mol. Cell. Biol.* **13:** 1983–1997.

Nieto-Sotelo, J., G. Wiederrecht, A. Okuda, and C.S. Parker. 1990. The yeast heat shock transcription factor contains a transcriptional activation domain whose activity is repressed under nonshock conditions. *Cell* **62:** 807–817.

Nishimura, R.N., B.E. Dwyer, K. Clegg, R. Cole, and J. de Vellis. 1991. Comparison of the heat shock response in cultured cortical neurons and astrocytes. *Mol. Brain Res.* **9:** 39–45.

Ohno, K., M. Fukushima, M. Fujiwara, and S. Narumiya. 1988. Induction of 68,000-dalton heat shock proteins by cyclopentenone prostaglandins. Its association with prostaglandin-induced G1 block in cell cycle progression. *J. Biol. Chem.* **263:** 19764–19770.

Perisic, O., H. Xiao, and J.T. Lis. 1989. Stable binding of *Drosophila* heat shock factor to head-to-head and tail-to-tail repeats of a conserved 5 bp recognition unit. *Cell.* **59:** 797–806.

Peteranderl, R. and H.C.M. Nelson. 1992. Trimerization of the heat shock transcription factor by a triple-stranded a-helical coiled-coil. *Biochemistry* **31:** 12272–12276.

Pirity, M., V.T. Nguyen, M.F. Dubois, O. Bensaude, A. Hever-Szabo, and A. Venetianer. 1992. Decreased stress inducibility of the HSP68 protein in a rat hepatoma variant clone. *Eur. J. Biochem.* **210:** 793–800.

Rabindran, S.K., G. Giorgi, J. Clos, and C. Wu. 1991. Molecular cloning and expression of a human heat shock factor, HSF1. *Proc. Natl. Acad. Sci.* **88:** 6906–6910.

Rabindran, S.K., R.I. Haroun, J. Clos, J. Wisniewski, and C. Wu. 1993. Regulation of heat shock factor trimer formation: Role of a conserved leucine zipper. *Science* **259:** 230–234.

Ron, D., A.R. Braiser, and J.F. Habener. 1991. Angiotensinogen gene-inducible enhancer-binding protein 1, a member of a new family of large nuclear proteins that recognize nuclear factor kappa B-binding sites through a zinc finger motif. *Mol. Cell. Biol.* **11:** 2887–2895.

Ron, D., A.R. Braiser, K.A. Wright, and J.F. Habener. 1990. The permissive role of glucocorticoids on interleukin-1 stimulation of angiotensinogen gene transcription is mediated by an interaction between inducible enhancers. *Mol. Cell. Biol.* **10:** 4389–4395.

Rotman, E.I., M.A. Brostrom, and C.O. Brostrom. 1992. Inhibition of protein synthesis in intact mammalian cells by arachidonic acid. *Biochem. J.* **282:** 487–494.

Sanchez, Y., J. Taulien, K.A. Borkovich, and S. Lindquist. 1992. Hsp104 is required for tolerance to many forms of stress. *EMBO J.* **11:** 2357–2364.

Santoro, M.G., E. Garaci, and C. Amici. 1989. Prostaglandins with antiproliferative activity induce the synthesis of a heat shock protein in human cells. *Proc. Natl. Acad. Sci.* **86:** 8407–8411.

Sarge, K.D., S.P. Murphy, and R.I. Morimoto. 1993. Activation of heat shock gene transcription by HSF1 involves oligomerization, acquisition of DNA binding activity, and nuclear localization and can occur in the absence of stress. *Mol. Cell. Biol.* **13:** 1392–1407.

Sarge, K.D., V. Zimarino, K. Holm, C. Wu, and R.I. Morimoto. 1991. Cloning and characterization of two mouse heat shock factors with distinct inducible and constitutive DNA-binding ability. *Genes Dev.* **5:** 1902–1911.

Sargent, C.A., I. Dunham, J. Trowsdale, and R.D. Campbell. 1989. Human major histocompatibility complex contains genes for the major heat shock protein HSP70. *Proc. Natl. Acad. Sci.* **86:** 1968–1972.

Scharf, K.-D., S. Rose, W. Zott, F. Schoff, and L. Nover. 1990. Three tomato genes code for heat stress transcription factors with a remarkable degree of homology to the DNA-binding domain of the yeast HSF. *EMBO J.* **9:** 4495–4501.

Schuetz, T.J., G.J. Gallo, L. Sheldon, P. Tempst, and R.E. Kingston. 1991. Isolation of a cDNA for HSF2: Evidence for two heat shock factor genes in humans. *Proc. Natl. Acad. Sci.* **88:** 6910–6915.

Singh, M.K. and J. Yu. 1984. Accumulation of a heat shock-like protein during differentiation of human erythroid cell line K562. *Nature* **309:** 631–633.

Sistonen, L., K.D. Sarge, and R.I. Morimoto. 1994. Human heat shock factors 1 and 2 are differentially activated and can synergistically induce HSP70 gene transcription. *Mol. Cell. Biol.* (in press).

Sistonen, L., K.D. Sarge, B. Phillips, K. Abravaya, and R. Morimoto. 1992. Activation of heat shock factor 2 during hemin-induced differentiation of human erythroleukemia cells. *Mol. Cell. Biol.* **12:** 4104–4111.

Sorger, P.K. and H.C.M. Nelson. 1989. Trimerization of a yeast transcriptional activator via a coiled-coil motif. *Cell* **59:** 807–813.

Taylor, I.C.A., J.L. Workman, T.J. Schuetz, and R.E. Kingston. 1991. Facilitated binding of GAL4 and heat shock factor to nucleosomal templates: Differential function of DNA-binding domains. *Genes Dev.* **5:** 1285–1298.

Theodorakis, N.G., D.J. Zand, P.T. Kotzbauer, G.T. Williams, and R.I. Morimoto. 1989. Hemin-induced transcriptional activation of the hsp70 gene during erythroid maturation in K562 cells is due to a heat shock factor-mediated stress response. *Mol. Cell. Biol.* **9:** 3166–3173.

Westwood, J.T. and C. Wu. 1993. Activation of *Drosophila* heat shock factor: Conformational change associated with a monomer-to-trimer transition. *Mol. Cell. Biol.* **13:** 3481–3486.

Westwood, J.T., J. Clos, and C. Wu. 1991. Stress-induced oligomerization and chromosomal relocalization of heat-shock factor. *Nature* **353:** 822–827.

Wiederrecht, G., D. Seto, and C.S. Parker. 1988. Isolation of the gene encoding the *S. cerevisiae* heat shock transcription factor. *Cell* **54:** 841–853.

Williams, G.T. and R.I. Morimoto. 1990. Maximal stress-induced transcription from the human hsp70 promoter requires interactions with the basal promoter elements independent of rotational alignment. *Mol. Cell. Biol.* **10:** 3125–3136.

Wilson, I.A., R.C. Ladner, J.J. Skehel, and D.C. Wiley. 1983. The structure and role of the carbohydrate moieties of influenza virus haemagglutinin. *Biochem. Soc. Trans.* **11:** 145–147.

Xiao, H. and J.T. Lis. 1988. Germline transformation used to define key features of the heat shock response element. *Science* **239:** 1139–1142.

Xiao, H., O. Perisic, and J. T. Lis. 1991. Cooperative binding of *Drosophila* heat shock factor to arrays of a conserved 5 bp unit. *Cell* **64:** 585–593.

Zimarino, V. and C. Wu. 1987. Induction of sequence-specific binding of *Drosophila* heat shock activator proteins without protein synthesis. *Nature* **327:** 727–730.

Zimarino, V., C. Tsai, and C. Wu. 1990. Complex modes of heat shock factor activation. *Mol. Cell. Biol.* **10:** 752–759.

18
Heat Shock Proteins and Stress Tolerance

Dawn A. Parsell and Susan Lindquist
Howard Hughes Medical Institute
The University of Chicago
Chicago, Illinois 60637

I. INTRODUCTION

The capacity of different individuals of the same species to survive short exposures to extreme temperatures (thermotolerance) varies over a remarkable range, commonly over several orders of magnitude. Both differences in growth conditions and differences in genetic background contribute. Although the contributions of genetic background are just beginning to be deciphered, the effects of growth conditions have been the subject of detailed and intense scrutiny. Nutrient availability, oxygen tension, diurnal rhythms, and a host of other variables exert highly reproducible effects on thermotolerance. By far the most closely studied phenomenon, however, is the tolerance afforded by short pretreatments at moderately elevated temperature. When whole organisms or cultured cells are given such pretreatments, their resistance to killing by extreme heat increases dramatically. This increased resistance, known as induced

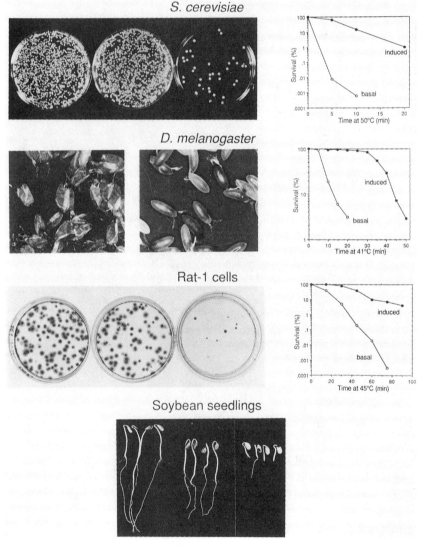

Figure 1 (See facing page for legend.)

thermotolerance (Fig. 1), is observed in virtually every organism studied. Remarkably, mild heat pretreatments elicit resistance not just to high temperatures, but to an extraordinary variety of other stresses. In addition, exposure to other mild stresses elicits protection not just against higher doses of those particular stresses, but against high temperatures as well.

Such tolerance-inducing treatments generally also induce the synthesis of a small number of proteins known as the heat shock proteins (hsps; Fig. 2) (Li and Laszlo 1985; Lindquist 1986; Nagao et al. 1986; Subjeck and Shyy 1986; Sanchez and Lindquist 1990; Nover 1991; Sanchez et al. 1992). Historically, many observations have suggested that hsps play a vital part in induced tolerance. For example, it is striking that hsps are induced at very different temperatures in different organisms, but in each case, induction occurs at a temperature that constitutes a stress for that particular organism. The rapidity and intensity of the response are also compelling. A wide variety of regulatory mechanisms, acting at both the transcriptional and posttranscriptional levels, are employed to ensure that

Figure 1 Induced thermotolerance in several common laboratory organisms. Induced thermotolerance is assayed in *S. cerevisiae* by shifting cells growing at 25°C to 50°C with (*middle plate*) or without (*right plate*) a preconditioning treatment at 37°C for 30 min. After the 50°C treatment, cultures were plated to assess colony-forming ability. *Left plate* shows an equivalent volume of a control culture maintained at 25°C (tenfold dilution). Induced thermotolerance in *D. melanogaster* embryos is assayed by exposing them to a killing temperature (41°C) with (*left plate*) or without (*right plate*) a preconditioning treatment at 36°C for 30 min. In this assay, embryos killed by high temperature do not hatch (dark egg cases), whereas surviving embryos hatch and crawl away (empty, light-colored egg cases). In mammalian tissue culture cells, induced thermotolerance is assayed by shifting cells growing at 37°C to 45°C for 30 min with (*middle plate*) or without (*right plate*) a tolerance-inducing 15-min incubation at 45°C followed by 16 hr at 37°C. A control dish that did not receive a heat treatment is shown at the left. Induced thermotolerance is assayed in soybeans by shifting seedlings growing at 28°C to 45°C for 2 hr with (*center*) or without (*right*) preincubation for 2 hr at 40°C. Heat treatment was followed by 72 hr of recovery at 28°C. Plants that did not receive a severe heat treatment are shown at the left. The *Drosophila* data were previously published in Welte et al. (1993), and the soybean data were previously published in Key et al. (1983). When thermotolerance is measured as a function of time at the killing temperatures, survival curves like those shown at the far right are produced. The curves labeled "induced" show survival of organisms given a tolerance-inducing mild heat treatment before the shift to the killing temperature, whereas the curves labeled "basal" show survival after a direct shift to the killing temperature.

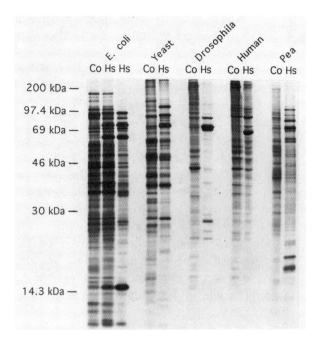

Figure 2 Induction of heat shock proteins. Logarithmically growing cells were pulse-labeled with ^{3}H amino acids either while growing at normal temperatures (*E. coli*, 37°C; *S. cerevisiae*, 25°C; *D. melanogaster*, 25°C; HeLa, 37°C; pea seedlings, 22°C) or following a shift to elevated temperatures (*E. coli*, 45°C, 10 min, 50°C, 10 min; *S. cerevisiae*, 39°C, 20 min; *D. melanogaster*, 36.5°C, 60 min; HeLa, 46°C, 5 min followed by 37°C, 120 min; pea seedlings, 40°C, 4 hr). Total cellular proteins were extracted, separated on a 10% SDS-polyacrylamide gel, and visualized by fluorography.

hsps are induced within minutes of exposure to high temperatures, giving hsp synthesis all the hallmarks of an emergency response (for review, see Lindquist 1986, 1993; Nover 1991; Yost et al. 1991).

During the past few years, the role of hsps in stress tolerance has moved from the realm of anecdote and correlation to convincing genetic proof. Moreover, new information on the biochemical functions of hsps has provided detailed insights about the mechanisms by which hsps contribute to stress tolerance. In general terms, hsps function by preventing the accumulation of stress-damaged proteins. Indeed, the artificial accumulation of "abnormal" proteins is sufficient to induce hsp synthesis (Pine 1967; Goldberg 1972; Goff and Goldberg 1985; Ananthan et al. 1986; Parsell and Sauer 1989), and many classic hsp-inducing treatments are now known to damage or denature proteins in vitro (Li and Laszlo 1985; Nover 1991). hsps prevent the accumulation of stress-damaged polypeptides in two ways: Some serve as "molecular chaperones," preventing the aggregation of denatured proteins and promoting their proper refolding, and others facilitate the degradation of abnormal proteins.

Here, we present an overview of current knowledge of hsp function in the cellular response to stress. After a discussion of the cellular consequences of heat stress, we examine in some detail the roles individual hsps play in ameliorating these effects. We also consider the role of other factors that influence stress tolerance and explore the evidence that differences in hsp expression or function may contribute to differences in basal stress tolerance. Finally, we consider possible explanations for a current enigma, that the relative importance of individual hsps for thermotolerance varies among organisms.

II. THE CELLULAR CONSEQUENCES OF STRESS

Heat damages a wide variety of cellular structures and metabolic processes (for detailed reviews, see Nover 1991; Laszlo 1992). Both the magnitude and duration of the stress determine the amount of damage incurred, and cells die when a threshold of damage is exceeded. A goal of current research is to differentiate, among the multitude of deleterious effects observed, those that are primary lethal lesions from those that are secondary consequences of other damage.

In higher eukaryotes, one of the most immediate effects of heat shock is extensive disruption of the cytoskeleton (Falkner et al. 1981; Coss et al. 1982; Glass et al. 1985; Welch and Suhan 1985; Iida et al. 1986). In most cell types, the intermediate filament (IF) network collapses from an intricate network of fine thread-like filaments to large aggregates that surround the nucleus. The effects of heat on other cytoskeletal elements are more variable. In some cells, heat induces the reorganization of the cytoplasmic microfilament network such that actin-containing fibers appear in the nucleus. In other cells, extreme heat destroys microfilaments and actin-containing contractile rings. Microtubules are also damaged by heat shock. Indeed, the extreme heat sensitivity of mitotic cells correlates with damage to the mitotic spindle, and the delicate microtubule network present in the cytoplasm of interphase cells is disrupted by heat in some cell types.

Although less visually dramatic, the damage heat produces in other cytoplasmic structures (Christiansen and Kvamme 1969; Welch and Suhan 1985) may be just as important in heat toxicity. Fragmentation of the Golgi is observed in some cells, and the number of lysosomes increases. Mitochondria swell slightly, their number decreases (perhaps due to aggregation around the nucleus), and there is a decrease in respiration and oxidative phosphorylation. Severe heat uncouples the two processes entirely (Patriarca and Maresca 1990).

The synthesis of many macromolecules is also profoundly disturbed by high temperatures, disruption of normal protein synthesis having received the most study. In many cell types, polysomes rapidly disappear (Lindquist 1986; Nover 1991). Two mechanisms have been proposed to account for their disappearance: the inactivation of initiation factors such as cap-binding protein and eIF2 (Panniers et al. 1985; Zapata et al. 1991) and the collapse of the IF network, to which many messages are attached (Heine et al. 1971; Howe and Hershey 1984). Many mRNAs remain bound to the collapsed IF network after polysome disassembly (Nover et al. 1989). This association may help protect them from degradation, although it also appears that mRNA degradative systems are inactivated by heat (Petersen and Lindquist 1988; Lindquist 1993). When temperatures decrease, these mRNAs do not need to be resynthesized, and normal protein synthesis may resume quite rapidly (Storti et al. 1980; Lindquist 1981).

The disruption of normal protein synthesis is not simply a toxic effect of high temperatures. In many cell types, it is part of a regulatory response that facilitates the expression of hsps; i.e., since heat shock messages can be translated under conditions when other messages are inactivated, the disruption of normal translation eliminates the competition for ribosomes and other synthetic factors, facilitating a rapid burst of hsp synthesis. Other stress-induced changes in gene expression show the same dichotomy: On the one hand, they may be viewed as toxic effects of heat, and on the other hand, they may be viewed as part of a sophisticated regulatory circuit that optimizes expression of hsps. For example, the transcription of normal messages decreases sharply at temperatures where transcription of heat-induced messages is at a maximum. Heat shock also disrupts message degradation mechanisms that act on *hsp70* and other short-lived mRNAs. This stabilizes the *hsp70* message and allows rapid accumulation of the protein at high temperatures (Theodorakis and Morimoto 1987; Petersen and Lindquist 1988). Finally, heat shock disrupts the splicing of mRNA precursors (Yost and Lindquist 1986), and heat-induced changes in the splicing apparatus have been observed (Bond 1988; Utans et al. 1992). This has little effect on the expression of hsps because most hsp mRNAs do not contain introns, and heat-induced changes in the splicing apparatus might promote preferential synthesis of hsps by blocking the maturation of competing species.

Many other effects of heat on gene expression do not confer any obvious advantage to hsp synthesis but appear simply to be toxic effects of heat. In the nucleolus, pre-rRNA processing halts immediately upon heat shock, and transcription by RNA polymerase I declines (Ellgaard and Clever 1971; Rubin and Hogness 1975; Bell et al. 1988). These metabolic effects coincide with drastic morphological changes in the nucleolus (Welch and Suhan 1985). Processing of RNA polymerase II transcripts is also disrupted by stress (Berger et al. 1985; Yost and Lindquist 1986).

DNA synthesis is also perturbed by heat, at the level of both initiation and elongation (Warters and Stone 1983). Furthermore, heat stress can inhibit chromatin assembly, leaving the DNA in an unstable, nuclease-sensitive conformation (Warters and Roti Roti 1981). The total protein content of the nucleus also increases with heat shock (Laszlo 1992).

Finally, heat shock induces a number of changes in cell membranes, including changes in surface morphology (surface smoothening, decreased number of villi), increased fluidity of the bilayer, and aggregation of integral membrane proteins (Bass et al. 1978; Kruuv et al. 1983; Lepock et al. 1983). The activities and abundance of membrane-localized receptors and transport proteins change as well.

It is thus clear that heat shock perturbs many aspects of normal cellular physiology. In many cases, mild heat pretreatments, which induce hsp synthesis, either reduce the extent of these perturbations or speed their repair. Unfortunately, although we know that hsps have a central role in thermotolerance and have learned a great deal about the general biochemical functions of these proteins, we still do not know which heat-induced cellular perturbations are responsible for lethality or which lethal lesions are the most susceptible to repair by hsps. The complexity of the phenomena makes these questions difficult to answer. Some investigators have attempted to address these issues by inducing isolated lesions with drugs or by comparing lesions in thermotolerant and thermosensitive cell lines. It is often difficult to integrate the results from different laboratories into a comprehensive picture, however, because experiments have been performed in widely different cell types using many different culture conditions and different levels of heat stress. Another strategy for discerning which cellular targets are the most vital is to develop a better understanding of induced protective mechanisms. Research in this area is currently focused on understanding the functions of hsps.

III. THE CHAPERONING FUNCTIONS OF HEAT SHOCK PROTEINS AND STRESS TOLERANCE

A. hsp70

The hsp70s are a highly conserved protein family (~50% amino acid identity among all species characterized) with distinct members in many cellular compartments. Some hsp70s are constitutively expressed, and others are induced by heat or cold; in multicellular organisms, there are tissue-specific versions (e.g., the testis-specific hsp70s) (Lindquist and Craig 1988; Craig et al., this volume). All hsp70 proteins contain a conserved amino-terminal ATP-binding domain (Flaherty et al. 1990; McKay 1991; McKay et al., this volume) and a less conserved carboxy-terminal substrate-binding domain (Hightower et al., this volume). This organization suggests that they share a common mechanism for utilizing the energy of ATP while recognizing a variety of substrates.

hsp70 proteins bind to a variety of substrates and participate in protein folding, unfolding, assembly, and disassembly processes (for review, see Gething and Sambrook 1992). The favored hypothesis to explain their role in these processes is that hsp70 proteins bind to hydrophobic surfaces, stabilizing target proteins in a fully or partially unfolded state, and prevent adventitious associations (Pelham 1986). ATP promotes substrate release, allowing folding, transport, or assembly to continue (Gething and Sambrook 1992). Other hsps potentiate the activities of hsp70s (Georgopoulos et al., this volume). Most notably, DnaJ and GrpE enhance interactions between DnaK (the *Escherichia coli* hsp70) and its targets and stimulate the ATPase activity of DnaK by promoting ATP hydrolysis and ADP/ATP exchange.

1. hsp70 Functions in Stress Tolerance

In most organisms, the hsp70s are among the most prominent proteins induced by heat (Fig. 2), and early experiments showed a close correlation between their induction and the induction of tolerance to high temperatures (Li and Werb 1982; Li and Laszlo 1985; Subjeck and Shyy 1986). When the role of hsp70 in chaperoning unfolded proteins at normal temperatures became apparent, a role in chaperoning proteins damaged by stress seemed to be a logical extension (Pelham 1986; Rothman 1989).

In vitro, two studies have demonstrated the biochemical ability of hsp70 to reactivate damaged proteins. In the pioneering study, DnaK was shown to protect RNA polymerase from inactivation at high temperatures (in an ATP-independent fashion) and to reactivate the previously heat-denatured enzyme (in an ATP-dependent fashion) (Skowyra et al. 1990). In a recent analysis, DnaK was shown to reactivate firefly luciferase both in vitro and in vivo. This reactivation was enhanced by the association of DnaJ with the denatured substrate (Schroder et al. 1993).

In vivo, the roles that heat-inducible forms of hsp70 play in stress tolerance have been examined in many organisms. Surprisingly, their importance varies among organisms to a large degree.

a. Escherichia coli. Deletion of the *E. coli dnaK* gene severely reduces growth at 30°C and eliminates it at both high (40°C) and low (11–16°C) temperatures (Bukau and Walker 1989; J.S. McCarty and G.C. Walker, in prep.). These *dnaK⁻* strains also strongly overexpress other hsps, consistent with the hypothesis that DnaK has a key role in hsp regulation. Most suppressors of the 30°C growth defect map to σ^{32}, suggesting that the major function of DnaK at 30°C is to down-regulate the expression of other hsps (Bukau and Walker 1990). Surprisingly, *dnaK⁻* mutations have only a very modest effect on tolerance to extreme temperatures (B. Bukau, pers. comm.). Instead, two newly described mechanisms appear to operate, one involving proteins under control of the stationary-phase regulator, σ^s, and the other involving other hsps (Squires et al. 1991; Hengge-Aronic 1993; Zambrano et al. 1993; B. Bukau, pers. comm.).

b. Saccharomyces cerevisiae. Yeast cells express four nuclear/cytosolic hsp70 proteins that form an essential subfamily (Craig et al., this volume). Cells with decreased constitutive hsp70 expression (*ssa1ssa2*) are temperature-sensitive for growth (Craig and Jacobsen 1985). Several suppressors of this defect map to the heat shock factor (HSF), decreasing its activity and suppressing the overexpression of other hsps seen in the *ssa1ssa2* mutant. Thus, as in *E. coli*, one of the major functions of hsp70 in yeast is the down-regulation of other hsps (Nelson et al. 1992; Craig et al., this volume).

Eliminating heat-inducible hsp70 expression prevents growth at moderately high temperatures (37°C) but does not cause any defect in survival at extreme temperatures (50°C) (Werner-Washburne et al. 1987). These mutations (*ssa1ssa3ssa4*) reduce survival at very high temperatures only when they are combined with mutations in the yeast *HSP104* gene, which encodes the major protein required for thermotolerance in yeast (Sanchez et al. 1993) (see below). Thus, as in *E. coli*, the heat-inducible hsp70 proteins of yeast help cells to cope with the stress of growth at moderately high temperatures but are not crucial for survival at extremes.

c. Drosophila melanogaster. Unlike *E. coli* and *S. cerevisiae*, *Drosophila* hsp70 appears to be the major protein involved in tolerance to extreme temperatures (40–42°C). As in other eukaryotes, *Drosophila* produces several hsp70 proteins constitutively (Craig et al. 1983), but the major heat-inducible hsp70 is virtually undetectable at normal temperatures (25°C). It is induced more than 1000-fold upon heat shock (37°C) (Velazquez et al. 1983). The accumulation of hsp70 is a rate-limiting factor in the development of thermotolerance both in cultured cells and in whole flies. Cultured cells transformed with extra copies of the *hsp70* gene acquire thermotolerance at a faster rate than wild-type cells, whereas cells transformed with *hsp70* antisense genes acquire it more slowly (Solomon et al. 1991). Moreover, when *hsp70* is expressed from a heterologous promoter at 25°C, without the induction of other hsps, it is sufficient to provide tolerance to direct exposure to 42°C (Solomon et al. 1991). The importance of hsp70 in thermotolerance was demonstrated at the organismal level in fly strains that carried 12 extra copies of the wild-type *hsp70* gene (Welte et al. 1993). The more rapid accumulation of hsp70 in embryos of these strains was paralleled by a more rapid acquisition of thermotolerance (Fig. 3). These studies suggest that it is possible to alter the stress tolerance of complex multicellular organisms by deliberate genetic intervention (Welte et al. 1993).

Although the heat-inducible form of hsp70 is beneficial for survival during exposure to high temperatures, it is detrimental to growth and/or cell division (Feder et al. 1992). Cells expressing the protein at normal temperatures stop growing, and when growth resumes after several days of continuous induction, it is accompanied by the sequestration of hsp70 into large granules. Similar granules appeared in wild-type cells approximately 12 hours after a standard heat shock. The sequestration of hsp70 is a regulated process that occurs with extraordinary speed at certain stages of development. In early embryos, it occurs within just a few minutes after return to normal temperatures, and thermotolerance is lost at a similar rate (M. Welte et al., unpubl.). This developmental stage is characterized by high rates of cell division. Thus, the extremely rapid inactivation of hsp70 by sequestration into granules may be essential for the resumption of normal development after heat shock in early embryos. Notably, preblastoderm embryos are not even capable of inducing hsp70. At this stage, it might be that the potential inhibiting effects of hsp70 outweigh the protection it affords upon heat shock.

d. Vertebrates. The role of hsp70 in induced thermotolerance has been studied extensively in vertebrate cells (Hahn 1982; Paliwal et al. 1988). Although a few reports contradict the general trend and suggest that hsp70 is not required for thermotolerance (for review, see Carper et al. 1987), other experiments provide a compelling argument for the importance of hsp70 in this phenomenon. Evidence includes (1) when several

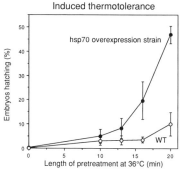

Figure 3   A *Drosophila* strain that overexpresses hsp70 acquires thermo-tolerance more rapidly than wild type. At 6 hr of development, embryos of the extra-copy and wild-type strains were collected, pretreated at 36°C, and heat-shocked for 25 min at 41°C. Hatching was scored as described in Fig. 1. The average and standard deviations of three independent experiments are shown. (Reprinted, with permission, from Welte et al. 1993.)

different cell types are exposed to a wide variety of tolerance-inducing treatments, induction of thermotolerance was reported to correlate more closely with accumulation of hsp70 than with accumulation of any other hsp (Li and Werb 1982; Li and Laszlo 1985); (2) the selection of heat-resistant cell lines by hyperthermic treatments yields cells that over-express hsp70 proteins (Laszlo and Li 1985; Anderson et al. 1989); (3) microinjection of anti-hsp70 antibodies prevents survival of fibroblasts at 45°C, whereas microinjection of control antibodies has no such effect (Riabowol et al. 1988); and (4) transformation of monkey cells (Angelidis et al. 1991) and rat cells (Li et al. 1991) with constitutively expressed human *hsp70* genes dramatically increases thermotolerance. In the latter case (Li et al. 1991), two-dimensional analysis of total cellular proteins demonstrated that increased basal expression of hsp70 was the only substantial alteration in the protein profile, making this study the most convincing demonstration of the importance of hsp70 in mam-malian thermotolerance. In a surprising extension of these studies, dele-tion of most of the highly conserved amino terminus of hsp70 did not im-pair its capacity to provide thermotolerance, suggesting that this domain is not required for thermotolerance (Li et al. 1992). However, in this case, the effects of the mutant protein on the expression of other proteins at normal temperatures were not examined. Because similar mutations in the *Drosophila* hsp70 protein up-regulate hsp expression (Solomon et al. 1991), the induction of endogenous hsps by the mutant hsp70 may ac-count for thermotolerance in this case.

2. Intracellular Targets of hsp70 Action

Although a great deal is known about the biochemical activities of hsp70 and its importance in protecting organisms at high temperatures, the critical biological processes that it protects from stress-induced damage have not yet been defined. Immunofluorescent localization studies show that hsp70 is concentrated at membranes, in nuclei, and in nucleoli (Pelham 1984; Pelham et al. 1984; Velazquez and Lindquist 1984; Welch and Feramisco 1984). This pattern is consistent with several observations suggesting that the protein is particularly important in these locations: (1) Members of the yeast hsp70 family that are both constitutive and heat-inducible promote the translocation of proteins across mitochondrial and endoplasmic reticulum (ER) membranes (Chirico et al. 1988; Deshaies et al. 1988; Vogel et al. 1990). (2) Purified hsp70 repairs heat-induced damage to some nuclear functions (e.g., mRNA splicing) in yeast cell-free systems (J.L. Vogel and S. Lindquist, unpubl.). (3) Constitutive expression of *Drosophila* hsp70 accelerates the recovery of nucleolar morphology after heat shock (Laszlo 1992; Pelham et al. 1984). It is not yet established which, if any, of these activities are relevant to the function of hsp70 in stress tolerance.

B. hsp100

The hsp100 proteins comprise a highly conserved family that includes proteins from bacteria, yeast, plants, trypanosomes, and mammals (Gottesman et al. 1990; Parsell et al. 1991; Squires et al. 1991; Squires and Squires 1992). The family has been divided into subfamilies (ClpA, ClpB, ClpC) on the basis of the organization of two highly conserved nucleotide-binding domains (Gottesman et al. 1990). The level of amino acid homology in these 200–300-amino-acid-long regions is remarkable: 50–80% identity and an additional 20–40% similarity. In the *S. cerevisiae* hsp104 protein, the two ATP-binding domains have different functions. One controls the assembly of the protein into a homoligomeric complex, and the other is responsible for most of the protein's ATPase activity (Parsell et al. 1994). Recently, more distantly related members of this family, which contain only one nucleotide-binding domain, have been identified in *E. coli* and mammals (Gottesman et al. 1990; C. Vandenberg, pers. comm.).

The proteins in the ClpB subfamily are all heat-inducible and have higher homology with each other than with the constitutive members of the family. This suggests that they share a conserved function in stress

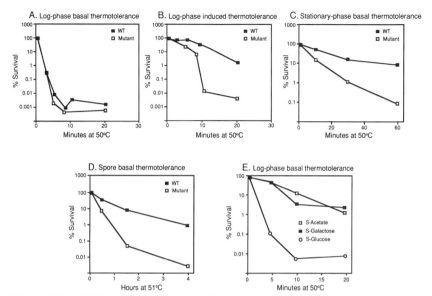

Figure 4 hsp104 is required for both basal and induced thermotolerance. Wild-type (WT) and hsp104 mutant (Mutant) cells were assayed for thermotolerance. To assay induced thermotolerance, cells were grown at 25°C to mid-log phase and either maintained at 25°C (*A*) or incubated at 37°C for 30 min (*B*) prior to exposure to 50°C. Survival at various times was assessed by colony-forming ability. To assay basal thermotolerance, stationary-phase cultures, grown for 72 hr from starter cultures inoculated with 2×10^4 cells/ml, were shifted directly from 25°C to 50°C (*C*). Spores were assayed for thermotolerance by heating at 51°C (*D*). Wild-type cells, grown at 25°C to mid-log phase in minimal acetate, galactose, or glucose media, were assayed for basal thermotolerance by shifting directly to 50°C (*E*). These data were taken from Sanchez et al. (1992).

tolerance (Gottesman et al. 1990; Squires and Squires 1992; E. Vierling, pers. comm.). Genetic analysis of heat-inducible family members from *S. cerevisiae* (hsp104) and *E. coli* (ClpB) confirms this hypothesis.

1. hsp100 Functions in Stress Tolerance

When *S. cerevisiae* cells grown at 25°C are pretreated at 37°C to induce tolerance, they survive exposure to 50°C 1000- to 10,000-fold better than nonpretreated cells (Fig. 4A,B). This tolerance is severely compromised by mutations in the hsp104 protein. Within a few minutes of exposure to 50°C, mutant cells begin to die at 100–1000 times the rate of wild-type cells (Fig. 4B) (Sanchez and Lindquist 1990). hsp104 seems to be specialized to function under extreme conditions. Mutations in the *HSP104* gene have no effect on growth on glucose at either low (25°C) or high (39°C) temperatures, and only a modest effect on survival at lower killing temperatures (e.g., at 44°C) (Sanchez et al. 1992). Like *hsp104* mutations, mutations in the *E. coli clpB* gene have no effect on growth at normal temperatures but cause cells to die more rapidly than wild-type cells at extreme temperatures (Squires et al. 1991). Thus, the tolerance functions of the heat-inducible members of this protein family have been evolutionarily conserved.

hsp100s, like other hsps, are induced by a wide variety of toxic conditions. However, their importance in providing tolerance to these stresses varies dramatically. For example, although hsp104 is strongly induced by arsenite and cadmium, mutations in the *HSP104* gene reduce survival during exposure to arsenite by only five- to tenfold and do not reduce survival during exposure to cadmium at all (Sanchez et al. 1992). This suggests that the damage caused by these agents is fundamentally different from that caused by heat. hsp104 must not repair this damage or, if it does, other proteins must be able to compensate for its absence. The hsp104 protein does have a major role in tolerance to ethanol. As is the case for heat, its function is most apparent under extreme conditions. Although the *hsp104* mutation reduces tolerance to 20% ethanol by 1000-fold, it reduces tolerance to 15% ethanol by only a fewfold (Sanchez et al. 1992).

2. Biochemical Function of hsp100 Proteins

The biochemical activities hsp100 proteins employ to provide cells with thermotolerance are not currently understood. Electron micrographs of wild-type and hsp104-deficient yeast cells exposed to high temperatures reveal the accumulation, just prior to killing, of large aggregates in the mutant cells (A. Kowal et al., unpubl.). Although it seems likely that this damage is responsible for the death of the mutant cells, it is not clear whether hsp104 acts to prevent these aggregates from forming or to dissolve them once they have formed.

A role for the heat-inducible hsp100 proteins in the proteolysis of stress-damaged proteins is suggested by the activity of one of the constitutive members of the family, the *E. coli* ClpA protein. ClpA functions as an ATP-dependent regulator of the ClpP protease. Experiments measuring the breakdown of amino acid analog-containing proteins indicate that the Clp complex (also known as protease Ti) is responsible for about 15% of the turnover of abnormal proteins in *E. coli* (Katayama et al. 1988). However, the role of the Clp protease in the turnover of stress-damaged proteins is unclear. Although the synthesis of ClpP increases about twofold upon heat shock (Kroh and Simon 1990), no stress-related phenotypes have been observed in *clpP* or *clpA* mutants (S. Gottesman, pers. comm.).

To date, attempts to demonstrate a proteolytic role for the heat-inducible members of the hsp100 family have failed. hsp104 and ClpB do not interact with ClpP or promote proteolysis in vitro (Woo et al. 1992; M. Maurizi, pers. comm.). Moreover, cells carrying *hsp104* and *clpB* mutations have no detectable proteolytic defects (Parsell et al. 1993; S. Gottesman pers. comm.).

3. Functional Relationship between hsp104 and hsp70

Analysis of *S. cerevisiae* strains carrying multiple mutations in *HSP104* and the heat-inducible hsp70 genes (*SSA1, SSA3, SSA4*) provide another clue to hsp104 function (Sanchez et al. 1993). When hsp70 levels are reduced, hsp104 becomes important for growth at normal and at moderately high temperatures. Conversely, when hsp104 levels are reduced, hsp70 becomes important in thermotolerance. These results strongly suggest that hsp70 and hsp104 function on the same pathway or on parallel pathways that partially overlap. For example, hsp104 might promote the proteolysis of damaged proteins that are not salvageable by hsp70. Alternatively, hsp104 might serve as another chaperone, binding to damaged proteins and promoting refolding.

C. hsp60

hsp60 proteins are found in the cytosol of bacteria (where they are known as GroEL proteins), in the matrix compartment of mitochondria, and in the stromal compartment of chloroplasts (where they are known as chaperonin-60 proteins) (for review, see Ellis and van der Vies 1991; Hartl et al. 1992; Lorimer 1992). They are among the most abundant cellular proteins in bacteria at normal temperatures. hsp60s share nearly 60% amino acid identity and a common oligomeric structure: a "double-doughnut" of two, seven-membered rings. hsp60 functions as a molecular chaperone, promoting the folding of newly made bacterial proteins (Horwich et al. 1993), as well as eukaryotic proteins translocated into mitochondria and chloroplasts (Frydman and Hartl, this volume). In vitro, it captures denatured substrates with high efficiency, and these substrates acquire elements of secondary structure while bound to hsp60. In most cases, further folding of the substrate protein and release from hsp60 requires ATP and another heat-inducible protein, hsp10 (also known as GroES or chaperonin-10). hsp10 also forms a homo-oligomeric ring-shaped particle that binds to one end of the hsp60 particle and regulates its ATPase activity and substrate associations.

In vitro studies show that hsp60 prevents the aggregation of proteins that denature at physiologically relevant temperatures. hsp60 suppresses the aggregation of dihydrofolate reductase and α-glucosidase when these proteins are heated to 40–45°C, and when the temperature is reduced, active enzymes are recovered upon addition of hsp10 and ATP (Holl-Neugebauer and Rudolph 1991; Martin et al. 1992). hsp60 becomes increasingly more important for protein folding as temperatures increase. For example, the refolding of chemically denatured Rubisco (ribulose bisphosphate carboxylase) and rhodanese requires hsp60 at high temperatures and high protein concentrations but is virtually independent of the chaperone at lower concentrations and temperatures (Goloubinoff et al. 1989; Viitanen et al. 1990; Mendoza et al. 1991a,b). In yeast mitochondria, hsp60 associates with a wide variety of proteins at high temperatures. These associations presumably prevent protein aggregation and promote protein refolding when cells are returned to lower temperatures (Martin et al. 1992).

Overexpression of hsp60 (GroEL) in *E. coli* suppresses a variety of conventional, temperature-sensitive mutations (Van Dyk et al. 1989). Presumably, these mutant proteins unfold at high temperatures, and GroEL prevents their aggregation and facilitates refolding. Overexpression of GroEL also suppresses some of the less conventional, temperature-sensitive for folding (TSF) mutations that have been isolated in bacteriophage P22 proteins. Proteins containing TSF mutations are stable at high temperatures when expressed and allowed to fold at low temperatures in *E. coli*, but they misfold and form inclusion bodies when expressed at 37–42°C (Yu and King 1988). Surprisingly, TSF mutations in different P22 proteins behave differently with respect to rescue by GroEL. Although TSF mutants in the tailspike protein can interact with the chaperone, GroEL overexpression does not suppress their temperature-sensitive phenotypes (Brunschier et al. 1993; S. Sather and J. King, pers. comm.). On the other hand, a large number of TSF mutations in the P22 coat protein are rescued by overexpression of GroEL and GroES (Gordon and King 1993; J. King, pers. comm.). Determining why only certain classes of TSF mutations are affected by GroEL may provide important insights into the specificity of hsp60 functions.

GroEL is strongly induced by heat, increasing to a remarkable 10–15% of total cellular protein (Neidhardt et al. 1984). This increase is essential for growth at high temperatures. An *E. coli* strain carrying a deletion in the gene encoding σ^{32} cannot induce hsps or grow above 20°C (Kusukawa and Yura 1988). Overexpression of GroEL alone permits growth at temperatures up to 40°C in these strains. Whether hsp60 has a significant role in protecting cells from short exposures to more extreme temperatures is unknown.

D. TF55/TCP

The central role that hsp60 proteins have in protein folding and stress tolerance in bacteria and eukaryotic organelles begs the question: What proteins provide these functions in the cytosolic and ER compartments of eukaryotes? A partial answer has come from an unexpected source. In the thermophilic archaebacteria *Sulfolobus shibatae*, high temperatures increase the expression of a single 55-kD protein, TF55 (Trent et al. 1990). The oligomeric structure of TF55 is similar to that of hsp60, consisting of two stacked nine-membered rings (Trent et al. 1991; Gao et al. 1992; Saibil et al. 1993). Other species of archaebacteria (Phipps et al. 1993) and eukaryotes (Frydman et al. 1992; Lewis et al. 1992; for review, see Willison and Kubota, this volume) synthesize proteins that form similar particles. In eukaryotes, these proteins (the T-complex proteins, or TCPs) are cytosolic and, surprisingly, not heat-inducible. TF55 and the TCPs show a low, but convincing, level of amino acid sequence homology with the hsp60s (Lewis et al. 1992).

The TF55 and TCP particles are also similar to hsp60 in that they bind to a wide variety of denatured proteins (Trent et al. 1991; J. Trent, pers. comm.; Willison and Kubota, this volume). TCP also possesses an hsp60-like chaperoning activity. Although TF55 has not yet been shown to facilitate the folding of any denatured substrate, it is almost certainly the central protein involved in thermotolerance in *Sulfolobus*. Unlike eubacteria and eukaryotes, which strongly induce 5–15 proteins in response to high temperatures, *Sulfolobus* produces only TF55. Furthermore, the induction of this protein is closely correlated with the induction of tolerance to otherwise lethal temperatures (Trent et al. 1990). Unfortunately, methods for creating site-directed mutations are not yet available in *Sulfolobus*, preventing a direct genetic test of this hypothesis. At this time, there is no evidence that TCP has a role in eukaryotic stress tolerance.

E. The Small Heat Shock Proteins

Of the major hsp families, the small hsps are the least conserved and, until recently, the most elusive in function (Arrigo and Landry, this volume). They are found in both prokaryotes (mycobacteria) and eukaryotes, and they are particularly abundant in plants, which encode small hsps localized to the endomembrane system and chloroplasts (for review, see Vierling 1991). The conservation of these proteins across prokaryotic and eukaryotic kingdoms and in different eukaryotic compartments strongly suggests that they have an important cellular function. One of the most interesting recent developments in the biochemical analysis of the small hsps is that they display elements of chaperone function in vitro. In one study, mouse hsp25 was shown to suppress the heat-induced aggregation of β-L-crystallin and α-glucosidase at molar ratios (hsp:substrate) of 1:20 and 1:10, respectively (Merck et al. 1993). In another report, small hsps were able to suppress aggregation of denatured substrates and also to promote their reactivation (Jakob et al. 1993). However, the formation of stable binary complexes between small hsps and their substrates has not yet been reported, and, in contrast to hsp70 and hsp60, the chaperoning activities of the small hsps are independent of ATP.

Genetic analysis of small hsp function in stress tolerance gives very different results in mammals and yeast. Hyperthermic selection of mutagenized Chinese hamster cells produced a family of thermotolerant cell lines in which the only detectable change in protein synthesis was the overproduction of hsp27 (Chretien and Landry 1988). More convincingly, transfection of naive CHO cells with constitutive hsp27 genes conferred constitutive thermotolerance, and transfection with metallothionein-regulated hsp27 genes conferred metal-regulated thermotolerance (Landry et al. 1989; Lavoie et al. 1993). To investigate the nature of this protection, several toxic effects of heat shock were examined in thermoresistant cells. Heat treatments disrupted protein synthesis, RNA synthesis, rRNA processing, and protein degradation to a similar extent in thermoresistant hsp27 transformants and wild-type cells (Landry et al. 1992). However, in the transformants, the microfilament (actin) network was dramatically protected from disruption by heat shock (Laszlo et al. 1993; Lavoie et al. 1993a,b). Studies in vitro suggest that the small hsps of avian cells can dramatically effect the polymerization of actin (Miron et al. 1991). Although the relationship between this observation and the protection of microfilament networks is not yet understood, it also suggests an important role for small hsps in microfilament dynamics (Arrigo and Landry, this volume).

Yeast cells contain only one major small hsp, hsp26 (Petko and Lindquist 1986; Susek and Lindquist 1989; Tuite et al. 1990). Overexpression of this protein provides, at best, only an extremely subtle increase in thermotolerance (Susek and Lindquist 1989; Tuite et al. 1990). More importantly, deletions of this gene have (1) no effect on growth at any temperature on a variety of carbon sources, under aerobic or anaerobic conditions, (2) no effect on tolerance to extreme temperatures or to ethanol in logarithmic and stationary-phase cells, and (3) no effect on spore morphogenesis, viability, germination, or thermotolerance (Petko and Lindquist 1986; J. Taulien et al., unpubl.). Furthermore, deletion of the *HSP26* gene in an *hsp104* mutant background has no effect on basal or induced thermotolerance (Sanchez et al. 1993). Thus, although the small hsps have an important role in thermotolerance in mammalian cells, they seem to have, at most, a minor role in yeast.

F. hsp90

Members of the hsp90 family are present in prokaryotes, in the cytosolic/nuclear compartment of eukaryotes, and in the ER of higher eukaryotes. Deletion of the *E. coli* hsp90 gene (*htpG*) has no effect on growth at normal temperatures and produces only a very subtle reduction in growth at high temperatures (Bardwell and Craig 1988). The *S. cerevisiae* hsp90 proteins, however, are essential at all temperatures. This suggests that the hsp90s have acquired novel functions in eukaryotes or that the functions they share with prokaryotes are now vital (Borkovich et al. 1989). Members of the hsp90 family interact with many other cellular proteins, including casein kinase II (Miyata and Yahara 1991), the heme-regulated eIF-2α kinase (Rose et al. 1989), several steroid hormone receptors (Pratt et al. 1992; Bohen and Yamamoto, this volume), oncogenic tyrosine kinases (Brugge 1986), calmodulin (Minami et al. 1993), actin (Nishida et al. 1986), and tubulin (Redmond et al. 1989; Fostinis et al. 1992).

In vitro, hsp90 functions as a molecular chaperone. Although the initial report demonstrated a relatively modest chaperoning activity (Wiech et al. 1992), recent results suggest it is considerably stronger (J. Buchner, pers. comm.). hsp90 may possess an inherent specificity that restricts its function to certain types of target proteins. Indeed, in vitro analysis with casein kinase II, a protein for which there is some evidence for an interaction with hsp90 in vivo, provides the strongest evidence for chaperone function (Miyata and Yahara 1992). Two characteristics appear to distinguish hsp90 from the other abundant chaperones, hsp70 and hsp60. First, hsp90 displays a high level of specificity in associating with particular target proteins. Second, the interactions between hsp90 and at least some target proteins are long-lived and have acquired important regulatory features (Brugge 1986; Picard et al. 1990; Pratt et al. 1992a,b; Xu and Lindquist 1993).

Genetic analysis in *S. cerevisiae* demonstrates that the amount of hsp90 required for growth increases as the temperature increases (Borkovich et al. 1989; L. Arwood and S. Lindquist, unpubl.). Its importance in helping cells survive extreme temperatures is less clear. hsp90 is the only protein whose expression is noticeably increased in one temperature-resistant CHO cell line (Yahara et al. 1986), and mammalian cells with reduced hsp90 levels (due to antisense RNA expression) are killed somewhat more rapidly than wild-type cells at extreme temperatures (Bansal et al. 1991). In yeast, however, mutational analysis in two different laboratories indicates that although increased expression of hsp90 is required for growth at the upper end of the normal growth range, the protein is not required for tolerance to extreme temperatures (Borkovich et al. 1989; Kimura et al. 1994).

G. Peptidyl Prolyl *Cis-Trans* Isomerases

Several peptidyl prolyl isomerases (PPIs) have recently been identified as hsps in widely divergent species. In *S. cerevisiae*, mutations in one class of PPIs, the cyclophilins, compromise the organism's tolerance to stress. For example, cells carrying mutations in the mitochondrial cyclophilin, Cpr3, cannot grow on lactose at 37°C (Davis et al. 1993). Two other yeast PPIs, the cyclophilins Cyp1 and Cyp2, are heat-inducible, and mutations in either or both of them cause cells to die five times more rapidly than wild-type cells at 48°C (Sykes et al. 1993). Critical substrates for these proteins have not yet been identified. An important recent discovery is that PPIs are present in complexes with two other hsps, hsp70 and hsp90, together with a variety of steroid hormone receptors (Tai et al. 1992, 1993; Nadeau et al. 1993). The PPI found in these complexes is heat-inducible and thus coregulated with hsp90 and hsp70.

Unlike the other hsps, PPIs function as true catalysts of protein folding, since the isomerization of specific Xaa-proline bonds is often a rate-limiting step in the folding process (Schmid et al. 1993). In vitro, PPIs also increase the yield of folded protein by reducing the formation of aggregates or by reducing adsorption of denatured proteins to the surface of the reaction vessel (Lilie et al. 1993). Thus, it is unclear whether it is the prolyl isomerase activity of PPIs or their anti-aggregation activities that are important in stress tolerance.

IV. PROTEOLYTIC FUNCTIONS OF HEAT SHOCK PROTEINS AND STRESS TOLERANCE

As discussed above, cells cope with proteins damaged by heat and other stresses in two general ways: They renature them or they degrade them. In the preceding sections, we have considered several hsps with chaperoning activities whose purpose is to salvage stress-damaged substrates. hsps also have an important role in the degradation of damaged proteins.

A. Ubiquitin and Other Enzymes of the Ubiquitin-dependent Degradation Pathway

Enzymes of the ubiquitin pathway target substrates for degradation by covalently attaching them to the highly conserved hsp, ubiquitin (for review, see Finley and Chau 1991; Hershko and Ciechanover 1992; Jentsch 1992). Work in both mammalian cells and yeast suggests that the ubiquitin system is responsible for much of the turnover of stress-damaged polypeptides in eukaryotes.

In mammalian cells, heat shock induces a burst of degradation of normally long-lived proteins that coincides with an increase in multi-ubiquitin-protein conjugates (Parag et al. 1987). Furthermore, a mutant cell line carrying a temperature-sensitive mutation in one of the enzymes in the ubiquitin pathway is unable to respond to temperature upshift with increased proteolytic activity and cannot grow at high temperatures (Parag et al. 1987).

In yeast, raising the intracellular concentration of ubiquitin appears to be an important part of the cellular response to stress. The *S. cerevisiae* polyubiquitin gene (*UBI4*) is heat-inducible, and although *ubi4* mutants grow at wild-type rates over the normal range of growth temperatures for yeast, they are hypersensitive to prolonged exposure to temperatures just beyond the growth range, to starvation, and to amino acid analogs (Finley et al. 1987). However, Ubi4p-deficient cells are not impaired in their ability to withstand short incubations at extreme temperatures.

The function of certain ubiquitin-conjugating enzymes (E2s), which catalyze the transfer of ubiquitin to proteolytic substrates (for review, see Jentsch 1992), is also crucial for yeast cell survival under conditions of stress (Seufert and Jentsch 1990). For example, cells carrying mutations in both of the heat-inducible *UBC4* and *UBC5* genes are inviable at elevated temperatures. These cells are also hypersensitive to cadmium, suggesting that Ubc4p and Ubc5p are important for the turnover of cadmium-damaged proteins as well (Jungmann et al. 1993). Ubc7p is another E2 enzyme that has a role in the degradation of cadmium-damaged proteins. Unlike Ubc4p and Ubc5p, however, Ubc7p does not appear to be important for the proteolysis of proteins damaged during heat shock (Jungmann et al. 1993). Thus, components of the ubiquitin-dependent degradation system appear to be specialized for the differential recognition of substrates damaged by different agents. This type of specialization was also observed in studies of the yeast hsp104 protein, which is crucial for protection against extreme heat, but provides no protection against cadmium (Sanchez et al. 1992).

The enzyme responsible for degrading ubiquitin-tagged substrates, the 26S protease, is composed of the 20S multi-catalytic proteasome and several other proteins (35–110 kD) (for review, see Hershko and Ciechanover 1992; Rechsteiner et al. 1993). Although most of its components are not heat-inducible, the 26S protease is clearly important for stress tolerance. Yeast strains carrying a missense mutation in the proteasome subunit, *pre1*, show increased sensitivity to both high temperatures and amino acid analogs (Heinemeyer et al. 1991).

B. The Lon and DegP Proteases

It is estimated from studies with canavanine-containing proteins that the Lon protease (also known as protease La) is responsible for roughly half of the turnover of abnormal proteins in the *E. coli* cytosol (for review, see Gottesman 1989; Gottesman and Maurizi 1992). This finding and the fact that Lon is induced by heat suggest that the protease is important for ridding the cell of stress-damaged polypeptides. However, since Lon is not essential for growth at high temperatures or for survival at extreme temperatures (S. Gottesman, pers. comm.), its importance in stress tolerance is not clear.

DegP (HtrA) is a 48-kD periplasmic protease that is strongly heat-inducible and essential for growth at high temperatures (Strauch et al. 1989; Lipinska et al. 1990). *degP* mutants prevent the turnover of periplasmic, but not cytoplasmic, fusion proteins (Strauch and Beckwith 1988; Lipinska et al. 1990), suggesting that DegP promotes the turnover of abnormal proteins in the periplasm. Thus, although little is known about the biochemical activities of DegP, it currently provides the best evidence for the biological importance of proteolysis in stress tolerance.

C. Role of Chaperones in Proteolysis

Genetic studies conducted in *E. coli* indicate that some of the "chaperoning" hsps (DnaK, DnaJ, GrpE, GroEL, and GroES) also have a role in the degradation of abnormal proteins (Baker et al. 1984; Keller and Simon 1988; Straus et al. 1988). Recent work has demonstrated physical associations between some of these hsps and certain proteolytic substrates (Sherman and Goldberg 1991, 1992), although it is unclear how these hsps facilitate the degradation of abnormal proteins. They might bind to proteases and use their chaperone functions either to maintain damaged substrates in an unfolded, proteolytically susceptible conformation or to stabilize a more active conformation of the protease itself. Alternatively, hsps could act independently of proteolytic complexes, employing their anti-aggregation activities to prevent damaged proteins from accumulating in a degradation-resistant form (A. Gragerov, pers. comm.).

Recent experiments suggest that, rather than cooperating with proteases, chaperones compete with them for stress-damaged substrates (see Craig et al., this volume). Eukaryotes encode a large family of enzymes that remove ubiquitin from proteins targeted for degradation. In *S. cerevisiae*, overexpression of one of these enzymes (Ubp3p) suppresses the growth defect of cells lacking the two constitutive cytosolic/nuclear hsp70 proteins (Ssa1p and Ssa2p). This suggests that the growth defect of *ssa1ssa2* mutant cells is due to the ubiquitin-dependent degradation of proteins normally salvaged by the hsp70 chaperone. Indeed, increasing the flux of substrates through the proteolytic pathway (by increasing the intracellular concentration of ubiquitin) reduces suppression by overexpressed Ubp3p.

V. OTHER FACTORS INFLUENCING STRESS TOLERANCE

Although in this chaper we have stressed the function of heat-inducible proteins in the cellular response to stress, it is important to remember that many other factors are also likely to have a role in stress tolerance. In fact, several reports suggest that there are two states of thermotolerance: The one that we have discussed thus far, which requires hsp synthesis, and another that functions under less severe conditions, decays more rapidly, and does not require hsp synthesis (Boon-Niermeijer et al. 1986; Hallberg 1986; Laszlo 1988; Lee and Dewey 1988). The latter tolerance mechanism might be mediated by small molecules like glycerol, betaine, and proline. These compounds accumulate to very high levels (100 mM to 4 M) under stressful conditions in a wide variety of evolutionarily diverse species (Yancey et al. 1982). In vitro, high concentrations of these solutes protect purified proteins from denaturation and aggregation (Yancey et al. 1982). Similarly, the disaccharide trehalose accumulates in several organisms during stationary-phase growth and during heat shock, two conditions that induce high levels of stress tolerance. In vitro, trehalose prevents the irreversible inactivation of enzymes during desiccation (Colaco et al. 1992). Recently, it has been proposed that the accumulation of adenosine 5'-phosphosulfate has an important role in tolerance (Jakubowshi and Goldman 1993). The presence of methionine in the media dramatically decreases the heat tolerance of yeast cells, presumably because it interferes with the accumulation of adenosine 5'-phosphosulfate (Jakubowshi and Goldman 1993). Finally, posttranslational modifications may promote the proper folding of proteins and reduce their susceptibility to stress damage. Most notably, glycosylation is suggested to protect ER proteins from aggregation, in effect mimicking the action of protein chaperones (Kern et al. 1992a,b).

VI. INTRINSIC FACTORS AFFECTING STRESS TOLERANCE

Although hsps are not the only factors involved, it is clear that they have key roles in the dramatic induction of tolerance by conditioning stress treatments. Are they also responsible for the dramatic differences in tolerance sometimes observed in the absence of stress conditioning? hsp expression changes during development and in response to physiological conditions, in a manner that correlates with changes in thermotolerance. For example, the steroid hormone ecdysone induces both the synthesis of small hsps and thermotolerance in *Drosophila* (Berger and Woodward 1983). hsps are also constitutively expressed in sporulating and stationary-phase fungi, which are characterized by very high levels of thermotolerance (Iida and Yahara 1984; Kurtz et al. 1986). A particularly interesting example is provided by the dimorphic pathogens, which induce hsp synthesis when they move from the environment into warm-blooded hosts. The degree to which they are capable of inducing hsp expression correlates with their pathogenicity (Young et al. 1990; Patriarca et al. 1992).

Genetic proof that changes in hsp expression in response to developmental and nutritional cues help to establish different levels of basal thermotolerance has recently been obtained in *S. cerevisiae*. hsp104 expression is high in both stationary-phase cells and spores and is required for the naturally high basal thermotolerance of these cell types (Fig. 4C,D) (Sanchez et al. 1992). Furthermore, constitutive levels of hsp104 are much higher in cells growing on acetate, which requires respiration, than in cells growing on glucose, which supports fermentation. Respiring cells have much higher basal thermotolerance than fermenting cells, and hsp104 is required for this thermotolerance (Fig. 4E) (Sanchez et al. 1992). Thus, the differences in thermotolerance observed between organisms at different stages of development and under different physiological conditions are, in at least some cases, directly related to differences in hsp expression.

A more enigmatic phenomenon is the large variation in basal thermotolerance sometimes observed between closely related organisms at the same stage in the life cycle, cultured under identical conditions. It is tempting to speculate that these naturally occurring, intrinsic differences in tolerance are due to polymorphisms in hsp expression or function. As pleasing as the hypothesis may be, however, the evidence to support it is, as yet, unconvincing.

As discussed above, thermoresistant mutants have been derived from a number of cultured cell lines by repeated cycles of growth and high-temperature killing. Different lines show elevated constitutive expression of hsp27, hsp90, or hsp70, and transformation of naive cells with genes constructed to increase the expression of these proteins demonstrates that their constitutive expression is sufficient to increase basal tolerance. However, although hsp overexpression is benefical in thermotolerance, other data suggest that it is detrimental to normal growth (Stone and Craig 1990; Feder et al. 1992). Thus, polymorphisms in hsp expression might not be a viable strategy for producing naturally occurring thermo-tolerance variations in whole organisms.

In several plant species, the importance of stress tolerance to agricultural productivity has led to the identification of thermosensitive and thermoresistant cultivars (for review, see Blum 1988). Breeding experiments suggest that a small number of genetic factors may control these difference in thermotolerance. In a few experiments, differences in thermotolerance among cultivars have been correlated with differences in hsp synthesis (Ougham and Stoddart 1986; Howarth 1989; Krishnan et al. 1989). In one case, however, where the cosegregation of hsp polymorphisms and thermotolerance polymorphisms was examined, no simple correlation was observed (Fender and O'Connell 1989). Unfortunately, thermotolerant cultivars have been characterized only in species with large genomes, and some of the hsps of these organisms are encoded by high-copy-number multigene families. There also appear to be regulatory variants in these species that influence the expression of different subsets of hsps in different tissues (Marmiroli et al. 1989; Fender and O'Connell 1990). The contributions that hsp polymorphisms make to thermotolerance in such species will be difficult to resolve.

Recently, naturally occurring thermotolerance polymorphisms have been identified in yeast cells, where the opportunities for genetic and molecular characterization of the phenomenon are better. A single polymorphic genetic locus, *HTT1*, makes a major contribution to thermotolerance in some backgrounds, and genetic analysis suggests it may do so through interaction with hsp104 (F. Rosenberg and S. Lindquist, in prep.). Most intriguingly, the *HTT1* locus appears to be genetically unstable, with high-thermotolerance variants arising in low-thermotolerance backgrounds at a high frequency. A mechanism for frequent production of genetic variants with different survival characteristics might provide an important strategy for coping with an unpredictable environment (F. Rosenberg and S. Lindquist, in prep.).

Mutations associated with oncogenic transformation may provide another source of genetic variation that contributes to differences in thermotolerance (for review, see Nover 1991). Hyperthermia has been employed in the treatment of tumors for many years, and early experiments suggested that intrinsic differences in the thermosensitivity of normal and transformed cells contributed to its efficacy. Unfortunately, current data provide no clear consensus. Some experiments support the conjecture, and others, including some of the more recent experiments, do not. Given the variety of oncogenic mechanisms and the multi-event character of transformation, it is not surprising that this problem still awaits resolution. There is no doubt, however, that the intrinsic thermosensitivity of transformed cells is only one of the factors that determine the therapeutic value of hyperthermic treatments. The degree of vascularization of the tumor, which in turn effects pH, nutrient availability, and oxygen tension, has profound effects on the ability of tumor cells to survive heat stress. Moreover, transformed cells have the capacity to synthesize hsps and acquire thermotolerance in response to heat pretreatments, complicating the scheduling of multiple hyperthermic treatments. The interplay between these different factors presents a serious challenge for clinicians.

VII. CONCLUDING REMARKS

A large body of evidence points to two general ways in which hsps prevent the accumulation of aberrant proteins generated as a result of exposure to stress. First, they function as molecular chaperones, preventing aggregation and restoring the native structures and activities of their substrates. Second, they enhance the flow of substrates through proteolytic pathways known to degrade structurally aberrant proteins. Identification of the features that usher substrates along one pathway or the other is an important problem for the future. Another outstanding question is whether chaperones function together with proteases to enhance the degradation of stress-damaged proteins or whether they compete with them in an effort to renature these substrates. The relative importance of refolding and degradation in the cellular response to stress is yet another major unresolved issue. Although mutations in many hsps with known chaperone functions have severe effects on growth and viability, the effects of mutations in proteolytic components are often much more subtle. This might indicate that degradation is less important than reactivation in coping with damaged proteins after a heat stress.

Organisms respond to different levels of stress by employing the activities of different hsps. For example, *E. coli* and *S. cerevisiae* require increased expression of hsp70 for growth at temperatures near the upper end of their normal growth ranges, and they require increased expression of hsp100 proteins for survival at extreme temperatures. hsps are also specialized to respond to different types of stress. In *S. cerevisiae*, hsp104 is crucial for tolerance to high temperatures and high concentrations of ethanol, moderately important for tolerance to arsenite, to lower temperatures, and to lower concentrations of ethanol, but it is of no importance at all for tolerance to copper and cadmium (Sanchez et al. 1992). Understanding how hsps differentially recognize substrates damaged by various types of stress is crucial for understanding how hsps function to provide stress tolerance.

One of the most perplexing aspects of hsp function in tolerance is that different organisms employ different hsps in response to what would appear to be similar levels of stress. For example, hsp70 is very important for tolerance to killing temperatures in mammals and *Drosophila*, whereas it seems to have only a minor role in yeast and *E. coli*. Similarly, members of the hsp100 family are very important for surviving extreme stress in yeast, somewhat important in *E. coli*, perhaps important in mammals, but apparently not important at all in *Drosophila*. In fact, *Drosophila* does not even synthesize a hsp100 protein in response to heat shock. Finally, small hsps promote stress tolerance in mammalian cells but appear to have little effect on tolerance in yeast.

To resolve this paradox, detailed biochemical knowledge of hsp function needs to be integrated with an understanding of the unique physiology of particular organisms. For example, hsps may have specialized roles in different cell types by protecting unique features of cellular architecture from disruption by stress. In certain mammalian cells, for example, one major effect of heat shock is the collapse of the cytoskeletal networks. Cells that overexpress small hsps are much less susceptible to this collapse, suggesting that small hsps may function in stress tolerance by preserving the integrity of cytoskeletal functions. In yeast, where small hsps are not especially important for tolerance, it is interesting that cytoskeletal networks are much less elaborate, and the defining feature for cell shape is the cell wall.

Metabolic constraints may also dictate which hsps will be employed for tolerance in a given organism. In *S. cerevisiae*, *hsp104* mutants have a subtle but surprising phenotype: When grown on carbon sources that force them into respiratory metabolism (yeast cells normally prefer to grow by fermentation), mutant cells grow slightly faster than wild-type cells (Sanchez et al. 1992). As discussed above, respiring cells constitutively express hsp104, and this expression greatly increases basal thermotolerance. The growth-rate phenotype suggests, however, that hsp100 proteins also have a slightly deleterious effect on respiring cells. In an organism like *Drosophila*, which is characterized by a very high rate of respiration (Wieser and Gnaiger 1989), such an effect of hsp100s might provide evolutionary pressure to find another avenue to stress tolerance: eliminate hsp100 expression and magnify hsp70 expression by amplification of the *hsp70* genes. The ability of hsp70 to fulfill a role in stress tolerance is clear in yeast, where overexpression of hsp70 partially compensates for the loss of hsp104 (Sanchez et al. 1993). However, amplification of the *hsp70* genes in *Drosophila* poses another problem: Constitutive expression of these proteins interferes with normal growth (Feder et al. 1992). *Drosophila* has coped with this problem by evolving multiple, stringent regulatory mechanisms that prevent hsp70 expression at normal temperatures, yet allow an enormous burst of expression in response to stress (Yost et al. 1991). Thus, the particular constellation of hsps induced in any given organism seems to represent a balance of beneficial and detrimental effects, suited both to the organism's own physiology and to the types of stress to which it is subject.

ACKNOWLEDGMENTS

We are very grateful to our many colleagues who shared their unpublished findings with us prior to publication. Specifically, we thank Y. Sanchez for the *S. cerevisiae* data, M. Welte for the *D. melanogaster* data, G. Li for the Rat-1 data, and J. Key for the soybean data in Figure 1; J. Taulien, Y. Sanchez, and M. Welte for the data presented in the other figures; and J. Feder and J. Vogel for critically reading the manuscript. We also thank the Howard Hughes Medical Institute for support. D.A.P. was supported by a postdoctoral fellowship from the Jane Coffin Childs Memorial Fund for Medical Research. Portions of this review previously appeared in the *Annual Review of Genetics*, Vol. 27, and are used with the permission of *Annual Reviews*.

REFERENCES

Ananthan, J., A.L. Goldberg, and R. Voellmy. 1986. Abnormal proteins serve as eukaryotic stress signals and trigger the activation of heat shock genes. *Science* **232:** 522–524.

Anderson, R.L., K.I. Van, P.E. Kraft, and G.M. Hahn. 1989. Biochemical analysis of heat-resistant mouse tumor cell strains: A new member of the HSP70 family. *Mol. Cell. Biol.* **9**: 3509–3516.

Angelidis, C.E., I. Lazaridis, and G.N. Pagoulatos. 1991. Constitutive expression of heat-shock protein 70 in mammalian cells confers thermoresistance. *Eur. J. Biochem.* **199**: 35–39.

Baker, T.A., A.D. Grossman, and C.A. Gross. 1984. A gene regulating the heat shock response in *Escherichia coli* also affects proteolysis. *Proc. Natl. Acad. Sci.* **81**: 6779–6783.

Bansal, G.S., P.M. Norton, and D.S. Latchman. 1991. The 90-kDa heat shock protein protects mammalian cells from thermal stress but not from viral infection. *Exp. Cell Res.* **195**: 303–306.

Bardwell, J.C. and E.A. Craig. 1988. Ancient heat shock gene is dispensable. (Published erratum appears in *J. Bacteriol.* 1988 Oct;170(10): 4999.) *J. Bacteriol.* **170**: 2977–2983.

Bass, H., J.L. Moore, and W.T. Coakley. 1978. Lethality in mammalian cells due to hyperthermia under oxic and hypoxic conditions. *Int. J. Radiat. Biol. Stud. Phys. Chem. Med.* **33**: 57–67.

Bell, J., L. Neilson, and M. Pellegrini. 1988. Effect of heat shock on ribosome synthesis in *Drosophila melanogaster*. *Mol. Cell. Biol.* **8**: 91–95.

Berger, E.M. and M.P. Woodward. 1983. Small heat shock proteins in *Drosophila* may confer thermal tolerance. *Exp. Cell Res.* **147**: 437–442.

Berger, E.M., M.P. Vitek, and C.M. Morganelli. 1985. Transcript length heterogeneity at the small heat shock protein genes of *Drosophila*. *J. Mol. Biol.* **186**: 137–148.

Blum, A. 1988. *Plant breeding for stress environments*. CRC Press, Boca Raton, Florida.

Bond, U. 1988. Heat shock but no other stress inducers leads to the disruption of a subset of snRNPs and inhibition of in vitro splicing in HeLa cells. *EMBO J.* **7**: 3509–3518.

Boon-Niermeijer, E.K., M. Tuyl, and H. van der Scheur. 1986. Evidence for two states of thermotolerance. *Int. J. Hyperthermia* **1**: 93–105.

Borkovich, K.A., F.W. Farrelly, D.B. Finkelstein, J. Taulien, and S. Lindquist. 1989. Hsp82 is an essential protein that is required in higher concentrations for growth of cells at higher temperatures. *Mol. Cell. Biol.* **9**: 3919–3930.

Brugge, J.S. 1986. Interaction of the Rous sarcoma virus protein pp60src with the cellular proteins pp50 and pp90. *Curr. Top. Microbiol. Immunol.* **123**: 1–22.

Brunschier, R., M. Danner, and R. Seckler. 1993. Interactions of phage P22 tailspike protein with GroE molecular chaperones during refolding in vitro. *J. Biol. Chem.* **268**: 2767–2772.

Bukau, B. and G.C. Walker. 1989. Cellular defects caused by deletion of the *Escherichia coli dnaK* gene indicate roles for heat shock protein in normal metabolism. *J. Bacteriol.* **171**: 2337–2346.

————. 1990. Mutations altering heat shock specific subunit of RNA polymerase suppress major cellular defects of *E. coli* mutants lacking the DnaK chaperone. *EMBO J.* **9**: 4027–4036.

Carper, S.W., J.J. Duffy, and E.W. Gerner. 1987. Heat shock proteins in thermotolerance and other cellular processes. *Cancer Res.* **47**: 5249–5255.

Chirico, W.J., M.G. Waters, and G. Blobel. 1988. 70K heat shock related proteins stimulate protein translocation into microsomes. *Nature* **332**: 805–810.

Chretien, P. and J. Landry. 1988. Enhanced constitutive expression of the 27-kDa heat shock proteins in heat-resistant variants from Chinese hamster cells. *J. Cell. Physiol.* **137**: 157–166.

Christiansen, E.N. and E. Kvamme. 1969. Effects of thermal treatment on mitochondria of brain, liver, and ascites cells. *Acta. Physiol. Scand.* **76:** 472–484.

Colaco, C., S. Sen, M. Thangavelu, S. Pinder, and B. Roser. 1992. Extraordinary stability of enzymes dried in trehalose: Simplified molecular biology. *Biotechnology* **10:** 1007–1011.

Coss, R.A., W.C. Dewey, and J.R. Bamburg. 1982. Effects of hyperthermia on dividing Chinese hamster ovary cells and on microtubules. *Cancer* **Res. 42:** 1059–1071.

Craig, E.A. and K. Jacobsen. 1985. Mutations in cognate genes of *Saccharomyces cerevisiae* hsp70 result in reduced growth rates at low temperatures. *Mol. Cell. Biol.* **5:** 3517–3524.

Craig, E.A., T.D. Ingolia, and L.J. Manseau. 1983. Expression of *Drosophila* heat-shock cognate genes during heat shock and development. *Dev. Biol.* **99:** 418–426.

Davis, E.S., A. Becker, J. Heitman, M.N. Hall, and M.B. Brennan. 1993. A novel yeast cyclophilin gene essential for lactate metabolism at high temperature. *Proc. Natl. Acad. Sci.* **89:** 11169–11173.

Deshaies, R.J., B.D. Koch, W.M. Werner, E.A. Craig, and R. Schekman. 1988. A subfamily of stress proteins facilitates translocation of secretory and mitochondrial precursor polypeptides. *Nature* **332:** 800–805.

Ellgaard, E.G. and U. Clever. 1971. RNA metabolism during puff induction in *D. melanogaster*. *Chromosoma* **36:** 60–78.

Ellis, R.J. and S.M. van der Vies. 1991. Molecular chaperones. *Annu. Rev. Biochem.* **60:** 321–347.

Falkner, F.G., H. Saumweber, and H. Biessmann. 1981. Two *Drosophila melanogaster* proteins related to intermediate filament proteins of vertebrate cells. *J. Cell. Biol.* **91:** 175–183.

Feder, J.H., J.M. Rossi, J. Solomon, N. Solomon, and S. Lindquist. 1992. The consequences of expressing hsp70 in *Drosophila* cells at normal temperatures. *Genes Dev.* **6:** 1402–1413.

Fender, S.E. and M.A. O'Connell. 1989. Heat shock protein expression in thermotolerant and thermosensitive lines of cotton. *Plant Cell Rep.* **8:** 37–40.

————. 1990. Expression of the heat shock response in a tomato interspecific hybrid is not intermediate between the two parental responses. *Plant Physiol.* **93:** 1140–1146.

Finley, D. and V. Chau. 1991. Ubiquitination. *Annu. Rev. Cell. Biol.* **7:** 25–69.

Finley, D., E. Ozkaynak, and A. Varshavsky. 1987. The yeast polyubiquitin gene is essential for resistance to high temperatures, starvation, and other stresses. *Ceil* **37:** 1035–1046.

Flaherty, K.M., F.C. DeLuca, and D.B. McKay. 1990. Three-dimensional structure of the ATPase fragment of a 70K heat-shock cognate protein (see comments). *Nature* **346:** 623–628.

Fostinis, Y., P.A. Theodoropoulos, A. Gravanis, and C. Stournaras. 1992. Heat shock protein HSP90 and its association with the cytoskeleton: A morphological study. *Biochem. Cell. Biol.* **70:** 779–786.

Frydman, J., E. Nimmesgern, B.H. Erdjument, J.S. Wall, P. Tempst, and F.-U. Hartl. 1992. Function in protein folding of TRiC, a cytosolic ring complex containing TCP-1 and structurally related subunits. *EMBO J.* **11:** 4767–4778.

Gao, Y., J.O. Thomas, R.L. Chow, G.H. Lee, and N.J. Cowan. 1992. A cytoplasmic chaperonin that catalyzes beta-actin folding. *Cell* **69:** 1043–1050.

Gething, M.J. and J. Sambrook. 1992. Protein folding in the cell. *Nature* **355:** 33–45.

Glass, J.R., R.G. DeWitt, and A.E. Cress. 1985. Rapid loss of stress fibers in Chinese hamster ovary cells after hyperthermia. *Cancer Res.* **45:** 258–262.

Goff, S.A. and A.L. Goldberg. 1985. Production of abnormal proteins in *E. coli* stimulates transcription of lon and other heat shock genes. *Cell* **41**: 587–595.

Goldberg, A.L. 1972. Degradation of abnormal proteins in *Escherichia coli. Proc. Natl. Acad. Sci.* **69**: 422–426.

Goloubinoff, P., A.A. Gatenby, and G.H. Lorimer. 1989. GroE heat-shock proteins promote assembly of foreign prokaryotic ribulose bisphosphate carboxylase oligomers in *Escherichia coli. Nature* **337**: 44–47.

Gordon, C.L. and J. King. 1993. Temperature-sensitive mutations in the phage P22 coat protein which interfere with polypeptide chain folding. *J. Biol. Chem.* **268**: 9358–9368.

Gottesman, S. 1989. Genetics of proteolysis in *Escherichia coli. Annu. Rev. Genet.* **23**: 163–198.

Gottesman, S. and M.R. Maurizi. 1992. Regulation by proteolysis: Energy-dependent proteases and their targets. *Microbiol. Rev.* **56**: 592–621.

Gottesman, S., W.P. Clark, and M.R. Maurizi. 1990. The ATP-dependent Clp protease of *Escherichia coli*: Sequence of *clpA* and identification of a Clp-specific substrate. *J. Biol. Chem.* **265**: 7886–7893.

Hahn, G.M. 1982. Hyperthermia and cancer. In *Thermotolerance: Thermotolerance and thermophily* (ed. K. Henle), vol. 1, pp. 97–112. Plenum Press, New York.

Hallberg, R.L. 1986. No heat shock protein synthesis is required for induced thermostabilization of translational machinery. *Mol. Cell. Biol.* **6**: 2267–2270.

Hartl, F.-U., J. Martin, and W. Neupert. 1992. Protein folding in the cell: The role of molecular chaperones Hsp70 and Hsp60. *Annu. Rev. Biophys. Biomol. Struct.* **21**: 293–322.

Heine, U., L. Sverak, J. Kondratick, and R.A. Bonar. 1971. The behavior of HeLa-S3 cells under the influence of supranormal temperatures. *J. Ultrastruct. Res.* **34**: 375–396.

Heinemeyer, W., J.A. Kleinschmidt, J. Saidowsky, C. Escher, and D.H. Wolf. 1991. Proteinase yscE, the yeast proteasome/multicatalytic-multifunctional proteinase: Mutants unravel its function in stress induced proteolysis and uncover its necessity for cell survival. *EMBO J.* **10**: 555–562.

Hengge-Aronic, R. 1993. Survival of hunger and stress: The role of rpoS in early stationary phase gene regulation in *E. coli. Cell* **72**: 165–168.

Hershko, A. and A. Ciechanover. 1992. The ubiquitin system for protein degradation. *Annu. Rev. Biochem.* **61**: 761–807.

Holl-Neugebauer, B. and R. Rudolph. 1991. Reconstitution of a heat shock effect in vitro: Influence of GroE on the thermal aggregation of a-glucosidase from yeast. *Biochemistry* **30**: 11609–11614.

Horwich, A.L., K.B. Low, W.A. Fenton, I.N. Hirshfield, and K. Furtak. 1993. Folding in vivo of bacterial cytoplasmic proteins: Role of GroEL. *Cell* **74**: 909–917.

Howarth, C. 1989. Heat shock proteins in sorghum bicolor and pennisetum americanum I. genotypic and developmental variation during seed germination. *Plant Cell Environ.* **12**: 471–477.

Howe, J.G. and J.W. Hershey. 1984. Translational initiation factor and ribosome association with the cytoskeletal framework fraction from HeLa cells. *Cell* **37**: 85–93.

Iida, H. and I. Yahara. 1984. Durable synthesis of high molecular weight heat shock proteins in G0 cells of the yeast and other eucaryotes. *J. Cell. Biol.* **99**: 199–207.

Iida, K., H. Iida, and I. Yahara. 1986. Heat shock induction of intranuclear actin rods in cultured mammalian cells. *Exp. Cell Res.* **165**: 207–215.

Jakob, U., M. Gaestel, K. Engel, and J. Buchner. 1993. Small heat shock proteins are molecular chaperones. *J. Biol. Chem.* **268**: 1517–1520.

Jakubowshi, H. and E. Goldman. 1993. Methionine-mediated lethality in yeast cells at elevated temperature. *J. Bacteriol.* **175:** 5469–5476.

Jentsch, S. 1992. The ubiquitin-conjugation system. *Annu. Rev. Genet.* **26:** 179–207.

Jungmann, J., H.-A. Reins, C. Schobert, and S. Jentsch. 1993. Resistance to cadmium mediated by ubiquitin-dependent proteolysis. *Nature* **361:** 369–371.

Katayama, Y., S. Gottesman, J. Pumphery, S. Rudikoff, W.P. Clark, and M.R. Maurizi. 1988. The two-component, ATP-dependent Clp protease of *Escherichia coli. J. Biol. Chem.* **263:** 15226–15236.

Keller, J.A. and L.D. Simon. 1988. Divergent effects of a dnaK mutation on abnormal protein degradation in *Escherichia coli. Mol. Microbiol.* **2:** 31–41.

Kern, G., M. Schmidt, J. Buchner, and R. Jaenicke. 1992a. Glycosylation inhibits the interaction of invertase with the chaperone GroEL. *FEBS Lett.* **305:** 203–205.

Kern, G., N. Schulke, F.X. Schmid, and R. Jaenicke. 1992b. Quaternary structure and stability of internal, external, and core-glycosylated invertase from yeast. *Protein Sci.* **1:** 120–131.

Key, J.L., C.Y. Lin, E. Ceglarz, and F. Schoffl. 1983. The heat shock response in soybean seedlings. In *Structure and function of plant genomes* (ed. O. Ciferri and L. Dure), vol. 63, pp. 25–36. Plenum Press, New York.

Kimura, Y., S. Matsumoto, and I. Yahara. 1994. Temperature-sensitive mutants of hsp82 of the budding yeast, *Saccharomyces cerevisiae. Mol. Gen. Genet.* (in press).

Krishnan, M., H.T. Nguyen, and J.J. Burke. 1989. Heat shock protein synthesis and thermal tolerance in wheat. *Plant Physiol.* **90:** 140–145.

Kroh, H.E. and L.D. Simon. 1990. The ClpP component of Clp protease is the sigma 32-dependent heat shock protein F21.5. *J. Bacteriol.* **172:** 6026–6034.

Kruuv, J., D. Glofcheski, K.H. Cheng, S.D. Campbell, H.M. Al-Qysi, W.T. Nolan, and J.R. Lepock. 1983. Factors influencing survival and growth of mammalian cells exposed to hypothermia. I. Effects of temperature and membrane lipid perturbers. *J. Cell Physiol.* **115:** 179–185.

Kurtz, S., J. Rossi, L. Petko, and S. Lindquist. 1986. An ancient developmental induction: Heat-shock proteins induced in sporulation and oogenesis. *Science* **231:** 1154–1157.

Kusukawa, N. and T. Yura. 1988. Heat shock protein GroE of *Escherichia coli*: Key protective roles against thermal stress. *Genes Dev.* **2:** 874–882.

Landry, J., P. Chretien, H. Lambert, E. Hickey, and L.A. Weber. 1989. Heat shock resistance conferred by expression of the human HSP27 gene in rodent cells. *J. Cell Biol.* **109:** 7–15.

Landry, J., J.N. Lavoie, E. Hickey, and L.A. Weber. 1992. HSP27 may couple signal transduction pathways to microfilament responses. In *The XXIth UOEH International Symposium on Stress Proteins*, 15, Supplement (pp. 111–121). Journal of UOEH, Kitakyushu, Japan.

Laszlo, A. 1988. Evidence for two states of thermotolerance in mammalian cells. *Int. J. Hyperthermia* **4:** 513–526.

———. 1992. The effects of hyperthermia on mammalian cell structure and function. *Cell Prolif.* **25:** 59–87.

Laszlo, A. and G.C. Li. 1985. Heat-resistant variants of Chinese hamster fibroblasts altered in expression of heat shock protein. *Proc. Natl. Acad. Sci.* **82:** 8029–8033.

Laszlo, A., T. Davidson, A. Hu, J. Landry, and J. Bedford. 1993. Putative determinants of the cellular response to hyperthermia. *Int. J. Radiat. Biol.* **63:** 569–581.

Lavoie, J.N., G. Gingras-Breton, R.M. Tanguay, and J. Landry. 1993a. Induction of Chinese hamster HSP27 gene expression in mouse cells confers resistance to heat shock. *J. Biol. Chem.* **268:** 3420–3429.

Lavoie, J.N., E. Hickey, L.A. Weber, and J. Landry. 1993b. Modulation of actin microfilament dynamics and fluid phase pinocytosis by phosphorylation of heat shock protein 27. *J. Biol. Chem.* **268:** 24210–24214.

Lee, Y.J. and W.C. Dewey. 1988. Thermotolerance induced by heat, sodium arsenite, or puromycin: Its inhibition and differences between 43 degrees C and 45 degrees C. *J. Cell. Physiol.* **135:** 397–406.

Lepock, J.R., K.H. Cheng, H. Al-Qysi, and J. Kruuv. 1983. Thermotropic lipid and protein transitions in Chinese hamster lung cell membranes: Relationship to hyperthermic cell killing. *Can. J. Biochem. Cell Biol.* **61:** 421–427.

Lewis, V.A., G.M. Hynes, D. Zheng, H. Saibil, and K. Willison. 1992. T-complex polypeptide-1 is a subunit of a heteromeric particle in the eukaryotic cytosol. *Nature* **358:** 249–252.

Li, G.C. and A. Laszlo. 1985. Thermotolerance in mammalian cells: A possible role for heat shock proteins. In *Changes in eukaryotic gene expression in response to environmental stress* (ed. B.G. Atkinson and D.B. Walden), pp. 227–254. Academic Press, Orlando, Florida.

Li, G.C. and Z. Werb. 1982. Correlation between synthesis of heat shock proteins and development of thermotolerance in Chinese hamster fibroblasts. *Proc. Natl. Acad. Sci.* **79:** 3218–3222.

Li, G.C., L. Li, R.Y. Liu, M. Rehman, and W.M. Lee. 1992. Heat shock protein hsp70 protects cells from thermal stress even after deletion of its ATP-binding domain. *Proc. Natl. Acad. Sci.* **89:** 2036–2040.

Li, G.C., L.G. Li, Y.K. Liu, J.Y. Mak, L.L. Chen, and W.M. Lee. 1991. Thermal response of rat fibroblasts stably transfected with the human 70-kDa heat shock protein-encoding gene. *Proc. Natl. Acad. Sci.* **88:** 1681–1685.

Lilie, H., K. Lang, R. Rudolph, and J. Buchner. 1993. Prolyl isomerases catalyze antibody folding in vitro. *Protein Sci.* **2:** 1490–1496.

Lindquist, S. 1981. Regulation of protein synthesis during heat shock. *Nature* **293:** 311–314.

———. 1986. The heat-shock response. *Annu. Rev. Biochem.* **55:** 1151–1191.

———. 1993. Autoregulation of the heat-shock response. In *Translational regulation of gene expression 2* (ed. J. Ilan), pp. 279–320. Plenum Press, New York.

Lindquist, S. and E.A. Craig. 1988. The heat-shock proteins. *Annu. Rev. Genet.* **22:** 631–677.

Lipinska, B., M. Zylicz, and C. Georgopoulos. 1990. The HtrA (DegP) protein, essential for *Escherichia coli* survival at high temperatures, is an endopeptidase. *J. Bacteriol.* **172:** 1791–1797.

Lorimer, G.H. 1992. Role of accessory proteins in protein folding. *Curr. Opin. Struct. Biol.* **2:** 26–34.

Marmiroli, N., C. Lorenzoni, A.M. Stanca, and V. Terzi. 1989. Preliminary study of the inheritance of temperature stress proteins in barley (hordeum vulgare L.). *Plant Sci.* **65:** 147–153.

Martin, J., A.L. Horwich, and F.-U. Hartl. 1992. Prevention of protein denaturation under heat stress by the chaperonin Hsp60. *Science* **258:** 995–998.

McKay, D.B. 1991. Structure of the 70-kilodalton heat-shock-related proteins. *Springer Semin. Immunopathol.* **13:** 1–9.

Mendoza, J.A., E. Rogers, G.H. Lorimer, and P.M. Horowitz. 1991a. Chaperonins facilitate the in vitro folding of monomeric mitochondrial rhodanese. *J. Biol. Chem.* **266:** 13044–10349.

———. 1991b. Unassisted refolding of urea unfolded rhodanese. *J. Biol. Chem.* **266:** 13587–13591.

Merck, K.B., P.J. Groenen, C.E. Voorter, H.H.W.A. de, J. Horwitz, H. Bloemendal, and J.W.W. de. 1993. Structural and functional similarities of bovine alpha-crystallin and mouse small heat-shock protein. A family of chaperones. *J. Biol. Chem.* **268:** 1046–1052.

Minami, Y., H. Kawasaki, K. Suzuki, and I. Yahara. 1993. The calmodulin-binding domain of the mouse 90-kDa heat shock protein. *J. Biol. Chem.* **268:** 9604–9610.

Miron, T., K. Vancompernolle, J. Vandekerckhove, M. Wilchek, and B. Geiger. 1991. A 25-kD inhibitor of actin polymerization is a low molecular mass heat shock protein. *J. Cell. Biol.* **114:** 255–261.

Miyata, Y. and I. Yahara. 1991. Cytoplasmic 8 S glucocorticoid receptor binds to actin filaments through the 90-kDa heat shock protein moiety. *J. Biol. Chem.* **266:** 8779–8783.

————. 1992. The 90-kDa heat shock protein, HSP90, binds and protects casein kinase II from self-aggregation and enhances its kinase activity. *J. Biol. Chem.* **267:** 7042–7047.

Nadeau, K., A. Das, and C.T. Walsh. 1993. Hsp90 chaperonins possess ATPase activity and bind heat shock transcription factors and peptidyl prolyl isomerases. *J. Biol. Chem.* **268:** 1479–1487.

Nagao, R.T., J.A. Kimpel, E. Vierling, and J.L. Key. 1986. The heat shock response: A comparative analysis. In *Oxford surveys of plant molecular and cell biology* (ed. J.B. Mifling) vol. 3, pp. 384–438. Oxford University Press, England.

Neidhardt, F.C., R.A. VanBogelen, and V. Vaughn. 1984. The genetics and regulation of heat-shock proteins. *Annu. Rev. Genet.* **18:** 295–329.

Nelson, R.J., M. Heschl, and E.A. Craig. 1992. Isolation and characterization of extragenic suppressors of mutations in the SSA hsp70 genes of *Saccharomyces cerevisiae. Genetics* **131:** 277–285.

Nishida, E., S. Koyasu, H. Sakai, and I. Yahara. 1986. Calmodulin-regulated binding of the 90-kDa heat shock protein to actin filaments. *J. Biol. Chem.* **261:** 16033–16036.

Nover, L. 1991. *Heat shock response.* CRC Press, Boca Raton, Florida.

Nover, L., K.D. Scharf, and D. Neumann. 1989. Cytoplasmic heat shock granules are formed from precursor particles and are associated with a specific set of mRNAs. *Mol. Cell Biol.* **9:** 1298–1308.

Ougham, H.J. and J.L. Stoddart. 1986. Synthesis of heat-shock protein and acquisition of thermotolerance in high-temperature tolerant and high-temperature susceptible lines of sorghum. *Plant. Sci.* **44:** 163–167.

Paliwal, B.R., F. Hetzel, and M.W. Dewhirst, eds. 1988. *Biological, physical and clinical aspects of hyperthermia.* American Institute of Physics, New York.

Panniers, R., E.B. Stewart, W.C. Merrick, and E.C. Henshaw. 1985. Mechanism of inhibition of polypeptide chain initiation in heat-shocked Ehrlich cells involved reduction of eukaryotic initiation factor 4F activity. *J. Biol. Chem.* **260:** 9648–9653.

Parag, H.A., B. Raboy, and R.G. Kulka. 1987. Effect of heat shock on protein degradation in mammalian cells: Involvement of the ubiquitin system. *EMBO J.* **6:** 55–61.

Parsell, D.A. and R.T. Sauer. 1989. Induction of a heat shock-like response by unfolded protein in *Escherichia coli:* Dependence on protein level not protein degradation. *Genes Dev.* **3:** 1226–1232.

Parsell, D.A., A.S. Kowal, and S. Lindquist. 1994. *Saccharomyces cerevisiae* Hsp104 protein: Purification and characterization of ATP-induced structural changes. *J. Biol. Chem.* **269:** (in press).

Parsell, D.A., J. Taulien, and S. Lindquist. 1993. The role of heat-shock proteins in thermotolerance. *Philos. Trans. R. Soc. Lond.* B **339:** 279–285.

Parsell, D.A., Y. Sanchez, J.D. Stitzel, and S. Lindquist. 1991. Hsp104 is a highly conserved protein with two essential nucleotide-binding sites. *Nature* **353:** 270–273.

Patriarca, E.J. and B. Maresca. 1990. Acquired thermotolerance following heat shock protein synthesis prevents impairment of mitochondrial ATPase activity at elevated temperatures in *Saccharomyces cerevisiae. Exp. Cell Res.* **190:** 57–64.

Patriarca, E.J., G.S. Kobayashi, and B. Maresca. 1992. Mitochondrial activity and heat-shock response during morphogenesis in the pathogenic fungus *Histoplasma capsulatum. Biochem. Cell. Biol.* **70:** 207–214.

Pelham, H.R. 1984. Hsp70 accelerates the recovery of nucleolar morphology after heat shock. *EMBO J.* **3:** 3095–3100.

———. 1986. Speculations on the functions of the major heat shock and glucose-regulated proteins. *Cell* **46:** 959–961.

Pelham, H., M. Lewis, and S. Lindquist. 1984. Expression of a *Drosophila* heat shock protein in mammalian cells: Transient association with nucleoli after heat shock. *Philos. Trans. R. Soc. Lond. B* **307:** 301–307.

Petersen, R. and S. Lindquist. 1988. The *Drosophila* hsp70 message is rapidly degraded at normal temperatures and stabilized by heat shock. *Gene* **72:** 161–168.

Petko, L. and S. Lindquist. 1986. Hsp26 is not required for growth at high temperatures, nor for thermotolerance, spore development, or germination. *Cell* **45:** 885–894.

Phipps, B.M., D. Typke, R. Heger, S. Volker, A. Hoffmann, K.O. Stettre, and W. Baumeister. 1993. Structure of a molecular chaperone from a thermophilic archaebacterium. *Nature* **361:** 475–477.

Picard, D., B. Khursheed, M.J. Garabedian, M.G. Fortin, S. Lindquist, and K.R. Yamamoto. 1990. Reduced levels of hsp90 compromise steroid receptor action in vivo. *Nature* **348:** 166–168.

Pine, M.J. 1967. Response of intracellular proteolysis to alteration of bacterial protein and the implications in metabolic regulation. *J. Bacteriol.* **93:** 1527–1533.

Pratt, W.B., K.A. Hutchison, and L.C. Scherrer. 1992a. Steroid receptor folding by heat-shock proteins and composition of the receptor heterocomplex. *Trends Endocrinol. Metab.* **3:** 326–333.

Pratt, W.B., L.C. Scherrer, K.A. Hutchison, and F.C. Dalman. 1992b. A model of glucocorticoid receptor unfolding and stabilization by a heat shock protein complex. *J. Steroid Biochem. Mol. Biol.* **41:** 223–229.

Rechsteiner, M., L. Hoffman, and W. Dubiel. 1993. The multicatalytic and 26S proteases. *J. Biol. Chem.* **265:** 6065–6068.

Redmond, T., E.R. Sanchez, E.H. Bresnick, M.J. Schlesinger, D.O. Toft, W.B. Pratt, and M.J. Welsh. 1989. Immunofluorescence colocalization of the 90-kDa heat-shock protein and microtubules in interphase and mitotic mammalian cells. *Eur. J. Cell. Biol.* **50:** 66–75.

Riabowol, K.T., L.A. Mizzen, and W.J. Welch. 1988. Heat shock is lethal to fibroblasts microinjected with antibodies against hsp70. *Science* **242:** 433–436.

Rose, D.W., W.J. Welch, G. Kramer, and B. Hardesty. 1989. Possible involvement of the 90-kDa heat shock protein in the regulation of protein synthesis. *J. Biol. Chem.* **264:** 6239–6244.

Rothman, J.E. 1989. Polypeptide chain binding proteins: Catalysts of protein folding and related processes in cells. *Cell* **59:** 591–601.

Rubin, G.M. and D.S. Hogness. 1975. Effect of heat shock on the synthesis of low molecular weight RNAs in *Drosophila*: Accumulation of a novel form of 5S RNA. *Cell* **6:** 207–213.

Saibil, H.R., D. Zheng, A.M. Roseman, A.S. Hunter, G.M.E. Watson, S. Chen, A. auf der

Mauer, B.P. O'Hara, S.P. Wood, N.H. Mann, L.K. Barnett, and R.J. Ellis. 1993. ATP induces large quaternary rearrangements in a cage-like chaperonin structure. *Curr. Biol.* **3:** 265–273.

Sanchez, Y. and S.L. Lindquist. 1990. HSP104 required for induced thermotolerance. *Science* **248:** 1112–1115.

Sanchez, Y., J. Taulien, K.A. Borkovich, and S. Lindquist. 1992. Hsp104 is required for tolerance to many forms of stress. *EMBO J.* **11:** 2357–2364.

Sanchez, Y., D.A. Parsell, J. Taulien, J.L. Vogel, E.A. Craig, and S. Lindquist. 1993. Genetic evidence for a functional relationship between Hsp104 and hsp70. *J. Bacteriol.* **175:** 6484–6491.

Schmid, F.X., L. Mayr, M. Mucke, and E.R. Schonbrunner. 1993. Prolyl isomerases: the role in protein folding. *Adv. Protein Chem.* **44:** 25–66.

Schroder, H.T., T. Langer, F.-U. Hartl, and B. Bukau. 1993. DnaK, DnaJ, GrpE form a cellular chaperone machinery capable of repairing heat-induced protein damage. *EMBO J.* **12:** 4137–4144.

Seufert, W. and S. Jentsch. 1990. Ubiquitin-conjugating enzymes UBC4 and UBC5 mediate selective degradation of short-lived and abnormal proteins. *EMBO J.* **9:** 543–550.

Sherman, M.Y. and A.L. Goldberg. 1991. Formation in vitro of complexes between an abnormal fusion protein and the heat shock proteins from *Escherichia coli* and yeast mitochondria. *J. Bacteriol.* **173:** 7249–7256.

———. 1992. Involvement of the chaperonin dnaK in the rapid degradation of a mutant protein in *Escherichia coli*. *EMBO J.* **11:** 71–77.

Skowyra, D., C. Georgopoulos, and M. Zylicz. 1990. The *E. coli dnaK* gene product, the hsp70 homolog, can reactivate heat-inactivated RNA polymerase in an ATP hydrolysis-dependent manner. *Cell* **62:** 939–944.

Solomon, J.M., J.M. Rossi, K. Golic, T. McGarry, and S. Lindquist. 1991. Changes in Hsp70 alter thermotolerance and heat-shock regulation in *Drosophila*. *New Biol.* **3:** 1106–1120.

Squires, C. and C.L. Squires. 1992. The Clp proteins: Proteolysis regulators or molecular chaperones? *J. Bacteriol.* **174:** 1081–1085.

Squires, C.L., S. Pedersen, B.M. Ross, and C. Squires. 1991. ClpB is the *Escherichia coli* heat shock protein F84.1. *J. Bacteriol.* **173:** 4254–4262.

Stone, D.E. and E.A. Craig. 1990. Self-regulation of 70-kilodalton heat shock proteins in *Saccharomyces cerevisiae*. *Mol. Cell. Biol.* **10:** 1622–1632.

Storti, R.V., M.P. Scott, A. Rich, and M.L. Pardue. 1980. Translational control of protein synthesis in response to heat shock in *D. melanogaster* cells. *Cell* **22:** 825–834.

Strauch, K.L. and J. Beckwith. 1988. An *Escherichia coli* mutation preventing degradation of abnormal periplasmic proteins. *Proc. Natl. Acad. Sci.* **85:** 1576–1580.

Strauch, K., K. Johnson, and J. Beckwith. 1989. Characterization of degP, a gene required for proteolysis in the cell envelope and essential for growth of *Escherichia coli* at high temperature. *J. Bacteriol.* **171:** 2689–2696.

Straus, D.B., W.A. Walter, and C.A. Gross. 1988. *Escherichia coli* heat shock gene mutants are defective in proteolysis. *Genes Dev.* **2:** 1851–1858.

Subjeck, J.R. and T.T. Shyy. 1986. Stress protein systems of mammalian cells. *Am. J. Physiol.* **17:** C1–17.

Susek, R.E. and S.L. Lindquist. 1989. hsp26 of *Saccharomyces cerevisiae* is related to the superfamily of small heat shock proteins but is without a demonstrable function. *Mol. Cell. Biol.* **9:** 5265–5271.

Sykes, K., M.J. Gething, and J. Sambrook. 1993. Proline isomerases function during heat shock. *Proc. Natl. Acad. Sci.* **90:** 5853–5857.

Tai, P.K., M.W. Albers, H. Chang, L.E. Faber, and S.L. Schreiber. 1992. Association of a 59-kilodalton immunophilin with the glucocorticoid receptor complex. *Science* **256:** 1315–1318.

Tai, P.K., H. Chang, M.W. Albers, S.L. Schreiber, D.O. Toft, and L.E. Faber. 1993. P59 (FKBP59) heat shock protein interaction is highly conserved and may involve proteins other than steroid receptors. *Biochemistry* **32:** 8842–8847.

Theodorakis, N.G. and R.I. Morimoto. 1987. Posttranscriptional regulation of hsp70 expression in human cells: Effects of heat shock, inhibition of protein synthesis, and adenovirus infection on translation and mRNA stability. *Mol. Cell. Biol.* **7:** 4357–4368.

Trent, J.D., J. Osipiuk, and T. Pinkau. 1990. Acquired thermotolerance and heat shock in the extremely thermophilic archaebacterium Sulfolobus sp. strain B12. *J. Bacteriol.* **172:** 1478–1484.

Trent, J.D., E. Nimmesgern, J.S. Wall, F.-U. Hartl, and A.L. Horwich. 1991. A molecular chaperone from a thermophilic archaebacterium is related to the eukaryotic protein t-complex polypeptide-1. *Nature* **354:** 490–493.

Tuite, M.F., N.J. Bentley, P. Bossier, and I.T. Fitch. 1990. The structure and function of small heat shock proteins: Analysis of the *Saccharomyces cerevisiae* Hsp26 protein. *Antonie van Leeuwenhoek* **58:** 147–154.

Utans, U., S.E. Behrens, R. Luhrmann, R. Kole, and A. Kramer. 1992. A splicing factor that is inactivated during in vivo heat shock is functionally equivalent to the (U4/U6.U5) triple snRNP-specific proteins. *Genes Dev.* **6:** 631–641.

Van Dyk, T.K., A.A. Gatenby, and R.A. LaRossa. 1989. Demonstration by genetic suppression of interaction of GroE products with many proteins. *Nature* **342:** 451–453.

Velazquez, J.M. and S. Lindquist. 1984. hsp70: Nuclear concentration during environmental stress and cytoplasmic storage during recovery. *Cell* **36:** 655–662.

Velazquez, J.M., S. Sonoda, G. Bugaisky, and S. Lindquist. 1983. Is the major *Drosophila* heat shock protein present in cells that have not been heat shocked? *J. Cell Biol* **96:** 286–290.

Vierling, E. 1991. The roles of heat shock proteins in plants. *Annu. Rev. Plant Physiol. Plant Mol. Biol.* **42:** 579–620.

Viitanen, P.V., T.H. Lubben, J. Reed, P. Goloubinoff, D.P. O'Keefe, and G.H. Lorimer. 1990. Chaperonin-facilitated refolding of ribulosebisphosphate carboxylase and ATP hydrolysis by chaperonin 60 (groEL) are K+ dependent. *Biochemistry* **29:** 5665–5671.

Vogel, J.P., L.M. Misra, and M.D. Rose. 1990. Loss of BiP/GRP78 function blocks translocation of secretory proteins in yeast. *J. Cell. Biol.* **110:** 1885–1895.

Warters, R.L. and J.L. Roti Roti. 1981. The effect of hyperthermia on replicating chromatin. *Radiat. Res.* **88:** 69–78.

Warters, R.L. and O.L. Stone. 1983. The effects of hyperthermia on DNA replication in HeLa cells. *Radiat. Res.* **93:** 71–84.

Welch, W.J. and J.R. Feramisco. 1984. Nuclear and nucleolar localization of the 72,000-dalton heat shock protein in heat-shocked mammalian cells. *J. Biol. Chem.* **259:** 4501–4513.

Welch, W.J. and J.P. Suhan. 1985. Morphological study of the mammalian stress response: Characterization of changes in cytoplasmic organelles, cytoskeleton, and nucleoli, and appearance of intranuclear actin filaments in rat fibroblasts after heat-shock treatment. *J. Cell. Biol.* **101:** 1198–1211.

Welte, M.A., J.M. Tetrault, R.P. Dellavalle, and S.L. Lindquist. 1993. A new method for manipulating transgenes: Engineering heat tolerance in a complex multicellular organism. *Current Biol.* **3:** (in press).

Werner-Washburne, M., D.E. Stone, and E.A. Craig. 1987. Complex interactions among members of an essential subfamily of hsp70 genes in *Saccharomyces cerevisiae. Mol. Cell. Biol.* **7:** 2568–2577.

Wiech, H., J. Buchner, R. Zimmermann, and U. Jakob. 1992. Hsp90 chaperones protein folding in vitro. *Nature* **358:** 169–170.

Wieser, W. and E. Gnaiger. 1989. *Energy transformations in cells and organisms.* Georg Thiem Verlag, Stuttgart.

Woo, K.M., K.I. Kim, A.L. Goldberg, D.B. Ha, and C.H. Chung. 1992. The heat-shock protein ClpB in *Escherichia coli* is a protein-activated ATPase. *J. Biol. Chem.* **267:** 20429–20434.

Xu, Y. and S. Lindquist. 1993. Heat-shock protein hsp90 governs the activity of p60[v-src] kinase. *Proc. Natl. Acad. Sci.* **90:** 7074–7078.

Yahara, I., H. Iida, and S. Koyasu. 1986. A heat shock-resistant variant of Chinese hamster cell line constitutively expressing heat shock protein of M_r 90,000 at high level. *Cell Struct. Funct.* **11:** 65–73.

Yancey, P.H., M.E. Clark, S.C. Hand, R.D. Bowlus, and G.N. Somero. 1982. Living with water stress: Evolution of osmolyte systems. *Science* **217:** 1214–1222.

Yost, H.J. and S. Lindquist. 1986. RNA splicing is interrupted by heat shock and is rescued by heat shock protein synthesis. *Cell* **45:** 185–193.

Yost, H.J., R.B. Petersen, and S. Lindquist. 1991. RNA metabolism: Strategies for regulation in the heat shock response. *Trends Genet.* **6:** 223–227.

Young, D.B., A. Mehlert, and D.F. Smith. 1990. Stress proteins and infectious diseases. In *Stress proteins in biology and medicine* (ed. R.I. Morimoto et al.), pp. 131–165. Cold Spring Harbor Laboratory Press, Cold Spring Harbor, New York.

Yu, M.-H. and J. King. 1988. Surface amino acids as sites of temperature-sensitive folding mutations in the P22 tailspike protein. *J. Biol. Chem.* **263:** 1424–1431.

Zambrano, M.M., D.A. Siegele, M. Almiron, A. Tormo, and R. Kolter. 1993. Microbial competition: *Escherichia coli* mutants that take over stationary phase cultures. *Science* **259:** 1757–1760.

Zapata, J.M., F.G. Maroto, and J.M. Sierra. 1991. Inactivation of mRNA cap-binding protein complex in *Drosophila melanogaster* embryos under heat shock. *J. Biol. Chem.* **266:** 16007–16014.

19

Heat Shock Proteins as Antigens in Immunity against Infection and Self

Stefan H.E. Kaufmann and Bernd Schoel
Department of Immunology
University of Ulm
D-89070 Ulm, Germany

I. INTRODUCTION

A major driving force for the evolution of the immune system is the continuous encounter with an almost infinite variety of invading pathogens. The immune system deals with this extreme diversity by generating an equally high number of receptors capable of specifically discriminating between these structures (Nossal 1993). The following are the two types of receptors: (1) antibodies that are produced by B lymphocytes and recognize their counterparts directly and (2) T-cell receptors that are expressed on the surface of T lymphocytes and recognize their counterparts indirectly. The diversity of both types of receptors is generated by similar mechanisms, in particular genetic recombination. The molecular entities recognized by these receptors are termed antigens. Antibodies recognize "their" antigens directly, comprising a stretch of six to seven amino acids or five to six carbohydrate residues on proteins or carbohydrates, respectively. Antibodies are therefore particularly suited to contest microbes and microbial secretory products present in the extracellular space. T

The Biology of Heat Shock Proteins and Molecular Chaperones
©1994 Cold Spring Harbor Laboratory Press 0-87969-427-0/94 $5 + .00

cells recognize their antigens indirectly, i.e., they interact with oligopep-
tides of about nine amino acids in length that are presented by self-
structures. Hence, antigen recognition by T cells requires intracellular
processing of microbial components and subsequent presentation on the
cell surface by specialized molecules that are the products of the major
histocompatibility gene complex (MHC) (Janeway 1993). T cells are
therefore particularly equipped for opposing microbes living inside host
cells. In a given setting, the immune system will contact certain invaders
more frequently than others. To focus the immune response to these
repeated encounters, the immune system develops a memory. In other
words, upon second contact with the same antigen, the specific immune
response to this very same antigen is qualitatively and quantitatively improved. This so-called booster effect is also the principal mechanism underlying vaccination.

To aggravate further the tasks of the immune system, microbial invaders and mammalian host cells are composed of many components
which are not that different at the molecular level. Therefore, the immune system had to devise a strategem that would allow distinction of
foreign from self-structures. This capacity was achieved by deletion or
functional inactivation ("silencing") of those lymphocytes that express
receptors for self-structures at an early stage of development (Marrack
and Kappler 1993). Frequently, B-lymphocyte responses depend on help
from T lymphocytes. Therefore, prevention of self-recognition is often
restricted to T cells. This silencing of autoreactive T cells is initiated in
the thymus and is maintained in the periphery through different
regulatory mechanisms (Schwartz 1989). Although these preventive
measures are generally successful, they sometimes are undermined and
recognition of self-structures occurs (Steinman 1993). Two major factors
are responsible: failure of certain self-antigens to be presented in the
thymus and extrathymic development of some T cells. As a potential
corollary of these events, host cells are recognized by T lymphocytes,
and subsequently effector mechanisms are mobilized: An autoimmune
disease may develop (Steinman 1993). It is, however, important to note
that autoreactivity is a more general phenomenon than originally thought
and that autoimmune disease is a possible, but not an inevitable, consequence of autoimmunity (Steinman 1993). In an extreme form, it has
been speculated that autoreactive lymphocytes form a network of
homeostasis, the disturbance of which may cause autoimmune disease
(Cohen 1992a,b).

Heat shock proteins (hsp) have several features that pose specific ob-
stacles to the immune system. First, hsp are among the most abundant
proteins in the biosphere, and their production is further elevated under

stress (Kaufmann 1990). Microbes that enter the host face various adversary conditions that result in increased hsp synthesis (Kaufmann 1990, 1991). Second, hsp are highly conserved. As a consequence of their abundance and high homology among various microbes, hsp are appropriate targets for the immune response. Focus on a small family of similar proteins already achieves its priming and boostering with a few antigens against the whole spectrum of infectious agents. hsp conservation, however, is not restricted to infectious agents but extends to mammalian host cells. This high conservation across all species barriers carries along the potential of inducing an autoimmune response. Therefore, the immune system must take care that those regions of hsp shared by microbial predators and mammalian prey are ignored. Ignorance of such cross-reactivity could culminate in autoimmune disease. Alternatively, ignorance of the whole hsp would provide an enormous advantage for the microbial invader since it would exclude major target antigens from the immune reaction.

II. MAJOR TYPES OF INFECTIOUS AGENTS

The various infectious agents have exploited different strategies for invasion and stable infection of the host. Some microbes preferentially reside in extracellular spaces where they frequently secrete toxic components that are harmful for the host. Antibodies are responsible both for attack of the infectious agents and for neutralization of their toxic products (Paul 1993). Typically, these microbes cause acute infectious diseases and are readily eradicated from the host once specific immunity has reached its maximum. Other bacteria and protozoa abuse host cells as habitat (Paul 1993; Kaufmann 1993a). These intracellular pathogens are shielded from antibody recognition. During their intracellular life, however, microbial proteins are degraded and then presented on the cell surface by MHC molecules. Thus, infected cells are recognized by T lymphocytes. Bacteria that enter the host are immediately confronted by various insults. In response to these changes, microbial pathogens produce increased hsp levels that thus become early targets of the immune response. Accordingly, hsp-specific antibody or T-cell responses have been frequently observed in infections with extracellular and intracellular pathogens. Viruses are unable to grow by themselves and are replicated by host cells. Free viral particles are potential targets for antibodies. During intracellular replication and assembly, however, viruses are unaccessible to antibodies. Yet, virus-infected host cells are detected by T cells because viral proteins undergo processing and presentation by MHC molecules (Kaufmann and Reddehase 1989). Although the viral genome does not seem to comprise genes that encode hsp, viral infection or infection with "prions" has been shown to elevate host hsp synthesis (Doherty et al. 1992; Diedrich et al. 1993). Therefore, viral infections can cause immune responses against self-hsp.

III. SUBDIVISION OF THE IMMUNE SYSTEM

The various survival strategies of infectious agents in the host demand highly adapted and specialized responses by the immune system. The first partitioning has been achieved by the already mentioned division of lymphocytes into B cells and T cells. For B cells, this is taken further by the subdivision of antibodies into different immunoglobulin classes (Nossal 1993). The T-cell system further segregates into phenotypically distinct subclasses. The vast majority of T lymphocytes in the periphery express a T-cell receptor composed of an α-chain and a β-chain (Janeway 1993). These α/β T cells in an almost mutually exclusive manner express either the CD4 or CD8 accessory molecule. CD4 α/β T cells are focused on antigenic peptides that are presented by MHC class II molecules expressed by a limited number of cell types, in particular by mononuclear phagocytes and B cells. Thus, the antigen-specific cross-talk between CD4 T cells and host cells is restricted to cells that express MHC class II. MHC class II molecules alternate between the endosomal compartment and the cell surface and hence serve as a transporter system of endosomal proteins. These are primarily bacterial and protozoan antigens (Kaufmann 1993b). CD4 T cells with specificity for various members of the hsp family have been frequently identified in numerous infectious diseases, suggesting their participation in antimicrobial immune responses. CD4 T cells perform helper functions, i.e., they produce different cytokines that, for example, induce antibody secretion in B cells and activate antimicrobial effector functions in macrophages (Janeway 1993). CD4 T cells also express cytolytic functions. Furthermore, evidence has accumulated for participation of CD4 T cells with specificity for self-hsp in autoimmune diseases.

CD8 T cells are focused on antigenic peptides presented by MHC class I molecules (Janeway 1993). The MHC class I gene products are expressed by virtually any host cell. Hence, CD8 T cells can communicate with host cells in an almost unrestricted fashion. MHC class I molecules transport newly synthesized proteins including viral and self-antigens from the cytoplasmic compartment to the cell surface. Although CD8 T cells with hsp specificity have only rarely been identified, evidence has been presented for participation of CD8 T cells with specificity for conserved regions of hsp in immunopathogenesis of infectious diseases. CD8 T cells typically perform cytolytic functions. Moreover, CD8 T cells are also capable of producing certain cytokines.

The minor T-cell population expresses a distinct T-cell receptor composed of a γ-chain and a δ-chain (Janeway 1993). These γ/δ T cells are generally double-negative, i.e., they lack both the CD4 and the CD8 accessory molecule. Many γ/δ T cells seem to perform functions similar to those of α/β T cells. Although the rules of antigen recognition by these γ/δ T cells are less well understood, peptide presentation by nonconventional MHC gene products has frequently been implicated in this event. Evidence has been presented that γ/δ T cells preferentially perform regulatory and scavenger functions (Haas et al. 1992; Kaufmann et al. 1993). For these purposes, γ/δ T cells are thought to recognize specific self-structures including hsp. The extrathymic development of a significant proportion of γ/δ T cells may promote the predilection of these T cells for autoantigens. Consistent with these notions, evidence has been presented that γ/δ T cells interact with activated or aberrant cells through hsp recognition (Haregewoin et al. 1989; Born et al. 1990; Fisch et al. 1990; Fu et al. 1993) and that MHC class Ib gene expression is elevated by hsp60 peptides (Imani and Soloski 1991).

IV. HEAT SHOCK PROTEINS AND INFECTIOUS DISEASES

The list of infectious diseases in which immune responses to hsp have been described is already extensive and continues to grow rapidly, ranging from various experimental animal models to human diseases. Since this topic has been reviewed extensively elsewhere (Young and Elliott 1989; Kaufmann 1990; Young 1990; Young and Mehlert 1990), we describe only representative examples and present major evidence for hsp involvement in infectious diseases in tabular form (Table 1). One of the first studies on this subject was the identification of hsp60-specific CD4 T-cell clones derived from leprosy patients and from BCG-vaccinated healthy individuals (Emmrich et al. 1986). Consistent with this, Kaufmann et al. (1987) isolated high frequencies of CD4 T cells with reactivity to hsp60 from mycobacterium-immunized mice. Furthermore, vaccination of young children with the trivalent vaccine against tetanus, diphtheria, and pertussis induced high anti-hsp antibody titers, further substantiating the dominance of this type of antigen (Del Guidice et al. 1993). These findings as well as the many others listed in Table 1 have been taken as evidence that hsp are dominant antigens of infections with various pathogens ranging from bacteria to protozoa to helminths as well as of antimicrobial vaccines. The identification of hsp-reactive T cells and antibodies in healthy individuals led to the extension that anti-hsp

Table 1 Heat Shock Proteins as Major Targets of Infectious Agents

Hsp	Infectious agent	Disease	References
hsp90	*Plasmodium falciparum*	malaria	Jendoubi and Bonnefoy (1988)
	Trypanosoma cruzi	Chagas' disease	Dragon et al. (1987)
	Trypanosoma brucei brucei	trypanosomiasis of cattle	Mottram et al. (1989)
	Leishmania amazonensis	leishmaniasis	Shapria and Pedraza (1990)
	Schistosoma mansoni	schistosomiasis	Johnson et al. 1989
	Candida albicans	candidiasis	Matthews et al. (1991)
hsp70	*Plasmodium falciparum*	malaria	Bianco et al. (1986);
			Ardeshir et al. (1987);
			Mattei et al. (1988);
			Engman et al. (1989)
	Trypanosoma cruzi	Chagas' disease	Glass et al. (1986)
	Trypanosoma brucei brucei	trypanosomiasis of cattle	MacFarlane et al. (1989)
	Leishmania donovani	visceral leishmaniasis	Lee et al. (1988)
	Leishmania major	leishmaniasis	Hedstrom et al. (1987, 1988)
	Schistosoma mansoni	schistosomiasis	Selkirk et al. (1989)
	Brugia malayi	filariasis	Rothstein et al. (1989)
	Onchocerca volvulus	onchocercosis	Young et al. (1988)
	Mycobacterium tuberculosis	tuberculosis	Garsia et al. (1989)
	Mycobacterium leprae	leprosy	

	Microorganism	Disease	Reference
	Chlamydia trachomatis	trachoma	Danilition et al. (1990)
	Borrelia burgdorferi	Lyme disease	Anzola et al. (1992)
hsp60	*Mycobacterium tuberculosis*	tuberculosis	Shinnick et al. (1988); Young et al. (1988)
	Mycobacterium leprae	leprosy	Shinnick et al. (1988); Young et al. (1988)
	Coxiella burnetii	Q-fever	Vodkin and Williams (1988)
	Treponema pallidum	syphilis	Hindersson et al. (1987)
	Legionella pneumophila	Legionnaire's disease	Hoffman et al. (1989)
	Chlamydia trachomatis	trachoma	Morrison et al. (1989)
	Borrelia burgdorferi	Lyme disease	Hansen et al. (1988)
	Brucella abortus	brucellosis	Roop et al. (1992)
	Bordetella pertussis	pertussis	Del Giudice et al. (1993)
	Helicobacter pylori	gastritis, ulcers (?)	Dunn et al. (1992)
Small hsp	*Schistosoma mansoni*	schistosomiasis	Nene et al. (1986)
	Mycobacterium leprae	leprosy	Nerland et al. (1988)
	Mycobacterium tuberculosis	tuberculosis	Verbon et al. (1992)
GroES	*Mycobacterium tuberculosis*	tuberculosis	Baird et al. (1988)
	Mycobacterium leprae	leprosy	Mehra et al. (1992)

immune responses are already induced by contact with low virulent microorganisms capable of invading and surviving in the host for a restricted time without causing clinical disease (Munk et al. 1988, 1989; Kaufmann et al. 1990). Frequent encounters with these microbes focus the immune response to hsp regions conserved in the microbial world. As a consequence of repeated boosters with such shared epitopes, these anti-hsp immune responses are elevated above those directed against species-specific antigens that are contacted less frequently. Hence, hsp are able to mobilize an immune response rapidly after infection with a highly virulent pathogen. In applying this concept, hsp have been successfully used as carriers for malaria-specific antigens to improve the antigenicity of the latter in BCG-immunized mice (Lussow et al. 1991; Barrios et al. 1992). These findings point to the feasibility of using hsp as carriers for specific antigens in novel subunit vaccines.

Microorganisms that invade the host are exposed to multiple adverse stimuli. First, microbes from the environment (such as food- and water-borne pathogens) or parasites transmitted by arthropode vectors must rapidly adapt their physiology to the novel situation inside the host, with alterations in temperature, pH, and pO_2, as well as other parameters (Kaufmann et al. 1990; Kaufmann 1991). In addition, microbes are rapidly confronted with natural host resistance mechanisms, particularly phagocytosis by professional phagocytes (Kaufmann 1993b). The pathogens inside these host defenders are confronted with highly toxic effector mechanisms including reactive oxygen and nitrogen intermediates, acidic pH, iron depletion, and attack by lysosomal enzymes. To survive in this adverse surrounding, microbial pathogens activate various evasion mechanisms, including hsp. Consistent with a role of hsp in microbial survival within professional phagocytes, it has been found that hsp gene deletion mutants of *Salmonella typhimurium* rapidly succumb to macrophage killing and that they lose their virulence in mice (Fields et al. 1986; Johnson et al. 1991). Conversely, these pathogens express elevated hsp levels following phagocytosis by macrophages or by treatment with reactive oxygen intermediates (Christman et al. 1985; Buchmeier and Heffron 1990).

The combative relationship between pathogens and host cells influences hsp expression not only in the pathogens, but also in the host cells, and evidence for increased hsp synthesis in human host cells upon microbial infection has been presented (Jäättelä 1990; Kaufmann et al. 1991; Steinhoff et al. 1991 and in prep.). The potentially causative stimuli are manifold and include direct disturbances of the intracellular metabolism by microbes, activation by microbial products, and stimulation and assault by host factors. Although increased hsp synthesis contributes to autoprotection of host cells (Kantengwa et al. 1991; Jäättelä and Wissing 1993), abundant production of self-hsp may transform these cells into targets of the anti-hsp immune response (Koga et al. 1989; Steinhoff et al. 1990 and in prep.; Kaufmann et al. 1991). In this way, the anti-infectious immune response is shifted from protection to pathogenesis. Indeed, evidence for hsp involvement in immunopathogenesis of infectious diseases has been presented (Steinhoff et al. 1990; Taylor et al. 1990; Dunn et al. 1992; Mistry et al. 1992; Ford et al. 1993). Moreover, elevated hsp production in both pathogen and host, paired with their high sequence homology, could form a bridge from infection to autoimmunity.

V. HEAT SHOCK PROTEINS AND AUTOIMMUNITY

As mentioned above, autoimmunity needs to be viewed in an unbiased form, i.e., it must be dissected from autoimmune disease (Cohen 1992a,b; Marrack and Kappler 1993; Steinman 1993). In this unbiased view, recognition and disposal of senescent or aberrant cells would be beneficial for the host and abnormal attack of physiologically active cells would be detrimental. hsp expression caused by various cellular insults could serve as a unique and universal indicator of stressed cells independent from causative insult, including inflammation, transformation, infection, and trauma. Finally, hsp are induced by cell activation and participate in protein neosynthesis in activated cells (Kaufmann et al. 1991). Recognition of activated immune cells by virtue of hsp expression could be an integral part of a regulatory immune network in which inhibitory hsp-reactive T lymphocytes promote down-regulation of ongoing immune responses and thus interfere with immune hyperreactivity (Cohen 1992a,b). Increased hsp synthesis in stressed or activated cells is followed by hsp degradation, thus possibly increasing cytoplasmic hsp peptide levels that can be disposed of through the MHC processing pathway. As a corollary, the proportion of hsp peptides among those presented by the MHC could be significantly increased. Finally, immunodominance of hsp may be related to their close functional association with the MHC processing machinery. In the tumor system, immunization with hsp gp96, hsp90, or hsp70 isolated from distinct tumors causes a specific immune response against the homologous tumor (Srivastava et al. 1986; Ullrich et al. 1986; Srivastava and Maki 1991; Srivastava 1993; Udono and Srivastava 1993). It was subsequently found that this was due to the firm association of antigenic tumor-specific peptides with these hsp cognates. These findings suggest that hsp are biased to antigenic peptides generated through MHC processing. Conversely, this close association could also be responsible for preferential introduction of hsp into the MHC processing pathway.

In taking these notions one step further, it has been speculated that the immune system is strongly biased toward hsp and other dominant self-antigens (Cohen and Young 1991). In this view, autoimmunity to self-hsp is the natural event that increases the rapidity and strength of the anti-infectious immune response by virtue of the similarity between microbial and self-antigens. Autoimmunity in this speculation is the consequence of improperly regulated immunity to self-hsp or other dominant autoantigens (Cohen and Young 1991). Recognition by the immune system of stressed cells could be achieved by two kinds of hsp expression:

1. Presentation of hsp peptides by MHC molecules on the cell surface. This allows recognition by CD4 or CD8 α/β T cells and at least some γ/δ T cells.
2. Surface expression of hsp proper. This allows recognition of stressed cells by antibodies and, perhaps, by some γ/δ T cells.

VI. MHC-RESTRICTED PRESENTATION OF SELF-HEAT SHOCK PROTEINS

The first evidence for recognition by CD4 T cells of self-hsp in the context of MHC class II gene products came from studies with synthetic peptides. In these experiments, peptides representing regions of self-hsp60 were added to cocultures of antigen-presenting cells and hsp-specific T lymphocytes and peptide recognition measured in appropriate readout systems (Lamb et al. 1989; Munk et al. 1989). Although these studies clearly demonstrated that T cells with specificity for self-hsp are not deleted during thymic development, they did not prove physiologic presentation of epitopes derived from endogenous hsp. Support for this assumption was derived from studies showing that hsp60-reactive T-cell lines and clones selectively recognize stressed host cells (Koga et al. 1989; Steinhoff et al. 1990 and in prep.). These T-cell clones were raised against mycobacterial hsp60 and were CD8[+]- and MHC-class-I-restricted. Increased levels of hsp60 were detected in these stressed cells, and the addition of hsp60-specific antisense oligonucleotides interfered with elevated hsp synthesis (Kaufmann et al. 1991; U. Steinhoff et al., in prep.). Importantly, treatment with antisense oligonucleotides specifically abolished recognition of stressed cells by the hsp60-reactive CD8 T lymphocytes (U. Steinhoff et al., in prep.). These findings therefore suggest that host cells present peptides derived from self-hsp under stress conditions. On the other hand, epitope mapping studies with synthetic peptides representing sequences of the mycobacterial hsp60 indicate that these cross-reactive T cells are specific for an epitope that is semi-conserved only in the mammalian hsp60 cognate (B. Schoel et al.; U. Zügel et al., both unpubl.). Therefore, the possibility cannot be formally excluded that mycobacterial hsp60-specific T cells recognize a structurally similar, but unrelated, peptide of host origin.

Born and co-workers (Born et al. 1990; O'Brien et al. 1992; Fu et al. 1993) have established a battery of γ/δ T-cell hybridomas from the thymuses of newborn mice and from the spleens of naive adult mice. These authors found that these γ/δ T cells recognize the mycobacterial hsp60 and, more precisely, that they are specific for amino acids 180–190 in the mycobacterial hsp60. Although this region shows only marginal sequence homology with its mammalian cognate, a peptide representing the homologous sequence of the mammalian hsp60 stimulated these γ/δ T cells as well. These findings suggest that a high frequency of γ/δ T cells recognize self-hsp60 peptides.

VII. HEAT SHOCK PROTEIN PEPTIDES ELUTED FROM MHC GENE PRODUCTS

The MHC class I and II gene products are responsible for the presentation of antigenic peptides to T lymphocytes (Janeway 1993; Nossal 1993). Both MHC molecules form a pocket in which peptides are accommodated and presented to T lymphocytes. The MHC class I surface complex is composed of an α heavy chain and the $β_2$-microglobulin and presents peptides to CD8 T cells. The α-chain consists of three extracellular domains, the outer two of which form a groove that fits peptides of 8–10 amino acids in length, with most peptides being nonamers (Rammensee et al. 1993). The MHC class II molecule that presents peptides to CD4 T cells is composed of an α-chain and β-chain. The peptide-accommodating groove is formed by the $α_1$ and $β_1$ domains of the α-chain and β-chain (Brown et al. 1993). The MHC class II pocket seems to be open at both ends, thus lodging longer peptides of more than 15 amino acids in length (Chicz et al. 1992; Hunt et al. 1992). However, nonamers are already efficiently recognized by CD4 T cells.

Recent progress has allowed the elution of naturally processed peptides from MHC molecules and their subsequent sequence determination by mass spectrometry and/or microsequencing (Jardetzky et al. 1991; Rammensee et al. 1993). These studies have formally proven that hsp-derived peptides are presented by cells in the context of MHC class I and II molecules. Because hsp molecules are cytosolic, presentation in the context of MHC class I gene products was not completely unexpected. However, hsp peptides as well as other cytosolic peptides were also eluted from MHC class II molecules, demonstrating that endogenous

Table 2 Similarities between Heat Shock Protein Peptides Eluted from MHC Gene Products and Those from Infectious Agents

Self-peptide					Homologous peptide		
sequence	origin	MHC	species/class	ref	sequence	origin	refs
RRIKEIVKK	human hsp89α	HLA B 27	man/I	1	**L**LI**	hsp 83 of *L. amazonensis*	5
					L*DLI	hsp85 of *T. cruzi*	6
					L*DLI	hsp83 of *T. brucei brucei*	7
RRVKEVVKK	human hsp89β	HLA B 27	man/I	1	**L*LI*	hsp83 of *L. amazonensis*	5
					L*DLI	hsp85 of *T. cruzi*	6
					L*DLI	hsp 83 of *T. brucei brucei*	7
TPSYVAFTDTERIGDA	human hsp70	HLA DR7	man/II	2	*****************	hsp70 of *P. falciparum*	8,9
					********E********	hsp70 of *L. major*	10
					*********S*******	hsp70 of *T. cruzi*	11
					*********S*******	hsp70 of *L. donovani*	12
					*********S*******	hsp70 of *L. major*	13
					*********S*******	hsp70 of *T. brucei brucei*	14

VNHFXAEFKRKHKKD	human hsp70	HLA DR11/w52	man/II	3	none	
XNHFIAEFKRKHK	human hsp70	HLA-DR11/w52	man/II	4	none	
IIANDQGNRTTPSY	murine hsp70	H-2I-A^k	mouse/II	4	************	hsp70 of *L. major* — 10
					************	hsp70 of *T. brucei brucei* — 14
					************	hsp70 of *L. donovani* — 12
					************	hsp70 of *P. falciparum* — 8,9
					************	hsp70 of *S. mansoni* — 15
					************	hsp70 of *L. major* — 13
					************	hsp70 of *T. cruzi* — 11

Sequences were compared from the Swiss-Prot Database (release 25.0) using the FASTA program (Pearson and Lipman 1988). The single-letter code of amino acids is given; asterisk indicates identical amino acids.

References: (1) Jardetzky et al. 1991; (2) Chicz et al. 1993; (3) Newcomb and Cresswell 1993; (4) Nelson et al. 1992; (5) Shapria and Pedraza 1990; (6) Dragon et al. 1987; (7) Mottram et al. 1989; (8) Yang et al. 1987; (9) Bianco et al. 1986; (10) Searle et al. 1989; (11) Requena et al. 1988; (12) MacFarlane et al. 1990; (13) Lee et al. 1988; (14) Glass et al. 1986; (15) Hedstrom et al. 1987.

Table 3 Similarities between Self-hsp Epitopes and Food Peptides

Self-peptide		Homologous peptide		refs
sequence	origin	sequence	origin	
RRIKEIVKK	human hsp89α	*********	hsp90 from chicken	1
RRVKEVVKK	human hsp89β	*I**I***	hsp90 from chicken	1
TPSYVAFTDTERIGDA	human hsp70	***************	hsp70 from soybean	2
		***************	hspSSA4 from baker's yeast	3
		*****G*********	hsp70 from maize	4
		*********D******	Grp78 from baker's yeast	5,6
VNHFXAEFKRKHKKD	human hsp70	****VS*****N***	hsp70 from soybean	2
		****VQ*****N***	hsp70 from maize	4
XNHFIAEFKRKHK	human hsp70	V***********	hsp70 cognate from bovine	7
		V***VS*****N*	hsp70 from soybean	2
IIANDQGNRTTPSY	murine hsp70	***********	hsp70 cognate protein 1 from tomato	8
		***********	hsp70 cognate protein 1 from maize	4
		*P*********	hsp70 from soybean	2
		*L**E**I****	Grp78 from baker's yeast	5,6

Sequences were compared from the Swiss-Prot Database (release 25.0) using FASTA program (Pearson and Lipman 1988). For further details, see Table 2.
References: (1) Binart et al. 1989; (2) Roberts and Key 1991; (3) Boorstein and Craig 1990; (4) Rochester et al. 1986; (5) Rose et al. 1989; (6) Normington et al. 1989; (7) Deluca-Flaherty and McKay 1990; (8) Lin et al. 1991.

cytosolic proteins are loaded into both pathways (Nelson et al. 1992; Newcomb and Cresswell 1993). These findings clearly demonstrate that hsp peptides are not excluded from antigen presentation to T cells.

Together with the identification of T-cell clones and lines capable of recognizing synthetic peptides representing amino acid stretches of self-hsp, this novel set of data suggests that self-hsp epitopes are targets of the T-cell response. Moreover, the identification of T cells that recognize regions conserved in the bacterial and mammalian hsp suggests that auto-hsp-reactive T cells can be activated by previous infection. To assess this notion further, we have searched for sequence homologies between peptides eluted from MHC molecules and proteins from infectious agents (see Table 2). This search revealed that an hsp70 peptide eluted from the murine MHC class II molecule was fully conserved in the hsp70 of numerous pathogens. Another hsp70 peptide derived from a human MHC class II molecule was almost identical with hsp70 of various pathogenic protozoa. In this latter case, a single amino acid of high similarity was altered (exchange of the negatively charged glutamic acid to aspartic acid or of the neutral threonine to serine). Although not listed in Table 2, the hsp peptides eluted from the MHC molecules also showed more than 55% similarity to unrelated proteins. However, in this case, major dissimilarities were observed including exchanges of positively charged amino acids by negatively charged or uncharged amino acids.

How is tolerance against such naturally presented self-epitopes including hsp peptides prevented? First, T-cell deletion in the thymus would guarantee absolute prevention including development of autoimmune disease. Second, those self-reactive T cells that evade thymic deletion must be kept silent through peripheral mechanisms. Since T cells with specificity for self-hsp have been described (see above), such peripheral tolerance mechanisms must be assumed for hsp. Although different mechanisms are possible, our search for sequence homologies reveals the interesting possibility that tolerance to the naturally presented hsp epitopes includes oral tolarization by cross-reactive hsp peptides from food (Mowat 1987; Schwartz 1989). As shown in Table 3, the naturally presented hsp-derived epitopes are fully to highly conserved in hsp of common foods, including chicken, beef, soybean, maize, tomato, and yeast. Thus, highly conserved hsp regions can be presented to the T-cell system through three different routes: (1) natural processing of self-hsp, (2) food uptake, and (3) infection. We propose that these different routes markedly influence the outcome of the immune response: Food-derived hsp contribute to tolerance, whereas infection promotes active immunization. The development of autoimmune disease could therefore depend on a balance between the two types of antigen deliveries. Autoimmune dis-

510   S.H.E. Kaufmann and B. Schoel

Table 4 Evidence for Surface Expression of hsp-related Proteins

Potential surface antigen	Cell/Organism	Method of detection	Major references
hsp60	PEER γ/δ T-cell leukemia/human	cytofluorimetry; surface iodination and immunoprecipitation	Jarjour et al. (1990)
	bone marrow macrophage/mouse	cytofluorimetry; surface iodination and immunoprecipitation	Wand-Württenberger et al. (1991; unpubl.)
	Daudi lymphoma/human	cytofluorimetry; surface iodination and immunoprecipitation	Fisch et al. (1990); Kaur et al. (1993)
	Daudi lymphoma/human	cytofluorimetry; immunotoxins	Poccia et al. (1992)
	H9 lymphoma/human	cytofluorimetry; immunotoxins	Poccia et al. (1992)
	U937 monocytic line/human	cytofluorimetry; immunotoxins	Poccia et al. (1992)
	pancreatic carcinoma cells/human	cytofluorimetry; immunotoxins	Poccia et al. (1992)
	oligodendrocytes/human	immunofluorescence; cytofluorimetry	Freedman et al. (1992)
	endothelial cells/rat	cytofluorimetry	Xu and Wick (1993)
hsp70	HL60 cells/human	immunofluorescence	Jarjour et al. (1989)
	tumor lines/human	cytofluorimetry; surface iodination and immunoprecipitation	Ferrarini et al. (1992)
	splenic B cells/mouse	cytofluorimetry	VanBuskirk et al. (1989)
	retro-ocular/human fibroblasts	immunofluorescence	Heufelder et al. (1991)
	retro-ocular/human fibroblasts	surface iodination and immunoprecipitation	Heufelder et al. (1992a,b)
hsp90	HL60/human	immunofluorescence	Jarjour et al. (1989)
	BHK21/hamster	immunofluorescence	La Thangue and Latchmann (1988)
	tumor lines/human	cytofluorimetry; surface iodination and immunoprecipitation	Ferrarini et al. (1992)
	blood monocytes/human	cytofluorimetry	Erkeller-Yueksel et al. (1992)

ease may be facilitated by frequent encounters through infection, and it may be retarded through continuous contacts through food uptake.

VIII.   SURFACE EXPRESSION OF SELF-HEAT SHOCK PROTEINS

Because of the intracellular location of hsp, their surface expression has been considered unlikely. More recently, however, evidence for surface localization of hsp (hsp-related proteins) has been obtained in experiments with hsp-specific antibodies (see Table 4). These experiments include analyses of labeled cells by microfluorimetry, in situ staining of tissue samples, and immunoprecipitation of surface-labeled molecules. In none of these cases, however, was the identity of the antibody-defined molecules confirmed by sequencing data. Cytofluorimetric analyses and immunoprecipitation demonstrated strong evidence for hsp60 expression by stressed mouse macrophages (Wand-Württenberger et al. 1991 and unpubl.). Similarly, evidence has been gathered for hsp60 expression by Daudi cells and for increased surface expression of an hsp70 cognate by murine spleen B cells (Fisch et al. 1990; Pierce et al. 1991; Kaur et al. 1993). In contrast, other studies failed to identify hsp60 expression on different human tumor cell lines and on human macrophages (Ferm et al. 1992).

Increased surface expression of hsp has also been reported for target cells of autoimmune lesions (Heufelder et al. 1991, 1992a; Erkeller-Yueksel et al. 1992; Freedman et al. 1992). These and other findings suggest that under certain conditions, cells expressing hsp on their surfaces become targets of autoreactive antibodies. In addition, evidence has been presented for target cell recognition by γ/δ T cells via hsp proper (Fisch et al. 1990; Rajasekar et al. 1990; Kaur et al. 1993). Specific interactions of γ/δ T cells with tumor cells could thus be blocked with anti-hsp60 antibodies (Fisch et al. 1990; Kaur et al. 1993). Moreover, circumstantial evidence suggests that γ/δ T cells from multiple sclerosis patients attack oligodendrocytes that express hsp60 on their cell surface (Freedman et al. 1992).

How can hsp be expressed on the cell surface if they are typical cytosolic proteins lacking appropriate leader sequences responsible for cell surface translocation? First, it is possible that the antibodies react with structurally similar epitopes of unrelated surface proteins; i.e., mimicry between hsp and unrelated proteins accounts for antibody recognition. Although several immunoprecipitation studies suggest that the precipitated surface molecules are indeed hsp, this possibility cannot be excluded as long as sequence data are not available. Alternatively, hsp could be translocated to the cell surface by unconventional mechanisms.

For example, because of their high peptide-binding activity, hsp60 and hsp70 could be translocated passively by unrelated cell surface proteins. Consistent with this assumption, coprecipitation of a 70-kD molecule with the hsp60 by rabbit anti-GroEL sera has been described recently (Kaur et al. 1993). Furthermore, evidence for a close association of hsp-like proteins with the MHC class I molecule has been reported (Ferrarini et al. 1992). Finally, circumstantial evidence has been presented for a peptide-acceptor/translocator function of hsp70 and gp96 in the course of antigen processing through the MHC class II and MHC class I pathway, respectively (Srivastava and Heike 1991).

IX. HEAT SHOCK PROTEINS AND IMMUNE SURVEILLANCE

T cells with reactivity to hsp have been frequently identified in normal individuals, in newborns, and in naive mice (Munk et al. 1988; Haregewoin et al. 1989; Born et al. 1990; Fischer et al. 1992; O'Brien et al. 1992; Fu et al. 1993). Recognition of stressed host cells by hsp-specific T cells would provide an ideal means of immune surveillance because it focuses on a single marker indicative for all types of cellular stress independent from the causative insult. Although both α/β and γ/δ T cells may contribute to hsp-directed surveillance, several studies have suggested a particular predilection of γ/δ T cells to hsp. A high percentage of γ/δ T cells develops in a thymus-independent manner and therefore evades thymic silencing of T lymphocytes with specificity for autoantigens including hsp (Haas et al. 1992). The type of antigen recognition by these γ/δ T cells remains elusive, and both conventional recognition of hsp-derived peptides in the context of MHC molecules and direct recognition of cell-surface-expressed hsp have been proposed (Haregewoin et al. 1989; Born et al. 1990; Fisch et al. 1990; Jarjour et al. 1990; Fisch et al. 1992; Fu et al. 1993; Kaur et al. 1993). For two reasons, recognition of autologous hsp by γ/δ T cells could represent a beneficial surveillance mechanism. First, γ/δ T cells have been shown to be activated rapidly and to precede α/β T cells at the site of inflammation or infection (Hiromatsu et al. 1992). Early identification of inflammatory or infectious processes through hsp specificity could therefore provide a universal response bridging nonspecific host resistance mechanisms with the highly antigen-specific reaction carried by α/β T cells (Kaufmann and Kabelitz 1991). Second, a high proportion of the intraepithelial lymphocytes that are fixed between epithelial cells in mucosal tissues express a γ/δ T-cell receptor (Brandtzaeg et al. 1988; Rajasekar et al. 1990; Allison and Havran 1991; Kaufmann 1993a). Focus of these immobilized γ/δ T cells to a few prevalent antigens would therefore be of greater

Table 5 Evidence for Increased hsp Expression in Inflammatory Diseases

Target cells	Disease	Major references
Synovial lining	rheumatoid juvenile arthritis	De Graeff-Meeder et al. (1990, 1991); Karlsson-Parra (1990); Boog et al. (1992)
Cartilage-pannus junction	rheumatoid arthritis	De Graeff-Meeder et al. (1990); Kleinau et al. (1991)
Oligodendrocytes	multiple sclerosis	Freedman et al. (1992); Selmaj et al. (1992)
Astrocytes	Alzheimer's disease	Hamos et al. (1991)
B cells	systemic lupus erythematosus	Norton et al. (1989); Erkeller-Yueksel et al. (1992); V.B. Dhillon et al. (in prep.); G.B. Faulds et al. (in prep.)
β-cell secretory granules	IDDM	Jones et al. (1990); Brudzynski et al. (1992a,b)
Muscle cells	inflammatory myopathy and polymyositis	Hohlfeld et al. (1991); Hohlfeld and Engel (1992)
Smooth muscle cells	atherosclerosis	Xu et al. (1992)
Thyroid cells, retro-ocular fibroblast	Graves' disease	Heufelder et al. (1991, 1992a,b)
Epidermal keratino-cytes	psoriasis	Rambukkana et al. (1992)
Gastric epithelial cells	chronic gastritis	Engstrand et al. (1991)

value than exclusive specificity for innumerable antigens. Recognition of stressed cells by γ/δ T cells in their near vicinity by means of hsp would promote rapid mobilization of host defense mechanisms at mucosal tissues that serve as major port of entry for a vast variety of infectious agents (Kaufmann 1993a). Although both assumptions are highly tempting, they remain speculative, and extensive studies will be required to prove or disprove their validity. The recently described accumulation of γ/δ T cells and increased hsp60 expression in inflammatory gastric epithelium are consistent with this notion (Engstrand et al. 1991).

Table 6 Major Evidence for hsp Involvement in Autoimmune Diseases

Disease	Evidence (pro and contra hsp involvement)	Major references
Experimental IDDM of NOD mice	*Pro*: hsp60-reactive T cells confer disease; prevention with peptide	Elias et al. (1990)
	Pro: shared epitope between mycobacterial and mammalian hsp60 amino acids 437–460	Elias et al. (1991)
	Contra: major inducing antigen: glutamic acid decarboxylase	Kaufman et al. (1993); Tisch et al. (1993)
Adjuvant arthritis of Lewis rats	*Pro*: hsp60-reactive T cells confer disease; epitope unique for mycobacterial hsp60 amino acids 180–188	van Eden et al. (1988)
	Pro: prevention by immunization with epitope amino acids 180–188	Yang et al. (1990, 1992); Wauben et al. (1992)
	Pro: prevention by immunization with recombinant vaccinia virus expressing mycobacterial hsp60	Hogervorst et al. (1991)
Pristane-induced arthritis of mice	*Pro*: prevention by immunization with mycobacterial hsp60	Thompson et al. (1990); Barker et al. (1992)
Adjuvant-induced acute inflammation of mice	*Pro*: T cells reactive to mammalian hsp60 amino acids 412–423	Anderton et al. (1993)
Reactive arthritis	*Pro*: hsp-reactive T cells; epitope unique for mycobacterial hsp60 amino acids 456–465	Gaston et al. (1990)
	Pro: hsp-reactive α/β T cells; epitope shared by mycobacterial and mammalian hsp60 cognate amino acids 370–540	Herrmann et al. (1991)
	Pro: hsp-reactive γ/δ T cells	Herrmann et al. (1992)
	Pro: cytolytic T-cell clones from sites of inflammation reactive to mycobacterial hsp60 and cross-reactive to human hsp60	Li et al. (1992)

	Contra: low frequency of hsp60-reactive T cells	Life et al. (1991)
	Contra: synovial fluid cells do not respond to human hsp60 in inflammatory synovitis	Pope et al. (1992)
Rheumatoid arthritis	*Pro*: hsp60-reactive T cells in synovial fluid of early chronic arthritis	Res et al. (1988)
	Pro: hsp60-reactive γ/δ T cells in synovial fluid of chronic arthritis	Holoshitz et al. (1989)
	Pro: hsp60-cross-reactive T cells in synovial fluid of juvenile chronic arthritis	De Graeff-Meeder et al. (1991)
	Pro: B- and T-cell responses to mycobacterial hsp60 and to 180–188 amino acids in juvenile, but not in adult, arthritis	Danieli et al. (1992a,b)
	Pro: increased serum levels of hsp60-specific antibodies	Tsoulfa et al. (1989)
	Contra: no hsp60-cross-reactive T cells in synovial fluid of adult chronic arthritis	De Graeff-Meeder et al. (1991)
	Contra: autoantibodies to human hsp infrequent	Jarjour et al. (1991)
	Contra: hsp60-reactive T-cell responses similar in synovial and peripheral samples from patients and in peripheral samples of patients and normal individuals	Fischer et al. (1991)
Multiple sclerosis	*Pro*: γ/δ T cells in acute plaques	Wucherpfennig et al. (1992)
	Pro: γ/δ T cells in chronically demyelinated lesions	Selmaj et al. (1991)
	Pro: colocalization of γ/δ T cells and hsp60 expression of oligodendrocytes of chronic lesions	Selmaj et al. (1991)
	Pro: hsp induction in oligodendrocytes in vitro	Freedman et al. (1992)
	Pro: hsp60-reactive spinal fluid T lymphocytes in multiple sclerosis	Birnbaum et al. (1993)
	Pro: T-cell responses to hsp70 in multiple sclerosis	Salvetti et al. (1992)
	Pro: hsp70 antibodies in serum and cerebrospinal fluid	Birnbaum and Kotilinek (1993)

(Continued on following page.)

Table 6 (Continued)

Disease	Evidence (pro and contra hsp involvement)	Major references
Systemic lupus erythematosus	*Pro:* hsp60-related B-cell stimulation by γ/δ T cells	Rajagopalan et al. (1990)
	Contra: no evidence for increased anti-hsp antibody levels	Kindas-Muegge et al. (1993); Jarjour et al. (1991)
	Contra: elevated IgM antibody levels to hsp60, but no consistent pattern	Panchapakesan et al. (1992)
Chronic gastritis	*Pro:* colocalization of γ/δ T cells and hsp60 expression in gastric epithelium during chronic gastritis	Engstrand et al. (1991)
Behcet's disease	*Pro:* T cells with specificity for human hsp60 peptides	Pervin et al. (1993)
Atherosclerosis	*Pro:* increased serum levels of hsp60-specific antibodies	Xu et al. (1993)
Systemic sclerosis	*Pro:* increased serum levels of hsp60-specific antibodies	Danieli et al. (1992a,b)
Psoriasis	*Pro:* increased serum levels of hsp60-specific antibodies	Rambukkana et al. (1993)
Kawasaki disease	*Pro:* increased serum levels of hsp60-specific antibodies	Yokota et al. (1993)

X. HEAT SHOCK PROTEINS AND AUTOIMMUNE DISEASE

The frequent identification of hsp-reactive T cells and antibodies in various autoimmune diseases has promoted speculations about a universal role of hsp as antigens responsible for autoaggression. This notion was further supported by the observation of increased hsp synthesis in cells present in inflammatory lesions (Table 5). Table 6 lists the major autoimmune diseases and arguments for or against hsp involvement. Convincing evidence for the participation of hsp in autoimmune diseases has been obtained in two experimental animal models: adjuvant arthritis of rats and insulin-dependent diabetes mellitus (IDDM) of nonobese diabetic (NOD) mice (van Eden et al. 1988; Elias et al. 1990, 1991; Cohen 1991).

Administration of complete Freund's adjuvant containing mycobacteria causes an autoimmune disease that closely resembles rheumatoid arthritis (Cohen 1991). T cells have been isolated from such mice that recognize a nonconserved region of the mycobacterial hsp60 (van Eden et al. 1988). These T cells are capable of adoptively transferring either induction of or protection against adjuvant arthritis. Moreover, preimmunization of rats with soluble hsp60 or the relevant peptide renders the animals resistant against subsequent attempts to induce adjuvant arthritis. In aging NOD mice, a diabetic disease develops spontaneously that closely resembles IDDM of humans (Elias et al. 1990, 1991; Cohen 1991). Again, T cells capable of adoptively transferring disease to recipients have been isolated from such mice that are specific for hsp60. In this case, however, the T cells recognize a conserved region shared by the mycobacterial and mammalian hsp60. In further support for hsp involvement in IDDM of NOD mice, evidence for increased hsp60 levels in pancreatic β cells has been obtained (Brudzynski et al. 1992a,b). Importantly, the spontaneous development of IDDM in NOD mice can be prevented by immunization with soluble hsp60 or the relevant peptide (Elias et al. 1990, 1991; Cohen 1991). Yet, the inducing antigen in IDDM seems to be the protein glutamic acid decarboxylase (Kaufman et al. 1993; Tisch et al. 1993). Although the mechanisms operative in these two experimental autoimmune diseases may differ as indicated by the involvement of different types of hsp epitopes (i.e., unique vs. conserved epitopes in adjuvant arthritis or IDDM, respectively), they provide strong evidence for a crucial role of hsp in the development of experimental autoimmune diseases. Yet, involvement of organ-specific autoantigens appears to be most likely.

In contrast, evidence supporting hsp involvement in human autoimmune diseases is far less convincing. Although it was originally proposed that hsp60 is central to human IDDM, more recent evidence suggests a minor (if any) role of hsp in this disease (Baekeshov et al. 1990; Jones et al. 1990). Most likely, as in the mouse, the major autoantigen is glutamic acid decarboxylase that besides having a similar molecular mass of 60 kD is totally unrelated to the hsp family. Recent identification of amino acid sequence similarities between hsp60 and glutamic acid decarboxylase may, however, be taken as circumstantial evidence for a relationship between these two antigens in IDDM (Jones et al. 1993). Similarly, it was originally claimed that hsp60-specific T cells are major players in the development of reactive arthritis (Hermann et al. 1991; Life et al. 1991). Although T cells with specificity for nonconserved mycobacterial epitopes, as well as for conserved epitopes shared by the bacterial and mammalian cognates, have been described, more recent findings indicate that the proportion of these hsp-specific T cells in lesions is small (Life et al. 1991). Original enthusiasm about hsp as responsible antigens for rheumatoid arthritis has also more recently been subdued (De Graeff-Meeder et al. 1991; Gaston et al. 1991; Res et al. 1991). Current evidence argues against a major role of hsp in rheumatoid arthritis of adults and favors contribution of hsp to the juvenile forms of arthritis only (De Graeff-Meeder et al. 1991). In chronic lesions and acute plaques of multiple sclerosis, accumulation of γ/δ T cells has been observed that was parallelled by increased hsp levels in intralesional oligodendrocytes (Selmaj et al. 1991). Evidence for a role of hsp in multiple sclerosis, although very appealing, still must await further confirmation (Georgopoulos and McFarland 1993).

In summary, the general role of hsp in the development of autoimmune diseases remains an interesting, but still unresolved, possibility (Fig. 1). The original proposal was that self-hsp-reactive T cells and antibodies are primarily stimulated by cross-reactive epitopes derived from infectious agents typically during chronic infections (Kaufmann 1990). The abundant presence of cross-reactive epitopes of microbial origin was considered sufficient to break tolerance against self. In this manner, the antimicrobial immune response would be deviated against host cells expressing self-hsp under the influence of various insults. In a slight modification of this assumption, T cells with specificity for epitopes shared by microbial and self-hsp would be attracted to a particular tissue site suffering from stress and then recognize cross-reactive epitopes derived from unrelated, tissue-specific antigens (Jones et al. 1993). These latter reactions would ultimately be responsible for the organ-specific autoimmune disease. The thought that hsp of microbial origin act as inducing antigens, however, may need revision. At the moment, it appears more likely that elevated hsp expression at the site of inflammation occurs as a result of tissue destruction and inflammation

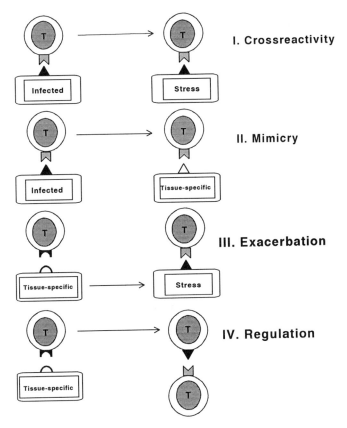

Figure 1 Heat shock protein and autoimmunity. (I) Heat shock protein cross-reactivity: T cells with specificity for microbial hsp are activated during infection. These T cells recognize a shared epitope of the mammalian hsp cognate. Stressed cells expressing such shared hsp regions become targets of T cells originally activated during infection. (II) Antigenic mimicry: T cells with specificity for microbial hsp may also cross-react with unrelated antigens that fortuitously contain regions resembling microbial hsp. Cells expressing such epitopes also become targets of T cells originally activated in response to infection. In both situations, hsp serve as inducing antigens. (III) Heat shock protein as exacerbating antigens: During an ongoing immune response, host cells become stressed and present epitopes derived from self hsp. Cells expressing elevated levels of hsp-derived epitopes are recognized by hsp-specific T cells that then exacerbate the inflammatory process. (IV) Heat shock proteins as target antigens of regulatory T cells: hsp may serve as target antigens of regulatory T cells that attempt to restrain an aggravated immune response at the site of inflammation. (*Closed triangles*) hsp epitope; (*open triangle*) similar, but unrelated epitope; (*half circles*) tissue-specific epitope.

caused by T cells recognizing tissue-specific antigens. In this scenario, hsp-reactive T cells would arrive secondarily and then further exacerbate local tissue destruction but would not be responsible for induction of the autoaggressive process. Alternatively, the possibility needs to be considered that hsp-reactive T cells act as down-regulators of exacerbated T-cell responses at the site of inflammation. Evidence consistent with such an assumption has been obtained in related experimental models (Doherty et al. 1992; Haas et al. 1992; Kaufmann et al. 1993). The identification of hsp-reactive T cells in autoimmune lesions would then primarily reflect the idle attempt of the immune response to restrain exacerbated immune responses.

XI. CONCLUDING REMARKS

We have attempted to briefly summarize evidence indicating a particularly intimate relationship of the immune system to hsp. Obviously, this is an ambivalent relationship with potentially beneficial or detrimental consequences for the host. On the positive side, hsp may serve to promote protective immunity against infectious agents; on the negative side, hsp may contribute to autoimmune disease. In the beginning, the detection of antibodies and T cells with specificity for hsp in either situation had promoted overenthusiastic speculations. It is now becoming more likely that hsp are neither the universal antigens of anti-infectious vaccines nor the central causes of autoimmune diseases. Rather, it seems more reasonable to assume that hsp perform auxiliary functions in either situation. Nevertheless, integration of hsp as vaccine carriers for pathogen-specific antigens to improve protective immunity against infection, on the one hand, and inhibition of exacerbated anti-hsp immune responses to interfere with ongoing autoaggression, on the other hand, will provide helpful control means as an adjunct to more specific modes of intervention.

ACKNOWLEDGMENTS

This work received financial support from SFB 322. Many thanks to Dr. Martin Munk for the Databank search and to Mrs. R. Mahmoudi for secretarial help.

REFERENCES

Allison, J.P. and W.L. Havran. 1991. The immunobiology of T cells with invariant γ/δ antigen receptors. *Annu. Rev. Immunol.* **9:** 679–705.
Anderton, S.M., R. Van der Zee, and J.A. Goodacre. 1993. Inflammation activates self hsp60-specific T cells. *Eur. J. Immunol.* **23:** 33–38.

Anzola, J., B.J. Luft, G. Gorgone, R.J. Battwyler, C Soderberg, R. Lahesmaa, and G. Peltz. 1992. *Borrelia burgdorferi* HSP70 homolog: Characterization of an immunoreactive stress protein. *Infect. Immun.* **60:** 3704–3713.

Ardeshir, F., J.E. Flint, S.J. Richman, and R.T. Reese. 1987. A 75kd merozoite surface protein of *Plasmodium falciparum* which is related to the 70kd heat-shock proteins. *EMBO J.* **6:** 493–499.

Baekeshov, S., H. Jan-Aanstoot, S. Christgau, A. Reetz, M. Solimena, M. Cascalho, F. Folli, H. Richter-Olesen, and P. De Camilli. 1990. Identification of the 65 K autoantigen in insulin-dependent diabetes as the GABA-synthesizing enzyme glutamic acid decarboxylase. *Nature* **347:** 151–156.

Baird, P.N., L.M.C. Hall, and A.R.M. Coates. 1988. A major antigen from *Mycobacterium tuberculosis* which is homologous to the heat shock proteins groES from *E. coli* and the htpA gene product of *Coxiella burneti. Nucleic Acids Res.* **16:** 9047.

Barker, R.N., G.R. Webb, S.J. Thompson, M. Ghoraishian, F.M. Ponsford, and C.J. Elson. 1992. Differential effects of immunisation with mycobacterial 65 kD heat shock protein on two models of autoimmunity. *Autoimmunity* **14:** 73–77.

Barrios, C., A.R. Lussow, J.D.A. Van Embden, R. Van der Zee, R. Rappuoli, P. Costantino, J.A. Louis, P.-H. Lambert, and G. Del Giudice. 1992. Mycobacterial heat-shock proteins as carrier molecules. II. The use of the 70-kDa mycobacterial heat-shock protein as carrier for conjugated vaccines can circumvent the need for adjuvants and Bacillus Calmette Guérin priming. *Eur. J. Immunol.* **22:** 1365–1372.

Bianco, A.E., J.M. Favaloro, T.R. Burkot, J.G. Culvenor, P.E. Crewther, G.V. Brown, R.F. Anders, R.L. Coppel, and D.J. Kemp. 1986. A repetitive antigen of *Plasmodium falciparum* that is homologous to heat shock protein 70 of *Drosophila melanogaster. Proc. Natl. Acad. Sci.* **83:** 8713–8717.

Binart, N., B. Chambraud, B. Dumas, D.A. Rowlands, C. Bigogne, J.M. Levin, J. Garnier, E.-E. Baulieu, and M.-G. Catelli. 1989. The cDNA-derived amino acid sequence of chick heat shock protein M_r 90,000 (hsp 90) reveals a "DNA-like" structure: Potential site of interaction with steroid receptors. *Biochem. Biophys. Res. Commun.* **159:** 140–147.

Birnbaum, G. and L. Kotilinek. 1993. Antibodies to 70-kD heat shock protein are present in CSF and sera from patients with multiple sclerosis. *Neurology* **43:** A162.

Birnbaum, G., L. Kotilinek, and L. Albrecht. 1993. Spinal fluid lymphocytes from a subgroup of multiple sclerosis patients respond to mycobacterial antigens. *Ann. Neurol.* **34:** 294–300.

Boog, C.J., E.R. De Graeff-Meeder, M.A. Lucassen, R. Van der Zee, M.M. Voorhorst-Ogink, P.J. van Kooten, H.J. Geuze, and W. van Eden. 1992. Two monoclonal antibodies generated against human hsp60 show reactivity with synovial membranes of patients with juvenile chronic arthritis. *J. Exp. Med.* **175:** 1805–1810.

Boorstein, W.R. and E.A. Craig. 1990. Structure and regulation of the SSA4 hsp70 gene of *Saccharomyces cervisiae. J. Biol. Chem.* **265:** 18912–18921.

Born, W., L. Hall, A. Dallas, J. Boymel, T. Shinnick, D. Young, P. Brennan, and R. O'Brien. 1990. Recognition of a peptide antigen by heat shock-reactive γ/δ T lymphocytes. *Science* **249:** 67–69.

Brandtzaeg, P., L.M. Sollid, P.S. Thrane, D. Kvale, K. Bjerke, H. Scott, K. Kett, and T.O. Rognum. 1988. Lymphoepithelial interactions in the mucosal immune system. *Gut* **29:** 1116–1130.

Brown, J.H., T.S. Jardetzky, J.C. Gorga, L.J. Stern, R.G. Urban, J.L. Strominger, and D.C. Wiley. 1993. Three-dimensional structure of the human class II histocompatibility antigen HLA-DR1. *Nature* **364:** 33–39.

Brudzynski, K., V. Martinez, and R.S. Gupta. 1992a. Secretory granule autoantigen in insulin-dependent diabetes mellitus is related to the 62 kDa heat-shock protein (hsp60). *J. Autoimmun.* **5:** 453–463.

————. 1992b. Immunocytochemical localization of heat-shock protein 60-related protein in beta-cell secretory granules and its altered distribution in non-obese diabetic mice. *Diabetologia* **35:** 316–324.

Buchmeier, N.A. and F. Heffron. 1990. Induction of *Salmonella* stress proteins upon infection of macrophages. *Science* **248:** 730–732.

Chicz, R.M., R.G. Urban, J.C. Gorga, D.A.A. Vignali, W.S. Lane, and J.L. Strominger. 1993. Specificity and promiscuity among naturally processed peptides bound to HLA-DR alleles. *J. Exp. Med.* **178:** 27–47.

Chicz, R.M., R.G. Urban, W.S. Lane, J.C. Gorga, L.J. Stern, D.A.A. Vignali, and J.L. Strominger. 1992. Predominant naturally processed peptides bound to HLA-DR1 are derived from MHC-related molecules and are heterogeneous in size. *Nature* **358:** 764–768.

Christman, M.F., R.W. Morgan, F.S. Jacobson, and B.N. Ames. 1985. Positive control of a regulon for defenses against oxidative stress and some heat shock proteins in *Salmonella typhimurium*. *Cell* **41:** 753–762.

Cohen, I.R. 1991. Autoimmunity to chaperonins in the pathogenesis of arthritis and diabetes. *Annu. Rev. Immunol.* **9:** 567–589.

————. 1992a. The cognitive paradigm and the immunological homunculus. *Immunol. Today* **13:** 490–494.

————. 1992b. The cognitive principle challenges clonal selection. *Immunol. Today* **13:** 441–444.

Cohen, I.R. and D.B. Young. 1991. Autoimmunity, microbial immunity and the immunological homunculus. *Immunol. Today* **12:** 105–110.

Danieli, M.G., M. Candela, A.M. Ricciatti, R. Reginelli, G. Danieli, I.R. Cohen, and A. Gabrielli. 1992a. Antibodies to mycobacterial 65 kDa heat shock protein in systemic sclerosis (scleroderma). *J. Autoimmun.* **5:** 443–452.

Danieli, M.G., D. Markovits, A. Gabrielli, A. Corvetta, P.L. Giorgi, R. Van der Zee, J.D. van Embden, G. Danieli, and I.R. Cohen. 1992b. Juvenile rheumatoid arthritis patients manifest immune reactivity to the mycobacterial 65-kDa heat shock protein, to its 180-188 peptide, and to a partially homologous peptide of the proteoglycan link protein. *Clin. Immunol. Immunopathol.* **64:** 121–128.

Danilition, S.L., I.W. Maclean, R. Peeling, S. Winston, and R.C. Brunham. 1990. The 75-kilodalton protein of *Chlamydia trachomatis*: A member of the heat shock protein 70 family. *Infect. Immun.* **58:** 189–196.

De Graeff-Meeder, E.R., R. Van der Zee, G.T. Rijkers, H.-J. Schuurman, W. Kuis, J.W.J. Bijlsma, B.J.M. Zegers, and W. van Eden. 1991. Recognition of human 60 kD heat shock protein by mononuclear cells from patients with juvenile chronic arthritis. *Lancet* **337:** 1368–1372.

De Graeff-Meeder, E.R., M. Voorhorst, W. van Eden, H.J. Schuurman, J. Huber, D. Barkley, R.N. Maini, W. Kuis, G.T. Rijkers, and B.J. Zegers. 1990. Antibodies to the mycobacterial 65-kd heat-shock protein are reactive with synovial tissue of adjuvant arthritic rats and patients with rheumatoid arthritis and osteoarthritis. *Am. J. Pathol.* **137:** 1013–1017.

Del Giudice, G., A. Gervaix, P. Costantino, C.-A. Wyler, C. Tougne, E.R. De Graeff-Meeder, J. van Embden, R. Van der Zee, L. Nencioni, R. Rappuoli, S. Suter, and P.-H. Lambert. 1993. Priming to heat shock proteins in infants vaccinated against pertussis. *J. Immunol.* **150:** 2025–2032.

DeLuca-Flaherty, C. and D.B. McKay. 1990. Nucleotide sequence of the cDNA of a bovine 70 kilodalton heat shock cognate protein. *Nucleic Acids Res.* **18:** 5569.

Diedrich, J.F., R.I. Carp, and A.T. Haase. 1993. Increased expression of heat shock protein, transferrin, and β2-microglobin in astrocytes during scrapie. *Microbial. Pathogen.* **15:** 1–6.

Doherty, P.C., W. Allan, M. Eichelberger, and S.R. Carding. 1992. Roles of α/β and γ/δ T cell subsets in viral immunity. *Annu. Rev. Immunol.* **10:** 123–151.

Dragon, E.A., S.R. Sias, E.A. Kato, and J.D. Gabe. 1987. The genome of *Trypanosoma cruzi* contains a constitutively expressed tandemly arranged multicopy gene homologous to a major heat shock protein. *Mol. Cell. Biol.* **7:** 1271–1275.

Dunn, B.E., R.M. Roop II, C.C. Sung, S.A. Sharma, G.I. Perez-Perez, and M.J. Blaser. 1992. Identification and purification of a cpn60 heat shock protein homolog from *Helicobacter pylori*. *Infect. Immun.* **60:** 1946–1951.

Elias, D., D. Markovits, T. Reshef, R. Van der Zee, and I.R. Cohen. 1990. Induction and therapy of autoimmune diabetes in the non-obese diabeitc (NOD/Lt) mouse by a 65-kDa heat shock protein. *Proc. Natl. Acad. Sci.* **87:** 1576–1580.

Elias, D., T. Reshef, O.S. Birk, R. Van der Zee, M.D. Walker, and I.R. Cohen. 1991. Vaccination against autoimmune mouse diabetes with a T-cell epitope of the human 65-kDa heat shock protein. *Proc. Natl. Acad. Sci.* **88:** 3088–3091.

Emmrich, F., J. Thole, J.D.A. Van Embden, and S.H.E. Kaufmann. 1986. A recombinant 64 kiloDalton protein of *Mycobacterium bovis* BCG specifically stimulates human T4 clones reactive to mycobacterial antigens. *J. Exp. Med.* **163:** 1024–1029.

Engman, D.M., L.V. Kirchhoff, and J.E. Donelson. 1989. Molecular cloning of mtp70, a mitochondrial member of the hsp70 family. *Mol. Cell. Biol.* **9:** 5163–5168.

Engstrand, L., A. Scheynius, and C. Pahlson. 1991. An increased number of γ/δ T cells and gastric epithelial cell expression of the groEL stress-protein homologue in *Helicobacter pylori*-associated chronic gastritis of the antrum. *Am. J. Gastroenterol.* **86:** 976–980.

Erkeller-Yueksel, F.M., D.A. Isenberg, V.B. Dhillon, D.S. Latchman, and P.M. Lydyard. 1992. Surface expression of heat shock protein 90 by blood mononuclear cells from patients with systemic lupus erythematosus. *J. Autoimmun.* **5:** 803–814.

Ferm, M.T., B. Soederstroem, S. Jindal, A. Groenberg, J. Ivanyi, R.A. Young, and R. Kiessling. 1992. Induction of human hsp60 expression in monocytic cell lines. *Int. Immunol.* **4:** 305–311.

Ferrarini, M., S. Heltai, M.R. Zocchi, and C. Rugarli. 1992. Unusual expression and localization of heat-shock proteins in human tumor cells. *Int. J. Cancer* **51:** 613–619.

Fields, P.I., R.V. Swanson, C.G. Haidaris, and F. Heffron. 1986. Mutants of *Salmonella typhimurium* that cannot survive within the macrophage are avirulent. *Proc. Natl. Acad. Sci.* **83:** 5189–5193.

Fisch, P., K. Oettel, N. Fudim, J.E. Surfus, M. Malkovsky, and P.M. Sondel. 1992. MHC-unrestricted cytotoxic and proliferative responses of two distinct human γ/δ T cell subsets to Daudi cells. *J. Immunol.* **148:** 2315–2323.

Fisch, P., M. Malkovsky, S. Kovats, E. Sturm, E. Braakman, B.S. Klein, S.D. Voss, L.W. Morrissey, R. DeMars, W.J. Welch, R.L.H. Bolhuis, and P.M. Sondel. 1990. Recognition by human Vγ9/Vδ2 T cells of a GroEL homolog on Daudi Burkitt's lymphoma cells. *Science* **250:** 1269–1273.

Fischer, H.P., C.E.M. Sharrock, and G.S. Panayi. 1992. High frequency of cord blood lymphocytes against mycobacterial 65-kDa heat shock protein. *Eur. J. Immunol.* **22:** 1667–1669.

Fischer, H.P., C.E. Charrock, M.J. Colston, and G.S. Panayi. 1991. Limiting dilution analysis of proliferative T cell responses to mycobacterial 65-kDa heat-shock protein fails to show significant frequency differences between synovial fluid and peripheral blood of patients with rheumatoid arthritis. *Eur. J. Immunol.* **21:** 2937–2941.

Ford, A.L., W.J. Britton, and P.J. Armati. 1993. Schwann cells are able to present exogenous mycobacterial hsp70 to antigen-specific T lymphocytes. *J. Neuroimmunol.* **43:** 151–159.

Freedman, M.S., N.N. Buu, T.C. Ruijs, K. Williams, and J.P. Antel. 1992. Differential expression of heat shock proteins by human glial cells. *J. Neuroimmunol.* **41:** 231–238.

Fu, Y.-X., R. Cranfill, M. Vollmer, R. Van der Zee, R.L. O'Brien, and W. Born. 1993. In vivo response of murine γ/δ T cells to heat shock protein-derived peptide. *Proc. Natl. Acad. Sci.* **90:** 322–326.

Garsia, R.J., L. Hellqvist, R.J. Booth, A.J. Radford, W.J. Britton, L. Astbury, R.J. Trent, and A. Basten. 1989. Homology of the 70-kilodalton antigens from *Mycobacterium leprae* and *Mycobacterium tuberculosis* 71-kilodalton antigen and with the conserved heat shock protein 70 of eucaryotes. *Infect. Immun.* **57:** 204–212.

Gaston, J.S., P.F. Life, P.J. Jenner, M.J. Colston, and P.A. Bacon. 1990. Recognition of a mycobacteria-specific epitope in the 65-kD heat-shock protein by synovial fluid-derived T cell clones. *J. Exp. Med.* **171:** 831–841.

Gaston, J.S., P.F. Life, R. Van der Zee, R.J. Jenner, M.J. Colston, S. Tonks, and P.A. Bacon. 1991. Epitope specificity and MHC restriction of rheumatoid arthritis synovial T cell clones which recognize a mycobacterial 65 kDa heat shock protein. *Int. Immunol.* **3:** 965–972.

Georgopoulos, C. and H. McFarland. 1993. Heat shock proteins in multiple sclerosis and other autoimmune diseases. *Immunol. Today* **14:** 373–375.

Glass, D.J., R.I. Polvere, and L.H.T. van der Ploeg. 1986. Conserved sequences and transcription of the hsp70 gene family in *Trypanosoma brucei. Mol. Cell. Biol.* **6:** 4657–4666.

Haas, W., P. Pereira, and S. Tonegawa. 1992. γ/δ T-cells. *Annu. Rev. Immunol.* **11:** 637–685.

Hamos, J.E., B. Oblas, D. Pulaski-Salo, W.J. Welch, D.G. Bole, and D.A. Drachman. 1991. Expression of heat shock proteins in Alzheimer's disease. *Neurology* **41:** 345–350.

Hansen, K., J.M. Bangsborg, H. Fjordvang, N.S. Pedersen, and P. Hindersson. 1988. Immunochemical characterization of, and isolation of the gene for a *Borrelia burgdorferi* immunodominant 60-kilodalton antigen common to a wide range of bacteria. *Infect. Immun.* **56:** 2047–2053.

Haregewoin, A., G. Soman, R.C. Hom, and R.W. Finberg. 1989. Human γ/δ[+] T cells respond to mycobacterial heat-shock protein. *Nature* **340:** 309–312.

Hedstrom, R., J. Culpepper, R.A. Harrison, N. Agabian, and G. Newport. 1987. A major immunogen in *S. mansoni* infections is homologous to the heat-shock protein hsp70. *J. Exp. Med.* **165:** 1430–1435.

Hedstrom, R., J. Culpepper, V. Schinski, N. Agabian, and G. Newport. 1988. Schistosome heat-shock proteins are immunologically distinct host-like antigens. *Mol. Biochem. Parasitol.* **29:** 275–282.

Hermann, E., A.W. Lohse, W.J. Mayet, R. Van der Zee, W. van Eden, P. Probst, T. Poralla, K.-H. Meyer zum Büschenfelde, and B. Fleischer. 1992. Stimulation of synovial fluid mononuclear cells with the human 65-kD heat shock protein or with live enterobacteria leads to preferential expansion of TCR-γ/δ[+] lymphocytes. *Clin. Exp. Immunol.* **89:** 427–433.

Hermann, E., A.W. Lohse, R. Van der Zee, W. van Eden, W.J. Mayet, P. Probst, T.

Poralla, K.-H. Meyer zum Büschenfelde, and B. Fleischer. 1991. Synovial fluid-derived Yersinia-reactive T cells responding to human 65-kDa heat-shock protein and heat-stressed antigen-presenting cells. *Eur. J. Immunol.* **21:** 2139–2143.

Heufelder, A.E., B.E. Wenzel, and R.S. Bahn. 1992a. Cell surface localization of a 72 kilodalton heat shock protein in retroocular fibroblasts from patients with Graves' ophthalmopathy. *J. Clin. Endocrinol. Metab.* **74:** 732–736.

Heufelder, A.E., J.R. Goellner, B.E. Wenzel, and R.S. Bahn. 1992b. Immunohistochemical detection and localization of a 72-kilodalton heat shock protein in autoimmune thyroid disease. *J. Clin. Endocrinol. Metab.* **74:** 724–731.

Heufelder, A.E., B.E. Wenzel, C.A. Gorman, and R.S. Bahn. 1991. Detection, cellular localization, and modulation of heat shock proteins in cultured fibroblasts from patients with extrathyroidal manifestations of Graves' disease. *J. Clin. Endocrinol. Metab.* **73:** 739–745.

Hindersson, P., J.D. Knudsen, and N.H. Axelsen. 1987. Cloning and expression of *Treponema pallidum* common antigen (Tp-4) in *E. coli* K-12. *J. Gen. Microbiol.* **133:** 587–596.

Hiromatsu, K., Y. Yoshikai, G. Matsuzaki, S. Ohga, K. Muramori, K. Matsumoto, J.A. Bluestone, and K. Nomoto. 1992. A protective role of γ/δ T cells in primary infection with *Listeria monocytogenes* in mice. *J. Exp. Med.* **175:** 49–56.

Hoffman, P.S., C.A. Butler, and F.D. Quinn. 1989. Cloning and temperature-dependent expression in *Escherichia coli* of a *Legionella pneumophila* gene coding for a genus-common 60-kilodalton antigen. *Infect. Immun.* **57:** 1731–1739.

Hogervorst, E.J., L. Schouls, J.P. Wagenaar, C.J. Boog, W.J. Spaan, J.D. van Embden, and W. van Eden. 1991. Modulation of experimental autoimmunity: Treatment of adjuvant arthritis by immunization with a recombinant vaccinia virus. *Infect. Immun.* **59:** 2029–2035.

Hohlfeld, R. and A.G. Engel. 1992. Expression of 65-kd heat shock proteins in the inflammatory myopathies. *Ann. Neurol.* **32:** 821–823.

Hohlfeld, R., A.G. Engel, K. Ii, and M.C. Harper. 1991. Polymyositis mediated by T lymphocytes that express the γ/δ receptor. *N. Engl. J. Med.* **324:** 877–881.

Holoshitz, J., F. Koning, J.E. Coligan, J. DeBruyn, and S. Strober. 1989. Isolation of CD4⁻ CD8⁻ mycobacterium-reactive T lymphocyte clones from rheumatoid arthritis synovial fluid. *Nature* **339:** 226–229.

Hunt, D.F., H. Michel, T.A. Dickinson, J. Shabanowitz, A.L. Cox, K. Sakaguchi, E. Appella, H.M. Grey, and A. Sette. 1992. Peptides presented to the immune system by the murine class II major histocompatibility complex molecule I-A^d. *Science* **256:** 1818–1820.

Imani, F. and M.J. Soloski. 1991. Heat shock proteins can regulate expression of the Tla region-encoded class Ib molecule Qa-1. *Proc. Natl. Acad. Sci.* **88:** 10475–10479.

Jäättelä, M. 1990. Effects of heat shock on cytolysis mediated by NK cells, LAK cells, activated monocytes and TNFs-α and β. *Scand. J. Immunol.* **31:** 175–182.

Jäättelä, M. and D. Wissing. 1993. Heat-shock proteins protect cells from monocyte cytotoxicity: Possible mechanism of self-protection. *J. Exp. Med.* **177:** 231–236.

Janeway, C.A. 1993. How the immune system recognizes invaders. *Sci. Am.* **269:** 72–79.

Jardetzky, T.S., W.S. Lane, R.A. Robinson, D.R. Madden, and D.C. Wiley. 1991. Identification of self peptides bound to purified HLA-B27. *Nature* **353:** 326–329.

Jarjour, W.N., B.D. Jeffries, J.S. Davis IV, W.J. Welch, T. Mimura, and J.B. Winfield. 1991. Autoantibodies to human stress proteins. A survey of various rheumatic and other inflammatory diseases. *Arthrit. Rheumat.* **34:** 1133–1138.

Jarjour, W., L.A. Mizzen, W.J. Welch, S. Denning, M. Shaw, T. Mimura, B.F. Haynes, and J.B. Winfield. 1990. Constitutive expression of a groEL-related protein on the surface of human cells. *J. Exp. Med.* **172:** 1857–1860.

Jarjour, W., V. Tsai, V. Woods, W. Welch, S. Pierce, M. Shaw, H. Mehta, W. Dillmann, N. Zvaifler, and J. Winfield. 1989. Cell surface expression of heat shock proteins. *Arthrit. Rheumat.* **32:** S44.

Jendoubi, M. and S. Bonnefoy. 1988. Identification of a heat shock-like antigen in *P. falciparum*, related to the heat shock protein 90 family. *Nucleic Acids Res.* **16:** 10928.

Johnson, K.S., K. Wells, J.V. Bock, V. Nene, D.W. Taylor, and J.S. Cordingley. 1989. The 86-kilodalton antigen from *Schistosoma mansoni* is a heat-shock protein homologous to yeast hsp-90. *Mol. Biochem. Parasitol.* **36:** 19–28.

Johnson, K., I. Charles, G. Dougan, D. Pickard, P. O'Gaora, G. Costa, T. Ali, I. Miller, and C. Hormaeche. 1991. The role of a stress-response protein in *Salmonella typhimurium* virulence. *Mol. Microbiol.* **5:** 401–407.

Jones, D.B., A.F.W. Coulson, and G.W. Duff. 1993. Sequence homologies between hsp60 and autoantigens. *Immunol. Today* **14:** 115–118.

Jones, D.B., N.R. Hunter, and G.W. Duff. 1990. Heat-shock protein 65 as a β-cell antigen of insulin-dependent diabetes. *Lancet* **336:** 583–585.

Kantengwa, S., Y.R.A. Donati, M. Clerget, I. Parini Maridonneau, F. Sinclair, E. Maréthoz, A.D.M. Rees, D.O. Slosman, and B.S. Polla. 1991. Heat shock proteins: An autoprotective mechanism for inflammatory cells? *Semin. Immunol.* **3:** 49–56.

Karlsson-Parra, A., K. Söderström, M. Ferm, J. Ivanyi, R. Kiessling, and L. Klareskog. 1990. Presence of human 65 kD heat shock protein (hsp) in inflamed joints and subcutaneous nodules of RA patients. *Scand. J. Immunol.* **31:** 283–288.

Kaufman, D.L., M. Clare-Salzler, J. Tian, T. Forsthuber, G.S.P. Ting, P. Robinson, M.A. Atkinson, E.E. Sercarz, A.J. Tobin, and P.V. Lehmann. 1993. Spontaneous loss of T-cell tolerance to glutamic acid decarboxylase in murine insulin-dependent diabetes. *Nature* **366:** 69–72.

Kaufmann, S.H.E. 1990. Heat shock proteins and the immune response. *Immunol. Today* **11:** 129–136.

———. 1991. Heat shock proteins and pathogenesis of bacterial infections. *Springer Semin. Immunopathol.* **13:** 25–36.

———. 1993a. Immunity to intracellular bacteria. In *Fundamental immunology* (ed. W.E. Paul), pp. 1251–1286. Raven Press, New York.

———. 1993b. Immunity to intracellular bacteria. *Annu. Rev. Immunol.* **11:** 129–163.

Kaufmann, S.H.E. and D. Kabelitz. 1991. Gamma/delta T lymphocytes and heat shock proteins. *Curr. Top. Microbiol. Immunol.* **167:** 191–207.

Kaufmann, S.H.E. and M.J. Reddehase. 1989. Infection of phagocytic cells. *Curr. Opin. Immunol.* **2:** 43–49.

Kaufmann, S.H.E., C. Blum, and S. Yamamoto. 1993. Crosstalk between α/β T cells and γ/δ T cells in vivo: Activation of α/β T cell responses after γ/δ T cell modulation with the monoclonal antibody GL3. *Proc. Natl. Acad. Sci.* **90:** 9620–9624.

Kaufmann, S.H.E., U. Väth, J.E.R. Thole, J.D.A. Van Embden, and F. Emmrich. 1987. Enumeration of T cells reactive with *Mycobacterium tuberculosis* organisms and specific for the recombinant mycobacterial 65 kiloDalton protein. *Eur. J. Immunol.* **178:** 351–357.

Kaufmann, S.H.E., B. Schoel, T. Koga, A. Wand-Württenberger, M.E. Munk, and U. Steinhoff. 1991. Heat shock protein 60: Implications for pathogenesis of and protection against bacterial infections. *Immunol. Rev.* **121:** 67–90.

Kaufmann, S.H.E., B. Schoel, A. Wand-Württenberger, U. Steinhoff, M.E. Munk, and T. Koga. 1990. T cells, stress proteins and pathogenesis of mycobacterial infections. *Curr.*

Top. Microbiol. Immunol. **155:** 125–141.

Kaur, I., S.D. Voss, R.S. Gupta, K. Schell, P. Fisch, and P.M. Sondel. 1993. Human peripheral gamma/delta T cells recognize hsp60 molecules on Daudi Burkitt's lymphoma cells. *J. Immunol.* **150:** 2046–2055.

Kindas-Muegge, I., G. Steiner, and J.S. Smolen. 1993. Similar frequency of autoantibodies against 70-kD class heat-shock proteins in healthy subjects and systemic lupus erythematosus patients. *Clin. Exp. Immunol.* **92:** 46–50.

Kleinau, S., K. Soederstroem, R. Kiessling, and L. Klareskog. 1991. A monoclonal antibody to the mycobacterial 65 kDa heat shock protein (ML30) binds to cells in normal and arthritic joints of rats. *Scand. J. Immunol.* **33:** 195–202.

Koga, T., A. Wand-Württenberger, J. DeBruyn, M.E. Munk, B. Schoel, and S.H.E. Kaufmann. 1989. T cells against a bacterial heat shock protein recognize stressed macrophages. *Science* **245:** 1112–1115.

Lamb, J.R., V. Bal, P. Mendez-Samperio, A. Mehlert, J. Rothbard, S. Jindal, R.A. Young, and D.B. Young. 1989. Stress proteins may provide a link between the immune response to infection and autoimmunity. *Int. Immunol.* **1:** 191–196.

La Thangue, N.B. and D.S. Latchman. 1988. A cellular protein related to heat-shock protein 90 accumulates during herpes simplex virus infection and is over-expressed in transformed cells. *Exp. Cell Res.* **178:** 169–179.

Lee, M.G., B.L. Atkinson, S.H. Giannini, and L.H.T. van der Ploeg. 1988. Structure and expression of the hsp70 gene family of *Leishmania major. Nucleic Acids Res.* **16:** 9567–9585.

Li, S.G., A.J. Quayle, Y. Shen, J. Kjeldsen-Kragh, F. Oftung, R.S. Gupta, J.B. Natvig, and O.T. Forre. 1992. Mycobacteria and human heat shock protein-specific cytotoxic T lymphocytes in rheumatoid synovial inflammation. *Arthrit. Rheumat.* **35:** 270–281.

Life, P.F., E.O.E. Bassey, and H.J.S. Gaston. 1991. T-cell recognition of bacterial heat shock proteins in inflammatory arthritis. *Immunol. Rev.* **121:** 113–135.

Lin, T.-Y., N.B. Duck, J. Winter, and W.R. Folk. 1991. Sequences of two hsc70 cDNAs from *Lycopersicon esculentum. Plant Mol. Biol.* **16:** 475–478.

Lussow, A.R., C. Barrios, J.D.A. Van Embden, R. Van der Zee, A.S. Verdini, A. Pessi, J.A. Louis, P.-H. Lambert, and G. Del Giudice. 1991. Mycobacterial heat-shock proteins as carrier molecules. *Eur. J. Immunol.* **21:** 2297–2302.

MacFarlane, J., M.L. Blaxter, R.P. Bishop, M.A. Miles, and J.M. Kelly. 1989. Characterization of a *Leishmania donovani* antigen similar to heat shock protein 70. *Biochem. Soc. Trans.* **17:** 168–169.

———. 1990. Identification and characterisation of a *Leishmania donovani* antigen belonging to the 70-kDa heat-shock protein family. *Eur. J. Biochem.* **190:** 377–384.

Marrack, P. and J.W. Kappler. 1993. How the immune system recognizes the body. *Sci. Am.* **269:** 80–89.

Mattei, D., L.S. Ozaki, and L. Pereira da Silva. 1988. A *Plasmodium falciparum* gene encoding a heat-shock-like antigen related to the rat 78 kD glucose-regulated protein. *Nucleic Acids Res.* **16:** 5204.

Matthews, R.C., J.P. Burnie, D. Howat, T. Rowland, and F. Walton. 1991. Autoantibody to heat shock protein 90 can mediate protection against systemic candidosis. *Immunology* **74:** 20–24.

Mehra, V., B.R. Bloom, A.C. Bajardi, C.L. Grisso, P.A. Sieling, D. Alland, J. Convit, X. Fan, S.W. Hunter, P.J. Brennan, T.H. Rea, and R.L. Modlin. 1992. A major T cell antigen of *Mycobacterium leprae* is a 10-kD heat-shock cognate protein. *J. Exp. Med.* **175:** 275–284.

Mistry, Y., D.B. Young, and R. Mukherjee. 1992. Hsp70 synthesis in Schwann cells in response to heat shock and infection with *Mycobacterium leprae. Infect. Immun.* **60:** 3105–3110.

Morrison, R.P., R.J. Belland, K. Lyng, and H.D. Caldwell. 1989. Chlamydial disease pathogenesis. The 57-kD chlamydial hypersensitivity antigen is a stress response protein. *J. Exp. Med.* **170:** 1271–1283.

Mottram, J., W. Murphy, and N. Agabian. 1989. A transcriptional analysis of the *Trypanosoma brucei* hsp83 gene cluster. *Mol. Biochem. Parasitol.* **37:** 115–128.

Mowat, A.M. 1987. The regulation of immune responses to dietary protein antigens. *Immunol. Today* **8:** 93–98.

Munk, M.E., B. Schoel, and S.H.E. Kaufmann. 1988. T cell responses of normal individuals towards recombinant protein antigens of *Mycobacterium tuberculosis. Eur. J. Immunol.* **18:** 1835–1838.

Munk, M.E., B. Schoel, S. Modrow, R.W. Karr, R.A. Young, and S.H.E. Kaufmann. 1989. Cytolytic T lymphocytes from healthy individuals with specificity to self epitopes shared by the mycobacterial and human 65 kDa heat shock protein. *J. Immunol.* **143:** 2844–2849.

Nelson, C.A., R.W. Roof, D.W. McCourt, and E.R. Unanue. 1992. Identification of the naturally processed form of hen egg white lysozyme bound to the murine major histocompatibility complex class II molecule I-A^k. *Proc. Natl. Acad. Sci.* **89:** 7380–7383.

Nene, V., D.W. Dunne, K.S. Johnson, D.W. Taylor, and J.S. Cordingley. 1986. Sequence and expression of a major egg antigen from *Schistosoma mansoni:* Homologies to heat shock proteins and alpha-crystallins. *Mol. Biochem. Parasitol.* **21:** 179–188.

Nerland, A.N., A.S. Mustafa, D. Sweetser, T. Godal, and R.A. Young. 1988. A protein antigen of *Mycobacterium leprae* is related to a family of small heat shock proteins. *J. Bacteriol.* **170:** 5919–5921.

Newcomb, J.R. and P. Cresswell. 1993. Characterization of endogenous peptides bound to purified HLA-DR molecules and their absence from invariant chain-associated α/β dimers. *J. Immunol.* **150:** 499–507.

Normington, K., K. Kohno, Y. Kozutsumi, M.-J. Gething, and J. Sambrook. 1989. S. cervisiae encodes an essential protein homologous in sequence and function to mammalian BiP. *Cell* **57:** 1223–1236.

Norton, P.M., D.A. Isenberg, and D.S. Latchman. 1989. Elevated levels of the 90 kd heat shock protein in a proportion of SLE patients with active disease. *J. Autoimmun.* **2:** 187–195.

Nossal, G.J.V. 1993. Life, death and the immune system. *Sci. Am.* **269:** 52–62.

O'Brien, R.L., Y.-X. Fu, R. Cranfill, A. Dallas, C. Ellis, C. Reardon, J. Lang, S.R. Carding, R. Kubo, and W. Born. 1992. Heat shock protein hsp60-reactive γ/δ cells: A large, diversified T-lymphocyte subset with highly focused specificity. *Proc. Natl. Acad. Sci.* **89:** 4348–4352.

Panchapakesan, J., M. Daglis, and P. Gatenby. 1992. Antibodies to 65 kDa and 70 kDa heat shock proteins in rheumatoid arthritis and systemic lupus erythematosus. *Immunol. Cell Biol.* **70:** 295–300.

Paul, W.E. 1993. Infectious diseases and the immune system. *Sci. Am.* **269:** 90–97.

Pearson, W.R. and D.J. Lipman. 1988. Improved tools for biological sequence comparison. *Proc. Natl. Acad. Sci.* **85:** 2444–2448.

Pervin, K., A. Childerstone, T. Shinnick, Y. Mizushima, R. van der Zee, R. Hasan, R. Vaughan, and T. Lehner. 1993. T cell epitope expression of mycobacterial and homologous human 65-kilodalton heat shock protein peptides in short term cell lines from patients with Behcet's disease. *J. Immunol.* **151:** 2273–2282.

Pierce, S.K., D.C. De Nagel, and A.M. van Buskirk. 1991. A role for heat shock proteins in antigen processing and presentation. *Curr. Top. Microbiol. Immunol.* **167**: 83–92.

Poccia, F., P. Piselli, S. Di Cesare, S. Bach, V. Colizzi, M. Mattei, A. Bolognesi, and F. Stirpe. 1992. Recognition and killing of tumour cells expressing heat shock protein 65 kD with immunotoxins containing saporin. *Brit. J. Cancer* **66**: 427–432.

Pope, R.M., R.M. Lovis, and R.S. Gupta. 1992. Activation of synovial fluid T lymphocytes by 60-kd heat-shock proteins in patients with inflammatory synovitis. *Arthrit. Rheumat.* **35**: 43–48.

Rajagopalan, S., T. Zordan, G.C. Tsokos, and S.K. Datta. 1990. Pathogenic anti-DNA autoantibody-inducing T helper cell lines from pateints with active lupus nephritis: Isolation of CD4-8- T helper cell lines that express the γ/δ T-cell antigen receptor. *Proc. Natl. Acad. Sci.* **87**: 7020–7024.

Rajasekar, R., G.-K. Sim, and A. Augustin. 1990. Self heat shock and γ/δ T-cell reactivity. *Proc. Natl. Acad. Sci.* **87**: 1767–1771.

Rambukkana, A., P.K. Das, L. Witkamp, S. Young, M.M. Meinardi, and J.D. Bos. 1993. Antibodies to mycobacterial 65-kDa heat shock protein and other immunodominant antigens in patients with psoriasis. *J. Invest. Dermatol.* **100**: 87–92.

Rambukkana, A., P.K. Das, S. Krieg, S. Young, I.C. Le Poole, and J.D. Bos. 1992. Mycobacterial 65,000 MW heat-shock protein shares a carboxy-terminal epitope with human epidermal cytokeratin 1/2. *Immunology* **77**: 267–276.

Rammensee, H.G., K. Falk, and O. Rötzschke. 1993. Peptides naturally presented by MHC class I molecules. *Annu. Rev. Immunol.* **11**: 213–244.

Requena, J.M., M.C. Lopez, A. Jimenez-Ruyiz, J.C. De la Torre, and C. Alonso. 1988. A head-to-tail organization of hsp70 genes in *Trypanosoma cruzi*. *Nucleic Acids Res.* **16**: 1393–1406.

Res, P., J. Thole, and R. De Vries. 1991. Heat-shock proteins and autoimmunity in humans. *Springer Semin. Immunopathol.* **13**: 81–98.

Res, P.C.M., C.G. Schaar, F.C. Breedveld, W. van Eden, J.D.A. van Embden, I.R. Cohen, and R.R.P. de Vries. 1988. Synovial fluid T cell reactivity against 65 kD heat shock protein of mycobacteria in early chronic arthritis. *Lancet* **II**: 478–480.

Robert, J.K. and J.L. Key. 1991. Isolation and characterization of a soybean hsp70 gene. *Plant Mol. Biol.* **16**: 671–683.

Rochester, D.E., J.A. Ainer, and D.M. Shah. 1986. The structure and expression of maize genes encoding the major heat shock protein, hsp70. *EMBO J.* **5**: 451–458.

Roop, R.M., III, M.L. Price, B.E. Dunn, S.M. Boyle, N. Sriranganathan, and G.G. Schurig. 1992. Molecular cloning and nucleotide sequence analysis of the gene encoding the immunoreactive *Brucella abortus* hsp60 protein, BA60K. *Microbial. Pathogen.* **12**: 47–62.

Rose, M.D., L.M. Misra, and J.P. Vogel. 1989. KAR2, a karyogamy gene, is the yeast homolog of the mammalian BiP/GRP78 gene. *Cell* **57**: 1211–1221.

Rothstein, N.M., G. Higashi, J. Yates, and T.V. Rajan. 1989. *Onchocerca volvulus* heat shock protein 70 is a major immunogen in amicrofilaremic individuals from a filariasis-endemic area. *Mol. Biochem. Parasitol.* **33**: 229–236.

Salvetti, M., C. Buttinelli, G. Ristori, M. Carbonari, M. Cherchi, M. Fiorelli, M.G. Grasso, L. Toma, and C. Pozzilli. 1992. T-lymphocyte reactivity to the recombinant mycobacterial 65- and 70-kDa heat shock proteins in multiple sclerosis. *J. Autoimmun.* **5**: 691–702.

Schwartz, R.H. 1989. Acquisition of immunologic self-tolerance. *Cell* **57**: 1073–1081.

Searle, S., A.J.R. Campos, R.M.R. Coulson, T.W. Spithill, and D.F. Smith. 1989. A family of heat shock protein 70-related genes are expressed in the promastigotes of *Leish-*

mania major. Nucleic Acids Res. **17:** 5081–5095.

Selkirk, M.E., D.A. Denham, F. Partono, and R.M. Maizels. 1989. Heat shock cognate 70 is a prominent immunogen in Brugian filariasis. *J. Immunol.* **143:** 299–308.

Selmaj, K., C.F. Brosnan, and C.S. Raine. 1991. Colocalization of lymphocytes bearing γ/δ T-cell receptor and heat shock protein hsp65+ oligodendrocytes in multiple sclerosis. *Proc. Natl. Acad. Sci.* **88:** 6452–6456.

———. 1992. Expression of heat shock protein-65 by oligodendrocytes in vivo and in vitro: Implications for multiple sclerosis. *Neurology* **42:** 795–800.

Shapria, M. and G. Pedraza. 1990. Sequence analysis and transcriptional activation of heat shock protein 83 of *Leishmania mexicana amazonensis. Mol. Biochem. Parasitol.* **42:** 247–256.

Shinnick, T.M., M.H. Vodkin, and J.C. Williams. 1988. The *Mycobacterium tuberculosis* 65-kilodalton antigen is a heat shock protein which corresponds to common antigen and to the *Escherichia coli* GroEl protein. *Infect. Immun.* **56:** 446–451.

Srivastava, P.K. 1993. Peptide-binding heat shock proteins in the endoplasmic reticulum: Role in immune response to cancer and in antigen presentation. *Adv. Cancer Res.* **62:** 153–177.

Srivastava, P.K. and M. Heike. 1991. Tumor-specific immunogenicity of stress-induced proteins: Convergence of two evolutionary pathways of antigen presentation. *Semin. Immunol.* **3:** 57–66.

Srivastava, P.K. and R.G. Maki. 1991. Stress-induced proteins in immune response to cancer. *Curr. Top. Microbiol. Immunol.* **167:** 109–123.

Srivastava, P.K., A.B. De Leo, and L.J. Old. 1986. Tumor rejection antigens of chemically induced sarcomas of inbred mice. *Proc. Natl. Acad. Sci.* **83:** 3407–3411.

Steinhoff, U., B. Schoel, and S.H.E. Kaufmann. 1990. Lysis of interferon-γ activated Schwann cells by crossreactive CD8 α/β T cells with specificity to the mycobacterial 65 kDa heat shock protein. *Int. Immunol.* **2:** 279–284.

Steinhoff, U., A. Wand-Württenberger, A. Bremerich, and S.H.E. Kaufmann. 1991. *Mycobacterium leprae* renders Schwann cells and mononuclear phagocytes susceptible or resistant against killer cells. *Infect. Immun.* **59:** 684–688.

Steinman, L. 1993. Autoimmune disease. *Sci. Am.* **269:** 106–115.

Taylor, H.R., I.W. Maclean, R.C. Brunham, S. Pal, and J. Wittum-Hudson. 1990. Chlamydial heat shock proteins and trachoma. *Infect. Immun.* **58:** 3061–3063.

Thompson, S.J., G.A. Rook, R.J. Brealey, R. Van der Zee, and C.J. Elson. 1990. Autoimmune reactions to heat-shock proteins in pristane-induced arthritis. *Eur. J. Immunol.* **20:** 2479–2484.

Tisch, R., X.-D. Yang, S.M. Singer, R.S. Liblau, L. Fugger, and H.O. McDevitt. 1993. Immune response to glutamic acid decarboxylase correlates with insulitis in non-obese diabetic mice. *Nature* **366:** 72–75.

Tsoulfa, G., G.A.W. Rook, G.M. Bahr, M.A. Sattar, K. Behbehani, D.B. Young, A. Mehlert, J.D.A. van Embden, F.C. Hay, D.A. Isenberg, and P.M. Lydyard. 1989. Elevated IgG antibody levels to the mycobacterial 65 kDa heat shock protein are characteristic of patients with rheumatoid arthritis. *Scand. J. Immunol.* **30:** 519–527.

Udono, H. and P.K. Srivastava. 1993. Heat shock protein 70-associated peptides elicit specific cancer immunity. *J. Exp. Med.* **178:** 1391–1396.

Ullrich, S.J., E.A. Robinson, L.W. Law, M. Willingham, and E. Appella. 1986. A mouse tumor-specific transplantation antigen is a heat-shock-related protein. *Proc. Natl. Acad. Sci.* **83:** 3121–3125.

VanBuskirk, A., B.L. Crump, E. Margoliash, and S.K. Pierce. 1989. A peptide binding protein having a role in antigen presentation is a member of the hsp 70 heat shock family. *J. Exp. Med.* **170:** 1799–1809.

van Eden, W., J.E. Thole, R. Van der Zee, A. Noordzij, J.D. van Embden, E.J. Hensen, and I.R. Cohen. 1988. Cloning of the mycobacterial epitope recognized by T lymphocytes in adjuvant arthritis. *Nature* **331:** 171–173.

Verbon, A., R.A. Hartskeerl, A. Schuitema, A.H.J. Kolk, D.B. Young, and R. Lathigra. 1992. The 14,000-molecular-weight antigen of *Mycobacterium tuberculosis* is related to the alpha-crystallin family of low-molecular-weight heat shock proteins. *J. Bacteriol.* **174:** 1352–1359.

Vodkin, M.H. and J.C. Williams. 1988. A heat shock operon in *Coxiella burnetii* produces a major antigen homologous to a protein in both *Mycobacteria* and *Escherichia coli. J. Bacteriol.* **170:** 1227–1234.

Wand-Württenberger, A., B. Schoel, J. Ivanyi, and S.H.E. Kaufmann. 1991. Surface expression by mononuclear phagocytes of an epitope shared with mycobacterial heat shock protein 60. *Eur. J. Immunol.* **21:** 1089–1092.

Wauben, M.H., C.J. Boog, R. Van der Zee, and W. van Eden. 1992. Towards peptide immunotherapy in rheumatoid arthritis: Competitor-modulator concept. *J. Autoimmun.* **5:** 205–208.

Wucherpfennig, K.W., J. Newcombe, H. Li, C. Keddy, M.L. Cuzner, and D.A. Hafler. 1992. γ/δ T-cell receptor repertoire in acute multiple sclerosis lesions. *Proc. Natl. Acad. Sci.* **89:** 4588–4592.

Xu, Q. and G. Wick. 1993. Surface expression of heat shock protein 60 on endothelial cells. *Immunobiology* **189:** 131–132.

Xu, Q., H. Dietrich, H.J. Steiner, A.M. Gown, B. Schoel, G. Mikuz, S.H.E. Kaufmann, and G. Wick. 1992. Induction of arteriosclerosis in normocholesterolemic rabbits by immunization with heat shock protein 65. *Arteriosclerosis Thrombosis* **12:** 789–799.

Xu, Q., J. Willeit, M. Marosi, R. Kleindienst, F. Oberhollenzer, S. Kiechl, T. Stulnig, G. Luef, and G. Wick. 1993. Association of serum antibodies to heat-shock protein 65 with carotid atherosclerosis. *Lancet* **341:** 225–259.

Yang, X.D., J. Gasser, and U. Feige. 1990. Prevention of adjuvant arthritis in rats by a nonapeptide from the 65-kD mycobacterial heat-shock protein. *Clin. Exp. Immunol.* **81:** 189–194.

―――. 1992. Prevention of adjuvant arthritis in rats by a nonapeptide from the 65-kD mycobacterial heat shock protein: Specificity and mechanism. *Clin. Exp. Immunol.* **87:** 99–104.

Yang, Y.-F., P. Tan-Ariya, Y.D. Sharma, and A. Kilejian. 1987. The primary structure of a *Plasmodium falciparum* polypeptide related to heat shock proteins. *Mol. Biochem. Parasitol.* **26:** 61–68.

Yokota, S., K. Tsubaki, T. Kuriyama, H. Shimizu, M. Ibe, T. Mitsuda, Y. Aihara, K. Kosuge, and H. Nomaguchi. 1993. Presence in Kawasaki disease of antibodies to mycobacterial heat shock protein hsp65 and autoantibodies to epitopes of human hsp65 cognate antigen. *Immunol. Immunopathol.* **2:** 163–170.

Young, D.B. and A. Mehlert. 1990. Stress proteins and infectious diseases. In *Stress proteins in biology and medicine* (ed. R. Morimoto et al.), pp. 131–165. Cold Spring Harbor Laboratory Press, Cold Spring Harbor, New York.

Young, D.B., R.B. Lathigra, R.W. Hendrix, D. Sweetser, and R.A. Young. 1988. Stress proteins are immune targets in leprosy and tuberculosis. *Proc. Natl. Acad. Sci.* **85:** 4267–4270.

Young, R.A. 1990. Stress proteins and immunology. *Annu. Rev. Immunol.* **8:** 401–420.

Young, R.A. and T.J. Elliott. 1989. Stress proteins, infection, and immune surveillance. *Cell* **59:** 5–8.

20

Expression and Function of Stress Proteins in the Ischemic Heart

Ivor J. Benjamin and R. Sanders Williams
University of Texas Southwestern Medical Center
Dallas, Texas 75235-8573

I. INTRODUCTION

The goal of this chapter is to summarize current understanding of the regulation and function of heat shock proteins in the ischemic heart. The clinical context in which this subject is discussed is an important one. Despite recent therapeutic advances, ischemic heart disease remains the leading cause of death and disability in industrialized nations. Coronary thrombosis, occurring against a backdrop of chronic atherosclerosis, leads rapidly to cell death within the heart (Reimer and Jennings 1988). If a sufficient mass of myocardium is injured, the patient will succumb either to inadequate pump function of the heart or to lethal dysrhythmias that accompany this condition. The progression to cellular necrosis is not, however, instantaneous, as shown by the beneficial effects of thrombolytic therapy delivered within the first few hours (Braunwald and Sobel 1988). Therefore, efforts to understand the mechanisms by which cells are damaged during ischemia and to identify compensatory or adaptive responses that may augment cell survival are of paramount importance (Williams and Benjamin 1991).

General and specific features of the molecular biology, genetics, and biochemistry of heat shock proteins are discussed in detail in other chap-

The Biology of Heat Shock Proteins and Molecular Chaperones
©1994 Cold Spring Harbor Laboratory Press 0-87969-427-0/94 $5 + .00

ters in this volume and will not be reiterated here. Rather, our discussion focuses explicitly on the heart and on efforts to understand the role of heat shock proteins in the complex series of events triggered by ischemia within the myocardium. Since terms used by cardiovascular investigators may be unfamiliar to molecular biologists and other scientists interested in heat shock proteins, some definitions are in order to enhance comprehension of the subsequent discussion.

A. Definition of Terms

Ischemia refers to the condition existing within a tissue in which blood flow is inadequate to meet metabolic demands. Although ischemic tissues experience *hypoxia*, the terms are not synonymous, and the consequences may be quite different (Allen and Orchard 1986). During ischemia, not only does oxygen become unavailable for mitochondrial respiration, but the cells are deprived of other blood-borne substrates, such as glucose, as well. Moreover, potentially toxic metabolites that normally would be removed by the circulation accumulate within the tissue.

Myocardial ischemia refers to ischemia within the wall of the heart and can have different consequences depending on several variables, most notably the severity and duration of the obstruction to blood flow (Reimer and Jennings 1979, 1988). Partial occlusion of a coronary artery may produce ischemia only during states of high metabolic demand, such as during exercise. Complete occlusion of a coronary artery also can produce a spectrum of abnormalities, depending on the magnitude of blood flow through unaffected arteries (collateral flow). Most commonly, however, complete occlusion produces severe ischemia that, if unrelieved, will lead rapidly and inexorably to necrosis of cells (*myocardial infarction*). In experimental models, myocardial ischemia can be *regional* (involving only a segment of the heart) or *global* (involving the entire heart).

Brief periods of myocardial ischemia (up to several minutes) produce profound changes in concentrations of intracellular metabolites and seriously compromise contractile function but may have no permanent sequelae. Interestingly, periods of brief ischemia induce effects within the myocardium that serve to limit the extent of injury produced by a subsequent period of prolonged ischemia (see Murry et al. 1986). This phenomenon, which is termed *ischemic preconditioning*, is reminiscent of the thermotolerance induced in many cell types by transient heat shock.

Reperfusion is the term used to describe the period that follows the restoration of blood flow to the ischemic heart. Current concepts suggest that many of the events that lead to lethal cellular injury are precipitated during reperfusion (Downing and Chen 1992; Yellon et al. 1992a,b). For example, restoration of oxygen generates superoxide radicals with resulting damage to both intracellular and extracellular structures. Leukocytes enter the tissue upon reperfusion and adhere to sites of endothelial damage in the microvasculature, releasing cytotoxic enzymes and clogging capillaries (no-reflow phenomenon), thereby extending the duration of ischemia at the cellular level (Entman et al. 1991). Even when reperfusion occurs in a timely manner so as to avoid irreversible cellular injury, certain consequences of ischemia may persist for hours or days. *Myocardial stunning* refers to transient post-ischemic contractile dysfunction of the heart seen after reperfusion in the absence of irreversible damage (Braunwald and Kloner 1982; for review, see Bolli 1990).

B. Experimental Models of Ischemia

Studies of stress protein expression and function have been examined in *cultured cells*, including skeletal myogenic cells and isolated, dispersed cardiomyocytes. These provide useful model systems in which many variables can be rigorously controlled within a relatively homogeneous cell population (for review, see Jennings and Morgan 1988). Conditions of ischemia can be simulated, although not entirely reproduced, by depriving cultured myocytes of substrates or by addition of metabolic inhibitors to the growth medium (Buja et al. 1985).

To examine consequences of ischemia under more authentic conditions, intact hearts can be excised from experimental animals and perfused ex vivo. Such *isolated heart preparations* can be maintained for several hours and permit simultaneous monitoring of physiological and biochemical variables under conditions in which blood flow, oxygen tension, and other components of the perfusate are controlled. Rat, hamster, or rabbit hearts are used most commonly for this purpose, and, with greater technical difficulty, mouse hearts also can be examined (Ng et al. 1991).

Open chest preparations involve anesthetized animals in which the heart is exposed but left in place within the chest (for review, see Reimer et al. 1985). The systemic circulation remains intact, but myocardial ischemia is produced by placing ligatures or snares around one or more coronary arteries. The animals are sacrificed at the end of the procedure. Rats, rabbits, or dogs serve as the usual experimental subjects, but other species also can be studied in this manner. *Closed chest preparations* require multistage procedures usually performed in dogs. Ameroid constrictors or snares that can be operated from an extrathoracic location are placed around a coronary vessel during an initial procedure. The surgical wounds are closed and the animals allowed to recover. Myocardial ischemia is induced at a later period after the animals are free of confounding effects associated with anesthesia and surgical stress. Open and closed chest preparations provide the closest experimental approximation to the clinical problem, but they present difficulties with respect to control of hemodynamic and metabolic variables that may influence the outcome of measurements.

All of these experimental systems have been employed productively over the past two decades to produce our current understanding of the effects of myocardial ischemia on metabolism, ultrastructure, electrophysiology, and contractile performance of the heart (Fozzard et al. 1988). Only in the last few years, however, have these models been used to assess changes in expression of heat shock proteins during ischemia and to attempt to draw conclusions regarding their functional role in the ischemic heart.

II. REGULATION OF STRESS PROTEIN EXPRESSION IN THE ISCHEMIC HEART

A. Induction of hsp70 by Metabolic Stresses

Studies from several laboratories using different experimental systems uniformly demonstrate increased expression of hsp70 mRNA and protein in response to simulated or authentic ischemia. The induction of hsp70 by ischemia is less rapid and usually of lesser magnitude than the effects of heat shock (Currie et al. 1987). Physiological stresses other than heat shock or ischemia also increase concentrations of hsp70 in the myocardial wall (Howard and Geoghegan 1986). Table 1 lists some of the reports that have examined expression of hsp70 in isolated myocytes or intact heart models of ischemia and other physiological stresses.

These studies demonstrate clearly that expression of hsp70 is induced by ischemia, and other investigations have sought to define the molecular mechanisms responsible for this induction. A series of studies from our laboratory have explored several features of this response. We observed that hypoxic induction of transcriptional activity of the hsp70 promoter in cultured myogenic cells required an intact heat shock element and was temporally linked to activation of the DNA-binding function of heat shock transcription factor (HSF). Thus, unlike other nonthermal stimuli known to activate expression of hsp70 (e.g., viral infection), hypoxia appeared to induce expression of hsp70 through the same final common pathway as heat shock (Benjamin et al. 1990).

Table 1 Induction of Expression of hsp70 Proteins in the Heart

Stimulus	Model	Species	Effect:Induction of (mRNA or protein)	References
Trauma	tissue slices	rat	hsp70	Currie and White (1981)
Hyperthermia	intact animal	rat	hsp70	Currie and White (1982); Hammond et al. (1982)
Aortic banding (1 hr)	open-chest	rat	hsp70	Hammond et al. (1982)
Myocardial ischemia	open-chest	dog	hsp70	Dillmann et al. (1986); Metha et al. (1988)
Isoproterenol	intact animal	rat	hsp70	White and White (1986)
Heterotopic transplant	isolated heart	rat	donor>recipient	Currie et al. (1987)
Myocardial stretch	isolated heart	rabbit	hsp70	Knowlton et al. (1991a,b)
Starvation	intact animal	rat	hsc70	Wing et al. (1991)
Aortic banding (2–4 days)	open-chest	rat	hsp70	Decayre et al. (1988)
Exhaustive exercise	intact animal	rat	hsp70	Salo et al. (1991)
Pulmonary artery banding	open-chest	rat	hsp70	Katayose et al. (1993)
Hypoxia (3 days)	intact animal	rat	no induction of hsp70	Katayose et al. (1993)

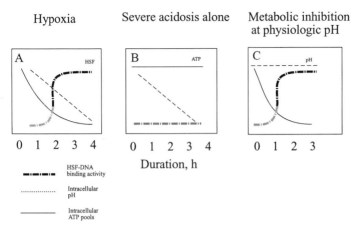

Figure 1 Summary of the relationships between intracellular pH, ATP pools, and HSF DNA-binding activity during conditions that simulate ischemia. (*A*) Pleiotropic effects of hypoxia in cultured cells at 37°C. In *B* and *C*, respectively, whereas severe acidosis (pH <6.7) alone failed, maneuvers that maintained physiological pH but lowered ATP pools are apparently sufficient to induce the HSF DNA-binding activity (Benjamin et al. 1992).

B. Metabolic Stimuli Leading the Activation of HSF

We went on to explore the relationships between specific metabolic perturbations that occur in ischemic cells and the DNA-binding activity of HSF (Sarge et al. 1991). Chemical inhibitors of mitochondrial respiration produced effects on HSF and expression of hsp70 similar to those resulting from hypoxia but with a more rapid time course (Benjamin et al. 1992). This result suggested that metabolic sequelae of hypoxia, rather than a direct oxygen-sensing mechanism, were responsible for activation of HSF in these cells. Such sequelae include a fall in ATP and other high-energy phosphate compounds within the cell, as well as a fall in pH, reflecting anaerobic glycolysis (Fig. 1A). In an attempt to discriminate between these two events as signals to activation of HSF, we employed conditions in which ATP concentrations and pH could be varied independently while expression of hsp70 and DNA-binding activity of HSF was monitored. Results from this set of experiments are summarized in Figure 1B–C.

Modest reductions in ATP (50% of control levels), as produced by glucose deprivation, were insufficient to activate HSF. In contrast, more severe reductions in ATP concentrations (<30% of control levels) resulting from glucose deprivation plus rotenone, a mitochondrial inhibitor, were associated with activation of HSE-binding activity within the cells. Under this latter condition, intracellular pH also fell, from 7.3 to 6.9, such that independent effects of acidosis and ATP depletion could not be discriminated. Further experiments were performed, however, in which intracellular pH was reduced by administration of amiloride and sodium propionate to 6.7 while ATP levels were preserved at normal levels. This maneuver was insufficient to activate HSF (Fig. 1B). In contrast, severe depletion of ATP remained linked to activation of HSF even under conditions in which pH was held constant at normal levels in the presence of high K^+ and nigericin (Fig. 1C). We concluded from these studies that the proximate stimulus to induction of hsp70 transcription in these cells was more closely associated with a fall in intracellular stores of high-energy phosphates than to perturbations of pH within the physiological range.

These results should be interpreted in light of additional findings from other laboratories. Earlier studies by Drummond et al. (1986) indicated that alterations of pH do not influence the induction of heat shock proteins. However, it has been possible to activate the DNA-binding function of HSF from HeLa cell extracts in vitro by reducing pH to unphysiologically low values (Mosser et al. 1990). The possibility therefore exists that under some conditions, severe acidosis may function as an inducing stimulus. With respect to the role of ATP as a stimulus to transcriptional activation of hsp70, recent studies of hypoxic cardiomyocytes have demonstrated induced expression of hsp70 at time points preceding a major decline in ATP concentrations (Iwaki et al. 1993). The same study described increased concentrations of hsp70 mRNA occurring in advance of activated HSF, as detected in gel mobility shift assays. These results suggest that alternative pathways for transcriptional activation of hsp70 during hypoxia or ischemia may exist as well. Further experiments are needed to determine whether an HSE is necessary for ischemia-induced transcriptional up-regulation of the hsp70 promoter in the myocardial wall and whether mechanisms distinct from activation of HSF may have a role in this induction.

If a fall in ATP constitutes a proximate stimulus to activation of HSF and subsequent induction of hsp70 transcription during ischemia, what intervening steps transduce this signal? Progress toward unraveling the events triggered by heat shock, as discussed elsewhere in this volume, suggests several possibilities. Unfolding of cytosolic proteins may provide a feature common to ischemia and other inducing stimuli that activate heat shock protein genes (Beckmann et al. 1992; Gething and Sambrook 1992; Craig 1993). Changes in cellular energy charge or redox potential due to diminished oxidative metabolism may destabilize the structure of certain proteins and trigger subsequent events identical to those encountered during heat shock. Such conformational changes also could occur directly through allosteric effects of nucleoside phosphates. ATP depletion may disrupt activity of membrane ion channels and result in changes in cytoplasmic or mitochondrial concentrations of divalent or monovalent cations. Any of these perturbations in the intracellular milieu may destabilize protein structure, but the events most important for activation of HSF remain to be determined. Future studies of the effects of ischemia on induction of heat shock genes also should address the phosphorylation status and *trans*-activation function of HSF, as distinct from DNA binding, since these functions are regulated independently in yeast (Sorger 1990) and perhaps also in mammalian cells (Morimoto 1993). In addition, future studies should recognize that multiple isoforms of HSF are present in mammalian cells (Morimoto 1993; Lis and Wu 1993) potentially adding to the complexity of the response mechanisms that are activated by ischemia.

Findings from the current literature leave these and other important questions unresolved. For example, no study has determined whether ischemic induction of hsp70 occurs contemporaneously or divergently within the many cell types (e.g., cardiomyocytes, Purkinje fibers, endothelial cells, and fibroblasts) that comprise the myocardium wall or whether spatial variation in expression is present within the ischemic region. Likewise, there is very limited information concerning the expression of specific isoforms of hsp70 or of other stress proteins (e.g., hsp60, low-molecular-weight heat shock proteins) in the ischemic and post-ischemic heart. One study reported modest elevations in expression of hsp60 in cardiac tissue 24 hours after an ischemic preconditioning protocol (Marber et al. 1993), but more detailed information is currently unavailable.

Compared to cardiomyocytes, endothelial cells may be relatively resistant to ischemic injury and may express a distinct set of hypoxia-associated proteins (Zimmerman et al. 1991). Augmented expression of hsp60 has been observed within atherosclerotic vessels (Kleindienst et al. 1993) and in inflammatory myopathies (Hohlfeld and Engel 1992) and has been proposed to stimulate recruitment of lymphocytes previously sensitized to highly conserved hsp60s of microbial organisms. Selective expression of hsp70 in vascular cells but not in cardiomyocytes occurs in response to neuroendocrine stimulation (Udelsman et al. 1993). In the ischemic brain, cell types differ in the timing and magnitude of induction of hsp70, and a relationship between restricted induction of stress proteins and the susceptibility of certain cells to lethal ischemic injury has been proposed (Abe et al. 1993; Aoki et al. 1993; Brown and Rush 1993). Such relationships, and their relevance to cellular viability, have not been examined in the ischemic heart. Thus, there is a current need to extend the types of analyses that have been performed with hsp70 to other stress proteins, as well as a need to define stress protein expression in the ischemic heart with methods that can distinguish among the different cell types resident within the complex architecture of the myocardial wall.

Besides the direct effects that evoke the heat shock response within ischemic tissues, other factors may indirectly influence the expression of stress proteins in nonischemic regions of the heart. The myocardial workload of both ventricles is obligatorily increased by the hemodynamic alterations necessary to sustain myocardial performance. Myocardial ischemia is also accompanied by release of metabolic by-products, catecholamines, and peptide growth factors. These factors are potential inducers of the heat shock response in myocardial regions remote from the ischemic stress. Hence, stress proteins not only may serve pivotal roles in cytoprotective agents to augment cellular survival (Williams et al. 1993), but may also participate in ventricular remodeling and hypertrophy following severe ischemia or infarction.

III. FUNCTIONS OF HEAT SHOCK PROTEINS DURING ISCHEMIA

Are heat shock proteins cytoprotective during myocardial ischemia? This fundamental question has occupied most of the attention of cardiovascular investigators to date. A putative cytoprotective function for heat shock proteins in the ischemic heart is suggested by the induction of expression of hsp70 during or following ischemia, and the apparent analogy between the ischemic preconditioning phenomenon and thermotolerance induced by prior heat shock.

A. Heat Shock Proteins and Ischemic Preconditioning

The potential role of heat shock proteins as determinants of the ischemic preconditioning phenomenon has been studied in several laboratories, and several of these studies are listed in Table 2. Certain preconditioning protocols are sufficient to limit myocardial damage during subsequent challenge with prolonged coronary occlusion in the absence of detectable increases in hsp70 expression induced by the preconditioning stimulus (for review, see Lawson and Downey 1993). In addition, events within the tissue that have no apparent relationship to heat shock proteins (e.g., release of adenosine) appear to be important determinants of infarct size in some models (Lawson and Downey 1993). On the other hand, an inverse correlation between expression of major stress proteins induced by ischemic preconditioning and the severity of subsequent injury resulting

Table 2 Association between Expression of hsp70 and Cellular Protection

Preconditioning stimulus	Model	Species	Outcome	References
Hyperthermia	isolated heart	rat	ventricular function improved; myocardial salvage	Currie et al. (1988); Karmazyn et al. (1990)
Hyperthermia	open–chest	rat	myocardial salvage at 35 min but not 45 min	Donnelly et al. (1992)
Hyperthermia	open–chest	rabbit	infarct size reduction 30 min but not 45 min	Currie et al. (1993)
Hyperthermia or ischemic preconditioning	intact animal	rabbit	equivalent reduction of infarct size	Marber et al. (1993)
Hyperthermia and brief ischemia	open–chest	rat	reduction of reperfusion arrhythmias	Steare and Yellon (1993)
Forced expression	dispersed cells	mouse	cytoprotection; growth inhibition	Williams et al. (1993)
Forced expression	dispersed cells	rat	cytoprotection	Mestril et al. (1994)

from prolonged ischemia has been reported from other groups (Currie et al. 1993; Marber et al. 1993). Two critical variables that may determine the relative importance of heat shock proteins in the ischemic preconditioning phenomenon are (1) the duration of the initial ischemic episode (the preconditioning stimulus) and (2) the time that elapses between the preconditioning stimulus and any subsequent ischemic event. Very brief initial periods of ischemia may protect the myocardium against subsequent ischemic episodes for a few hours through mechanisms that are largely independent of heat shock proteins. A more prolonged initial ischemic episode may generate protective effects that are more long-lasting (24 hr or longer) and dependent, at least in part, on induction of heat shock proteins.

B. Heat Shock as a Preconditioning Stimulus

Several reports listed in Table 2 describe preconditioning with thermal stress as a different approach to limit infarct size in animal models of coronary occlusion. In an early study, Currie (1988) subjected rats to whole body hyperthermia and then examined contractile function of isolated perfused hearts subjected to 10 minutes of global ischemia (10% of normal coronary flow rate). During ischemia, the heat-shocked hearts demonstrated declines in developed pressure that were similar to those of controls. However, upon restoration of normal flow, hearts pretreated by heat shock recovered function more rapidly than controls. These data indicate that hyperthermia, sufficient to induce hsp70 expression in the heart, provides cross-tolerance to ischemic injury, manifested by more rapid recovery of cardiac function after reversible ischemia (Currie et al. 1988; Karmazyn et al. 1990). Similar findings have been observed in other studies using different models (Donnelly et al. 1992; Marber et al. 1993). As illustrated in Figure 2, only modest increases in expression of hsp70 (approximately threefold) induced by ischemic or thermal preconditioning are associated with distinct effects on the magnitude of irreversible injury following prolonged ischemia.

Several conclusions may reasonably be drawn from these studies, viewed in aggregate and in the context of current concepts of ischemic cell injury. Although induction of heat shock proteins may, at least under some conditions, contribute to the ischemic preconditioning phenomenon, this stimulus triggers a complex array of responses that influence cell survival, and the role played by currently known heat shock proteins may be relatively minor. With respect to thermal preconditioning, there seems to be a reasonable consensus that prior heat shock sufficient to induce expression of major heat shock proteins serves to limit the sub-

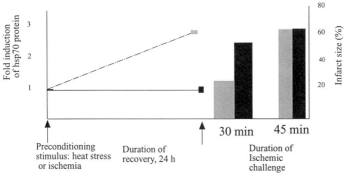

Figure 2   Effects of preconditioning on myocardial protection. Animals were subjected to either whole-body hyperthermia or ischemic preconditioning. Analyses were performed 24 hr later when modest increases in hsp70 expression were observed. The ischemic protection afforded by both forms of preconditioning are roughly equivalent following 30-min occlusion, but this beneficial effect is apparently abolished when more prolonged periods of myocardial ischemia are exceeded (Currie et al. 1993; Marber et al. 1993).

sequent injury resulting from a severe ischemic episode. As discussed in other chapters in this volume and elsewhere (Currie and Tanguay 1991), however, thermal stress has many sequelae in addition to induction of major heat shock proteins that may influence cellular viability. For example, preconditioning by thermal stress is associated with increased catalytic activity of the antioxidant, catalase, for which a protective role is postulated during ischemic reperfusion (Currie and Tanguay 1991). Hence, the beneficial effects of prior heat shock on the extent of ischemic injury in the heart suggest a cytoprotective function for heat shock proteins but do not establish a cause and effect relationship.

To circumvent some of the limitations inherent to the correlative analysis employed by the studies cited in Table 2, we recently used a genetic approach to address the functional significance of hsp70 as a determinant of cell viability during simulated ischemia in cultured cells (Williams et al. 1993). Our major hypothesis was that loading cells with high concentrations of hsp70 prior to the onset of a severe metabolic stress would extend the period of time before irreversible injury ensued. A human *hsp70* gene (Hunt and Morimoto 1985) was placed under the control of the constitutively active human β-actin promoter and transfected into mouse 10T1/2 cells. Viability was monitored in transfected cells as the intracellular retention of an enzyme marker (luciferase) expressed from a second plasmid cotransfected with the hsp70 expression vector. The human *hsp70* transgene was expressed to high levels in these transiently transfected cells, as assessed by RNase protection assays, Western blots, and immunohistochemical analysis. By comparison to cells transfected with a control plasmid, cells genetically engineered in this manner to overexpress human hsp70 were resistant to injury resulting from glucose deprivation and blockade of mitochondrial respiration. This study provides direct evidence that hsp70 exerts a cytoprotective function during simulated ischemia in mammalian cells. An elegantly detailed study (see Mestril et al. 1994) reaches similar conclusions from analysis of H9c2(2-1) cells (a line derived from the embryonic rat heart) stably transformed to overexpress an inducible hsp70 isoform.

The molecular mechanisms of this cytoprotective effect of hsp70 are unknown, but they are presumably related to the well-established functions of this class of proteins in normal, unstressed cells and to the role of heat shock proteins in thermotolerance. The activity of hsp70 and other heat shock proteins as molecular chaperones is discussed in detail in this volume. We presume that binding of hsp70 to proteins denatured during ischemia promotes more efficient refolding to functional configurations so that normal cellular processes can resume more rapidly in the post-ischemic state. Apparently, the endogenous pool of hsp70 available during metabolic stress is inadequate to saturate all sites at which this function of hsp70 can influence cell viability, since increasing this pool by forced overexpression results in additional cytoprotective effects.

The relevance of such findings in cultured cell models to bona fide myocardial ischemia in the intact heart remains to be determined but may be addressed in the future by analysis of transgenic mice engineered to overexpress human hsp70 in cardiomyocytes or endothelial cells of the myocardium. The results observed in cultured cells strongly support the hypothesis that hsp70 is cytoprotective during metabolic and hypoxic stresses encountered by ischemic cells and demonstrate further that the endogenous defense mechanisms engaged by cells during energy deprivation do not constitute a biological limit. It is possible, at least under some conditions, to employ genetic approaches to improve cell survival in the face of severe metabolic perturbations.

Although transient elevations of hsp70 appear to exert a cytoprotective function during metabolic stress, deleterious effects may ensue from sustained, long-term overexpression. In stably transformed lines of insect or mammalian cells, chronic elevation of hsp70 is accompanied by constraints on proliferative growth of cells under nonstressed conditions (Solomon et al. 1991; Williams et al. 1993). It remains to be determined whether postmitotic cardiac myocytes or quiescent endothelial cells of the adult heart also would suffer deleterious effects from sustained elevations of hsp70.

IV. SUMMARY AND FUTURE DIRECTIONS

This volume is devoted to a description of the burgeoning supply of new information concerning the structure, regulation, and function of the ancient and interesting families of heat shock proteins. The potential relevance of this knowledge to the formidable clinical problem of ischemic heart disease has not escaped the attention of cardiovascular investigators. Although relatively few studies have been pursued to explore the regulation and function of heat shock proteins in the ischemic myocardium, the following conclusions seem to be warranted.

Expression of hsp70 is induced in ischemic cells through mechanisms that include, but may not be entirely limited to, activation of heat shock transcription factor(s) in a manner similar to the effects of thermal stress. The proximate stimulus generated within ischemic cells that leads to activation of HSF is currently unknown. There is a general association with abnormalities in intracellular metabolites, such as ATP, but no firm evidence that any single metabolic abnormality is either necessary or sufficient to trigger transcriptional activation of genes encoding major heat shock proteins. The time course in which hsp70 is up-regulated within the ischemic heart indicates that this induction is not a premorbid event reflecting irrevocable cell injury, but rather a process activated in stressed but viable cells. Limited data suggest that certain other stress proteins such as hsp60 also may be induced in the ischemic heart, but a comprehensive examination of known heat shock proteins other than hsp70 has not been performed, and the possibility that novel members of heat shock protein families are induced by metabolic stresses in the ischemic heart cannot be excluded. Indeed, an hypoxia-inducible DNA-binding factor recently identified from studies of an erythropoietin gene enhancer (Wang and Semenza 1993) appears to be expressed in the heart, although the downstream genes that may be controlled by this factor are not currently known.

Several studies describe an inverse relationship between expression of major heat shock proteins in the heart and the severity of injury resulting from coronary occlusion. This correlation seems to be valid and is supported by studies in which either heat shock or transient ischemia was employed as a preconditioning stimulus. These studies are inadequate, however, to establish a causal relationship between expression of heat shock proteins and the magnitude of ischemic injury. More direct evidence for a cytoprotective function of heat shock proteins during ischemia comes from analysis of cells transfected in culture in which expression of a human *hsp70* transgene delayed the onset of irreversible injury resulting from metabolic stress simulating ischemia. Additional studies are required to determine whether this apparent cytoprotective effect of *hsp70* is pertinent to the more complex environment of the intact, ischemic heart.

The increasing interest of cardiovascular scientists in heat shock proteins coupled with technological advances to facilitate analysis and manipulation of gene expression in intact animals should lead to rapid progress in the near future with respect to several major questions (Black and Lucchesi 1993; Chien 1993). More detailed studies of heat shock protein expression in the ischemic and post-ischemic heart should identify differences, if any, between this stress response in cardiomyocytes and other cell types of the myocardial wall. The effects of ischemia on hsp60, mitochondrial hsp70 (Grp75), hsc70, and proteins of the hsp26 class will be defined in a systematic manner. Studies performed primarily in isolated, dispersed cells have now firmly established the synthesis of the glucose-regulated class of stress proteins (GRPs) during anoxia or glucose deprivation (for review, see Black and Subject 1991). An important future challenge will be studies to define the molecular signals that induce expression of these classes of stress proteins in the intact heart during ischemia and following reperfusion.

Analysis of mitochondrial stress proteins may be of particular interest in the context of ischemic heart disease, because of the enormous respiratory demand placed on cardiomyocytes and the necessity for rapid restoration of mitochondrial function for viability and function of these cells in the post-ischemic period. As discussed in another chapter, proteins encoded by nuclear genes are translated on cytoplasmic ribosomes and subsequently imported into mitochondria by one of several mechanisms, the molecular details of which have been studied extensively in unicellular eukaryotes (Neupert, this volume) and are known to require participation of several members of the heat shock protein gene family. Cytoplasmic hsp70 binds nascent polypeptides destined for mitochondrial import and maintains them in a form competent for recognition by components of the import machinery on the surface of the mitochondrial outer membrane (Craig 1993). Other members of the heat shock family, the mammalian forms of which are termed hsp60 and Grp75, reside within the mitochondrial matrix and are essential for transposition of newly synthesized proteins across mitochondrial membranes and assembly of mitochondrial enzyme complexes (Kang et al. 1990; Lubben et al. 1990; Mizzen et al. 1991). Mitochondrial hsp60 prevents thermal denaturation and irreversible aggregation of preexisting mitochondrial enzymes in vitro (Martin et al. 1992; Hartman et al. 1993), a function that may be of special importance in the heart because of the strict dependence of cardiomyocytes on mitochondrial respiration to meet the high-energy demands of continuous contractile work. Defective expression of hsp60 in humans has been described in an infant dying of severe multisystem failure (Agsteribbe et al. 1993).

The application of transgenic technology (Hanahan 1989) to the study of heat shock proteins in the myocardium should provide more definitive answers to questions concerning the effects of specific heat shock proteins on cell survival following myocardial ischemia. The experimental approaches to this end may not, however, be entirely straightforward. Gain-of-function mutations to overexpress specific heat shock proteins in hearts of transgenic mice, for example, may result in developmental abnormalities, including embryonic lethality, that would preclude assessment of the response to ischemia. The same concerns cloud prospects for analyzing loss-of-function mutations produced by homologous recombination in embryonic stem cells (Capecchi 1989). If such problems occur, investigators will have to resort to more complex genetic engineering strategies or rely on somatic cell gene transfer methods to produce the desired phenotype. Prospects for productive application of transgenic technology to assess the role of heat shock proteins in the ischemic heart have been enhanced by recent efforts to miniaturize experimental systems previously applied to larger animals for application to mice (Ng et al. 1991; Chien 1993) or to modify transgenic methods for application to larger animals (Graves and Moreadith 1993).

From the clinician's point of view, what are the prospects that knowledge of the biology of heat shock proteins may be exploited for the benefit of patients with ischemic heart disease or other cardiac disorders? Additional research on the effects of preconditioning the heart by sublethal thermal stress prior to cardiopulmonary bypass could lead to changes in peri-operative management (Udelsman et al. 1991; Liu et al. 1992). Further research on the mechanisms of induction of stress proteins in the heart could result in new pharmaceuticals to manipulate expression of heat shock proteins in patients undergoing cardiac surgery or experiencing myocardial infarction. If future experiments in experimental animals confirm the putative cytoprotective function of heat shock proteins during myocardial ischemia, one can envisage the development of gene therapy strategies in which native or modified heat shock proteins are manipulated in cells of the heart. Gene therapy protocols using heat shock protein genes will, however, require technical advances to permit targeted delivery of transgenes to cardiomyocytes or endothelial cells within the heart (Gerard and Meidell 1993). Finally, ischemia-inducible promoter elements from heat shock protein genes may be employed to regulate expression of other gene products to be used in genetic strategies to prolong life and limit morbidity in patients with coronary artery disease.

REFERENCES

Abe, K., J.I. Kawagoe, M. Aoki, and K. Kogure. 1993. Dissociation of HSP70 and HSC70 heat shock mRNA inductions as an early biochemical marker of ischemic neuronatal death. *Neurosci. Lett.* **149:** 165–168.

Agsteribbe, E., A. Huckriede, M. Veenhuis, M.H.J. Ruiters, K.E. Neizen-Koning, K. Skjeldal, R.S. Gupta, R. Hallbert, O.P. vanDiggelen, and H.R. Scholte. 1993. A fatal systemic mitochondrial disease with decreased mitochondrial enzyme activities, abnormal ultra structure of the mitochondria and deficiency of heat shock protein 60. *Biochem. Biophys. Res. Commun.* **193:** 146–154.

Allen, D.G. and C.H. Orchard. 1986. Myocardial contractile function during ischemia and hypoxia. *Circ. Res.* **60:** 153–167.

Aoki, M., K. Abe, J.I. Kawagoe, S. Sato, S. Nakamura, and K. Kogure. 1993. Temporal profile of the induction of heat shock protein 70 and heat shock cognate protein 70 mRNAs after transient ischemia in gerbil brain. *Brain Res.* **601:** 185–192.

Beckmann, R.P., M. Lovett, and W.J. Welch. 1992. Examining the function and regulation of hsp70 in cells subjected to metabolic stress. *J. Cell Biol.* **117:** 1137–1150.

Benjamin, I.J., B. Kroger, and R.S. Williams. 1990. Activation of the heat shock transcription factor by hypoxia in mammalian cells. *Proc. Natl. Acad. Sci.* **87:** 6263–6267.

———. 1992. Induction of stress proteins in cultured myogenic cells: Molecular signals for the activation of heat shock transcription factor during ischemia. *J. Clin. Invest.* **89:** 1685–1689.

Black, A.R. and J.R. Subjeck. 1991. The biology and physiology of the heat shock and glucose-regulated stress protein systems. *Methods Achiev. Exp. Pathol.* **15:** 126–166.

Black, S.C. and B.R. Lucchesi. 1993. Heat shock proteins and the ischemic heart. An endogenous protective mechanism. *Circulation* **87:** 1048–1051.

Bolli, R. 1990. Mechanism of a myocardial "stunning." *Circulation* **82:** 724–738.

Braunwald, E. and R.A. Kolner. 1982. The stunned myocardium: Prolonged post-ischemic ventricular dysfunction. *Circulation* **66:** 1146–1149.

Braunwald, E. and B.E. Sobel. 1988. Coronary blood flow and myocardial ischemia. In *Heart disease: A textbook of cardiovascular medicine* (ed. E. Braunwald), pp. 1191–1221. Saunders, Philadelphia.

Brown, I.R. and S.J. Rush. 1993. Expression of heat shock genes (hsp70) in the mammalian brain: Distinguishing constitutively expressed and hyperthermia-inducible mRNA species. *J. Neurosci. Res.* **25:** 14–19.

Buja, L.M., H.K. Hagler, D. Parsons, K. Chien, R.C. Reynolds, and J.T. Willerson. 1985. Alteration of ultrastructure and elemental composition by cultured neonatal rat cardiac myocytes after metabolic inhibition with iodoacetic acid. *Lab. Invest.* **53:** 397–412.

Capecchi, M. 1989. Altering the genome by homologous recombination. *Science* **222:** 1288–1292

Chien, K.R. 1993. Molecular advances in cardiovascular biology. *Science* **260:** 916–917.

Craig, E.A. 1993. Chaperones: Helpers along the pathway to protein folding. *Science* **260:** 1902–1903.

Currie, R.W. 1987. Protein synthesis in heterotopically transplanted rat hearts. *Exp. Cell Biol.* **55:** 46–56.

Currie, R.W. and R.M. Tanguay. 1991. Analysis of RNA for transcripts for catalase and SP71 in rat hearts after *in vivo* hyperthermia. *Biochem. Cell Biol.* **69:** 375–382.

Currie, R.W. and F.P. White. 1981. Trauma-induced protein in rat tissues: A physiological role for a "heat shock" protein? *Science* **214:** 72–73.

———. 1982. Characterization of the synthesis and accumulation of a 71-kilodalton protein induced in rat tissues after hyperthermia. *Can. J. Biochem. Cell Biol.* **61:** 438–446.

Currie, R.W., B.M. Ross, and T.A. Davis. 1988. Heat-shock response is associated with enhanced postischemic ventricular recovery. *Circ. Res.* **63:** 543–549.

Currie, R.W., R.M. Tanguay, and J.G. Kingma, Jr. 1993. Heat-shock response and limitation of tissue necrosis during occlusion/reperfusion in rabbit hearts. *Circulation* **87:** 963–971.

Currie, R.W., V.K. Sharma, S.M. Stepkowski, and R.F. Payce. 1987. Effects of ischaemia and perfusion temperature on the synthesis of stress induced (heat shock) proteins in isolated and perfused rat hearts. *J. Mol. Cell Cardiol.* **19:** 795–808.

Delcayre, C., J.-L. Samuel, F. Marotte, M. Best-Belpomme, J.J. Mercadier, and L. Rappaport. 1988. Synthesis of stress proteins in rat cardiac myocytes 2-4 days after imposition of hemodynamic overload. *J. Clin. Invest.* **82:** 460–468.

Dillmann, W.H., H. Mehta, A. Barrieux, B.D. Guth, W.E. Neeley, and J. Ross. 1986. Ischemia of the dog heart induces the appearance of a cardiac in RNA coding for a protein with migration characteristics similar to heat-shock/stress protein 71. *Circ. Res.* **59:** 110–114.

Donnelly, T.J., R.E. Sievers, F.I.J. Vissern, W.J. Welch, and C.L. Wolfe. 1992. Heat shock protein induction in rat hearts: A role for improved myocardial salvage after ischemia and reperfusion?. *Circ. Res.* **85:** 769–778.

Downing, S.E. and V. Chen. 1992. Acute hibernation and reperfusion of the ischemic heart. *Circulation* **85:** 699–707.

Drummond, I.A.S., S.A. McClure, M. Phoenie, R.Y. Tsien, and R.A. Steinhardt. 1986. Large changes in intracellular pH and calcium observed during heat shock are not responsible for the induction of heat shock proteins in *Drosophila melanogaster*. *Mol. Cell Biol.* **6:** 1767–1775.

Entman, M.L., L. Michael, R.D. Rossen, W.J. Dreyer, D.C. Anderson, and A.A. Taylor. 1991. Inflammation in the course of early myocardial ischemia. *FASEB J.* **5:** 2529–2537.

Fozzard, H.A., E. Haber, R.B. Jennings, A.M. Katz, and H.E. Morgan, eds. 1988. *The heat and carciovascular system.* Raven Press, New York.

Gerard, R.D. and R.S. Meidell. 1993. Adenovirus mediated gene transfer. *Trends Cardiovasc. Med.* **3:** 9–105.

Gething, M.J. and J. Sambrook. 1992. Protein folding in the cell. *Nature* **355:** 33–45.

Graves, K.H. and R.W. Moreadith. 1993. Derivation and characterization of putative pluripotent ES cell lines from preimplantation rabbit embryos. *Mol. Reprod. Dev.* **36:** 424–433.

Hanahan, D. 1989. Transgenic mice as probes into complex systems. *Science* **246:** 1263–1275.

Hartman, D.J., B.P. Surin, N.E. Dixon, N.J. Hoogenraad, and P.B. Hoj. 1993. Substoichiometric amounts of the molecular chaperones GroEL and GroES prevent thermal denaturation and aggregation of mammalian mitochondrial malate dehydrogenase *in vitro. Proc. Natl. Acad. Sci.* **90:** 2276–2280.

Hohlfield, R. and A.G. Engel. 1992. Expression of 65-kd heat shock proteins in the inflammatory myopathies. *Annu. Neurol.* **32:** 821–823.

Howard, G. and T.E. Geoghegan. 1986. Altered cardiac tissue gene expression during acute hypoxic exposure. *Mol. Cell Biochem.* **69:** 155–160.

Hunt, C. and R.I. Morimoto. 1985. Conserved features of eukaryotic hsp70 genes revealed by comparison with the nucleotide sequence of human hsp70. *Proc. Natl. Acad. Sci.* **82:** 6455–6459.

Iwaki, K., S.-H. Chi, W.H. Dillmann, and R. Mestril. 1993. Induction of HSP70 in cultured rat neonatal cardiomyocytes by hypoxia and metabolic stress. *Circulation* **87:** 2023-2032.

Jennings, R.B. and H. E. Morgan. 1988. Strategy of experimental design in metabolic experiments. In *The heart and cardiovascular system* (ed. H.A. Fozzard et al.), pp. 1133-1201. Raven Press, New York.

Kang P-J, J. Ostermann, J. Shilling, W. Neupert, E.A. Craig, and N. Pfanner. 1990. Requirement for hsp70 in the mitochondrial matrix for translocation and folding of precursor proteins. *Nature* **348:** 137-142.

Karmazyn, M., K. Mailer, and R.W. Currie. 1990. Acquisition and decay of heat-shock-enhanced postischemic ventricular recovery. *Am. Physiol. Soc.* **259:** H424-H431.

Katayose, D., I. Shogen, F. Hiroyoshi, and S. Shibahara. 1993. Separate regulation of heme oxygenase and heat shock protein 70 mRNA expression in the rat heart by hemodynamic stress. *Biochem. Biophys. Res. Commun.* **191:** 587-594.

Kleindienst, R., Q. Xu, J. Willeit, F.R. Waldenberger, S.Weimann, and G. Wick. 1993. Immunology of artherosclerosis: Demonstration of heat shock protein 60 expression and T lymphocytes bearing μ/b or gamma/d receptor in human atherosclerotic lesions. *Am. J. Pathol.* **142:** 1927-1937.

Knowlton, A.A., P. Brecher, and C.S. Apstein. 1991a. Rapid expression of heat shock protein in the rabbit after brief cardiac ischemia. *J. Clin. Invest.* **87:** 139-147.

―――. 1991b. A single myocardial stretch or decreased systolic fiber shortening stimulates the expression of heat shock protein 70 in the isolated, erythrocyte-perfused rabbit heart. *J. Clin. Invest.* **88:** 2018-2025.

Lawson, C.S. and J.M. Downey. 1993. Preconditioning: State of the art myocardial protection. *Cardiovasc. Res.* **27:** 542-550.

Lis, J. and C. Wu. 1993. Protein traffic on the heat shock promoter: Parking, stalling, trucking along. *Cell* **74:** 1-4.

Liu, X., R.M. Engelman, I.I. Moraru, J.A. Rousou, J.E. Flack III, D.W. Deaton, N. Maulik, and D.K. Das. 1992. Heat shock: A new approach for myocardial preservation in cardiac surgery. *Circulation* **86:** 358-363.

Lubben, T..H., A.A. Gatenby, G.K. Donaldson, G.H. Lorimer, and P.V. Viitanen. 1990. Identification of a groES-like chaperonin in mitochondria that facilitates protein folding. *Proc. Natl. Acad. Sci.* **87:** 7683-7687.

Marber, M.S., D.S. Latchman, J.M. Walker, and D.M. Yellon. 1993. Cardiac stress protein elevation 24 hours after brief ischemia or heat stress is associated with resistance to myocardial infarction. *Circulation* **88:** 1264-1272.

Martin, J., A.L. Horwich, and F.U. Hartl. 1992. Prevention of protein denaturation under heat stress by the chaperonin Hsp60. *Science* **258:** 995-998.

Mehta, H.B., B.K. Popovich, and W.H. Dillmann. 1988. Ischemia induces changes in the level of mRNAs coding for stress protein 71 and creatine kinase M. *Circ. Res.* **63:** 512-517.

Mestril, R., S.-H Chi, M.R. Sayen, K. O'Reilly, and W.H. Dillmann. 1994. Expression of inducible stress protein 70 in rat heart myogenic cells confers protection against simulated ischemia-induced injury. *J. Clin. Invest.* **93:** (in press).

Mizzen L.A., A.N. Kabiling, and W.J. Welch. 1991. The two mammalian mitochondrial stress proteins, grp75 and hsp58, transiently interact with newly synthesized mitochondrial proteins. *Cell Reg.* **2:** 165-179.

Mosser, D.D., P.T. Kotzbauer, K.D. Sarge, and R.I. Morimoto. 1990. In vitro activation of the heat shock transcription factor DNA-binding by calcium and biochemical conditions that affect protein conformation. *Proc. Natl. Acad. Sci.* **87:** 3748-3752.

Morimoto, R.I. 1993. Cells in stress: Transcriptional activation of heat shock genes. *Science* **259:** 1409-1410.

Murry, C.E., R.B. Jennings, and K.A. Reimer. 1986. Preconditioning with ischemia: A delay of lethal cell injury in ischemic myocardium. *Circulation* 74: 1124–1136.

Ng, W.A., I.L. Grupp, A. Subramanian, and J. Robbins. 1991. Cardiac myosin heavy chain mRNA expression and myocardial function in the mouse heart. *Circ. Res.* 69: 1742–1750.

Reimer, K.A. and R.B. Jennings. 1979. The changing anatomic reference base of evolving myocardial infarction. *Circulation* 4: 866–876.

———. 1985. Animal models for protecting ischemic myocardium: Results of the NHLBI cooperative study. Comparison of unconscious and conscious dog models. *Circ. Res.* 56: 651–665.

———. 1988. Myocardial ischemia hypoxia and infarction. In *The heart and cardiovascular system* (ed. H.A. Fozzard et al.), pp. 1133–1201. Raven Press, New York.

Salo, D.C., C.M. Donovan, and K.J.A. Davies. 1991. Hsp70 and other possible heat shock or oxidative stress proteins are induced in skeletal muscle, heart, and liver during exercise. *Radical Biol. Med.* 11: 239–246.

Sarge, K.D., V. Zimarino, K. Holm, C. Wu, and R.I. Morimoto. 1991. Cloning and characterization of two mouse heat shock factors with distinct inducible and constitutive DNA-binding ability. *Genes Dev.* 5: 1902–1911.

Solomon. J.M., J.M. Rossi, K. Golic, T. McCarry, and S. Lindquist. 1991. Changes in hsp70 alter thermotolerance and heat-shock regulation in *Drosophila*. *New Biol.* 3: 1106–1120.

Sorger, P.K. 1990. Yeast heat shock factor contains separable transient and sustained response transcriptional activators. *Cell* 62: 793–805.

Steare, S.E. and D.M. Yellon. 1993. The protective effect of heat stress against reperfusion arrhythmias in the rat. *J. Mol. Cell Cardiol.* 25: (in press).

Udelsman, R., M.J. Blake, and N.J. Holbrook. 1991. Molecular response to surgical stress: Specific and simultaneous heat-shock protein induction in the adrenal cortex, aorta, and vena cava. *Surgery* 110: 1125–1131.

———. 1993. Vascular heat shock protein expression in response to stress. Endocrine and autonomic regulation of this age-dependent response. *J. Clin. Invest.* 91: 465–473.

Wang, G.L. and G.L. Semenza. 1993. General involvement of hypoxia-inducible factor 1 in transcriptional response to hypoxia. *Proc. Natl. Acad. Sci.* 90: 4303–4308.

White, F.P. and S.R. White. 1986. Isoproterenol induced myocardial necrosis is associated with stress protein synthesis in rat heart and thoracic aorta. *Cardiovasc. Res.* 20: 512–515.

Williams, R.S. and I.J. Benjamin. 1991. Stress proteins and cardiovascular disease. *Mol. Biol. Med.* 8: 197–206.

Williams, R.S., J.A. Thomas, M. Fina, Z. German, and I.J. Benjamin. 1993. Human Hsp70 protects murine cells from injury during metabolic stress. *J. Clin. Invest.* 93: 503–508.

Wing, S.S., H.-L. Chiang, A.L. Goldberg, and J.F. Dice. 1991. Proteins containing peptide sequences related to Lys-Phe-Glu-Arg-Gln are selectively depleted in liver and heart, but not skeletal muscle, of fasted rats. *Biochem. J.* 275: 165–169.

Yellon, D.M., E. Iliodromitis, D.S. Latchman, D.M. VanWinkle, J.M. Downey, F.M. Williams, and T.J. Williams. 1992a. Whole body heat stress fails to limit infarct size in the reperfused rabbit heart. *Cardiovasc. Res.* 26: 342–346.

———. 1992b. The protective role of heat stress in the ischaemic and reperfused rabbit myocardium. *J. Mol. Cell Cardiol.* 24: 895–907.

Zimmermann, L..H., R.A. Levine, and H.W. Farber. 1991. Hypoxia induces a specific set of stress proteins in cultured endothelial cells. *J. Clin. Invest.* 87: 908–914.

21

Postischemic Stress Response in Brain

Thaddeus S. Nowak, Jr.
Department of Neurology
University of Tennessee
Memphis, Tennessee 38163

Hiroshi Abe
Department of Neurosurgery
Brain Research Institute, Niigata University
Niigata City 951, Japan

I. INTRODUCTION

Cerebral ischemia occurs in the context either of focal reductions in blood flow to brain regions (stroke) or as a global deficit after cardiac arrest. In the United States alone, these conditions annually affect 500,000 and more than 1 million individuals, respectively (Wolf et al. 1992). As summarized below, studies in animal models have identified a prominent heat shock/stress response in the brain following such insults. It can be inferred that ischemic injury constitutes one of the primary settings in which the stress response impacts human brain pathophysiology. Parallel studies have documented a postischemic stress response in other tissues including, for example, recent studies in liver (Tacchini et al. 1993) and kidney (Van Why et al. 1992). Substantial work in cardiac ischemia is considered elsewhere in this volume (Benjamin and Williams).

In view of the anatomical complexity of the brain, the distribution of the stress response after ischemia and other insults is a particularly important issue. Most work has relied on immunocytochemistry and in situ hybridization to map the expression of a highly inducible member of the 70-kD heat shock protein family, hsp72, as an index of the stress response in the brain. Studies of other members of this family, as well as of ubiquitin and other heat shock responsive genes, are also beginning to emerge. As detailed below, striking differences in the distribution and cell-type specificity of pathology observed after global and focal ischemia are well correlated with the pattern of hsp72 expression after these insults, confirming a close association between the stress response and cell damage. Apart from its utility as a marker of pathology, it may be presumed that insight into the mechanism of the postischemic stress response may help to identify mechanisms of cellular injury. This is especially significant in view of the hours to days over which brain damage evolves after ischemic insults, during which there may be an opportunity for intervention should such mechanisms be understood. Mechanistic aspects of the postischemic stress response are therefore considered in the context of its relationship to induction of the proto-oncogene, c-*fos*, and other genomic responses to ischemia. Finally, evidence of a robust stress response in cell populations that are destined to survive ischemia and reports of induced tolerance after mild insults suggest that the stress response, together with other changes in gene expression observed after such insults, is likely to be of functional importance in the process of cell survival after ischemia.

II. GLOBAL ISCHEMIA

An early report suggested that circulatory insufficiency increased the expression of a 70-kD protein in diverse tissues, including the brain (Currie and White 1981). This response was later characterized in established models of global cerebral ischemia in the gerbil and rat (Nowak 1985; Dienel et al. 1986; Kiessling et al. 1986), employing in vitro translation or in vivo labeling and two-dimensional gel electrophoresis to demonstrate the increased expression of a protein having properties of an inducible member of the hsp70 family. This induced protein has been designated "hsp72" in many subsequent studies based on its comigration and immunological identity with the previously characterized stress protein (Welch and Suhan 1986; Vass et al. 1988; Watowich and Morimoto 1988). Blot hybridizations demonstrated increased levels of a corresponding mRNA detected either with an oligonucleotide selective for an inducible hsp70 species (Nowak et al. 1990a; Miller et al. 1991) or with less selective probes (Abe et al. 1991). Several studies have also identified a moderate increase in the level of heat shock cognate (hsc70) mRNA expression after ischemia (Abe et al. 1991; Kawagoe et al. 1992c; Aoki et al. 1993b), although early studies did not indicate significant changes in its expression at the translational level (Nowak 1985).

The characteristic feature of brain pathology after brief global ischemia, both in humans and in animal models, is discrete neuron loss involving selectively vulnerable cell populations (Ito et al. 1975; Kirino
1982; Pulsinelli et al. 1982; Smith et al. 1984; Ross and Duhaime 1989;
Ross and Graham 1993). Most extensively studied is the delayed death of
CA1 pyramidal neurons in hippocampus that occurs during the interval
of 2–4 days recirculation after the initial insult (Kirino 1982; Pulsinelli et
al. 1982). hsp72 expression after global ischemia is also selectively distributed in neurons, with the duration of mRNA induction directly correlated with the relative vulnerability of a given population (Nowak 1991;
Kawagoe et al. 1992c). This is illustrated in Figure 1, showing the time
course of hsp72 mRNA expression after 5 minutes ischemia in the gerbil
characterized by in situ hybridization with a ^{35}S-labeled oligonucleotide
probe. hsp72 is initially expressed in all of the major hippocampal
neuron populations but is sequentially lost from the less vulnerable

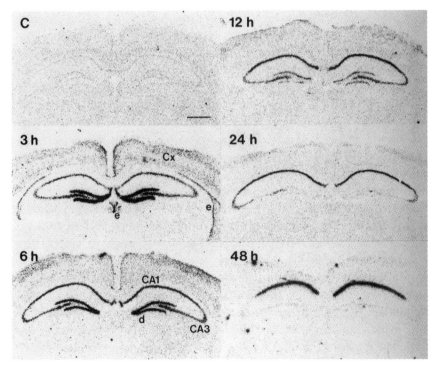

Figure 1 In situ hybridization of hsp72 mRNA expression in gerbil hippocampus after transient ischemia, illustrating prolonged expression in the selectively vulnerable CA1 pyramidal neurons. Signal is progressively lost from
dentate granule cells (d) and CA3 neurons that survive the insult. Ependymal
cells (e) that constitutively express hsp72 in the gerbil also show transient induction after ischemia. (Reprinted, with permission, from Nowak 1991.)

dentate granule cells and CA3 pyramidal neurons, whereas the signal is persistently expressed in CA1 neurons essentially until their eventual disappearance. Ependymal cells lining the ventricles constitutively express hsp72 immunoreactivity in the gerbil (Vass et al. 1988) but also show transiently elevated hsp72 mRNA levels after ischemia. hsp72 mRNA is expressed in other brain regions where neuronal damage can occur after ischemia (not shown), notably including dorsolateral striatum and cortex (Nowak 1991). The precise distribution and time course of mRNA expression varies with the severity of the insult in a particular model, with more lasting and more diffusely distributed hybridization generally observed under conditions resulting in increasing damage (Blumenfeld et al. 1992; Kawagoe et al. 1992c; Nowak et al. 1992).

Detection of postischemic hsp72 expression at the protein level is often more restricted after global ischemia than would be expected from the above hybridization studies. In the gerbil model, in particular, there is negligible accumulation of hsp72 immunoreactivity in CA1 after ischemia of a duration that injures this neuron population (Fig. 2a) (Vass et al. 1988). This discrepancy is somewhat less pronounced in most rat ischemia models in which significant hsp72 immunoreactivity accumulates in CA1 neurons (Chopp et al. 1991; Simon et al. 1991), although its expression in CA1 is usually attenuated and is less pronounced than in the adjacent CA3. A primary factor contributing to this difference between hsp72 mRNA and protein expression in vulnerable neurons is likely to be the well-documented prolonged deficit in translational activity that is particularly pronounced in the gerbil (Thilmann et al. 1986) and also documented in the rat (Dienel et al. 1980; Widmann et al. 1991). We have recently demonstrated that ischemia in an established rat model results in quantitatively less pronounced deficits in overall translational activity in CA1 neurons than observed in the gerbil, consistent with the better translational expression of hsp72 (M. Okawa et al., in prep.). As a practical issue, it is evident that there are species differences in the ischemic thresholds for severe translational deficits, since increasing ischemic duration eventually results in a further attenuation in CA1 hsp72 immunoreactivity in the rat (Simon et al. 1991). Conversely, short ischemic insults in both rat and gerbil result in preferential expression of immunoreactive hsp72 in CA1 (Kirino et al. 1991; Simon et al. 1991), shown in Figure 2b for the gerbil. This is considered more fully below in the context of induced ischemic tolerance. Recent evidence indicates that pharmacological interventions that reduce ischemic damage are associated with a shift toward the pattern of expression seen after shorter periods of ischemia (Bergstedt 1993).

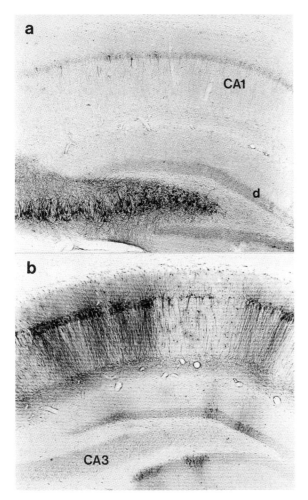

Figure 2 Immunocytochemical localization of hsp72 at 48 hr of recirculation after transient ischemia in the gerbil. (*a*) Ischemia of 5-min duration results in strong CA3 staining but minimal accumulation of hsp72 in CA1 neurons that show prolonged expression of the encoding mRNA (see Fig. 1). More transient expression in dentate granule cells (d) is no longer evident at this relatively late time point. (*b*) Brief 2-min ischemia gives rise to selective hsp72 staining in CA1 neurons, with patches of persistent expression in the dentate gyrus. Areas of reduced staining in CA1 correspond to regions of early neuron loss in which scattered, small hsp72-positive glia are detected.

It is likely that other factors must influence the detection of hsp72 protein expression after ischemia. There is a striking lag in the detection of hsp72 immunoreactivity, even in hippocampal neuron populations such as dentate and CA3 that eventually express detectable levels of the protein (Vass et al. 1988). This delay in detection extends well beyond the recirculation intervals at which translational recovery has been documented to occur (Thilmann et al. 1986) and during which other induced proteins such as Fos and Jun transcription factors have been shown to be expressed (Nowak et al. 1993; S. Suga and T.S. Nowak, Jr., in prep.). Cryptic expression of 70-kD stress proteins also has been documented in cell cultures, depending on the specific antibodies used for detection (Milarski et al. 1989), and it has been suggested that this may reflect a masking of epitopes involved in protein-protein interactions of functional importance in the stress response. Difficulties in hippocampal hsp72 detection have also been reported in blot studies under some conditions (Dwyer et al. 1989).

hsp72 expression in nonneuronal cells is a major component of the response to focal ischemia (see below) but has also been detected under some conditions after global ischemia. Usually, this has been observed under conditions of more severe global insults that result in focal areas of damage involving multiple cell types (Gonzalez et al. 1991; Sharp et al. 1991). However, hsp72-positive glia can sometimes be detected in regions of selective neuronal damage after relatively mild ischemia (Fig. 2b) and other insults (Sloviter and Lowenstein 1992). Whether this relatively infrequently reported finding reflects a potential for injury to these cells or occurs transiently as a component of the transition to "reactive" astrocytes or microglia remains to be determined.

From the above observations, it can be concluded that hsp72 is not a selective marker for neuron populations that will necessarily be lost after global ischemia. Rather, most cells that initially express hsp72 transcripts eventually also express detectable immunoreactive protein and go on to survive. Robust, delayed expression of hsp72 immunoreactivity in surviving neurons after ischemia may therefore be suggestive of recovery from a period of active stress (Vass et al. 1988). However, even neurons that accumulate significant levels of hsp72 protein can be lost (Chopp et al. 1991; Simon et al. 1991; Deshpande et al. 1992). Recent results in an electrical stimulation paradigm demonstrate directly that hsp72 expression may best be considered as an indicator of potential injury, recovery from which is dependent on the timely cessation of the inducing insult (Sloviter and Lowenstein 1992). Conversely, not all postischemic injury mechanisms are associated with a detectable stress response. Neurons of the reticular thalamic nucleus are very rapidly lost in a rat cardiac arrest model (Kawai et al. 1992; Mies et al. 1993), and under these conditions, even hsp72 hybridization fails to be detected (N. Saito et al., in prep.). Together, these results suggest that hsp72 expression preferentially indicates those conditions resulting in delayed rather than acute cell damage after global ischemia.

Relatively few studies have examined other components of the stress response after ischemia. Early two-dimensional gels of in vivo translation products identified several additional species showing increased expression (Kiessling et al. 1986). Ubiquitin mRNA is also induced after global ischemia, although the increases are modest relative to control expression (Nowak et al. 1990a). The pattern of ubiquitin-conjugate immunoreactivity demonstrates a striking depletion of the control staining in all major hippocampal neuron populations, with a progressive recovery except in the vulnerable CA1 region (Magnusson and Wieloch 1989), generally paralleling the time course observed for hsp72 detection in most models. Blot studies of ubiquitin conjugates in a particulate brain fraction demonstrated a prolonged increase in the levels of detectable ubiquitin conjugates, indicating activation of the ubiquitination pathway after ischemia (Hayashi et al. 1992a,b). Recent studies have demonstrated induction of heme oxygenase after ischemia/reperfusion in the kidney (Maines et al. 1993) as well as in the liver (Tacchini et al. 1993) and have identified a predominantly glial response to oxidative stress in brain (Ewing and Maines 1993). Effects of transient ischemia on heme oxygenase in the brain remain to be examined. Clearly, there is more to be done to characterize the spectrum of heat shock proteins associated with the postischemic stress response.

III. FOCAL ISCHEMIA

In contrast to the selective neuronal injury characteristic of transient global ischemia, long-lasting focal ischemic insults lead to infarction, with injury to essentially all cell types within the affected territory. Around the core of severely deprived tissue is a "penumbra" of intermediate flow values in which a graded cellular response is observed (Astrup et al. 1981). The distribution of hsp72 expression in focal ischemia models correlates well with this distribution of pathology. At relatively early intervals following occlusion of a major vessel, hsp72 hybridization is concentrated in the border zone surrounding the ischemic core, with relatively little mRNA expressed within regions of severe flow disruption (Welsh et al. 1992). This is consistent with the relatively complete failure of substrate availability for energy metabolism within this territory, precluding significant transcriptional activity. Upon recirculation after prolonged focal ischemia, there is a pronounced increase in hybridization signal within this ischemic core (Kawagoe et al. 1992b; Welsh et al. 1992; Kinouchi et al. 1993), but even under conditions of permanent

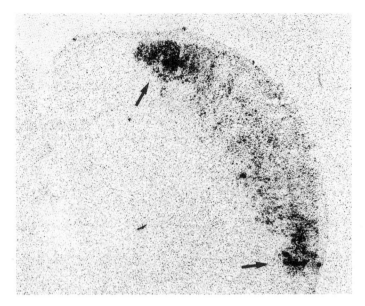

Figure 3 Localization of hsp72 mRNA expression in a model of focal ischemia in the rat. The hybridization signal is detected in a circumscribed region of the ischemic cortex. The most pronounced expression is evident in areas of significant residual perfusion at the periphery of the infarct (arrows), with greatly attenuated expression in the ischemic core.

arterial occlusion, there can be a progressive but modest increase in hybridization signal in such a region (Welsh et al. 1992), indicating the presence of at least some blood flow via collateral circulation. The pattern of hsp72 hybridization observed after 24 hours of middle cerebral artery occlusion in the spontaneously hypertensive rat (SHR) model is illustrated in Figure 3 (M. Jacewicz et al., unpubl.). Parallel increases in the constitutively expressed hsc70 have also been described after focal ischemia (Kawagoe et al. 1992b). It should be mentioned that in some models, significant hsp72 mRNA can be expressed in well-perfused regions remote from the ischemic focus (e.g., hippocampus) (Welsh et al. 1992). The possibility that there may be distinct mechanisms responsible for local versus *trans*-synaptic induction of the stress response after focal ischemia is considered in more detail below. Following either permanent or temporary focal insults, the hybridization signal is gradually lost during an interval of several days (Welsh et al. 1992; Kinouchi et al. 1993).

There is a striking variation in the cellular localization of the stress response in the different regions of an evolving infarct as evidenced by hsp72 immunocytochemistry (Gonzalez et al. 1989; Li et al. 1992; Kinouchi et al. 1993). Within the ischemic core, immunoreactive hsp72 is found almost exclusively within vascular endothelial cells, which constitute the major surviving cell type in this region. At the edge of an infarct is a rim of hsp72-positive glia that have been considered by some to be astrocytes (Kinouchi et al. 1993), but that others have identified as microglia (Li et al. 1992). The bulk of expression in the surrounding region is found in neurons. Shorter periods of focal ischemia that are compatible with the subsequent survival of cells within the injury focus result in more striking hsp72 expression in neurons throughout the reperfused territory (Kinouchi et al. 1993). It should be noted that a generally comparable pattern of hsp72 expression in diverse cell types occurs following a number of other focal injuries in nervous tissue (Brown et al. 1989; Gower et al. 1989; Tanno et al. 1993). The above observations indicate that immunoreactive hsp72 preferentially accumulates in surviving cell populations after focal insults, as shown previously for global ischemia. A region of impaired protein synthesis extends beyond the zone of severe focal blood flow deficit and appears to define the range of the final infarct (Jacewicz et al. 1986; Xie et al. 1989; Mies et al. 1991), suggesting that translational impairment is also a significant factor limiting accumulation of hsp72 and other induced proteins after focal ischemia. The continued increase in neuronal hsp72 staining intensity noted between 8 and 24 hours in the region surrounding a focal insult (Kinouchi et al. 1993) suggests that there may be a lag in hsp72 detectability in recovering neurons, again comparable to that seen after global ischemia, although this has not been sytematically evaluated. Similarly, it remains to be explicitly determined whether a subpopulation of stressed neurons expressing hsp72 in the periphery of the lesion is eventually lost as the focal injury evolves.

IV. REGULATION OF THE POSTISCHEMIC STRESS RESPONSE

Mechanisms regulating the postischemic induction of heat shock genes remain to be fully defined, but as detailed below, a number of relevant observations have been made in the case of neuronal expression after global ischemia. Recent experiments confirm the activation of heat shock factor (HSF) in hippocampus in the gerbil model, indicating the proximal involvement of this established mechanism in the postischemic stress response. Further information regarding signal transduction pathways that are potentially involved either directly in HSF activation or in synergistic effects on hsp72 expression may be inferred from character-

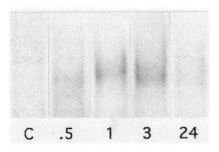

C   .5   1   3   24

Figure 4 Time course of HSF activation following transient ischemia. Gel-shift assays of nuclear extracts from gerbil hippocampus demonstrate a detectable increase in heat shock element binding by 30 min, with significant activity remaining at 24 hr.

izations of induction thresholds and coordinate regulation of other responsive genes.

Given the multiple regulatory elements that have been demonstrated upstream of characterized *hsp70* genes (Wu et al. 1986; Hunt and Calderwood 1990; Choi et al. 1991), it is conceivable that mechanisms distinct from HSF activation could be involved in the postischemic stress response. However, as shown in Figure 4, preliminary studies employing electrophoretic mobility shift assays of gerbil hippocampal extracts demonstrate a prolonged increase in the activity of a factor binding a synthetic heat shock consensus sequence (W. Valentine et al., unpubl.), paralleling the long time course of hsp72 mRNA expression in this tissue illustrated above. This is consistent with the HSF activation previously demonstrated following hypoxia/ischemia in cultured muscle cells (Benjamin et al. 1990) and implicates HSF in the induction of hsp72 and other heat shock genes after ischemia in vivo.

There is strong circumstantial evidence for the involvement of receptor-mediated signaling pathways in injury mechanisms and the stress response in the brain and other organs. The observation that hsp72 induction and HSF activation can occur in response to behavioral stress (Blake et al. 1991; Holbrook and Udelsman, this volume) suggests that the heat shock response may be generated in several tissues by signaling mechanisms that are not associated with a direct environmental challenge to these cells. The hsp72 induction that occurs in neurons of cingulate cortex in response to the *N*-methyl-D-aspartate (NMDA) antagonist, MK-801, is attenuated by pharmacological intervention with other classes of receptor blockers, indicating the potential involvement of complex neuronal input in the inducing stimulus (Olney et al. 1991; Sharp et al. 1992). Prolonged seizure activity in response to kainic acid or other agents results in neuronal injury and accompanying expression of hsp72

(Uney et al. 1988; Gonzalez et al. 1989; Vass et al. 1989; Lowenstein et al. 1990). Potentiating electrical stimulation of hippocampal slices is associated with hsp72 induction, although inconsistently so (Mackler et al. 1992), suggesting that it may arise secondary to stimulation-induced injury as described in vivo (Sloviter and Lowenstein 1992). All of these observations are consistent with proposed "excitotoxic" mechanisms of neuronal injury after ischemia (Choi and Rothman 1990), although the nature of the transition from physiological to pathological stimulus remains to be defined.

hsp72 is coexpressed with c-*fos* after global ischemia and in many other stimulation paradigms in which both have been examined (Andrews et al. 1987; Gubits and Fairhurst 1988; Nowak et al. 1990b; Schiaffonati et al. 1990; Nowak 1991; Wessel et al. 1991). Although there is considerable dissociation of c-*fos* and hsp72 expression after focal ischemia (Welsh et al. 1992), the region of overlap with hsp72 expression in the lesion periphery constitutes a zone of c-*fos* induction that appears to be mechanistically dictinct from that observed in areas remote from the lesion (Uemura et al. 1991b; Gass et al. 1992). Together, these observations demonstrate significant overlap in the signals mediating c-*fos* and hsp72 induction in neurons after both global and focal ischemia. As noted above, the presence of both $Ca^{++}/cAMP$ and serum response elements in c-*fos* and in those hsp70 genes that have been characterized (Voellmy et al. 1985; Wu et al. 1986; Visvader et al. 1988; Hunt and Calderwood 1990; Leung et al. 1990; Sheng and Greenberg 1990) provides common mechanisms by which this coexpression could be mediated. Coordinate changes in the expression of a growing number of other genes have been described after ischemia in the brain and other tissues (Dempsey et al. 1991; Müller et al. 1991; Lindvall et al. 1992; Matsuyama et al. 1993b; Saito and Nowak 1993), which may eventually assist in the identification of regulatory pathways.

Progress in the understanding of HSF activation in other systems also suggests direct mechanisms by which HSF itself could be regulated by signaling pathways of particular relevance to the postischemic stress response. It has been shown that HSF can be phosphorylated (Larson et al. 1988; Sorger and Pelham 1988) and that DNA-binding activity can be dissociated from transcriptional activation (Sorger and Pelham 1988; Price and Calderwood 1991; Sarge et al. 1993). There is evidence for involvement of a calcium-dependent step in acquisition and maintenance of DNA-binding capability, although transcriptional activation apparently requires ATP and tyrosine kinase activity (Price and Calderwood 1991). Increased calcium influx after ischemia is well established and has long been considered a factor in ischemic injury (Siesjö and Beng-

tsson 1989). Recent studies directly demonstrate tyrosine phosphorylation of microtubule-associated protein kinase after ischemia (Campos-
González and Kindy 1992), and there is a suggestion that tyrosine kinase
inhibitors may be neuroprotective (Kindy 1993), although the mechanisms of tyrosine kinase activation and its distribution relative to ischemic
injury and the postischemic stress response remain to be established. In
view of the identified interactions between the heat shock and Ca^{++}/
cAMP response elements in a human hsp70 promoter (Choi et al. 1991),
it remains possible that such interactions also contribute to the postischemic stress response. Although several distinct HSF activities have now
been defined (Scharf et al. 1990; Rabindran et al. 1991; Schuetz et al.
1991; Nakai and Morimoto 1993), their possible differential roles have
yet to be investigated in the context of ischemia.

V. ISCHEMIC TOLERANCE

A number of studies have now demonstrated a phenomenon of induced
ischemic tolerance comparable to the thermotolerance observed in many
experimental systems after mild heat shock. The most striking effects
have been observed following brief intervals of global ischemia, resulting in significant protection of CA1 neurons after subsequent longer insults (Kitagawa et al. 1990; Kato et al. 1991; Kirino et al. 1991; Liu et al.
1992). Injury in other brain regions can also be protected (Kitagawa et al.
1991b). Some reports have also described induction of ischemic tolerance after hyperthermic stresses in vivo (Chopp et al. 1989; Kitagawa et
al. 1991a) or after systemic oxidative stress (Ohtsuki et al. 1992). In the
case of prior ischemic challenges, the available data convincingly
demonstrate expression of hsp72 in the otherwise vulnerable neurons that
are protected following such treatments (Kirino et al. 1991; Simon et al.
1991). hsp72 was shown to be induced in hippocampus following protective oxidative stress, although a cellular localization was not established
(Ohtsuki et al. 1992). In these studies, the period of effective tolerance
spans an interval of 1–7 days following the initial challenge, corresponding roughly to the duration of hsp72 expression in the various models. It
is somewhat more difficult to invoke a direct role for the stress response
in heat-induced ischemic tolerance since hsp72 induction appears to be
preferentially localized in glial and vascular cells in the brain after hyperthermia in vivo (Sprang and Brown 1987; Marini et al. 1990), although
modest expression can be detected in hippocampal neurons under some
conditions (Blake et al. 1990; Pardue et al. 1992). There is precedent for
cross-tolerance in other in vivo injury models, as observed in the case of

hyperthermia-induced protection against light damage in the retina (Barbe et al. 1988). In addition, hyperthermia has been reported to induce tolerance to glutamate toxicity in cultured neurons (Lowenstein et al. 1991; Rordorf et al. 1991), with clear relevance to potential excitotoxic mechanisms of ischemic injury, and there is also evidence that excito-toxic agents can induce self-tolerance (Marini and Paul 1992). Several studies now indicate that hsc70 mRNA is increased at lower thresholds than required for hsp72 induction following either hyperthermic (Miller et al. 1991) or ischemic stresses (Kawagoe et al. 1992a), raising the pos-sibility that its expression may be of particular importance in the context of induced tolerance in some of these models. Recent results have noted more rapid postischemic induction of both hsp70 (hsp72) and hsc70 mRNAs in tolerant versus naive animals, accompanied by more rapid detection of hsp72 immunoreactivity (Aoki et al. 1993a).

The generation of threshold ischemic insults sufficient to induce a stress response but not of a severity that results in significant injury presents a considerable challenge in an animal model. Ischemia of even 2 minutes duration can result in some injury in CA1 (Fig. 2b), although this appears to be more evident in the gerbil than in the rat, consistent with the differences in injury thresholds established in these models. Recent studies of hsp72 expression after brief ischemia have contributed to a better understanding of a number of variables that influence isch-emic severity, with a particular focus on temperature. It is well estab-lished that temperature during ischemia is a critical determinant of the resulting damage (Churn et al. 1990; Dietrich et al. 1990; Welsh et al. 1990), and reductions in temperature during recirculation can slow the rate of cell loss (Morse and Davis 1990; Dietrich et al. 1993). There is also evidence that moderate, transient increases in temperature that occur spontaneously during early recirculation in some models can affect the extent of injury (Kuroiwa et al. 1990), although conflicting results have also been obtained (Welsh and Harris 1991; Dietrich et al. 1993). We have found that postischemic hyperthermia is a major factor affecting the stress response after brief ischemia in the gerbil, with temperatures above 39°C during the interval between the 30- and 90-minute recirculation re-quired to induce detectable hsp72 mRNA expression after threshold in-sults (Suga and Nowak 1993). We have also examined hsp72 immuno-reactivity after brief ischemia and confirmed a requirement for similar temperature elevation to produce strong hsp72 staining in the gerbil (H. Abe and T. S. Nowak, Jr., unpubl.). The response illustrated in Figure 2b was produced with a peak postischemic temperature of 39.8°C. This temperature is below that usually required for robust thermal activation of the heat shock response in brain (Nowak et al. 1990a; Miller et al.

1991) and results in the selective neuronal distribution characteristic of ischemic insults, rather than the glial and vascular expression that predominates after hyperthermia (Sprang and Brown 1987; Marini et al. 1990). Although the temperature dependence of hsp72 induction after brief ischemia appears more likely to reflect the thermal sensitivity of mechanisms leading to neuronal damage, a selective lowering of the set point of neuronal HSF activation after ischemia remains a possible interpretation. As a practical issue, these results should provide a basis for better controlled models of ischemic tolerance and further emphasize the importance of the early recirculation period as an interval during which ischemic outcome may be significantly influenced.

In addition to the stress response, changes in apparent neuronal vulnerability may potentially involve complex interactions of other mechanisms within cells and within the organism as a whole. In terms of responses intrinsic to the affected neurons, there is good evidence that many other changes in gene expression occur in CA1 neurons following mild ischemic insults. As discussed above, a number of immediate early genes show increased transcription following ischemia (Jørgensen et al. 1989; Nowak et al. 1990b; Wessel et al. 1991; Gubits et al. 1993). Whereas expression of Fos protein is limited in vulnerable neurons after ischemia (Uemura et al. 1991a), Jun transcription factor immunoreactivity has been shown to be strikingly increased in CA1 neurons after brief, but not after severe, insults in the gerbil (Suga and Nowak 1991; S. Suga and T.S. Nowak, Jr., in prep.), consistent with the possible involvement of further cascades of altered gene expression under conditions that produce tolerance. There is evidence that tolerance can be blocked independent of effects on hsp72 expression (Kato et al. 1992). Further studies are clearly necessary to identify the range of transcriptional and translational responses that may characterize the tolerant state.

As an additional caution, there may be mechanisms distinct from direct effects of priming challenges on vulnerable neurons by which brief ischemia could influence the outcome of later insults. For example, it is well known that a subpopulation of hippocampal neurons in dentate hilus is also highly vulnerable to ischemic injury (Johansen et al. 1987). A recent study suggests that these neurons are lost even after brief 2-minute insults that leave CA1 neurons largely unaffected (Matsuyama et al. 1993a). Although the effect of this particular lesion in the hippocampal circuitry remains to be fully characterized, a number of other selective ablations of neuronal circuitry have been reported to result in protection of CA1 neurons (Johansen et al. 1986; Jørgensen et al. 1987; Benveniste et al. 1989). Removal of stimulating inputs that may be involved in triggering postischemic injury may therefore also contribute to tolerance.

Finally, as noted above, a range of physiological variables including temperature can critically influence the impact of an ischemic insult, and for the most part, these have yet to be critically examined in tolerance studies. Therefore, although it is to be expected that the stress response and other changes in gene expression are likely to have significant roles in affecting the survival and function of cells after ischemia, the interactions between these and other variables affecting ischemic severity remain to be fully evaluated.

VI. SUMMARY

It is apparent that the heat shock/stress response, for the most part using hsp72 as the representative indicator, is closely associated with cellular injury in the brain after both global and focal ischemia. Particularly in the case of global ischemia, the mapping of stressed cells with this technique identifies not only those highly vulnerable neurons that are likely to be lost, but also a number of cell populations that generally survive the insult. The stress response therefore appears to be a sensitive indicator of cells at risk. In both global and focal insults, there is generally better detection of mRNA than protein in injured cells, consistent in part with the involvement of translational deficits as a component of injury, but possibly also reflecting masking of hsp72 immunoreactivity during a period of active stress.

A number of studies have shown that tolerance to ischemic insults can result following prior challenges that also induce the stress response. Many other transcriptional and translational responses are known to occur under such conditions, and it is premature to assign specific importance to any given component of these responses, but it is clear that expression of hsp72 and other stress proteins is one characteristic of the tolerant state.

Heat shock factor activation has been demonstrated after ischemia, although it remains a possibility that factors involved in other postischemic transcriptional responses could also interact with the stress response. Available evidence suggests a good correlation between identified steps in heat shock factor activation and known changes in signal transduction pathways after ischemia, but these connections remain to be clearly established.

Future studies may be expected to examine more fully the range of proteins that constitute the postischemic stress response, to identify the signal transduction mechanisms that are responsible for its induction, and to clarify the functional roles of stress proteins within the ever broadening context of diverse changes in gene expression that occur in response to injury.

REFERENCES

Abe, K., R.E. Tanzi, and K. Kogure. 1991. Induction of HSP70 mRNA after transient ischemia in gerbil brain. *Neurosci. Lett.* **125:** 166–168.

Andrews, G.K., M.A. Harding, J.P. Calvet, and E.D. Adamson. 1987. The heat shock response in HeLa cells is accompanied by elevated expression of the c-*fos* proto-oncogene. *Mol. Cell Biol.* **7:** 3452–3458.

Aoki, M., K. Abe, J. Kawagoe, S. Nakamura, and K. Kogure. 1993a. Acceleration of HSP70 and HSC70 heat shock gene expression following transient ischemia in the preconditioned gerbil hippocampus. *J. Cereb. Blood Flow Metab.* **13:** 781–788.

Aoki, M., K. Abe, J. Kawagoe, S. Sato, S. Nakamura, and K. Kogure. 1993b. Temporal profile of the induction of heat shock protein 70 and heat shock cognate protein 70 mRNAs after transient ischemia in gerbil brain. *Brain Res.* **601:** 185–192.

Astrup, J., B.K. Siesjö, and L. Symon. 1981. Thresholds in cerebral ischemia. The ischemic penumbra. *Stroke* **12:** 723–725.

Barbe, M.F., M. Tytell, D.J. Gower, and W.J. Welch. 1988. Hyperthermia protects against light damage in the rat retina. *Science* **241:** 1817–1820.

Benjamin, I.J., B. Kröger, and R.S. Williams. 1990. Activation of the heat shock transcription factor by hypoxia in mammalian cells. *Proc. Natl. Acad. Sci.* **87:** 6263–6267.

Benveniste, H., M.B. Jørgensen, M. Sandberg, T. Christensen, H. Hagberg, and N.H. Diemer. 1989. Ischemic damage in hippocampal CA1 is dependent on glutamate release and intact innervation from CA3. *J. Cereb. Blood Flow Metab.* **9:** 629–639.

Bergstedt, K. 1993. "Ischemic and hypoglycemic brain damage. Studies on protein synthesis and heat-shock protein expression in the rat brain." Ph.D. thesis, Lund University, Sweden.

Blake, M.J., T.S. Nowak, Jr., and N.J. Holbrook. 1990. In vivo hyperthermia induces expression of HSP70 mRNA in brain regions controlling the neuroendocrine response to stress. *Mol. Brain Res.* **8:** 89–92.

Blake, M.J., R. Udelsman, G.J. Feulner, D.D. Norton, and N.J. Holbrook. 1991. Stress-induced heat shock protein 70 expression in adrenal cortex: An adrenocorticotropic hormone-sensitive, age-dependent response. *Proc. Natl. Acad. Sci.* **88:** 9873–9877.

Blumenfeld, K.S., F.A. Welsh, V.A. Harris, and M.A. Pesenson. 1992. Regional expression of c-*fos* and heat shock protein-70 mRNA following hypoxia-ischemia in immature rat brain. *J. Cereb. Blood Flow Metab.* **12:** 987–995.

Brown, I.R., S. Rush, and G.O. Ivy. 1989. Induction of a heat shock gene at the site of tissue injury in the rat brain. *Neuron* **2:** 1559–1564.

Campos-González, R. and M. Kindy. 1992. Tyrosine phosphorylation of microtubule-associated protein kinase after transient ischemia in the gerbil brain. *J. Neurochem.* **59:** 1955–1958.

Choi, D. and S.M. Rothman. 1990. The role of glutamate neurotoxicity in hypoxic-ischemic neuronal death. *Annu. Rev. Neurosci.* **13:** 171–182.

Choi, H.-K., B. Li, Z. Lin, L.E. Huang, and A.Y.-C. Liu. 1991. cAMP and cAMP-dependent protein kinase regulate the human heat shock protein 70 gene promoter activity. *J. Biol. Chem.* **266:** 11858–11865.

Chopp, M., Y. Li, M.O. Dereski, S.R. Levine, Y. Yoshida, and J.H. Garcia. 1991. Neuronal injury and expression of 72-kDa heat-shock protein after forebrain ischemia in the rat. *Acta Neuropathol.* **83:** 66–71.

Chopp, M., H. Chen, K.-L. Ho, M.O. Dereski, E. Brown, F.W. Hetzel, and K.M.A. Welch. 1989. Transient hyperthermia protects against subsequent forebrain ischemic cell damage in the rat. *Neurology* **39:** 1396–1398.

Churn, S.B., W.C. Taft, M.S. Billingsley, R.E. Blair, and R.J. DeLorenzo. 1990. Temperature modulation of ischemic neuronal death and inhibition of calcium/calmodulin-dependent protein kinase II in gerbils. *Stroke* **21**: 1715–1721.

Currie, R.W. and F.P. White. 1981. Trauma-induced protein in rat tissues: A physiological role for a "heat shock" protein? *Science* **214**: 72–73.

Dempsey, R.J., J.M. Carney, and M.S. Kindy. 1991. Modulation of ornithine decarboxylase mRNA following transient ischemia in the gerbil. *J. Cereb. Blood Flow Metab.* **11**: 979–985.

Deshpande, J., K. Bergstedt, T. Lindén, H. Kalimo, and T. Wieloch. 1992. Ultrastructural changes in the hippocampal CA1 region following transient cerebral ischemia: Evidence against programmed cell death. *Exp. Brain Res.* **88**: 91–105.

Dienel, G.A., W.A. Pulsinelli, and T.E. Duffy. 1980. Regional protein synthesis in rat brain following acute hemispheric ischemia. *J. Neurochem.* **35**: 1216–1226.

Dienel, G.A., M. Kiessling, M. Jacewicz, and W.A. Pulsinelli. 1986. Synthesis of heat shock proteins in rat brain cortex after transient ischemia. *J. Cereb. Blood Flow Metab.* **6**: 505–510.

Dietrich, W.D., R. Busto, I. Valdes, and Y. Loor. 1990. Effects of normothermic versus mild hyperthermic forebrain ischemia in rats. *Stroke* **21**: 1318–1325.

Dietrich, W.D., R. Busto, O. Alonso, M.Y.-T. Globus, and M.D. Ginsberg. 1993. Intraischemic but not postischemic brain hypothermia protects chronically following global forebrain ischemia in rats. *J. Cereb. Blood Flow Metab.* **13**: 541–549.

Dwyer, B.E., R.N. Nishimura, and I.R. Brown. 1989. Synthesis of the major inducible heat shock protein in rat hippocampus after neonatal hypoxia-ischemia. *Exp. Neurol.* **104**: 28–31.

Ewing, J.F. and M.D. Maines. 1993. Glutathione depletion induces heme oxygenase-1 (HSP32) mRNA and protein in rat brain. *J. Neurochem.* **60**: 1512–1519.

Gass, P., M. Spranger, T. Herdegen, R. Bravo, P. Köck, W. Hacke, and M. Kiessling. 1992. Induction of FOS and JUN proteins following focal ischemia in the rat cortex: Differential effect of MK-801. *Acta Neuropathol.* **84**: 545–553.

Gonzalez, M.F., D. Lowenstein, S. Fernyak, K. Hisanaga, R. Simon, and F.R. Sharp. 1991. Induction of heat shock protein 72-like immunoreactivity in the hippocampal formation following transient global ischemia. *Brain Res. Bull.* **26**: 241–250.

Gonzalez, M.F., K. Shiraishi, K. Hisanaga, S.M. Sagar, M. Mandabach, and F.R. Sharp. 1989. Heat shock proteins as markers of neural injury. *Mol. Brain Res.* **6**: 93–100.

Gower, D.J., C. Hollman, S. Lee, and M. Tytell. 1989. Spinal cord injury and the stress protein response. *J. Neurosurg.* **70**: 605–611.

Gubits, R.M. and J.L. Fairhurst. 1988. c-*fos* mRNA levels are increased by the cellular stressors, heat shock and sodium arsenite. *Oncogene* **3**: 163–168.

Gubits, R.M., R.E. Burke, G. Casey-McIntosh, A. Bandele, and F. Munell. 1993. Immediate early gene induction after neonatal hypoxia-ischemia. *Mol. Brain Res.* **18**: 228–238.

Hayashi, T., K. Takada, and M. Matsuda. 1992a. Post-transient ischemia increase in ubiquitin conjugates in the early reperfusion. *NeuroReport* **3**: 519–520.

———. 1992b. Subcellular distribution of ubiquitin-protein conjugates in the hippocampus following transient ischemia. *J. Neurosci. Res.* **31**: 561–564.

Hunt, C. and S. Calderwood. 1990. Characterization and sequence of a mouse *hsp70* gene and its expression in mouse cell lines. *Gene* **87**: 199–204.

Ito, U., M. Spatz, J.T.J. Walker, and I. Klatzo. 1975. Experimental cerebral ischaemia in Mongolian gerbils. I. Light microscopic observations. *Acta Neuropathol.* **32**: 209–223.

Jacewicz, M., M. Kiessling, and W.A. Pulsinelli. 1986. Selective gene expression in focal cerebral ischemia. *J. Cereb. Blood Flow Metab.* **6:** 263–272.

Johansen, F.F., M.B. Jørgensen, and N.H. Diemer. 1986. Ischemic CA-1 pyramidal cell loss is prevented by preischemic colchicine destruction of dentate granule cells. *Brain Res.* **377:** 344–347.

Johansen, F.F., J. Zimmer, and N.H. Diemer. 1987. Early loss of somatostatin neurons in dentate hilus after cerebral ischemia in the rat precedes CA1 pyramidal loss. *Acta Neuropathol.* **73:** 110–114.

Jørgensen, M.B., F.F. Johansen, and N.H. Diemer. 1987. Removal of the entorhinal cortex protects hippocampal CA-1 neurons from ischemic damage. *Acta Neuropathol.* **73:** 189–194.

Jørgensen, M.B., J. Deckert, D.C. Wright, and D.R. Gehlert. 1989. Delayed c-*fos* proto-oncogene expression in the rat hippocampus induced by transient global cerebral ischemia: An in situ hybridization study. *Brain Res.* **484:** 393–398.

Kato, H., Y. Liu, T. Araki, and K. Kogure. 1991. Temporal profile of the effects of pretreatment with brief cerebral ischemia on the neuronal damage following secondary ischemic insult in the gerbil: Cumulative damage and protective effects. *Brain Res.* **553:** 238–242.

Kato, H., Y. Liu, T. Araki, and K. Kogure. 1992. MK-801, but not anisomycin, inhibits the induction of tolerance to ischemia in the gerbil hippocampus. *Neurosci. Lett.* **139:** 118–121.

Kawagoe, J., K. Abe, and K. Kogure. 1992a. Different thresholds of HSP70 and HSC70 heat shock mRNA induction in post-ischemic gerbil brain. *Brain Res.* **599:** 197–203.

Kawagoe, J., K. Abe, S. Sato, I. Nagano, S. Nakamura, and K. Kogure. 1992b. Distributions of heat shock protein (HSP) 70 and heat shock cognate protein (HSC) 70 mRNAs after transient focal ischemia in rat brain. *Brain Res.* **587:** 195–202.

――――. 1992c. Distributions of heat shock protein-70 mRNAs and heat shock cognate protein-70 mRNAs after transient global ischemia in gerbil brain. *J. Cereb. Blood Flow Metab.* **12:** 794–801.

Kawai, K., L. Nitecka, C.A. Ruetzler, G. Nagashima, F. Joó, G. Mies, T.S. Nowak, Jr., N. Saito, J. Lohr, and I. Klatzo. 1992. Global cerebral ischemia associated with cardiac arrest in the rat: I. Dynamics of early neuronal changes. *J. Cereb. Blood Flow Metab.* **12:** 238–249.

Kiessling, M., G.A. Dienel, M. Jacewicz, and W.A. Pulsinelli. 1986. Protein synthesis in postischemic rat brain: A two-dimensional electrophoretic analysis. *J. Cereb. Blood Flow Metab.* **6:** 642–649.

Kindy, M.S. 1993. Inhibition of tyrosine phosphorylation prevents delayed neuronal death following cerebral ischemia. *J. Cereb. Blood Flow Metab.* **13:** 372–377.

Kinouchi, H., F.R. Sharp, M.P. Hill, J. Koistinaho, S.M. Sagar, and P.H. Chan. 1993. Induction of 70-kDa heat shock protein and hsp70 mRNA following transient focal cerebral ischemia in the rat. *J. Cereb. Blood Flow Metab.* **13:** 105–115.

Kirino, T. 1982. Delayed neuronal death in the gerbil hippocampus following ischemia. *Brain Res.* **239:** 57–69.

Kirino, T., Y. Tsujita, and A. Tamura. 1991. Induced tolerance to ischemia in gerbil hippocampal neurons. *J. Cereb Blood Flow Metab* **11:** 299–307.

Kitagawa, K., M. Matsumoto, M. Tagaya, K. Kuwabara, R. Hata, N. Handa, R. Fukunaga, K. Kimura, and T. Kamada. 1991a. Hyperthermia-induced neuronal protection against ischemic injury in gerbils. *J. Cereb. Blood Flow Metab.* **11:** 449–452.

Kitagawa, K., M. Matsumoto, K. Kuwabara, M. Tagaya, T. Ohtsuki, R. Hata, H. Ueda, N. Handa, K. Kimura, and T. Kamada. 1991b. "Ischemic tolerance" phenomenon detected in various brain regions. *Brain Res.* **561:** 203–211.

Kitagawa, K., M. Matsumoto, M. Tagaya, R. Hata, H. Ueda, M. Niinobe, N. Handa, R. Fukunaga, K. Kimura, K. Mikoshiba, and T. Kamada. 1990. "Ischemic tolerance" phenomenon found in brain. *Brain Res.* **528**: 21–24.

Kuroiwa, T., P. Bonnekoh, and K.-A. Hossmann. 1990. Prevention of postischemic hyperthermia prevents ischemic injury of CA_1 neurons in gerbils. *J. Cereb. Blood Flow Metab.* **10**: 550–556.

Larson, J.S., T.J. Schuetz, and R.E. Kingston. 1988. Activation in vitro of sequence-specific DNA binding by a human regulatory factor. *Nature* **335**: 372–375.

Leung, T.K.C., M.Y. Rajendran, C. Monfries, C. Hall, and L. Lim. 1990. The human heat-shock protein family. Expression of a novel heat-inducible HSP70 (HSP70B') and isolation of its cDNA and genomic DNA. *Biochem. J.* **267**: 125–132.

Lindvall, O., P. Ernfors, J. Bengzon, Z. Kokaia, M.-L. Smith, B.K. Siesjö, and H. Persson. 1992. Differential regulation of mRNAs for nerve growth factor, brain-derived neurotrophic factor, and neurotrophin 3 in the adult rat brain following cerebral ischemia and hypoglycemic coma. *Proc. Natl. Acad. Sci.* **89**: 648–652.

Li, Y., M. Chopp, J.H. Garcia, Y. Yoshida, Z.G. Zhang, and S.R. Levine. 1992. Distribution of the 72-kd heat-shock protein as a function of transient focal cerebral ischemia in rats. *Stroke* **23**: 1292–1298.

Liu, Y., H. Kato, N. Nakata, and K. Kogure. 1992. Protection of rat hippocampus against ischemic neuronal damage by pretreatment with sublethal ischemia. *Brain Res.* **586**: 121–124.

Lowenstein, D.H., P.H. Chan, and M.F. Miles. 1991. The stress protein response in cultured neurons: Characterization and evidence for a protective role in excitotoxicity. *Neuron* **7**: 1053–1060.

Lowenstein, D.H., R.P. Simon, and F.R. Sharp. 1990. The pattern of 72-kDa heat shock protein-like immunoreactivity in the rat brain following flurothyl-induced status epilepticus. *Brain Res.* **531**: 173–182.

Mackler, S.A., B.P. Brooks, and J.H. Eberwine. 1992. Stimulus-induced coordinate changes in mRNA abundance in single postsynaptic hippocampal CA1 neurons. *Neuron* **9**: 539–548.

Magnusson, K. and T. Wieloch. 1989. Impairment of protein ubiquitination may cause delayed neuronal death. *Neurosci. Lett.* **96**: 264–270.

Maines, M.D., R.D. Mayer, J.F. Ewing, and W.K. McCoubrey, Jr. 1993. Induction of kidney heme oxygenase-1 (HSP32) mRNA and protein by ischemia/reperfusion: Possible role of heme as both promoter of tissue damage and regulator of HSP32. *J. Pharmacol. Exp. Ther.* **264**: 457–462.

Marini, A.M. and S.M. Paul. 1992. N-methyl-D-aspartate receptor-mediated neuroprotection in cerebellar granule cells requires new RNA and protein synthesis. *Proc. Natl. Acad. Sci.* **89**: 6555–6559.

Marini, A.M., M. Kozuka, R.L. Lipsky, and T.S. Nowak, Jr. 1990. 70-Kilodalton heat shock protein induction in cerebellar astrocytes and cerebellar granule cells in vitro: Comparison with immunocytochemical localization after hyperthermia in vivo. *J. Neurochem.* **54**: 1509–1516.

Matsuyama, T., M. Tsuchiyama, H. Nakamura, M. Matsumoto, and M. Sugita. 1993a. Hilar somatostatin neurons are more vulnerable to an ischemic insult than CA1 pyramidal neurons. *J. Cereb. Blood Flow Metab.* **13**: 229–234.

Matsuyama, T., H. Michishita, H. Nakamura, M. Tsuchiyama, S. Shimizu, K. Watanabe, and M. Sugita. 1993b. Induction of copper-zinc superoxide dismutase in gerbil hippocampus after ischemia. *J. Cereb. Blood Flow Metab.* **13**: 135–144.

Mies, G., S. Ishimaru, Y. Xie, K. Seo, and K.-A. Hossmann. 1991. Ischemic thresholds of cerebral protein synthesis and energy state following middle cerebral artery occlusion in rat. *J. Cereb. Blood Flow Metab.* **11:** 753–761.

Mies, G., K. Kawai, N. Saito, G. Nagashima, T.S. Nowak Jr., C.A. Ruetzler, and I. Klatzo. 1993. Cardiac arrest-induced complete cerebral ischaemia in the rat: Dynamics of postischaemic *in vivo* calcium uptake and protein synthesis. *Neurol. Res.* **15:** 253–263.

Milarski, K.L., W.J. Welch, and R.I. Morimoto. 1989. Cell cycle-dependent association of HSP70 with specific cellular proteins. *J. Cell Biol.* **108:** 413–423.

Miller, E.K., J.D. Raese, and M. Morrison-Bogorad. 1991. Expression of heat shock protein 70 and heat shock cognate 70 messenger RNAs in rat cortex and cerebellum after heat shock or amphetamine treatment. *J. Neurochem.* **56:** 2060–2071.

Morse, J.K. and J.N. Davis. 1990. Regulation of ischemic hippocampal damage in the gerbil: Adrenalectomy alters the rate of CA_1 cell disappearance. *Exp. Neurol.* **110:** 86–92.

Müller, M., M. Cleef, G. Röhn, P. Bonnekoh, A.E.I. Pajunen, H.-G. Bernstein, and W. Paschen. 1991. Ornithine decarboxylase in reversible cerebral ischemia: An immunohistochemical study. *Acta Neuropathol.* **83:** 39–45.

Nakai, A. and R. Morimoto. 1993. Characterization of a novel chicken heat shock transcription factor, heat shock factor 3, suggests a new regulatory pathway. *Mol. Cell Biol.* **13:** 1983–1997.

Nowak, T.S., Jr. 1985. Synthesis of a stress protein following transient ischemia in the gerbil. *J. Neurochem.* **45:** 1635–1641.

———. 1991. Localization of 70 kDa stress protein mRNA induction in gerbil brain after ischemia. *J. Cereb. Blood Flow Metab.* **11:** 432–439.

Nowak, T.S., Jr., U. Bond, and M.J. Schlesinger. 1990a. Heat shock RNA levels in brain and other tissues after hyperthermia and transient ischemia. *J. Neurochem.* **54:** 451–458.

Nowak, T.S., Jr., J. Ikeda, and T. Nakajima. 1990b. 70 Kilodalton heat shock protein and c-*fos* gene expression following transient ischemia. *Stroke* (suppl. III) **21:** 107–111.

Nowak, T.S., Jr., O.C. Osborne and J. Ikeda. 1992. Role of altered gene expression in development of neuronal changes after ischemia. In *Maturation phenomenon in cerebral ischemia* (ed. U. Ito et al.), pp. 121–128. Springer-Verlag, Berlin.

Nowak, T.S., Jr., O.C. Osborne, and S. Suga. 1993. Stress protein and proto-oncogene expression as indicators of neuronal pathophysiology after ischemia. In *Neurobiology of ischemic brain damage—Progress in brain research* (ed. K. Kogure et al.), vol. 96, pp. 195–208. Elsevier Science Publishers, Amsterdam.

Ohtsuki, T., M. Matsumoto, K. Kuwabara, K. Kitagawa, K. Suzuki, N. Taniguchi, and T. Kamada. 1992. Influence of oxidative stress on induced tolerance to ischemia in gerbil hippocampal neurons. *Brain Res.* **599:** 246–252.

Olney, J.W., J. Labruyere, G. Wang, D.F. Wozniak, M.T. Price, and M.A. Sesma. 1991. NMDA antagonist neurotoxicity: Mechanism and prevention. *Science* **254:** 1515–1518.

Pardue, S., K. Groshan, J.D. Raese, and M. Morrison-Bogorad. 1992. Hsp70 mRNA induction is reduced in neurons of aged rat hippocampus after thermal stress. *Neurobiol. Aging* **13:** 661–672.

Price, B.D. and S.K. Calderwood. 1991. Ca^{2+} is essential for multistep activation of the heat shock factor in permeabilized cells. *Mol. Cell. Biol.* **11:** 3365–3368.

Pulsinelli, W.A., J.B. Brierley, and F. Plum. 1982. Temporal profile of neuronal damage in a model of transient forebrain ischemia. *Ann. Neurol.* **11:** 491–498.

Rabindran, S.K., G. Giorgi, J. Clos, and C. Wu. 1991. Molecular cloning and expression of a human heat shock factor, HSF1. *Proc. Natl. Acad. Sci.* **88:** 6906–6910.

Rordorf, G., W.J. Koroshetz, and J.V. Bonventre. 1991. Heat shock protects cultured neurons from glutamate toxicity. *Neuron* **7**: 1043–1051.

Ross, D.T. and A.C. Duhaime. 1989. Degeneration of neurons in the thalamic reticular nucleus following transient ischemia due to raised intracranial pressure: Excitotoxic degeneration mediated via non-NMDA receptors? *Brain Res.* **501**: 129–143.

Ross, D.T. and D.I. Graham. 1993. Selective loss and selective sparing of neurons in the thalamic reticular nucleus following human cardiac arrest. *J. Cereb. Blood Flow Metab.* **13**: 558–567.

Saito, N. and T.S. Nowak, Jr. 1993. Delayed expression of mRNA encoding microtubule-associated protein 2c (MAP2c) in vulnerable neuron populations after ischemia. *J. Cereb. Blood Flow Metab.* **13**: S467 (abstr.).

Sarge, K.D., S. Murphy, and R.I. Morimoto. 1993. Activation of heat shock gene transcription by heat shock factor 1 involves oligomerization, acquisition of DNA-binding activity, and nuclear localization and can occur in the absence of stress. *Mol. Cell Biol.* **13**: 1392–1407.

Scharf, K.-D., S. Rose, W. Zott, F. Schöff, and L. Nover. 1990. Three tomato genes code for heat stress transcription factors with a region of remarkable homology to the DNA-binding domain of the yeast HSF. *EMBO J.* **9**: 4495–4501.

Schiaffonati, L., E. Rappocciolo, L. Tacchini, G. Cairo, and A. Bernelli-Zazzera. 1990. Reprogramming of gene expression in postischemic rat liver: Induction of proto-oncogenes and hsp70 gene family. *J. Cell. Physiol.* **143**: 79–87.

Schuetz, T.J., G.J. Gallo, L. Sheldon, P. Tempst, and R.E. Kingston. 1991. Isolation of a cDNA for HSF2: Evidence for two heat shock factor genes in humans. *Proc. Natl. Acad. Sci.* **88**: 6911–6915.

Sharp, F.R., D. Lowenstein, R. Simon, and K. Hisanaga. 1991. Heat shock protein hsp72 induction in cortical and striatal astrocytes and neurons following infarction. *J. Cereb. Blood Flow Metab.* **11**: 621–627.

Sharp, F.R., M. Butman, S. Wang, J. Koistinaho, S.H. Graham, S.M. Sagar, L. Noble, P. Berger, and F.M. Longo. 1992. Haloperidol prevents induction of the hsp70 heat shock gene in neurons injured by phencyclidine (PCP), MK801, and ketamine. *J. Neurosci. Res.* **33**: 605–616.

Sheng, M. and M.E. Greenberg. 1990. The regulation and function of c-*fos* and other immediate early genes in the nervous system. *Neuron* **4**: 477–485.

Siesjö, B.K. and F. Bengtsson. 1989. Calcium fluxes, calcium antagonists, and calcium-related pathology in brain ischemia, hypoglycemia, and spreading depression: A unifying hypothesis. *J. Cereb. Blood Flow Metab.* **9**: 127–140.

Simon, R.P., H. Cho, R. Gwinn, and D.H. Lowenstein. 1991. The temporal profile of 72-kDa heat-shock protein expression following global ischemia. *J. Neurosci.* **11**: 881–889.

Sloviter, R.S. and D.H. Lowenstein. 1992. Heat shock protein expression in vulnerable cells of the rat hippocampus as an indicator of excitation induced neuronal stress. *J. Neurosci.* **12**: 3004–3009.

Smith, M.-L., R.N. Auer, and B.K. Siesjö. 1984. The density and distribution of ischemic brain injury in the rat following 2-10 min of forebrain ischemia. *Acta Neuropathol.* **64**: 319–332.

Sorger, P.K. and H.R.B. Pelham. 1988. Yeast heat shock factor is an essential DNA-binding protein that exhibits temperature-dependent phosphorylation. *Cell* **54**: 855–864.

Sprang, G.K. and I.R. Brown. 1987. Selective induction of a heat shock gene in fiber tracts and cerebellar neurons of the rabbit brain detected by in situ hybridization. *Mol.*

Brain Res. **3:** 89–93.

Suga, S. and T.S. Nowak, Jr. 1991. Localization of immunoreactive Fos and Jun proteins in gerbil brain: Effect of transient ischemia. *J. Cereb. Blood Flow Metab.* **11:** S352 (abstr.).

―――――. 1993. Pharmacology of the postischemic stress response: Effects of temperature on hsp70 expression after transient ischemia and hypothermic action of NBQX in the gerbil. In *Pharmacology of cerebral ischemia 1992* (ed. J. Krieglstein and H. Oberpichler-Schwenk), pp. 279–286. Wissenschaftliche Verlag, Stuttgart.

Tacchini, L., L. Schiaffonati, C. Pappalardo, S. Gatti, and A. Bernelli-Zazzera. 1993. Expression of HSP 70, immediate-early response and heme oxygenase genes in ischemic-reperfused rat liver. *Lab. Invest.* **68:** 465–471.

Tanno, H., R.P. Nockels, L.H. Pitts, and L.J. Noble. 1993. Immunolocalization of heat shock protein after fluid percussive brain injury and relationship to breakdown of the blood-brain barrier. *J. Cereb. Blood Flow Metab.* **13:** 116–124.

Thilmann, R., Y. Xie, P. Kleihues, and M. Kiessling. 1986. Persistent inhibition of protein synthesis precedes delayed neuronal death in postischemic gerbil hippocampus. *Acta Neuropathol.* **71:** 88–93.

Uemura, Y., N.W. Kowall, and M.F. Beal. 1991a. Global ischemia induces NMDA receptor-mediated c-*fos* expression in neurons resistant to injury in gerbil hippocampus. *Brain Res.* **542:** 343–347.

Uemura, Y., N.W. Kowall, and M.A. Moskowitz. 1991b. Focal ischemia in rats causes time-dependent expression of c-*fos* protein immunoreactivity in widespread regions of ipsilateral cortex. *Brain Res.* **552:** 99–105.

Uney, J.B., P.N. Leigh, C.D. Marsden, A. Lees, and B.H. Anderton. 1988. Stereotaxic injection of kainic acid into the striatum of rats induces synthesis of mRNA for heat shock protein 70. *FEBS Lett.* **235:** 215–218.

Van Why, S.K., F. Hildebrandt, T. Ardito, A.S. Mann, N.J. Siegel, and M. Kashgarian. 1992. Induction and intracellular localization of HSP-72 after renal ischemia. *Am. J. Physiol.* **263:** F769–F775.

Vass, K., W.J. Welch, and T.S. Nowak, Jr. 1988. Localization of 70 kDa stress protein induction in gerbil brain after ischemia. *Acta Neuropathol.* **77:** 128–135.

Vass, K., M.L. Berger, T.S. Nowak, Jr., W.J. Welch, and H. Lassmann. 1989. Induction of stress protein HSP70 in nerve cells after status epilepticus in the rat. *Neurosci. Lett.* **100:** 259–264.

Visvader, J., P. Sassone-Corsi, and I.M. Verma. 1988. Two adjacent promoter elements mediate nerve growth factor activation of the c-*fos* gene and bind distinct nuclear complexes. *Proc. Natl. Acad. Sci.* **85:** 9474–9478.

Voellmy, R., A. Ahmed, P. Schiller, P. Bromley, and D. Rungger. 1985. Isolation and functional analysis of a human 70,000-dalton heat shock protein gene segment. *Proc. Natl. Acad. Sci.* **82:** 4949–4953.

Watowich, S.S. and R.I. Morimoto. 1988. Complex regulation of heat shock- and glucose-responsive genes in human cells. *Mol. Cell. Biol.* **8:** 393–405.

Welch, W.J. and J.P. Suhan. 1986. Cellular and biochemical events in mammalian cells during and after recovery from physiological stress. *J. Cell Biol.* **103:** 2035–2052.

Welsh, F.A. and V.A. Harris. 1991. Postischemic hypothermia fails to reduce ischemic injury in gerbil hippocampus. *J. Cereb. Blood Flow Metab.* **11:** 617–620.

Welsh, F.A., R.E. Sims, and V.A. Harris. 1990. Mild hypothermia prevents ischemic injury in gerbil hippocampus. *J. Cereb. Blood Flow Metab.* **10:** 557–563.

Welsh, F.A., D.J. Moyer, and V.A. Harris. 1992. Regional expression of heat shock protein-70 mRNA and c-*fos* mRNA following focal ischemia in rat brain. *J. Cereb.*

Blood Flow Metab. **12:** 204–212.

Wessel, T.C., T.H. Joh, and B.T. Volpe. 1991. In situ hybridization analysis of c-*fos* and c-*jun* expression in the rat brain following transient forebrain ischemia. *Brain Res.* **567:** 231–240.

Widmann, R., T. Kuroiwa, P. Bonnekoh, and K.-A. Hossmann. 1991. [^{14}C]Leucine incorporation into brain proteins in gerbils after transient ischemia: Relationship to selective vulnerability of hippocampus. *J. Neurochem.* **56:** 789–796.

Wolf, P.A., J.L. Cobb, and R.B. D'Agostino. 1992. Epidemiology of stroke. In *Stroke. Pathophysiology, diagnosis and management*, 2nd ed. (ed. H.J.M. Barnett et al.), pp. 3–27. Churchill Livingstone, New York.

Wu, B.J., R.E. Kingston, and R.I. Morimoto. 1986. Human hsp70 promoter contains at least two distinct regulatory domains. *Proc. Natl. Acad. Sci.* **83:** 629–633.

Xie, Y., G. Mies, and K.-A. Hossmann. 1989. Ischemic threshold of brain protein synthesis after unilateral carotid artery occlusion in gerbils. *Stroke* **20:** 620–626.

22

Heat Shock Protein Gene Expression in Response to Physiologic Stress and Aging

Nikki J. Holbrook
Laboratory of Molecular Genetics
National Institute on Aging
Baltimore, Maryland 21224

Robert Udelsman
Laboratory of Endocrine Surgery
The Johns Hopkins Hospital
Baltimore, Maryland 21287

I. INTRODUCTION

Human aging is accompanied by a progressive decline in physiologic processes and in particular by a decreased ability to maintain homeostasis when faced with the stresses of life (Shock 1961; Shock et al. 1984). Current concepts suggest that aging results at least in part from damage to molecules, cells, and tissues by a variety of toxic factors that either are endogenously produced or occur as a result of environmental exposure (Martin et al. 1993). Genetic systems have evolved to detect specific forms of damage and to activate the expression of genes whose products are presumed to increase the resistance of cells to such damage and/or to aid in its repair. Believing that the continued effectiveness of these genetic responses to stress may be a major factor in resistance to disease and aging, we initiated a series of studies to examine the expression of various stress response genes as a function of aging. Because the expression of heat shock proteins is the most highly conserved and best understood of these genetic stress responses, and because the beneficial effects of heat shock proteins during stress have been clearly documented, we have focused efforts on this group of proteins.

Most of our knowledge concerning the homeostatic role of heat shock proteins has come from studies using cultured cells. Although much less is known about their expression in vivo, heat shock proteins are induced acutely in the intact animal in response to localized tissue injury, as well as systemically following heat stress (Vass et al. 1988; Gower et al. 1989; Blake et al. 1990; Morimoto et al. 1990). Furthermore, elevated levels of heat shock proteins also occur in certain chronic disease states including Hashimoto's thyroiditis, Graves' disease, arthritis, and atherosclerosis (Morimoto et al. 1990; Heufelder et al. 1992). Our own studies have examined the expression of heat shock proteins in physiologically relevant stress models. We have demonstrated in the rat that expression of the major heat shock protein, hsp70, is induced in vivo in response to a variety of stresses, including mild elevations in body temperature (<1.5°C rise), ether anesthesia, surgery, and the stresses engendered by restraint (Blake et al. 1990, 1991; Udelsman et al. 1991, 1993). We have concentrated our efforts on the response to restraint that is markedly attenuated as a function of age. We review here our progress in studies aimed at understanding the molecular mechanisms involved in controlling this response and the causes for its age-related decline.

II. RESTRAINT-INDUCED HSP70 EXPRESSION

We have shown that simply placing a rat in a partial restraint device such as that shown in Figure 1 results in the induction of hsp70 expression (Blake et al. 1991; Udelsman et al. 1993). This type of restraint apparatus limits the mobility of the rat and is frequently used in research for many other purposes including drug delivery and bloodletting. It does not cause direct damage to cells or tissues and in this respect differs from most stress paradigms used to elicit the heat shock response. Viewed more as a behavioral stress, this model has been utilized extensively to induce physiologic stress response syndromes in animals and has been employed to examine central and peripheral mechanisms associated with stress-related disorders such as gastric ulcers (for review, see Pare and Glavin 1986).

Figure 2 shows a representative Northern blot that demonstrates the magnitude of the hsp70 mRNA induction in adrenals and thoracic aortas of rats restrained for 60 minutes. A 2.3-kb transcript corresponding to the

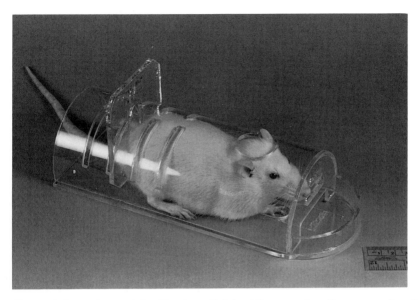

Figure 1 Restraint model utilized to elicit hsp70 expression. Rats were placed in the ventilated Plexiglas restraint device as shown. Stress conditions were approved by the Animal Care and Use Committee of the National Institute on Aging and were in accordance with the National Institutes of Health guidelines on the care and use of laboratory of animals.

constitutively expressed hsp70-related gene (hsc73; Fig. 2C) in the rat is seen in the adrenals of both control and stressed animals. It is present in lower amounts in the aorta and is therefore not evident in the autoradiographic exposures shown in Figure 2. Restraint results in the rapid induction of two stress-inducible transcripts at about 3.1 and 3.3 kb (I), with maximum mRNA expression achieved within 30–60 minutes of placement of the animal in the restraint device. Longer periods of restraint lead to a waning in the response, presumably the result of adaptation. Although the magnitude of the induction varies between individual animals, vascular expression is invariably greater than adrenal expression. Induction of mRNA is followed by an elevation in hsp70 protein, with maximum expression occurring between 3 and 6 hours after restraint (Udelsman et al. 1993).

Restraint-induced hsp70 expression is unique in several respects. First, it is a tissue-restricted response to whole body stress; the response is present in the adrenal gland and vasculature and absent in all other organ tissues examined, including brain, heart, lung, kidney, liver, small bowel, testes, pituitary, thyroid, ovary, skeletal muscle, and spleen. Vascular expression is not limited to the aorta but also occurs to a lesser de-

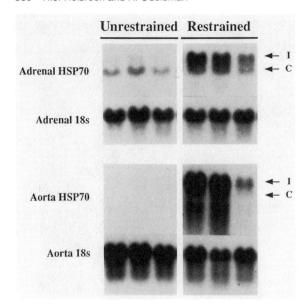

Figure 2 Representative Northern blot showing restraint-induced expression of hsp70 mRNA in adrenal and vascular tissues of rats. (C) Constitutive (hsc70) expression; (I) inducible transcripts.

gree in the vena cava. A second unique feature of this in vivo stress response is that it is limited to specific heat shock proteins. Examination of the expression of two members of the hsp90 family, hsp89 α and hsp89 β, and the low-molecular-weight heat shock protein, hsp27, by Northern analysis demonstrated that in the adrenal gland, none of these three heat shock proteins were induced by restraint. In the vasculature, however, hsp27 expression was also markedly increased in restrained rats.

To determine whether the expression was specific to certain regions or cell types within the adrenal and vascular tissues, in situ hybridization was performed. These studies utilized a 30-bp oligonucleotide that recognizes transcripts corresponding to the inducible hsp70 genes only (Blake et al. 1991). The adrenal gland is composed of two distinct regions: the outer cortex responsible for synthesis of mineralocorticoids, glucocorticoids, and sex steroids and the inner medulla where catecholamines are synthesized. In situ hybridization in the adrenal revealed that hsp70 mRNA expression was limited to the inner cortical regions but absent in the medulla (Blake et al. 1991). In the aorta, expression occurred throughout the smooth muscle cells of the media, with little expression evident in either the luminal endothelial cell layer or external adventitia (Udelsman et al. 1993).

III. ENDOCRINE CONTROL OF THE RESPONSE

The fact that immobilization alone could activate the heat shock response suggested that an endogenous factor(s) modulated by restraint might be responsible for the induction of hsp70 in these tissues. In mammals, restraint has been shown to result in the rapid activation of both the hypothalamic-pituitary-adrenal (HPA) axis and the sympathetic nervous system (Pare and Glavin 1986). This fact, coupled with the anatomic localization of hsp70 expression in the adrenal cortex and vascular smooth muscle, led us to examine whether these neuroendocrine systems were linked to restraint-induced hsp70 expression.

Stress induces the secretion of corticotropin-releasing hormone (CRH) from the hypothalamus, which in turn results in secretion of adrenocorticotropic hormone (ACTH) from the anterior pituitary gland. ACTH then stimulates the adrenal cortex, increasing both the synthesis and release of glucocorticoids into the peripheral circulation. Glucocorticoids influence a myriad of homeostatic processes including gluconeogenesis, the acute phase response, immune function, and cardiovascular stability (Munck et al. 1984; Udelsman et al. 1986). Furthermore, they feed back to both the hypothalamus and pituitary to inhibit CRH and ACTH secretion. Activation of the sympathetic nervous system results in the synthesis and release of catecholamines from both peripheral ganglia and the adrenal medulla. These catecholamines act locally in a paracrine-like fashion as well as systemically, particularly through their binding to specific adrenergic receptors in the cardiovascular system. We found that restraint-induced hsp70 expression is linked to activation of both the HPA axis and the sympathetic nervous system. However, the adrenal cortical and vascular responses are differentially regulated.

A. ACTH Mediates the Adrenal Response

The fact that restraint-induced hsp70 expression was localized to the cortex of the adrenal gland suggested that there could be a link between this expression and the HPA axis. To test the role of the HPA axis in mediating restaint-induced hsp70 expression, we examined the chronic effects of pharmacologic doses of agents capable of perturbing normal HPA axis activity on restraint-induced hsp70 induction. These included dexamethasone (a potent synthetic glucocorticoid), CRH, and the antiglucocorticoid/antiprogestin agent RU486. Chronic exposure to dexamethasone caused a marked attenuation of the adrenal response to restraint compared to placebo-treated controls, whereas CRH and RU486 had minimal effects. Importantly, dexamethasone had no effect on aortic hsp70 expression (Udelsman et al. 1994).

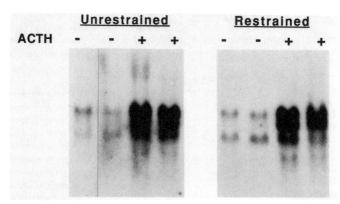

Figure 3 Influence of the HPA axis on the adrenal response. ACTH induces hsp70 mRNA expression in adrenals of hypophysectomized rats. (Reprinted from Blake et al. 1991.)

To delineate further the regulatory role of the HPA axis, we utilized a hypophysectomized rat model that eliminates endogenous ACTH. Removal of the pituitary gland abolished the induction of hsp70 mRNA in the adrenal gland (Blake et al. 1991), suggesting that either ACTH, glucocorticoid, or some other pituitary or adrenal-derived hormone was required for the response. Additional experiments revealed that acute injection of ACTH alone into hypophysectomized animals resulted in induction of hsp70 in the adrenal gland (Fig. 3). Dexamethasone, on the other hand, was without effect (Udelsman et al. 1993). From these studies, we concluded that stress-induced adrenal hsp70 expression is dependent on an intact HPA axis and specifically requires ACTH.

B. The α_1 Adrenergic Receptor Mediates the Vascular Response

Elevated hsp70 mRNA levels were seen in the aortas of hypophysectomized animals after restraint regardless of the presence or absence of ACTH or dexamethasone (Udelsman et al. 1993). Thus, in contrast to the adrenal response, vascular hsp70 expression is not dependent on the HPA axis. We speculated that adrenergic hormones might regulate this response.

Initial experiments to test the role of adrenergic agents in mediating the vascular response to restraint relied on specific adrenergic receptor antagonists to block the effects of endogenous compounds. As shown in Figure 4 (top), we demonstrated that acute injection of the specific α_1 adrenergic-blocking agent, prazosin, prior to restraint virtually elimi-

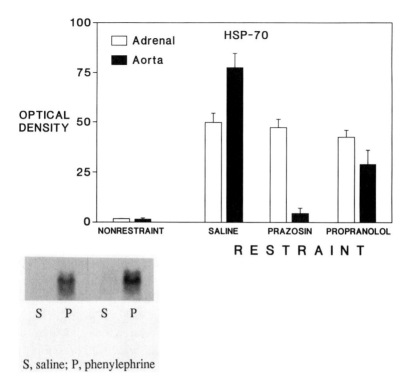

S, saline; P, phenylephrine

Figure 4 Influence of α_1 adrenergic stimulation on the vascular response. (*Top*) The α_1 adrenergic antagonist prazosin blocks vascular hsp70 expression in response to restraint. (*Bottom*) The α_1 adrenergic agonist phenylephrine induces hsp70 expression in the aorta. (Reprinted, with permission, from Udelsman et al. 1993.)

nated the induction of hsp70 in the vasculature (Udelsman et al. 1993). In contrast, the β adrenergic receptor antagonist, propranolol, had a lesser effect, even at very high doses. Importantly, these adrenergic receptor antagonists had no effect on the adrenal response. To test directly the effect of α_1 adrenergic stimulation on hsp70 expression in the aorta, we developed a chronic central venous catheter model to inject specific adrenergic agents into conscious nonstressed rodents. As shown in Figure 4 (bottom), the specific α_1 adrenergic agonist, phenylephrine, induced the expression of hsp70 in the aorta, whereas saline had no effect. Consistent with the inability of propranolol to block the effect significantly, administration of the β adrenergic agonist, isoproterenol, failed to induce hsp70 mRNA expression. Thus, we concluded that the vascular response to restraint is dependent on activation of the sympathetic nervous system, specifically via the α_1 adrenergic receptor.

IV. TRANSCRIPTIONAL CONTROL OF THE RESPONSE

As discussed in greater detail in this monograph and elsewhere (Morimoto 1993), expression of heat shock proteins in eukaryotes is regulated primarily through the activity of one or more of a family of heat shock transcription factors (HSFs). HSFs interact with a specific nucleotide sequence, the heat shock element (HSE), present in the promoter regions of heat shock protein genes, resulting in an increase in their rates of transcription. At least two different HSFs are present in mammalian cells (Rabindran et al. 1991; Schuetz et al. 1991). Current evidence suggests that functional differences exist between these HSFs and that the signals that activate the DNA-binding properties of each are distinct (Kroeger et al. 1993; Morimoto 1993). HSF1 has been shown to be the mediator of induction of heat shock protein gene expression in response to heat and other classical stressors known to elicit the heat shock response (Sarge et al. 1993). Another HSF family member, HSF2, appears to mediate induction of hsp70 transcription during hemin-induced differentiation of human K562 erythroleukemia cells and may have a role in mouse spermatogenesis (Sistonen et al. 1992; Morimoto 1993). It was therefore of great interest to determine which, if either, of these HSFs are involved in the activation of hsp70 expression in response to restraint.

Using gel mobility shift analysis and a consensus HSE oligonucleo-tide, we provided evidence to suggest that expression of hsp70 in the adrenal gland of restrained rats is mediated through an HSF (Blake et al. 1991). We demonstrated that there was constitutive HSE-binding activity present in extracts prepared from the adrenals of Wistar rats and that the binding activity increased following 15 minutes of restraint. Although we could not entirely rule out the possibility that the constitutive binding ac-tivity was due to activation of non-DNA-binding complexes during the preparation of extracts, this seemed unlikely since no appreciable DNA-binding activity was evident in extracts prepared from other tissues (i.e., the brain and liver) which did not show restraint-induced hsp70 expres-sion.

More recently, we have examined the nature of this binding activity in greater depth in Fischer 344 rats. As previously noted in Wistar rats, we observed that HSE-binding activity is present constitutively in the adrenal glands of unrestrained Fischer 344 rats. However, following restraint, we noted little change in the actual amount of HSE-binding ac-tivity in this rat strain, but rather, we observed an alteration in the pattern of DNA-binding complexes seen in gel mobility shift assays (T.W. Faw-cett et al., in prep.). As shown in Figure 5, two distinct DNA-binding complexes designated C1 and C2 are apparent in adrenal extracts. Both

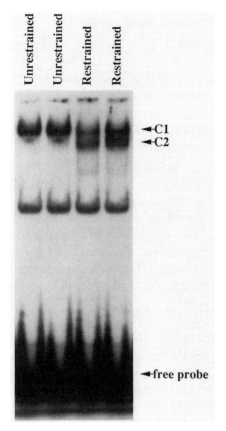

Figure 5 HSE-binding activities in adrenal extracts from unrestrained and restrained Fischer 344 rats. Gel-shift assays were performed as described previously (Blake et al. 1991).

of these complexes represent specific binding to the HSE as determined via competition analysis with cold HSE oligonucleotide and nonspecific competitors. In the absence of restraint, C1 predominates with little C2 evident. Following restraint, however, C2 becomes the predominant complex with a diminution in the amount of C1.

To define the nature of these complexes and determine which, if either, HSF is present in the complexes, we performed binding reactions in the presence of antibodies specific for either HSF1 or HSF2 (Sarge et al. 1993). As shown in Figure 6, treatment of adrenal extracts with antibody specific to HSF1 eliminated binding of both the C1 and C2 complexes to the HSE, whereas antibody specific to HSF2 had no effect. Thus, HSF1 appears to mediate the response to restraint.

V. AGE-RELATED DECLINE IN RESTRAINT-INDUCED HSP70 EXPRESSION

Using a variety of in vitro model systems, we and other investigators have shown that the induction of heat shock proteins in response to heat stress declines with aging (Lui et al. 1989; Fargnoli et al. 1990; Heydari et al. 1993). In our initial studies, we examined the effect of aging on hsp70 expression in cultures of lung- and skin-derived fibroblasts obtained from young (5-month) and old (24-month) male Wistar rats. We observed lower levels of heat-induced hsp70 mRNA and protein expression in cultures derived from the aged animals following their exposure to elevated temperatures (42.5°C) (Fargnoli et al. 1990). Additional experiments with freshly excised lung tissue from old and young rats showed a similar age-related decline in heat-induced hsp70 expression. Recent studies by Heydari et al. (1993), in which they examined hsp70 expression following heat stress in freshly isolated hepatocytes of young and old rats, have supported our earlier findings. These investigators further showed that the reduced expression of hsp70 in response to heat was associated with reduced HSE-binding activity.

Several additional studies have examined hsp70 expression in response to heat stress as a function of in vitro cellular senescence (Liu et al. 1989; Luce and Cristafalo 1993). Cells nearing the end of their in vitro proliferative life span showed a reduced response to heat relative to early-passage cells. Again, the age-related decline in the induction of hsp70 was associated with reduced levels of HSE-binding activity (Choi et al. 1990). Taken together, these in vitro findings provided considerable evidence for an age-related decline in the heat shock response, but this remained to be shown in the intact animal. Our findings with restraint-induced hsp70 expression have provided such evidence.

Figure 7 shows a comparison of hsp70 mRNA levels in mature adult (6-month) and aged (24-month) rats following 60 minutes of restraint. As can be seen, even though the adrenal and vascular responses are differentially regulated, aged animals show a significant decline in the induction of hsp70 expression in both tissues in response to restraint (see Udelsman et al. 1993). We have found such age-related differences in both inbred and outbred Wistar strains and in both male and female Fischer 344 rats.

What is the cause of this age-related decline? One simple possibility is that aged rats do not perceive restraint to be as stressful as do young rats. A second possibility is that the lower induction in aged rats occurs secondary to alterations in endocrine function. This is an important consideration since both the HPA axis and sympathetic nervous system have been reported to undergo alterations with aging (Roberts and Tumer 1987; Sapolsky 1990; Nielsen et al. 1992). We addressed these pos-

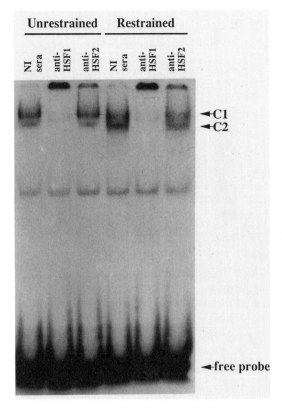

Figure 6 Polyclonal antibodies specific to HSF1 but not HSF2 eliminate HSE-binding activity in adrenal extracts of rats. (NI) Nonimmune sera.

sibilities with respect to adrenal hsp70 expression in several ways. First, we measured plasma ACTH and corticosterone (the major glucocorticoid in rats) levels in young and old restrained rats. We found that both young and old animals respond to restraint with an elevation in the levels of ACTH and corticosterone, indicating that both age groups viewed the restraint as stressful. However, no difference in the magnitude of the hormonal responses was found comparing young and old animals (Udelsman et al. 1993). In other experiments, we examined the relative level of expression of hsp70 mRNA in young and aged rats injected with a dose of ACTH sufficient to cause maximal elevation in glucocorticoid levels. In keeping with our findings depicted in Figure 3, ACTH treatment led to a significant induction (greater than fivefold) of hsp70 mRNA in young rats but resulted in less than a twofold increase in aged rats (D.J. Putney et al., unpubl.).

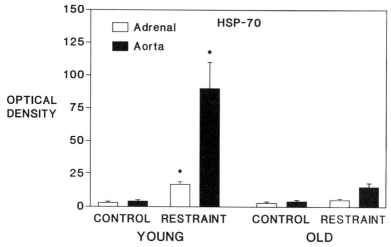

Figure 7 Aging is associated with a decrease in both the adrenal and vascular tissue responses to restraint. (Reprinted, with permission, from Udelsman et al. 1993.)

A. HSE-binding Activity in Young and Old Rats

To investigate further the possible cause for the age-related decline in hsp70 mRNA expression, we examined levels of HSE-binding activity in young and old rats. As shown in Figure 8, we observed significantly lower levels of HSE-binding activity in adrenal extracts of aged rats relative to their younger cohorts. Although these results show levels of binding activity only following restraint, constitutive binding activity in unrestrained animals was also lower in the older age group (T.W. Fawcett et al., in prep.).

The fact that HSE-binding activity is lower in aged rats suggests either that there are lower amounts of the HSF1 protein in aged rats relative to young animals or that there is a deficit in some other cellular component necessary to generate DNA-binding complexes. To address the first possibility, we utilized anti-HSF1 antibodies to determine levels of HSF1 protein in adrenal and aorta extracts from young and aged rats by Western analysis. We observed no age-related differences in the quantity of HSF1 protein in either of these tissues, and no difference in the amount of HSF1 protein was found in unrestrained versus restrained rats (T.W. Fawcett et al., in prep.). From these results, it appears that either HSF1 itself is altered with age or there is an alteration in the activity or amount of some other factor or component of the signal transduction pathway necessary for the activation of HSF to a DNA-binding form.

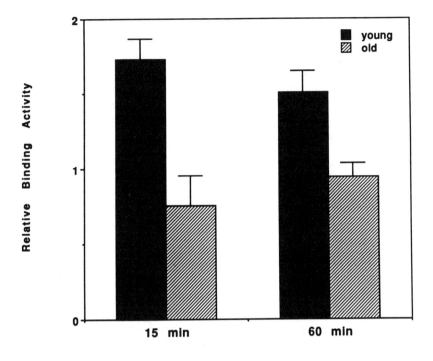

Time of Restraint

Figure 8 Adrenal extracts from aged rats have reduced levels of HSE-binding activity. Results are the mean ± SEM binding activity derived from six individual rats of each age group for each stress condition.

B. Transplantation Model for Studying Age-related Alterations in the Vascular Response

To study the mechanisms responsible for the age-related decline in vascular heat shock protein expression, we have developed a cross-transplantation model in which thoracic aortas harvested from young rats are transplanted into aged rats in an infrarenal position, and vice versa (Li et al. 1994). Using this model, we have begun to address a fundamental question: Is the aging process inherent to the aortic tissue or is it the environment in which the aorta resides that is responsible for the age-associated decline in stress-induced hsp70 expression?

Preliminary experiments performed during the developmental phase of this model have yielded two interesting observations. First, there is a gradient of the magnitude of restraint-induced hsp70 expression extending from the proximal to distal aorta. This gradient parallels the density of α_1 adrenergic receptors reported by others (Griendling et al. 1984), such that the greatest response occurs in the thoracic aorta, whereas a diminished response was noted in the abdominal aorta. Second, regardless of age, the process of transplanting a donor thoracic aorta to an infrarenal position in the recipient results in an attenuated hsp70 response in the transplanted vessel when the recipient is subsequently stressed. Although the cause for this decline is not yet understood, several possibilities exist. Injury and/or denervation secondary to transplantation might impede subsequent aortic smooth muscle responsiveness. Alternatively, since our model necessitates transplantation from the thoracic to the infrarenal abdominal position, an alteration in flow characteristics may explain this observation. Finally, it is possible that the transplanted aorta undergoes an alteration in its α_1 receptor density or function, which secondarily affects its ability to mount this response. Regardless of the cause for this transplantation effect, even after taking into account such positional differences, our preliminary results indicate that the heat shock protein response in aged aortas is partially rejuvenated by transplanting them into young animals. In contrast, young aortas behave more like old vessels when they are transplanted into old hosts (R. Udelsman and N.J. Holbrook, unpubl.). These data indicate that the environment in which the aorta resides is a significant factor in determining the level of hsp70 expression seen in this stressed tissue. It is possible that some negative, presumably circulating, factor is present in the old host that depresses responsiveness. Alternatively, aging could lead to the loss of a positive factor necessary for maintaining the response. Ongoing studies are addressing these possibilities.

VI. CONCLUSIONS AND PERSPECTIVES

The physiologic response to stress has been a major area of interest to investigators since at least the time of Hippocrates. Origins of the word "stress" can be traced back to the ancient Greek term "*strangale*," which means a halter, and the ancient Latin term "*stringere*," which means to draw tight (for review, see Chrousos et al. 1988). For modern scientists, stress congers up variable images dependent on one's particular field of interest encompassing physical, chemical, and emotional stimuli capable of causing distress to a cell, tissue, or whole organism. Although the expression of heat shock proteins in response to physical and chemical stresses has long been appreciated, prior to our studies, the possibility that an emotional stress could elicit the heat shock response had not been considered. Our finding that restraint, a behavioral stress akin to stresses encountered by humans, results in hsp70 expression in rodents provides such evidence. Indeed, we have demonstrated that this "primitive" cellular response to stress is intimately linked to neuro-hormonal stress responses in mammals and is therefore likely to play an important homeostatic role in coping with the stresses of everyday life. This could have important implications with respect to general health and well-being as a variety of physiological and emotional disorders are associated with perturbations in HPA axis and sympathetic nervous system function (Gold et al. 1988). It is of particular interest that the expression of hsp70 declines as a function of age, as does the ability of the elderly to maintain homeostasis in response to various environmental, metabolic, and emotional disturbances. We believe that a reduced ability to mount this response renders the aged more vulnerable to the deleterious effects of stress, thereby contributing to the aging process.

Although we have described the phenomenon of restraint-induced heat shock protein expression in detail, a number of important questions remain to be clarified: (1) What is the functional significance of this localized response to restraint? (2) How do ACTH and adrenergic agents act on cells to result in HSF activation? (3) Why is the response diminished in aged hosts? (4) What are the consequences of the reduced heat shock protein expression in the aged animals? The search for answers to these questions poses current and future challenges for us as well as other researchers in the field.

REFERENCES

Blake, M.J., D. Gershon, J. Fargnoli, and N.J. Holbrook. 1990. Discordant expression of heat shock protein mRNAs in tissues of heat-stressed rats. *J. Biol. Chem.* **265:** 15275–15279.

Blake, M.J., R. Udelsman, G.J. Feulner, D.D. Norton, and N.J. Holbrook. 1991. Stress-induced heat shock protein 70 expression in adrenal cortex: An adrenocorticotropic hormone-sensitive, age-dependent response. *Proc. Natl. Acad. Sci.* **88:** 9873–9877.

Choi, H.S., Z. Lin, B. Li, and A.Y.-C. Liu. 1990. Age-dependent decrease in the heat-inducible DNA sequence-specific binding activity in human diploid fibroblasts. *J. Biol. Chem.* **265:** 18005–18011.

Chrousos, G.P., D.L. Loriaux, and P.W. Gold. 1988. The concept of stress and its historical development. *Adv. Exp. Biol. Med.* **245:** 3–7.

Fargnoli, J., T. Kunisada, A.J. Fornace, Jr., E.L. Schneider, and N.J. Holbrook. 1990. Decreased expression of heat shock protein 70 mRNA and protein after heat treatment in cells of aged rats. *Proc. Natl. Acad. Sci.* **87:** 846–850.

Gold, P.W., S.K. Goodwin, and G.P. Chrousos. 1988. Clinical and biochemical manifestations of depression: Relation to the neurobiology of stress. *N. Engl. J. Med.* **319:** 348–353.

Gower, D.J., C. Hollman, K.S. Lee, and M.T. Tytell. 1989. Spinal cord injury and the stress protein response. *J. Neurosurg.* **70:** 605–611.

Griendling, K.K., A. Sastre, and W.R. Milnor. 1984. Regional differences in alpha 1-adrenoceptor numbers and responses in canine aorta. *Am. J. Physiol.* **247:** H928–H935.

Heufelder, A.E., J.R. Goellner, B.E. Wenzel, and R.S. Bahn. 1992. Immunohistochemical detection and localization of a 72-kilodalton heat shock protein in autoimmune thyroid disease. *J. Clin. Endocrinol. Metab.* **74:** 724–731.

Heydari, A.R., B. Wu, R. Takahashi, R. Strong, and A. Richardson. 1993. Expression of heat shock protein 70 is altered by age and diet at the level of transcription. *Mol. Cell. Biol.* **13:** 2909–2918.

Kroeger, P.E., K.D. Sarge, and R.I. Morimoto. 1993. Mouse heat shock transcription factors 1 and 2 prefer a trimeric binding site but interact differently with the HSP70 heat shock element. *Mol. Cell. Biol.* **13:** 3370–3383.

Li, D., C.A. Stagg, C. Gordon, and R. Udelsman. 1994. Vascular heat shock portein 70: Effect of aortic transplantation. In *Vascular surgery* (ed. Z.G. Wang). International Academic Publishers, Beijing. (In press.)

Liu, A.Y.-C., Z. Lin, H.S. Choi, F. Sorhage, and B. Li. 1989. Attenuated induction of heat shock gene expression in aging diploid fibroblasts. *J. Biol. Chem.* **164:** 12037–12045.

Luce, M.C. and V.J. Cristafalo. 1992. Reduction in heat shock gene expression correlates with increased thermosensitivity in senescent human fibroblasts. *Exp. Cell Res.* **202:** 9–16.

Martin, G.R., D.B. Danner, and N.J. Holbrook. 1993. Aging—Causes and defenses. *Annu. Rev. Med.* **44:** 419–429.

Morimoto, R.I. 1993. Cells in stress: Transcriptional activation of heat shock genes. *Science* **259:** 1409–1410.

Morimoto, R.I., A. Tissières, and C. Georgopoulous. 1990. *Stress proteins in biology and medicine.* Cold Spring Harbor Laboratory Press, Cold Spring Harbor, New York.

Munck, A., P.M. Guyre, and N.J. Holbrook. 1984. Physiological functions of glucocorticoids in stress and their relation to pharmacological actions. *Endocrine Rev.* **5:** 25–44.

Nielson, H., J.M. Hasenkam, H.K. Pilegaard, C. Aalkjaer, and F.V. Mortensen. 1992. Age-dependent changes in alpha- adrenoceptor-mediated contractility of isolated human resistance arteries. *Am. J. Physiol.* **263:** H1190–H1196.

Pare, W.P. and G.B. Glavin. 1986. Restraint stress in biomedical research: A review. *Neurosci. Biobehav. Rev.* **10:** 339–370.

Rabindran, S.K., G. Giorgi, J. Clos, and C. Wu. 1991. Molecular cloning and expression of a human heat shock factor, HSF1. *Proc. Natl. Acad. Sci.* **88:** 6906–6910.

Roberts, J. and N. Tumer. 1987. Age-related changes in autonomic function of catecholamines. *Rev. Biol. Res. Aging* **3:** 257–298.

Sapolsky, R.M. 1990. The adrenocortical axis. In *The handbook of the biology of aging,* 3rd ed. (ed. E.L. Schneider and J.W. Rowe), pp. 330–346. Academic Press, New York.

Sarge, K.D., S.P. Murphy, and R.I. Morimoto. 1993. Actvation of heat shock gene transcription by HSF1 involves oligomerization, acquisition of DNA binding activity, and nuclear localization and can occur in the absence of stress. *Mol. Cell. Biol.* **13:** 1392–1407.

Schuetz, T.J., G.J. Gallo, L. Sheldon, P. Tempst, and R.E. Kingston. 1991. Isolation of a cDNA for HSF2: Evidence for two heat shock factor genes in humans. *Proc. Natl. Acad. Sci.* **88:** 6910–6915.

Shock, N.W. 1961. Physiological aspects of aging in man. *Proc. Natl. Acad. Sci.* **23:** 97–122.

Shock, N.W., R.C. Greulich, R.A. Andres, D. Arenberg, P.T. Costa, Jr., E.G. Lakatta, and J.D. Tobin. 1984. *Normal human aging. The Baltimore Longitudinal Study of Aging,* vol. 84, p. 2450. U.S. Government Printing Office, Washington, D.C.

Sistonen, L., K.D. Sarge, B. Phillips, K. Abravaya, and R. Morimoto. 1992. Activation of heat shock factor 2 during hemin-induced differentiation of human erythroleukemia cells. *Mol. Cell. Biol.* **12:** 4104–4111.

Udelsman, R., M.J. Blake, and N.J. Holbrook. 1991. Molecular response to surgical stress: Specific and simultaneous heat shock protein induction in the adrenal cortex, aorta, and vena cava. *Surgery* **110:** 1125–1131

———. 1994. Endocrine control of stress-induced HSP-70 expression in vivo. *Surgery* (in press).

Udelsman, R., M.J. Blake, C.A. Stagg, D. Li, D.J. Putney, and N.J. Holbrook. 1993. Vascular heat shock protein expression in response to stress. *J. Clin. Invest.* **91:** 465–473.

Udelsman, R., J. Ramp, W.T. Gallucci, A. Gordon, E. Lipford, J.A. Norton, D.L. Loriaux, and G.P. Chrousos. 1986. Adaptation during surgical stress: A reevaluation of the role of glucocorticoids. *J. Clin. Invest.* **77:** 1377–1381.

Vass, K., W.J. Welch, and T.S. Nowak, Jr. 1988. Localization of the 70-K stress protein induction in gerbil brain after ischemia. *Acta Neuropathol.* **77:** 128–135.

Index

Molten globule state, ANS dye affinity,
268
mt-hsp70
ATP binding, 63–67
hsp60 interaction, 72, 74
inner membrane localization, 61
mechanism of action, 61–63
role
membrane translocation, 61–63, 72
precursor protein unfolding, 66–67
protein folding, 72, 74
sequence homology, 155–156
Multidrug resistance promoter, HSF
binding, 441
Multiple sclerosis, heat shock protein
role, 518

Octylglucoside yeast microsome
solubilization, 95, 97
Oligomerization. *See also* Heat shock
transcription factor
BiP, 114–115, 163, 170–171, 185
effect
ADP ribosylation, 172
ATP, 171, 185
GroEL requirement for complex assembly, 267
hsc70, 185
Ornithine transcarbamylase, GroE protein
folding, 267
Oxidant injury, stress protein response,
20
Oxyradicals, hsp27 phosphorylation induction, 353

P1 RepA, Dna protein activation, 220
p53
binding
DNA, 23, 216
heat shock protein, 23, 196
purification from *E. coli*, 216
p60, aporeceptor
complex component, 317
cycling role, 325
p88, binding motif, 202
PapD
activity, 12
crystal structure of complex, 12, 122–
123
Pathogen
heat shock protein

autoimmunity, 503–504
disease role, 499–503
expression, 22, 497, 502
induction in host, 497, 502–503
host
adaptation, 502
entry, 497
virus
heat shock protein induction, 498
protein processing by MHC
molecules, 497
Pbp74, peptide-binding affinity, 197
Peptidyl prolyl *cis-trans* isomerases. *See
also* hsp56
activity, 14, 476
distribution, 14
ninaA-encoded, 14–15
regulation, 476
role
bacteria, 15
thermotolerance, 476
yeast, 15
substrate conformation, 15
pH, heat shock response role, 538–539
Potassium, ATP-binding effect, 181
pp60^{src}, hsp90 association, 7
Prepro-α factor
hsc70 interaction, 90–91
translocation in yeast, 89–91
Presequence-binding factor
mechanism of action, 59
structure, 59
translocation competence role, 59
Progesterone aporeceptor
cellular localization, 315, 325
components, 317
domains, 318
hsp70 association, 320
Proline, isomerization in protein folding,
14
Promoter, heat shock gene
association
GAGA factor, 383–384, 413
RNA polymerase II, 383–384
TATA-binding protein, 383–385, 413
configuration in chromatin, 383
heat-shock-induced changes, 385–386
transcription factors, 383
Prostaglandins
activation of HSF, 446
heat shock protein induction, 445–446
Protease La. *See* Lon protease
Protease Ti. *See* ClpP
Protein

aggregation, 4
intracellular concentration, 4
Protein disulfide isomerase
activity, 15–16, 111
Dsb proteins in bacteria, 16–17
intracellular distribution, 15
role in PPIase activity, 15
role in yeast, 16
Protein folding
activation energy, 292
aggregation prevention, 252
kinetic partitioning, 293
mechanism, 252–253, 263–264, 268–272
pathways, 252
rate constants, 292–293
rate-determining steps, 111, 290, 476
Protein kinase C
arachidonic acid induction, 449
HSF1 phosphorylation, 449
Proteolysis. *See* Lysosomes; Ubiquitin system
prp73
binding specificity, 142
lysosomal proteolysis role, 142

RepA, DnaJ binding, 217, 258
Reperfusion, cellular damage, 535
Restraint. *See* Stress, behavioral
Rheumatoid arthritis, heat shock protein role, 518
Rhodanese
DnaJ binding, 216
refolding
model, 262, 275
rate constant, 305
Ribonuclease A
lysosomal proteolysis, 139–140, 143
prp73 affinity, 142
SecB protein binding, 288
Ribose-binding protein, SecB protein binding, 288
Ribulose bisphosphate carboxylase. *See* Rubisco-binding protein
RNA polymerase
domains, 384
heat shock promoter association, 383–385
paused in heat shock gene transcription, 383–385, 387, 413
phosphorylation, 384–385
protection from heat inactivation
DnaK, 222, 260

GroEL, 222–223, 226
refolding by DnaK chaperone machine, 211–212, 464
rpoH gene
bacterial distribution, 241
promoters, 234
regulation
cis-acting regulatory regions, 235–237
posttranslational, 235–237
temperature, 235–236
transcriptional, 234–235
σ^{32} transcription factor encoding, 19, 233
rpoN gene, transcription, 240–241
Rubisco-binding protein. *See* cpn60

Saccharomyces cerevisiae. See also individual proteins
DnaJ homolog, 49, 222
GrpE homolog
cloning, 221
purification, 221
sequence homology with GrpE, 222
hsp70
cytosolic. *See also* Ssa protein; Ssb protein
evolution, 34
functions, 31, 49–50
subfamilies, 34
genes, 31, 35
protein translocation
ATP role, 98–100
mechanism, 98–101
Salicylic acid, effects
heat shock protein induction, 23, 446
HSF1 DNA binding, 423, 446
Scj1p, homology with DnaJ, 74–75, 114, 260
Sec61p
interaction
BiP, 99–100
Sss1p, 96
role in yeast protein translocation, 94–95, 100
Sec62p, role in yeast protein translocation, 94–95, 98–100
Sec63p
complex with other Sec proteins, 96–97
cross-linking studies, 96
DnaJ similarity, 114, 254
role in yeast protein translocation, 94–95, 98–99, 101, 260

Ssc1p, protein translocation role, 97, 254, 261
Sss1p
  role in yeast protein translocation, 96
  Sec61p interaction, 96
Steroid receptors. *See also individual receptors*
  cross-talk of signaling networks, 326–328
  heat shock protein roles, 326–328
  hsp90 interactions, 315–317, 322–324, 327–328
  segments, 317–318
Stop-transfer model, intermembrane space protein sorting, 65
Stress, behavioral
  heat shock protein response
    aging effects, 586–587
    aorta environment effects, 589–590
    endocrine control, 581–583
    restraint, 21–22, 578–580
    significance, 590–591
    transcriptional control, 584–585
  role in aging, 591
Stress-70 proteins. *See also* ATPase; BiP; DnaK; hsp70; mt-hsp70
  activity
    clathrin uncoating ATPase, 154
    protein folding, 154
    translocation, 154
    transmembrane targeting, 154
  affinity chromatography, 154
  ATP affinity, 154, 163–166
  discovery, 153
  endoplasmic reticulum retention signal, 157
  functional differences between species, 197
  ligand-induced conformational changes, 170–171
  localization, 154
  self-peptide binding, 172
  sequence homology, 154–158, 196–197
  thermal stability, 159, 161–162
Stroke. *See* Cerebral ischemia
Synthetic lethality, yeast, 96–97

T cells
  autoimmimune disease role, 496
  function
    CD4 cells, 498
    CD8 cells, 498–499
  γ/δ, 499, 512
  heat shock protein
    antigenicity in disease, 499–502
    autoimmunity, 503–505
  receptor
    components, 498
    expression, 495
  recognition
    antigen, 496
    heat shock protein recognition, 22
  silencing of autoreactive cells, 496
  subclasses, 498–499
T-complex polypeptide 1. *See* TCP-1
TATA-binding protein. *See* Promoter
TCP-1. *See also* TF55
  activity, 13, 272, 274, 304, 309, 473
  ATPase activity, 274, 308
  CCT subunits, 300–304, 309
  chaperone function, 274
  cofactors, 304
  evolution, 307–309
  gene, 272, 300, 302–303
    cloning, 299
    evolution, 306–307
    expression, 307
    mapping, 306
    yeast, 306
  importance in yeast, 274–275
  purification, 300, 306
  rate constants
    binding, 304
    folding, 304
  sequence homology between species, 274, 300, 303, 305, 307–308
  structure, 13, 272, 304
  substrate specificity, 303–305
  symmetry, 300–301
  TRiC ring complex, 272
  yeast genetics, 305–306
TF55. *See also* TCP-1
  sequence homology with hsp60, 473
  structure, 13, 473
  symmetry, 301
  thermotolerance role, 473
TFIID, chromatin disruption, 441, 444
Thermosome
  archaebacteria chaperonin, 301
  symmetry, 301
Thermotolerance. *See also* Hyperthermia
  effect
    development stage, 480–481
    growth conditions, 458
    hyperthermia pretreatment, 458–459
    protein glycosylation, 480

DATE DUE	
FEB 1 7 2006	

GAYLORD   PRINTED IN U.S.A.